ELEVENTH EDITION

Choices in Relationships

An Introduction to Marriage and the Family

David Knox

East Carolina University

Caroline Schacht

East Carolina University

WADSWORTH
CENGAGE Learning

Australia • Brazil • Japan • Korea • Mexico • Singapore • Spain • United Kingdom • United States

Choices in Relationships: An Introduction to Marriage and the Family, Eleventh Edition
David Knox and Caroline Schacht

Senior Publisher: Linda Schreiber-Ganster

Acquiring Sponsoring Editor: Erin Mitchell

Assistant Editor: Linda Stewart

Editorial Assistant: Mallory Ortberg

Associate Media Editor: Melanie Cregger

Marketing Program Manager: Tami Strang

Project Management: PreMediaGlobal

Design Director: Rob Hugel

Senior Art Director: Caryl Gorska

Print Buyer: Karen Hunt

Senior Rights Specialist: Dean Dauphinais

Cover Design: Brie Hattey/Riezebos/Holzbauer

Cover Image: Coco Marlet/Photolibrary

Compositor: PreMediaGlobal

> For product information and technology assistance, contact us at
> **Cengage Learning Customer & Sales Support, 1-800-354-9706.**
> For permission to use material from this text or product,
> submit all requests online at **www.cengage.com/permissions.**
> Further permissions questions can be e-mailed to
> **permissionrequest@cengage.com.**

Library of Congress Control Number: 2011932999

Student Edition:
ISBN-13: 978-1-111-83322-0

ISBN-10: 1-111-83322-2

Loose-leaf Edition:
ISBN-13: 978-1-111-83950-5

ISBN-10: 1-111-83950-6

Wadsworth
20 Davis Drive
Belmont, CA 94002-3098
USA

Cengage Learning is a leading provider of customized learning solutions with office locations around the globe, including Singapore, the United Kingdom, Australia, Mexico, Brazil, and Japan. Locate your local office at **www.cengage.com/global.**

Cengage Learning products are represented in Canada by Nelson Education, Ltd.

To learn more about Wadsworth, visit **www.cengage.com/wadsworth.**

Purchase any of our products at your local college store or at our preferred online store **www.CengageBrain.com.**

Printed in the United States of America
1 2 3 4 5 6 7 15 14 13 12 11

To Karin and Dave who have made a wonderful and wise choice in each other for a life together.

Contents in Brief

Contents

CHAPTER **5**

Singlehood, Hanging Out, Hooking Up, and Cohabitation 117

CHAPTER **6**

Selecting a Partner 153

CHAPTER **16**

Relationships in the Later Years 474

Relationships, marriages, families, and the choices we make in reference to them are undergoing rapid change in our society. This new edition includes awareness of these choices in the context of our accelerated digital age but remains grounded in one focus: the choices we make in our relationships have consequences for the happiness, health, and well-being of ourselves, our partners, our marriage, our parents, and our children. By making deliberate informed choices, everyone wins. Not to take our relationship choices seriously is to limit our ability to enjoy deep, emotionally fulfilling relationships—the only game in town.

Over 600 new research studies, new Self-Assessment scales (e.g., Communication Danger Signs Scale), and new Applying Social Research boxes (e.g., "Sexting: Sexual Content/Images in Romantic Relationships") are featured in this new edition, which continues the cutting edge platform for which *Choices in Relationships* is known. Additionally, a new section has been added to the end of every chapter—"The Future of…"—where we predict, based on trends found in the latest research studies, what the future is likely to hold for marriage, singlehood, parenting, divorce/remarriage, and other chapter topics. Other new content added to each chapter includes the following.

New to the Eleventh Edition: Chapter by Chapter

Chapter 1 Choices in Relationships: An Introduction

34 new research articles

Benefits of marriage

When do people define themselves as a couple?

Self-Assessment: Family of Origin Expressiveness Scale

When families are destroyed—the Australian Aboriginal example

Applying Social Research: The "Relationship Talk"—Assessing Partner Commitment

Comparison of a web based marriage education class and a traditional face to face class

Chapter 2 Gender

47 new research articles

Gender roles across 24 highly developed countries

Benevolent sexism as a negative consequence of the traditional female role

Personal Choices: Choosing Gender Behavior that Fits

Chapter 12 Work and Family Life

34 new research articles

Effects of recession on marriages and families

Health care reform of 2010 as it relates to the family

Online financial management programs for undergraduates

Effects of money on emotional well being

Research Application: Undergraduate Love and Sex on the Job

Wife's contribution to the household—an economist speaks

Chapter 13 Violence and Abuse in Relationships

55 new research articles

Strategies abused women use in coping with abuse

How parent training programs can improve parenting practices

Spiritual abuse

Demographics of men who abuse

Cybercontrol as abuse

Chapter 14 Stress and Crisis in Relationships

42 new research articles

Scaling back and restructuring family roles as a stress-management strategy

Technology and therapy—computerized Internet therapy and telerelationship therapy

Alienation of affection laws in divorce proceedings

Chapter 15 Divorce and Remarriage

46 new research articles

Occupations at high and low risk for divorce

Al and Tipper Gore's divorce—why there should be no surprise

Revoking one's marriage license?

Living Apart Together as a structural solution to problems of stepfamily living

Value of stepfamily education programs for parents and children

Value of self-help books on stepfamily living

Divorce insurance

Chapter 16 Relationships in the Later Years

42 new research articles

Siblings who share (or refuse to share) elder parental care

Sexuality changes in men and women as they age

Use of technology to maintain relationships

Divorce as a behavior in the last year of life

Unique Features of the Text

Choices in Relationships has several unique features that are a part of every chapter.

Self-Assessment Scales

Each chapter features one or more self-assessment scales that allow students to measure a particular aspect of themselves or their relationships. Scales new to this edition include "Communication Danger Signs Scale," "Satisfaction with Married Life Scale," and "Childfree Lifestyle Scale."

Applying Social Research

To emphasize that *Choices in Relationships* is not merely a self-help trade book but a college textbook, we present a research application feature in every chapter and specify how new research may be applied to one's interpersonal relationships. Examples of new Applying Social Research features in this edition include "Secrets in Romantic Relationships," "The Relationship Talk: Assessing Partner Commitment," and "On Your Knees: Prayer in Romantic Relationships."

Social Policy

Congress is concerned about enacting into law "social policies that work." In each chapter we review social policies relevant to marriage and the family. Examples include abstinence-only vs. comprehensive sex education in the public school system, marriage education in public schools, and divorce mediation before litigation in divorce proceedings.

Personal Choices

An enduring popular feature of the text is the Personal Choices—detailed discussions of personal choice dilemmas. Examples include "Who is the Best Person for You to Marry?" "Should I Get Involved in a Long-Distance Relationship?" and "Deciding to Have Intercourse in a New Relationship."

What if?

To further personalize the focus of choices in relationships, we present "what ifs?" in every chapter. Examples include "What if you have made a commitment to marry someone but feel it is a mistake?" "What if you are in love with two people at the same time?" and "What if an old lover contacts you?"

Diversity in the United States and Diversity in Other Countries

The "one-size-fits-all" model of relationships and marriage is nonexistent. Individuals may be described as existing on a continuum from heterosexuality to homosexuality, from rural to urban dwellers, and from being single and living alone to being married and living in communes. Emotional relationships range from being close and loving to being distant and violent. Family diversity includes two parents (other or same-sex), single-parent families, blended families, families with adopted children, multigenerational families, extended families, and families representing different racial, religious, and ethnic backgrounds. Diversity is the term that accurately describes marriage and family relationships today.

Offset in various paragraphs in each chapter, "Diversity in the United States" reveals racial, religious, same-sex, economic, and educational differences in regard to relationship phenomena. In addition, to reveal courtship, marriage, and family patterns in other societies, "Diversity in Other Countries" paragraphs are presented throughout the text. For example, after years of having a One

Child Policy backed up by forced abortions, sterilizations, and economic policies, China's need for young workers has resulted in a revised policy for some provinces to allow for a couple to have two children. Both of these features have been thoroughly updated and include new discussions of walking marriages in the Mosuo culture of China and the prevalence of Hispanic same-sex couples raising children in the United States.

National and International Data

To replace speculation and guessing with facts, we provide data from national samples as well as data from around the world. For example, divorce rates in the United States are not soaring but have stabilized. Likewise, around the world we are seeing relatively high divorce rates.

Photo Essays

A new photo essay on the roles of parenting and new photos for the photo essay on weddings in other cultures are provided in Chapters 7 and 11.

Special Topics

This edition features three Special Topics, highlighting marriage and family careers, contraception, and additional resources and organizations. Each topic has been fully revised and updated.

Chapter Summaries

Each chapter ends with a summary, formatted as questions and answers, with each question relating back to the Learning Objectives listed at the beginning of every chapter.

Key Terms

Boldface type indicates key terms, which are defined and featured in the margin of the text as well as the glossary at the end of the text.

Web Links

The Internet is an enormous relationship resource. Internet addresses are provided at the end of each chapter. These have been checked at the time of publication to ensure that they are "live."

Quotes

New quotes are scattered throughout the text to give unique perspectives and to generate discussion of various key topics.

Supplements and Resources

Choices in Relationships, Eleventh Edition, is accompanied by a wide array of supplements prepared by the authors for both instructors and students. Some new resources have been created specifically to accompany the eleventh edition, and all of the continuing supplements have been thoroughly revised and updated.

Supplements for the Instructor

Instructor's Resource Manual with Test Bank   This manual provides instructors with learning objectives, a list of key concepts and terms (with page references), detailed lecture outlines, extensive student projects and classroom activities, identification of how current movies and television programs can be used in the classroom, updated InfoTrac article questions, current Internet resources, and self-assessment handouts for each chapter. Also included is a concise user guide for InfoTrac and a table of contents for the *ABC News* Marriage and Family Video Series. The Test Bank contains fifty multiple-choice, ten true/false, ten short answer/discussion questions, and five essay questions per chapter. In this revision, we have focused on including more conceptual/application questions and minimizing the number of factual questions in the multiple-choice and true/false sections. Our goal in making this change is to create a more well-rounded test bank to encourage student application of content rather than memorizing facts. The Test Bank items are also available electronically on the Power Lecture with Examview DVD.

PowerLecture with ExamView®   Available on DVD, this one-stop class preparation tool contains ready-to-use Microsoft PowerPoint® slides, enabling you to assemble, edit, publish, and present custom lectures with ease. PowerLecture helps you bring together text-specific lecture outlines and photos/art from the text along with videos and your own materials—culminating in powerful, personalized, media-enhanced presentations. Featuring automatic grading, **ExamView®** is also available within PowerLecture, allowing you to create, deliver, and customize tests and study guides (both print and online) in minutes. See assessments onscreen exactly as they will print or display online. Build tests of up to 250 questions using up to 12 question types and enter an unlimited number of new questions or edit existing questions. PowerLecture also includes the text's Instructor's Resource Manual and Test Bank as Word documents.

Classroom Activities for Marriage and Family   Created from contributions by instructors who teach Marriage and Family courses, this book will add new life to your lectures. Includes group exercises, lecture ideas, and homework assignments.

The Wadsworth Sociology Video Library, Vol. 1 & 2   This video library drives home the relevance of course topics through short, provocative clips of current and historical events. Perfect for enriching lectures and engaging students in discussion, many of the segments on this volume have been gathered from BBC Motion Gallery. Through an agreement with BBC Motion Gallery, Cengage Learning selects content from the BBC archive to enhance students' knowledge and understanding of significant events that relate to the concepts in your course. Clips are drawn from the BBC's vast library of award-winning news, science, business, humanities and social science-related programming, including biographies of notable artists, authors, scientists, and inventors from history through today. Ask your Cengage Learning representative for complete details on this amazing resource.

WebTutor with eBook   Jumpstart your course with customizable, rich, text-specific content within your Course Management System.

- **Jumpstart** – Simply load a WebTutor cartridge into your Course Management System
- **Customizable** – Easily blend, add, edit, reorganize, or delete content.
- **Content** – Rich, text-specific content, media assets, ebook, quizzing, web links, discussion topics, interactive games and exercises, and more

Supplements for the Student

CourseMate   The book's CourseMate website includes chapter-specific resources for instructors and students. For instructors, the site offers a password-protected instructor's manual, Microsoft PowerPoint presentation slides, and more. For students, there are a multitude of text-specific study aids: tutorial practice quizzes that can be scored and e-mailed to the instructor, web links, InfoTrac

College Edition exercises, flash cards, MicroCase Online data exercises, cross-word puzzles, Virtual Explorations, and much more!

Relationship Skills Exercises This newly updated supplement, full of assessments and questionnaires, will make students think more reflectively on important topics related to marriage, such as finances and intimacy. Assignments can be done in-class or at home, alone or with a partner.

InfoTrac® College Edition Give your students anytime, anywhere access to current reliable resources with InfoTrac College Edition, the online library. This fully searchable database offers twenty years' worth of full-text articles from thousands of diverse sources, such as academic journals, newsletters, and up-to-the minute periodicals including *Time, Newsweek, Science, Forbes,* and *USA Today.* The incredible depth and breadth of material—available twenty-four hours a day from any computer with Internet access—makes conducting research so easy, your students will want to use it to enhance their work in every course! Through InfoTrac's InfoWrite, students now also have instant access to critical thinking and paper writing tools. Both adopters and their students receive unlimited access for four months.

Acknowledgments

Texts are always a collaborative and collective product. This eleventh edition reflects the commitment and vision of Erin Mitchell. We thank Erin for her state-of-the-art content guidance throughout this new edition. We would also like to thank Linda Stewart, our developmental editor who was vital in moving the revision forward; Tami Strang, the marketing program manager; Melanie Cregger, the media editor; Rathi Thirumalai, the production project manager; Katrina Wilbur, the copy editor; and Susan Buschhorn and Shawn DeJong, the permissions researchers. All were superb, and we appreciate their professionalism and attention to detail. We would also like to thank Chelsea E. Curry for her superb photographic skill, evident in her various photos throughout the text; Alora Brackett, who assisted with various aspects of each chapter; and Twyanna Purkett for updating the information on contraception/sexually transmitted infections.

Reviewers for the Eleventh Edition
Sheldon Helfing, College of the Canyons; Alissa King, Iowa Central Community College; Michallene McDaniel, Gainesville State College; Jodi McKnight; Mid Continent University; Nancy Reeves, Glouchester County College; Linda Stone, Towson University

Reviewers for Previous Editions
Grace Auyang, University of Cincinnati; Rosemary Bahr, Eastern New Mexico University; Von Bakanic, College of Charleston; Mary Beaubien, Youngstown State University; Sampson Lee Blair, Arizona State University; Mary Blair-Loy, Washington State University; David Daniel Bogumil, Wright State University; Elisabeth O. Burgess, Georgia State University; Craig Campbell, Weber State University; Michael Capece, University of South Florida; Lynn Christie, Baldwin-Wallace College; Laura Cobb, Purdue University and Illinois State University; Jean Cobbs, Virginia State University; Donna Crossman, Ohio State University; Karen Dawes, Wake Technical Community College; Susan Brown Donahue, Pearl River Community College; Doug Dowell, Heartland Community College; John Engel, University of Hawaii; Kim Farmer, Martin Community College; Mary Ann Gallagher, El Camino College; Shawn Gardner, Genesee Community College;

Ted Greenstein, North Carolina State University; Heidi Goar, St. Cloud State University; Norman Goodman, State University of New York at Stony Brook; Jerry Ann Harrel-Smith, California State University, Northridge; Gerald Harris, University of Houston; Rudy Harris, Des Moines Area Community College; Terry Hatkoff, California State University, Northridge; Christina Hawkey, Arizona Western College; Sheldon Helfing, College of the Canyons; Tonya Hilligoss, Sacramento City College; Rick Jenks, Indiana University; Richard Jolliff, El Camino College; Diane Keithly, Louisiana State University; Steve Long, Northern Iowa Area Community College; Patricia B. Maxwell, University of Hawaii; Carol May, Illinois Central College; Tina Mougouris, San Jacinto College; Jane A. Nielsen, College of Charleston; Lloyd Pickering, University of Montevallo; Scott Potter, Marion Technical College; Janice Purk, Mansfield University; Cherylon Robinson, University of Texas at San Antonio; Patricia J. Sawyer, Middlesex Community College; Cynthia Schmiege, University of Idaho; Susan Schuller Friedman, California State University Los Angeles; Eileen Shiff, Paradise Valley Community College; Scott Smith, Stanly Community College Beverly Stiles; Tommy Smith, Auburn University; Beverly Stiles, Midwestern State University; Dawood H. Sultan, Louisiana State University; Elsie Takeguchi, Sacramento City College; Myrna Thompson, Southside Virginia Community College; Teresa Tsushima, Iowa State University; Janice Weber-Breaux, University of Southwestern Louisiana; Kathleen Wells, University of Arizona South; Loreen Wolfer, University of Scranton

We love the study, writing, and teaching of marriage and the family and recognize that no one has a corner on relationships. We welcome your insights, stories, and suggestions for improvement in the next edition of this text. We check our e-mail frequently and invite you to send us your thoughts. We will respond.

David Knox, *Knoxd@ecu.edu*
Caroline Schacht, *Schachtc@ecu.edu*

About the Authors

David Knox, Ph.D., is Professor of Sociology at East Carolina University, where he teaches courtship and marriage, marriage and the family, and sociology of human sexuality. He is a marriage and family therapist and the author or coauthor of twelve books and over 100 professional articles. He and Caroline Schacht are married.

Caroline Schacht, M.A. in Sociology and M.A. in Family Relations, is instructor of sociology at East Carolina University and teaches courtship and marriage, introduction to sociology, and the sociology of food. Her clinical work includes marriage and family relationships. She is also a divorce mediator and the coauthor of several books, including *Understanding Social Problems* (Wadsworth 2012).

CHAPTER 1

Choices in Relationships: An Introduction

It's choice—not chance—that determines your destiny.

Jean Nidetch, Founder of Weight Watchers

Sybrand Cillie

Learning Objectives

Review the various elements of marriage.

List and define the various types of families.

Identify the theme of this text.

Summarize how marriage and the family have changed in the last 60 years.

Name the various theoretical frameworks for viewing marriage and the family.

Describe the steps in the research process and the caveats to be kept in mind.

Predict the future of marriage.

1

1. Having no close relationship connections (e.g. marriage, family, kinship) can be compared to smoking 15 cigarettes a day and being an alcoholic in terms of the effect on one's health.

2. Marriage is no longer a dominant lifestyle goal. By age 65 only about half of Americans have ever married.

3. Sharing funds or exchanging financial support is a "marker" that two individuals are becoming a "couple."

4. A web-based marriage education course is just as effective as a traditional face to face class and both are more valuable than no course at all.

5. Marriage education classes are helpful only for individuals BEFORE marriage (once a couple is married, involvement in these types of programs shows little to no effect).

Answers: 1. T   2. F   3. T   4. T   5. F

A n aging professor of marriage and family revealed a pattern of always giving the same tests each year. Colleagues thought the teacher both lazy and unfair since the old tests would get out and be used by new students. But the teacher justified giving the same exams every year on the premise that while the questions were the same, the answers kept changing. For example, in 1960, two-thirds (68%) of all twenty-somethings were married. More recently, just over a fourth (26%) in this age category was married (Pew Research Center 2010a). The rush to marry soon after high school has been replaced by great caution and delay.

What has not changed is the importance of family. All of us were born into a family and will end up in a family (however one defines this concept) of our own. Indeed, "raising a family" remains one of the top values in life for undergraduates. In a nationwide study of 201,818 undergraduates in 279 colleges and universities, 73% identified this as an essential objective (77% for financial success) (Pryor et al. 2011). In a study of 2,922 undergraduates, a "happy marriage" (50%) was the top value in life over "having a career I love" (28%) and financial security (17%) (Knox and Hall 2010). In this chapter we review the meaning of marriage and family and the choices we make in regard to these and other relationship options.

Getting married is a way to show family and friends that you have a successful personal life. It's the ultimate merit badge.

Andrew Cherlin, sociologist

Marriage a legal relationship that binds a couple together for the reproduction, physical care, and socialization of children.

Marriage

The federal government regards **marriage** as a legal relationship that binds a couple together for the reproduction, physical care, and socialization of children. Each society works out its own details of what marriage is. In the United States, marriage is a legal contract between a heterosexual couple (we discuss same-sex marriage later in the chapter) and the state in which they reside that specifies the economic relationship between the couple (they become joint owners of their income and debt) and encourages sexual fidelity. Various elements implicit in the marriage relationship in the United States are discussed in the following.

National **Data**

Over 95% of U.S. adult women (96%) and men (95.6%) aged 75 and older report having been married at least once (*Statistical Abstract of the United States* 2011, Table 34).

Elements of Marriage

No one definition of marriage can adequately capture its meaning. Rather, marriage might best be understood in terms of its various elements. Some of these include the following.

Legal Contract   Marriage in our society is a legal contract into which two people of different sexes and legal age (usually eighteen or older) may enter when they are not already married to someone else. The marriage license certifies that a legally empowered representative of the state married the individuals, often with two witnesses present. The marriage contract actually gives more power to the government and its control over the couple (Aulette 2010). The government will dictate not only who may marry (e.g. heterosexuals in most states, age 18 or above, not already married) but also the conditions of divorce (e.g. alimony and child support).

Under the laws of the state, the license means that spouses will jointly own all future property acquired and that each will share in the estate of the other. In most states, whatever the deceased spouse owns is legally transferred to the surviving spouse at the time of death. In the event of divorce and unless the couple had a prenuptial agreement, the property is usually divided equally regardless of the contribution of each partner. The license also implies the expectation of sexual fidelity in the marriage. Though less frequent because of no-fault divorce, infidelity is a legal ground for both divorce and alimony in some states.

The marriage license is also an economic license that entitles a spouse to receive payment for medical bills by a health insurance company if the partner is insured, to collect Social Security benefits at the death of the other spouse, and to inherit from the estate of the deceased. One of the goals of gay rights advocates who seek the legalization of marriage between homosexuals is that the couple will have the same rights and benefits as heterosexuals.

Though the courts are reconsidering the definition of what constitutes a "family," the law is currently designed to protect spouses, not lovers or cohabitants. An exception is **common-law marriage**, in which a heterosexual couple cohabits and presents themselves as married; they will be regarded as legally married in those states that recognize such marriages. Common-law marriages exist in fourteen states (Alabama, Colorado, Georgia, Idaho, Iowa, Kansas, Montana, New Hampshire, Ohio, Oklahoma, Pennsylvania, Rhode Island, South Carolina, and Texas) and the District of Columbia. Some states have restrictions—e g. Pennsylvania only recognizes common law marriages created before 2005.

Common-law marriage a heterosexual cohabiting couple presenting themselves as married.

Emotional Relationship   Most people in the United States regard being in love with the person they marry as an important reason for staying married. Forty percent of 2,922 undergraduates reported that they would divorce if they no longer loved their spouse (Knox and Hall 2010). This emphasis on love is not shared throughout the world. Individuals in other cultures (for example, India and Iran) do not require love feelings to marry—love is expected to follow, not precede, marriage. In these countries, parental approval and similarity of religion, culture, and education are considered more important criteria for marriage than love.

The heart has its reasons that reason knows nothing of.

Blaise Pascal, *Pensées*, 1670

National **Data**

In regard to reasons to get married, adults point to love (93%), making a lifelong commitment (87%) and companionship (81%) rather than having children (59%) or financial stability (31%) (Pew Research Center 2010a).

Eighty percent of first marriages occur in a religious context. The ceremony is a joyous occasion attended by friends, family, and well wishers. No one attends an individual's divorce.

Colleen Breen

Love: a temporary insanity, curable by marriage.

Ambrose Bierce, journalist

Sexual Monogamy   Marital partners expect sexual fidelity. Over half (54%) of 2,922 undergraduates agreed, "I would divorce a spouse who had an affair," and over two thirds (67%) agreed that they would end a relationship with a partner who cheated on them (Knox and Hall 2010).

Legal Responsibility for Children   Although individuals marry for love and companionship, one of the most important reasons for the existence of marriage from the viewpoint of society is to legally bind a male and a female for the nurture and support of any children they may have. In our society, child rearing is the primary responsibility of the family, not the state.

Marriage is a relatively stable relationship that helps to ensure that children will have adequate care and protection, will be socialized for productive roles in society, and will not become the burden of those who did not conceive them. Even at divorce, the legal obligation of the noncustodial parent to the child is maintained through child-support payments.

Announcement/Ceremony   The legal binding of a couple is often preceded by an announcement in the local newspaper and then followed by a formal ceremony in a church or synagogue. Such a ceremony reflects the cultural importance of the event. Telling parents, siblings, and friends about wedding plans helps to verify the commitment of the partners, and signifies that the marriage is a social event. The custom of gift giving after the ceremony also helps to marshal the economic support to launch the couple into married life.

Benefits of Marriage

Most people in our society get married, which has enormous benefits. Researchers Holt-Lunstad et al. (2010) emphasized the value of marriage as a social context conducive to one's mental as well as physical health and mortality. Their conclusion is based on a review of 148 studies involving 308,849 participants, which showed a 50% increase in the likelihood of survival for both women and men who had high social connections with family, friends, neighbors, and colleagues. Low social interaction/involvement had negative consequences and could be compared to smoking 15 cigarettes a day, being an alcoholic, not exercising, and being obese.

TABLE 1.1 Benefits of Marriage and the Liabilities of Singlehood

	Benefits of Marriage	Liabilities of Singlehood
Health	Spouses have fewer hospital admissions, see a physician more regularly, and are "sick" less often.	Single people are hospitalized more often, have fewer medical checkups, and are "sick" more often.
Longevity	Spouses live longer than single people.	Single people die sooner than married people.
Happiness	Spouses report being happier than single people.	Single people report less happiness than married people.
Sexual satisfaction	Spouses report being more satisfied with their sex lives, both physically and emotionally.	Single people report being less satisfied with their sex lives, both physically and emotionally.
Money	Spouses have more economic resources than single people.	Single people have fewer economic resources than married people.
Lower expenses	Two can live more cheaply together than separately.	Cost is greater for two singles than one couple.
Drug use	Spouses have lower rates of drug use and abuse.	Single people have higher rates of drug use and abuse.
Connected	Spouses are connected to more individuals who provide a support system—partner, in-laws, etc.	Single people have fewer individuals upon whom they can rely for help.
Children	Rates of high school dropouts, teen pregnancies, and poverty are lower among children reared in two-parent homes.	Rates of high school dropouts, teen pregnancies, and poverty are higher among children reared by single parents.
History	Spouses develop a shared history across time with significant others.	Single people may lack continuity and commitment across time with significant others.
Crime	Spouses are less likely to be involved in crime.	Single people are more likely to be involved in crime.
Loneliness	Spouses are less likely to report loneliness.	Single people are more likely to report being lonely.

When married people are compared with singles, the differences are strikingly in favor of the married (see Table 1.1). The advantages of marriage over singlehood have been referred to as the **marriage benefit** and are true for first as well as subsequent marriages. Explanations for the marriage benefit include economic resources (e.g. higher income/can afford health care), social control (e.g. spouses ensure partner moderates alcohol/drug use, does not ride motorcycle) and psychosocial support and strain (e.g. in-resident counselor, caring partner) (Carr and Springer 2010). However, just being married is not beneficial to all individuals. In fact, being in a stressful marriage is associated with elevated risk of cardiovascular disease and diabetes, which collectively have been shown to increase risk for heart attack, diabetes, stroke, and mortality (Whisman et al. 2010).

Marriage benefit when compared to being single, married persons are healthier, happier, live longer, less drug use, etc.

Types of Marriage

Although we think of marriage in the United States as involving one man and one woman, other societies view marriage differently. **Polygamy** is a generic term for marriage involving more than two spouses. Polygamy occurs "throughout the world . . . and is found on all continents and among adherents of all world religions" (Zeitzen 2008). There are three forms of polygamy: polygyny, polyandry, and pantagamy.

Polygamy a generic term for marriage involving more than two spouses.

Joe Jessop, an elder of the FLDS, has 5 wives, 46 children and 239 grandchildren.

Polygyny in the United States

Polygyny involves one husband and two or more wives and is practiced illegally in the United States by some religious fundamentalist groups. These groups are primarily in Arizona, New Mexico, and Utah (as well as Canada), and have splintered off from the Church of Jesus Christ of Latter-day Saints (commonly known as the Mormon Church). To be clear, the Mormon Church does not practice or condone polygyny. Those that split off from the Mormon Church represent only about 5% of Mormons in Utah. The largest offshoot is called the Fundamentalist Church of Jesus Christ of the Latter-day Saints (FLDS). Members of the group feel that the practice of polygyny is God's will. Joe Jessop, age 88, is an elder of the FLDS. He has 5 wives, 46 children, and 239 grandchildren. Although it is illegal, polygynous individuals are rarely prosecuted since a husband will have only one legal wife and the others will just live with and be a part of the larger family. Women are socialized to bear as many children as possible to build up the "celestial family" that will remain together for eternity (Anderson 2010a).

National **Data**

There are an estimated 38,000 breakaway Mormon fundamentalists who practice plural marriage in North America. FLDS is the largest group which was founded in Hildale and Colorado City astride the Utah-Arizona border. There are about 10,000 FLDS members who live in the western U.S. and Canada (Anderson 2010a).

It is often assumed that polygyny in FLDS marriages exists to satisfy the sexual desires of the man, that the women are treated like slaves, and that jealousy among the wives is common. In most polygynous societies, however, polygyny has a political and economic rather than a sexual function. Polygyny, for members of the FLDS, is a means of having many children to produce a celestial family. In other societies, a man with many wives can produce a greater number of children for domestic or farm labor. Wives are not treated like slaves (although women have less status than men in general), all household work is evenly distributed among the wives, and each wife is given her own house or own sleeping quarters. Jealousy is minimal because the husband often has a rotational system for conjugal visits, which ensures that each wife has equal access to sexual encounters.

Polygyny type of marriage involving one husband and two or more wives.

Polyandry type of marriage in which one wife has two or more husbands.

DIVERSITY IN OTHER COUNTRIES

Jacob Zuma, the president of South Africa, has 5 wives, 19 children, and a fiancée. In the West African country of Mali, 43% of women live in polygamous marriages. Infertility is the number one reason that women allow co-wives in their marriages—they are necessary to protect the marriage rather than divorce. Though the majority of the women disapprove of polygyny unions, divorce is not an option (Tabi et al. 2010).

Polyandry The Buddhist Tibetans foster yet another brand of polygamy, referred to as **polyandry**, in which one wife has two or more (up to five) husbands. These husbands, who may be brothers, pool their resources to support one wife. Polyandry is a much less common form of polygamy than polygyny. The major reason for polyandry is economic. A family that cannot afford wives or marriages for each of its sons may find a wife for the eldest son only. Polyandry allows the younger brothers to also have sexual access to the one wife that the family is able to afford.

Chapter 1 Choices in Relationships: An Introduction

Pantagamy   Pantagamy describes a group marriage in which each member of the group is "married" to the others. Pantagamy is a formal arrangement that was practiced in communes (for example, Oneida) of the nineteenth and twentieth centuries. Pantagamy is, of course, illegal in the United States.

Our culture emphasizes monogamous marriage and values stable marriages. One expression of this value is the concern for marriage education (see the Social Policy feature on pages 8–9).

Pantagamy a group marriage in which each member of the group is "married" to the others.

Family

Most people who marry choose to have children and become a family. However, the definition of what constitutes a family is sometimes unclear. This section examines how families are defined, their numerous types, and how marriages and families have changed in the past sixty years.

Call it a clan, call it a network, call it a tribe, call it a family. Whatever you call it, whoever you are, you need one.

Jane Howard, journalist

Definitions of Family

The U.S. Census Bureau defines **family** as a group of two or more people related by blood, marriage, or adoption. This definition has been challenged because it does not include foster families or long-term couples (heterosexual or homosexual) that live together. The answer to "who is family?" is important because access to resources such as health care, social security, and retirement benefits is involved. Cohabitants are typically not viewed as "family" and are not accorded health benefits, social security, and retirement benefits of the partner. Indeed, the "live-in partner" or a partner who is gay (although a long-term significant other) may not be allowed to see the beloved in the hospital, which limits visitation to "family only." Nevertheless, the definition of who counts as family is being challenged. In some cases, families are being defined by function rather than by structure—what is the level of emotional and financial commitment and interdependence? How long have they lived together? Do the partners view themselves as a family?

Family a group of two or more people related by blood, marriage, or adoption.

National **Data**

Eighty six percent of U.S adults say a single parent and child constitute a family; nearly as many (80%) say an unmarried couple living together with a child is a family. Sixty three percent say a gay or lesbian couple raising a child is a family (Pew Research Center 2010a).

Friends sometimes become family. Due to mobility, spouses may live several states away from their respective families. Although they may visit their families for holidays, they often develop close friendships with others on whom they rely locally for emotional and physical support on a daily basis.

Sociologically, a family is defined as a kinship system of all relatives living together or recognized as a social unit, including adopted people. The family is regarded as the basic social institution because of its important functions of procreation and socialization, and because it is found in some form in all societies.

Same-sex couples (for example, Ellen DeGeneres and her partner) certainly define themselves as family. New Hampshire, Massachusetts, Connecticut, Vermont, and Iowa recognize marriages between same-sex individuals. Short of marriage, New Jersey recognizes committed gay relationships as **civil unions** (pair-bonded relationship given legal significance in terms of rights and privileges).

Civil union a pair-bonded relationship given legal significance in terms of rights and privileges.

Social policies are purposive courses of action that individuals or groups of individuals take regarding a particular issue or problem of concern. The U.S. government actively supports and encourages marriage via its National Health Marriage Resource Center (see website listing at end of chapter). Public schools are a major source for socialization of values and skills in American youth. With almost half of new marriages ending in divorce, politicians ask whether social policies designed to educate youth about the realities of marriage might be beneficial in promoting marital quality and stability. Might students profit from education about marriage, before they get married, if such relationship skill training is made mandatory in the school curriculum? Such marriage education courses might occur in a traditional face to face class or by means of a web-based self-directed Internet course. Both have been found to be equally effective over no instruction (Duncan et al. 2009). In addition, 1,409 individuals were exposed to marriage preparation information in four contexts: in a seated class, at a workshop sponsored by a community or church, as premarital counseling, and self-directed. While all contexts were helpful, the seated class and self-directed learning rated more highly than counseling and workshops, independent of gender.

The philosophy behind premarital education is that building a fence at the top of a cliff is preferable to putting an ambulance at the bottom. Over 2,000 public schools nationwide offer a marriage education course. In Florida, all public high school seniors are required to take a marriage and relationship skills course. Kerpelman et al. (2010) examined the effectiveness of a relationship education curriculum in a sample of 1,430 adolescents attending health classes across 39 public high schools. Results revealed that those who took the class, compared to those who did not, could recognize faulty relationship beliefs.

Adler-Baeder et al. (2010) studied 1293 ethnically and economically diverse adults participating in relationship and marriage education (RME) programs and noted that both men and women showed gains (e.g. couple functioning/confidence in one's relationship) from such exposure. Women reported the greatest change if their partners attended and persons who were married showed more gains that those who were single. Morris et al. (2011) compared the backgrounds of participants who do and do not attend marriage education workshops and found that those who do attend report lower levels of self-esteem, marital communication quality, marital commitment, marital satisfaction and increased levels of marital conflict. Hence, these programs seem to be reaching the intended audience. Finally, Halpern-Meekin (2011) compared mandated and voluntary programs and found greater gains by students in mandated programs and students from two parent homes. Students from severely economically disadvantaged homes showed no gains. The federal government has a vested interest in marriage education programs. One motivation is economic. The estimated cost for divorce to U.S. society is almost $35 billion (Wilmoth et al. 2010) because divorce often plunges individuals into poverty, making them dependent on public resources. Hence, to the degree that people enter marriages that turn out to be stable, there is greater economic stability for the family and less drain on social services in the U.S. for single-parent mothers.

Another motivation for concern about marriage education courses in the public schools is control of the content. The School Textbook Marriage Protection Act, considered in the House education committee in Arkansas, would require that marriage textbooks define marriage as between one man and one woman. While the bill was defeated in the senate, its existence emphasizes the political incentives behind interest in these courses and the desire to normalize/influence youth about marriage (and exclude/deny same sex relationships).

There is also opposition to marriage education programs. Opponents question using school time for relationship courses. Teachers are viewed as already overworked, and an

Domestic partnerships
relationships in which cohabiting individuals are given some kind of official recognition by a city or corporation so as to receive partner benefits (for example, health insurance).

Although other states typically do not recognize same-sex marriages or civil unions (and thus people moving from these states to another state lose the privileges associated with marriage), over twenty-four cities and countries (including Canada) recognize some form of domestic partnership. **Domestic partnerships** are relationships in which cohabiting individuals are given some kind of official recognition by a city or corporation so as to receive partner benefits (for example, health insurance). Disney employees recognizes domestic partnerships.

Some individuals view their pets as part of their family. In a Harris Poll (2007) survey of 2,455 adults, 88% regarded their pets as family members—more women (93%) than men (84%), and more dog owners (93%) than cat

additional course on marriage seems to press the system to the breaking point. In addition, some teachers lack the training to provide relationship courses. Although training teachers would stretch already-thin budgets, many schools already have programs in family and consumer sciences, and teachers in these programs are trained in teaching about marriage and the family. A related concern with teaching about marriage and the family in high school is the fear on the part of some parents that the course content may be too liberal. Some parents who oppose teaching sex education in the public schools fear that such courses lead to increased sexual activity.

Marriage education is also offered in various communities and through various churches. Wilmont et al. (2010) observed that 86% of weddings are performed by clergy, who often provide marriage education prior to the wedding. Such clergy may take various courses in counseling and use standardized marriage material in their programs.

Regardless of the source—school, communication, or clergy—marriage education has positive outcomes at any time in a couple's relationship. Stanley et al. (2010) conducted a study on married U.S. Army couples who were assigned to either PREP for Strong Bonds (n = 248) delivered by U.S. Army chaplains or to a no-treatment control group (n = 228). One year after the intervention, couples who received PREP for Strong Bonds had one-third the rate of divorce of the control group. Specifically, 6.20% of the control group divorced, while 2.03% of the intervention group divorced. These findings suggest that couple education can reduce the risk of divorce. Other researchers have confirmed the value of marriage education programs on participants (Calligas et al. 2010).

Your Opinion?

1. To what degree do you believe marriage education belongs in the public school system?

2. What evidence reveals that marriage education is effective?

3. Should marriage be encouraged by the federal government?

Sources

Adler-Baeder, F., A. Bradford, E. Skuban, M. Lucier-Greer, S. Ketring, and T. Smith. 2010. Demographic predictors of relationship and marriage education participants' pre- and post-program relational and individual functioning. *Journal of Couple & Relationship Therapy* 9: 113–132.

Calligas, A., F. Adler-Baeder, M. Keiley, T. Smith, and S. Ketring. 2010. Examining change in parenting dimensions in relation to change in couple dimensions. Poster, National Council on Family Relations annual meeting, November 3–5. Minneapolis, MN.

Duncan, S. F., G. R. Childs, and J. H. Larson. 2010. Perceived helpfulness of four different types of marriage preparation interventions. *Family Relations* 59: 623–636.

Duncan, S., A. Steed, and C. M. Needham. 2009. A comparison evaluation study of web-based and traditional marriage and relationship education. *Journal of Couple & Relationship Therapy* 8: 162–180.

Halpern-Meekin. 2011. High school relationship and marriage education: A comparison of mandated and self-selected treatment detail. *Journal of Family Issues* 32: 394–419.

Kerpelman, J., J. F. Pittman, F. Adler-Baeder, K. Stringer, S. Eryigit, H. Cadely, and M. Harrell-Levy. 2010. What adolescents bring to and learn from relationship education classes: Does social address matter? *Journal of Couple & Relationship Therapy* 9: 95–112.

Morris, M. L., H. McMillan, S. D. Duncan, and J. Larson. 2011. Who will attend? Characteristics of couples and individuals in marriage education. *Marriage & Family* 47: 1–22.

Stanley, S. M., E. S. Allen, H. J. Markman, G. K. Rhoades, and D. L. Prentice. 2010. Decreasing divorce in U.S. Army Couples: Results from a randomized controlled trial using PREP for strong bonds. *Journal of Couple & Relationship Therapy* 9: 149–160.

Wilmont, J. D., S. L. Smyser, T. Staier, and T. M. Phillips. 2010. Influence of a statewide marriage initiative on clergy involvement in marriage preparation. *Marriage and Family Review* 46: 278–299.

owners (89%). Weinstein and Alexander (2010) found that cat owners report that they have personalities similar to their cats. Mooney et al. (2010) studied attachment as the benchmark of the human-companion animal bond and found that, in a sample of 250 adults who had an animal, women, those not married, those over 65, and dog-only owners reported the greatest attachment.

Examples of treating pets like children include buying presents for the pet at birthdays, buying "clothes" for one's pet, and leaving money in one's will for the care of a pet. Some pet owners buy accident insurance—Progressive© insurance covers pets. Gregory (2010) noted that pets are now the legal subject of divorce—that parents are granted custody and visitation rights.

This Labrador retriever is part of the family. Recently he ate a six foot dog leash, which required surgery and hospitalization. This "mother" took care of the "baby" during the dog's recovery.

David Knox

Types of Families

There are various types of families.

Family of Origin Also referred to as the **family of orientation**, this is the family into which you were born or the family in which you were reared. It involves you, your parents, and your siblings. When you go to your parents' home for the holidays, you return to your *family of origin*. Your experiences in your family of origin have an impact on your own relationships. Chiu and Busby (2010) found that individuals reared in homes where their parents had a high-quality relationship were likely to have a similar high-quality marriage and to be familisticly oriented. Similarly, Reczek et al. (2010) documented how the relationship with one's parents affects the quality of one's own marriage.

The degree to which your family of origin was open about emotional expression also has an effect on your openness. See the Self-Assessment on Family of Origin Expressiveness Scale to assess the degree to which the family in which you were reared encouraged open emotional expression. Twenty-six undergraduate males and 85 undergraduate females (in the authors' classes) completed the scale with average scores of 79.73 and 85.40, respectively. Since the lowest possible score was 22 (reflecting low expressiveness) and the highest possible score was 110 (reflecting high permissiveness) with a midpoint of 66, both sexes tended to come from emotionally expressive homes, women more than men.

Your family of origin also affects your access to resources. Fairlie et al. (2010) found that teenagers who have access to home computers are 6 to 8 percentage points more likely to graduate from high school than teenagers who do not have home computers. Home computers are associated with graduating from high school in that they make completing homework assignments easier and students doing homework on computers are not engaging in criminal activitity.

Chapter 1 Choices in Relationships: An Introduction

Directions: The family of origin is the family with which you spent most or all of your childhood years. Each family is unique and has its own ways of doing things. What is important is you respond as honestly as you can. Apply the following statements to your family or origin.

Key:

1 = Strongly Disagree that it describes my family of origin.
2 = Disagree that it describes my family of origin.
3 = Neutral: sometimes yes, other times no.
4 = Agree that it describes my family of origin.
5 = Strongly Agree that it describes my family of origin.

_____ 1. My family members usually were sensitive to one another's feelings.

_____ 2. In my family, I felt that I could talk things out.

_____ 3. I found it easy in my family to express how I thought and felt.

_____ 4. My attitudes and my feelings frequently were ignored or criticized in my family.

_____ 5. In my family, I felt free to express my own opinions.

_____ 6. The atmosphere in my family was cold and negative.

_____ 7. The members of my family were very receptive to each other's feelings.

_____ 8. I found it difficult to express my own opinions in my family.

_____ 9. I remember my family as being warm and supportive.

_____ 10. My family had an unwritten rule: Don't express your feelings.

_____ 11. My parents encouraged me to express my views openly.

_____ 12. We usually were able to work out conflicts in my family.

_____ 13. Sometimes in my family, I did not have to say anything, but I felt understood.

_____ 14. In my family, certain feelings were not allowed to be expressed.

_____ 15. Conflicts in my family seldom got resolved.

_____ 16. The atmosphere in my family was usually unpleasant.

_____ 17. In my family, no one cared about the feelings of other family members.

_____ 18. My parents discouraged us from expressing views different from theirs.

_____ 19. In my family, people took responsibility for what they did.

_____ 20. My parents encouraged family members to listen to one another.

_____ 21. Mealtimes in my home usually were friendly and pleasant.

_____ 22. My parents openly admitted it when they were wrong.

The new 22 item measure "FAMILY-OF-ORIGIN EXPRESSIVE ATMOSPHERE SCALE"

13 Positive items are: 1, 2, 3, 5, 7, 9, 11, 12, 13, 19, 20, 21, 22
9 Reversed items are: 4, 6, 8, 10, 14, 15, 16, 17, 18
Change the reversed items (1=5, 5=1, etc.) and add all 22 items for the score.
High score means expressive atmosphere.
Low score means lack of expressive atmosphere.

Source

The Family of Origin Expressiveness Scale by Yelsma, P., Hovestadt, A. J., Anderson, W. T., & Nilsson, J. E. (2000). Family of origin expressiveness: Measurement, meaning, and relationship to alexithymia. *Journal of Marital and Family Therapy* 26: 353–363. Used by permission of Blackwell Publishing Co.

Siblings represent an important part of one's family of origin. Meinhold et al. (2006) noted that the relationship with one's siblings, particularly sister-sister relationships, represent the most enduring relationship in a person's lifetime. Sisters who lived near one another and who did not have children reported the greatest amount of intimacy and contact. Myers (2011) studied 124 adults who provided an explanation for why they maintained their relationship with one of their siblings. "We are family" was one of the major categories of reasons.

Family of Procreation   The **family of procreation** represents the family that you will begin should you marry and have children. Of U.S. citizens living in the United States, 95.83% marry and establish their own family of procreation (*Statistical Abstract of the United States 2011*, Table 34). Across the life cycle, individuals move from the family of orientation to the family of procreation.

Nuclear Family   The **nuclear family** refers to either a family of origin or a family of procreation. In practice, this means that your nuclear family consists of you, your parents, and your siblings; or you, your spouse, and your children.

To our children, we give two things, one is roots, the other is wings.

Unknown

Family of procreation the family a person begins typically by getting married and having children.

Nuclear family consists of you, your parents, and your siblings or you, your spouse, and your children.

The relationship with one's sibling, particularly sister-sister (as in this photo) is likely to be the most enduring of all relationships, longer than with one's parents, spouse, or children.

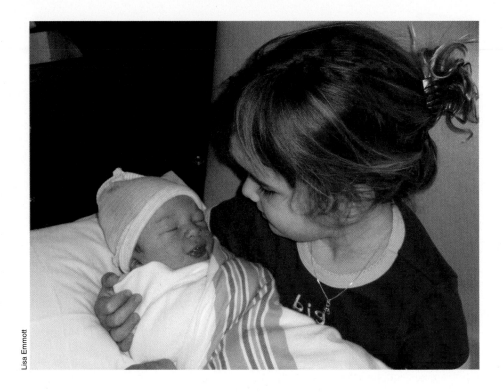

Lisa Emmott

Generally, one-parent households are not referred to as nuclear families. They are binuclear families if both parents are involved in the child's life, or single-parent families if one parent is involved in the child's life and the other parent is totally out of the picture.

Traditional, Modern, and Postmodern Family
Silverstein and Auerbach (2005) distinguished between three central concepts of the family. The traditional family is the two-parent nuclear family, with the husband as breadwinner and wife as homemaker. The modern family is the dual-earner family, where both spouses work outside the home. Postmodern families represent a departure from these models, such as lesbian or gay couples and mothers who are single by choice, which emphasizes that a healthy family need not be heterosexual or include two parents.

Binuclear family a family in which the members live in two households.

Binuclear Family   A **binuclear family** is a family in which the members live in two separate households. This family type is created when the parents of the children divorce and live separately, setting up two separate units, with the children remaining a part of each unit. Each of these units may also change again when the parents remarry and bring additional children into the respective units **(blended family)**. Hence, the children may go from a nuclear family with both parents, to a binuclear unit with parents living in separate homes, to a blended family when parents remarry and bring additional children into the respective units.

Blended family a family created when two individuals marry and at least one of them brings a child or children from a previous relationship or marriage. Also referred to as a stepfamily.

Extended family the nuclear family or parts of it plus other relatives such as grandparents, aunts, uncles, and cousins.

Extended Family   The **extended family** includes not only the nuclear family (or parts of it) but other relatives as well. These relatives include grandparents, aunts, uncles, and cousins. An example of an extended family living together would be a husband and wife, their children, and the husband's parents (the children's grandparents). The extended family is particularly important for African-American married couples. Marks et al. (2010) observed that the willingness to help extended kin financially makes it very difficult for the African-American married couple to get ahead financially. "Knocks of need" come often and African-American couples often respond as generously as they can.

When Families Are Destroyed— The Australian Aboriginal Example

In Australia, between 1885 and 1969, between 50,000 and 100,000 half caste (one white parent) Aboriginal children were taken by force from their parents by the Australian police. The rationale by the white society was that it wanted to convert these children to Christianity and to destroy their Aboriginal culture which was viewed as primitive and without value. The children walked or were taken by camel hundreds of miles away from their parents to church missions.

Bob Randall (2008) is one of the children who was taken by force from his parents at age 7, never to see them again. Of his experience, he wrote,

> Instead of the wide open spaces of my desert home, we were housed in corrugated iron dormitories with rows and rows of bunk beds. After dinner we were bathed by the older women, put in clothing they called pajamas, and then tucked into one of the iron beds between the sheets. This was a horrible experience for me. I couldn't stand the feel of the cloth touching my skin (p. 35).

The Australian government subsequently apologized for the laws and policies of successive parliaments and governments that inflicted profound grief, suffering, and loss on the Aborigines. He noted that the Aborigines continue to be marginalized and nothing has been done to compensate for the horror of taking children from their families.

Bob Randall is an Australian Aboriginal and was taken at age 7 from his parents by the government and brought up in missionary camps. He never saw his parents again.

David Knox

Differences between Marriage and Family

The concepts of marriage and the family are often used in tandem. Marriage can be thought of as a set of social processes that lead to the establishment of family. Indeed, every society or culture has mechanisms (from "free" dating to arranged marriages) of guiding their youth into permanent emotionally, legally, or socially bonded heterosexual relationships that are designed to result in procreation and care of offspring. Although the concepts of marriage and the family are closely related, they are distinct. Sociologist Dr. Lee Axelson identified some of these differences in Table 1.2.

Changes in Marriage and the Family in the Last Sixty Years

Enormous changes have occurred in marriage and the family since the 1950s. Among these changes, divorce has replaced death as the endpoint for the majority of marriages, marriage and intimate relations have become legitimate objects of scientific study, feminism and changes in gender roles in marriage have risen, and remarriages have declined (Amato et al. 2007). Other changes include a delay in age at marriage, increased acceptance of singlehood, cohabitation, and childfree marriages. Even the definition of what constitutes a family is being revised, with some emphasizing that durable emotional bonds between individuals is the core of "family," whereas others insist on a more legalistic view, emphasizing connections by blood marriage or adoption mechanisms. Table 1.3 reflects some of the changes from the 1950s until now.

In spite of the persistent and dramatic changes in marriage and the family, marriage and the family continue to be resilient. Using this marriage-resilience perspective, changes in the institution of marriage are not viewed negatively nor are they indicative that marriage is in a state of decline. Indeed, these changes are thought to have "few negative consequences for adults, children, or the wider society" (Amato et al. 2007, p. 6).

TABLE 1.2 Differences between Marriage and the Family in the United States	
Marriage	**Family**
Usually initiated by a formal ceremony	Formal ceremony not essential
Involves two people	Usually involves more than two people
Ages of the individuals tend to be similar	Individuals represent more than one generation
Individuals usually choose each other	Members are born or adopted into the family
Ends when spouse dies or is divorced	Continues beyond the life of the individual
Sex between spouses is expected and approved	Sex between near kin is neither expected nor approved
Requires a license	No license needed to become a parent
Procreation expected	Consequence of procreation
Spouses are focused on each other	Focus changes with addition of children
Spouses can voluntarily withdraw from marriage	Parents cannot divorce themselves from obligations via divorce to children
Money in unit is spent on the couple	Money is used for the needs of children
Recreation revolves around adults	Recreation revolves around children

Reprinted by permission of Dr. Lee Axelson.

TABLE 1.3 Changes in Marriages and Families—1950 and 2012		
	1950	**2012**
Family Relationship Values	Strong values for marriage and the family. Individuals who wanted to remain single or childless were considered deviant, even pathological. Husband and wife should not be separated by jobs or careers.	Individuals who remain single or childfree experience social understanding and sometimes encouragement. Single and childfree people are no longer considered deviant or pathological but are seen as self-actuating individuals with strong job or career commitments. Husbands and wives can be separated for reasons of job or career and live in a commuter marriage. Married women in large numbers have left the role of full-time mother and housewife to join the labor market.
Gender Roles	Rigid gender roles, with men dominant and earning income while wives stay home, taking care of children.	Egalitarian gender roles with both spouses earning income. Greater involvement of men in fatherhood.
Sexual Values	Marriage was regarded as the only appropriate context for intercourse in middle-class America. Living together was unacceptable, and a child born out of wedlock was stigmatized. Virginity was sometimes exchanged for marital commitment.	For many, concerns about safer sex have taken precedence over the marital context for sex. Virginity is no longer exchanged for anything. Living together is regarded as not only acceptable but sometimes preferable to marriage. For some, unmarried single parenthood is regarded as a lifestyle option. Hooking up is new courtship norm.
Homogamous Mating	Strong social pressure existed to date and marry within one's own racial, ethnic, religious, and social class group. Emotional and legal attachments were heavily influenced by obligation to parents and kin.	Dating and mating have become more heterogamous, with more freedom to select a partner outside one's own racial, ethnic, religious, and social class group. Attachments are more often by choice.
Cultural Silence on Intimate Relationships	Intimate relationships were not an appropriate subject for the media.	Talk shows, interviews, and magazine surveys are open about sexuality and relationships behind closed doors.
Divorce	Society strongly disapproved of divorce. Familistic values encouraged spouses to stay married for the children. Strong legal constraints kept couples together. Marriage was forever.	Divorce has replaced death as the endpoint of a majority of marriages. Less stigma is associated with divorce. Individualistic values lead spouses to seek personal happiness. No-fault divorce allows for easy divorce. Marriage is tenuous. Increasing numbers of children are being reared in single-parent households apart from other relatives.
Familism versus Individualism	Families were focused on the needs of children. Mothers stayed home to ensure that the needs of their children were met. Adult concerns were less important.	Adult agenda of work and recreation has taken on increased importance, with less attention being given to children. Children are viewed as more sophisticated and capable of thinking as adults, which frees adults to pursue their own interests. Day care is used regularly.
Homosexuality	Same-sex emotional and sexual relationships were a culturally hidden phenomenon. Gay relationships were not socially recognized.	Gay relationships are increasingly a culturally open phenomenon. Some definitions of the family include same-sex partners. Domestic partnerships are increasingly given legal status in some states. Same-sex marriage is a hot social and political issue. More states legalizing same-sex marriage. "Don't ask, don't tell" repealed.
Scientific Scrutiny	Aside from Kinsey, few studies were conducted on intimate relationships.	Acceptance of scientific study of marriage and intimate relationships.
Family Housing	Husbands and wives lived in same house.	Husbands and wives may "live apart together" (LAT), which means that, although they are emotionally and economically connected, they (by choice) maintain two households, houses, condos, or apartments. They may be separated for reasons of career, or mutually desire the freedom and independence of having a separate domicile.

Choices in Relationships—View of the Text

Whatever your relationship goal, in this text we encourage a proactive approach of taking charge of your life and making wise relationship choices. The World Health Organization defines *health* as a state of complete physical, mental, and social well-being, and not merely the absence of disease or infirmity. This definition underscores the importance of social relationships as an important element in our individual and national health. Making the right choices in our relationships, including marriage and family relationships, is critical to our health, happiness, and sense of well-being. Our times of greatest elation and sadness are in reference to our love relationships.

The central theme of this text is choices in relationships. Although we have many such choices to make, among the most important in our society are whether to marry, whom to marry, when to marry, whether to have children, whether to remain emotionally and sexually faithful to one's partner, and whether to use a condom. Though structural and cultural influences are operative, a choices framework emphasizes that individuals have some control over their relationship destiny by making deliberate choices to initiate, respond to, nurture, or terminate intimate relationships.

When Do Two People Define Themselves as a Couple?

Chaney and Marsh (2009) interviewed 62 married and 60 cohabiting couples to find out when they first identified themselves as a couple. There were four "markers": relationship events, affection/sex, having or rearing children, and time and money.

1. Relationship events included a specific event such as visiting the parents of one's partner, becoming engaged, or moving in together.

2. Affection/sexual events such as the first time the couple had sex. Losing one's virginity was a salient event.

3. Children—becoming pregnant, having a child together, or the first time the partner assumed a parenting role.

4. Time/money—spending a lot of time together, sharing funds, or exchanging financial support.

The Applying Social Research feature (on pages 18–19) for this chapter reveals how partners in an ongoing relationship go about assessing the degree to which the partner is committed.

Facts about Choices in Relationships

The facts to keep in mind when making relationship choices include the following.

> *If you don't design your own life plan, chances are you'll fall into someone else's plan. And guess what they have planned for you? Not much.*
>
> Jim Rohn, business philosopher

Not to Decide Is to Decide Not making a decision is a decision by default. If you are sexually active and decide not to use a condom, you have made a decision to increase your risk for contracting a sexually transmissible infection, including HIV. If you don't make a deliberate choice to end a relationship that is unfulfilling, abusive, or going nowhere, you have made a choice to continue in that relationship and have little chance of getting into a more positive and satisfying relationship. If you don't make a decision to be faithful to your partner, you have made a decision to be vulnerable to cheating.

Some Choices Require Correction Some of our choices, although appearing correct at the time that we make them, turn out to be disasters. Once we realize that a choice is having consistently negative consequences, we need to stop defending the old choice, reverse the position, make new choices, and move for-

What if You Have Made a Commitment to Marry but Feel it is a Mistake?

We know three individuals who reported the following: "On my wedding day, I knew it was a mistake to marry this person." Although all had their own reasons for going through with marrying, the basic reason was social pressure. All three are now divorced and clearly regret marrying. The take-home message is to "listen to your senses" and to act accordingly. To avoid acting on the feelings that the relationship is doomed may be to delay the inevitable. The price of ending a marriage, particularly with children, is much higher than ending a relationship before marriage. If your moving toward marriage with a particularly person does not "feel" right, stop the process.

ward. Otherwise, one remains consistently locked into continued negative outcomes of "bad" choices. For example, choosing a partner who was loving and kind but who turns out to be abusive and dangerous requires correcting that choice. To stay in the abusive relationship will have predictable disastrous consequences. To make the decision to disengage and to move on opens the opportunity for a loving relationship with another partner. In the meantime, living alone may be a better alternative than living in a relationship in which you are abused and may end up dead. Other examples of making corrections involve ending dead or loveless relationships (perhaps after investing time and effort to improve the relationship or love feelings), changing jobs or career, and changing friends.

The greatest mistake you can make in life is to be continually fearing you will make one.

Elbert Hubbard, writer

Choices Involve Trade-Offs   By making one choice, you relinquish others. Every relationship choice you make will have a downside and an upside. If you decide to stay in a relationship that becomes a long-distance relationship, you are continuing involvement in a relationship that is obviously important to you. However, you may spend a lot of time alone and wonder if you made the right decision to continue the relationship. If you decide to marry, you will give up your freedom to pursue other emotional and/or sexual relationships, and you will also give up some of your control over how you spend your money—but you may also get a wonderful companion with whom to share life.

The doors we open and close each day decide the lives we live.

Flora Whittenmore, author

Choices Include Selecting a Positive or Negative View   As Thomas Edison progressed toward inventing the light bulb, he said, "I have not failed. I have found ten thousand ways that won't work." In spite of an unfortunate event in your life, you can choose to see the bright side. Regardless of your circumstances, you can choose to view a situation in positive terms. A breakup with a partner you have loved can be viewed as the end of your happiness or an opportunity to become involved in a new, more fulfilling relationship. The discovery of your partner cheating on you can be viewed as the end of the relationship or an opportunity to examine your relationship, to open up communication channels with your partner, and to develop a stronger relationship. Finally, discovering that one is infertile can be viewed as a catastrophe or as a challenge to face adversity with one's partner. One's point of view does make a difference—it is the one thing we have control over.

It's so hard when I have to, and so easy when I want to.

Annie Gottlier, writer

Choices Involve Different Decision-Making Styles   Allen et al. (2008a) identified four patterns in the decision-making process of 148 college students. These patterns and the percentage using each pattern included

For dating couples, "the talk" is culturally understood to mean a discussion whereby both parties reveal their feelings about each other and their commitment to the future together. Typically, one partner feels a greater need to clarify the future and instigates "the talk." The goal of "the talk" is to confirm that the partner is interested in and committed to a future. This study examined "the talk" in terms of how long partners are involved before they have "the talk," specific words/strategies used in having "the talk," and the context (e.g. during sex?, after sex?, while watching TV?, during dinner?) of "the talk." Other research questions included how the partner responded and the effect of "the talk" on the couple's relationship.

Sample

Data for the study was based on a sample of 211 undergraduate student volunteers at a large southeastern university who completed a 15 item in-class questionnaire. A majority of respondents were female (77.7%, with 22.3% male) and white (78.0%, with 11.4% African American).

Context of "The Talk"

Mealtime was the context in which "the talk" was most often initiated. Over thirty percent (30.5%) of the respondents reported that they had "the talk" during a meal. Other contexts were "after sexual intimacy" (16.4%), "before sexual intimacy" (11.3%), and "while we were on a trip" (10.0%). Alcohol was usually not involved. Three percent of those initiating "the talk" said that they had been drinking alcohol; 0.6% reported that their partner was drinking when they initiated "the talk." In regard to being anxious about initiating "the talk," on a ten point scale with 1 being minimal anxiety and 10 indicting extreme anxiety, the mean level of anxiety was 4.6. Hence the respondents were moderately anxious when they initiated "the talk."

Strategies involved in "The Talk"

Various strategies were used to initiate the talk and assess the level of commitment of the partner. The top ten are presented below.

Direct Question about Future (30%) – "What do you see as far as the future of this relationship? Where do you plan for this to go?"

Questioning of Motives (15%) – "What do you want out of this relationship?"

Direct Question about Marriage (8%) – "I asked if he ever saw a future in us and if he ever thought we would get married."

Assessment of Level of Interest in Relationship (7%) – "I just asked how serious he was about this relationship."

Assumption of Marriage (7%) – "We would just randomly be talking and say things like, 'when we get married.'

Modeling (6%) – "I mentioned to my partner what my feelings were towards him and looked to see how he responded to what I said. He told me he actually felt the same way."

Question and Evaluate (6%) – "Asked if there was a future for us and we talked about the pros and cons."

Assessment of View of Partner (5%) – "Am I your soul mate?"

Soft Cotton Approach (4%) – "I mentioned something about the future without putting any pressure on the situation. Tried to avoid awkwardness and help him feel comfortable."

> *The most decisive actions in our life—I mean those that are most likely to decide the whole course of our future—are, more often than not, unconsidered.*
>
> Andre Gide, French author

(1) "I am in control" (45%), (2) "I am experimenting and learning" (33%), (3) "I am struggling but growing" (14%), and (4) "I have been irresponsible" (3%). Of those who reported that they were in control, about a third said that they were "taking it slow," and about 11% reported that they were "waiting it out." Men were more likely to report that they were "in control." Hence, these college students could conceptualize their decision-making style; they knew what they were doing. Of note, only 3% labeled themselves as being irresponsible.

Choices Produce Ambivalence Choosing among options and trade-offs often creates ambivalence—conflicting feelings that produce uncertainty or indecisiveness as to a course of action to take. There are two forms of ambivalence: sequential and simultaneous. In sequential ambivalence, the individual experiences one wish and then another. For example, a person may vacillate between wanting to stay in a less-than-fulfilling relationship or to end it. In simultaneous

Ultimatum (4%) – "I just asked what he wanted out of this relationship and if it wasn't the same thing I wanted, I would then end the relationship."

Reactions of the Partner to "The Talk"

Reactions by the partner who was asked about the future of the relationship fell into four categories.

Commitment to the future – The most frequent response to "the talk" was a clear statement that he or she wanted a future with the partner. Over half (50.5%) were quick to confirm that they wanted a future with the partner. While the word "marriage" was often not spoken, the assumption on the part of both partners was that the partners would eventually get married.

Uncertain – The second most frequent response (32.3%) to "the talk" was uncertainty. The partner simply said that he or she did not know about the future and kept the partner and the relationship in limbo.

No future – The third most frequent response (9%) was a "subtle revelation that there was no future." The partner was not brutal but implied that the relationship would not go anywhere.

Other responses – Aside from "yes, "not sure," and "no," other responses included "made a joke out of it" (4%) or was "brutal in making clear that there was no future" (1%).

Relationship Effects of Having "The Talk"

The effect of having "the talk" on the relationship was variable. Over a third (34.7%) of the respondents reported that the effect was "to move us closer and forward." Fourteen percent ended the relationship and 11.7% said that their relationship became strained since they had different feelings about the future. Another 10% said they were still talking about the issue.

Summary and Implications

This study reported the strategies and outcomes of individuals in a romantic relationship who initiated "the talk" with their partner about the future of the relationship. Most felt moderate anxiety about bringing up the future and did so during a meal. Their primary strategy was to be direct—to ask the partner if he or she saw a future for the relationship. While marriage was implicated, it was rarely made explicit.

Over half of the partners were quick to confirm that, indeed, they did see a future for the couple. Another third gave no indication about the future—they were uncertain. And almost 10% of the partners said that there was no future.

In evaluating "the talk," over a third of the respondents who initiated the talk reported that it brought them closer together with another 10% saying that they were still talking. Perhaps clarity tends to be preferred over ambiguity. Since about 10% reported that their partner said there was no future, this allowed the person asking the question either to try to change the person's mind or to move on.

Source

Abridged and adapted from Nelms, B. J., D. Knox, and B. Easterling. Poster, "THE RELATIONSHIP TALK": Assessing Partner Commitment. Annual Eastern Sociological Society Meeting, Philadelphia Feb 24–27, 2011.

ambivalence, the person experiences two conflicting wishes at the same time. For example, the individual may feel both the desire to stay with the partner and the desire to break up at the same time. The latter dilemma is reflected in the saying, "You can't live with them, and you can't live without them." Some anxiety about choices is normative and should be embraced.

Most Choices Are Revocable; Some Are Not Most choices can be changed. For example, a person who has chosen to be sexually active with multiple partners can later decide to be monogamous or to abstain from sexual relations. Individuals who have in the past chosen to emphasize career, money, or advancement over marriage and family can choose to prioritize relationships over economic and career-climbing behaviors. People who have been unfaithful in the past can elect to be emotionally and sexually committed to a new partner.

Sometimes it's the smallest decisions that can change your life forever.

Keri Russell, actress

Generation Y children of the baby boomers (typically born between 1979 and 1984).

This is the generation that won't commit to going to a party on Saturday night because something better might come along— someone better might come along.

Shannon Fox, psychotherapist

For you and I are past our dancing days.

William Shakespeare

Once I had made the decision, I didn't worry about it. If I made a wrong decision, I made another one to correct it.

Harry Truman, President

Other choices are less revocable. For example, backing out of the role of spouse is much easier than backing out of the role of parent. Whereas the law permits disengagement from the role of spouse (formal divorce decree), the law ties parents to dependent offspring (for example, through child support). Hence, the decision to have a child is usually irrevocable. Choosing to have unprotected sex can also result in a lifetime of coping with sexually transmitted infections.

Choices of Generation Y Those in **Generation Y** (typically born between 1979 and 1984) are the children of the baby boomers. About 40 million of them, these Generation Yers (also known as the Millennial or Internet Generation) have been the focus of their parents' attention. They have been nurtured, coddled, and scheduled into day-care centers for getting ahead. The result is a generation of high self-esteem, self-absorbed individuals who believe they "are the best." Unlike their parents who believe in paying one's dues, getting credentials, and sacrifice through hard work to achieve economic stability, Generation Yers focus on fun, enjoyment, and flexibility. They might choose a summer job at the beach if it buys a burger and a room with six friends over an internship at IBM that smacks of the corporate America sellout. Generation Yers know college graduates who work at McDonald's, so they may wonder, *why bother?* and instead, seek innovative ways of drifting through life; they may continue to live with their parents, live communally, or get food by "dumpster diving." In effect, they are the generation of immediate gratification; they focus only on the here and now. The need for social security is too far off, and health care is available, so they say, "for free at the local hospital emergency room."

Generation Yers are also relaxed about relationship choices. Rather than pair-bond, they "hang out," "hook up," and "live together." They are in no hurry to find "the one," to marry, or to begin a family. To be sure, not all youth fit this characterization. Some have internalized their parents' values, and are focused on education, credentials, a stable job, a retirement plan, and health care. They may view education as the ticket to a good job and expect their college to provide a credential they can market. However, increasingly, Generation Yers are taking their time getting to the altar and focusing on education, career, and enjoying their freedom in the meantime.

Choices are Influenced by the Stage in the Family Life Cycle The choices a person makes tend to be individualistic or familistic, depending on the stage of the family life cycle that the person is in. Before marriage, individualism characterizes the thinking and choices of most individuals. Individuals need only be concerned with their own needs. Most people delay marriage in favor of completing school, becoming established in a career, and enjoying the freedom of singlehood.

Once married, and particularly after having children, the person's familistic values and choices ensue as the needs of a spouse and children begin to influence. Evidence of familistic choices is reflected in the fact that spouses with children are less likely to divorce than spouses without children.

Making Wise Choices Is Facilitated by Learning Decision-Making Skills Choices occur at the individual, couple, and family level. Deciding to transfer to another school or take a job out of state may involve all three levels, whereas the decision to lose weight is more likely to be an individual decision. Regardless of the level, the steps in decision making include setting aside enough time to evaluate the issues involved in making a choice, identifying alternative courses of action, carefully weighing the consequences for each choice, and being attentive to your own inner voice (listen to your senses). The goal of most people is to make relationship choices that result in the most positive and least negative consequences.

TABLE 1.4 "Best" and "Worst" Choices Identified by University Students

Best Choice	Worst Choice
Waiting to have sex until I was older and involved.	Cheating on my partner.
Ending a relationship with someone I did not love.	Getting involved with someone on the rebound.
Insisting on using a condom with a new partner.	Making decisions about sex when drunk.
Ending a relationship with an abusive partner.	Staying in an abusive relationship.
Forgiving my partner and getting over cheating.	Changing schools to be near my partner.
Getting out of a relationship with an alcoholic.	Not going after someone I really wanted.

All of us are proud of some of the choices we have made. We also regret other choices. We asked our students to identify their "best" and "worst" relationship choices (see Table 1.4).

Global, Structural/Cultural, and Media Influences on Choices

Choices are influenced by global, structural/cultural, and media factors. This section reviews the ways in which globalization, social structure, and culture impact choices in relationships. Although a major theme of this book is the importance of taking active control of your life in making relationship choices, it is important to be aware that the social world in which you live restricts and channels such choices. For example, enormous social disapproval for marrying someone of another race is part of the reason that most individuals in the United States marry someone of the same race.

Globalization   Families exist in the context of world globalization. Indeed, "globalization is the critical driving force that is fundamentally restricting the social order around the world, and families are at the center of this change" (Trask 2010, p. v). Economic, political, and religious happenings throughout the world affect what happens in your marriage and family in the United States. When the price of oil per barrel increases in the Middle East, gasoline costs more, leaving fewer dollars to spend on other items. When the stock market in Hong Kong drops 500 points, Wall Street reacts, and U.S. stocks drop. The politics of the Middle East (for example, terrorist or nuclear threats from Iran) impact Homeland Security measures so that getting through airport security to board a plane may take longer. The tsunami in Japan put U.S. people/resources on the plane to render aid in their country.

The country in which you live also affects your happiness and well-being. For example, in a study, citizens of thirteen countries were asked to indicate their level of life satisfaction on a scale from 1 (dissatisfied) to 10 (satisfied): citizens in Switzerland averaged 8.3, those in Zimbabwe averaged 3.3, and those in the United States averaged 7.4 (Veenhoven 2007). The Internet, CNN, and mass communications provide global awareness so that families are no longer isolated units.

Social Structure   The social structure of a society consists of institutions, social groups, statuses, and roles.

   1. *Institutions.* The largest elements of society are social **institutions**, which may be defined as established and enduring patterns of social relationships. The institution of the family in the U.S. is definitely biased toward heterosexism. For example, there is no federal recognition of gay marriage. Indeed, in the face of

Experience is not what happens to someone but what that person does with what happens to them.

Aldous Huxley, philosopher

Institution established and enduring patterns of social relationships (e.g. the family).

traditional terms in marriage such as husband and wife, gay and lesbian individuals struggle with which term to use (lover, partner, significant other, companion, spouse, life mate, and "wife").

In addition to the family, major institutions of society include the economy, education, and religion. Institutions affect individual decision making. For example, you live in a capitalistic society where economic security is valued—the number-one value held by college students (Pryor et al. 2011). In effect, the more time you spend focused on obtaining money, the less time you have for relationships. You are now involved in the educational institution that will impact your choice of a mate (college-educated people tend to select and marry one another). Religion also affects sexual and relationship choices (for example, religion may result in delaying first intercourse, not using a condom, or marrying someone of the same faith). The family is a universal institution. Spouses who "believe in the institution of the family" are highly committed to maintaining their marriage and do not regard divorce as an option.

2. *Social groups.* Institutions are made up of social groups, defined as two or more people who share a common identity, interact, and form a social relationship. Most individuals spend their day going between social groups. You may awaken in the context (social group) of a roommate, partner, or spouse. From there you go to class with other students, lunch with friends, work with the boss, and talk on the phone to your parents. So, within twenty-four hours you have been in at least five social groups. These social groups have varying influences on your choices. Your roommate influences what other people you can have in your room for how long, your friends may want to eat at a particular place, your boss will assign you certain duties, and your parents may want you to come home for the weekend.

> *Students sometimes argue that they—as individuals—make choices. In reality, the choices they make are only the ones the social context permits. Iranian women have almost no choice to be "childfree." Similarly, a Mormon woman married to a Mormon man in the Mormon Church has almost no choice to be "childfree." Although the Mormon woman will tell you she is "choosing" to have seven children, her illusion of choice is context controlled. Change her context (so that she is no longer a member of the Mormon Church and is married to a man who wants to be childfree) and she is now free to be childfree (but only because her context has changed) (Zusman 2011).*

The Blind Side (a 2009 film for which Sandra Bullock won the Best Actress Oscar) illustrates the enormous impact of family on one's choices. A major question of the movie is whether Mike was "free" to select any college to play football for or whether he was socialized to "select" the college where his adoptive parents went as undergraduates. Further, Mike never considered being an ice hockey player since he grew up in the American South were ice hockey was not played. Clearly, one's family has an enormous influence on one's choices.

Your interpersonal choices are influenced mostly by your partner and peers (for example, your sexual values, use of condoms, and the amount of alcohol you consume). Thus, selecting a partner and peers carefully is important. As Falstaff, one of Shakespeare's characters, said, "Company, villainous company, hath been the spoil of me."

The age of the partner you select to become romantically involved with is influenced by social context. If you are a woman, your parents and peers will probably approve of you dating and marrying "someone a little older." Likewise, if you are a man, your parents and peers will probably approve of you dating and marrying someone "a little younger."

As a first year female student, you will be in demand by freshmen, sophomore, and senior men. But because of the **mating gradient** which gives social approval to men who seek out younger, less educated,

Child marriage marriages in which females as young as 8 to 12 are required by their parents to marry an older man.

Mating gradient norms which give social approval to men who seek out younger, less educated, less financially secure women and vice versa.

DIVERSITY IN OTHER COUNTRIES

In countries such as Nepal and Afghanistain, **child marriage** occurs whereby young females (ages 8 to 12) are required to marry an older man identified by their parents. Suicide is the only alternative "choice" for these children.

Chapter 1 Choices in Relationships: An Introduction

less financially secure women, as a senior female you are no longer "younger" and only senior men are typically available. Hence, your pool of eligible men shrinks each year.

As a first year male student, just the opposite occurs. During your first year, only other freshmen women are potential partners (and of course, women from your high school). But as you become a senior, all women on campus become options. In effect, your pool of eligibles increases every year. In effect, social structure (your place in student class rank) affects the number of partners you can date, not your "personality."

Social groups may be categorized as primary or secondary. **Primary groups**, which tend to involve small numbers of individuals, are characterized by interaction that is intimate and informal. Parents are members of one's primary group and may exercise enormous influence over one's mate choice. Lehmiller and Agnew (2007) found that romantically involved couples that perceived that their parents and friends did not approve of their relationship were more likely to break up than couples that viewed approval from these social networks.

Although parents may register direct disapproval, there may also be a less direct way in how they influence the mate choices of their children. By living in the "right" neighborhood, joining a particular church, and enrolling their children in a college or university, parents influence the context in which their children are likely to meet and select a "suitable" marriage partner.

In contrast to primary groups, **secondary groups** may involve small or large numbers of individuals and are characterized by interaction that is impersonal and formal. Being in a context of classmates, coworkers, or fellow students in the library are examples of secondary groups. Members of secondary groups have much less influence over one's relationship choices than members of one's primary groups.

Most people regard primary groups as crucial for their personal happiness and feel adrift if they have only secondary relationships. Indeed, in the absence of close primary ties, they may seek meaning in secondary group relationships. Comedian George Carlin said that his "fans were his family" because he was in a different town performing more than half the weekends a year, implying he had no "real" family.

3. Statuses. Just as institutions consist of social groups, social groups consist of statuses. A status is a position a person occupies within a social group. The statuses we occupy largely define our social identity. The statuses in a family may consist of mother, father, child, sibling, stepparent, and so on. In discussing family issues, we refer to statuses such as teenager, cohabitant, and spouse. Statuses are relevant to choices in that many choices can significantly change one's status. Making decisions that change one's status from single person to spouse to divorced person can influence how people feel about themselves and how others treat them.

4. Roles. Every status is associated with many roles, or sets of rights, obligations, and expectations associated with a status. Our social statuses identify who we are; our roles identify what we are expected to do. Roles guide our behavior and allow us to predict the behavior of others. Spouses adopt a set of obligations and expectations associated with their status. By doing so, they are better able to influence and predict each other's behavior.

Because individuals occupy a number of statuses and roles simultaneously, they may experience role conflict. For example, the role of the parent may conflict with the role of the spouse, employee, or student. If your child needs to be driven to the math tutor, your spouse needs to be picked up at the airport, your employer wants you to work late, and you have a final exam all at the same time, you are experiencing role conflict.

Culture   Just as social structure refers to the parts of society, culture refers to the meanings and ways of living that characterize people in a society. Two central elements of culture are beliefs and values.

Primary groups small numbers of individuals whereby the interaction is intimate and informal.

Secondary groups groups in which the interaction is impersonal and formal.

1. *Beliefs.* Beliefs refer to definitions and explanations about what is true. The beliefs of an individual or couple influence the choices they make. Couples that believe that young children flourish best with a full-time parent in the home make different job/child-care decisions than do couples who believe that day care offers opportunities for enrichment. Hall (2010a) noted that individuals have different beliefs about marriage (e.g. successful marriages are destined to succeed or that successful marriages grow naturally) and that these respective views impact choices.

2. *Values.* Values are standards regarding what is good and bad, right and wrong, desirable and undesirable. J. D. Salinger, author of *The Catcher in the Rye*, noted that he felt powerless NOT to buy into society's value for fame. "I'm sick of not having the courage to be an absolute nobody. I'm sick of myself and everybody else that wants to make some kind of a splash" (Lacayo 2010, p. 66).

Values influence choices. **Individualism** is making decisions that serve the individual's interests rather than the family's interests (**familism**). "What makes me happy" not "what makes my family happy" is the focus of the individualist. Americans are characteristically individualistic, whereas Hispanics are characteristically familistic. Cherlin (2009) noted a paradox in American values—they value marriage and abhor divorce, but they are also individualistic, which ensures a high divorce rate. Forty percent of 2,922 undergraduates reported that they would divorce their spouse if they fell out of love (Knox and Hall 2010). Allowing one's personal love feelings to dictate the stability of a marriage is a highly individualistic value. Aside from individualism is **collectivism,** which emphasizes doing what is best for the group (not specific to the family group); collectivism is characteristic of traditional Chinese families.

Those who remain single, who live together, who seek a childfree lifestyle, and who divorce are more likely to be operating from an individualistic philosophical perspective than those who marry, do not live together before marriage, rear children, and stay married (a familistic value). Collectivistic values would be illustrated by an Asian child on a swim team who would work for the good of the team, not for personal acclaim.

These elements of social structure and culture play a central role in making interpersonal choices and decisions. One of the goals of this text is to emphasize the influence of social structure and culture on your interpersonal decisions. Sociologists refer to this awareness as the **sociological imagination** (or sociological mindfulness). For example, though most people in the United States assume that they are free to select their own sex partner, this choice (or lack of it) is in fact heavily influenced by structural and cultural factors. Most people date, have sex with, and marry a person of the same racial background. Structural forces influencing race relations include segregation in housing, religion, and education. The fact that African Americans and European Americans live in different neighborhoods, worship in different churches, and often attend different schools makes meeting a person of a different race unlikely. When such encounters occur, prejudices and bias may influence these interactions so that individuals are hardly "free" to act as they choose. Hence, cultural values (transmitted by and through parents and peers) generally do not support or promote mixed racial interaction, relationship formation, and marriage. In a study of college students, DeCuzzi et al. (2006) found that both European Americans and African Americans tended to view their respective groups more positively. Consider the last three relationships in which you were involved, the level of racial similarity, and the structural and cultural influences on those relationships.

Media A Kaiser Family Foundation (2010) study of 8- to 18-year-olds revealed that daily consumption of watching television averages 270 minutes, music = 151 minutes, and playing video games = 73 minutes. Media in all of its forms (television, music, video games, print) influences how we think about and make our relationship choices. For example, media exposure transmits the "acceptability"

Individualism making decisions that serve the individual's rather than the family's interests.

Familism value that decisions are made in reference to what is best for the family.

Collectivism pattern in which one regards group values and goals as more important than one's own values and goals.

Sociological imagination the influence of social structure and culture on interpersonal decisions.

Chapter 1 Choices in Relationships: An Introduction

of various sexual values and norms. At the 2011 Academy Awards, unmarried and pregnant Natalie Portman accepted her award for best actress.

Other Influences on Relationship Choices

Aside from structural and cultural influences on relationship choices, other influences include family of origin (the family in which you were reared), unconscious motivations, habit patterns, individual personality, and previous experiences.

Family of Origin Your family of origin is a major influence on your subsequent choices and relationships. Trotter (2010) found that college students who were from an intact family viewed their parents' relationship more positively than those from a non-intact family. In addition, Cui and Fincham (2010) studied the impact of parental divorce and marital conflict on a young couple's romantic relationship. They found that individuals with divorced parents reported lower marital satisfaction in that they were less committed to their partner. In addition, marital conflict of one's parents was associated with conflict in one's own romantic relationships.

Habits Habit patterns also influence choices. People who are accustomed to and enjoy spending a great deal of time alone may be reluctant to make a commitment to live with people who make demands on their time. A person who has workaholic tendencies is unlikely to allocate enough time to a relationship to make it flourish. Alcohol abuse is associated with a higher number of sexual partners and not using condoms to avoid pregnancy and contraction of sexually transmitted infections.

Personality One's personality (for example, introverted, extroverted; passive, assertive) also influences choices. For example, people who are assertive are more likely than those who are passive to initiate conversations with someone they are attracted to at a party. People who are very quiet and withdrawn may never choose to initiate a conversation even though they are attracted to someone. People with a bipolar disorder who are manic one part of the semester (or relationship) and depressed the other part are likely to make different choices when each phase is operative.

Friends, Relationships, and Life Experiences Busse et al. (2010) observed that adolescent virgins talking with friends about sex was associated with choosing to have one's first sexual experience. In effect, the approval/acceptance of one's friends paves the way for engaging in sexual behavior.

Current and past relationship experiences also influence one's perceptions and choices. Individuals in a current relationship are more likely to hold relativistic sexual values (choose intercourse over abstinence). Similarly, people who have been cheated on in a previous relationship are vulnerable to not trusting a new partner.

The life of Ann Landers illustrates how experiences over time change one's views and choices. After Landers's death, her only child, Margo Howard (2003), noted that her mother was once against premarital intercourse, divorce, and involvement with a married man. However, when Margo told her that unmarried youth were having intercourse, Landers shifted her focus to the use of contraception. When Margo divorced, Landers began to say that ending an unfulfilling marriage is an option. And, when Landers was divorced and fell in love with a married man, she said that you can't control who you fall in love with.

The effect of one relationship on another is illustrated by the life of Chet Baker, legendary trumpet player. His biographer details Baker's innocent love for a woman who dumped him to marry his buddy. ". . . [T]hat woman to whom he

Maybe our girlfriends are our soul mates, and guys are just people to have fun with.

Carrie, *Sex and the City*

had opened his heart and who turned out to be a liar and a phony, permanently marred his attitude toward love. In future relationships he would veer between a need for mothering and paranoid mistrust" (Gavin 2003, p. 28).

While being aware of the numerous influences on one's choices, the theme of this text is to make deliberate relationship choices. See the Personal Choices section for examples of taking charge of your life.

PERSONAL CHOICES

Relationship Choices—Deliberately or by Default?

It is a myth that you can avoid making relationship decisions, because not to make a decision is to make a decision by default. Some examples follow:

- If you don't make a decision to pursue a relationship with a particular person, you have made a decision (by default) to let that person drift out of your life.
- If you do not decide to do the things that are necessary to keep or improve your current relationship, you have made a decision to let the relationship slowly disintegrate.
- If we do not make a decision to be faithful to your partner, you have made a decision to be open to situations and relationships in which you are vulnerable to being unfaithful.
- If you do not make a decision to delay having intercourse early in a new relationship, you have made a decision to have intercourse soon in a relationship (which typically has a negative outcome).
- If you are sexually active and do not make a decision to use birth control or a condom, you have made a decision to expose yourself to risk for pregnancy or a sexually transmitted infection.
- If you do not make a decision to break up with an emotionally/physically abusive partner or spouse, you have made a decision to continue the relationship and incur further damage to your self-esteem and being hurt emotionally/physically.

Throughout the text, as we discuss various relationship choices, consider that you automatically make a choice by being inactive—that not to make a choice is to make one. We encourage a proactive style whereby you make deliberate relationship choices.

Theoretical Frameworks for Viewing Marriage and the Family

Although we emphasize choices in relationships as the framework for viewing marriage and the family, other conceptual theoretical frameworks are helpful in understanding the context of relationship decisions. All **theoretical frameworks** are the same in that they provide a set of interrelated principles designed to explain a particular phenomenon and provide a point of view. In essence, theories are explanations.

Theoretical framework a set of interrelated principles designed to explain a particular phenomenon.

Social Exchange Framework

In a review of 673 empirical articles, exchange theory was most common among those using a theoretical perspective (Taylor and Bagd 2005). The **social exchange framework** views interaction and choices in terms of cost and profit.

The social exchange framework also operates from a premise of **utilitarianism**—that individuals rationally weigh the rewards and costs associated with behavioral choices. Each interaction between spouses, parents, and children can be understood in terms of each individual seeking the most benefits at the least cost so as to have the highest "profit" and avoid a "loss" (White and

Social exchange framework views interaction and choices in terms of cost and profit.

Utilitarianism individuals rationally weigh the rewards and costs associated with behavioral choices.

Klein 2002). Both men and women marry because they perceive more benefits than costs for doing so. Similarly, those who remain single or who divorce perceive fewer benefits and more costs for marriage.

A social exchange view of marital roles emphasizes that spouses negotiate the division of labor on the basis of exchange. For example, a man participates in child care in exchange for his wife earning an income, which relieves him of the total financial responsibility. Social exchange theorists also emphasize that power in relationships is the ability to influence, and avoid being influenced by, the partner.

The various bases of power, such as money, the need for a partner, and brute force, may be expressed in various ways, including withholding resources, decreasing investment in the relationship, and violence.

Family Life Course Development Framework

The **family life course development framework** is the second most frequently used theory in empirical family studies (Taylor and Bagd 2005), and emphasizes the important role transitions of individuals that occur in different periods of life and in different social contexts. For example, young married couples become young parents, which changes the interaction between the couple. As spouses age and retire, their new roles impact not only each of them but also their partners. Cherlin (2010) noted that the family life cycle should be viewed as a basic set of stages through which not all individuals pass and that there is great diversity particularly in regard to race and education (e.g. African-Americans less likely to marry, highly educated less likely to divorce). For example, Pelton and Hertlein (2011) emphasized that voluntary childfree couples do not progress through the traditional family life cycle. Indeed they pass through a different set of stages specific to these couples. The family life course developmental framework has a basis in sociology (for example, role transitions), whereas the **family life cycle** has a basis in psychology, with its emphasis on stages that identify the various developmental tasks family members face across time (for example, marriage, childbearing, preschool, school age children, teenagers, and so on). If developmental tasks at one stage are not accomplished, functioning in subsequent stages will be impaired. For example, one of the developmental tasks of early marriage is to emotionally and financially separate from one's family of origin. If such separation does not take place, independence as individuals and as a couple is impaired.

The family life course development framework may help to identify the choices with which many individuals are confronted throughout life. Each family stage presents choices. For example, never-married people are struggling with educational, career, and partner choices; newly married people are making choices about housing and when to begin a family; soon-to-be-divorced people are making decisions about custody, child support, and division of property; and remarried people are making choices with regard to stepchildren and ex-spouses. Grandparents are making choices about how much time they want to commit to child care of their grandchildren, and widows or widowers are concerned with where to live (with children, a friend, alone, or in a retirement home). A specific example of the effect on one's stage in the family life cycle on choices is the choice Johnny Depp made to make *Pirates of the Caribbean*. He noted that he made the movie specifically in reference to his young children—a movie that they could see (Goodall 2010).

Structure-Function Framework

The **structure-function framework** emphasizes how marriage and family contribute to society. Just as the human body is made up of different parts that work together for the good of the individual, society is made up of different institutions (family, education, economics, and so on) that work together for the good

Family life course development framework the stages and process of how families change over time.

Family life cycle stages that identify the various developmental tasks family members face across time.

Structure-function framework emphasizes how marriage and family contribute to society.

of society. Functionalists view the family as an institution with values, norms, and activities meant to provide stability for the larger society. Such stability is dependent on families performing various functions for society.

First, families serve to replenish society with socialized members. Because our society cannot continue to exist without new members, we must have some way of ensuring a continuing supply. However, just having new members is not enough. We need socialized members—those who can speak our language and know the norms and roles of our society. So-called **feral** (meaning wild, not domesticated) **children** are those who are thought to have been reared by animals. Newton (2002) detailed nine such children, the most famous of which was Peter the Wild Boy found in the Germanic woods at the age of 12 and brought to London in 1726. He could not speak; growling and howling were his modes of expression. He lived until the age of 70 and never learned to talk. Feral children emphasize that social interaction and family context make us human. Girgis et al. (2011) emphasized that "societies rely on families to produce upright people who make for conscientious, law-abiding citizens lessening the demand for governmental policing and social services" (p. 245).

Real-life Genie is a young girl who was discovered in the 1970s who had been kept in isolation in one room in her California home for twelve years by her abusive father (James 2008). She could barely walk and could not talk. Although provided intensive therapy at UCLA and the object of thousands of dollars of funded research, Genie progressed only briefly. Today, she is in her mid fifties, institutionalized, and speechless. Her story illustrates the need for socialization; the legal bond of marriage and the obligation to nurture and socialize offspring help to ensure that this socialization will occur.

Second, marriage and the family promote the emotional stability of the respective spouses. Society cannot provide enough counselors to help us whenever we have problems. Marriage ideally provides in-residence counselors who are loving and caring partners with whom people share their most difficult experiences.

Children also need people to love them and to give them a sense of belonging. This need can be fulfilled in a variety of family contexts (two-parent families, single-parent families, extended families). The affective function of the family is one of its major offerings. No other institution focuses so completely on meeting the emotional needs of its members as marriage and the family.

Third, families provide economic support for their members. Although modern families are no longer self-sufficient economic units, they provide food, shelter, and clothing for their members. One need only consider the homeless in our society to be reminded of this important function of the family.

In addition to the primary functions of replacement, emotional stability, and economic support, other functions of the family include the following:

• *Physical care*—families provide the primary care for their infants, children, and aging parents. Other agencies (neonatal units, day care centers, assisted-living residences) may help, but the family remains the primary and recurring caretaker. Spouses are also concerned about the physical health of one another by encouraging the partner to take medications and to see a health care professional for regular checkups.

• *Regulation of sexual behavior*—spouses are expected to confine their sexual behavior to each other, which reduces the risk of having children who do not have socially and legally bonded parents, and of contracting or spreading sexually transmitted infections.

• *Status placement*—being born in a family provides social placement of the individual in society. One's family of origin largely determines one's social class, religious affiliation, and future occupation. Prince William, the son of Prince Charles and the late Princess Diana, was automatically in the upper class and destined to be in politics by virtue of being born into a political family. His wedding to Catherine Middleton in 2011 was a media frenzy.

Feral children "wild, not domesticated" children thought to have been reared by animals.

- *Social control*—spouses in high-quality, durable marriages provide social control for each other that results in less criminal behavior. Parole boards often note that the best guarantee that a person released from prison will not continue a life of crime is a spouse who expects the partner to get a job and avoid criminal behavior and who reinforces these goals.

Conflict Framework

Conflict framework views individuals in relationships as competing for valuable resources. Conflict theorists recognize that family members have different goals and values that result in conflict. Conflict is inevitable between social groups (such as parents and children). Conflict theory provides a lens through which to view these differences. Whereas functionalists look at family practices as good for the whole, conflict theorists recognize that not all family decisions are good for every member of the family. Indeed, some activities that are good for one member are not good for others. For example, a woman who has devoted her life to staying home and taking care of the children may decide to return to school or to seek full-time employment. This may be a good decision for her personally, but her husband and children may not like it. Similarly, divorce may have a positive outcome for spouses in turmoil but a negative outcome for children, whose standard of living and access to the noncustodial parent are likely to decrease.

Conflict theorists also view conflict not as good or bad but as a natural and normal part of relationships. They regard conflict as necessary for change and growth of individuals, marriages, and families. Cohabitation relationships, marriages, and families all have the potential for conflict. Cohabitants are in conflict about commitment to marry, spouses are in conflict about the division of labor, and parents are in conflict with their children over rules such as curfew, chores, and homework. These three units may also be in conflict with other systems. For example, cohabitants are in conflict with the economic institution for health benefits for their partners. Similarly, employed parents are in conflict with their employers for flexible work hours, maternity or paternity benefits, and day-care or eldercare facilities.

Karl Marx emphasized that conflict emanates from struggles over scarce resources and for power. Though Marxist theorists viewed these sources in terms of the conflict between the owners of the means of production (bourgeoisie) and the workers (proletariat), they are also relevant to conflicts within relationships. The first of these concepts, conflict over scarce resources, reflects the fact that spouses, parents, and children compete for scarce resources such as time, affection, and space. Spouses may fight with each other over how much time should be allocated to one's job, friends, or hobbies. Parents are sometimes in conflict with each other over who will do what housework or child care. Children are in conflict with their parents and with one another over time, affection, what programs to watch on television, and money.

Conflict theory is also helpful in understanding choices in relationships with regard to mate selection and jealousy. Unmarried individuals in search of a partner are in competition with other unmarried individuals for the scarce resources of a desirable mate. Such conflict is particularly evident in the case of older women in competition for men. At age 85 and older, there are twice as many women (3.9 million) as there are men (1.9 million) (*Statistical Abstract of the United States 2011*, Table 8). Jealousy is also sometimes about scarce resources. People fear that their "one and only" will be stolen by someone else who has no partner.

Conflict theorists also emphasize conflict over power in relationships. Premarital partners, spouses, parents, and teenagers also use power to control one another. The reluctance of some courtship partners to make a marital commitment is an expression of wanting to maintain their autonomy because marriage implies a relinquishment of power from each partner to the other. Spouse abuse is sometimes the expression of one partner trying to control the other through fear, intimidation, or force. Divorce may also illustrate control. The person who

Conflict framework the view that individuals in relationships compete for valuable resources.

When we are no longer able to change a situation, we are challenged to change ourselves.

Victor Frankl, *Man's Search for Meaning*

executes the divorce is often the person with the least interest in the relationship. Having the least interest gives that person power in the relationship. Parents and adolescents are also in a continuous struggle over power. Parents attempt to use privileges and resources as power tactics to bring compliance in their adolescent. However, adolescents may use the threat of suicide as their ultimate power ploy to bring their parents under control.

Symbolic Interaction Framework

Symbolic interaction framework views marriages and families as symbolic worlds in which the various members give meaning to each other's behavior.

Symbolic interaction framework views marriages and families as symbolic worlds in which the various members give meaning to each other's behavior. Human behavior can be understood only by the meaning attributed to behavior (White and Klein 2002). Curran et al. (2010) assessed the meaning of marriage for 31 African Americans of different ages and found that the two most common meanings were commitment and love. Herbert Blumer (1969) used the term *symbolic interaction* to refer to the process of interpersonal interaction. Concepts inherent in this framework include the definition of the situation, the looking-glass self, and the self-fulfilling prophecy.

Definition of the Situation   Two people who have just spotted each other at a party are constantly defining the situation and responding to those definitions. Is the glance from the other person (1) an invitation to approach, (2) an approach, or (3) a misinterpretation—the other person was looking at someone behind the person? The definition a person arrives at will affect subsequent interaction.

Looking-Glass Self   The image people have of themselves is a reflection of what other people tell them about themselves (Cooley 1964). People may develop an idea of who they are by the way others act toward them. If no one looks at or speaks to them, they will begin to feel unsettled, according to Cooley. Similarly, family members constantly hold up social mirrors for one another into which the respective members look for definitions of self.

The importance of the family (and other caregivers) as an influence on the development and maintenance of a positive self-concept cannot be overemphasized (Brown et al. 2009). Orson Welles, known especially for his film *Citizen Kane,* once said that he was taught that he was wonderful and that everything he did was perfect. He never suffered from a negative self-concept. Cole Porter, known for creating such memorable songs as "I Get a Kick out of You," had a mother who held up social mirrors offering nothing but praise and adoration. Because children spend their formative years surrounded by their family, the self-concept they develop in that setting is important to their feelings about themselves and their positive interaction with others.

G. H. Mead (1934) believed that people are not passive sponges but evaluate the perceived appraisals of others, accepting some opinions and not others. Although some parents teach their children that they are worthless, they typically overcome this definition when they are loved/valued by others.

It goes far toward thinking a man faithful to let him understand that you think him so; and he that does but suspect I will deceive him gives me a sort of right to do it.

Seneca

Self-Fulfilling Prophecy   Once people define situations and the behaviors in which they are expected to engage, they are able to behave toward one another in predictable ways. Such predictability of behavior also tends to exert influence on subsequent behavior. If you feel that your partner expects you to be faithful, your behavior is likely to conform to these expectations. The expectations thus create a self-fulfilling prophecy.

Symbolic interactionism as a theoretical framework helps to explain various choices in relationships. Individuals who decide to marry have defined their situation as a committed reciprocal love relationship. This choice is supported by the belief that the partners will view each other positively (looking-glass self) and be faithful spouses and cooperative parents (self-fulfilling prophecies).

Chapter 1 Choices in Relationships: An Introduction

Later we will discuss the negative emotion of violence. Turner (2007) noted how symbolic interactionism may be used to explain the dynamics of intense emotions (such as violence). He suggested that extreme violence (which depends on intense negative emotions) has its genesis when negative emotions about the self and identity are repressed. All of this happens through the manipulation of symbols (negative self-reference statements) inside one's head.

Family Systems Framework

Systems theory is the most recent of all the theories for understanding family interaction (White and Klein 2002). The **family systems framework** views each member of the family as part of a system and the family as a unit that develops norms of interacting, which may be explicit (for example, parents specify chores for the children) or implied (for example, spouses expect fidelity from each other). These rules serve various functions, such as allocating the resources (money for vacation), specifying the division of power (who decides how money is spent), and defining closeness and distance between systems (seeing or avoiding parents or grandparents).

Rules are most efficient if they are flexible. For example, they should be adjusted over time in response to children's growing competence. A rule about not leaving the yard when playing may be appropriate for a 4-year-old but inappropriate for a 15-year-old. The rules and individuals can be understood only by recognizing that "all parts of the system are interconnected" (White and Klein 2002, p. 122).

Family members also develop boundaries that define the individual and the group and separate one system or subsystem from another. A boundary is a "border between the system and its environment that affects the flow of information and energy between the environment and the system" (White and Klein 2002, p. 124). A boundary may be physical, such as a closed bedroom door, or social, such as expectations that family problems will not be aired in public. Boundaries may also be emotional, such as communication, which maintains closeness or distance in a relationship. Some family systems are cold and abusive; others are warm and nurturing.

In addition to rules and boundaries, family systems have roles (leader, follower, scapegoat) for the respective family members. These roles may be shared by more than one person or may shift from person to person during an interaction or across time. In healthy families, individuals are allowed to alternate roles rather than being locked into one role. In problem families, one family member is often allocated the role of scapegoat, or the cause of all the family's problems (for example, an alcoholic spouse).

Family systems may be open, in that they are open to information and interaction with the outside world, or closed, in that they feel threatened by such contact. The Amish have closed family systems and minimize contact with the outside world. Some communes also encourage minimal outside exposure. Twin Oaks Intentional Community of Louisa, Virginia, does not permit any of its almost 100 members to own or keep televisions in their rooms. Exposure to the negative drumbeat of the evening news is seen as harmful.

Family systems theory also emphasizes how one system affects another. Malinen et al. (2010) found that spouses who were happy with each other were also happy with their children. In effect, the marital system had a positive effect on the family system.

Feminist Framework

Although a **feminist framework** views marriage and family as contexts of inequality and oppression for women, there are eleven feminist perspectives, including lesbian feminism (oppressive heterosexuality and men's domination of social spaces), psychoanalytic feminism (cultural domination of men's phallic-oriented ideas and repressed emotions), and standpoint feminism (neglect of women's perspective and experiences in the production of knowledge) (Lorber 1998). Regardless of which feminist framework is being discussed, all feminist

It's not what you call me, but what I answer to.

African Proverb

Family systems framework views each member of the family as part of a system and the family as a unit that develops norms of interaction.

Feminist framework views marriage and family as contexts of inequality and oppression for women.

frameworks have the themes of inequality and oppression. According to feminist theory, gender structures our experiences (for example, women and men will experience life differently because there are different expectations for the respective genders) (White and Klein 2002). Feminists seek equality in their relationships with their partners.

The major theoretical frameworks for viewing marriage and the family are summarized in Table 1.5.

TABLE 1.5 **Theoretical Frameworks for Marriage and the Family**					
Theory	**Description**	**Concepts**	**Level of Analysis**	**Strengths**	**Weaknesses**
Social Exchange	In their relationships, individuals seek to maximize their benefits and minimize their costs.	Benefits Costs Profit Loss	Individual Couple Family	Provides explanations of human behavior based on outcome.	Assumes that people always act rationally and all behavior is calculated.
Family Life Course Development	All families have a life course that is composed of all the stages and events that have occurred within the family.	Stages Transitions Timing	Institution Individual Couple Family	Families are seen as dynamic rather than static. Useful in working with families who are facing transitions in their life courses.	Difficult to adequately test the theory through research.
Structure-Function	The family has several important functions within society; within the family, individual members have certain functions.	Structure Function	Institution	Emphasizes the relation of family to society, noting how families affect and are affected by the larger society.	Families with nontraditional structures (single-parent, same-sex couples) are seen as dysfunctional.
Conflict	Conflict in relationships is inevitable, due to competition over resources and power.	Conflict Resources Power	Institution	Views conflict as a normal part of relationships and as necessary for change and growth.	Sees all relationships as conflictual, and does not acknowledge cooperation.
Symbolic Interaction	People communicate through symbols and interpret the words and actions of others.	Definition of the situation Looking-glass self Self-fulfilling prophecy	Couple	Emphasizes the perceptions of individuals, not just objective reality or the viewpoint of outsiders.	Ignores the larger social interaction context and minimizes the influence of external forces.
Family Systems	The family is a system of interrelated parts that function together to maintain the unit.	Subsystem Roles Rules Boundaries Open system Closed system	Couple Family	Very useful in working with families who are having serious problems (violence, alcoholism). Describes the effect family members have on each other.	Based on work with systems, troubled families, and may not apply to nonproblem families.
Feminism	Women's experience is central and different from man's experience of social reality.	Inequality Power Oppression	Institution Individual Couple Family	Exposes inequality and oppression as explanations for frustrations women experience.	Multiple branches of feminism may inhibit central accomplishment of increased equality.

Research Process and Caveats

Research is valuable since it helps to provide evidence for or against a hypothesis. For example, there is a stigma associated with persons who have tattoos and it is often assumed that students who have tattoos make lower grades than those who do not have tattoos. But Martin and Dula (2010) compared the GPA of persons who had tattoos and those who did not and found no significant differences.

Researchers follow a standard sequence when conducting a research project and there are certain caveats to be aware of when reading any research finding.

Steps in the Research Process

Several steps are used in conducting research.

1. *Identify the topic or focus of research.* Select a focus about which you are passionate. For example, are you interested in studying cohabitation of college students? Give your projected study a title in the form of a question—"Do People Who Cohabit before Marriage Have Happier Marriages than Those Who Do Not?"

2. *Review the literature.* Go online to the various databases of your college or university and read research that has already been published on cohabitation. Not only will this prevent you from "reinventing the wheel" (you might find a research study has already been conducted on exactly what you want to study), but it will give you ideas for your study.

3. *Develop hypotheses.* A **hypothesis** is a suggested explanation for a phenomenon. For example, you might suggest that cohabitation results in greater marital happiness and less divorce because the partners have a chance to "test-drive" each other and their relationship.

Hypothesis a suggested explanation for a phenomenon.

4. *Decide on a method of data collection.* To test your hypothesis, do you want to interview college students, give them a questionnaire, or ask them to complete an online questionnaire? Of course, you will also need to develop a list of questions for your interview or survey questionnaire.

5. *Get IRB approval.* To ensure the protection of people who agree to be interviewed or who complete questionnaires, researchers must submit a summary

Research posters are presented at professional meetings and provide cutting edge research for marriage and family texts (the content of this poster is included in this text).

TABLE 1.6	Potential Research Problems in Marriage and Family	
Weakness	**Consequences**	**Example**
Sample not random	Cannot generalize findings	Opinions of college students do not reflect opinions of other adults.
No control group	Inaccurate conclusions	Study on the effect of divorce on children needs control group of children whose parents are still together.
Age differences between groups of respondents	Inaccurate conclusions	Effect may be due to passage of time or to cohort differences.
Unclear terminology	Inability to measure what is not clearly defined	What is living together, marital happiness, sexual fulfillment, good communication, quality time?
Researcher bias	Slanted conclusions	A researcher studying the value of a product (e.g. The Atkins Diet) should not be funded by the organization being studied.
Time lag	Outdated conclusions	Often-quoted Kinsey sex research is over sixty years old.
Distortion	Invalid conclusions	Research subjects exaggerate, omit information, and/or recall facts or events inaccurately. Respondents may remember what they wish had happened.

of their proposed research to the Institutional Review Board (IRB) of their institution who reviews the research to ensure that the project is consistent with research ethics and poses no undue harm to participants. Important considerations include collecting data from individuals who are told that their participation is completely voluntary, that maintains their anonymity, and that is confidential. Such approval ensures that research studies such as those conducted in the 1940s in Guatemala where American scientists deliberately infected prisoners and patients with syphilis does not recur.

6. *Collect and analyze data.* There are various statistical packages designed to analyze data to discover if your hypotheses are true or false.

7. *Write up and publish results.* Writing up and submitting your findings for publication is important so that your study becomes part of the academic literature.

Caveats to Consider in Research Quality

"New Research Study" is a frequent headline in popular magazines such as *Cosmopolitan, Glamour,* and *Redbook* promising accurate information about "hooking up," "what women want," "what men want," or other relationship, marriage, and family issues. As you read such articles, as well as the research in such texts as this, be alert to their potential flaws. The various issues to keep in mind when evaluating research are identified in Table 1.6.

The Future of Marriage

While marriage as a lifestyle choice is declining somewhat among all groups in the U.S. (Fincham and Beach 2010), and particularly among middle class Americans (high school education/no four year degree), it remains the dominant choice for most Americans, particularly for college educated individuals with a good income (Wilcox 2010). Though these individuals will increasingly delay getting married until their late twenties/early thirties (to complete their educations, launch their careers, and/or become economically independent), there is no evidence that marriage will cease to be a life goal. Indeed six in ten never married adults say they want to get married. Even those on the lower rungs of the socio-economic ladder have the goal of marriage but place a higher premium on economic security as a condition for marriage. Almost 70% say they are optimistic about marriage (Pew Research Center 2010a). The almost 5 million who marry annually and the over 125 million already married are among the happiest, healthiest, and most sexually fulfilled in our society.

Summary

What is marriage?

Marriage is a system of binding a man and a woman together for the reproduction, care (physical and emotional), and socialization of offspring. The federal government regards marriage as a legal contract between a heterosexual couple and the state in which they reside that regulates their economic and sexual relationship. Other elements of marriage involve emotion, sexual monogamy, and a formal ceremony. Types of marriage include monogamy and polygamy. Various forms of polygamy are polygyny, polyandry, and pantagamy.

What is family?

The U.S. Census Bureau defines family as a group of two or more people related by blood, marriage, or adoption. In recognition of the diversity of families, the definition of family is increasingly becoming two adult partners whose interdependent relationship is long-term and characterized by an emotional and financial commitment. The family of origin is the family into which you were born or the family in which you were reared. The family of procreation represents the family that you will begin should you marry and have children. Types of family include nuclear, extended, and blended.

Changes in the family in the last sixty plus years include: divorce has replaced death as the endpoint for the majority of marriages, marriage and intimate relations have emerged as legitimate objects of scientific study, feminism and changes in gender roles in marriage have risen, the age at marriage has increased, some spouses may now live apart (LAT), and the acceptance of singlehood, cohabitation, and childfree marriages has increased.

What is the view/theme of this text?

A central theme of this text is to encourage you to be proactive—to make conscious, deliberate relationship choices to enhance your own well-being and the well-being of those in your intimate groups. Although there are over a hundred such choices, among the most important are whether to marry, whom to marry, when to marry, whether to have children, whether to remain emotionally and sexually faithful to one's partner, and whether to use a condom. Important issues to keep in mind about a choices framework for viewing marriage and the family are that (1) not to decide is to decide, (2) some choices require correcting, (3) all choices involve trade-offs, (4) choices include selecting a positive or negative view, (5) making choices produces ambivalence, and (6) some choices are not revocable. For today's youth, they are in no hurry to find "the one," to marry, and to begin a family. Individuals tend to define themselves as a couple when they spend a lot of time together/share money, visit each other's parents, first have sex together, and have a child.

What are the theoretical frameworks for viewing marriage and the family?

Theoretical frameworks used to study the family include the (1) social exchange framework (spouses exchange resources, and decisions are made on the basis of perceived profit and loss), (2) structural-functional framework (how the family functions to serve society), (3) conflict framework (family members are in conflict over scarce resources of time and money), (4) symbolic interaction framework (symbolic worlds in which the various family members give meaning to each other's behavior), (5) family systems framework (each member of the family is part of a system and the family as a unit develops norms of interaction), (6) family life course development framework (the stages and process of how families change over time), (7) feminist framework (inequality and oppression). The framework used in most empirical family studies is the social exchange framework.

What are steps in the research process and what caveats should be kept in mind?

Steps in the research process include identifying a topic, reviewing the literature, deciding on methods and data collection procedures, ensuring protection of subjects, analyzing the data, and submitting the results to a journal for publication.

Caveats that are factors to be used in evaluating research include a random sample (the respondents providing the data reflect those who were not in the sample), a control group (the group not subjected to the experimental design for a basis of comparison), terminology (the phenomenon being studied should be objectively defined), researcher bias (present in all studies), time lag (takes two years from study to print), and distortion or deception (although rare, some researchers distort their data). Few studies avoid all research problems.

What is the future of marriage?

Marriage will continue to be the lifestyle of choice for the majority (85–90%) of U.S. adults. Though individuals will increasingly delay getting married until their late twenties to early thirties (to complete their educations, launch their careers, and/or become economically independent) and there will be an increase in those who never marry, there is no evidence that marriage will cease to be a life goal for most.

Key Terms

Binuclear family	Family life cycle	Nuclear family
Blended family	Family of orientation	Pantagamy
Child marriage	Family of procreation	Polyandry
Civil union	Family systems framework	Polygamy
Collectivism	Feminist framework	Polygyny
Common-law marriage	Feral children	Primary groups
Conflict framework	Generation Y	Secondary groups
Domestic partnerships	Hypothesis	Social exchange framework
Extended family	Individualism	Sociological imagination
Familism	Institution	Structure-function framework
Family	Marriage	Symbolic interaction framework
Family life course development framework	Marriage benefit	Theoretical frameworks
	Mating gradient	Utilitarianism

Web Links

Family Process
   http://www.trinity.edu/~mkearl/family.html

Gilder Lehrman Institute of American History—History of the Family
   http://www.digitalhistory.uh.edu/historyonline/familyhistory.cfm

National Council on Family Relations
   http://www.ncfr.org/

National Healthy Marriage Resource Center
   http://twoofus.org/index.aspx

National Marriage Project University of Virginia
   http://www.virginia.edu/marriageproject/
   http://www.stateofourunions.org

U.S. Census Bureau
   http://www.census.gov/

Gender

Newsflash: You're the boy. I'm the girl. You text me first or we don't talk today.

Unknown, University Newspaper

Lisa Emmott

Learning Objectives

Learn the various terms used to discuss gender roles.

Know the theories of gender role development.

Identify the agents of gender roles.

Compare gender roles in other societies.

Summarize the consequences of traditional roles for women and men.

Reflect on how gender roles are changing.

Predict the future of gender roles.

1. In spite of women having equal education and pay as men, women do more housework.

2. Persons who ignore cultural gender role expectations typically have a profound sense of well being.

3. Greater gender equality has resulted in women having equal power in negotiating condom use with men.

4. Men now have as much incentive to enter female jobs as women do for male jobs.

5. Women are aware of gender inequities and expect to be paid less at the beginning and end of their careers.

Answers: 1. T   2. F   3. F   4. F   5. T

Caster Semenya won gold in the women's 800-m race at the World Athletics Championship in 2009. But some colleagues questioned if the 18 year old South African was, in fact, a woman. Tests showed that she had three times the level of testosterone of a typical woman and that she was intersexed, which blurred her gender and gave her an unfair advantage. Her right to keep her gold medal has been controversial and her case emphasizes the ambiguity of gender.

Sociologists note that one of the defining moments in an individual's life is when the sex of a fetus (in the case of an ultrasound) or infant (in the case of a birth) is announced. "It's a boy" or "It's a girl" immediately summons an

Caster Semenya won gold in the women's 800-m race at the World Athletics Championship in 2009, but colleagues questioned if the 18 year old South African was, in fact, a woman.

Gallo Images / Alamy

Chapter 2 Gender

onslaught of cultural programming affecting the color of the nursery (e.g. blue for a boy and pink for a girl), name of the baby (there are few gender-free names such as Chris), and occupational choices (in spite of Nancy Pelosi and Hillary Clinton, few women are in politics). In this chapter, we examine variations in gender roles and the way they express themselves in various relationships. We begin by looking at the terms used to discuss gender issues.

Terminology of Gender Roles

In common usage, the terms *sex* and *gender* are often used interchangeably, but social scientists do not regard these terms as synonymous. After clarifying the distinction between *sex* and *gender*, we clarify other relevant terminology, including *gender identity*, *gender role*, and *gender role ideology*.

Sex

Sex refers to the biological distinction between females and males. Hence, to be assigned as a female or male, several factors are used to determine the biological sex of an individual:

- *Chromosomes:* XX for females; XY for males
- *Gonads:* Ovaries for females; testes for males
- *Hormones:* Greater proportion of estrogen and progesterone than testosterone in females; greater proportion of testosterone than estrogen and progesterone in males
- *Internal sex organs:* Fallopian tubes, uterus, and vagina for females; epididymis, vas deferens, and seminal vesicles for males
- *External genitals:* Vulva for females; penis and scrotum for males

Even though we commonly think of biological sex as consisting of two dichotomous categories (female and male), biological sex exists on a continuum. Sometimes not all of the five bulleted items just listed are found neatly in one person (who would be labeled as a female or a male). Rather, items typically associated with females or males might be found together in one person, resulting in mixed or ambiguous genitals; such persons are called **intersexed individuals**. Indeed, the genitals in these intersexed (or middlesexed) individuals (about 2% of all births) are not clearly male or female (Crawley et al. 2008). Intersex development refers to congenital variations in the reproductive system, sometimes resulting in ambiguous genitals.

Even if chromosomal makeup is XX or XY, too much or too little of the wrong kind of hormone during gestation can also cause variations in sex development. Meyer-Bahlburg (2005) suggested that the genesis of the ambiguity may be the brain anatomy, whereby the wiring is different for those with gender identity disorder (GID). Genetics, hormones, and brain mechanisms might all underlie neuroanatomic changes inducing intersexuality. Dysart-Gale (2010) included intersexed in her discussion of nurses providing treatment for lesbian, gay, bisexual, transgendered, intersexed, and queer (henceforth LGBTIQ) patients' health needs. She noted that these individuals often feel a sense of exclusion, isolation, and fear due to negative social attitudes.

Gender

The term **gender** is a social construct and refers to the social and psychological characteristics associated with being female or male. For example, women usually see themselves (and men often agree) as moody and easily embarrassed; men usually see themselves (and women often agree) as competitive, sarcastic, and sexual (Knox et al. 2004). Robnett and Susskind (2010) found that 3rd and 4th grade children valued their own sex's personality traits, but that boys with a lot of same sex friends tended to denigrate female traits. Boys who acted like girls were criticized.

He is half of a blessed man. Left to be finished by such as she; and she a fair divided excellence, whose fullness of perfection lies in him.

William Shakespeare

Sex the biological distinction between females and males.

When a man opens a car door for his wife, it's either a new car or a new wife.

Prince Philip, Duke of Edinburgh

Intersexed individuals those with mixed or ambiguous genitals.

Gender social construct which refers to the social and psychological characteristics associated with being female or male.

What if You Have a Child Who Is Intersexed?

If you had an intersexed child, would you rear the child as an intersexed child or would you dress and socialize the child as a girl or as a boy? How would you feel about arranging for your child to have surgery to alter the genitals to match the gender on which you had decided? How would you regard your child's intersexed status—as an example of diversity, a disability, a challenge, or a unique opportunity for the child to be able to experience both sides of the gender world? What would you tell your child about the intersex "condition"? Do you think acceptance of an intersexed child would vary by gender of the parent (for example, would mothers be more accepting than fathers)? How would you respond to a spouse who did not accept your intersexed child?

In popular usage, gender is dichotomized as an either/or concept (feminine or masculine). Each gender has some characteristics of the other. However, gender may also be viewed as existing along a continuum of femininity and masculinity.

The Self-Assessment of this chapter examines various beliefs about women.

SELF-ASSESSMENT | The Beliefs about Women Scale (BAWS)

The following statements describe different attitudes toward men and women. There are no right or wrong answers, only opinions. Indicate how much you agree or disagree with each statement, using the following scale: (A) strongly disagree, (B) slightly disagree, (C) neither agree nor disagree, (D) slightly agree, or (E) strongly agree.

_____ 1. Women are more passive than men.

_____ 2. Women are less career-motivated than men.

_____ 3. Women don't generally like to be active in their sexual relationships.

_____ 4. Women are more concerned about their physical appearance than are men.

_____ 5. Women comply more often than men.

_____ 6. Women care as much as men do about developing a job or career.

_____ 7. Most women don't like to express their sexuality.

_____ 8. Men are as conceited about their appearance as are women.

_____ 9. Men are as submissive as women.

_____ 10. Women are as skillful in business-related activities as are men.

_____ 11. Most women want their partner to take the initiative in their sexual relationships.

_____ 12. Women spend more time attending to their physical appearance than men do.

_____ 13. Women tend to give up more easily than men.

_____ 14. Women dislike being in leadership positions more than men.

_____ 15. Women are as interested in sex as are men.

_____ 16. Women pay more attention to their looks than most men do.

_____ 17. Women are more easily influenced than men.

_____ 18. Women don't like responsibility as much as men.

_____ 19. Women's sexual desires are less intense than men's.

_____ 20. Women gain more status from their physical appearance than do men.

The Beliefs about Women Scale (BAWS) consists of fifteen separate subscales; only four are used here. The items for these four subscales and coding instructions are as follows:

1. Women are more passive than men (items 1, 5, 9, 13, 17).
2. Women are interested in careers less than men (items 2, 6, 10, 14, 18).
3. Women are less sexual than men (items 3, 7, 11, 15, 19).
4. Women are more appearance conscious than men (items 4, 8, 12, 16, 20).

Score the items as follows: strongly agree = +2; slightly agree = +1; neither agree nor disagree = 0; slightly disagree = −1; strongly disagree = −2.

Scores range from 0 to 40; subscale scores range from 0 to 10. The higher your score, the more traditional your gender beliefs about men and women.

Source

William E. Snell, Jr., PhD. 1997. College of Liberal Arts, Department of Psychology, Southeast Missouri State University. Reprinted with permission. Contact Dr. Snell for further use: wesnell@semo.edu.

Gender differences are a consequence of biological (e.g. chromosomes and hormones) and social factors (e.g. male/female models such as parents, siblings, peers). The biological provides a profound foundation for gender role development. As evidence for this biological influence, is the experiment of the late John Money, psychologist and former director of the now-defunct Gender Identity Clinic at Johns Hopkins University School of Medicine, who encouraged the parents of a boy (Bruce) to rear him as a girl (Brenda) because of a botched circumcision that rendered the infant without a penis. Money argued that social mirrors dictate one's gender identity, and thus, if the parents treated the child as a girl (for example, name, dress, toys), the child would adopt the role of a girl and later that of a woman. The child was castrated and sex reassignment began.

However, the experiment failed miserably; the child as an adult (David Reimer—his real name) reported that he never felt comfortable in the role of a girl and had always viewed himself as a boy. He later married and adopted his wife's three children.

In the past, David's situation was used as a textbook example of how "nurture" is the more important influence in gender identity, if a reassignment is done early enough. Today, his case makes the point that one's biological wiring dictates gender outcome (Colapinto 2000). Indeed, David Reimer noted in a television interview, "I was scammed," referring to the absurdity of trying to rear him as a girl. Distraught with the ordeal of his upbringing and beset with financial difficulties, he committed suicide in May 2004 via a gunshot to the head.

The story of David Reimer is a landmark in terms of the power of biology in determining gender identity. Other research supports the critical role of biology. Cohen-Kettenis (2005) emphasized that biological influences in the form of androgens in the prenatal brain are very much at work in creating one's gender identity.

Nevertheless, **socialization** (the process through which we learn attitudes, values, beliefs, and behaviors appropriate to the social positions we occupy) does impact gender role behaviors, and social scientists tend to emphasize the role of social influences in gender differences.

Gender Identity

Gender identity is the psychological state of viewing oneself as a girl or a boy, and later as a woman or a man. Such identity is largely learned and is a reflection of society's conceptions of femininity and masculinity. Some individuals experience gender dysphoria, a condition in which one's gender identity does not match one's biological sex. An example of gender dysphoria is transsexualism (discussed in the next section).

Transgender

The word **transgender** is a generic term for a person of one biological sex who displays characteristics of the other sex. **Cross-dresser** is a broad term for individuals who may dress or present themselves in the gender of the other sex. Some cross-dressers are heterosexual adult males who enjoy dressing and presenting themselves as women. Cross-dressers may also be women who dress as men and present themselves as men. Cross-dressers may be heterosexual, homosexual, or bisexual.

Transsexuals are individuals with the biological and anatomical sex of one gender (for example, male) but the self-concept of the other sex (that is, female).

When I look in the mirror, I don't see a female. I see a soldier.

Teresa King, drill sergeant
U.S. Army

Socialization the process through which we learn attitudes, values, beliefs, and behaviors appropriate to the social positions we occupy.

Gender identity the psychological state of viewing oneself as a girl or a boy, and later as a woman or a man.

Transgender a generic term for a person of one biological sex who displays characteristics of the other sex.

Cross-dresser individuals who dress or present themselves in the gender of the other sex.

Transsexual an individual with the biological and anatomical sex of one gender (for example, male) but the self-concept of the other sex (that is, female).

Some female-to-male transsexuals say, "I am a man trapped in a woman's body."

Lauren Lemmond

"I am a woman trapped in a man's body" reflects the feelings of the male-to-female transsexual (MtF), who may take hormones to develop breasts and reduce facial hair and may have surgery to artificially construct a vagina. Such a person lives full-time as a woman. Lawrence (2010) observed that nonhomosexual MtF transsexuals are more prevalent in individualistic countries than collectivist countries.

The female-to-male transsexual (FtM) is one who is a biological and anatomical female but feels "I am a man trapped in a woman's body." This person may take male hormones to grow facial hair and deepen her voice and may have surgery to create an artificial penis. This person lives full-time as a man. Thomas Beatie, born a biological woman, viewed himself as a man and transitioned from living as a woman to living as a man. His wife, Nancy, had two adult children before having a hysterectomy (which rendered her unable to bear more children) so her husband agreed to be artificially inseminated (because he had ovaries), became pregnant, and delivered a child in the summer of 2008. Because he was legally a male, the media referred to him as "The Pregnant Man"; they asked him: if he gave birth to the child, could he be the father? Technically, Oregon law defines birth as an expulsion or extraction from the mother so Tom is the technical mother. However, the new parents could petition the courts and have "him" declared as the father and his wife as the mother (Heller 2008).

Individuals need not take hormones or have surgery to be regarded as transsexuals. The distinguishing variable is living full-time in the role of the gender opposite one's biological sex. A man or woman who presents full-time as the other gender is a transsexual by definition.

The woman most in need of liberation is the woman in every man and the man in every woman.

William Sloan Coffin, clergyman and political activist

Gender Roles

Gender roles are social norms which specify the socially appropriate behavior for females and males in a society. All societies have expectations of how boys and girls, men and women "should" behave. Gender roles influence women and men in virtually every sphere of life, including family and occupation. In spite of women having equal education and pay than men, they still do more housework (Lachance-Grzela and Bouchard 2010).

Gender roles social norms which specify the socially appropriate behavior for females and males in a society.

Chapter 2 Gender

And, even with transgender male-to-female individuals, women end up doing more domestic work. Pfeffer (2010) interviewed 50 women partners of transgender and transsexual men to assess the division of labor. Women often spoke of an inegalitarian division of labor, but rationalized the reasons for this division. Ani stated: "I do the dishes; but I'm so neurotic about having a clean house and he is not . . . . I definitely do more than he does but, again, I'm the one that happens to be a neat freak."

Another prevalent norm in family life is that women end up devoting more time to child rearing and child care. Lareau and Weininger (2008) studied the division of labor of parents getting their children to various leisure activities and found that traditional gender roles were the norm and that mothers "are the ones who must satisfy these demands" (p. 450). The sheer number of activities forces some mothers to cut back. One mother said:

> I know that all these things are good for my child and they develop all those things but time-wise, I just don't have time for all that stuff. I mean I have to live up to my label as the meanest Mom in the world and I try to do a real good job of it [laughs]. I told her she could be either on soccer or softball. She'll be on one of them. (p. 450)

Walzer (2008) noted that marriage is a place where men and women "do gender" in the sense that roles tend to be identified as breadwinning, housework, parenting, and emotional expression and are gender-differentiated. She also noted that divorce generates "redoing" gender in the sense that it changes the expectations for masculine and feminine behavior in families (for example, women become breadwinners, men become single parents, and so on).

The term sex roles is often confused with and used interchangeably with the term *gender roles*. However, whereas gender roles are socially defined and can be enacted by either women or men, **sex roles** are defined by biological constraints and can be enacted by members of one biological sex only—for example, wet nurse, sperm donor, child bearer.

Gender Role Ideology

Gender role ideology refers to beliefs about the proper role relationships between women and men in a society (e.g., man more likely than a woman to initiate first interaction). Where there is gender equality, there is enhanced relationship satisfaction (Walker and Luszcz 2009). Egalitarian wives are also most happy with their marriages if their husbands share both the work and the emotions of managing the home and caring for the children (Wilcox and Nock 2006).

In spite of the rhetoric regarding the entrenchment of egalitarian interaction between women and men in the United States, there is evidence of traditional gender roles in mate selection with men in the role of initiating relationships. When 692 undergraduate females at a large southeastern university were asked if they had ever asked a new guy to go out, 60.1% responded, "no" (Ross et al. 2008). In another study, 30% of the female respondents reported a preference of marrying a traditional man (one who saw his role as provider and who was supportive of his wife staying home to rear children) (Abowitz et al. 2011). Some undergraduate men also prefer a traditional wife (see Applying Social Research feature).

Traditional American gender role ideology has perpetuated and reflected patriarchal male dominance and male bias in almost every sphere of life. Even our language reflects this male bias. For example, the words *man* and *mankind* have traditionally been used to refer to all humans. There has been a growing trend away from using male-biased language. Dictionaries have begun to replace *chairman* with *chairperson* and *mankind* with *humankind*.

Gender Differences in Viewing Romantic Relationships

Abowitz et al. (2009) assessed the respective gender views of a sample of 326 undergraduates in regard to romantic relationships and found that men were significantly more likely to believe that bars are good places to meet a potential

Men are so made that they can resist sound argument, and yet yield to a glance.

Balzac, French novelist

Sex roles roles defined by biological constraints and can be enacted by members of one biological sex only—for example, wet nurse, sperm donor, child-bearer.

Gender role ideology the proper role relationships between women and men in a society.

Women always worry about the things that men forget; men always worry about the things women remember.

Author Unknown

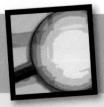

When the young lovers (Christine and Raoul) of *The Phantom of the Opera* are finally alone, Raoul sings his song of love and protection:

> *No more talk of darkness, forget these wide-eyed fears.*
>
> *I'm here, nothing can harm you; my words will warm and calm you.*
>
> *Let me be your shelter, let me be your life . . .*

Christine responds with:

> *Promise me everything you say is true, that's all I ask of you.*

This scene reflects the traditional relationship where the man is the protector and the woman is being taken care of. Like the lovers in *Phantom,* some university students also seek a traditional relationship whereby the wife stays at home and the husband is the breadwinner. This research focused on the degree to which undergraduate men at a large southeastern university sought a traditional wife and the various background characteristics of these men.

Methods

The data were taken from a larger nonrandom sample of 1,027 undergraduates at a large southeastern university. Of these undergraduates, 335 men or 30.8% agreed with the statement, "I prefer to marry a traditional wife who sees her role as staying home and rearing children." In contrast, 69.2% of the undergraduate men disagreed with the statement. Cross-classification was conducted to determine any relationships, with chi-square utilized to assess statistical significance.

Findings and Discussion

Analysis of the data revealed the following five statistically significant differences in regard to the characteristics of those who wanted a traditional wife in contrast to those who did not:

1. Financial security valued over happiness. Almost three-fourths (73.6%) of the men who valued financial security wanted a traditional wife compared to 42.8% who wanted happiness over money.

   Two factors may explain this association: One reason men who valued financial security sought a traditional wife could be the belief that a man would be better able to earn money and to focus on his career if he had a traditional wife who supported them (for example, prepared meals and took care of the children). Indeed, career wives often say that they need a "wife" to take care of them (for example, have dinner ready, take care of the children, do the grocery shopping, take care of laundry, and so on).

   A second reason men who valued financial security sought a traditional wife may be that a traditional wife might actually be very cost-effective. Sefton (1998) calculated that the value of the stay-at-home mother in terms of what it would cost to pay for all services that she provides (domestic cleaning, laundry, meal planning and preparation, shopping, providing transportation to child's activities, taking the children to the doctor, and running errands) and, adjusting for changes in the consumer price index, the figure was around $50,000 in 2011. A study on the marital satisfaction of wives who did a lot of housework revealed that they felt appreciated to the

partner, that cohabitation improves marriage, that men control relationships, and that people will "cheat" if they feel they will not be caught. In contrast, women were significantly more likely to believe that love is more important than factors like age and race in choosing a mate, that couples stop "trying" after they marry, and that women "know" when their men are lying.

Theories of Gender Role Development

Various theories attempt to explain why women and men exhibit different characteristics and behaviors.

degree that their husbands spent a lot of time with them (Lee and Waite 2010).

2. A wife's higher income believed to weaken the marriage. Of the respondents who believed that a woman's higher income weakens one's marriage, 61.9% wanted a nontraditional wife. Indeed, men preferring a traditional wife may have felt that a wife earning more money threatened their masculinity, head of the household, and chief breadwinner roles.

3. Belief that children turn out better when one parent stays home. Men who believed that children benefit from staying at home with their mother were significantly more likely than those who rejected this belief to want to marry a traditional woman. Of the respondents who believed that children turn out better when one parent stays home, 51.3% preferred a traditional wife.

4. Religion. Of the male respondents who viewed themselves as "religious," 33.6% reported a preference for marrying a traditional woman who wanted to stay home and rear children. In contrast, 18.6% of those who did not regard themselves as religious reported a preference for a traditional wife.

5. Divorce not considered an option. The undergraduates who did not believe divorce to be an option were 25.9% more likely to prefer a traditional wife than men who regarded divorce as an alternative. Being against divorce and having a traditional family with only one breadwinner (the husband) are both traditional values.

Implications

The study has three implications. The finding that men who want a traditional wife value money over happiness has implications for the professions they choose. Men who seek such a wife will tend to seek those professions such as business, medicine, engineering, and those associated with high incomes. Similarly, men who are not intent on seeking a traditional wife are more likely to seek those professions such as elementary school education, art, social work, and those associated with relatively low incomes.

A second implication is in reference to the finding that men who seek a traditional wife tend to believe that the marriage will be weakened by a wife's earning a high income. Such men may have a need to be in control because money is considered power. These women might be sensitive to the need of such a husband to exercise control in the relationship.

A third implication is in reference to the finding that men who seek a traditional wife believe that children benefit from the wife staying at home. Such a belief reveals that men may view parenting as the primary responsibility of the mother with a limited role for the father. Indeed, these men may view earning money as their role with little interest in hands-on parenting. This view may be consistent with the traditional wife's view. Or she may reject this view and want a more involved father.

*Abridged and updated from Knox, D., and M. Zusman. 2007. Traditional wife? Characteristics of college men who want one. *Journal of Indiana Academy of Social Sciences* 11: 27–32.

Sources

Denton, M. L. 2004. Gender and marital decision making: Negotiating religious ideology and practice. *Social Forces* 82: 1151–80.

Lee, Y., and L. J. Waite. 2010. How appreciated do wives feel for the housework they do? *Social Science Quarterly* 91: 476–492.

Sefton, B. W. 1998. The market value of the stay-at-home mother. *Mothering* 86: 26–29.

Biosocial

In the discussion of gender at the beginning of the chapter, we noted the profound influence of biology on one's gender. **Biosocial theory** emphasizes that social behaviors (for example, gender roles) are biologically based and have an evolutionary survival function. For example, women tend to select and mate with men whom they deem will provide the maximum parental investment in their offspring. The term **parental investment** refers to any investment by a parent that increases the offspring's chance of surviving and thus increases reproductive success. Parental investments require time and energy. Women have a great deal of parental investment in their offspring (including nine months of gestation), and they tend to mate with men who have high status, economic resources, and a willingness to share those economic resources.

Biosocial theory also referred to as sociobiology, social behaviors (for example, gender roles) are biologically based and have an evolutionary survival function.

Parental investment any investment by a parent that increases the offspring's chance of surviving and thus increases reproductive success.

The biosocial explanation (also referred to as sociobiology) for mate selection is extremely controversial. Critics argue that women may show concern for the earning capacity of a potential mate because they have been systematically denied access to similar economic resources, and selecting a mate with these resources is one of their remaining options. In addition, it is argued that both women and men, when selecting a mate, think more about their partners as lovers and companions than as future parents of their offspring.

Social Learning

Derived from the school of behavioral psychology, the social learning theory emphasizes the roles of reward and punishment in explaining how a child learns gender role behavior. This is in contrast to the biological explanation for gender roles. For example, consider two young brothers who enjoy playing "lady"; each of them puts on a dress, wears high-heeled shoes, and carries a pocketbook. Their father came home early one day and angrily demanded, "Take those clothes off and never put them on again. Those things are for women." The boys were punished for "playing lady" but rewarded with their father's approval for playing sports and being aggressive.

Reward and punishment alone are not sufficient to account for the way in which children learn gender roles. Another way children learn is when parents or peers offer direct instruction (for example, "girls wear dresses" or "a man stands up and shakes hands"). In addition, many of society's gender rules are learned through modeling. In modeling, children observe and imitate another's behavior. Gender role models include parents, peers, siblings, and characters portrayed in the media. Parents may fear some media models—Lady GaGa.

The impact of modeling on the development of gender role behavior is controversial. For example, a modeling perspective implies that children will tend to imitate the parent of the same sex, but children in all cultures are usually reared mainly by women. Yet this persistent female model does not seem to interfere with the male's development of the behavior that is considered appropriate for his gender. One explanation suggests that boys learn early that our society generally grants boys and men more status and privileges than girls and women. Therefore, boys devalue the feminine and emphasize the masculine aspects of gender role learning.

Regardless of the source, Witt and Wood (2010) found that expectations for following gender role standards are so powerful that one's well being is related to the degree that one reflects these standards. Persons who accept the standards but who do not live up to them feel an impaired sense of well being.

Identification

Freud was one of the first researchers to study gender role development. He suggested that children acquire the characteristics and behaviors of their same-sex parent through a process of identification. Girls identify with their mothers; boys identify with their fathers. For example, girls are more likely to become involved in taking care of children because they see women as the primary caregivers of young children. In effect, they identify with their mothers and will see their own primary identity and role as those of a mother. Likewise, boys will observe their fathers and engage in similar behaviors to lock in their own gender identity. The classic example is the son who observes his father shaving and wants to do likewise (be a man too).

Cognitive-Developmental

The cognitive-developmental theory of gender role development reflects a blend of biological and social learning views. According to this theory, the biological readiness of the child, in terms of cognitive development, influences how the child responds to gender cues in the environment (Kohlberg 1966). For example,

Chapter 2 Gender

gender discrimination (the ability to identify social and psychological characteristics associated with being female or male) begins at about age 30 months. However, at this age, children do not view gender as a permanent characteristic. Thus, even though young children may define people who wear long hair as girls and those who never wear dresses as boys, they also believe they can change their gender by altering their hair or changing clothes.

Not until age 6 or 7 do children view gender as permanent (Kohlberg 1966; 1969). In Kohlberg's view, this cognitive understanding involves the development of a specific mental ability to grasp the idea that certain basic characteristics of people do not change. Once children learn the concept of gender permanence, they seek to become competent and proper members of their gender group. For example, a child standing on the edge of a school playground may observe one group of children jumping rope while another group is playing football. That child's gender identity as either a girl or a boy connects with the observed gender-typed behavior, and the child joins one of the two groups. Once in the group, the child seeks to develop behaviors that are socially defined as gender-appropriate.

I wanted to be more of a business woman. I wanted to take this name I had made for myself and take it to the next level.

Tyra Banks, model & business woman

Agents of Socialization

Three of the four theories discussed in the preceding section emphasize that gender roles are learned through interaction with the environment. Indeed, though biology may provide a basis for one's gender identity, cultural influences in the form of various socialization agents (parents, peers, religion, and the media) shape the individual toward various gender roles. These powerful influences in large part dictate what people think, feel, and do in their roles as man or woman. In the next section, we look at the different sources influencing gender socialization.

Family

The family is a gendered institution with female and male roles highly structured by gender. The names parents assign to their children, the clothes they dress them in, and the toys they buy them all reflect gender. Parents (particularly African-American mothers) may also be stricter on female children—determining the age they are allowed to leave the house at night, the time of curfew, and using directives such as "call your mamma when you get to the party." Boys are also required to perform fewer chores (Mandara et al. 2010).

Siblings also influence gender role learning. As noted in Chapter 1, the relationship with one's sibling (particularly in sister-sister relationships) is likely to be the most enduring of all relationships (Meinhold et al. 2006). Also, growing up in a family of all sisters or all brothers intensifies social learning experiences toward femininity or masculinity. A male reared with five sisters and a single-parent mother is likely to reflect more feminine characteristics than a male reared in a home with six brothers and a stay-at-home dad.

Other things may change us, but we start and end with family.

Anthony Brandt, author

Race/Ethnicity

The race and ethnicity of one's family also influence gender roles. Although African American families are often stereotyped as being matriarchal, the more common pattern of authority in these families is egalitarian (Taylor 2002). Both President Barack and First Lady Michelle Obama have law degrees from Harvard, and their relationship appears to be very egalitarian.

The fact that African American women have increased economic independence provides a powerful role model for young African American women. A similar situation exists among Hispanics, who represent the fastest-growing segment of the U.S. population. Mexican American marriages have great variability, but

I am the son of a black man from Kenya and a white woman from Kansas.

Barack Obama

where the Hispanic woman works outside the home, her power may increase inside the home. However, because Hispanic men are much more likely to be in the labor force than Hispanic women, traditional role relationships in the family are more likely to be the norm.

Life is partly what we make it, and partly what it is made by the friends we choose.

Tennessee Williams, author

Peers

Though parents are usually the first socializing agents that influence a child's gender role development, peers become increasingly important during the school years. Haynie and Osgood (2005) analyzed data from the National Longitudinal Study of Adolescents reflecting responses from adolescents in grades 7 through 12 at 132 schools over an eleven-year period and confirmed the influence of peers in delinquent behavior. If friends drank, smoked cigarettes, skipped school without an excuse, and became involved in serious fights, then the adolescents had an increased likelihood that they would also engage in delinquent acts. Regarding gender, the gender role messages from adolescent peers are primarily traditional. Boys are expected to play sports and be career-oriented. Female adolescents are under tremendous pressure to be physically attractive and thin, popular, and achievement-oriented.

The religion of one age is the literary entertainment of the next.

Ralph Waldo Emerson, poet

Religion

Because women (particularly white women) are "socialized to be submissive, passive, and nurturing," they may be predisposed to greater levels of religion and religious influence (Miller and Stark 2002). Such exposure includes a traditional framing of gender roles. Male dominance is indisputable in the hierarchy of religious organizations, where power and status have been accorded mostly to men. Mormons, particularly, adhere to traditional roles in marriage where men are regarded as the undisputed head of the household. Maltby et al. (2010) observed that the stronger the religiosity for men, the more traditional and sexist their view of women. This association was not found for women.

National **Data**

Of Americans age 18 and older, 56% of adults in the United States say that religion is "very important" in their lives; 48% of those 18 to 29 report this level of importance for religion (Pew Research 2010b).

To be truly ignorant, be content with your own knowledge.

Chuang Tzu, Chinese philosopher

Education

The educational institution serves as an additional socialization agent for gender role ideology. However, such an effect must be considered in the context of the society or culture in which the "school" exists and of the school itself. Schools are basic cultures of transmission in that they make deliberate efforts to reproduce the culture from one generation to the next.

Occupational sex segregation the concentration of women and men in different occupations.

Economy

The economy of the society influences the roles of the individuals in the society. The economy is a very gendered institution. **Occupational sex segregation** is the concentration of men and women in different occupations which has grown out of traditional gender roles. Men dominate as airline pilots, architects, and auto mechanics; women dominate as elementary school teachers, florists, and hair stylists (Weisgram et al. 2010). Female-dominated occupations tend to require less education and have lower status. England (2010) noted that gains of women have been uneven because women have had a strong incentive to enter male jobs, but men have had little incentive to take on female activities or jobs. The salaries in these occupations (e. g. elementary education) have remained relatively low. Women are aware that there are gender inequities in pay and they expect to be paid less than men both at the beginning and the peak of their careers (Hogue, 2010).

Increasingly, occupations are becoming less segregated on the basis of gender, and social acceptance of nontraditional career choices has increased. In 1960, 98% of persons entering veterinary medicine were male; today only 49% are male (Lincoln 2010). The U.S. government is committed to opening occupations to both genders.

Mass Media

Mass media, such as movies, television, magazines, newspapers, books, music, computer games, and music television videos, both reflect and shape gender roles. Media images of women and men typically conform to traditional gender stereotypes, and media portrayals depicting the exploitation, victimization, and sexual objectification of women are common. In regard to the influence of television on gender roles, Rivadeneyraa and Lebob (2008) studied ninth grade students and found that watching "romantic" television (for example, soaps, Lifetime movies) was associated with having more traditional gender role attitudes in dating situations. As for music, Ter Gogt et al. (2010) studied 410 13- to 16-year-old students and found that a preference for hip-hop music was associated with gender stereotypes (e.g. men are sex driven and tough, women are sex objects). In regard to music television videos, Wallis (2011) conducted a content analysis of 34 music videos and found that significant gender displays reinforced stereotypical notions of women as sexual objects, females as subordinate, and males as aggressive.

Basow (2010) emphasized the new trend by authors of gender textbooks to move beyond gender role stereotypes. But there are cultural limits. Basow (2010) noted that she wanted to change the title of one of her textbooks to *A Feminist Perspective* but this idea was rejected by the publisher for fear of losing adoptions.

Pompper (2010) noted that media may project gender images that become unsettling. The researcher conducted a series of interviews and noted how the media may threaten traditional conceptions of masculinity for some males. No longer is "the man" the tough, rugged cowboy. Real men can also be clean shaven-metrosexual "pretty boys." But there is caution. "Views such as Ray's, an African-American/ Black father/uncle, were common: "A real man is a man, not a woman. He's a pillar of strength, not no soft dude . . ."

The cumulative effects of family, peers, religion, education, the economy, and mass media perpetuate gender stereotypes. Each agent of socialization reinforces gender roles that are learned from other agents of socialization, thereby creating a gender role system that is deeply embedded in our culture. All of these influences effect relationship choices (see Table 2.1).

Gender Roles in Other Societies

Because culture largely influences gender roles, individuals reared in different societies typically display the gender role patterns of those societies. The following subsections discuss how gender roles differ in various societies.

TABLE 2.1 Effects of Gender Role Socialization on Relationship Choices

Women

1. Women who are socialized to invest in relationships, to be in love, and to be nurturing may be unable to leave an emotionally or physically abusive relationship.

2. Women who are socialized to play a passive role and not initiate relationships are limiting interactions that could develop into valued relationships.

3. Women who are socialized to accept that they are less valuable and important than men are less likely to seek, achieve, or require egalitarian relationships with men.

4. Women who internalize society's standards of beauty and view their worth in terms of their age and appearance are likely to feel bad about themselves as they age. Their negative self-concept, more than their age or appearance, may interfere with their relationships.

5. Women who are socialized to accept that they are solely responsible for taking care of their parents, children, and husband are likely to experience role overload. In this regard, some women feel angry and resentful which has a negative impact on their relationships.

6. Women who are socialized to emphasize the importance of relationships in their lives will continue to seek relationships that are emotionally satisfying.

Men

1. Men who are socialized to define themselves more in terms of their occupational success and income and less in terms of positive individual qualities leave their self-esteem and masculinity vulnerable should they become unemployed or work in a low-status job.

2. Men who are socialized to restrict their experience and expression of emotions are denied the opportunity to discover the rewards of emotional interpersonal sharing.

3. Men who are socialized to believe it is not their role to participate in domestic activities (child rearing, food preparation, house cleaning) will not develop competencies in these life skills. Potential partners often view domestic skills as desirable.

4. Heterosexual men who focus on cultural definitions of female beauty overlook potential partners who might not fit the cultural beauty ideal but who would nevertheless be wonderful life companions.

5. Men who are socialized to have a negative view of women who initiate relationships will be restricted in the long term relationships they are willing to develop.

6. Men who are socialized to be in control of relationship encounters may alienate their partners, who may desire an egalitarian relationship.

International **Data**

When education, politics, health, and economics are considered, progress in closing the global gender inequality gap reveals that the U.S. is number 19 of 134 countries. Iceland, Norway, and Finland lead with Pakistan, Chad, and Yemen being the worst countries (*Time Magazine*, 2010, p. 19).

Some men spend a lifetime in an attempt to comprehend the complexities of women. Others pre-occupy themselves with somewhat simpler tasks, such as understanding the theory of relativity!

Albert Einstein, brilliant Nobel Prize winner in physics

Gender Roles in Latino/Hispanic Families

Although there is no one Latino/Hispanic family and the gender roles differ across families, there seems to be a drift toward less role rigidity. Royo-Vela et al. (2008) emphasized that, although the traditional family model in Spain calls for men as providers and women as homemakers and mothers, a new feminine culture and a social reality is moving steadily toward gender equality and complementariness between genders.

Parra-Cardona et al. (2008) compared the gender role views of sixty-four foreign and U.S.-born parents who identified themselves as Latino or Hispanic. They found that foreign-born parents tended to regard the role of the man as provider and the role of the woman as primary caretaker and the main source of emotional nurturance for children and the family. However, 30% of foreign-born parents "expressed the need to challenge predetermined gender roles."

One father felt that mothers are often blamed for the outcome of their children when fathers should be more involved. ". . . . [I]t's the mother who gets blamed for [this] because we say 'Where were you?, What were you doing?, Did you spoil them too much?' . . . I think we should not delegate so much on the mother" (p. 165).

U.S.-born Latinos were even more (half) egalitarian. A mother expressed, "I grew up in a family in which girls were supposed to do the cleaning and the cooking . . . boys were supposed to take care of the garbage and fix the bicycles . . . I got divorced because my ex-husband was like that" (p. 167). A U.S.-born Latino reported that her husband not only participated in domestic chores but remained attentive to her personal needs: "My husband is really involved with the children . . . He also cooks and cleans . . . He goes to all my OBGYN appointments . . . No matter what I ask from him, he always says to me 'all right, I'm here'" (p. 168).

Finally a male talked about his interpretation of "machismo," which is usually associated with negative stereotypes. "Being a macho . . . a lot of people take the macho thing the wrong way . . . Being a macho is about being a man . . . being strong for your family . . . not giving up when you feel down . . . to know what you are doing . . . being a man about it" (p. 169).

Gender Roles in Afghanistan under the Taliban

U.S. military involvement in Afghanistan continues. It is a country about the size of Texas, with an estimated population of 32 million. The Taliban reached the peak of their dominance there in 1996. The Revolutionary Association of the Women of Afghanistan (RAWA) compiled an "abbreviated" list of restrictions against women (including "creating noise when they walk" and "wearing white socks"). The life for many women and children was often cruel, demeaning, and fatal. Some women drank household bleach rather than continue to endure their plight. They were not allowed to go to school or work and thus were completely dependent economically. Indeed, they were required to stay in the house, to paint the windows black, and to leave the house only if they were fully clothed (wearing a burka) and accompanied by a male relative. Some could not afford burkas and had no living male relatives. According to Skaine, "There are two places for women: one is the husband's bed and the other is the graveyard" (2002, 64). One mother reported that the Taliban came to her house, dragged her 19-year-old daughter from it, and drove away with her. "They sell them," she lamented (p. 116).

Although the plight of Afghan women may seem horrible to an outsider, some Afghan women who are thoroughly socialized in the culture and tradition may not feel oppressed but rather are accepting of their role. They may feel love, protection, and security inside the context of their marriage and family. Some may relish tradition and be against those who wish to change their sacred traditions. In 2010 the French passed a law outlawing the burqa—"no one can, in the public space, wear clothing intended to hide the face" No one asked the 2000 veiled women in France how they felt about the burqa.

Nevertheless, subsequent to 9/11, the United States attacked the Taliban in Afghanistan with the goal of removing them. Success is debatable and Afghan women are still oppressed. The August 9, 2010 cover of *Time* magazine featured the photo of 18-year-old Aisha whose nose had been cut off by her husband because she had tried to run away (Baker 2010). The role of women in Afghanistan, particularly in the rural areas, has typically been one of submissiveness. "Most of the women in rural areas (which comprise over 80% of Afghanistan) have never had the opportunity to get out of their own little house, little village, little province" (Consolatore 2002, 13).

Outside of Kabul, Afghan women go uneducated, become child brides, produce children, and rarely expect their daughters' lives to be different. The patriarchal social structure and the absence of a centralized and modernized state in Afghanistan predict that changes will be limited for Afghan women (Moghadam 2002). In spite of some changes, *self-immolation* continues "at a steady rate" (Raj et al. 2008). Over 100 Afghan women (Raj et al. 2008) have set themselves on fire as a means of escaping their mistreatment and to use the only voice they have in protest (public suicide). While gains have been made since the U.S presence began in 2001, a return of the Taliban will mean a return to unbridled oppression (Baker 2010).

Gender Roles in Caribbean Families

For spring break, college students sometimes go to the Bahamas, Jamaica, or other English-speaking islands in the Caribbean (for example, Barbados, Trinidad, Guyana) and may wonder about the family patterns and role relationships of the people they encounter. The natives of the Caribbean represent more than 30 million, with a majority being of African ancestry. Their family patterns are diverse but are often characterized by women and their children as the primary family unit—the fathers of these children rarely live in the home (29% in St. Kitts and 55% in Jamaica) (Roopnarine et al. 2005). Hence, men may have children with different women and be psychologically and physically absent from their children's lives. When they do live with a woman, traditional division of labor prevails, with women taking care of domestic and child-care tasks.

About half of all female household heads in the Caribbean have never been married. Women view motherhood, not marriage, as the symbol of their womanhood. Hence, their focus is on taking care of their children and the children of others. Caribbean fathers vary from showing negligible levels of involvement with their children (e.g., in Belize) to showing levels comparable to those in other societies (e.g., in Trinidad and Guyana) (ibid.).

Gender Roles in East and South African Families

Africa is a diverse continent with more than fifty nations. The cultures range from Islamic and Arab cultures of northern Africa to industrial and European influences in South Africa. In some parts of East Africa (for example, in Kenya), gender roles are in flux.

Meredith Kennedy (2007) has lived in East Africa and makes the following observations of gender roles:

> *The roles of men and women in most African societies tend to be very separate and proscribed, with most authority and power in the men's domain. For instance, Maasai wives of East Africa do not travel much, since when a husband comes home he expects to find his wife (or wives) waiting for him with a gourd of sour milk. If she is not, he has the right to beat her when she shows up. As attempts are made to soften these boundaries and equalize the roles, the impacts are very visible and cause a lot of reverberations throughout these communal societies. Many African women who believe in and desire better lives will not call themselves "feminists" for fear of social censure. Change for people whose lives are based on tradition and "fitting in" can be very traumatic.*

> *Similarly, young East African men (for example, Kenyans) view white American coeds who visit their country as students to be very forward and available for sex. These perceptions are sometimes due to the way female students dress and require an adjustment of female dress to avoid inaccurate perceptions.*

In South Africa, where more than 75% of the racial population is African and around 10% are white, the African family is also known for its traditional role relationships and patriarchy. African men were socialized by the Dutch with firm patriarchal norms and adopted this style in their own families. In addition, women were subordinate to men "within a wider kinship system, with the chief as the controlling male."

Consequences of Traditional Gender Role Socialization

This section discusses different consequences, both negative and positive, for women and men, of traditional female and male socialization in the United States.

Consequences of Traditional Female Role Socialization

Table 2.2 summarizes some of the negative and positive consequences of being socialized as a woman in U.S. society. Each consequence may or may not be true for a specific woman. For example, although women in general have less education and income, a particular woman may have more education and a higher income than a particular man.

Negative Consequences of Traditional Female Role Socialization There are several negative consequences of being socialized as a woman in our society.

1. *Less Income.* Although women now earn about half of PhDs (47% to be exact) (Welch 2009), they have lower academic rank and earn less money. The lower academic rank is because women give priority to the care of their children and family. In addition, women tend to be more concerned about the nonmonetary aspects of work. In a study of 102 seniors and 504 alumni from a mid-sized Midwestern public university that rated 48 job characteristics, women gave significantly higher ratings to family life accommodations, pleasant working conditions, travel, and interpersonal relationships. While the economy of 2010–2012 reflected more women as breadwinners, women still earn about two-thirds of what men earn, even when the level of educational achievement is identical (see Table 2.3). Their visibility in the ranks of high corporate America is also still low. Of Fortune 500 companies, women run only thirteen as CEO. Angela Braly is one of them. She heads WellPoint and earns 9.1 million annually (Fortune 500, 2008). However, while women are, increasingly, more likely than men to complete a four year college degree (Wells et al. 2011), their incomes have not caught up.

2. *Feminization of Poverty.* Another reason many women are relegated to a lower income status is the **feminization of poverty**. This term refers to the disproportionate percentage of poverty experienced by women living alone or with their children. Single mothers are particularly associated with poverty.

Feminization of poverty the idea that women (particularly those who live alone or with their children) disproportionately experience poverty.

TABLE 2.2 Consequences of Traditional Female Role Socialization

Negative Consequences	Positive Consequences
Less income (more dependent)	Longer life (five years)
Feminization of poverty	Stronger relationship focus
Higher STD/HIV infection risk	Keep relationships on track
Negative body image	Bonding with children
Less personal/marital satisfaction	Identity not tied to job

TABLE 2.3 Women's and Men's Median Income with Similar Education

	Bachelor's	Master's	Doctoral Degree
Men	$57,278	$70,973	$90,575
Women	$36,294	$48,000	$60,619

Source: *Statistical Abstract of the United States*, 2011. 130th ed. Washington, DC: U.S. Bureau of the Census, Table 701.

When head-of-household women are compared with married-couple households, the median income is $33,073 versus $73,010 (*Statistical Abstract of the United States, 2011*, Table 691). The process is cyclical—poverty contributes to teenage pregnancy because teens have limited supervision and few alternatives to parenthood. (The median income for head-of-household men is $49,186.) Such early childbearing interferes with educational advancement and restricts women's earning capacity, which keeps them in poverty. Their offspring are born into poverty, and the cycle begins anew.

Low pay for women is also related to the fact that they tend to work in occupations that pay relatively low incomes. Indeed, women's lack of economic power stems from the relative dispensability of women's labor (it is easy to replace) and how work is organized (men control positions of power). Women also live longer than men, and poverty is associated with being elderly.

When women move into certain occupations, such as teaching, there is a tendency in the marketplace to segregate these occupations from men's, and the result is a concentration of women in lower-paid occupations. The salaries of women in these occupational roles increase at slower rates. For example, salaries in the elementary and secondary teaching profession, which is predominately female, have not kept pace with inflation.

Conflict theorists assert that men are in more powerful roles than women and use this power to dictate incomes and salaries of women and "female professions." Functionalists also note that keeping salaries low for women keeps women dependent and in child-care roles so as to keep equilibrium in the family. Hence, for both conflict and structural reasons, poverty is primarily a feminine issue. One of the consequences of being a woman is to have an increased chance of feeling economic strain throughout life.

3. *Higher Risk for Sexually Transmitted Infections.* Gender roles influence a woman's vulnerability to sexually transmitted infection and HIV, not only because women receive more bodily fluids from men, who have a greater number of partners (and are therefore more likely to be infected), but also because some women feel limited power to influence their partners to wear condoms (East et al. 2011). Sassler and Miller (2011) also revealed that women are less likely than men to be assertive in intimate relationships.

4. *Negative Body Image.* Just as young girls tend to have less positive self concepts than boys (Yu and Zie 2010), they also feel more negatively about their bodies due to the cultural emphasis on being thin and trim (Grogan 2010). Darlow and Lobel (2010) noted that overweight women who endorse cultural values of thinness have lower self-esteem. There are more than 3,800 beauty pageants annually in the United States. The effect for many women who do not match the cultural ideal is to have a negative body image. Hollander (2010) reported on a study whereby teenage girls who viewed themselves as overweight had an increased risk of having first sex by age 13.

American women also live in a society that devalues them in a larger sense. Their lives and experiences are not taken as seriously. **Sexism** is an attitude, action, or institutional structure that subordinates or discriminates against individuals or groups because of their sex. Sexism against women reflects the tradition of male dominance and presumed male superiority in American society.

Benevolent sexism (reviewed by Maltby et al. 2010) is a related term and reflects the belief that women are innocent creatures who should be protected and supported. While such a view has positive aspects, it assumes that women are best suited for domestic roles and need to be taken care of by a man since they are not capable of taking care of themselves.

Sexism an attitude, action, or institutional structure that subordinates or discriminates against individuals or groups because of their biological sex.

Benevolent sexism the belief that women are innocent creatures who should be protected and supported.

DIVERSITY IN OTHER COUNTRIES

Gender differences in body dissatisfaction do not exist throughout the world. Mellor et al. (2010) studied a sample of 513 Malay, Indian, and Chinese adolescent boys and girls in Malaysia and found no differences between the genders. The researchers noted that most Asian females tend to have trim bodies which approximate the Western cultural ideal. However, the researchers did observe that boys reported greater engagement in strategies to increase muscles. Again, Asian males tend to be smaller and may be more motivated to work out to achieve the male Western cultural ideal of a muscular body.

5. *Less Personal/Marital Satisfaction.* Read and Grundy (2011) surveyed 2,511 married couples and found that wives had poorer mental health than their husbands. Simon and Lively (2010) also found that women were more angry and more depressed than men. The researchers noted that one interpretation is that such anger is related to the sense of powerlessness that women feel in America—inequitable division of labor, lower status/lower wage jobs they are more likely to hold, less power in relationships, etc. Corra et al. (2009) analyzed General Social Survey data over a thirty-year period (1972–2002), controlled for socioeconomic factors such as income and education, and found that women reported less marital satisfaction than men. Similarly, twice as many husbands as wives among 105 late-life couples (average age, 69 years) reported that they had "no disappointments in the marriage" (15% versus 7%), suggesting greater dissatisfaction among wives (Henry et al. 2005). Bulanda (2011) also found lower marital satisfaction among wives. The lower marital satisfaction of wives is attributed to power differentials in the marriage. Traditional husbands expect to be dominant, which translates into their earning an income and the expectation that the wife not only will earn an income but also will take care of the house and children. The latter expectation results in a feeling of unfairness.

Before ending this section on negative consequences of being socialized as a woman, look at the Social Policy feature on **female genital alteration**. This is more of an issue for females born in some African, Middle Eastern, and Asian countries than for women in the United States. However, the practice continues even here.

Female genital alteration cutting off the clitoris or excising (partially or totally) the labia minora.

Positive Consequences of Traditional Female Role Socialization We have discussed the negative consequences of being born and socialized as a woman. However, there are also decided benefits.

1. *Longer Life Expectancy.* Women have a longer life expectancy than men. It is not clear if their greater longevity is related to biological or to social factors.

National **Data**

Females born in the year 2015 are expected to live to the age of 81.4, in contrast to men, who are expected to live to the age of 76.4 (*Statistical Abstract of the United States, 2011*, Table 102).

2. *Stronger Relationship Focus.* Women continue to prioritize family relationships over work relationships (Stone 2007) and spend more time taking care of children than men (Craig and Mullan 2010). Female family members, in contrast to male family members, are viewed as more nurturing and responsive (Monin et al. 2008). And, they provide more "emotion work" for children in the sense of attending to their emotional needs, helping them with what they are struggling with, etc. (Minnottea et al. 2010).

3. *Keep Relationships on Track.* Because women evidence more concern for relationships, they are more likely to be motivated to keep them on track and to initiate conversation when there is a problem. In a study of 203 undergraduates, two-thirds of the women, in contrast to 60% of the men, reported that they were likely to start a discussion about a problem in their relationship (Knox and Hatfield et al. 1998). In a study of almost a thousand high school students in Taiwan, female students were significantly more likely to use the Internet to connect and communicate with others than were males (who used the Internet to "explore") (Tsai and Tsai 2010).

4. *Bonding with Children.* Another advantage of being socialized as a woman is the potential to have a closer bond with children. In general, women tend to be more emotionally bonded with their children than men do. Although the new cultural image of the father is of one who engages emotionally with his

Female genital alteration, more commonly known as FGC (female genital cutting), female genital mutilation, or female circumcision, involves cutting off the clitoris or excising (partially or totally) the labia minora. "World-wide about 130 million women have undergone FGC. In the USA, more than 168,000 females have had or are at risk for this procedure and the number may be increasing as the admission ceiling for African refugees is raised. Federal law criminalizes the performance of FGC on females under 18 in the USA; however, the procedure is not unknown in this country. More commonly, young women are sent back to their country of origin for the procedure. Over 90% of women from Djibouti, Egypt, Eritrea, Ethiopia, Mali, Sierra Leone, and Northern Sudan have had the procedure" Nicoletti, 2007). Ninety-eight percent of the women in Somali have had the procedure (Simister, 2010). The American Academy of Pediatrics condemns all types of female genital cutting (Policy Statement, 2010).

The practice of FGC is not confined to a particular religion. The reasons for the practice include:

a. Sociological/cultural—parents believe that female circumcision makes their daughters lose their desire for sex, which helps them maintain their virginity and helps to ensure their marriageability and fidelity to their husbands. Hence, the "circumcised" female is seen as one whom males will desire as a wife. FGC is seen as a "rite of passage" that initiates a girl into womanhood and increases her bonding and social cohesion with other females.

b. Hygiene/aesthetics—female genitalia are considered dirty and unsightly so that their removal promotes hygiene and provides aesthetic appeal.

c. Religion—some Muslim communities practice FGC in the belief that the Islamic faith demands it, but it is not mentioned in the Qur'an.

d. Myths: FGC is thought to enhance fertility and promote child survival (Nicoletti 2007).

Elnashar and Abdelhady (2007) compared the sexuality of married women who had been circumcised with those who had not. The researchers found statistically significant differences such that the former were more likely to report pain during intercourse, loss of libido, and failure to orgasm. The wives who had been circumcised also reported more physical complaints, anxiety, and phobias.

Changing a country's deeply held beliefs and values concerning this practice cannot be achieved by denigration. More effective approaches to discouraging the practice include the following:

1. Respect the beliefs and values of countries that practice female genital operations. Calling the practice "genital mutilation" and "a barbaric practice" and referring to it as a form of "child abuse" and "torture" convey disregard for the beliefs and values of the cultures where it is practiced. In essence, we might adopt a culturally relativistic point of view (without moral acceptance of the practice).

---

children, many fathers continue to be content for their wives to take care of their children, with the result that mothers, not fathers, become more emotionally bonded with their children.

> *Therapists report that the most common complaint of women in distressed marriages is that their husbands are too withdrawn and don't openly share enough.*
>
> Howard Markman, marriage researcher

Consequences of Traditional Male Role Socialization

Male socialization in American society is associated with its own set of consequences. Both the negative and positive consequences are summarized in Table 2.4. As with women, each consequence may or may not be true for a specific man.

Negative Consequences of Traditional Male Role Socialization   There are several negative consequences associated with being socialized as a man in U.S. society.

   1. *Identity Synonymous with Occupation.* Ask men who they are, and many will tell you what they do. Society tends to equate a man's identity with his occupational role. Male socialization toward greater involvement in the labor force is evident in governmental statistics.

Maume (2006) analyzed national data on taking vacation time and found that men were much less likely to do so. They cited fear that doing so would affect

2. Remember that genital operations are arranged and paid for by loving parents who deeply believe that the surgeries are for their daughters' welfare.

3. It is important to be culturally sensitive to the meaning of being a woman. Indeed, genital cutting is mixed up with how a woman sees herself.

4. Simister (2010) studied national samples of genital alteration in Kenya and found that the higher the education of the mother, the lower the incidence of genital alteration in their daughters. Hence, increasing the educational level of women in a community is a structural way to reduce genital alteration in females.

Braun (2009) discussed the practice of female genital cosmetic surgery (FGCS) where Western women voluntarily have their genitals altered for aesthetic reasons. Such surgeries include those designed to reduce and make symmetrical the labia minora, those designed to 'augment' the labia majora, and those designed to tighten the vagina (so called 'vaginal rejuvenation' or vaginoplasty). These operations are sometimes called "designer vaginas" and have been critiqued by the American College of OBGYN. Researcher Braun finds it interesting that surgery on the genitals can be both a cause for public outcry and an issue of personal cosmetics.

Your Opinion?

1. To what degree do you feel the United States should become involved in the practice of female genital alterations of U.S. citizens?

2. To what degree can you regard the practice from the view of traditional parents and daughters?

3. How could not having the operation be a liability and a benefit for the woman whose culture supports the practice?

4. How does education influence the probability of genital alteration?

5. How might women who have female genital cutting view women who elect to have vaginal surgeries?

Sources

Braun, V. 2009. 'The women are doing it for themselves': The rhetoric of choice and agency around female genital 'cosmetic surgery'. *Australian Feminist Studies* 24: 233–249.

Elnashar, A., and R. Abdelhady. 2007. The impact of female genital cutting on health of newly married women. *International Journal of Gynecology & Obstetrics* 97: 238–244.

Nicoletti, A. 2007. Female genital cutting. *Journal of Pediatric & Adolescent Gynecology* 20: 261–262.

Policy Statement—Ritual Genital Cutting of Female Minors. 2010. *Pediatrics* 125: 1088–1093.

Simister, J. G. 2010. Domestic violence and female genital mutilation in Kenya: Effects of ethnicity and education. *Journal of Family Violence* 25: 247–257.

TABLE 2.4 Consequences of Traditional Male Role Socialization	
Negative Consequences	**Positive Consequences**
Identity tied to work role	Higher income and occupational status
Limited emotionality	More positive self-concept
Fear of intimacy; more lonely	Less job discrimination
Disadvantaged in getting custody	Freedom of movement; more partners to select from; more normative to initiate relationships
Shorter life (by five years)	Happier marriage

National **Data**

Seventy two percent of men, compared with 59.23% of women, were in the civilian workforce in 2009 (*Statistical Abstract of the United States, 2011*, Table 585).

Congressman John Boehner's being tearful/emotional has been commented on in the media (a congresswoman being emotional would go without notice).

their job or career performance evaluation. Women, on the other hand, were much more likely to use all of their vacation time. However, the "work equals identity" equation for men may be changing. Increasingly, as women are more present in the labor force and become co-providers, men become co-nurturers and co-homemakers. In addition, more stay-at-home dads and fathers are seeking full custody in divorce litigation. These changes challenge cultural notions of masculinity.

That men work more and play less may translate into fewer friendships and relationships. In a study of 377 university students, 25.9% of the men compared to 16.7% of the women reported feeling a "deep sense of loneliness" (Vail-Smith et al. 2007). Similarly, Grief (2006) reported that a quarter of 386 adult men reported that they did not have enough friends. Grief also suggested some possible reasons for men having few friends—homophobia, lack of role models, fear of being vulnerable, and competition between men. McPherson et al. (2006) also found that men reported fewer confidantes than women.

2. *Limited Expression of Emotions.* Some men feel caught between society's expectations that they be competitive, aggressive, and unemotional and their own desire to be more cooperative, passive, and emotional. Indeed, men see themselves as less emotional and loving than women (Hill 2007), and are pressured to disavow any expression that could be interpreted as feminine (for example, be emotional). Congressman John Boehner's being tearful/emotional has been commented on in the media (a congresswoman being emotional would go without notice). Also, 55% of the soldiers serving in Iraq or Afghanistan reported that they feared they would appear "weak" if they expressed feelings of fear or symptoms of post-traumatic stress disorder (Thompson 2008).

Lease et al. (2010) confirmed that the socialization of men (particularly white men) involves men "being more emotionally isolated and competitive in relationships and less competent at providing support to others, disclosing their own feelings, or managing conflict effectively. Indeed, adhering to norms of emotional self-control, the use of physical force as a viable response, and avoidance of anything deemed feminine or 'weak' is likely to reduce one's ability to be emotionally available and open when frustrated with romantic partners."

Garfield (2010) reviewed men's difficulty with emotional intimacy and noted that their emotional detachment stems from the provider role which requires them to stay in control. Being emotional is seen as weakness. Men's groups where men learn to access their feelings and express them have been helpful in increasing men's emotionality. Murphy et al. (2010) emphasized that men profit from becoming involved in the emotional labor of maintaining a relationship— their satisfaction increases.

3. *Fear of Intimacy.* Men may be socialized to withhold information about themselves that encourages the development of intimacy. Giordano et al. (2005) analyzed data from the National Longitudinal Study of Adolescent Health, consisting of more than 9,000 interviews, and found that adolescent boys reported less willingness to disclose than adolescent girls.

4. *Custody Disadvantages.* Courts are sometimes biased against divorced men who want custody of their children. Because divorced fathers are typically regarded as career-focused and uninvolved in child care, some are relegated to seeing their children on a limited basis, such as every other weekend or four evenings a month.

5. Shorter Life Expectancy. Men typically die five years sooner (at age 76) than women (*Statistical Abstract of the United States, 2011,* Table 102). One explanation is that the traditional male role emphasizes achievement, competition, and suppression of feelings, all of which may produce stress. Not only is stress itself harmful to physical health, but it may lead to compensatory behaviors such as smoking, alcohol, and other drug abuse, and dangerous risk-taking behavior (all of which is higher in males). In effect, the traditional male gender role is hazardous to men's physical health. Kimmel (2008) noted that

Males often know how to fish together. It is one of the few cultural rituals which allows for male bonding.

the socialization of white middle class males ages 16 to mid twenties can also be confusing. He notes that men are socialized in "guyland" where "real guys" tend to love girls but hate women, give acceptance to having sex with women/being sexually aggressive, and express disdain for gays by calling them "faggot." As guys emerge from guyland they are thrown into an uncertain world—women may have more education and be their supervisor. In addition, a male may not be able to find or keep a job which may wreck his sense of masculinity.

Benefits of Traditional Male Socialization   As a result of higher status and power in society, men tend to have a more positive self-concept and greater confidence in themselves. In a sample of 288, almost half (48%) of undergraduate/graduate men, in contrast to 30% of undergraduate women, agreed that "we determine whatever happens to us, and nothing is predestined" (Dotson-Blake et al. 2008). Men also enjoy higher incomes and an easier climb up the good-old-boy corporate ladder; they are rarely stalked or targets of sexual harassment. Other benefits are the following:

1. Freedom of Movement. Men typically have no fear of going anywhere, anytime. Their freedom of movement is unlimited. Unlike women, who are taught to fear rape and to be aware of their surroundings, walk in well-lit places, and who do not walk alone after dark; men are oblivious to these fears and perceptions. They can be alone in public and be anxiety-free about something ominous happening to them.

2. Greater Available Pool of Potential Partners. Because of the mating gradient (men marry "down" in age and education whereas women marry "up"), men tend to marry younger women so that a 35-year-old man may view women from 20 years to 40 years as possible mates. However, a woman of age 35 is more likely to view men her same age or older as potential mates. As she ages, fewer men are available; less so for men.

3. Norm of Initiating a Relationship. Men are advantaged because traditional norms allow men to be aggressive in initiating relationships with women. In contrast, women are less often aggressive in initiating a relationship. In a study of 1,027 undergraduates, 61.1% of the female respondents reported that they had not "asked a guy to go out" (Ross et al. 2008).

Changing Gender Roles

Gender roles are changing. The next subsections discuss androgyny, gender role transcendence, and gender postmodernism.

Androgyny

Adam Lambert of *American Idol* fame personifies **androgyny**, a blend of traits that are stereotypically associated with masculinity and femininity. Other androgynous celebrities include David Bowie, Boy George, Patti Smith, Annie Lennox, and k.d. Lang. Two forms of androgyny are:

1. Physiological androgyny refers to intersexed individuals, discussed earlier in the chapter. The genitals are neither clearly male nor female, and there is a mixing of "female" and "male" chromosomes and hormones.

2. Behavioral androgyny refers to the blending or reversal of traditional male and female behavior, so that a biological male may be very passive, gentle, and nurturing and a biological female may be very assertive, rough, and selfish. While not androgynous in the traditional use of the term, Nancy Pelosi has both masculine and feminine qualities (Dabbous and Ladley 2010). She is portrayed as 'very strong' and 'decisive' with a 'spine of steel' but she also possesses a 'heart of gold'—she is a mother of five and a grandmother of six.

Androgyny may also imply flexibility of traits; for example, an androgynous individual may be emotional in one situation, logical in another, assertive in another, and so forth. Gender role identity (androgyny, masculinity, femininity) was assessed in a sample of Korean and American college students with androgyny emerging as the largest proportion in the American sample and femininity in the Korean sample (Shin et al. 2010). Masculine stereotypes emerge early and may have negative outcomes. Granie (2010) found that boys' and girls' injury-risk behaviors were predicted by masculine stereotype conformity.

Woodhill and Samuels (2003) emphasized the need to differentiate between positive and negative androgyny. **Positive androgyny** views androgyny devoid of the negative traits associated with masculinity (aggression, hardheartedness, indifference, selfishness, showing off, and vindictiveness) and femininity (being passive, submissive, temperamental, and fragile). The researchers also found that positive androgyny is associated with psychological health and well-being.

Gender Role Transcendence

Beyond the concept of androgyny is that of gender role transcendence. We associate many aspects of our world, including colors, foods, social or occupational roles, and personality traits, with either masculinity or femininity. The concept of **gender role transcendence** means abandoning gender frameworks and looking at phenomena independent of traditional gender categories. DiDonato and Berenbaum (2011) studied 401 undergraduate students and

Gregorio T. Binuya/Everett Collection

Adam Lambert personifies androgyny.

Androgyny a blend of traits that are stereotypically associated with masculinity and femininity.

Positive androgyny a view of androgyny that is devoid of the negative traits associated with masculinity (e.g. aggression) and femininity (e.g. being passive).

Gender role transcendence abandoning gender frameworks and looking at phenomena independent of traditional gender categories.

found that psychological adjustment was positively associated with flexible gender attitudes.

Transcendence is not equal for women and men. Although females are becoming more masculine, in part because our society values whatever is masculine, men are not becoming more feminine. Indeed, adolescent boys may be described as very gender-entrenched. Beyond gender role transcendence is gender postmodernism.

Gender Postmodernism

Gender postmodernism abandons the notion of gender as natural and emphasizes that gender is socially constructed. Monro (2000) noted that people in the post modern society would no longer be categorized as male or female but be recognized as capable of many identities—"a third sex" (p. 37). A new conceptualization of "trans" people calls for new social structures, "based on the principles of equality, diversity and the right to self determination" (p. 42). No longer would our society telegraph transphobia but embrace pluralization "as an indication of social evolution, allowing greater choice and means of self-expression concerning gender" (p. 42).

Gender postmodernism abandons the notion of gender as natural and emphasizes that gender is socially constructed.

PERSONAL CHOICES

Choosing Gender Behavior that Fits

Being aware that gender role behavior is socially constructed gives one the freedom to engage in whatever occupational and relationship gender role behavior that seems a natural fit for one's personality. Occupational choices traditionally reserved for women (e. g. elementary school teacher) or men (e.g. athletic coach) need no longer be off the table for the other sex. Similarly, dating roles whereby the woman initiates and the man is passive or marital roles whereby the woman is the primary breadwinner and the man is the child-focused homemaker become options.

The Future of Gender Roles

Imagine a society in which women and men each develop characteristics, lifestyles, and values that are independent of gender role stereotypes. Characteristics such as strength, independence, logical thinking, and aggressiveness are no longer associated with maleness, just as passivity, dependence, emotions, intuitiveness, and nurturance are no longer associated with femaleness. Both sexes are considered equal, and women and men may pursue the same occupational, political, and domestic roles. These changes are occurring. . . slowly (Department of Commerce et al. 2011). Lucier-Greer and Adler-Baeder (2010) provided data that, compared to 2000, gender role attitudes of today are becoming more egalitarian. They observed the trend specifically among those who divorced and remarried with the latter being more likely to express egalitarian attitudes. Fisher (2010) emphasized that peer marriage or marriage between equals is the most profound change in marriage in recent years.

Another change in gender roles is the independence and ascendency of women. Hymowitz (2011) predicted that women will less often require marriage for fulfillment, will increasingly take care of themselves economically, and will opt for having children via adoption or donor sperm rather than foregoing motherhood. That women are slowly outstretching men in terms of education will provide the impetus for these changes.

Summary

What are the important terms related to gender?

Sex refers to the biological distinction between females and males. One's biological sex is identified on the basis of one's chromosomes, gonads, hormones, internal sex organs, and external genitals, and exists on a continuum rather than being a dichotomy. *Gender* is a social construct and refers to the social and psychological characteristics associated with being female or male. Other terms related to gender include *gender identity* (one's self-concept as a girl or boy), *gender role* (social norms of what a girl or boy "should" do), *gender role ideology* (how women and men "should" interact), and *transgender* (presenting one's self different from one's biological sex).

What theories explain gender role development?

Biosocial theory emphasizes that social behaviors (for example, gender roles) are biologically based and have an evolutionary survival function. Social learning theory emphasizes the roles of reward and punishment in explaining how children learn gender role behavior. Identification theory says that children acquire the characteristics and behaviors of their same-sex parent through a process of identification. Cognitive-developmental theory emphasizes biological readiness, in terms of cognitive development, of the child's responses to gender cues in the environment. Once children learn the concept of gender permanence, they seek to become competent and proper members of their gender group.

What are the various agents of socialization?

Various socialization influences include parents and siblings (representing different races and ethnicities), peers, religion, the economy, education, and mass media. These shape individuals toward various gender roles and influence what people think, feel, and do in their roles as woman or man. For example, the family is a gendered institution with female and male roles highly structured by gender. The names parents assign to their children, the clothes they dress them in, and the toys they buy them all reflect gender. Parents may also be stricter on female children, determining the age they are allowed to leave the house at night, time of curfew, and directives such as "text your mamma when you get to the party."

How are gender roles expressed in other societies?

Although there is no one Latino or Hispanic family and the gender roles differ across families, there seems to be a drift from the male provider or family homemaker toward less role rigidity. However, foreign-born Latino or Hispanic parents tend to regard the role of the man as provider and the role of the woman as primary caretaker and the main source of emotional nurturance for children and the family.

Women under Taliban rule in Afghanistan have been very oppressed. Some women drank household bleach rather than continue to endure their plight. They were not allowed to go to school or work and thus were completely dependent economically.

In the Caribbean, family patterns are diverse but are often characterized by women and their children as the primary family unit, with men often not living in the home.

What are the consequences of traditional gender role socialization?

Traditional female role socialization sometimes results in negative outcomes such as, less income, negative body image, and lower marital satisfaction but positive outcomes such as a longer life, a stronger relationship focus, keeping relationships on track, and a closer emotional bond with children. Traditional male role socialization may result in the fusion of self and occupation, a more limited expression of emotion, disadvantages in child custody disputes, and a shorter life but higher income, greater freedom of movement, a greater available pool of potential partners, and greater acceptance in initiating relationships. About 30% of college men in one study reported their preference for marrying a traditional wife (one who would stay at home to take care of children).

How are gender roles changing?

Androgyny refers to a blend of traits that are stereotypically associated with both masculinity and femininity. It may also imply flexibility of traits; for example, an androgynous individual may be emotional in one situation, logical in another, assertive in another, and so forth. The concept of gender role transcendence involves abandoning gender schema (for example, becoming "gender aschematic"), so that personality traits, social and occupational roles, and other aspects of our lives become divorced from gender categories. However, such transcendence is not equal for women and men. Although females are becoming more masculine partly because our society values whatever is masculine, men are not becoming more feminine.

What is the future of gender roles?

Imagine a society in which women and men each develop characteristics, lifestyles, and values that are independent of gender role stereotypes. Characteristics such as strength, independence, logical thinking, and aggressiveness are no longer associated with maleness, just as passivity, dependence, emotions, intuitiveness, and nurturance are no longer associated with female-ness. Both sexes are considered equal, and women and men may pursue the same occupational, political, and domestic roles. These changes are occurring, albeit slowly.

Women are also becoming more ascendant—they are earning more college degrees than men, they increasingly can take care of themselves economically, and they less often require marriage for their personal fulfillment.

Key Terms

Androgyny

Benevolent sexism

Biosocial theory

Cross-dresser

Female genital alteration

Feminization of poverty

Gender

Gender identity

Gender postmodernism

Gender role ideology

Gender role transcendence

Gender roles

Intersexed individuals

Occupational sex segregation

Parental investment

Positive androgyny

Sex

Sex roles

Sexism

Socialization

Transgender

Transsexual

Web Links

American Men's Studies Association
http://www.mensstudies.org/

Equal Employment Opportunity Commission
http://www.eeoc.gov/

Intersex Society of North America
http://www.isna.org/

National Organization for Women (NOW)
http://www.now.org/

Transgender Forum
http://www.tgforum.com/

Transsexuality
http://www.transsexual.org/

CHAPTER 3

Love

*Love is merely a madness; and,
I tell you, deserves a dark house
and a whip as madmen do: and
the reason why they are not so punished and cured is that lunacy is so
ordinary that the whippers are in
love too.*

William Shakespeare

Victoria Oliver

Learning Objectives

Review some of the ways love is conceptualized.

Know why and how love is subject to social control, especially from parents.

Summarize the various theories of love.

Identify the process of falling in love.

Review the ways in which love is a context for problems.

Specify the definition and types of jealousy.

Compare and contrast compersion, polyamory, and open relationships.

Predict the future of love relationships.

TRUE OR FALSE?

1. The most common love style of college students is "playing games" rather than passionate romantic love.

2. Moving from dating to living together to getting married is associated with putting on weight.

3. Parents in 40% of the world's population (e.g. China, India, and Indonesia) select the mate for their son or daughter.

4. Most companies/corporations have policies designed to control office romances.

5. Romantic love is losing its appeal among adolescents, particularly among males.

Answers: 1. F   2. T   3. T   4. F   5. F

Anyone who has been in love knows how mad and bizarre it can be (see opening quote). The love story that takes the cake for being bizarre is about Burt and Linda. Burt Pugach, 31, was married when he fell in love with 21-year-old Linda Riss. This New York love story of the fifties featured Linda finding out that Burt was married, breaking off the relationship, and getting involved with someone new (Larry Schwartz). Burt asked her to end the relationship with Larry and threatened that if she did not, he would "fix" it so no one else would want her. After Burt found out that she was engaged to Larry, he hired three men to throw lye in her face which blinded her in one eye (she eventually lost sight in the other eye). Burt was sent to prison for 14 years and wrote to Linda throughout his incarceration.

Meanwhile, Linda's fiancée Larry left her and other men found it difficult to cope with her blindness and eye disfigurement. At the time of Burt's release from prison, she was alone, unemployed, and financially destitute. She redefined Burt's insane abusive act as intense love for her and decided to let him support her, which he was willing to do. They married shortly after his release from prison in 1974 and were still married in 2011. In 1997, Burt was once again accused of threatening another woman with whom he was having an affair. Linda appeared at the trial as a character witness on behalf of her husband. *Crazy Love* is a documentary on the couple's relationship (http://www.crazylovefilm.com/home.html).

Love, and its romantic ideals of love forever, faithfulness, etc., remain firmly entrenched in American culture. A team of researchers surveyed 14,121 adolescents and found that the romantic ideal continues regardless of gender or sexual identity—an overwhelming proportion of the respondents rated love, faithfulness, and lifelong commitment as extremely important for marriage or long-term relationships (Meier et al. 2009).

J. H. Newman, cardinal and philosopher, noted that we tend to fear less that life will end but, rather, that life will never begin. His point targets the importance of love in one's life that provides an unparalleled richness, meaning, and happiness. Demir (2008) emphasized that involvement in a romantic relationship moves one to a new level of happiness independent of one's personality. In other words, although some individuals have personalities that tend to be happy anyway, love moves them to an even higher level.

For many, being in love is a prerequisite for getting and remaining married. Lovers affect each other. Schoebi (2008) identified hard (angry) and soft (depressed) emotions and the degree to which the emotions of one spouse affected another. When one spouse was experiencing hard emotions, the partner tended to mirror those, particularly when feeling interpersonal insecurity.

Love is very much a part of student life. More than half (51%) of 2,922 undergraduates reported that they were emotionally involved with one person; 6% were engaged (Knox and Hall 2010). This chapter reviews various conceptions, theories, and problems associated with love. Along the way we discover how love develops and what makes it last. Because jealousy in love relationships is common, we also examine its causes and consequences.

Ways of Conceptualizing Love

The word love is polysemous—the meanings are numerous (Berscheid 2010). Love is elusive and incapable of being defined by those caught in its spell. Hegi and Bergner (2010) provide a standard definition—*"Investment in the well-being of the other for his or her own sake."* Love is often confused with lust and infatuation (Jefson 2006). Love is about deep, abiding feelings with a focus on the long term (Foster 2010); **lust** is about sexual desire and the present, and **infatuation** is about emotional feelings based on little actual exposure to the love object. In the following section, we look at the various ways of conceptualizing love.

Lust sexual desire.

Infatuation intense emotional feelings based on little actual exposure to the love object.

Love Styles

Theorist John Lee (1973; 1988) identified a number of styles of love that describe the way lovers relate to each other. Keep in mind that the same individual may view love in more than one way at a time or may view love in different ways at different times. These love styles are also independent of one's sexual orientation—no one love style is characteristic of heterosexuals or homosexuals.

1. *Ludic.* The **ludic love style** views love as a game in which the player has no intention of getting seriously involved. The ludic lover refuses to become dependent on any one person and does not encourage another's intimacy. Two essential skills of the ludic lover are to juggle several partners at the same time and to manage each relationship so that no one partner is seen too often.

Ludic love style views love as a game. The ludic lover has no intention of getting involved.

These strategies help to ensure that the relationship does not deepen into an all-consuming love. Don Juan represented the classic ludic lover. "Love 'em and leave 'em" is the motto of the ludic lover. Tzeng et al. (2003) found that, whereas men were more likely than women to be ludic lovers, ludic love characterized the love style of college students the least.

While ludic lovers may sometimes be characterized as manipulative and uncaring (Jonason and Kavanagh 2010), they may also be compassionate and very protective of another's feelings. For example, some uninvolved, soon-to-graduate seniors avoid involvement with anyone new and become ludic lovers so as not to encourage anyone to fall in love whom they would soon leave.

2. *Pragma.* The **pragma love style** is the love of the pragmatic—that which is logical and rational. Pragma lovers assess their partners on the basis of assets and liabilities. Economic security may be a guiding consideration. One single parent mother of four married a neurosurgeon who paid for the expensive college educations of her children at Ivy League schools. The day the last child graduated, she divorced her physician husband. One wonders if this mother was operating from a pragma love style perspective. Pragma lovers do not become involved with interracial, long-distance, or age-discrepant partners because logic argues against doing so. See the Personal Choices section about making decisions with one's heart or head.

Pragma love style love style that is logical and rational. The love partner is evaluated in terms of pluses and minuses and is regarded as a good or bad "deal."

Do You Make Relationship Choices with Your Heart or Head?

Lovers are frequently confronted with the need to make decisions about their relationships, but they are divided on whether to let their heart or head rule in such decisions. Some evidence suggests that the heart rules. Almost half (49.8%) of 2,922 undergraduates agreed with the statement, "I make relationship decisions more with my heart than my head" (Knox and Hall 2010), suggesting that the heart tends to rule in relationship matters. We asked students in our classes on marriage and family to fill in the details about deciding with their heart or head. Some of their answers follow:

Heart

Those who relied on their hearts for making decisions (women more than men) felt that emotions were more important than logic and that listening to their heart made them happier. One woman said:

> In deciding on a mate, my heart would rule because my heart has reasons to cry and my head doesn't. My heart knows what I want, what would make me most happy. My head tells me what is best for me. But I would rather have something that makes me happy than something that is good for me.

Some men also agreed that the heart should rule. One said:

> I went with my heart in a situation, and I'm glad I did. I had been dating a girl for two years when I decided she was not the one I wanted and that my present girlfriend was. My heart was saying to go for the one I loved, but my head was telling me not to because if I broke up with the first girl, it would hurt her, her parents, and my parents. But I decided I had to make myself happy and went with the feelings in my heart and started dating the girl who is now my fiancée.

Relying on one's emotions does not always have a positive outcome, as the following experience illustrates:

> Last semester, I was dating a guy I felt more for than he did for me. Despite that, I wanted to spend any opportunity I could with him when he asked me to go somewhere with him. One day he had no classes, and he asked me to go to the park by the river for a picnic. I had four classes that day and exams in two of them. I let my heart rule and went with him. Nothing ever came of the relationship and I didn't do well in those classes.

Head

In contrast to making relationship decisions with one's heart, 16% of 2,922 undergraduates reported that they made such decisions with their head (Knox and Hall 2010). Some student comments about making relationship decisions rationally follow from a class on marriage and family:

> In deciding on a mate, I feel my head should rule because you have to choose someone that you can get along with after the new wears off. If you follow your heart solely, you may not look deep enough into a person to see what it is that you really like. Is it just a pretty face or a nice body? Or is it deeper than that, such as common interests and values? After the new wears off, it's the person inside the body that you're going to have to live with. The "heart" sometimes can fog up this picture of the true person and distort reality into a fairy tale.

Another student said:

> Love is blind and can play tricks on you. Two years ago, I fell in love with a man who I later found out was married. Although my heart had learned to love this

man, my mind knew the consequences and told me to stop seeing him. My heart said, "Maybe he'll leave her for me," but my mind said, "If he cheated on her, he'll cheat on you." I got out and am glad that I listened to my head.

Some individuals feel that both the head and the heart should rule when making relationship decisions.

When you really love someone, your heart rules in most of the situations. But if you don't keep your head in some matters, then you risk losing the love that you feel in your heart. I think that we should find a way to let our heads and hearts work together.

There is an adage, "Don't wait until you find the person you can live with; wait and find the person that you can't live without!" Upon hearing this quote, one of our students said, "I think both are important. I want my head to let me know it 'feels' right." In your own decisions you might consider the relative merits of listening to your heart or head and moving forward recognizing there is not one "right" answer for all individuals on all issues.

Eros love style also known as romantic love, the love of passion and sexual desire.

Mania love style the out-of-control love whereby the person "must have" the love object. Obsessive jealousy and controlling behavior are symptoms of manic love.

Storge love style also known as companionate love, a calm, soothing, nonsexual love devoid of intense passion.

Agape love style also known as compassionate love, characterized by a focus on the well-being of the love object, with little regard for reciprocation.

Romantic love an intense love whereby the lover believes in love at first sight, only one true love, and that love conquers all.

We sometimes encounter people, even perfect strangers, who begin to interest us at first sight, somehow suddenly, all at once, before a word has been spoken.

Fyodor Dostoevsky, Russian author

3. *Eros.* Just the opposite of the pragmatic love style, the **eros love style**, which is also known as romantic love, is imbued with passion and sexual desire. Eros is the most common love style of college women and men (Tzeng et al. 2003). The idea of a soul mate is part of American culture. In a study of over 1000 adults, 66% reported "belief in the idea of a soul mate;" 34% disagreed (Cary and Gellers 2010).

4. *Mania.* The **mania love style** is the out-of-control love whereby the person "must have" the love object. Obsessive jealousy and controlling behavior are symptoms of manic love. The lover is possessive, dependent, and "must have" the beloved. People who are extremely jealous and controlling reflect manic love. "If I can't have you, no one else will" is sometimes the mantra of the manic lover.

5. *Storge.* The **storge love style**, also known as companionate love, is a calm, soothing, nonsexual love devoid of intense passion. Respect, friendship, commitment, and familiarity are characteristics that help to define the relationship. The partners care deeply about each other but not in a romantic or lustful sense. Their love is also more likely to endure than fleeting romance. One's grandparents who have been married fifty years and who still love and enjoy each other are likely to have a storge type of love.

6. *Agape.* **Agape love style** (also known as compassionate love) is characterized by a focus on the well-being of the love object, with little regard for reciprocation. Key qualities of agape love are not responding to one's negativity and not expecting an exchange for positives but believing that the other means well and will respond kindly in time. Berscheid (2010) suggested that compassionate love may be the most enduring of all loves since it does not depend on immediate reciprocation. The love parents have for their children is often described as agape love.

Romantic versus Realistic Love

Love may also be described as being on a continuum from romanticism to realism (see Self Assessment Scale on Love Attitudes). For some people, love is romantic; for others, it is realistic. **Romantic love**, said to have appeared in all human groups at all times in human history (Berscheid 2010), is characterized in modern America by such beliefs as "love at first sight," and "If I were really in love, I would marry someone I had known for only a short time."

Regarding these beliefs, 21% of 2,922 undergraduates reported that they had experienced love at first sight; over a quarter (27%) reported that they would marry someone they had known for only a short time if they were in love (Knox and Hall 2010). Men were more likely than women to report having experienced love at first sight (38.3% versus 32.7%). One explanation is that men must be visually attracted to young, healthy females to inseminate them. This biologically-based reproductive attraction is interpreted as a love attraction so that the male

SELF-ASSESSMENT | The Love Attitudes Scale

This scale is designed to assess the degree to which you are romantic or realistic in your attitudes toward love. There are no right or wrong answers.

Directions

After reading each sentence carefully, circle the number that best represents the degree to which you agree or disagree with the sentence.

1	2	3	4	5
Strongly agree	Mildly agree	Undecided	Mildly disagree	Strongly disagree

_____ 1. Love doesn't make sense. It just is.

_____ 2. When you fall "head over heels" in love, it's sure to be the real thing.

_____ 3. To be in love with someone you would like to marry but can't is a tragedy.

_____ 4. When love hits, you know it.

_____ 5. Common interests are really unimportant; as long as each of you is truly in love, you will adjust.

_____ 6. It doesn't matter if you marry after you have known your partner for only a short time as long as you know you are in love.

_____ 7. If you are going to love a person, you will "know" after a short time.

_____ 8. As long as two people love each other, the educational differences they have really do not matter.

_____ 9. You can love someone even though you do not like any of that person's friends.

_____ 10. When you are in love, you are usually in a daze.

_____ 11. Love "at first sight" is often the deepest and most enduring type of love.

_____ 12. When you are in love, it really does not matter what your partner does because you will love him or her anyway.

_____ 13. As long as you really love a person, you will be able to solve the problems you have with the person.

_____ 14. Usually you can really love and be happy with only one or two people in the world.

_____ 15. Regardless of other factors, if you truly love another person, that is a good enough reason to marry that person.

_____ 16. It is necessary to be in love with the one you marry to be happy.

_____ 17. Love is more of a feeling than a relationship.

_____ 18. People should not get married unless they are in love.

_____ 19. Most people truly love only once during their lives.

_____ 20. Somewhere there is an ideal mate for most people.

_____ 21. In most cases, you will "know it" when you meet the right partner.

_____ 22. Jealousy usually varies directly with love; that is, the more you are in love, the greater your tendency to become jealous will be.

_____ 23. When you are in love, you are motivated by what you feel rather than by what you think.

_____ 24. Love is best described as an exciting rather than a calm thing.

_____ 25. Most divorces probably result from falling out of love rather than failing to adjust.

_____ 26. When you are in love, your judgment is usually not too clear.

_____ 27. Love comes only once in a lifetime.

_____ 28. Love is often a violent and uncontrollable emotion.

_____ 29. When selecting a marriage partner, differences in social class and religion are of small importance compared with love.

_____ 30. No matter what anyone says, love cannot be understood.

Scoring

Add the numbers you circled. 1 (strongly agree) is the most romantic response and 5 (strongly disagree) is the most realistic response. The lower your total score (30 is the lowest possible score), the more romantic your attitudes toward love. The higher your total score (150 is the highest possible score), the more realistic your attitudes toward love. A score of 90 places you at the midpoint between being an extreme romantic and an extreme realist. Both men and women undergraduates typically score above 90, with men scoring closer to 90 than women.

Reference

Medora, N. P., J. H. Larson, N. Hortacsu, and P. Dave. 2002. Perceived attitudes towards romanticism: A cross-cultural study of American, Asian-Indian, and Turkish young adults. _Journal of Comparative Family Studies_ 33:155–178.

Source

Knox, D. "Conceptions of Love at Three Developmental Levels" Dissertation, Florida State University, 1969. Permission to use the scale for research available from David Knox at davidknox2@yahoo.com or by contacting Dr. Knox, Department of Sociology, East Carolina University, Greenville, NC 27858.

feels immediately drawn to the female, but he may actually see an egg needing fertilization. Dotson-Blake et al. (2008) found further evidence that males are more romantic than females in that men were significantly more likely (85% versus 73%) than women to believe that they could solve any relationship problem as long as they were in love. Raley and Sullivan (2010) found that African-American black adolescent males were more romantic than white males.

Data from a survey of 641 young adults at three international universities indicated that young American adults are the most romantic, followed by Turkish students, with Indians having the lowest romanticism scores (Medora et al. 2002).

Romantic love provides an even stronger motivation than the sex drive.

Helen Fisher, anthropologist

DIVERSITY IN OTHER COUNTRIES

The theme of American culture is individualism, which translates into personal fulfillment, emotional intimacy, and love as the reason for marriage. In Asian cultures (e.g. China) the theme is collectivism, which focuses on family, comradeship, obligations to others, and altruismî with love as secondary (Riela et al. 2010).

There's nothing half so sweet in life

As love's young dream.

Thomas Moore, *Love's Young Dream*

The heart has reasons that reason does not understand.

Jacques Benigne Boussuet, French bishop and theologian

In regard to love at first sight, Barelds and Barelds-Dijkstra (2007) studied the relationships of 137 married couples or cohabitants (together for an average of twenty-five years) and found that those who fell in love at first sight had similar relationship quality to those couples who came to know each other more gradually. Huston et al. (2001) found that, after two years of marriage, the couples that had fallen in love more slowly were just as happy as couples that fell in love at first sight.

The symptoms of romantic love include drastic mood swings, palpitations of the heart, and intrusive thoughts about the partner. F. Scott Fitzgerald immortalized the concept of romantic obsession in *The Great Gatsby*. Of Daisy Buchanan, he wrote, "She was the first girl I ever loved and I have faithfully avoided seeing her . . . to keep that illusion perfect." He actually was writing about a real-life true love, Ginevra King, whom he had met when she was 16; she eventually married another man (West 2005).

Infatuation comes from the same root word as *fatuous*, meaning "silly" or "foolish," and refers to a state of passion or attraction that is not based on reason. Infatuation is characterized by the tendency to idealize the love partner. People who are infatuated magnify their lovers' positive qualities ("My partner is always happy") and overlook or minimize their negative qualities ("My partner doesn't have a problem with alcohol; he just likes to have a good time").

David Knox

This romantically in love couple was married 30 minutes before this photo was taken. Over time, as spouses, their love will move toward being conjugal, companionate love.

In contrast to romantic love, which is characterized by excitement and passion, is realistic love, also known as conjugal love. This type of love is characterized by companionship, calmness, comfort and security. **Conjugal love** is the love between married people. The Love Attitudes Scale (see page 69) provides a way for you to assess the degree to which you tend to be romantic or realistic (conjugal) in your view of love. When you determine your score from the Love Attitudes Scale, be aware that your tendency to be a romantic or a realist is neither good nor bad. Both romantics and realists can be happy individuals and successful relationship partners. Love also conveys enormous benefits, including positive mental health. Plant et al. (2010) emphasized that these benefits are so pronounced for the individual that he or she will protect these benefits by diverting one's self when he or she feels attracted to an alternative.

Triangular View of Love

Sternberg (1986) developed the "triangular" view of love, consisting of three basic elements: intimacy, passion, and commitment. The presence or absence of these three elements creates various types of love experienced between individuals, regardless of their sexual orientation. These various types include:

1. *Nonlove*—the absence of intimacy, passion, and commitment. Two strangers looking at each other from afar have a nonlove.

2. *Liking*—intimacy without passion or commitment. A new friendship may be described in these terms of the partners liking each other.

3. *Infatuation*—passion without intimacy or commitment. Two people flirting with each other in a bar may be infatuated with each other.

4. *Romantic love*—intimacy and passion without commitment. Love at first sight reflects this type of love.

5. *Conjugal love (also known as married love)*—intimacy and commitment without passion. A couple that has been married for fifty years is said to illustrate conjugal love.

6. *Fatuous love*—passion and commitment without intimacy. Couples who are passionately wild about each other and talk of the future but do not have an intimate connection with each other have a fatuous love.

7. *Empty love*—commitment without passion or intimacy. A couple who stay together for social and legal reasons but who have no spark or emotional sharing between them have an empty love.

8. *Consummate love*—combination of intimacy, passion, and commitment; Sternberg's view of the ultimate, all-consuming love.

Individuals bring different combinations of the elements of intimacy, passion, and commitment (the triangle) to the table of love. One lover may bring a predominance of passion, with some intimacy but no commitment (romantic love), whereas the other person brings commitment but no passion or intimacy (empty love). The triangular theory of love allows lovers to see the degree to which they are matched in terms of passion, intimacy, and commitment in their relationship (see Figure 3.1).

A common class exercise among professors who teach about marriage and the family is to randomly ask class members to identify one word they most closely associate with love. Invariably, students identify different words (commitment, feeling, trust, altruism, and so on), suggesting great variability in the way we think about love. Indeed, just the words "I love you" have different meanings, depending on whether they are said by a man or a woman. In a study of 147 undergraduates (72% female, 28% male), men (more than women) reported that saying "I love you" was a ploy to get a partner to have sex, whereas women (more than men) reported that saying

Conjugal love the love between married people characterized by companionship, calmness, comfort, and security.

"Beyond the Physical"

Physical touch has evaded us, yet I feel you.

We know not what we are, yet the thought of losing you tears at my very being.

We share no quarters, yet you sleep in my heart each night.

We take no walks, yet joy overwhelms me as I imagine you near.

We have not shared the night, yet I see your likeness in the stars.

Until we can finally touch, I dwell with you in a world of splendor beyond the physical.

Jeff Johnson, poet, scholar

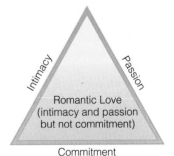

Figure 3.1 Romantic love is characterized by eros— intimacy and passion without commitment—and is the type of love most prevalent among college students.

"I love you" was a reflection of their feelings for the partner (with no sexual motive) (Brantley et al. 2002).

Why a Love Relationship Lasts

Caryl Rusbult's investment model of commitment has been used to identify why relationships last. While love is important, there are other factors involved....

> *people become dependent on their relationships because they (a) are satisfied with the relationship—it gratifies important needs, including companionship, intimacy, and sexuality; (b) believe their alternatives are poor—their most important needs could not be gratified independent of the particular relationship (e.g., in an alternative relationship, by friends and kin); and (c) have invested many important resources in the relationship (e.g., time, effort, shared friendship network, and material possessions). The model thereby includes not only internal factors (i.e., satisfaction) to explain why partners stick with each other but also external, structural factors to capture individuals in their interpersonal context (Finkenauer 2010, 162).*

Hence, people stay in a relationship because their needs are being met, they have no place to go, and they have made considerable investment in getting where they are and don't want to give it all up.

Social Control of Love

Because romantic love is such a powerful emotion and marriage such an important relationship, mate selection is not left to chance in connecting an outsider into an existing family. Parents inadvertently influence the mate choice of their children by moving to certain neighborhoods, joining certain churches, and enrolling their children in certain schools. Doing so increases the chance that their offspring will "hang out" with, fall in love with, and marry people who are similar in race, education, and social class. Although twenty-first-century parents normally do not have large estates and are not concerned about the transfer of wealth, they usually want their offspring to meet someone who will "fit in" and with whom they will feel comfortable. Peers exert a similar influence on homogenous mating by approving or disapproving certain partners. Their motive is similar to that of parents—they want to feel comfortable around the people their peers bring with them to social encounters. Both parents and peers are influential, as most offspring and friends end up falling in love with and marrying people of the same race, education, and social class.

Social approval of one's partner is normally important for a love relationship to proceed on course. If all of your friends disapprove of the person you are involved with, it is difficult for you to continue the relationship.

The ultimate social control of love is **arranged marriage**—mate selection pattern whereby parents select the spouse of their offspring. A matchmaker may be used but the final selection is someone the parents approve of. Parents arrange 80% of marriages in China, India, and Indonesia (three countries representing 40% of the world's population).

Parents in these familistic countries select the mate for their child in an effort to prevent any potential love relationship from forming with the "wrong" person and to ensure that the child marries the "right" person. Such a person must belong to the desired social class, have the education/economic resources that the parents desire, and share the same cultural/religious values. Marriage is regarded as the linking of two families; the love feelings of the respective partners are irrelevant. Love is expected to follow marriage, not precede it.

In the traditional arranged marriage the couple may get a fifteen-minute meeting followed by a few brief encounters (with parents nearby) before the wedding. However, love marriages—where the individuals meet, fall in love, and then convince and cajole their parents to approve of a wedding—are slowly

No disguise can long conceal where love exists, or long feign it where it is lacking.

La Rochfoucauld, French author

Arranged marriage mate selection pattern whereby parents select the spouse of their offspring. A matchmaker may be used but the selection is someone of which the parents approve.

becoming more common in Eastern societies (D. Jones 2006). Similarly, to accommodate the needs of traditional parents in a small village in Western Turkey who want to arrange the marriage of their children but appear "modern," anthropologist Hart (2007) observed that they now allow a period of time for the couple to develop romantic love feelings for each other.

America is a country that prides itself on the value of individualism. We proclaim that we are free to make our own choices. Not so fast, however. Love may be blind, but it knows what color a person's skin is. The data are clear—potential spouses seem to see and select people of similar color, as about 90% of people marry someone of their own racial background (*Statistical Abstract of the United States 2011*, Table 60). Hence, parents and peers may approve of their offsprings' and friends' love choice when the partner is of the same race and disapprove of the selection when the partner is not. These approval and disapproval mechanisms illustrate social control of love.

National **Data**

Fewer than 1% of the almost 60 million married couples in the United States include an African American spouse and a white spouse (*Statistical Abstract of the United States 2011*, Table 60).

> *Love is an ideal thing, marriage a real thing; a confusion of the real with the ideal never goes unpunished.*
>
> Goethe, German writer

> *We can cure physical diseases with medicine, but the only cure for loneliness, despair, and hopelessness is love. There are many in the world who are dying for a piece of bread, but there are many more dying for a little love.*
>
> Mother Teresa, Nobel Peace Prize winner

Another example of the social control of love is that individuals attracted to someone of the same sex quickly feel the social and cultural disapproval of this attraction. Although we discuss same-sex relationships later in the text, these relationships are challenged by the lack of institutional support. Even though upwards of ten states legally sanction same-sex marriage, these couples are no longer "married" once they cross the border into another state. Nevertheless, regardless of the law, same-sex love is common. Diamond (2003) emphasized that individuals are biologically wired and capable of falling in love and establishing intense emotional bonds with members of their own or opposite sex (hence, one's partners for love desire and for sexual desire can be different).

The social control of love may also occur in the workplace (see the Social Policy section).

Theories on the Origins of Love

Various theories have been suggested with regard to the origins of love.

Evolutionary Theory

Gillath et al. (2008) provided evidence that sexual interest and arousal are associated with motives to form and maintain a close relationship, to fall in love. They suggested these motives are hardwired to ensure a stable relationship for producing offspring. Although these motives are subject to distraction of new sexual opportunities, they nevertheless suggest a broader relationship

Evolutionary theory says that this woman and man are wired to seek each other, to fall in love, to have sex, and to rear children.

With women being almost half of the American workforce, people delaying marriage until they are older, and workers being around each other for 8 or more hours a day, the workplace has become a common place for romantic relationships to develop. More future spouses may meet at work than in academic, social, or religious settings. Barack and Michelle Obama, Bill and Melinda Gates, and Brad Pitt and Angelina Jolie all met "on the job." Almost 60% of a national survey of workers revealed that they had become involved in a romance with someone at work (Stott 2010).

Although such relationships are most often between peers, sometimes a love relationship develops between individuals occupying different status positions. And it can get ugly. Such was the case of Harry Stonecipher (a 68-year-old married man and head of Boeing) and a female employee, which resulted in Stonecipher being fired. His dismissal was not because of the affair (there were no company rules) but because of the negative publicity he brought to Boeing when his steamy e-mails became public. These types of love relationships are sometimes problematic in the workplace.

Advantages of an Office Romance

The energy that both fuels and results from intense love feelings can also fuel productivity on the job. If the coworkers eventually marry or enter a nonmarital but committed, long-term relationship, they may be more satisfied with and committed to their jobs than spouses whose partners work elsewhere. Working at the same location enables married couples to commute together, go to company-sponsored events together, and talk shop together. Workplaces such as academia often try to hire both spouses since they are likely to become more permanent workers.

Recognizing the potential benefits of increased job satisfaction, morale, productivity, creativity, and commitment, some companies encourage love relationships among employees. Aware that their single employees are interested in relationships, in Tokyo, Japan, Hitachi Insurance Service provides a dating service for its 400,000 employees (many of whom are unmarried) called Tie the Knot. Those interested in finding a partner complete an application and a meeting or lunch is arranged with a suitable candidate through the Wedding Commander. In America, some companies hire two employees who are married, reflecting a focus on the value of each employee to the firm rather than on their love relationship outside work.

Disadvantages of an Office Romance

However, workplace romances can also be problematic for the individuals involved as well as for their employers. When a workplace romance involves a supervisor/subordinate relationship, other employees might make claims of favoritism or differential treatment. In a typical differential-treatment allegation, an employee (usually a woman) claims that the company denied her a job benefit because her supervisor favored a female coworker—who happens to be the supervisor's girlfriend.

Across the gateway of my heart

I wrote, "No Thoroughfare,"

But love came laughing by, and cried:

"I enter everywhere."

Herbert Shipman, *No Thoroughfare*

motivation to sex. In effect, love has an evolutionary purpose by providing a bonding mechanism between the parents during the time their offspring are dependent infants.

Love's strongest bonding lasts about four years, the time when children are most dependent and when two parents can cooperate in handling their new infant. "If a woman was carrying the equivalent of a twelve-pound bowling ball in one arm and a pile of sticks in the other, it was ecologically critical to pair up with a mate to rear the young," observed anthropologist Helen Fisher (Toufexis 1993). The "four-year itch" is Fisher's term for the time at which parents with one child are most likely to divorce—the time when the woman can more easily survive without parenting help from the male. If the couple has a second child, doing so resets the clock, and "the seven-year itch" is the next most vulnerable time.

Love may also have a physiological basis. Dopamine is a chemical in the brain that is associated with the development of love. The prairie vole is a monogamous rodent. If a drug that reduces the effect of dopamine is injected into the brain of a monogamous vole, it loses the preference to "commit" to one mate. If the animal is then placed with a new mate and given a drug that increases dopamine levels, it develops the equivalent of romantic feelings for the new partner (Fisher et al. 2006).

If a workplace relationship breaks up, it may be difficult to continue to work in the same environment (and others at work may experience the fallout). A breakup that is less than amicable may result in efforts by partners to sabotage each other's work relationships and performance, incidents of workplace violence, harassment, and/or allegations of sexual harassment. In a survey of 774 respondents who had experience in the workplace, over a third (36.3%) recommended avoiding involvement in an office romance (Merrill and Knox 2010). Raso (2008) noted that such love relationships may also become a problem because coworkers do not want to be subjected to the open display of a roller coaster love affair at work.

Company Policies on Office Romances

Some companies such as Disney, Universal, and Columbia have "anti-fraternization" clauses that impose a cap on workers talking about private issues or sending personal e-mails. Some British firms have "love contracts" that require workers to tell their managers if they are involved with anyone from the office.

Most companies (Wal-Mart is an example) do not prohibit romantic relationships among employees. However, the company may have a policy prohibiting open displays of affection between employees in the workplace and romantic relationships between supervisor and a subordinate. Most companies have no policy regarding love relationships at work and generally regard romances between coworkers as "none of their business." There are some exceptions to the general permissive policies regarding workplace romances. Many companies have written policies prohibiting intimate relationships when one member of the couple is in a direct supervisory position over the other. These policies may be enforced by transferring or dismissing employees who are discovered in romantic relationships.

Your Opinion?

1. To what degree do you believe corporations should develop policies in regard to workplace romances?
2. What are the advantages and disadvantages of a workplace romance for a business?
3. What are the advantages and disadvantages for individuals involved in a workplace romance?
4. How might an office romance of peers affect coworkers?

Sources

Merrill, J., and D. Knox. 2010. *Finding Love from 9 to 5: Trade Secrets of Office Romance*. Santa Barbara, CA: Praeger.

Raso, R. 2008. How to handle workplace romance, foster interpersonal skills. *Nursing Management* 39:56–57.

Stott, P. 2010. No recession for workplace romance. Vault.Com Office Romance Survey 2010. Posted on Vault.com's website February 14 and retrieved February 18 at http://vaultcareers.wordpress.com/2010/02/17/no-recession-for-workplace-romance/

Learning Theory

Unlike evolutionary theory, which views the experience of love as innate, learning theory emphasizes that love feelings develop in response to certain behaviors in which a partner engages. Individuals in a new relationship who look at each other, smile at each other, compliment each other, touch each other endearingly, do things for each other, and do enjoyable things together are engaging in behaviors that make love feelings develop easily. In effect, love can be viewed as a feeling that results from a high frequency of positive behavior and a low frequency of negative behavior. One high-frequency behavior is positive labeling whereby the partners flood each other with positive statements. We asked one of our students who reported that she was deliriously in love to identify the positive statements her partner had said to her. She kept a list for a week, which included:

> *Angel, Sweetie, Cinderella, Sleeping Beauty.*
> *You understand me so well.*
> *Precious jewel, Sugar bear, Snow White, Honey.*
> *You are my best friend.*
> *I never thought you were out there.*

Love looks not with the eyes, but with the mind,

And therefore is winged Cupid painted blind.

William Shakespeare, *A Midsummer Night's Dream*

You know me better than anyone in my whole life.
You're always safe in my arms.
I would sell my guitar for you if you need money.

People who "fall out of love" may note the high frequency of negatives on the part of their partner and the low frequency of positives. People who say, "this is not the person I married," are saying the ratio of positives to negatives has changed dramatically.

Cunningham et al. (2005) used the term **social allergy** to refer to being annoyed and disgusted by a repeated behavior on the part of the partner. Examples are uncouth habits (for example, picking one's teeth), inconsiderate acts (not offering to get something from the kitchen when going oneself), intrusive behaviors (opening one's mail or e-mail or checking one's cell phone to see the listing on speed-dial), and norm violations (drinking out of someone else's glass). The researchers found that these types of behaviors increased over time and were associated with both decreased relationship satisfaction and termination of the relationship.

Sociological Theory

Fifty years ago, Ira Reiss (1960) suggested the wheel model as an explanation for how love develops. Basically, the wheel has four stages—rapport, self-revelation, mutual dependency, and fulfillment of personality needs. In the rapport stage, each partner has the feeling of having known the partner before, feels comfortable with the partner, and wants to deepen the relationship.

Such desire leads to self-revelation or self-disclosure, whereby each reveals intimate thoughts to the other about oneself, the partner, and the relationship. Such revelations deepen the relationship because it is assumed that the confidences are shared only with special people, and each partner feels special when listening to the revelations of the other. Indeed, Shelon et al. (2010) confirmed that intimacy develops when individuals disclose personal information about themselves and perceive that the listener understands/validates and cares about their disclosure. As the level of self-disclosure becomes more intimate, a feeling of mutual dependency develops. Each partner is happiest in the presence of the other and begins to depend on the other for creating the context of these euphoric feelings. "I am happiest when I am with you" is the theme of this stage.

The feeling of mutual dependency involves the fulfillment of personality needs. The desires to love and be loved, to trust and be trusted, and to support and be supported are met in the developing love relationship.

Psychosexual Theory

According to psychosexual theory, love results from blocked biological sexual desires. In the sexually repressive mood of his time, Sigmund Freud (1905/1938) referred to love as "aim-inhibited sex." Love was viewed as a function of the sexual desire a person was not allowed to express because of social restraints. In Freud's era, people would meet, fall in love, get married, and have sex. Freud felt that the socially required delay from first meeting to having sex resulted in the development of "love feelings." By extrapolation, Freud's theory of love suggests that love dies with marriage (access to one's sexual partner).

Biochemical Theory

There may be a biochemical basis for love feelings. **Oxytocin** is a hormone released from the pituitary gland during the expulsive stage of labor that has been associated with the onset of maternal behavior in lower animals. It has been referred to as the "cuddle chemical" because of its significance in

Social allergy being annoyed and disgusted by a repeated behavior on the part of the partner.

Oxytocin a hormone released from the pituitary gland during the expulsive stage of labor that has been associated with the onset of maternal behavior in lower animals.

Chapter 3 Love

bonding. Later in life, oxytocin seems operative in the development of love feelings between lovers during sexual arousal. Oxytocin may be responsible for the fact that more women than men prefer to continue cuddling after intercourse.

Phenylethylamine (PEA) is a natural, amphetamine-like substance that makes lovers feel euphoric and energized. The high that they report feeling just by being with each other is from the PEA that the brain releases in their bloodstream. The natural chemical high associated with love may explain why the intensity of passionate love decreases over time. As with any amphetamine, the body builds up a tolerance to PEA, and it takes more and more to produce the special kick. Hence, lovers develop a tolerance for each other. "Love junkies" are those who go from one love affair to the next in rapid succession to maintain the high. Alternatively, some lovers break up and get back together frequently as a way of making the relationship new again and keeping the high going.

Zeki (2007) emphasized the neurobiology of love in that both romantic love and maternal love are linked to the perpetuation of the species. Romantic love bonds the male and female together to reproduce, take care of, and socialize new societal members, whereas maternal love ensures that the mother will prioritize the care of her baby over other needs. Because of the social functions of these love states, neurobiologists have learned via brain imaging techniques that both types of attachment activate regions of the brain that access the brain's reward system (areas rich in oxytocin and vasopressin receptors). At the same time, negative cognitions or emotions about these relationships are shut down to allow the positive to predominate. No wonder both lovers and mothers seem very happy and focused. They are on a biological mission, and the reward center of their brain keeps them on track.

Meyer (2007) noted that taking selective serotonin reuptake inhibitor (SSRI) medications commonly used for depression and anxiety can affect relationship satisfaction (for example, blunt emotions and decrease sexual interest); Meyer also emphasized the importance of checking with one's physician. In some cases, other medications can be used to block these negative side effects.

Attachment Theory

The attachment theory of love emphasizes that a primary motivation in life is to be connected with other people. Monteoliva et al. (2005) confirmed that the attachment style an individual has with one's parents is associated with the quality of one's later romantic relationships. Specifically, a secure emotional attachment with loving adults as a child is associated with later involvement in a satisfying, loving, communicative relationship. This finding was true regardless of ethnic or racial background. Berscheid (2010) noted that individuals transition their love from parents to others and feel most secure when they are connected. Thus individuals struggle to maintain an intimate relationship since being attached to one creates a sense of security. Mohr et al. (2010) also used attachment theory to note that undergraduates who measured high on anxiety (presumably from lack of secure attachments in their family) predicted challenges in long term romantic relationship. Men and singles were particularly more likely to predict difficulties over time with a romantic partner.

Attachment theory has its basis in the work of Rene Spitz and Harry Harlow. The former emphasized the importance of infants being held and nurtured for their physical and emotional development. Dr. Harlow studied infant rhesus monkeys and found that they preferred soft mother-like dummies that offered no food over dummies that provided a food source but were made of wire and were less pleasant to the touch.

Each of the theories of love presented in this section has critics (see Table 3.1).

TABLE 3.1   Love Theories and Criticisms

Theory	Criticism
Evolutionary—love is the social glue that bonds parents with dependent children and spouses with each other to care for offspring.	The assumption that women and children need men for survival is not necessarily true today. Women can have and rear children without male partners.
Learning—positive experiences create love feelings.	This theory does not account for (1) why some people will share positive experiences yet will not fall in love, and (2) why some people stay in love despite negative behavior.
Psychosexual—love results from blocked biological drive.	The theory does not account for people who report intense love feelings yet are having sex regularly.
Sociological—the wheel theory whereby love develops from rapport, self-revelation, mutual dependency, and personality need fulfillment.	Not all people are capable of rapport, revealing oneself, and so on.
Biochemical—love is chemical. Oxytocin is an amphetamine-like chemical that bonds mother to child and produces a giddy high in young lovers.	This theory does not specify how much of what chemicals result in the feeling of love. Chemicals alone cannot create the state of love; cognitions are also important.
Attachment—primary motivation in life is to be connected to others. Children bond with parents and spouses to each other.	Not all people feel the need to be emotionally attached to others. Some prefer to be detached.

Love grants in a moment

What toil can hardly achieve in an age.

Goethe

How delicious is the winning,

Of a kiss at Love's beginning.

Thomas Campbell, Scottish poet

Falling in Love

Various social, physical, psychological, physiological, and cognitive conditions affect the development of love relationships.

Social Conditions for Love

Love is a social label given to an internal feeling. Our society promotes love through popular music, movies, and novels. These media convey the message that love is an experience to pursue, enjoy, and maintain. People who fall out of love are encouraged to try again: "love is lovelier the second time you fall." Love also takes place in a cultural context.

Body Type Condition for Love

The probability of being involved in a love relationship is influenced by approximating the cultural ideal of physical appearance. Halpern et al. (2005) analyzed data on a nationally representative sample of 5,487 African American, white, and Hispanic adolescent females and found that, for each one-point increase in body mass index (BMI), the probability of involvement in a romantic relationship dropped by 6%. Hence, to the degree that a woman approximates the cultural ideal of being trim and "not being fat," she increases the chance of attracting a partner and becoming involved in a romantic love relationship. One of our former students dropped from 225 pounds to 125 pounds and noted, "You wouldn't believe the dramatic difference in the way guys noticed and talked to me [between] when I was beefed up and when I was trim. I was engaged within three months of getting the weight off and am now married."

Ambwani and Strauss (2007) found that body image has an effect on sexual relations and that relationships

DIVERSITY IN OTHER COUNTRIES

Madathil and Benshoff (2008) compared the importance attributed to love of Asian and Indian couples living in the United States who had arranged marriages versus American couples who selected their own partners. The researchers found greater importance attributed to love by Asian and Indian couples than American couples. One explanation was the greater joy of Asian and Indian couples who were experiencing freedom since they were not living in their home countries and being restricted by their culture as compared to American couples who took freedom to love and choice of their mate as a given.

affect their self-image. Hence, women who felt positively about their body were more likely to report having sexual relations with a partner. The fact that they were in a relationship was associated with positive feelings about themselves.

Psychological Conditions for Love

Five psychological conditions associated with falling in love are perception of reciprocal liking, personality, high self-esteem, self-disclosure and gratitude.

Perception of Reciprocal Liking   Riela et al. (2010) conducted two studies on falling in love using both American and Chinese samples. The researchers found that one of the most important psychological factors associated with falling in love was the perception of reciprocal liking. When one perceives that he or she is desired by someone else, this has the effect of increasing the attraction toward that person. Such an increase is particularly strong if the person is very physically attractive (Greitemeyer 2010).

Personality Qualities   Riela et al. (2010) also found that the personal qualities of the love object has an important effect on falling in love. Viewing the partner as intelligent or having a sense of humor are examples of qualities that make the lover want to be with the beloved.

Self-Esteem   High self-esteem is also important for falling in love since it enables individuals to feel worthy of being loved. Feeling good about yourself allows you to believe that others are capable of loving you. Individuals with low self-esteem doubt that someone else can love and accept them. Rill et al. (2009) found that one's positive self esteem was not only positively related to one's level of commitment but to the perception that one's partner was also committed. Having high self-esteem provides other benefits:

1. It allows one to be open and honest with others about both strengths and weaknesses.

2. It allows one to feel generally equal to others.

3. It allows one to take responsibility for one's own feelings, ideas, mistakes, and failings.

4. It allows for the acceptance of both strengths and weaknesses in oneself and others.

5. It allows one to validate oneself and not to expect the partner to do this.

6. It permits one to feel empathy—a very important skill in relationships.

7. It allows separateness and interdependence, as opposed to fusion and dependence.

Positive physiological outcomes also follow from high self-esteem. People who feel good about themselves are less likely to develop ulcers and are likely to cope with anxiety better than those who don't. In contrast, low self-esteem has devastating consequences for individuals and the relationships in which they become involved. Not feeling loved as a child and, worse, feeling rejected and abandoned creates the context for the development of a negative self-concept and mistrust of others. People who have never felt loved and wanted may require constant affirmation from a partner as to their worth, and may cling desperately to that person out of fear of being abandoned. Such dependence (the modern term is *codependency*) may also encourage staying in unhealthy relationships (for example, abusive and alcoholic relationships) because the person may feel "this is all I deserve." Fuller and Warner (2000) studied 257 college students and observed that women had higher codependency scores than men. Codependency was also associated with being reared in families that were stressful and alcoholic.

One characteristic of individuals with low self-esteem is that they may love too much and be addicted to unhealthy love relationships. Petrie et al. (1992) studied fifty-two women who reported that they were involved in unhealthy love relationships in which they had selected men with problems (such as alcohol or other drug addiction) that they attempted to solve at the expense of neglecting themselves. "Their preoccupation with correcting the problems of others may be an attempt to achieve self-esteem," the researchers noted (p. 17). "I know I can help this man" is the motif of these women. Although having positive feelings about oneself when entering into a love relationship is helpful, sometimes these develop after one becomes involved in the relationship. "I've always felt like an ugly duckling," said one woman. "But once I fell in love with him and him with me, I felt very different. I felt very good about myself then because I knew that I was somebody that someone else loved." High self-esteem, then, is not necessarily a prerequisite for falling in love. People who have low self-esteem may fall in love with someone else as a result of feeling deficient. The love they perceive the other person has for them may compensate for the perceived deficiency and improve their self-esteem. This phenomenon can happen with two individuals with low self-esteem—love can elevate the self concepts of both of them.

Self-Disclosure Disclosing oneself is necessary if one is to fall in love—to feel invested in another. Ross (2006) identified eight dimensions of self-disclosure: (1) background and history, (2) feelings toward the partner, (3) feelings toward self, (4) feelings about one's body, (5) attitudes toward social issues, (6) tastes and interests, (7) money and work, and (8) feelings about friends. Disclosed feelings about the partner included "how much I like the partner," "my feelings about our sexual relationship," "how much I trust my partner," "things I dislike about my partner," and "my thoughts about the future of our relationship"—all of which were associated with relationship satisfaction. Of interest in Ross's findings is that disclosing one's tastes and interests was negatively associated with relationship satisfaction. By telling a partner too much detail about what one likes, partners may discover something that turns them off and lowers relationship satisfaction.

It is not easy for some people to let others know who they are, what they feel, or what they think. They may fear that, if others really know them, they will be rejected as a friend or lover. To guard against this possibility, they may protect themselves and their relationships by allowing only limited information about their past behaviors and present thoughts and feelings. Some people keep others at a distance—they do not want psychological intimacy. Audrey Hepburn, Academy Award-winning screen actress of the 1950s, is said to have been wary of being close. Stanley Donen, with whom she was involved in the making of three movies, noted:

> I longed to get closer, to get behind whatever was the invisible, but decidedly present barrier between her and the rest of us, but I never got to the deepest part of Audrey. I don't mean to imply that I thought she was playing a game with me. But she always kept a little of herself in reserve, which was hers alone, and I couldn't ever find out what it was, let alone share it with her. She was the pot of gold at the end of the rainbow. (Spoto 2006, 251)

Trust is the condition under which people are most willing to disclose themselves. When people trust someone, they tend to feel that whatever feelings or information they share will not be judged and will be kept safe with that person. If trust is betrayed, people may become bitterly resentful and vow never to disclose themselves again. One woman said, "After I told my partner that I had had an abortion, he told me that I was a murderer and he never wanted to see me again. I was devastated and felt I had made a mistake telling him about my past. You can bet I'll be careful before I disclose myself to someone else" (personal communication).

Gallmeier et al. (1997) studied the communication patterns of 360 undergraduates at two universities and found that women were significantly more likely to disclose information about themselves. Specific areas of disclosure included previous love relationships, what they wanted for the future of the relationship, and what their partners did that they did not like.

Gratitude Algoe et al. (2010) studied cohabitating couples and found that partners who received "thoughtful benefits" (e.g. positive behaviors) experienced feelings of both gratitude and indebtedness to the other. Gratitude particularly was associated with increases in relationship connection and a feeling of satisfaction for both the recipient and benefactor. The authors concluded that gratitude acted like a booster shot of love for the partner and for the relationship.

Physiological and Cognitive Conditions for Love

Riela et al. (2010) noted that arousal (strong physiological reactions when in the presence of the other) is associated with falling in love. This physical chemistry is powerful. Another physiological change is feeling anxious and interpreting this stirred up state as love (Walster and Walster 1978).

In the absence of one's cognitive functioning, love feelings are impossible. Individuals with brain cancer who have had the front part of their brain (between the eyebrows) removed are incapable of love. Indeed, emotions are not present in them at all (Ackerman 1994). The social, physical, psychological, physiological, and cognitive conditions are not the only factors important for the development of love feelings. The timing must also be right. There are only certain times in life (for example, when educational and career goals are met or within sight) when people seek a love relationship. When those times occur, a person is likely to fall in love with another person who is there and who is also seeking a love relationship. Hence, many love pairings exist because each of the individuals is available to the other at the right time—not because they are particularly suited for each other.

Other Factors Associated with Falling in Love

In addition to the factors listed above, Riela et al. (2010) identified others associated with falling in love. These include:

Appearance—delight in seeing a nice-looking person with a nice body

Similarity—having common interests, views, backgrounds

Familiarity—enjoyment in spending time together

Social influence—approval expressed by the self's or the other's friends/family

Meeting one's needs—feeling happy, being cared for

Specific cues—unique characteristics of the partner that produce strong attractions (normal fetish), e.g. blue eyes, Asian features

Isolation—being alone with the other

Mysteriousness—wanting to know about the partner

How fast a person falls in love has also been studied by researchers. Falling in love in America is intense and quick. In a study of 239 undergraduates from both America and China who wrote about their experience of falling in love, 14% reported that they fell in love very fast, 42% fell in love fast, 36% fell in love slow, and 8% fell in love very slow. There were no significant differences between women and men in regard to speed of falling in love (Riela et al. 2010).

The Applying Social Research feature on page 82 reviews how love is experienced by black and white adolescents.

Love as a Context for Problems

Though love may bring great joy, it also creates a context for problems—the sadness of its ending, unrequited love, dangerous decisions, etc.

If love be good, from whennes comth my wo?

Geoffrey Chaucer, English poet

To what degree are African American and white adolescent experiences with love similar or different?

Sample and Methods

To find out, a team of researchers analyzed data collected during interviews with adolescents in grades 7 through 11 at more than 80 high schools. The sample consisted of 575 African American girls, 379 African American boys, 1,528 white girls, and 985 white boys. Each of these respondents reported having a "current" or "recent" relationship. The data are part of Add Health, a longitudinal study of a nationally representative sample of adolescents.

Selected Findings and Conclusions

1. *European Americans valued romantic love relationships more than African Americans.* When the respondents were asked, "How much would you like to have a romantic relationship in the next year?" they selected a number on a continuum from 1 to 7 (the higher the number, the greater the importance):

White respondents, compared with African American respondents, rated having a romantic relationship in the next year as significantly more important (mean = 3.47 and 3.24, respectively). Interestingly, a similar percentage (36% of European Americans and 34% of African Americans) reported current involvement in a love relationship.

2. *European Americans engaged in more romantic behaviors than African Americans.* When "romantic behaviors" were defined as "told other people that we were a couple," "went out together alone," "kissed," "held hands," "gave each other presents," "told each other we loved each other," and "thought of ourselves as a couple," white adolescents, on average, reported significantly more romantic behaviors with the current partner than did African American teens.

3. *African Americans were less likely to report involvement in an exclusive relationship.* Although more than 90% of both European Americans and African Americans reported current involvement in only one relationship,

---

Take love away from life and you take away its pleasures.

Moliere, French playwright

Profound Sadness/Depression when a Love Relationship Ends

Fisher et al. (2010) noted that "romantic rejection causes a profound sense of loss and negative affect. It can induce clinical depression and in extreme cases lead to suicide and/or homicide." The researchers studied brain changes via magnetic resonance imaging of 10 women and 5 men who had recently been rejected by a partner but reported they were still intensely "in love." Participants alternately viewed a photograph of their rejecting beloved and a photograph of a familiar individual interspersed with a distraction-attention task. Their responses while looking at their rejecter included love, despair, good, and bad memories, and wondering why this happened. Compared with data from happily-in-love individuals, brain activation suggested that reward/survival systems are involved in romantic passion regardless of whether one is happily or unhappily in love. Brain reactions to being rejected by a lover are similar to withdrawal from cocaine.

There is also physical pain. "The pain one experiences in response to an unwanted breakup is identical to the pain one experiences when physically hurt," noted Dr. Steven Richeimer (2011) of the University of Southern California Pain Center. He was speaking in reference to a University of Michigan study where researchers asked people who recently had an unwanted romantic breakup to look at a picture of their ex or hold a hot cup of coffee. The pain reaction in the brain was exactly the same.

While there is no recognized definition or diagnostic criteria for "love addiction," some similarities to substance dependence include: euphoria and unrestrained desire in the presence of the love object or associated stimuli (drug intoxication); negative mood and sleep disturbance when separated from the love object (drug withdrawal); intrusive thoughts about the love object; and

African American respondents were less likely than white respondents to report exclusive involvement in their current relationship (94% versus 98%).

4. *African Americans were less likely to report intimate self-disclosure than European Americans.* Self-disclosure was identified in terms of telling a partner about a problem. Females, older individuals, and those who had been in a relationship for a considerable amount of time were also more likely to self-disclose than males, younger adolescents, and those who had known each other for only a short period of time.

5. *African Americans reported longer current relationships and more inclusion of sexual intercourse.* The duration of African American relationships was about a month longer than those of white respondents. African Americans also more often reported the inclusion of sexual intercourse in their current relationships, a circumstance that might be related to the fact that sexual involvement increases with relationship duration.

What are the implications of this study? One, love remains an important experience for adolescents, even as young as the seventh grade. Two, although there are statistically "significant" differences in some of the variables studied, the actual experienced differences may be irrelevant. For example, the mean scores of 3.47 and 3.24, respectively, for European Americans and African Americans on a scale of 1 to 7 (reflecting the importance of wanting to be involved in a love relationship) may actually reflect more similarities than differences between the races. Other findings reflect the similarity of the races, rather than the differences (for example, more than 90% of both races reported exclusive involvement in a relationship).

Source

Giordano, P. C., W. D. Manning, and M. A. Longmore. 2005. The Romantic Relationships of African–American and White Adolescents. *The Sociological Quarterly* 46:545, p. 68. Used by permission of John Wiley and Sons.

problems associated with love which may lead to clinically significant impairment or distress (Reynaud et al. 2011).

Berscheid (2010) noted that the intensity of romantic love decreases across time in a relationship. It is not unusual for lovers to vary in the intensity of their love for each other. The interesting question is whether being the person who loves more in a relationship is better than the person who loves less. The person who loves more may suffer more anguish. Some regard the anguish as worth the price as captured in the quote, "Tis better to have loved and lost, than never to have loved at all" (Alfred Lord Tennyson).

WHAT IF?

What if You Fall out of Love with the Person You Had Planned to Marry?

In a sample of 2,922 undergraduate students, exactly half agreed that, "I would divorce my spouse if I no longer loved him or her" (Knox and Hall 2010). Although Asians typically view love as a feeling that is to follow marriage, Americans have been socialized to expect being in love with their spouse-to-be, and to feel embarrassed about and to hide it if they are not. Although falling in love with someone after marriage is more than possible, someone socialized in America with individualistic values might be very cautious in proceeding with a wedding (because such an enormous value is placed on love in U.S. society). To do so is to run the risk of being in a "loveless" marriage that may make one vulnerable to "falling in love" outside the marriage.

The hottest love has the coldest end.

Socrates, philosopher

Unrequited Love

Sometimes love does not end, but is never returned in the first place. Such a love is typically referred to as **unrequited love**. The result was referred to in the sixteenth century as "lovesickness" and today is called erotomania or erotic melancholy (Kem 2010). An example is the case of Dean (six year husband of Cindy) in the movie *Blue Valentine* (2011). He is confronted with the fact that she no longer loves him and unable to accept her rejection.

Love Involving Risky, Dangerous, or Questionable Choices

Plato said that "love is a grave mental illness," and some research suggests that individuals in love make risky, dangerous, or questionable decisions. In a study on "what I did for love," college students reported that "driving drunk," "dropping out of school to be with my partner," and "having sex without protection" were among the more dubious choices they had made while they were under the spell of love (Knox and Zusman et al. 1998). Similarly, non-smokers who become romantically involved with a smoker are more likely to begin smoking (Kennedy et al. 2011). And, a team of researchers examined the relationship between having a romantic love partner and engaging in minor acts of delinquency (for example, getting drunk, skipping school); they found that females were particularly influenced by their "delinquent" boyfriends (Haynie et al. 2005). Their data source was the National Longitudinal Study of Adolescent Health. Furthermore, researchers have found that women who are "romantically in love" are less likely to use condoms with their partners. Doing so isn't regarded as very romantic, and they elect not to inject realism into a love context (East et al. 2007; Cousins et al. 2010).

Ending the Relationship with One's Parents

Some parents disapprove of the partner their son or daughter is involved with (e.g. issues of race, age, religion, ethnicity) to the point that their offspring will end the relationship with their parents. "They told me I couldn't come home if I kept dating this guy, so I stopped going home" said one of our students who was involved with a partner of a different race. Choosing to end a relationship with one's parents is a definite downside of love.

Simultaneous Loves

Sometimes an individual is in love with two or more people at the same time. While this is acceptable in open relationships where the partners agree on multiple relationships, simultaneous loves become a major problem unless there has been a previous understanding/agreement. Only 2.6% of 2922 undergraduates agreed "I can feel good about my partner having an emotional/sexual relationship with someone else" (Knox and Hall 2010).

Putting on Weight after Moving In

The and Gordon-Larsen (2009) studied 1,293 dating, cohabiting, and married romantic couples and found that those who transitioned from single/dating to cohabiting or married were more likely to become obese than those who were dating. The strongest obesity-related behaviors (e. g. sedentary behavior, no exercise) were strongest for married couples and couples who had lived together 2 years. Explanations include that cohabitants and marrieds are more likely to plan their time so that they eat together (hence focus on food) and the tendency to "let oneself go" (e.g. stop exercising) after getting the partner committed. Previous research has also shown that if one partner is obese, the other has an increased chance of obesity.

What if You Are in Love with Two People at the Same Time?

One answer to the dilemma of simultaneous loves is to let the clock run. Most love relationships do not have a steady course. Time has a way of changing them. If you maintain both relationships, one is likely to emerge as more powerful, and you will have your answer. Alternatively, if you feel "guilty" for having two loves, you may make the conscious choice to spend your time and attention with one partner and let the other relationship go in terms of actual time spent with the partner. Although you can have emotions for two people at the same time, you generally cannot psychically be with more than one person at a time. The person with whom you choose to spend your time is likely to be the person you love "a little bit more" and with whom your love feelings are likely to increase. Indeed, Lundstrom and Jones-Gotman (2009) noted that being romantically in love with one partner helps one to deflect the development of love with a potential new partner.

Abusive/Stalking Relationships

Thirty percent of 2922 undergraduates reported that they had been involved in an emotionally abusive relationship with a partner. As for physical abuse, 7.5% reported such previous involvement. Some stay in these relationships because they are "in love" (Knox and Hall 2010).

Rejected lovers (more often men) may stalk (repeated pursuit of a target victim that threatens the victim's safety) a partner because of anger and jealousy and try to win the partner back. Twenty-five percent of 2922 undergraduates reported that they had been stalked. We discuss both abusive relationships and stalking in detail in Chapter 13.

Jealousy in Relationships

Jealousy can be defined as an emotional response to a perceived or real threat to an important or valued relationship. People experiencing jealousy fear being abandoned and feel anger toward the partner or the perceived competition. People become jealous when they fear replacement. Although jealousy does not occur in all cultures (polyandrous societies value cooperation, not sexual exclusivity; Cassidy and Lee 1989), it does occur in our society and among both heterosexuals and homosexuals. Of 2,922 university students, 41% reported, "I am a jealous person" (Knox and Hall 2010). In another study, 185 students gave information about their experience with jealousy (Knox et al. 1999). On a continuum of 0 ("no jealousy") to 10 ("extreme jealousy"), with 5 representing "average jealousy," these students reported feeling jealous at a mean level of 5.3 in their current or last relationship. Students who had been dating a partner for a year or less were significantly more likely to report higher levels of jealousy (mean = 4.7) than those who had dated 13 months or more (mean = 3.3). Hence, jealously is more likely to occur early in a couple's relationship. Consistent with this theme, Gatzeva and Paik (2011) compared noncohabiting, cohabiting, and married couples in regard to satisfaction as related to jealous conflict. They found that married couples were less likely to report jealous conflict and that satisfaction was higher in the absence of such conflict. Finally, the sexual orientation to the

Jealousy an emotional response to a perceived or real threat to an important or valued relationship.

> *Jealousy in romance is like salt in food. A little can enhance the savor, but too much can spoil the pleasure and, under certain circumstances, can be life-threatening.*
>
> Maya Angelou, poet

Reactive jealousy jealous feelings that are a reaction to something the partner is doing.

Anxious jealousy obsessive ruminations about the partner's alleged infidelity that can make one's life a miserable emotional torment.

Possessive jealousy involves attacking the partner or the alleged person to whom the partner is showing attention.

affair object affects the level of jealousy. Cox et al. (2011) identified how sexual orientation of the affair object affected willingness to continue a relationship. The researchers found that heterosexual men were less likely to continue an imagined long-term relationship following a partner's heterosexual affair compared to homosexual affair. For women, both affair types resulted in a low willingness to continue the relationship, but especially so for homosexual affairs.

Types of Jealousy

Barelds-Dijkstra and Barelds (2007) identified three types of jealousy as reactive jealousy, anxious jealousy, and possessive jealousy. **Reactive jealousy** consists of feelings that are a reaction to something the partner is doing (for example, texting a former lover). **Anxious jealousy** is obsessive ruminations about the partner's alleged infidelity that make one's life a miserable emotional torment. **Possessive jealousy** involves an attack at the partner or the alleged person to whom the partner is showing attention. Jealousy is a frequent motive when one romantic partner kills another. Less dramatic, note Fisher and Cox (2011), is "competitor derogation" whereby the person talks negatively about the alleged suitor.

Causes of Jealousy

Jealousy can be triggered by external or internal factors.

External Causes   External factors refer to behaviors a partner engages in that are interpreted as (1) an emotional and/or sexual interest in someone (or something) else, or (2) a lack of emotional and/or sexual interest in the primary partner. In the study of 185 students previously referred to, the respondents identified "actually talking to a previous partner" (34%) and "talking about a previous partner" (19%) as the most common sources of their jealousy. Also, men were more likely than women to report feeling jealous when their partner talked to a previous partner, whereas women were more likely than men to report feeling jealous when their partner danced with someone else. Buunk et al. (2010) found that what makes women jealous is the physical attractiveness of a rival; what makes a man jealous is the rival's physical dominance.

Internal Causes   Internal causes of jealousy refer to characteristics of individuals that predispose them to jealous feelings, independent of their partner's behavior. Examples include being mistrustful, having low self-esteem, being highly involved in and dependent on the relationship, and having no perceived alternative partners available (Pines 1992). The following are explanations of these internal causes of jealousy:

1. *Mistrust.* If an individual has been deceived or cheated on in a previous relationship, that individual may learn to be mistrustful in subsequent relationships. Such mistrust may manifest itself in jealousy. Mistrust and jealousy may be intertwined.

2. *Low self-esteem.* Individuals who have low self-esteem tend to be jealous because they lack a sense of self-worth and hence find it difficult to believe anyone can value and love them. Feelings of worthlessness may contribute to suspicions that someone else is valued more.

3. *Lack of perceived alternatives.* Individuals who have no alternative person or who feel inadequate in attracting others may be particularly vulnerable to jealousy. They feel that, if they do not keep the person they have, they will be alone.

4. *Insecurity.* Individuals who feel insecure in a relationship with their partner may experience higher levels of jealousy. Khanchandani (2005) found that individuals who had been in relationships for a shorter time, who were in less committed relationships, and who were less satisfied with their relationships were more likely to be jealous.

86   **Chapter 3** Love

5. *Physiology.* There may be chemical contexts conducive to jealousy. Cobey et al. (2011) found in a sample of 275 women that higher levels of ethinyl estradiol (an estrogen found in combined oral contraceptive pills) were associated with reported feelings of jealousy.

Consequences of Jealousy

Jealousy can have both desirable and undesirable consequences.

Desirable Outcomes   Barelds-Dijkstra and Barelds (2007) studied 961 couples and found that reactive jealousy is associated with a positive effect on the relationship. Not only may reactive jealousy signify that the partner is cared for (the implied message is "I love you and don't want to lose you to someone else"), but also the partner may learn that the development of other romantic and sexual relationships is unacceptable.

One wife said:

> When I started spending extra time with this guy at the office, my husband got jealous and told me he thought I was getting in over my head and asked me to cut back on the relationship because it was "tearing him up." I felt he really loved me when he told me this, and I chose to stop having lunch with the guy at work. (personal communication)

The researchers noted that making the partner jealous may also have the positive function of assessing the partner's commitment and of alerting the partner that one could leave for greener mating pastures. Hence, one partner may deliberately evoke jealousy to solidify commitment and ward off being taken for granted. In addition, sexual passion may be reignited if one partner perceives that another would take the love object away. That people want what others want is an adage that may underlie the evocation of jealousy.

Undesirable Outcomes   Shakespeare referred to jealousy as the "green-eyed monster," suggesting that it sometimes leads to undesirable outcomes for relationships. Anxious jealousy with its obsessive ruminations about the partner's alleged infidelity can make an individual miserable, and such jealousy spills over into one's evaluation or experience of the relationship as negative. If the anxious jealousy results in repeated unwarranted accusations, a partner can tire of such attacks and end the relationship.

In its extreme form, jealousy may have devastating consequences. In the name of love, people have stalked or shot the beloved and killed themselves in reaction to rejected love. Barelds-Dijkstra and Barelds (2007) noted that possessive jealousy involves an attack on a partner or an alleged person to whom the partner is showing attention. Possessive jealousy definitely may have negative consequences for a relationship. The next section details the different ways women and men cope with jealousy.

Gender Differences in Coping with Jealousy

Jealousy is a theme of popular movies and part of our cultural language. Defined as "one's emotional reaction to the perception that one's love relationship may end because of a third party," jealousy was the topic of a study of 291 undergraduates where 51.9% "agreed" or "strongly agreed" that "jealousy is normal" (Knox et al. 2007).

Analysis of the data on women's and men's reactions to jealousy revealed four significant differences:

1. *Food.* Women were significantly more likely than men to report that they turned to food when they felt jealous: 30.3% of women, in contrast to 22% of men, said that they "always, often, or sometimes" looked to food when they felt jealous. One coed said, "When I feel jealous, my favorite guys are "Ben and Jerry."

> *The jealous are troublesome to others, but a torment to themselves.*
>
> William Penn, *Some Fruits of Solitude*, 1693

> *Jealousy is... a tiger that tears not only its prey but also its own raging heart.*
>
> Michael Beer, professor

2. *Alcohol.* Men were significantly more likely than women to report that they drank alcohol or used drugs when they felt jealous: 46.9% of men, in contrast to 27.1% of women, said that they "always, often, or sometimes" would drink or use drugs to make the pain of jealousy go away.

3. *Friends.* Women were significantly more likely than men to report that they turned to friends when they felt jealous: 37.9% of women, in contrast to 13.5% of men, said that they "always" turned to friends for support when feeling jealous.

4. *Nonbelief that "jealousy shows love."* Women were significantly more likely than men to disagree or to strongly disagree that "jealousy shows how much your partner loves you": 63.2% of women, in contrast to 42.6% of men, disagreed with the statement. This difference may reflect the fact that jealous males are more likely to abuse women.

Coping Strategies   Implications of the data may be relevant to women who may be alert to the "extra urge" to eat in reaction to jealousy, and turn instead to vigorous exercise as a way of reducing stress. Not only might exercise better reduce the stress; it will do so without adding pounds, which could lead to further self-deprecation and depression. Similarly, men might consider talking with a buddy rather than turning to the bottle; strengthening friendships would be more productive than having a hangover or risking a fatal car wreck.

Compersion, Polyamory, and Open Relationships

Compersion sometimes thought of as the opposite of jealousy, the approval of a partner's emotional and sexual involvement with an-other person.

Polyamory a lifestyle in which two lovers embrace the idea of having multiple lovers. By agreement, each partner may have numerous emotional and sexual relationships.

Compersion, sometimes thought of as the opposite of jealousy, is the approval (indeed embracing) of a partner's emotional and sexual involvement with another person. **Polyamory** means multiple loves (poly = many; amor = love). **Polyamory** is a lifestyle in which two lovers embrace the idea of having multiple lovers. By agreement, each partner may have numerous emotional and sexual relationships. About 25% of the 100 members of Twin Oaks Intentional Community in Louisa, Virginia, are polyamorous in that each partner may have several emotional or physical relationships with others at the same time. Although not legally married, these adults view themselves as emotionally bonded to each other and may even rear children together. Polyamory is not swinging, as polyamorous lovers are concerned about enduring, intimate relationships that include sex. People in polyamorous relationships seek to rid themselves of jealous feelings and to increase their level of compersion. To feel happy for a partner who delights in the attention and affection of—and sexual involvement with—another person is the goal of polyamorous couples.

Advantages and Disadvantages of Polyamory

Embracing polyamory has both advantages and disadvantages (Wilson and Rodrigous 2010). Advantages of polyamory include greater variety in one's emotional and sexual life; the avoidance of hidden affairs and the attendant feelings of deception, mistrust, or betrayal; and the opportunity to have different needs met by different people. The disadvantages of polyamory involve having to manage one's feelings of jealousy, greater exposure of oneself and partners to human immunodeficiency virus and other sexually transmitted infections, and limited time with each partner. Of the latter, one polyamorous partner said, "With three relationships and a full-time job, I just don't have much time to spend with each partner so I'm frustrated about who I'll be with next. And managing the feelings of the other partners who want to spend time with me is a challenge."

Rules of an Open Relationship

While polyamorous relationships involve emotional as well as sexual involvement with others, **open relationships** are more sexual and recreation focused. These relationships involve individuals agreeing that they may have sexual encounters with others. Most undergraduates are not comfortable with open relationships (only 2.6% of 2,922 said they were comfortable, Knox and Hall 2010). Nevertheless, couples who have open relationships often develop certain rules. The following are rules of a couple in an open relationship ((author's files):

1. Honesty—we tell each other everything we do with someone outside the relationship. If we flirt, we even tell that. Openness about our feelings is a must—if we get uncomfortable or jealous, we must talk about it.

2. Recreational sex—sex with the other person will be purely recreational—it is not love and the relationship with the other person is going nowhere. The people we select to have sex with must know that we have a loving committed relationship with someone else.

3. Condom—a requirement every time.

4. Approval—every person we have sex with must be approved by the partner in advance. Each partner has the right to veto a selection. The person in question must not be into partner snatching, looking for romance, or jealous. Persons off the list are co-workers, family (he can't have sex with her sister or she with his brother), old lovers, and old friends.

5. Online hunting—prohibited. Each agrees not to go looking on the Internet for sex partners.

Open relationship relationship in which the partners agree that each may have sexual relationships with those outside the dyad.

The Future of Love Relationships

Love will continue to be one of the most treasured experiences in life. Love will be sought, treasured, and when lost or ended, will be met with despair and sadness. After a period of recovery, a new search will begin. As our society becomes more diverse, the range of potential love partners will widen to include those with demographic characteristics different from oneself. Romantic love will continue and love will maintain its innocence as those getting remarried will love just as deeply and intensely as those in love for the first time.

Summary

What are some ways of conceptualizing love?

Love remains an elusive and variable phenomenon. Researchers have conceptualized love as a style (e.g. ludic, eros, storge, mania), as a continuum from romanticism to realism (e.g. from belief in love at first sight/one true love/love conquers all to these beliefs being nonsense), and as a triangle consisting of three basic elements (intimacy, passion, and commitment; e.g. infatuation is passion without intimacy). People stay in a relationship because it meets important emotional needs (e.g. satisfaction), they have few alternatives (e.g. no place to go), and they have already invested resources (e.g. time, money, friendship networks).

Why and how do parents attempt to control who their offspring fall in love with?

All parents attempt to influence and control the person their children fall in love with. Love may be blind, but offspring are socialized to know what color a person's skin is (about 90% of Americans fall in love with/marry someone of their same racial background). Because romantic love is such a powerful emotion and marriage such an important relationship, mate selection is not left to chance when connecting an outsider with an existing family and peer network. Parents inadvertently influence the mate choice of their children by moving to certain neighborhoods, joining certain churches, and enrolling their children in certain schools. Doing so increases the chance that their offspring will "hang out" with, fall in love with, and marry people who are similar in race, education, and social class.

What are the various theories of love?

Theories of love include evolutionary (love provides the social glue needed to bond parents with their dependent children and spouses with each other to care for their dependent offspring), learning (positive experiences create love feelings), sociological (Reiss's "wheel" theory), psychosexual (love results from a blocked biological drive), and biochemical (love involves feelings produced by biochemical events). For example, the neurobiology of love emphasizes that, because romantic love and maternal love are linked to the perpetuation of the species, biological wiring locks in the bonding of the male and female to rear offspring and to ensure the bonding of the mother to the infant. Finally, attachment theory focuses on the fact that a primary motivation in life is to be connected with other people.

What is the process of "falling in love?"

Love occurs under certain conditions. Social conditions include a society that promotes the pursuit of love, peers who enjoy it, and a set of norms that link love and marriage. Psychological conditions involve high self-esteem, a willingness to disclose oneself to others, perception that the other person has a reciprocal interest, and gratitude. Physiological and cognitive conditions imply that the individual experiences a stirred-up state and labels it "love." Body type is related to falling in love in that the closer one's body type matches the cultural ideal, the more likely the person is to fall in love.

How is love a context for problems?

For all of its joy, love is associated with problems. Sadness/depression often follow the end of an intense love relationship, love that is not returned (unrequited love) is painful, and lovers sometimes make unwise decisions—drop out of school to be with a partner in another state or have sexual intercourse without using protection (it isn't romantic). Individuals in love may lie to or end the relationship with their parents if their parents don't like who they are dating, become involved in more than one love relationship at the same time, and let themselves go when they begin to cohabit or marry.

What is the definition of jealousy? What types of jealousy are there?

Jealousy is an emotional response to a perceived or real threat to a valued relationship. Types of jealousy are reactive (partner shows interest in another), anxious (ruminations about partner's unfaithfulness), and possessive (striking back at a partner or another). Jealous feelings may have both internal and external causes and may have both positive and negative consequences for a couple's relationship.

What is comperson?; polyamopry and open relationships?

Compersion is the opposite of jealousy and involves feeling positive about a partner's emotional and physical relationship with another person. Polyamory ("many loves") is an arrangement whereby lovers agree to have numerous emotional and/or sexual relationships with others at the same time. An open relationship is similar to what polyamorous individuals have except their other relationships are sexual, not emotional.

What is the future of love relationships?

Love will continue to be one of the most treasured experiences in life. As our society becomes more diverse, the range of potential love partners will widen to include those demographic characteristics different from oneself. Romantics will continue and love will maintain its innocence as those getting remarried will love just as deeply as those falling in love for the first time.

Key Terms

Agape love style	Jealousy	Possessive jealousy
Anxious jealousy	Ludic love style	Pragma love style
Arranged marriage	Lust	Reactive jealousy
Compersion	Mania love style	Romantic love
Conjugal love	Open relationship	Social allergy
Eros love style	Oxytocin	Storge love style
Infatuation	Polyamory	Unrequited love

Web Links

Love Quotes
  http://library.lovingyou.com/quotes/
The Polyamory Society
  http://www.polyamorysociety.org/

Love at Answers.com
  http://www.answers.com/topic/love

CHAPTER 4

Communication

Communication is the Vaseline of marital intercourse.

Neil Emmott, entrepreneur

Chelsea E. Curry

Learning Objectives

Know the principles of communication as well as gender/cultural differences in communication styles.

Review the effects of self-disclosure, dishonesty, and secrets on relationships.

Discuss the effects of lying and cheating on relationships.

Identify sociological theories relevant to communication.

Summarize the steps involved in conflict resolution.

Predict the future of communication.

1. In a study of over 1000 partners in a romantic relationship the most common text message they sent to each other was about making plans for the evening.

2. Spouses mirror each other in communication—when one spouse swears, the other swears back.

3. Some teenagers may be "addicted" to IM (instant messaging) to the point that their relationships and grades suffer.

4. In a study on sexting, while there were no gender differences in sexting behavior, couples who sent sex messages were happier.

5. Females, Blacks, and homosexuals keep more secrets from their romantic partners than males, Whites, and heterosexuals.

Answers: 1. F   2. T   3. T   4. F   5. T

Famed magician Harry Houdini and his new bride Bess had a fight which ended in his putting her on a train back to her home in Bridgeport, Connecticut. He had told her if she "disobeyed" him he would send her home. "I always keep my word. Good bye, Mrs. Houdini," he said mockingly as he watched the train depart.

In Bridgeport, Bess was distraught, but at 2:00 A.M. the doorbell rang. She rushed to the door where Houdini stood. "See, darling, I told you I would send you away if you disobeyed, but I didn't say I wouldn't fly after you and bring you back" (Kalush, and Sloman 2006, 37). The Houdini spat and resolution reflect communication and conflict in relationships—the subject of this chapter.

"Good communication" is regarded as the primary factor responsible for a good relationship. Individuals report that communication confirms the quality of their relationship ("We can talk all night about anything and everything") or condemns their relationship ("We have nothing to say to each other; we are getting a divorce"). Braun et al. (2010) noted the positive effect of positive marital communication on mental health. Caregivers of spouses with dementia who reported positive communication with spouses were not depressed in contrast to those reporting negative communication.

Communication between romantic partners is influenced by a technological explosion including cell phones, texting, Facebook, Skype, etc. Lovers now "stay connected all day". This chapter acknowledges the effect of technology on relationships today. We begin by looking at the nature of interpersonal communication.

We tried to talk it over but the words got in the way.

"This Masquerade," Leon Russell, musician and songwriter

Communication the process of exchanging information and feelings between two or more people.

Nonverbal communication the "message about the message," using gestures, eye contact, body posture, tone, volume, and rapidity of speech.

Interpersonal Communication

Communication can be defined as the process of exchanging information and feelings between two or more people. Wickrama et al. (2010) studied 540 couples and found that spouses mirror each other in terms of how they relate to each other. When one spouse insults, swears, or shouts, the other is likely to engage in the same behavior. Communication is both verbal and nonverbal. Although most communication is focused on verbal content, most (estimated to be as high as 80%) interpersonal communication is nonverbal. **Nonverbal communication** is the "message about the message," the gestures, eye contact, body posture, tone, volume, and rapidity of speech. Even though a person says, "I love you and am faithful to you," crossed arms and lack of eye contact will convey a very different meaning than the same words accompanied by a tender embrace and sustained eye-to-eye contact. Bos et al. (2007) found that the greater the congruence between verbal and nonverbal communication, the better. Indeed, when the two are congruent, a person experiences fewer stressful events and lowered depression.

Moore (2010) studied how the sexes communicate with each other nonverbally and, at the same time, move the relationship forward. Moore (2010) focused on **flirting** defined as playing at courtship—showing interest without serious intent. Her review included basic courtship behaviors: keeping one's body in shape to provide a positive stimulus, preening such as stroking one's hair/adjusting one's clothing, and positional cues such as leaning toward or away from the target. Individuals go through a series of steps on their way to sexual intercourse and the female is responsible for signaling the male that she is interested so that he moves the interaction forward. At each stage, she must signal readiness to move to the next stage. For example, if the male holds the hand of the female, she must squeeze his hand before he can entwine their fingers.

Having a supportive communicative partner, one who listens and is engaged is a plus for any relationship. The Self-Assessment allows you to assess the degree to which your relationship is characterized by supportive communication.

Words versus Action

A great deal of social discourse depends on saying things that sound good but which have no meaning in terms of behavioral impact. "Let's get together" or "let's hang out" sounds good since they imply an interest in spending time with each other. But the phrase has no specific plan so that the 'intent' is likely never to happen—just the opposite. So "let's hang out" really means "we won't be spending any time together." Similarly, "come over anytime" means "never come."

Where there is behavioral intent, a phrase with meaning is "Let's meet Thursday night at seven for dinner" (rather than "let's hang out") and "Can you

Flirting to show interest without serious intent.

The single biggest problem in communication is the illusion that it has taken place.

George Bernard Shaw, playwright

There's a language in her eye, her cheek, her lip;

Nay, her foot speaks. Her wanton spirits look out at every joint and motive of her body.

Shakespeare, *Troilus and Cressida*

What nonverbal message are these females conveying to their partners?

Chelsea E. Curry

come over Sunday afternoon at four to play video games?" (rather than "come over any time"). With one's partner, "let's go camping sometime" means we won't ever go camping. "Let's go camping at the river this coming weekend" means you are serious about camping. A major theme of all communication is how emotionally closed the partners want to be with each other. This issue is addressed in the Personal Choices section.

PERSONAL CHOICES

How Close Do You Want to Be?

Individuals differ in their capacity for and interest in an emotionally close and disclosing relationship. These preferences may vary over time; the partners may want closeness at some times and distance at other times. Individuals frequently choose partners according to an "emotional fit"—agreement about the amount of closeness they desire in their relationship.

In addition to emotional closeness, some partners prefer a pattern of physical presence and complete togetherness (the current buzzword is *codependency*), in which they spend all of their leisure time together. Others enjoy time alone and time with other friends and do not want to feel burdened by the demands of a partner with high companionship needs. Partners might consider their own preferences and those of their partners in regard to emotional and spatial closeness.

SELF-ASSESSMENT | Supportive Communication Scale

This scale is designed to assess the degree to which partners experience supportive communication in their relationships. After reading each item, circle the number that best approximates your answer.

0 = strongly disagree (SD)
1 = disagree (D)
2 = undecided (UN)
3 = agree (A)
4 = strongly agree (SA)

	SD	D	UN	A	SA
1. My partner listens to me when I need someone to talk to.	0	1	2	3	4
2. My partner helps me clarify my thoughts.	0	1	2	3	4
3. I can state my feelings without my partner getting defensive.	0	1	2	3	4
4. When it comes to having a serious discussion, it seems we have little in common (reverse scored).	0	1	2	3	4
5. I feel put down in a serious conversation with my partner (reverse scored).	0	1	2	3	4
6. I feel discussing some things with my partner is useless (reverse scored).	0	1	2	3	4
7. My partner and I understand each other completely.	0	1	2	3	4
8. We have an endless number of things to talk about.	0	1	2	3	4

Scoring

Look at the numbers you circled. Reverse score the numbers for questions 4, 5, and 6. For example, if you circled a 0, give yourself a 4; if you circled a 3, give yourself a 1, and so on. Add the numbers and divide by 8, the total number of items. The lowest possible score would be 0, reflecting the complete absence of supportive communication; the highest score would be 4, reflecting complete supportive communication. The average score of 94 male partners who took the scale was 3.01; the average score of 94 female partners was 3.07. Thirty-nine percent of the couples were married, 38% were single, and 23% were living together. The average age was just over 24.

Source

Sprecher, S., S. Metts, B. Burleson, E. Hatfield, and A. Thompson. 1995. Domains of expressive interaction in intimate relationships: Associations with satisfaction and commitment. *Family Relations 44*: 203–210. Copyright © 1995 by the National Council on Family Relations.

Texting, IM, MySpace, and Interpersonal Communication

Personal, portable, wirelessly networked technologies in the form of iPhones, Blackberries, Droids, iPads etc. have become commonplace in the lives of individuals (Looi et al. 2010). Bauerlein (2010) noted that today's youth are being socialized in a hyper-digital age where traditional modes of communication will be replaced by gadgets and texting will become the primary mode of communication, with an average of 2,272 messages a month. This shift to greater use of technology affects relationships in both positive and negative ways. On the positive side, they allow an instant and unabated connection—individuals can text each other throughout the day so that they are "in effect, together all the time." One of our students noted that on a "regular" day, she and her boyfriend will exchange 50 to 60 text messages.

Coyne et al. (2011) examined the use of technology by 1039 individuals in sending messages to their romantic partner. The respondents were more likely to use their cell phones to send text messages than any other technology. The most common reason was to express affection (75%) followed by discussing serious issues (25%) and apologizing (12%). There were no significant differences in use by gender, ethnicity, or religion. Pettigrew (2009) emphasized that **texting** or text messaging (short typewritten messages sent via mobile telephones) is used to "commence, advance, maintain" interpersonal relationships and is viewed as more constant and private than talking on cell phones. Women text more than men. In a Nielsen State of the Media Survey in the third quarter of 2010, women spent 818 minutes texting 640 messages; men, 716 minutes texting 555 messages (Carey and Salazar 2011). On the negative side, these devices encourage the continued interruption of face to face communication between individuals and encourage the intrusion of one's work/job into the emotional intimacy of a couple and the family.

Huang and Leung (2010) studied instant messaging and identified four characteristics of "addiction" in teenagers: preoccupation with IM, loss of relationships due to overuse, loss of control, and escape. Results also showed that shyness and alienation from family, peers, and school were significantly and positively associated with levels of IM addiction. As expected, both the level of IM use and level of IM addiction are significantly linked to teenagers' academic performance decrement.

Texting short typewritten messages sent via a cell phone that are used to "commence, advance, and maintain" interpersonal relationships.

Chelsea E. Curry

College students are focused on texting throughout the day, so much so that they sometimes ignore potential face-to-face interactions.

MySpace is the largest social networking site on the Internet that is available to the public. Unlike Facebook, MySpace does not require membership within a particular network to view public profiles. A team of researchers (Walker et al. 2009) examined 44 profiles of undergraduates at two universities in the Midwest. Users were 18- to 28-years-old, with a mean age of 21.6 years. Most students (n = 28) self-identified as "White/Caucasian." The number of friends for each student ranged from 1 to 623 (average 128). Almost 80% (77%) had logged into MySpace within 2 days of the date of data collection. The number of comments students received ranged from 0 to 766 (the average was approximately 173).

The researchers identify six types of comments to and from undergraduate students on MySpace: (1) friendly greetings or inquiries (e.g. "Hey baby—how'd you do on the test?"), (2) expressions of affection and encouragement (e.g. "Good luck on your chemistry exam."), (3) suggestion and confirmation of plans (e.g. "We're meeting for drinks at 10:00?."), (4) personal asides and inside jokes (e.g. "Maria can't stay away from the donuts."), (5) exchanges of information and news (e.g. "Lydia is pregnant."), and (6) entertainment. An example of the latter is pictures with humorous captions. Some forwarded jokes about sex, alcohol, etc. The value of MySpace is that it connects the individual to others, confirms his or her identity/acceptance with others, and provides support as he or she navigates their developing interpersonal world.

Another way in which technology affects communication, particularly in romantic relationships is **sexting** (sending erotic text and photo images via a cell phone). See the section on Applying Social Research on pages 98–99 for new data on sexting in romantic relationships.

Sexting sending erotic text and photo images via a cell phone.

Conflicts in Relationships

Conflict is the context in which the perceptions or behavior of one person are in contrast to or interfere with the other. A professor in a Marriage and Family class said, "If you haven't had a conflict with your partner, you haven't been going together long enough." This section explores the inevitability, benefits/drawbacks, and sources of conflict in relationships.

Conflict the context in which the perceptions or behavior of one person are in contrast to or interfere with the other.

Inevitability of Conflict

If you are alone this Saturday evening from six o'clock until midnight, you are assured of six conflict-free hours. But if you plan to be with your partner, roommate, or spouse during that time, the potential for conflict exists. Whether you eat out, where you eat, where you go after dinner, and how long you stay must be negotiated. Although it may be relatively easy for you and your companion to agree on one evening's agenda, marriage involves the meshing of desires on an array of issues for potentially 60 years or more. Indeed, conflict is inevitable in any intimate relationship.

Benefits and Drawbacks of Conflict

Conflict can be healthy and productive for a couple's relationship. Ignoring an issue may result in the partners becoming increasingly resentful and dissatisfied with the relationship. Indeed, not talking about a concern may do more damage to a relationship (since resentment may build) than bringing up the issue and discussing it. Couples in trouble are not those who disagree but those who never discuss their disagreements. However, sustained conflict over chronic problems contributes to poor mental and physical health of the partners in the relationship (Wickrama et al. 2010).

Sources of Conflict

Conflict has numerous sources, some of which are easily recognized, whereas others are hidden inside the web of marital interaction.

1. *Behavior.* Money is a frequent issue over which couples conflict. The behavioral expression of a money issue might include how the partner spends money (excessively), the lack of communication about spending (i.e. does not consult the partner), and the target (i.e., items considered unnecessary by the partner).

2. *Cognitions and perceptions.* Aside from your partner's actual behavior, your cognitions and perceptions of a behavior can be a source of satisfaction or dissatisfaction. Claffey and Mickelson (2010) noted that it is not the actual division of labor but how the partners perceive it which has an impact on their lives and relationship.

3. *Value differences.* Because you and your partner have had different socialization experiences, you may also have different values—about religion (one feels religion is a central part of life; the other does not), money (one feels uncomfortable being in debt; the other has the buy-now-pay-later philosophy), in-laws (one feels responsible for parents when they are old; the other does not), and children (number, timing, discipline). The effect of value differences depends less on the degree of the difference than on the degree of rigidity with which each partner holds his or her values. Dogmatic and rigid thinkers, feeling threatened by value disagreement, may try to eliminate alternative views and thus produce more conflict. Partners who recognize the inevitability of difference may consider the positives of an alternative view and move toward acceptance. When both partners do this, the relationship takes priority and the value differences suddenly become less important.

4. *Inconsistent rules.* Partners in all relationships develop a set of rules to help them function smoothly. These unwritten but mutually understood rules include what time you are supposed to be home after work, whether you should call if you are going to be late, how often you can see friends alone, and when and how to make love. Conflict results when the partners disagree on the rules or when inconsistent rules develop in the relationship. For example, one wife expected her husband to take a second job so they could afford a new car, but she also expected him to spend more time at home with the family.

5. *Leadership ambiguity.* Unless a couple has an understanding about which partner will make decisions in which area (for example, the wife may make decisions about money management; the husband will make decisions about rearing the children), unnecessary conflict may result. Whereas some couples may want to discuss certain issues, others may want to develop a clear specification of roles.

Styles of Conflict Resolution

Conflict can be resolved by one or both partners changing positions more in line with the partner, compromising, agreeing to disagree (no resolution), or forcing one's position. All are likely to be used over the course of a relationship with modifying one's position and compromise having the most positive outcome.

Fifteen Principles and Techniques of Effective Communication

People who want effective communication in their relationship follow various principles and techniques, including the following:

1. *Make communication a priority.* Communicating effectively implies making communication an important priority in a couple's relationship. When communication is a priority, partners make time for it to occur in a setting

"The picture was of me, and I sent it." These words by Anthony Weiner, Democratic Representative from New York gave sexting nationwide visibility in the summer of 2011. He had been discovered sending sex text messages and explicit photos to various women. But sexting is also routine in romantic relationship—the focus of this research study. Sexting is defined as the sending of sexually explicit messages, photos and videos to each other using a cell phone. Use of technology has become common in the initiation, development, and maintenance of romantic relationships among undergraduates. The type of technology, frequency of use, gender/racial differences, and outcomes of sexting were among the research foci of this study.

Research Questions

The purpose of this research was to assess the degree to which technology was being used in a sample of undergraduates to send sexual messages, photos, and videos to a romantic partner. Several research questions guided this investigation. These included: What types of sexually explicit technology are the most common? Are males or females more likely to use sexually explicit technology? Are males or females more likely to be the first to initiate sexually explicit technology? What is the effect of the relationship of sending sexual content to a romantic partner?

Questionnaire

A 25 item questionnaire was posted on the Internet. In addition to specific types of technology and frequency of use, respondents were asked about sending erotic messages, photo, or video to a romantic partner.

Sample

The sample consisted of 483 undergraduate volunteer respondents in lower division sociology classes at a large southeastern university. Three-fourths of the respondents were female; 25% were male. The mean age was 19.7 with almost three-fourths (72%) sophomores or freshmen (18% were juniors, 9% seniors). Three-fourths self identified as white, 15% black, and 5% "other" (Hispanic, Asian, biracial). In regard to relationship status, 45% were dating one person, 38% were not dating anyone, and 12% were dating different people. Six percent were engaged/partnered/married. Of those who were involved in a relationship, the mean number of months was 10.4. Over ninety percent (91.5%) identified their sexual orientation as heterosexual.

Method

Regression analysis was performed to determine if there were gender differences in using sexually explicit technology and its perceived effects on a relationship. Regression analysis was also utilized to explore impacts of sexting on relationship happiness.

Good communication is as stimulating as black coffee, and just as hard to sleep after.

Anne Morrow Lindbergh,
author

without interruptions: they are alone; they are not answering e-mail/surfing/texting, do not answer the phone; and they turn the television off. Making communication a priority results in the exchange of more information between partners, which increases the knowledge each partner has about the other.

Negative relationship outcomes occur when partners do not prioritize communication with each other but are passionately and obsessively interacting with others via the Internet. Seguin-Levesque et al. (2003) found that the use of the Internet is not destructive per se but the obsessive passion of involvement with the Internet. Playing video games for hours on end also has a negative effect on a couple's communication and relationship.

2. *Avoid negative/hurtful statements to your partner.* Because intimate partners are capable of hurting each other so intensely, it is important to avoid brutal statements to the partner. Indeed, be very careful how you give negative feedback or communicate disapproval to your partner. Markman and Rhoades et al. (2010) noted that couples in marriage counseling often will report "it was a bad week" based on one negative comment made by the partner.

Results

Analysis revealed six primary findings:

1. Use of technology—Almost two thirds (64%) of the respondents reported sending a sexual text message, over forty percent (42.9%) a sexual photo, and over ten percent (12.7%) a sex video to a romantic partner.

2. Frequency of use—The frequency of sending sexual content varied from daily (4%) to less than once a month (26%).

3. Sex differences—There were no significant differences between women and men in sexting behavior. However, men were the first to initiate sending sexual content. Men were also more likely to perceive that sexting would have a positive effect on the couple's relationship (no such relationship was found).

4. Racial differences—Nonwhites compared to whites were more likely to report ever having used technology to send sexual content to a romantic partner.

5. Class in school—Sophomores, juniors, and seniors when compared to first year students were more likely to report ever having used technology to send sexual content to a romantic partner. Hence, the older the undergraduate, the more likely to report engaging in sexting behavior.

6. Type of technology—Text messaging was the technology used most frequently to send sexual content to a current or frequent partner.

Implications of the Study

The data revealed that it is not unusual for undergraduates to send sexual text messages, photos, and videos to their romantic partners. Since one's private cell phone is the technology of choice for sending this erotic content, employers are not involved. However, should an individual use their e-mail account at work to send sexual content, there could be trouble. The Supreme Court ruled in 2010 that employers can monitor e-mails of employees if there is reasonable grounds to assume that e-mails are personal and not work related. Hence, employees can get fired for sending sexual messages.

While undergraduates are not at risk as long as the parties are age 18 or older, sending erotic photos of individuals younger than 18 can be problematic. Sexting is considered by many countries to be child pornography, and laws related to child pornography have been applied in cases of sexting. Six high school students in Greensburg, Pennsylvania, were charged with child pornography after three teenage girls allegedly took nude or semi-nude photos of themselves and shared them with male classmates via their cell phones. Some undergraduates are under age 18. Having or sending nude images of underage individuals is a felony which can result in fines, imprisonment, and a criminal record.

Source

Parker, M., D. Knox, and Easterling. 2011. SEXTING: Sexual content/images in romantic relationships. Poster, Eastern Sociological Society, Philadelphia, Feb 24–26. Abridged.

Finkenauer et al. (2010) used signal detection theory to identify the degree to which an individual detects behaviors the partner does or does not engage in and how this detection affects their feelings about their partner and the relationship. The researchers focused on **correct rejections** where both partners agree that one of them did not engage in a behavior. In effect both partners are correct that one of them did not engage in a rejecting behavior. For example, Jose is late but Maria does not say anything about it. In this case, both Jose and Maria are aware that Maria chose not to comment on Jose being late. Such "correct rejections" are associated with higher relationship quality, greater ease of coordination in everyday life, and lower levels of conflict. In effect, when a partner elects not to make a negative comment, the other is aware of this and the outcome is positive for the couple.

Correct rejections both partners are aware that one of them chose not to engage in a negative comment.

3. *Say positive things about your partner.* Markman and Rhoades et al. (2010) also emphasized the need for partners to make positive comments to each other and that doing so was associated with more stable relationships.

People like to hear others say positive things about them. These positive statements may be in the form of compliments ("You look terrific!") or appreciation

("Thanks for putting gas in the car."). Gable et al. (2003) asked fifty-eight heterosexual dating couples to monitor their interaction with one another. The respondents observed that they were overwhelmingly positive, at a five-to-one ratio.

4. *Establish and maintain eye contact.* Shakespeare noted that a person's eyes are the "mirrors to the soul." Partners who look at each other when they are talking not only communicate an interest in each other but also are able to gain information about the partner's feelings and responses to what is being said. Not looking at your partner may be interpreted as lack of interest and prevents you from observing nonverbal cues.

5. *Ask open-ended questions.* When your goal is to find out your partner's thoughts and feelings about an issue, using **open-ended questions** is best. Such questions (for example, "How do you feel about me?") encourage your partner to give an answer that contains a lot of information. **Closed-ended questions** (for example, "Do you love me?"), which elicit a one-word answer such as yes or no, do not provide the opportunity for the partner to express a range of thoughts and feelings.

6. *Use reflective listening.* Effective communication requires being a good listener. One of the skills of a good listener is the ability to use the technique of **reflective listening**, which involves paraphrasing or restating what the person has said to you while being sensitive to what the partner is feeling. For example, suppose you ask your partner, "How was your day?" and your partner responds, "I felt exploited today at work because I went in early and stayed late and a memo from my new boss said that future bonuses would be eliminated because of a company takeover." Listening to what your partner is both saying and feeling, you might respond, "You feel frustrated because you really worked hard and felt unappreciated . . . and it's going to get worse."

Reflective listening serves the following functions: (1) it creates the feeling for speakers that they are being listened to and are being understood; and (2) it increases the accuracy of the listener's understanding of what the speaker is saying. If a reflective statement does not accurately reflect what a speaker thinks and feels, the speaker can correct the inaccuracy by restating the thoughts and feelings.

An important quality of reflective statements is that they are nonjudgmental. For example, suppose two lovers are arguing about spending time with their respective friends and one says, "I'd like to spend one night each week with my friends and not feel guilty about it." The partner may respond by making a statement that is judgmental (critical or evaluative), such as those exemplified in Table 4.1. Judgmental responses serve to punish or criticize people for what they think, feel, or want and often result in frustration and resentment.

Table 4.1 also provides several examples of nonjudgmental reflective statements.

7. *Use "I" statements.* **"I" statements** focus on the feelings and thoughts of the communicator without making a judgment on others. Because "I" statements are a clear and nonthreatening way of expressing what you want and how you feel, they are likely to result in a positive change in the listener's behavior. Making "I" statements reflects being authentic. Impett et al. (2010) emphasized the need to be "authentic" when communicating. Being **authentic** means speaking and acting in a manner according to what one feels. Being authentic in a relationship means being open with the partner about one's preferences and feelings about the partner's behavior. Being authentic has positive consequences for the relationship in that one's thoughts and feelings are out in the open (in contrast to being withdrawn and resentful).

In contrast, **"you" statements** blame or criticize the listener and often result in increasing negative feelings and behavior in the relationship. For example, suppose you are angry with your partner for being late. Rather

---

Open-ended question question which elicits a lot of information.

Closed-ended question question that allows for a one-word answer and does not elicit much information.

Reflective listening paraphrasing or restating what the person has said to you while being sensitive to what the partner is feeling.

"I" statements statements which focus on the feelings and thoughts of the communicator without making a judgment on others.

Authentic being who one is and saying what one feels.

"You" statement statement that blames or criticizes the listener and often results in increasing negative feelings and behavior in the relationship.

TABLE 4.1 Judgmental and Nonjudgmental Responses to a Partner's Saying, "I'd like to go out with my friends one night a week."

Nonjudgmental, Reflective Statements	Judgmental Statements
You value your friends and want to maintain good relationships with them.	You only think about what you want.
You think it is healthy for us to be with our friends some of the time.	Your friends are more important to you than I am.
You really enjoy your friends and want to spend some time with them.	You just want a night out so that you can meet someone new.
You feel it is important that we not abandon our friends just because we are involved.	You just want to get away so you can drink.
You feel that our being apart one night each week will make us even closer.	You are selfish.

than say, "You are always late and irresponsible" (which is a "you" statement), you might respond with, "I get upset when you are late and will feel better if you call me when you will be delayed." The latter focuses on your feelings and a desirable future behavior rather than blaming the partner for being late.

8. *Touch.* Hertenstein et al. (2007) identified the various meanings of touch such as conveying emotion, attachment, bonding, compliance, power, and intimacy. The researchers also emphasized the importance of using touch as a mechanism of nonverbal communication to emphasize one's point or meaning. Thompson and Hampton (2011) compared the ability of romantic couples and strangers to communicate emotions solely via touch. Results showed that both strangers and romantic couples were able to communicate universal and prosocial emotions through touch, whereas only romantic couples were able to communicate the self-focused emotions of envy and pride.

9. *Use "soft" emotions.* Sanford and Grace (2011) identified "hard" emotions (for example, angry or outraged) or "soft" emotions (sad or hurt) displayed during conflict. Examples of flat emotions are being bored or indifferent. The use of hard emotions resulted in an escalation of negative communication, whereas the display of "soft" emotions resulted in more benign communication and an increased feeling regarding the importance of resolving interpersonal conflict.

10. *Tell your partner what you want.* Focus on what you want rather than on what you don't want. Rather than say, "You spend too much time in playing video games," an alternative might be "When I come home, please stop playing video games, help me with dinner, ask me about my day and play video games after 10 P.M.). Rather than say, "You never call me when you are going to be late," say "Please call me when you are going to be late."

11. *Stay focused on the issue.* Wheeler et al. (2010) studied conflict resolution strategies and found that couples who have a solution-oriented style in contrast to a confrontational (attack) or nonconfrontational (ignore) style report higher levels of marital satisfaction. Hence, couples who focus on the issue to get it resolved handle conflict with minimal marital fallout. **Branching** refers to going out on different limbs of an issue rather than staying focused on the issue. If you are discussing the overdrawn checkbook, stay focused

There is a sacredness in tears. They are not the mark of weakness, but of power. They speak more eloquently than ten thousand tongues. They are messengers of overwhelming grief . . . and unspeakable love.

Washington Irving, American author

Kind words can be short and easy to speak, but their echoes are truly endless.

Mother Teresa, Nobel Peace Prize winner

Branching in communication, going out on different limbs of an issue rather than staying focused on the issue.

DIVERSITY IN OTHER COUNTRIES

The culture in which one is reared will influence the meaning of various words. An American woman was dating a man from Iceland. When she asked him, "Would you like to go out to dinner?" he responded, "Yes, maybe." She felt confused by this response and was uncertain whether he wanted to eat out. It was not until she visited his home in Iceland and asked his mother, "Would you like me to set the table?"—to which his mother replied, "Yes, maybe" —that she discovered that "Yes, maybe" means "Yes, definitely."

I didn't say that I didn't say it. I said that I didn't say that I said it. I want to make that very clear.

George Romney, former Michigan governor

Power the ability to impose one's will on the partner and to avoid being influenced by the partner.

on the checkbook. To remind your partner that he or she is equally irresponsible when it comes to getting things repaired or doing housework is to get off the issue of the checkbook. Stay focused.

12. *Make specific resolutions to disagreements.* To prevent the same issues or problems from recurring, agreeing on what each partner will do in similar circumstances in the future is important. For example, if going to a party together results in one partner's drinking too much and drifting off with someone else, what needs to be done in the future to ensure an enjoyable evening together? In this example, a specific resolution would be to decide how many drinks the partner will have within a given time period and to agree to stay together.

13. *Give congruent messages.* **Congruent messages** are those in which the verbal and nonverbal behaviors match. A person who says, "Okay, you're right" and smiles while embracing the partner is communicating a congruent message. In contrast, the same words accompanied by leaving the room and slamming the door communicate a very different message of disagreement and outrage.

14. *Share power.* One of the greatest sources of dissatisfaction in a relationship is a power imbalance and conflict over power (Kurdek 1994a). **Power** is the ability to impose one's will on the partner and to avoid being influenced by the partner. Oyamot et al. (2010) examined the relationships of 227 undergraduates and found that most viewed their relationships to be those of equal power. Over half (53%) of 2,922 undergraduates from two universities reported that they had the same amount of power in the relationship as their partner. Thirteen percent felt that they had less power; 17% more power (Knox and Hall 2010).

Expressions of power in a relationship are numerous and include the following:

Withdrawal (not speaking to the partner)

Guilt induction ("How could you ask me to do this?")

Being pleasant ("Kiss me and help me move the sofa.")

Negotiation ("We can go to the movie if we study for a couple of hours before we go.")

Deception (running up credit card debts of which the partner is unaware)

Blackmail ("I'll find someone else if you won't have sex with me.")

Physical abuse or verbal threats ("You will be sorry if you try to leave me.")

Criticism ("I can't think of anything good about you.")

Dominance ("I make more money than you so I will decide where we go.")

Power may also take the form of love and sex. The person in the relationship who loves less and who needs sex less has enormous power over the partner who is very much in love and who is dependent on the partner for sex. This pattern reflects the principle of least interest.

15. *Keep the process of communication going.* Communication includes both content (verbal and nonverbal information) and process (interaction). It is important not to allow difficult content to shut down the communication process (Turner 2005). To ensure that the process continues, the partners should focus on the fact that sharing information is essential and reinforce each other for keeping the process alive. For example, if your partner tells you something that you do that bothers him or her, it is important to thank him or her for telling you that rather than becoming defensive. In this way, your partner's feelings about you stay out in the open rather than hidden behind a wall of resentment. Otherwise, if you punish such disclosure because you don't like the content, subsequent disclosure will stop.

Although effective communication skills can be learned, Robbins (2005) noted that physiological capacities may enhance or impede the acquisition of these skills. She noted that people with attention-deficit/hyperactivity disorder (ADHD) might have deficiencies in basic communication and social skills.

Gender, Culture, and Communication

Gender Differences in Communication

Numerous jokes address the differences between how women and men communicate. One anonymous quote on the Internet follows:

> When a woman says, "Sure . . . go ahead," what she means is "I don't want you to." When a woman says, "I'm sorry," what she means is "You'll be sorry." When a woman says, "I'll be ready in a minute," what she means is "Kick off your shoes and start watching a football game on TV."

Women and men differ in their approach to and patterns of communication. Women are more communicative about relationship issues, view a situation emotionally, and initiate discussions about relationship problems. Deborah Tannen (1990; 2006) is a specialist in communication. She observed that, to women, conversations are negotiations for closeness in which they try "to seek and give confirmations and support, and to reach consensus" (1990, 25). A woman's goal is to preserve intimacy and avoid isolation. To men, conversations are about winning and achieving the upper hand.

The genders differ in regard to emotionality. Garfield (2010) reviewed men's difficulty with emotional intimacy. He noted that their emotional detachment stems from the provider role which requires them to stay in control. Being emotional is seen as weakness. Men's groups where men learn to access and express their feelings have been helpful in increasing men's emotionality.

In contrast, women tend to approach situations emotionally. For example, if a child is seriously ill, wives will want their husbands to be emotional, to cry, to show that they really care that their child is ill. But a husband might react to a seriously ill child by putting pressure on the wife to be mature about the situation and by encouraging stoicism. Mothers and fathers also speak differently to their children. Shinn and O'Brien (2008) observed the interactions between parents and their third grade children and found that mothers used more affiliative (relationship) speech than fathers, and fathers used more assertive speech than mothers. No sex differences in children's speech were found, suggesting that these differences do not emerge until later.

Women disclose more in their relationships than men do (Gallmeier et al. 1997). In a study of 360 undergraduates, women were more likely to disclose information about previous love relationships, previous sexual relationships, their love feelings for the partner, and what they wanted for the future of the relationship. They also wanted their partners to reciprocate this level of disclosure, but such disclosure was not forthcoming.

Men and women also differ in the degree to which they use tentative speech—words which convey uncertainty and a lack of confidence for a communicator. Examples include hedges (e.g., sort of, maybe), disclaimers (e.g., I'm not sure), and tag questions (e.g., don't you think?). This is particularly true when the topic is "masculine" or "feminine". Hence women are more tentative in speech when talking about sports (Palomares 2009).

Cultural Differences in Communication

Communication styles vary by country. Individuals reared in France, Germany, Italy, or Greece regard arguing as a sign of closeness—to be blunt and argumentative is to keep the interaction alive and dynamic; to have a tone of agreement is boring. Asian cultures (e.g. Japanese, Chinese, Thai, and Pilipino) place a high value on avoiding open expression of disagreement and emphasizing harmony. Deborah Tannen (1998) observed the different perceptions of a Japanese woman married to a Frenchman:

> "He frequently started arguments with her, which she found so upsetting that she did her best to agree and be conciliatory. This only led him to seek another point on which

to argue. Finally she lost her self-control and began to yell back. Rather than being angered, he was overjoyed. Provoking arguments was his way of showing interest in her, letting her know how much he respected her intelligence. To him, being able to engage in spirited disagreement was a sign of a good relationship" (p. 211).

Self-Disclosure, Dishonesty, and Secrets

Shakespeare noted in Macbeth that "the false face must hide what the false heart doth know," suggesting that withholding information and being dishonest may affect the way one feels about oneself and relationships with others. All of us make choices, consciously or unconsciously, about the degree to which we disclose, are honest, and/or keep secrets.

Self-Disclosure in Intimate Relationships

One aspect of intimacy in relationships is self-disclosure, which involves revealing personal information and feelings about oneself to another person.

Relationships become more stable when individuals disclose themselves—their formative years, previous relationships (positive and negative), experiences of elation and sadness or depression, and goals (achieved and thwarted). We noted in the discussion of love in Chapter 3 that self-disclosure is a psychological condition necessary for the development of love. To the degree that you disclose yourself to another, you invest yourself in and feel closer to that person. People who disclose nothing are investing nothing and remain aloof. One way to encourage disclosure in one's partner is to make disclosures about one's own life and then ask about the partner's life.

PERSONAL CHOICES

How Much Do I Tell My Partner about My Past?

Because of the fear of HIV infection and other sexually transmitted infections (STIs), some partners want to know the details of each other's previous sex life, including how many partners they have had sex with and in what contexts. Those who are asked will need to decide whether to disclose the requested information, which may include one's sexual orientation, present or past sexually transmitted diseases, and any sexual proclivities or preferences the partner might find bizarre (for example, bondage and discipline). Ample evidence suggests that individuals are sometimes dishonest with regard to the sexual information about their past they provide to their partners. The "number of previous sexual partners" is the most frequent lie undergraduates report telling each other.

In deciding whether or not to talk honestly about your past to your partner, you may want to consider the following questions: How important is it to your partner to know about your past? Do you want your partner to tell you (honestly) about her or his past? What impact on your relationship will open disclosure have? What impact will withholding such information have on the level of intimacy you have with your partner?

Forms of Dishonesty and Deception

Dishonesty and deception take various forms. *An Education* (nominated for Best Picture of 2009) featured Jenny falling in love with an older man only to discover that he was married with children. In addition to telling an outright lie (e.g., the older man presented himself as single and available), people may exaggerate or conceal the truth, pretend, or withhold information. Regarding the latter, in virtually every relationship, partners may not share things with each other about themselves or their past. We often withhold information or keep secrets in our intimate relationships for what we believe are good reasons—we believe that we are protecting our partners from anxiety or hurt feelings, protecting ourselves from criticism and rejection, and protecting our relationships from conflict and disintegration. Information posted on social networking sites is often not true. Only 31% of a large sample of social network site users reported that they were "totally honest" about what they put on Facebook, Twitter, etc. (Carey and Trap 2010a).

Finkenauer and Hazam (2000) found that happy relationships depend on withholding information. The researchers contend, "Nobody wants to be criticized or talk about topics that are known to be conflictive." Ennis et al. (2008) noted three types of lies: (1) self-centered to protect oneself ("I didn't do it."); (2) oriented to protect another ("Your hair looks good today."); or (3) altruistic to protect a third party ("She didn't do it."). The Applying Social Research feature on page 106 reveals what romantic partners keep secrets from each other and why.

Find the grain of truth in criticism—chew it and swallow it.

D. Sutton

Family Secrets

Some romantic secrets are also family secrets. Table 4.2 reflects various family secrets identified by 70 students.

TABLE 4.2 Family Secrets*

(n = 70)

Abuse in the Family (5)
- Cousin molested by paternal uncle
- Sexual assault by parent
- Abusive grandfather
- Chopped off brother's finger on purpose but pretended it was an accident
- My brother threw a rock at me and knocked out a tooth

Abuse from a partner in a relationship (3)
- Hidden abuse in marriage
- Sister's boyfriend beat her
- Dad's parents were abusive to each other in front of him

Substance abuse (9)
- Dad smokes weed
- Aunt has marijuana plant
- Drug/alcohol abuse
- Cousin is a drug addict and a thief
- Best friend died of drug overdose and family claims it was a car accident
- Alcoholism
- Family member does heavy drugs

Cheating/Adultery (11)
- Infidelities
- Dad may have siblings in France from an affair when grandfather was in the war
- Friend's uncle had an affair, divorced her aunt after 20 years of marriage and is dating the other lady
- Dad has 3 kids outside of marriage
- Husband is unfaithful
- Cheating on wife
- Boyfriend's grandma left her fiancé for his grandfather
- Dad cheats on Mom frequently
- My dad cheated on my mom
- Cheating
- Aunt is pregnant with another man's baby who is not her husband

Homosexuality (5)
- Sister is a lesbian
- Gay cousin currently not out to family
- Neighbors are lesbians
- My cousin is a lesbian
- My uncle is gay

Adoption (9)
- Child unaware that they were adopted
- Uncle gave child up for adoption because he wasn't married
- Mother and Uncle were adopted
- I'm adopted
- My cousin is adopted
- My aunt thinks she's adopted, but I don't think she is
- My aunt is adopted
- One of the children is adopted in the family
- Mom and her sisters were adopted

Heritage (3)
- Some of my relatives were Nazis
- I'm Polish and Cherokee
- Grandmother changed the spelling of our last name

Illness/death (5)
- Grandmother has a serious mental illness
- Mother attempted suicide and is in therapy
- Best friend has an eating disorder
- My great grandfather committed suicide
- Grandpa committed suicide in 1995

Other (5)
- Grandfather ran moonshine
- Brother's engaged
- A recipe
- A wreck I was in
- Family was placed under the Witness Protection Agency

No Answer 16

*Seventy students were asked to identify "A family secret (not necessarily my own)". Unpublished data collected for this text. Department of Sociology, East Carolina University.

Arnold Schwarzenegger's disclosure that he fathered a child with his housekeeper while Governor of California and the Tiger Woods sex scandal (including 13 alleged mistresses) gave renewed visibility to the fact that keeping secrets in romantic relationships occurs often.

Vangelisti (1994) identified three categories of secrets: taboo secrets, rule violations, and conventional secrets. Taboo secrets, if disclosed, would bring stigma to the individual. Examples include having a sexually transmitted infection, being gay, or having been sexually abused. Rule violations refer to activities which break rules established and enforced by one's family (e.g., adolescents keeping drug use and sexual behavior from parents). Conventional secrets refer to content regarded as no one's business outside the family such as finances, abortion, and mental illness. Of these three types, taboo secrets are the most common, with more than half of all secrets being sexual, including unwanted pregnancy, rape, sexually transmitted infections, abortion, homosexuality, and promiscuity. Both the Schwarzenegger and Woods secrets fall into the category of "taboo" secrets.

The purpose of the current research was to identify the social correlates of keeping secrets in a romantic relationship. Specifically, do women or men, people dating or people married, blacks or whites, and gays or straights keep more secrets?

Description of the Sample

The sample consisted of 431 undergraduates. A majority of respondents were female (75.2%), heterosexual (77.3%), and white (70.5%). The mean age of the respondents was slightly over 22. Class standing in college was distributed across categories, with the largest category (29.2%) being sophomores. Relationship status was also distributed across categories with dating one person and not dating anyone representing the largest respondent categories. For those in a relationship, the mean amount of time was about 17 months. Individual happiness was a mean of 7.07 on a scale of 1–10 and relationship happiness was a mean of 6.48 on a scale from 1–10, with higher numbers indicating higher levels of happiness. Response categories for type of secret most likely to be kept varied, with the largest category being their own secret (41.6%). Reasons for keeping secrets also varied with the largest response category being "to avoid hurting the partner" (38.9%).

Results and Discussion

1. *Percent Keeping a Secret.* Over 60% of respondents reported ever having kept a secret from a romantic partner and over one-quarter of respondents reported currently keeping a secret.

2. *Females More Likely to Keep Secrets than Males.* Sensitivity to the partner's reaction, desire to avoid hurting the partner, and desire to avoid damaging the relationship may be the primary motives for why females are more likely than males to keep a secret from a romantic partner. Abundant research has demonstrated that females are more invested in relationships than males (Monin et al. 2008; Stone 1007). This relationship value may translate into keeping secrets to minimize relationship discord/ pain.

Lying and Cheating

> *I'm not upset that you lied to me; I'm upset that from now on I can't believe you.*
>
> Friedrich Nietzsche, German philosopher

Relationships are compromised by lying and cheating.

Lying in American Society

Lying, a deliberate attempt to mislead, is pervasive in American society. Presidential candidate John Edwards repeatedly lied about his affair with Rielle Hunter, a member of his campaign staff, until he was caught meeting her at a hotel. Investment consultant Bernie Madoff lied to 4,800 clients over twenty-five years and stole over $50 billion from them. Baseball pitcher Roger Clemons and hitter Alex Rodriguez ("A-Rod") lied to investigators about steroid use. Politicians routinely lie to citizens ("Lobbyists can't buy my vote."), and citizens lie to the government (via cheating on taxes). Teachers lie to students ("The test will be easy."), and students lie to teachers ("I studied all night."). Parents lie to their children ("It won't hurt."), and children lie to their parents about where they

3. **Spouses More Likely to Keep Secrets than Dating Partners.** The cultural script for spouses is that they have "open communication" with fidelity a given. The Schwarzenegger and Tiger Woods mistress scandals became international news because they were MARRIED. Had Schwarzenegger or Woods been single and cheating on their girlfriends, there would be much less interest by the media.

   Given higher cultural standards for fidelity, the need to avoid disapproval from a spouse requires keeping secret anything of which the partner would disapprove. Even the husband who eats a candy bar on the way home and destroys the wrapper is keeping a secret. Hence, spouses keep more secrets than singles because the cultural script demands a higher level of honesty (which is unrealistic but the illusion of openness is maintained by secrecy).

4. **Blacks More Likely to Keep Secrets than Whites.** Blacks are a minority who may still be victimized by the white majority. One way to avoid such disapproval is to keep one's thoughts to oneself—to keep secrets. This skill of deception may generalize to one's romantic relationships.

5. **Homosexuals More Likely to Keep Secrets.** One explanation is that homosexuals live in a heterosexist context which encourages regular deception by hiding their sexual orientation to avoid prejudice and discrimination. Indeed the phrase "in the closet" implies "keeping a secret." Since being deceptive may become a part of one's personality, its expression may generalize to one's romantic relationship. Everyone has secrets and everyone decides on a daily basis the degree to which they will be open and disclosing about each issue or issues. Homosexuals may keep a secret from a romantic partner more often because they can develop a pattern of keeping secrets.

6. **Reasons for Keeping Secrets.** Respondents were asked why they kept a personal secret from a romantic partner. "To avoid hurting the partner" was the top reason reported by almost forty percent (38.9%) of the respondents. "It would alter our relationship" and "I feel so ashamed for what I did" were reported by 17.7% and 10.7% of the respondents respectively. Afifi and Steuber (2010) noted that if one partner views another as having the potential to respond negatively to the revelation of a secret, information is withheld.

References

Afifi, T. D. and K. Steuber 2010. The cycle of concealment model. *Journal of Social and Personal Relationships* 27: 1019–1034.

Monin, J. K., M. S. Clark, and E. P. Lemay. (2008). Communal responsiveness in relationships with female versus male family members. *Journal Sex Roles* 59:176–88.

Stone, P. (2007). *Opting out?* Berkley: University of California Press.

Vangelisti, A. (1994). Family secrets: Forms, functions and correlates. *Journal of Social and Personal Relationships* 1: 113–135.

* Abridged and updated from Brackett, A., D. Knox, and B. Easterling. 2010. Secrets in romantic relationships: Sexual orientation and other correlates. Poster, Southern Sociological Society, Atlanta, April.

have been, whom they were with, and what they did. Dating partners lie to each other ("I've had a couple of previous sex partners."), women lie to men ("I had an orgasm."), and men lie to women ("I'll call."). The price of lying is high—distrust and alienation. A student in class wrote:

> At this moment in my life I do not have any love relationship. I find college dating to be very hard. The guys here lie to you about anything and you wouldn't know the truth. I find it's mostly about sex here and having a good time before you really have to get serious. That is fine, but that is just not what I am all about.

Lying in College Student Relationships

Lying is epidemic in college student relationships. In response to the statement, "I have lied to a person I was involved with," 57% of 2,922 undergraduates reported "yes." Fourteen percent of 2,922 undergraduates reported having lied to a partner about their previous number of sexual partners (Knox and Hall 2010).

SOCIAL POLICY | Should One Partner Disclose Human Immunodeficiency Virus (HIV)/Sexually Transmitted Infection (STI) Status to Another?

An estimated 25% of undergraduates report that they have or have had an STI. Individuals often struggle over whether or how to tell a partner if they have an STI, including HIV infection. If a person in a committed relationship acquires an STI, then that individual, or the partner, may have been unfaithful and have had sex with someone outside the relationship. Thus, disclosure about an STI may also mean confessing one's own infidelity or confronting the partner about the possible infidelity. (However, the infection may have occurred prior to the current relationship but gone undetected.) Individuals who have an STI and who are beginning a new relationship face a different set of concerns. Will their new partner view them negatively? Will they want to continue the relationship? One Internet ad began, "I have herpes—Now that that is out of the way. . . ."

Although telling a partner about having an STI may be difficult and embarrassing, avoiding disclosure or lying about having an STI represents a serious ethical violation. The responsibility to inform a partner that one has an STI—before having sex with that partner—is a moral one. But there are also legal reasons for disclosing one's sexual health condition to a partner. If you have an STI and you do not tell your partner, you may be liable for damages if you transmit it to your partner. Ayres and Baker (2004) proposed a new crime, reckless sexual conduct, of which a person would be guilty for not using a condom the first time of intercourse with a person. The penalty for the perpetrator would be three months in prison.

Khalsa (2006) noted that reporting HIV infection and acquired immunodeficiency syndrome (AIDS) is mandatory

Cheating in College Student Relationships

Cheating may be defined as having sex with someone else while involved in a relationship with a romantic partner. When 2,922 undergraduates were asked if they had cheated on a partner they were involved with, almost a quarter (24%) reported that they had done so (Knox and Hall 2010). McAlister et al. (2005) noted that extradyadic activity (defined as kissing or "sexual activity") among young adults who were dating could be predicted. Those young adults who had a high number of previous sexual partners, who were impulsive, who were not satisfied in their current relationship, and who had attractive alternatives were more vulnerable to being unfaithful.

Even in "monogamous" relationships, there is considerable cheating. Vail-Smith et al. (2010) found that 27.2% of the males and 19.8% of the females of 1,341 undergraduates reported having oral, vaginal, or anal sex outside of a relationship that their partner considered monogamous. People most likely to cheat

WHAT IF?

What if an Old Lover Contacts You?

Partners may differ in terms of how they would respond to an old lover who contacts them. Although some may not respond at all to an e-mail or letter, others may e-mail, call, and meet without the current partner's knowledge. Still others may make the partner aware of the contact and negotiate an outcome. One scenario is to meet the person in a public place with a time frame (for example, lunch at McDonalds). Another is that the old lover may be made aware of the new relationship and given the choice to meet the partner (for example, come to the apartment for lunch). To increase the security and strength of the new relationship, include the current partner in what is happening with the "old love surfacing." To keep the contact with the old love secret is to invite escalation of the relationship, deception, and eventual relationship disaster.

in most states, although partner notification laws vary from state to state. New York has a strong partner notification law that requires health care providers to either notify any partners the infected person names or to forward the information about partners to the Department of Health, where public health officers notify the partners that they have been exposed to an STI and to schedule an appointment for STI testing. The privacy of the infected individual is protected by not revealing names to the partner being notified of potential infection. In cases where the infected person refuses to identify partners, standard partner notification laws require doctors to undertake notification without cooperation if they know of the sexual partner or spouse.

Your Opinion?

1. What percentage of undergraduates (who knowingly has an STI) would have sex with another undergraduate and not tell the partner?

2. What do you think the penalty should be for deliberately exposing a person to an STI?

3. What partner notification law do you recommend?

Sources

Ayres, I., and K. Baker. 2004. A separate crime of reckless sex. *Yale Law School, Public Law orking Paper No. 80.*

Khalsa, A. M. 2006. Preventive counseling, screening, and therapy for the patient with newly diagnosed HIV infection. *American Family Physician* 73: 271–280.

in these "monogamous" relationships were men over the age of 20, those who were binge drinkers, members of a fraternity, male NCAA athletes, and those who reported that they were "nonreligious." White et al. (2010) also studied 217 couples where both partners reported on their own risk behaviors and their perceptions of their partner's behavior. Three percent of women and 14% of men were unaware that their partner had recently had a concurrent partner. Eleven percent and 12%, respectively, were unaware that their partner had ever injected drugs; 10% and 12% were unaware that their partner had recently received an STD diagnosis; and 2% and 4% were unaware that their partner was HIV-positive. These data suggest a need for people in "committed" relationships to reconsider their risk of sexually transmitted infections and to protect themselves via condom usage. In addition, one of the ways in which college students deceive their partners is by failing to disclose that they have an STI. Approximately 25% of college students will contract an STI while they are in college (Purkett 2010). Because the potential to harm an unsuspecting partner is considerable, should we have a national social policy regarding such disclosure (see Social Policy feature)?

Strickler and Hans (2010) conceptualized infidelity (cheating) as both sexual and nonsexual. Sexual cheating was intercourse, oral sex, and kissing. Nonsexual cheating could be interpersonal (secret time together, flirting), electronic (text messaging, e-mailing), or alone (sexual fantasies with others, pornography, masturbation). Of 400 undergraduates, 74% of the males and 67% of the females in a committed relationship reported that they had cheated according to their own criteria. Hence, in the survey, they identified a specific behavior as cheating and later reported that they had engaged in that behavior.

Theories Applied to Relationship Communication

Symbolic interactionism and social exchange are theories that help to explain the communication process.

What if You Discover Your Partner Is Cheating on You?

Because cheating does occur, to deny that this will ever happen in one's own relationship may be unrealistic. Reactions will vary from immediate termination of the relationship, to taking a break from the relationship, to revenge by cheating also. One scenario is to discuss the dishonesty with the partner to discover any relationship deficits that may be corrected. Another is to discuss the acceptability of the behavior in terms of frequency. Does everyone make a mistake sometimes and this is to be overlooked, or is the dishonesty of a chronic variety that will continue? Most individuals are devastated to discover a betrayal but find a way to continue the relationship. This reaction is functional only if the dishonesty is not chronic. Chronic unfaithfulness takes advantage of the forgiveness of the partner and permanently infuses the relationship with distrust and deceit—a recipe for disaster. Hannon et al. (2010) emphasized that getting beyond an act of betrayal requires the cooperation of the perpetrator who expresses sorrow/apologizes/makes amends in exchange for the partner who forgives. Both work together to get beyond the impasse.

Symbolic Interactionism

Interactionists examine the process of communication between two actors in terms of the meanings each attaches to the actions of the other. Definition of the situation, the looking-glass self, and taking the role of the other (discussed in Chapter 1) are all relevant to understanding how partners communicate. With regard to resolving a conflict over how to spend the semester break (for example, vacation alone or go to see parents), the respective partners must negotiate their definitions of the situation (is it about their time together as a couple or their loyalty to their parents?). The looking-glass self involves looking at each other and seeing the reflected image of someone who is loved and cared for and someone with whom a productive resolution is sought. Taking the role of the other involves each partner's understanding the other's logic and feelings about how to spend the break.

Social Exchange

Exchange theorists suggest that the partners' communication can be described as a ratio of rewards to costs. Rewards are positive exchanges, such as compliments, compromises, and agreements. Costs refer to negative exchanges, such as critical remarks, complaints, and attacks. When the rewards are high and the costs are low, the outcome is likely to be positive for both partners (profit). When the costs are high and the rewards low, neither may be satisfied with the outcome (loss).

When discussing how to spend the semester break, the partners are continually in the process of exchange—not only in the words they use but also in the way they use them. If the communication is to continue, both partners need to feel acknowledged for their points of view and to feel a sense of legitimacy and respect. Communication in abusive relationships is characterized by the parties criticizing and denigrating each other, which usually results in a shutdown of the communication process.

Fighting Fair: Seven Steps in Conflict Resolution

Before reading about how to resolve conflict, you may wish to take the Communication Danger Signs Scale to assess the degree to which you may have a problem in this area.

The art of being wise is the art of knowing what to overlook.

William James, philosopher and psychologist

SELF-ASSESSMENT | Communication Danger Signs Scale

This scale is designed to assess the degree to which there is communication trouble in your relationship.

Directions

After reading each sentence carefully, circle the number that best represents how often this happens in your relationship.

1 Never/almost never	2 Occasionally	3 Frequently

	N	O	F
1. Little arguments escalate into ugly fights with accusations, criticisms, name calling, or bringing up past hurts.	1	2	3
2. My partner criticizes or belittles my opinions, feelings, or desires.	1	2	3
3. My partner seems to view my words or actions more negatively than I mean them to be.	1	2	3
4. When we have a problem to solve, it is like we are on opposite teams.	1	2	3
5. I hold back from telling my partner what I really think and feel.	1	2	3
6. I feel lonely in this relationship.	1	2	3
7. When we argue, one of us withdraws, that is, doesn't want to talk about it anymore; or leaves the scene.	1	2	3

Scoring

Add the numbers you circled. 1 (never) is the response reflecting the ultimate safe context in which you communicate with your partner and a 3 (frequently) is the most toxic of communication contexts. The lower your total score (7 is the lowest possible score), the greater your communication context comfort. The higher your total score (21 is the highest possible score), the more you are likely to feel anxious when around or communicating with your partner. A score of 14 places you at the midpoint between being extremely comfortable communicating with your partner (a score of 7) and being extremely uncomfortable (a score of 21).

Source

Dr. Howard Markman, Director Center for Marital and Family Studies at the University of Denver. Used by permission. Contact Dr. Markman at http://loveyourrelationship.com/ for information about Communication/ Relationship Retreats. Also see, Markman, H. J., S. M. Stanley, and S. L. Blumberg (2010). *Fighting for your marriage* (3rd ed.) San Francisco, CA: Jossey-Bass.

When a disagreement ensues, it is important to establish rules for fighting that will leave the partners and their relationship undamaged after the disagreement. Indeed, Lavner and Bradbury (2010) studied 464 newlyweds over a 4 year period, noticed the precariousness of relationships (even those reporting considerable satisfaction divorced) and recommended that couples "impose and regularly maintain ground rules for safe and nonthreatening communication." Such guidelines for fair fighting/effective communication include not calling each other names, not bringing up past misdeeds, and not attacking each other.

Indeed, being positive about the partner is essential. Gottman (2004) found that partners who say positive things to each other at a ratio of 5:1 (positives to negatives) seem to stay together (Gottman 1994). And, not beginning a heated discussion late at night is important. In some cases, a good night's sleep has a way of altering how a situation is viewed and may even result in the problem no longer being an issue.

Fighting fairly also involves keeping the interaction focused, respective, and moving toward a win-win outcome. If recurring issues are not discussed and resolved, conflict may create tension and distance in the relationship, with the result that the partners stop talking, stop spending time together, and stop being intimate. A conflictual, unsatisfactory marriage is similar to divorce in terms of its impact on the diminished psychological, social, and physical well-being of the partners (Hetherington 2003). Developing and using skills for fair fighting and conflict resolution are critical for the maintenance of a good relationship. Resolving issues via communication is not easy. Rhoades et al. (2009) noted that relationship partners who decided to live together to test their relationship noted that communication was difficult or negative.

Howard Markman is head of the Center for Marital and Family Studies at the University of Denver. He and his colleagues have been studying 150 couples at yearly intervals (beginning before marriage) to determine those factors most responsible for marital success. They have found that communication skills that reflect the ability to handle conflict, which they call "constructive arguing," are the single biggest predictor of marital success over time (Marano 1992). According to Markman, "Many people believe that the causes of marital problems are the differences between people and problem areas such as money, sex, and children. However, our findings indicate it is not the differences that are important, but how these differences and problems are handled, particularly early in marriage" (Marano 1992, 53). Markman and Rhoades et al. (2010) provide details for constructive communication in their new book *Fighting for Your Marriage*. The following sections identify standard steps for resolving interpersonal conflict.

Success in marriage is only partly attributable to compatibility. It's about how you manage those differences and whether you have a style for doing so that is successful.

Gregory A. Kuhlman, psychologist

Address Recurring, Disturbing Issues

Addressing issues in a relationship is important. As noted earlier, couples who stack resentments rather than discuss conflictual issues do no service to their relationship. Indeed, the healthiest response to feeling upset about a partner's behavior is to engage the partner in a discussion about the behavior. Not to do so is to let the negative feelings fester, which will result in emotional and physical withdrawal from the relationship.

Identify New Desired Behaviors

Dealing with conflict is more likely to result in resolution if the partners focus on what they *want* rather than what they *don't want*. Tell your partner specifically what you want him or her to do. For example if your partner is chronically late, rather than say, "Stop being late" you might ask him or her to "please be on time or call me if you are going to be late." (say with a smile)

Identify Perceptions to Change

Rather than change behavior, changing one's perception of a behavior may be easier and quicker. Rather than expect one's partner to always be "on time," it may be easier to drop the expectation that one's partner be on time and to stop being mad about something that doesn't matter. South et al. (2010) emphasized the importance of perception of behavior in regard to marital satisfaction.

Summarize Your Partner's Perspective

We often assume that we know what our partner thinks and why he or she does things. Sometimes we are wrong. Rather than assume how our partner thinks and feels about a particular issue, we might ask open-ended questions in an effort to learn our partner's thoughts and feelings about a particular situation.

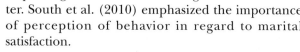

Generate Alternative Win-Win Solutions

Looking for win-win solutions to conflicts is imperative. Solutions in which one person wins means that one person is not getting needs met. As a result, the person who loses may develop feelings of resentment, anger, hurt, and hostility toward the winner and may even look for ways to get even. In this way, the winner is also a loser. In intimate relationships, one winner really means two losers.

Chelsea E. Curry

These partners are so worked up and angry they are unlikely to be able to summarize each other's point of view.

Generating win-win solutions to interpersonal conflict often requires **brainstorming**. The technique of brainstorming involves suggesting as many alternatives as possible without evaluating them. Brainstorming is crucial to conflict resolution because it shifts the partners' focus from criticizing each other's perspective to working together to develop alternative solutions.

With our colleagues (Knox et al. 1995), we studied the degree to which 200 college students who were involved in ongoing relationships were involved in win-win, win-lose, and lose-lose relationships. Descriptions of the various relationships follow:

Win-win relationships are those in which conflict is resolved so that each partner derives benefits from the resolution. For example, suppose a couple have a limited amount of money and disagree on whether to spend it on eating out or on seeing a current movie. One possible win-win solution might be for the couple to eat a relatively inexpensive dinner and rent a movie.

An example of a **win-lose solution** would be for one of the partners to get what he or she wanted (eat out or go to a movie), with the other partner getting nothing of what he or she wanted. Caughlin and Ramey (2005) studied demand-and-withdraw patterns in parent–adolescent dyads and found that the demand on the part of one of them was usually met by withdrawal on the part of the other partner. Such demand may reflect a win-lose interaction.

A **lose-lose solution** is one in which both partners get nothing that they want—in the scenario presented, the partners would neither go out to eat nor see a movie, be mad a each other and not talk.

More than three-quarters 77.1% of the students reported being involved in a win-win relationship, with men and women reporting similar percentages. Of the respondents, 20% were involved in win-lose relationships. Only 2% reported that they were involved in lose-lose relationships. Of the students in win-win relationships, 85% reported that they expected to continue their relationship, in contrast to only 15% of students in win-lose relationships. No student in a lose-lose relationship expected the relationship to last.

After a number of solutions are generated, each solution should be evaluated and the best one selected. In evaluating solutions to conflicts, it may be helpful to ask the following questions:

1. Does the solution satisfy both individuals? Is it a win-win solution?
2. Is the solution specific? Does it specify exactly who is to do what, how, and when?
3. Is the solution realistic? Can both parties realistically follow through with what they have agreed to do?
4. Does the solution prevent the problem from recurring?
5. Does the solution specify what is to happen if the problem recurs?

Kurdek (1995) emphasized that conflict-resolution styles that stress agreement, compromise, and humor are associated with marital satisfaction, whereas conflict engagement, withdrawal, and defensiveness styles are associated with lower marital satisfaction. In his own study of 155 married couples, the style in which the wife engaged the husband in conflict and the husband withdrew was particularly associated with low marital satisfaction for both spouses.

Communicating effectively and creating a context of win-win in one's relationship contributes to a high-quality marital relationship, which is good for one's health.

Forgive

Too little emphasis is placed on forgiveness as an emotional behavior that can move a couple from a deadlock to resolution. Merolla and Zhang (2011) noted that offender remorse positively predicted forgiveness and that such forgiveness was associated with helping to resolve the damange. Hill (2010) studied forgiveness and emphasized that it is less helpful to try to "will" one's self to forgive the transgressions of another than to engage a process of self-reflection—that one

Brainstorming suggesting as many alternatives as possible without evaluating them.

Win-win relationships relationship in which conflict is resolved so that each partner derives benefits from the resolution.

Win-lose solution outcome of a conflict in which one partner wins and the other loses.

Lose-lose solution a solution to a conflict in which neither partner benefits.

Remember not only to say the right thing in the right place, but to leave unsaid the wrong thing at the tempting moment.

Benjamin Franklin, founding father

Marriage is three parts love and seven parts forgiveness of sins.

Langdon Mitchell, American playwright

Amae expecting a close other's indulgence when one behaves inappropriately.

Defense mechanism unconscious techniques that function to protect individuals from anxiety and to minimize emotional hurt.

Escapism the simultaneous denial of and avoidance of dealing with a problem.

Rationalization the cognitive justification for one's own behavior that unconsciously conceals one's true motives.

Projection attributing one's own thoughts, feelings, and desires to someone else while avoiding recognition that these are one's own thoughts, feelings, and desires.

Displacement shifting one's feelings, thoughts, or behaviors from the person who evokes them onto someone else.

has also made mistakes, hurt others, is guilty and to empathize with the fact that we are all fallible and need forgiveness. In addition, forgiveness ultimately means "letting go" of one's anger, resentment, hurt and its power comes from offering forgiveness as an expression of love to the person who has betrayed him or her. Forgiveness also has a personal benefit—it releases one from feeling hurt, reduces hypertension and feelings of stress. To forgive is to restore the relationship—to pump life back into it. Of course, forgiveness given too quickly may be foolish. A person who has deliberately hurt their partner without remorse may not deserve forgiveness.

It takes more energy to hold on to resentment than to move beyond it. One reason some people do not forgive a partner for a transgression is that one can use the fault to control the relationship. "I wasn't going to let him forget," said one woman of her husband's infidelity.

A related concept to forgiveness is amae. Marshall et al. (2011) studied the concept of **amae** in Japanese romantic relationships. The term means expecting a close other's indulgence when one behaves inappropriately. Thirty Japanese undergraduate romantic couples kept a diary for two weeks that assessed their amae behavior (requesting, receiving, providing amae). Results revealed that amae behavior was associated with greater relationship quality and less conflict. "Cutting one some slack" may be another way of expressing amae.

Be Alert to Defense Mechanisms

Effective conflict resolution is sometimes blocked by **defense mechanisms**—unconscious techniques that function to protect individuals from anxiety and to minimize emotional hurt. The following paragraphs discuss some common defense mechanisms.

Escapism is the simultaneous denial of and avoidance of dealing with a problem. The usual form of escape is avoidance. The spouse becomes "busy" and "doesn't have time" to think about or deal with the problem, or the partner may escape into playing video games, sleep, alcohol, marijuana, or work. Denying and withdrawing from problems in relationships offer no possibility for confronting and resolving the problems.

Rationalization is the cognitive justification for one's own behavior that unconsciously conceals one's true motives. For example, one wife complained that her husband spent too much time at the health club in the evenings. The underlying reason for the husband's going to the health club was to escape an unsatisfying home life. However, the idea that he was in a dead marriage was too painful and difficult for the husband to face, so he rationalized to himself and his wife that he spent so much time at the health club because he made a lot of important business contacts there. Thus, the husband concealed his own true motives from himself (and his wife).

Projection is attributing one's own thoughts, feelings, and desires to someone else while avoiding recognition that these are one's own thoughts, feelings, and desires. For example, the wife who desires to have an affair may accuse her husband of being unfaithful to her. Projection may be seen in such statements as "You spend too much money" (projection for "I spend too much money") and "You want to break up" (projection for "I want to break up"). Projection interferes with conflict resolution by creating a mood of hostility and defensiveness in both partners. The issues to be resolved in the relationship remain unchanged and become more difficult to discuss.

Displacement involves shifting your feelings, thoughts, or behaviors from the person who evokes them onto someone else. The wife who is turned down for a promotion and the husband who is driven to exhaustion by his boss may direct their hostilities (displace them) onto each other rather than toward their respective employers. Similarly, spouses who are angry at each other may displace this anger onto someone else, such as the children.

By knowing about defense mechanisms and their negative impact on resolving conflict, you can be alert to them in your own relationships. When a conflict continues without resolution, one or more defense mechanisms may be operating.

The best answer to anger is silence.

Unknown

The Future of Communication

The future of communication will increasingly involve technology in the form of texting, smart phones, Facebook, etc. Such technology will be used to initiate, enhance, and maintain relationships. Indeed, intimates today may text each other 60 times a day. Over 2,000 messages a month are not unusual. Parental communication with children will also be altered. Aponte and Pessagno (2010) noted that technology may have positive and negative effects on the family. A positive effect is that parents will be able to use technology to monitor content as their children surf the Internet, send text messages, and send/receive photos on their cell phone. Parents may also use technology to know where their children are by global tracking systems embedded in their cell phones. The downside is that children can use this same technology to establish relationships external to the family which may be nefarious (e.g. child predators).

Summary

What is communication and what are some principles of effective communication?

What is communication and what are some principles of effective communication? Communication is the exchange of information and feelings by two individuals. It involves both verbal and nonverbal messages. The nonverbal part of a message often carries more weight than the verbal part. "Good communication" is regarded as the primary factor responsible for a good relationship. Individuals report that communication confirms the quality of their relationship or condemns their relationship. Words have no meaning unless they are backed up by actions. Being told that one is loved is meaningless unless the person saying the words engages in behavior which shows love—being available for time together, being faithful, supporting the partner's interests, etc.

Some basic principles and techniques of effective communication include making communication a priority, maintaining eye contact, asking open-ended questions, using reflective listening, using "I" statements, complimenting each other, and sharing power. Partners must also be alert to keeping the dialogue (process) going even when they don't like what is being said (content).

Men and women tend to focus on different content in their conversations. Men tend to focus on activities, information, logic, and negotiation and "to achieve and maintain the upper hand." To women, communication focuses on emotion, relationships, interaction, and maintaining closeness. A woman's goal is to preserve intimacy and avoid isolation. Women are also more likely than men to initiate discussion of relationship problems, and women disclose more than men.

Communication styles also vary by country. Individuals reared in France, Germany, Italy, or Greece regard arguing as a sign of closeness—to be blunt and argumentative is to keep the interaction alive and dynamic; to have a tone of agreement is boring. In contrast, Asian cultures (e.g. Japanese, Chinese, Thai, Pilipino) place a high value on avoiding open expression of disagreement and emphasize harmony.

How does self-disclosure, dishonesty, and secrets impact relationship?

High levels of self-disclosure are associated with increased intimacy. Most individuals value honesty in their relationships. Dishonesty and deception take various forms. In addition to telling an outright lie, people may exaggerate or conceal the truth, pretend, or withhold information. Regarding the latter, in virtually every relationship, partners may not share things with each other about themselves or their past. We often withhold information or keep secrets in our intimate relationships for what we believe are good reasons—we believe that we are protecting our partners from anxiety or hurt feelings, protecting ourselves from criticism and rejection, and protecting our relationships from conflict and disintegration.

What is the extent of lying and cheating in relationships?

Almost 60% of almost 3000 undergraduates reported that they had lied to their partner. One in four reported that they had cheated on a partner. Even those in "monogamous" relationships had cheated—27% of males and 20% of females reported having had sex (oral, vaginal, or anal) with someone other than their partner.

How are interactionist and exchange theories applied to relationship communication?

Symbolic interactionists examine the process of communication between two actors in terms of the meanings each attaches to the actions of the other. Definition of the situation, the looking-glass self, and taking the role of the other are all relevant to understanding how partners communicate.

Exchange theorists suggest that the partners' communication can be described as a ratio of rewards to costs. Rewards are positive exchanges, such as compliments, compromises, and agreements. Costs refer to negative exchanges, such as critical remarks, complaints, and attacks. When the rewards are high and the costs are low, the outcome is likely to be positive for both partners (profit). When the costs are high and the rewards low, neither may be satisfied with the outcome (loss).

What are the steps involved in conflict resolution?

The sequence of resolving conflict includes deciding to address recurring issues rather than suppressing them, asking the partner for help in resolving issues, finding out the partner's point of view, summarizing in a non-judgmental way the partner's perspective, and finding alternative win-win solutions. Defense mechanisms that interfere with conflict resolution include escapism, rationalization, projection, and displacement.

What is the future of communication?

The future of communication will increasingly involve technology to initiate, enhance, and maintain relationships. Parental communication with children will also be altered. Technology will have both positive and negative effects on communication in relationships.

Key Terms

Amae

Authentic

Brainstorming

Branching

Closed-ended question

Communication

Conflict

Congruent messages

Correct rejections

Defense mechanism

Displacement

Escapism

Flirting

"I" statements

Lose-lose solution

Nonverbal communication

Open-ended question

Power

Projection

Rationalization

Reflective listening

Sexting

Texting

Win-lose solution

Win-win relationships

"You" statement

Web Links

Association for Couples in Marriage Enrichment
http://www.bettermarriages.org/

Guidelines on Effective Communication, Healthy Relationships & Successful Living
http://www.drnadig.com/

Episcopal Marriage Encounter
http://www.episcopalme.com/

Love Your Relationship
http://loveyourrelationship.com/

CHAPTER 5

Singlehood, Hanging Out, Hooking Up, and Cohabitation

Individuals today select from a veritable smorgasbord of romantic options, including entering into casual, short-term sexual relationships; dating as an end toward finding a long-term partner; entering into shared living with a romantic partner (cohabitation) as an alternative to living alone; forming a cohabiting union as a precursor to marriage; or living with a partner as a substitute for formal marriage.

Sharon Sassler, researcher

David Knox

Learning Objectives

Describe the status of singlehood today.

Identify the categories of singles.

Discuss the functions of and changes in dating.

Review the ways individuals find a partner.

Summarize the types and consequences of cohabitation.

Reflect on the phenomenon of living apart together.

Predict the future of singlehood, long-term relationships, and cohabitation.

1. By 2030 over half of U.S. adults will choose to remain single rather than marry as their permanent lifestyle.

2. In general, women who are cohabitating are just as happy as women who are married.

3. In searching for a partner on the Internet, men are more likely to lie about their relationship goals and women are more likely to lie about their weight.

4. Cohabitants in the United States live together a much shorter period of time than cohabitants in Canada.

5. When romantic partners disagree about whether to cohabit, the preferences of the one who wants to cohabit win out and the couple end up living together.

Answers: 1. F   2. F   3. T   4. T   5. F

No, I don't have a boyfriend. No, I don't need a boyfriend. I am enough. And I am complete just the way I am. I choose to be single, just like I choose to not listen to people who make marriage seem like the only possible pinnacle a life can have.

Lauren Rohrer

Singlehood lifestyle of not being legally married.

Susan Boyle is the single British singer who, at age 50, soared to fame after appearing on the TV show *Britain's Got Talent*, capturing audiences with her beautiful voice and album *I Dreamed a Dream*. But her manager and siblings report that after her performances worldwide she returns alone to her hotel room. "Susan would be the first to admit that sometimes she feels lonely," says her manager, Jerry. Susan is not alone in her sentiments. While all of us are lonely from time to time, we are socialized to seek and enjoy being in a relationship.

Though most end up married, youth today are in no hurry. They enjoy the freedom of singlehood and most put off marriage until their late twenties. Young American adults reflect the pattern of youth in other countries: individuals in France, Germany, and Italy are engaging in a similar pattern of delaying marriage. In the meantime, the process of courtship has evolved, with various labels and patterns, including hanging out (undergraduates less often use the term "dating" but say they are "seeing someone"), hooking up (the new term for "one-night stand"), and pairing off for seeing each other exclusively, which may include cohabitation as a prelude to marriage. We begin with examining singlehood versus marriage.

Singlehood

In this section, we discuss the various categories of singles, why individuals are delaying marriage, the choice to be permanently unmarried, and the legal blurring of the married and unmarried.

Categories of Singles

The term **singlehood** is most often associated with young unmarried individuals. However, there are three categories of single people: the never-married, the divorced, and the widowed. See Figure 5.1 for the distribution of the American adult population by relationship status.

Never-Married Singles   Oprah Winfrey, Bill Maher, Sean "P.Diddy" Combs, and Jack Kevorkian are examples of heterosexuals who have never married. It is rare for people to remain unmarried their entire life. One reason is stigma. DePaulo (2006) asked 950 undergraduate college students to describe single people. In contrast to married people, who were described as "happy, loving, stable," single people were described as "lonely, unhappy, and insecure." Wienke and Hill (2009) also compared single people with married people and cohabitants (both heterosexual and homosexual) and found that single people were less happy regardless of sexual orientation. Also, just as some single men are viewed as having characteristics which decrease their attractiveness as a married partner, Manning et al. (2010) indentified variables which are liabilities for females who want to find a mate: being poor, having mental/physical health issues, using drugs, and having children with multiple partners.

Indeed, there is a cultural norm that disapproves of singlehood as a permanent lifestyle. And, the data comparing young never-married females with married women reflects difficulty with the lifestyle for women. In a national sample of 9,507, never-married women reported more symptoms of depression than married women, but only at younger ages when they are most likely to feel frustrated at not having achieved the culturally prescribed marital status of spouse (LaPierre 2009). The difference disappears in middle age when never married women are likely to have adapted to being single. However, never married women who are living together with a partner are less happy/more depressed than married women, perhaps because the former had wanted to transition into marriage but did not.

The stigma of remaining single in one's thirties weighs heavily on some individuals. Sharp and Ganong (2007) interviewed thirty-two white never-married college-educated women ages 28 to 34 who revealed a sense of uncertainty about their lives. One reported:

Like all or nothing, it is either—you assume it [your life] is either going to be great or horrible. You just have to get better at accepting the fact that you don't know, it is probably somewhere in between and you are just going to have to wait and see.

Although some were despondent that they would ever meet a man and have children, others (particularly when they became older) reminded themselves of the advantages of being single: freedom, financial independence, ability to travel, and so on. However, the overriding theme of these respondents was that they were "running out of time to marry and to have children." The researchers emphasized the enormous cultural expectation to follow age-graded life transitions and to stay on time and on course. People outside the norm struggle with managing their difference, with varying degrees of success. See Table 5.1 for a review of the issues identified by a single woman.

Divorced Singles   The divorced are also regarded as single. For many divorced, the return to singlehood is a difficult transition. Knox and Corte (2007) studied a sample of individuals going through divorce, many of whom reported their unhappiness and a desire to reunite with their partner.

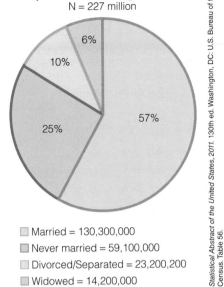

U.S. Population over the age of 18 in 2009
N = 227 million

6%
10%
25%
57%

☐ Married = 130,300,000
☐ Never married = 59,100,000
☐ Divorced/Separated = 23,200,200
☐ Widowed = 14,200,000

Statistical Abstract of the United States, 2011. 130th ed. Washington, DC: U.S. Bureau of the Census. Table 56.

Figure 5.1 Marital status of U.S. population over the age of 18 in 2009.

The dread of loneliness is greater than the fear of bondage, so we get married.

Cyril Connolly, *The Unquiet Grave*

TABLE 5.1 A Never-Married Single Woman's View of Singlehood

A never-married woman, 40 years of age, spoke to our marriage and family class about her experience as a single woman. The following is from the outline she developed and the points she made about each topic.

Stereotypes about Never-Married Women

Various assumptions are made about the never-married woman and why she is single. These include the following:

Unattractive—She's either overweight or homely, or else she would have a man.

Lesbian—She has no real interest in men and marriage because she is homosexual.

Workaholic—She's career-driven and doesn't make time for relationships.

Poor interpersonal skills—She has no social skills, and she embarrasses men.

History of abuse—She has been turned off to men by the sexual abuse of, for example, her father, a relative, or a date.

Negative previous relationships—She's been rejected again and again and can't hold a man.

Man-hater—Deep down, she hates men.

Frigid—She hates sex and avoids men and intimacy.

Promiscuous—She is indiscriminate in her sexuality so that no man respects or wants her.

Too picky—She always finds something wrong with each partner and is never satisfied.

Too weird—She would win the Miss Weird contest, and no man wants her.

Positive Aspects of Being Single

1. Freedom to define self in reference to own accomplishments, not in terms of attachments (for example, spouse).
2. Freedom to pursue own personal and career goals and advance without the time restrictions posed by a spouse and children.
3. Freedom to come and go as you please and to do what you want, when you want.
4. Freedom to establish relationships with members of both sexes at the desired level of intensity.
5. Freedom to travel and explore new cultures, ideas, values.

Negative Aspects of Being Single

1. Increased extended-family responsibilities. The unmarried sibling is assumed to have the time to care for elderly parents.
2. Increased job expectations. The single employee does not have marital or family obligations and consequently can be expected to work at night, on weekends, and holidays.
3. Isolation. Too much time alone does not allow others to give feedback such as "Are you drinking too much?", "Have you had a checkup lately?", or "Are you working too much?"
4. Decreased privacy. Others assume the single person is always at home and always available. They may call late at night or drop in whenever they feel like it. They tend to ask personal questions freely.
5. Less safety. A single woman living alone is more vulnerable than a married woman with a man in the house.
6. Feeling different. Many work-related events are for couples, husbands, and wives. A single woman sticks out.
7. Lower income. Single women have much lower incomes than married couples.
8. Less psychological intimacy. The single woman may not have an emotionally intimate partner at the end of the day.
9. Negotiation skills lie dormant. Because single people do not negotiate issues with someone on a regular basis, they may become deficient in compromise and negotiation skills.
10. Patterns become entrenched. Because no other person is around to express preferences, the single person may establish a very repetitive lifestyle.

Maximizing One's Life as a Single Person

1. Frank discussion. Talk with parents about your commitment to and enjoyment of the single lifestyle and request that they drop marriage references. Talk with siblings about joint responsibility for aging parents and your willingness to do your part. Talk with employers about spreading workload among all workers, not just those who are unmarried and childfree.
2. Relationships. Develop and nurture close relationships with parents, siblings, extended family, and friends to have a strong and continuing support system.
3. Participate in social activities. Go to social events with or without a friend. Avoid becoming a social isolate.
4. Be cautious. Be selective in sharing personal information such as your name, address, and phone number.
5. Money. Pursue education to maximize income; set up a retirement plan.
6. Health. Exercise, have regular checkups, and eat healthy food. Take care of yourself.

Chapter 5 Singlehood, Hanging Out, Hooking Up, and Cohabitation

Most divorced individuals have children. Most of these are single mothers, but increasingly, single fathers have sole custody of their children. Most single parents prioritize their roles of "single parent" as a parent first and as a single adult second. One newly divorced single parent said, "My kids come first. I don't have time for anything else now." We discuss the topic of single parenthood in greater detail in Chapter 10 on planning children.

The divorced have a higher suicide risk. Denney (2010) examined the living arrangements of over 800,000 adults and found that being married or living with children decreases one's risk of suicide. One explanation for greater longevity among marrieds is the protective aspect of marriage. "The protection against diseases and mortality that marriage provides may take the form of easier access to social support, social control, and integration, which leads to risk avoidance, healthier lifestyles, and reduced vulnerability" (p. 375). Married people also look out for the health of the other. Spouses often prod each other to "go to the doctor," "have that rash on your skin looked at," and "remember to take your medication." Single people often have no one in their life to nudge them toward regular health maintenance. Regarding the insulation against suicide, marrieds are more likely to be "connected" to intimates; this "connection" seems to insulate a person from suicide. And these connections can occur outside of marriage.

Widowed Singles Although divorced people often choose to leave their spouses and be single again, the widowed are forced into singlehood. The stereotype of the widow and widower is utter loneliness, even though there are compensations (for example, escape from an unhappy marriage, social security). Ha (2008) compared widowed people with married couples and found the former less likely to have a confidant, but they received greater support from children, friends, and relatives. Hence, the widowed may have a broader array of relationships than those who are married. Nevertheless, widowhood is a difficult time for most individuals. Widowed men, in contrast to widowed women, may feel particularly disconnected from their children (Kalmijn 2007).

National **Data**

There are 11.4 million widowed females and 2.8 widowed males in the United States (*Statistical Abstract of the United States 2011*, Table 57).

In a study conducted by the American Association of Retired Persons (AARP), most widows were between the ages of 40 and 69. Almost a third of the respondents (31%) were in an exclusive relationship and another 32% were dating nonexclusively. Of the remaining 37%, only 13% reported that they were actively looking; 10% said that they had no desire to date; and the rest said they were open to meeting someone but not obsessive about it (Mahoney 2006).

Individuals Are Delaying Marriage Longer

Having identified the three categories of singles as the never-married, divorced, and widowed, we now focus on the never-married, the largest group. A primary question is whether or not more people are choosing to remain single. We do not know the answer. What we do know is that individuals are *delaying marriage.* As evidence, in 1960, two-thirds (68%) of all twenty-somethings were married. More recently, just over a fourth (26%) in this age category was married (Pew Research Center 2010a). Will those who are delaying getting married, eventually marry? We suspect most will but don't know. We need to wait till the current

Many divorced or widowed people do with their singleness what they should have done before they married for the first time: live alone, find their own rhythms, date a variety of people, go into therapy, develop new friends and interests, learn how to live with and care for themselves.

Harville Hendrix, French author

Marriage itself has become increasingly optional as a context for intimate partnerships and parenthood.

Megan Sweeney, demographer

TABLE 5.2 Reasons to Remain Single	
Benefits of Singlehood	**Limitations of Marriage**
Freedom to do as one wishes	Restricted by spouse or children
Variety of lovers	One sexual partner
Spontaneous lifestyle	Routine, predictable lifestyle
Close friends of both sexes	Pressure to avoid close other-sex friendships
Responsible for one person only	Responsible for spouse and children
Spend money as one wishes	Expenditures influenced by needs of spouse and children
Freedom to move as career dictates	Restrictions on career mobility
Avoid being controlled by spouse	Potential to be controlled by spouse
Avoid emotional and financial stress of divorce	Possibility of divorce

cohort of youth reach age 75 and beyond to see if as high a percentage (96%) eventually marry as is true of the current percent of those age 75 and older (*Statistical Abstract of the United States 2011*).

Table 5.2 lists the standard reasons people give for remaining single. The primary advantage of remaining single is freedom and control over one's life. Once a decision has been made to involve another in one's life, one's choices become vulnerable to the influence of that other person. The person who chooses to remain single may view the needs and influence of another person as something to avoid.

Though white never-married single females are delaying marriage to pursue educational and career opportunities, never married African American women note a lack of potential marriage partners. Educated black women report a particularly difficult time finding eligible men from which to choose.

National **Data**

Among adults 18 years and older, about 41% of black women and 43% of black men have never married, in contrast to about 20% of white women and 27% of white men (*Statistical Abstract of the United States 2011*, Table 56).

As noted in the national data, there is a great racial divide in terms of remaining single. In her article, "Marriage is for White People," J. Jones (2006, B5) notes:

Sex, love, and childbearing have become a la carte choices rather than a package deal that comes with marriage. Moreover, in an era of brothers on the "down low," the spread of sexually transmitted diseases and the decline of the stable blue-collar jobs that black men used to hold, linking one's fate to a man makes marriage a risky business for a black woman.

But men in general, not just black men, may be scarce or less desirable. Rosin (2010) in her article on "The End of Men" noted that (due to the recession) women now hold the majority of the nation's jobs. The result is that men are being marginalized and bring less to the table economically. Rosin also states that modern industrial society no longer values men's size and strength, that "social intelligence, open communication, the ability to sit still and focus are, at a minimum, not predominantly male"—yet these are precisely the qualities now demanded in the global economy.

Some people do not set out to be single but drift into singlehood longer than they anticipated—and discover that they like it. Meredith Kennedy (shown

I never married because I have three pets at home that serve the same purpose as a husband. I have a dog that growls every morning, a parrot that sleeps all afternoon, and a cat that comes home late at night.

Marie Corelli, British novelist

This never married female "drifted" into being single and now thoroughly enjoys her life. She is a veterinarian, singer, author, teacher, and playwright.

in the above photo) is a never-married veterinarian who has found that single-hood works very well for her. She writes:

> As a little girl, I always assumed I'd grow up to be swept off my feet, get married, and live happily ever after. Then I hit my 20s, became acquainted with reality, and discovered that I had a lot of growing up to do. I'm still working on it now, in my late fourties.
>
> I've gotten a lot out of my relationships with men over the years, some serious and some not so serious, and they've all left their impression on me. But gradually I've moved away from considering myself "between boyfriends" to getting very comfortable with being alone, and finding myself good company. The thought of remaining single for the rest of my life doesn't bother me, and the freedom that comes with it is very precious. I've worked and traveled all over the world, and my schedule is my own. I don't think this could have come about with the responsibilities of marriage and a family, and the time and space I have as a single woman have allowed me to really explore who I am in this life.
>
> It's not always easy to explain why I'm single in a culture that expects women to get married and to have children, but the freedom and independence I have allow me to lead a unique and interesting life.

Some delay getting married because they are afraid. Manning et al. (2010) interviewed single women, some of whom reported that they were afraid to get married. One respondent, Tania, a 24-year-old cohabiting mother of three, reported

> I'm scared to get married ... and being committed to somebody for so long ... cause I know some people do get married and there are a lot of arguments in the house. It really depends on the relationship and on the person. And I know there can be a lot of arguing ... And I think that's what scares me.

Manning et al. (2010) also interviewed some who did not want to marry because of fear of divorce. Helen, a 36-year-old single mother of one child, explained

> It's like, I mean I do [want to get married] and then I don't ... Because ... here every time I turn around somebody's marriage is ending, and people are getting divorced. And that just bothers me.

Alternatives to Marriage Project

The Alternatives to Marriage Project is an organization which supports those who do not marry. According to the mission statement identified on the website of the Alternatives to Marriage Project (http://www.unmarried.org/aboutus.php), the emphasis is to advocate "for equality and fairness for unmarried people, including people who are single, choose not to marry, cannot marry, or live together before marriage." The nonprofit organization is not against marriage but provides support and information for the unmarried and "fights discrimination on the basis of marital status. . . . We believe that marriage is only one of many acceptable family forms, and that society should recognize and support healthy relationships in all their diversity."

The Alternatives to Marriage Project is open to everyone, "including singles, couples, married people, individuals in relationships with more than two people, and people of all genders and sexual orientations. We welcome our married supporters, who are among the many friends, relatives, and allies of unmarried people."

He travels fastest who travels alone.

Kipling, *The Winners*

Legal Blurring of the Married and Unmarried

One factor involved in more individuals delaying marriage is that the legal distinction between married and unmarried couples is blurring. Whether it is called the deregulation of marriage or the deinstitutionalization of marriage, the result is the same—more of the privileges previously reserved for the married are now available to unmarried and/or same-sex couples. For example, individuals who define themselves as being in a **domestic partnership** may have many of the rights and privileges (for example, health benefits for a partner) previously available only to married people.

Domestic partnership two adults who have chosen to share each other's lives in an intimate and committed relationship of mutual caring. These relationships are given some kind of official recognition by a city or corporation so as to receive partner benefits (for example, health insurance).

PERSONAL CHOICES

Is Singlehood for You?

Singlehood is not a one-dimensional concept. Whereas some are committed to singlehood, others enjoy it for now but intend to eventually marry, and still others are conflicted about it. There are many styles of singlehood from which to choose. As a single person, you may devote your time and energy to career, travel, privacy, heterosexual or homosexual relationships, living together, communal living, or a combination of these experiences over time. An essential difference between traditional marriage and singlehood is the personal, legal, and social freedom to do as you wish. Although singlehood offers freedom, issues such as loneliness, less money, and establishing an identity sometimes challenge single people.

1. *Loneliness.* For some singles, being alone is a desirable and enjoyable experience. Ellen Burstyn, actress, said of solitude, "I just revel in it, indulge in it, bathe in it...I desire loneliness" (Rountree 1993, 18). In an American Association of Retired Persons (AARP) survey of single women aged 40 to 69, 93% noted that their "independence" was important for their quality of life (and that this overshadowed the occasional feelings of loneliness) (Mahoney 2006). Henry David Thoreau, who never married, spent two years alone on fourteen acres bordering Walden Pond in Massachusetts. He said of his experience, "I love to be alone. I have never found the companion that was as companionable as solitude."

Nevertheless, some singles are lonely. In the AARP study of single women, 28% reported that they had felt lonely occasionally in the past two weeks or most of the time (13% of married women reported these same feelings) (Mahoney 2006). Some research suggests that single men may be lonelier than single women. In a study of 377 undergraduates, 25.9% of the men compared to 16.7% of the women agreed that they felt a "deep sense of loneliness" (Knox et al. 2007). Research on dog ownership and a sense of psychological well-being revealed that dog ownership was associated with a sense of well being, particularly for single individuals and particularly for women (Cline and Marie 2010).

2. *Less Money.* Single people living alone typically have less income than married people. The median income of the single female householder with no husband is $33,073 (for the male single householder with no wife, $49,186) compared to a married couple—$73,010 (*Statistical Abstract of the United States 2011*, Table 691). Married couples have higher incomes since there are typically two incomes.

3. *Social Identity.* Single people must establish a social identity—a role—that defines who they are and what they do, independent of the role of spouse. Married couples find housing together and share economic resources—they are spouses to each other.

Single people find other roles and avenues to identity. A meaningful career is the avenue most singles pursue. A career provides structure, relationships with others, and a strong sense of identity ("I am a veterinarian—I love my work," said Meredith Kennedy).

4. *Children.* Some individuals want to have a child but not a spouse. We will examine this issue in Chapter 11 in a section on Single Mothers by Choice. There are no data on single men seeking the role of parent. Although some custodial divorced men are single fathers, this is not the same as never having married and having a child.

In evaluating the single lifestyle, to what degree, if any, do you feel that loneliness is or would be a problem for you? What is your social identity, your work role satisfaction? What are your emotional and structural needs for "marriage"? For children? The old idea that you can't be happy unless you are married is no longer credible. Whereas marriage will be the first option for some, it will be the last option for others. As one 76-year-old single person by choice said, "A spouse would have to be very special to be better than no spouse at all."

Sources

Knox, D., K. Vail-Smith, and M. Zusman. 2007. The lonely college male. *International Journal of Men's Health.* 6: 273–279.

Statistical Abstract of the United States 2011. 130th ed. Washington, DC: U.S. Census Bureau.

About age 30 most women think about having children, most men think about dating them.

Judy Carter, actress and comedian

Functions, History, and Changes in "Dating"

Dating in the traditional sense of the male calling the female several days in advance and asking her to a specific event has given way to hanging out and meeting someone. Sometimes one individual will send a message to another on

WHAT IF?

What if You Are Afraid of Being Alone and Just Settle for Someone?

Fear of loneliness is a powerful motivator to pair-bond with anyone who may be available. It is possible to blend your life with the person who is available and have a good life. A larger question is your level of attraction ("chemistry") and similarity. If you don't feel emotionally/physically attracted to the person and have nothing in common, the ending may be sad because these are basic prerequisites for an enjoyable and enduring relationship.

Facebook such as "Hey—wanna hang out Thursday night at the ......?" The person will text back "sure" and typically show up. After the partners "hang out" on several occasions they will get more specific about when they will see each other again. In effect, they slide into the structure of seeing each other at predictable times rather than impose the structure initially. Whether it is called "hanging out" or "seeing someone" (the current term for dating), the couple typically spend increasing amounts of time together and become involved. Such involvement has various functions.

Functions of Involvement with a Partner

Meeting and becoming involved with someone has at least seven functions: (1) confirmation of a social self; (2) recreation; (3) companionship, intimacy, and sex; (4) anticipatory socialization; (5) status achievement; (6) mate selection; and (7) health enhancement.

    1. *Confirmation of a social self.* In Chapter 1, we noted that symbolic interactionists emphasize the development of the self. Parents are usually the first social mirrors in which we see ourselves and receive feedback about who we are; new partners continue the process. When we are hanging out with a person, we are continually trying to assess how that person sees us: What does the person think of me? Does the person like me? Will the person want to be with me again? When the person gives us positive feedback through speech and gesture, we feel good about ourselves and tend to view ourselves in positive terms. "Seeing someone" or dating provides a context for the confirmation of a strong self-concept in terms of how we perceive our effect on other people.

    2. *Recreation.* The focus of hanging out and pairing off is fun. Reality television programs such as *Jersey Shore* use recreational activities as a context to help participants interact. Being "fun" is a factor in being selected. Individuals who just met make only small talk and learn very little about each other—what seems important is not that they have common interests, values, or goals but that they "have fun" together.

    3. *Companionship, intimacy, and sex.* Beyond fun, major motivations for finding a new person and pairing off are companionship, intimacy, and sex. The impersonal environment of a large university makes a secure relationship very appealing. "My last two years have been the happiest ever," remarked a

This couple simply enjoys being together and telling jokes or funny stories—having fun together is what they do.

Chelsea E. Curry

senior in interior design. "But it's because of the involvement with my partner. During my freshman and sophomore years, I felt alone. Now I feel loved, needed, and secure."

4. *Anticipatory socialization.* Before puberty, boys and girls interact primarily with same-sex peers. A fifth grader may be laughed at if showing an interest in someone of the other sex. Even when boy-girl interaction becomes the norm at puberty, neither sex may know what is expected of the other. "Seeing someone" provides a context for individuals to learn how to interact with potential partners. Though the manifest function of "seeing someone" is to teach partners how to negotiate differences (for example, how much sex and how soon), the latent function is to help them learn the skills necessary to maintain long-term relationships (e. g. empathy, communication, and negotiation). In effect, pairing off involves a form of socialization that anticipates a more permanent union in one's life. Individuals may also try out different role patterns, like dominance or passivity, and try to assess the feel and comfort level of each.

5. *Status achievement.* Being involved with someone is usually associated with more status than being unattached and alone. In a couple world, sometimes there is embarrassment with "I'm not seeing anyone." Some may seek such involvement because of the associated higher status. Others may become involved for peer acceptance and conformity to gender roles, not for emotional reasons.

6. *Mate selection.* Finally, "seeing someone" may eventually lead to marriage, which remains a major goal in our society. Selecting a mate has become big business. Barnes & Noble, one of the largest bookstore chains in the United States, lists more than 127,000 books on "relationships" and 966 books on "mate selection" on their online catalog.

7. *Health enhancement.* In Chapter 1, we reviewed the benefits of marriage. Not the least of marital benefits is health. Specifically, there is a direct relationship between getting married for the first time and the cessation of smoking (Weden and Kimbro 2007). Such influence may begin in the early stages of "seeing someone"—the partner may discourage smoking.

Changes in "Dating" in the Past Sixty Years

There have been numerous changes to dating and marriage patterns since 1950. The changes include an increase in the age at marriage. Marrying at age 28 rather than 22 provides more time and opportunity to hang out with more people.

The dating pool today also includes an increasing number of individuals in their thirties who have been married before. These individuals often have children, which changes the nature of a date from two adults going to a movie alone to watching a DVD and taking care of the kids in the home of one of the partners.

As we will note later in this chapter, cohabitation has become more normative. For some couples, the sequence of dating, falling in love, and getting married has been replaced by dating, falling in love, and living together. Such a sequence results in the marriage of couples that are more relationship-savvy than those who dated and married just after high school.

Not only do individuals now "see" more partners and live together more often, but also gender role relationships have become more egalitarian. Though the double standard still exists, women today are more likely than women in the 1950s to ask men out (see Applying Social Research), to have sex without being in love or requiring a commitment, and to postpone marriage until meeting their own educational and career goals. Women no longer feel desperate to marry but consider marriage one of many goals they have for themselves.

Mae West, film actress of the 1930s and 1940s, is remembered for being very forward with men. Two classic phrases of hers are "Is that a pistol in your pocket or are you glad to see me?" and "Why don't you come up and see me sometime, I'm home every night?" As a woman who went after what she wanted, West was not alone—then or now. There have always been women not bound by traditional gender role restrictions. This study was concerned with women who initiate relationships with men—women who, like Mae West, have ventured beyond the traditional gender role expectations of the passive female.

Sample and Methodology

Data for this study consisted of 692 undergraduate women who answered "yes" or "no" on a questionnaire to the question, "I have asked a new guy to go out with me"—a nontraditional gender role behavior.

Findings and Discussion

Of the 692 women surveyed, 39.1% reported that they had asked a new guy out on a date; 60.9% had not done so. Analysis of the data revealed ten statistically significant findings in regard to the characteristics of those women who had initiated a relationship with a man and those who had not done so.

1. *Nonbeliever in "one true love."* Of the women who asked men out, 42.3% did not believe in "one true love" in contrast to 31.7% who believed in one true love—a statistically significant difference (p < .01). These assertive women felt they had a menu of men from which to choose.

2. *Experienced "love at first sight."* Of the women who had asked a man out, 45.7% had experienced falling in love at first sight in contrast to 28.2% who had not had this experience. Hence, women who let a man know they were interested in him were likely to have already had a "sighting" of a man with whom they fell in love.

3. *Sought partner on the Internet.* Only a small number (59 of 692; 8.5%) of women reported that they had searched for a partner using the Internet. However, 54.2% of those who had done so (in contrast to only 37.5% who had not used the Internet to search for a partner) reported that they had asked a man to go out (p < .02). Because both seeking a partner on the Internet and asking a partner to go out verbally are reflective of nontraditional gender role behavior, these women were clearly in charge of their lives and moved the relationship forward rather than waiting for the man to make the first move.

4. *Nonreligious.* Respondents who were not religious were more likely to ask a guy out than those who were religious (74.6% versus 64.2%; p < .001). This finding comes as no surprise, as previous research suggests that being nonreligious is associated with having nontraditional values or roles (McCready and McCready 1973; Miller and Stark 2002).

5. *Nontraditional sexual values.* Consistent with the idea that women who had asked a guy out were also nonreligious (a nontraditional value) is the finding that these

---

> *My knight in shining armor turned out to be a loser in aluminum foil.*
>
> Anonymous

Couples today are also more aware of the impermanence of marriage. However, most couples continue to feel that divorce will not happen to them, and they remain committed to domestic goals. Almost three-quarters (73%) of all first-year college students in the United States reported that "raising a family" was "an essential or very important goal" for them (Pryor et al. 2011).

Unlike during the 1950s, both sexes today are more aware of and cautious of becoming HIV-infected. Sex has become potentially deadly, and condoms are being used more frequently. The 1950s fear of asking a druggist for a condom has been replaced by the confidence and mundaneness of buying condoms along with one's groceries.

Dating after Divorce

The pool of potential partners also includes over 2 million newly divorced Americans each year. As evidenced by the fact that more than three-quarters of divorced people remarry within five years, most divorced people are open to a new relationship. But this single-again population differs from those becoming involved for the first time.

same women tended to have nontraditional sexual values. Of those women who had initiated a relationship with a guy, 44.4% reported having a hedonistic sexual value ("If it feels good, do it.") compared to 25.7% who regarded themselves as having absolutist sexual values ("Wait until marriage to have intercourse."). Hence, women with nontraditional sexual values were much more likely to be aggressive in initiating a new relationship with a man.

6. *Open to cohabitation.* Of the women who reported that they had asked a man to go out, 44.6% reported that they would cohabit with a man compared to 26.0% who would not cohabit. This finding is supported by the research of Michael et al. (1994), who confirmed that cohabiting men and women are more likely to have nontraditional sexual values.

7. *White.* Over 40% (41.4%) of the white women, compared to 28.2% of the black women, in the sample reported that they had asked a guy out. This finding is not surprising in that black people are traditionally more conservative than white people in religion (Sherkat 2002) and sexual values (Michael et al. 1994).

Implications

Analysis of these data revealed that 39.1% of the undergraduate women at a large southeastern university had asked a guy to go out (a nontraditional gender role behavior). There are implications of this finding for both women and men. Women who feel uncomfortable asking a man out, who fear rejection for doing so, or who lack the social skills to do so ("Hey big boy! Wanna get a pizza?") may be less likely to find the man who will be whisked away by women who have such comfort, overcome their fear of rejection, and who make their interest in a partner known.

The implication of this study for men is not to be surprised when a woman makes a direct request to go out—relationship norms are changing. For some men, this comes as a welcome trend in that they feel burdened that they must always be the first one to indicate interest in a partner and to move the relationship forward. Men might also reevaluate their negative stereotypical notions of women who initiate relationships ("they are loose") and be reminded that the women in this study who had asked men out were *more* likely to have been faithful in previous relationships than those who had not.

Sources

McCready, W., and N. McCready. 1973. "Socialization and the Persistence of Religion." In *The Persistence of Religion,* ed. M. Sussman and S. Steinmetz, 58–68. New York: Macmillan.

Michael, R. T., J. H. Gagnon, E. O. Laumann, and G. Kolata. 1994. *Sex in America: A Definitive Survey.* Boston, MA: Little, Brown and Company.

Miller, A. S., and R. Stark. 2002. "Gender and religiousness: Can socialization explanations be saved?" *American Journal of Sociology* 107: 1399–1423.

Sherkat, D. E. 2002. "African American Religious Affiliation in the Late 20th Century: Cohort Variations and Patterns of Switching 1973–1998." *Journal of the Scientific Study of Religion* 41: 485–493.

*Abridged from C. Ross, D. Knox, and M. Zusman. 2008. "Hey Big Boy": Characteristics of university women who initiate relationships with men. Poster, Southern Sociological Society Annual Meeting, April, Richmond, VA.

1. *Internet for new partner.* Aside from traditional ways to meet new partners (through friends, at work, at health or exercise clubs, or religious services), the Internet has become a valuable tool for divorced people in finding someone new. Divorced people are very busy. With full-time jobs and single parenthood, they may have little time for traditional dating. The Internet provides a quick way to sift through hundreds of people after work/at home alone and start up an e-mail relationship on the home computer.

2. *Older population.* Divorced individuals are, on the average, ten years older than people in the marriage market who have never been married before. Hence, divorced people tend to be in their mid-to-late thirties. Widows and widowers are usually 40 and 30 years older, respectively (hence, around ages 65 and 55), when they begin to date the second time around. Most divorced people date and marry others who are divorced. Since they are older they may also be learning the new norms of hanging out and seeing each other; some will miss the structure and comfort of traditional dating.

3. *Fewer potential partners for the divorced.* Most men and women who are dating the second time around find fewer partners from whom to choose than

when they were dating before their first marriage. The large pools of never-married people (26% of the population) and currently married people (57% of the population) are usually not considered an option (*Statistical Abstract of the United States 2011*, Table 56). Most divorced people (10% of the population) date and marry others who have also been married before.

4. *Increased HIV risk.* As might be assumed, the older unmarried people are, the greater the likelihood that they have had multiple sexual partners, which is associated with increased risk of contracting HIV and other STIs. Therefore, individuals entering the dating market for the second time are advised to *always* use a condom and to assume that one's partner has been sexually active and is at risk.

5. *Children.* More than half of the divorced people who are dating again have children from a previous marriage. How these children feel about their parents dating, how the partners feel about each other's children, and how the partners' children feel about each other are challenging issues. Deciding whether to have intercourse when one's children are in the house, when a new partner should be introduced to the children, and what the children should call the new partner are other issues familiar to parents dating for the second time.

6. *Ex-spouse issues.* Ex-spouses may be uncomfortable with their former partner's involvement in a new relationship. They may not only create anxiety on the part of their former spouse but may also directly attack the new partner. Dealing with an ex-spouse may be challenging. Ties to an ex-spouse, in the form of child support or alimony, and phone calls may also have an influence on the new dating relationship. Some individuals remain psychologically and sexually involved with their exes. In other cases, if the divorce was bitter, partners may be preoccupied or frustrated in their attempts to cope with a harassing ex-spouse (and feel emotionally distant).

7. *Brief courtship.* Divorced people who are dating again tend to have a shorter courtship period than people married for the first time. In a study of 248 individuals who remarried, the median length of courtship was nine months, as opposed to seventeen months the first time around (O'Flaherty and Eells 1988). A shorter courtship may mean that sexual decisions are confronted more quickly—timing of first intercourse, discussing the use of condoms and contraceptives, and clarifying whether the relationship is to be monogamous. People who have been previously divorced or widowed derive a new sense of well-being for spending time with or becoming involved with a new partner—"I'm getting used to being loved agan" is the lyric and title to an old country western song by Gene Watson. Caution may be prudent because the old rules still apply. Knowing a partner at least two years before marrying that person is predictive of a happier and more durable relationship than marriage after a short courtship. Divorced people pursuing a new relationship might consider slowing down their relationship.

Aside from the various issues to keep in mind regarding a new partner, Krumrei et al. (2007) studied unhappiness and maladjustment of divorced people in twenty-one studies in reference to social relationships. Divorced people who had a network of relationships reported higher levels of positive adjustment. One take-home message is that dating again and creating a new network of friendships and relationships is functional for one's divorce adjustment.

Singlehood and HIV Infection Risk

Individuals who are not married or not living with someone are at greater risk for contracting human immunodeficiency virus (HIV) and other sexually transmitted infections (STIs). Though women typically report having had fewer sexual partners than men, the men they have sex with have usually had multiple sexual partners. Hence, women are more likely to get infected from men than men are from women. In addition to the social reason for increased risk of infection, there is a biological reason—sperm that may be HIV-infected is deposited into the woman's body.

Finding a Partner

Whether an individual has the goal of remaining unmarried or eventually getting married, most have the goal of finding a partner with whom they can have fun with/share their lives. One of the unique qualities of the college or university environment is that it provides a context in which to meet thousands of potential partners of similar age, education, and social class. This context will likely never recur following graduation. Although people often meet through friends (including through Facebook) or on their own in school, work, or recreation contexts, an increasing number are open to a range of alternatives, from hanging out to the Internet.

Hanging Out

Hanging out, also referred to as getting together, refers to going out in groups where the agenda is to meet others and have fun. The individuals may watch television, rent a DVD, go to a club or party, and/or eat out. Of 2,922 undergraduates, 85% reported that "hanging out for me is basically about meeting people and having fun" (Knox and Hall 2010). Hanging out may occur in group settings such as at a bar, a sorority or fraternity party, or a small gathering of friends that keeps expanding. Friends may introduce individuals, or they may meet someone "cold," as in initiating a conversation. There is usually no agenda beyond meeting and having fun. Of the 2,922 respondents, only 4% said that hanging out was about beginning a relationship that may lead to marriage (Knox and Hall 2010).

Hooking Up

Hooking up is a sexual encounter that occurs between individuals who have no relationship commitment. There is generally no expectation of seeing one another again and alcohol is often involved. Data on the frequency of college students having experienced a hookup varies. Reporting on more than 17,000 students at 20 colleges and universities (from Stanford University's Social Life Survey), sociologist Paula England revealed that 72% of both women and men reported having hooked up (40% intercourse, 35% kiss, touch; 12% hand genital; 12% oral sex). Men reported having ten hookups, women seven (Jayson 2011), Owen et. al. (2010) surveyed 832 undergraduates and found that 60% of Caucasian and 35% of African Americans reported having "hooked up." More alcohol use, more favorable attitudes toward hooking up, and higher parental income were associated with a higher likelihood of having hooked up. Fieldera and Careya (2010) surveyed 118 first semester female college students on their hook up experiences (60% had done so). Most hookups involved friends (47%) or acquaintances (23%) rather than strangers.

These data reflect that hooking up is becoming the norm on college and university campuses. Not only do female students outnumber men students (60 to 40) (which means women are less able to bargain sex for commitment since men have a lot of options), individuals want to remain free for summer internships, study abroad, and marriage later (Uecker and Regnerus 2011). Kalish and Kimmel (2011) suggested that hooking up is a way of confirming one's heterosexuality.

Being single used to mean that nobody wanted you. Now it means you're pretty sexy and you're taking your time deciding how you want your life to be and who you want to spend it with.

Carrie, *Sex and the City*

Hanging out refers to going out in groups where the agenda is to meet others and have fun.

DIVERSITY IN OTHER COUNTRIES

Americans are not alone in looking for a partner. There are 180 million single individuals in China looking for a partner. While there are more males than females, men view women who reach age 28 as "old" (Macleod 2010a).

Eye Contact: A method utilized by a single woman to communicate to a man that she is interested in him. Despite being advised to do so, many women have difficulty looking a man directly in the eyes, not necessarily due to the shyness, but usually due to the fact that a woman's eyes are not located in her chest.

The Dictionary of Dating

Hooking up a sexual encounter that occurs between individuals who have no relationship commitment.

Alcohol use is common in most hookup encounters.

Chelsea E. Curry

The hooking up experience is also variable. Bradshaw et al. (2010) compared the experiences of women and men who hooked up. Men benefited more since they were able to have casual sex with a willing partner and no commitment. Women were more at risk for feeling guilty, becoming depressed, and defining the experience negatively. Both experienced the hook up in a context of deception (neither being open about their relationship goals) and may have exposed themselves to an STI.

Bogle (2008) interviewed fifty-seven college students and alumni at two universities in the eastern United States about their experiences with dating and sex. She found that hooking up had become the *primary* means for heterosexuals to get together on campus. About half (47%) of hookups started at a party and involved alcohol, with men averaging five drinks and women three drinks (England and Thomas 2006).

Researchers (Bogle 2008; Eshbaugh and Gute 2008) noted that, although hooking up may be an exciting sexual adventure, it is fraught with feelings of regret. Some of the women in their studies were particularly disheartened to discover that hooking up usually did not result in the development of a relationship that went beyond a one-night encounter. Eshbaugh and Gute (2008) examined hooking up as a predictor of sexual regret in 152 sexually active college women and identified two sexual behaviors that were particularly predictive of participants' regret: (a) engaging in sexual intercourse with someone once and only once, and (b) engaging in intercourse with someone known for less than twenty-four hours. Noncoital hookups (performing and receiving oral sex) were not significantly related to regret.

Bogle (2008) noted three outcomes of hooking up. In the first, previously described, nothing results from the first-night sexual encounter. In the second, the college students will repeatedly hook up with each other on subsequent occasions of "hanging out." However, a low level of commitment characterizes this type of relationship, in that each person is still open to hooking up with someone else. A third outcome of hooking up, and the least likely, is that the two people begin going out or spending time together in an exclusive relationship. Hence, hooking up is most often a sexual adventure that rarely results in the development of a relationship. However, there are exceptions. When 2,922

Chapter 5 Singlehood, Hanging Out, Hooking Up, and Cohabitation undergraduates were presented with the statement "People who 'hook up' and have sex the first night don't end up in a stable relationship," 14% disagreed (Knox and Hall 2010).

Men and Emotional Hookups

While previous research has identified men as preferring and benefitting from traditional casual sex themes of hooking up and friends with benefits, a team of researchers (Epstein et al. 2009) interviewed 19 men about their recent experiences and found that about half (47%) reported that their real life experiences did not match the traditional scripts. "Some men engaged in hookups, but found it difficult to remain detached from the experience. Others used the term hookup in relational situations, such as the beginning and ending stages of dating, or even an unsuccessful attempt at a dating relationship. For the group as a whole, it appeared that hookups were not simply casual" (p. 420). These respondents were specific that "the choice of a hookup partner is not necessarily limited to strangers, and that there may be an emotional connection between the two" (p. 422).

The Internet—Meeting Online and After

Heino et al. (2010) noted that online dating services have become clear in their mission—to provide a place where people go to "shop" for potential romantic partners and to "sell" themselves in hopes of creating a successful romantic relationship. Persons (34) in interviews who had used online dating services, revealed that they used economic metaphors to describe the experience— "supermarket," "catalog," etc. Indeed there were five themes: assessing others' market worth, determining one's own market worth, shopping for perfect parts, maximizing inventory, and calibrating selectivity. The latter referred to assessing one's own market worth by the number of e-mails received and changing their presentation of self to elicit more interest, e.g. changing photos or putting up more photos (Heino et al. 2010). Such "calibration" also involved comparing what one had to offer with what one could ask for. Some online daters said that since they were older and had put on weight they had to be willing to accept a greater age range and weight in a potential partner.

Better to be quirky alone than unhappy together.

Anita Hamilton

Where else can you go in a matter of 20 minutes, look at 200 people who are single and want to go on a date?

College student seeking online partner

David Knox

This couple set their Internet search for "anywhere in the U.S." and found each other. He lived in Chicago, and she in a small town in North Carolina. After meeting online, they met in person, and married a year later.

Although some individuals feel "We could never let anyone know we met over the Internet," such stigmatization is decreasing. Almost three-fourths (74%) of single Americans have used the Internet to find a romantic partner, and 15% report that they know someone who met their spouse or significant other online (Madden and Lenhart 2006). Over 16 million U.S. adults report that they have been online looking for a partner (Heino et al. 2010). Three to six percent of marriages or long-term relationships began online (Sprecher 2009). eHarmony founder, Dr. Neil Clark Warren, has advertised that their broad-based compatibility system, in which they claim to match highly compatible singles, will possibly reduce the divorce rate in the United States by 1% (Sprecher 2009). There is no evidence to support this claim.

Online meetings will continue to increase as people delay getting married and move beyond contexts where hundreds or thousands of potential partners are available (the undergraduate coed classroom filled with same-age potential mates is rarely equaled in the workplace after college). Persons who are busy, don't have time for traditional dating, or who are shy may also be attracted to finding a partner online. The profiles that individuals construct or provide for others are an example of impression management—presenting an image that is perceived to be what the target audience wants. In this regard, men tend to emphasize their status characteristics (for example, income, education, career), whereas women tend to emphasize their youth, trim body, and beauty (Spitzberg and Cupach 2007).

DIVERSITY IN OTHER COUNTRIES

Chih-Chien and Ya-Ting (2010) interviewed 36 Internet users and surveyed 248 students in Taiwan to discover their motives for using the Internet to create a new relationship (cyber relationship motives). Factors included the desire to meet new people, the ease of finding someone new with a click, and the psychological comfort of doing so (less anxiety over the Internet than in person). Respondents also mentioned that they wanted to escape reality into the virtual world, were curious about the experience of making an Internet friend, and were interested in finding emotional support for a specific interest (e.g. a person who likes computer games might find others who share the interest). Finding a romantic or sexual partner were also motives.

As noted previously, a primary attraction of meeting someone online is its efficiency. It takes time and effort to meet someone at a coffee shop for an hour, only to discover that the person has habits (for example, does or does not smoke) or values (too religious or too agnostic) that would eliminate them as a potential partner. One can spend a short period of time and literally scan hundreds of potential partners without leaving the house. For noncollege people who are busy in their job or career, the Internet offers the chance to meet someone outside their immediate social circle. "There are only six guys in my office," noted one Internet user. "Four are married and the other two are alcoholics. I don't go to church and don't like bars so the Internet has become my guy store." An example is a woman who devoted a month to doing nothing but finding a mate on the Internet. She sifted through hundreds of guys, ended up seeing eight of them, settled on two of them and ended up marrying one of them. She said of the experience, "I was exhausted. But I found my man."

Another advantage of looking for a partner online is that it removes emotion/chemistry/first meeting magic from the mating equation so that individuals can focus on finding someone with common interests, background, values, and goals. In real life, you can "fall in love at first sight" and have zero in common (Heino et al. 2010). Right Mate at Heartchoice.com not only provides a way to meet others but a free "Right Mate Checkup" to evaluate similarities with a potential partner. Some websites exist to target specific interests such as black singles (BlackPlanet.com), Jewish singles (Jdate.com), and gay people (Gay.com). In one study on online dating, women received an average of fifty-five replies compared to men who reported receiving thirty-nine replies. Younger women (average age of 35), attractive women, and those who wrote longer profiles were more successful in generating replies (Whitty 2007).

Internet Use: The Downside

Lying occurs in Internet dating (as it does in non-Internet dating). Hall et al. (2010) identified seven categories of misrepresentation used by 5,020 individuals who posted profiles in search of an Internet date. These included personal assets

("I own a house at the beach"), relationship goals ("I want to get married'), personal interests ("I love to exercise"), personal attributes ("I am religious"), past relationships ("I have only been married once"), weight, and age. Men were most likely to misrepresent personal assets, relationship goals, personal interests, and personal attributes, whereas women were more likely to misrepresent weight. Heino et al. (2010) interviewed 34 online dating users and found that there is the assumption of exaggeration and a compensation for such exaggeration. The female respondents noted that men exaggerate how tall they are, so the women downplay their height. If a man said he was 5'11" the woman would assume he was 5'9".

Toma and Hancock (2010) examined the role of an online daters' physical attractiveness in the probability that they would be deceptive in their profile self-presentation. Sixty-nine online daters had their photograph taken in the lab. Independent judges rated the online daters' physical attractiveness. Results showed that the lower the online daters' attractiveness, the more likely they were to enhance their profile photographs and lie about their physical descriptors (height, weight, age).

Some online users also lie about being single. "Saleh" was married yet maintained fifty simultaneous online relationships with other women. He allegedly wrote intoxicating love letters, many of which were cut and paste jobs, to various women. He made marriage proposals to several and some bought wedding gowns in anticipation of the wedding (Albright 2007). Although Saleh is an "Internet guy," it is important to keep in mind that people not on the Internet may also be very deceptive and cunning. To suggest that the Internet is the only place where deceivers lurk is to turn a blind eye to those people met through traditional channels.

A theme of the website WildXAngel.com is that it is important to be cautious of meeting someone online. The website features horror stories of some people who met online. Although the Internet is a good place to meet new people, it also allows someone you rejected or an old lover to monitor your online behavior. Most sites note when you have been online last, so if you reject someone online by saying, "I'm really not ready for a relationship," that same person can log on and see that you are still looking. Some individuals become obsessed with a person they meet online and turn into a cyberstalker when rejected. Cyberstalking will be discussed in Chapter 13 on violence and abuse. Some people use the Internet to try on new identities. For example a person who feels he or she is attracted to same sex individuals may present a gay identity online.

Other disadvantages of online meeting include the potential to fall in love too quickly as a result of intense mutual disclosure; not being able to assess "chemistry" or to observe nonverbal cues and gestures or how a person interacts with your friends or family; and the tendency to move too quickly (from e-mail to phone to meeting to first date) to marriage, without spending much time to get to know each other.

Another disadvantage of using the Internet to find a partner is that having an unlimited number of options sometimes results in not looking carefully at the options one has. Wu and Chiou (2009) studied undergraduates looking for romantic partners on the Internet who had 30, 60, and 90 people to review and found that the more options the person had, the less time the undergraduate spent carefully considering each profile. The researchers concluded that it was better to examine a small number of potential online partners carefully than to be distracted by a large pool of applicants, which does not permit the time for close scrutiny.

McGinty (2010) noted the importance of using Internet dating sites safely, including not giving out home or business phone numbers or addresses, always meeting the person in one's own town with a friend, and not posting photos that are "too revealing," as these can be copied and posted elsewhere. It is also important to take it slow—after connecting in an e-mail through the dating site,

If you are looking for love on the Internet, you better look here first.

Lisa, founder of
WildXAngel.com

move to instant messages, phone calls, texting, Skypeing, then meet in a public place with friends near. She recommends being open and honest: "Let them know who you are and who you are looking for," she suggests.

The Internet may also be used to find out information about a partner. Argali.com can be used to find out where the Internet mystery person lives, Zabasearch.com for how long the person has lived there, and Zoominfo.com for where the person works. The person's birth date can be found at Birthdatabase. com. Women might want to see if any red flags have been posted on the Internet at Dontdatehimgirl.com.

For individuals who learn about each other online, what is it like to finally meet? Baker (2007) is clear: "If they have presented themselves accurately and honestly online, they encounter few or minor surprises at the first meeting offline or later on in further encounters" (p. 108). About 7% end up marrying someone they met online (Albright 2007).

Speed-Dating

Dating innovations that involve the concept of speed include the eight-minute date. The website http://www.8minutedating.com/ identifies these "Eight-Minute Dating Events" throughout the country, where a person has eight one-on-one dates at a bar that last eight minutes each. If both parties are interested in seeing each other again, the organizer provides contact information so that the individuals can set up another date. Speed-dating is time-effective because it allows daters to meet face-to-face without burning up a whole evening. Adams et al. (2008) interviewed participants who had experienced speed-dating to assess how they conceptualized the event. They found that women were more likely to view speed-dating as an investment of time and energy to find someone (58% versus 25%), whereas men were more likely to see the event as one of exploration (75% versus 17%). Wilson et al. (2006) collected data on nineteen young men who had three-minute social exchanges with nineteen young women and found that those partners who wanted to see each other again had more in common than those who did not want to see each other again. Common interests were assessed using the compatibility quotient (CQ).

A rural variation of speed dating is **weed dating** whereby individuals who want to meet someone new show up at a farm, pay ten dollars and begin to hoe a row/pull weeds with a potential partner. During the weeding the partners chit-chat and discover if they have "chemistry" between the rows. After seven minutes, a bell rings and one of the partners moves one row over to weed a new row with a new potential partner. The idea began in rural Turnbridge, Vermont where opportunities to meet potential partners are more limited (Pasanen and Silverman 2010).

Weed dating alternative to speed dating where people in a rural area meet on a farm and get to know each other while weeding between rows.

High End Matchmaking

Wealthy busy clients looking for marriage partners pay Selective Search (www. selectivesearch.com) $20,000 to find them a mate. The Chicago-based service personally interviews the client and then searches the data base (for men there are 140,000 women in the database) for a partner. The women selected may also be personally interviewed again to ensure a match. Barbie Adler is the CEO of Selective Search and notes that 1,221 marriages and 417 babies have occurred to date. Almost ninety percent (88%) of clients meet their eventual spouse within the first nine months (Stein 2011).

International Dating

Go to Google.com and type in "international brides," and you will see an array of sites dedicated to finding foreign women for Americans. Not listed is Ivan Thompson, who specializes in finding Mexican women for his American clients.

As documented in the movie *Cupid Cowboy*, Ivan takes males (one at a time) to Mexico (Torreón is his favorite place). For $3,000, Ivan places an ad in a local newspaper for a young (age 20 to 35), trim (less than 130 pounds), single woman "interested in meeting an American male for romance and eventual marriage" and waits in a hotel for the phone to ring. They then meet and interview "candidates" in the hotel lobby. Ivan says his work is done when his client finds a woman he likes.

PERSONAL CHOICES

Should I Get Involved in a Long-Distance Relationship?

One outcome of online dating is that you may meet someone who does not live close to you so that you end up in a long-distance relationship. Alternatively, you may already be in a relationship and the separation occurs. Whatever the reason, the primary advantages of **long-distance relationships** (defined here as being separated from a romantic partner by 300 or more miles which precludes regular weekly face-to-face contact) include: positive labeling ("even though we are separated, we care about each other enough to maintain our relationship"), keeping the relationship "high" because constant togetherness may dull it, having time to devote to school or a career, and having a lot of one's own personal time and space. Pistole (2010) in a review of the literature on long distance relationships noted that they are as satisfying and stable as geographically close relationships.

People suited for such relationships have developed their own autonomy or independence for the times they are apart, have a focus for their time such as school or a job, have developed open communication with their partner to talk about the difficulty of being separated, and have learned to trust each other because they spend a lot of time away from each other. Another advantage is that the partner may actually look better from afar than up close. One respondent noted that he and his partner could not wait to live together after they had been separated—but "when we did, I found out I liked her better when she wasn't there."

The primary disadvantages of long-distance relationships include being frustrated over not being able to be with the partner and loneliness. Pistole et al. (2010) noted that being attached to someone is the strongest of social behaviors and that there is a "separation protest" when the two are separated. This longing for each other may be stronger than either anticipated. When the long distance lovers are reunited, they spend higher order quality time together.

Other disadvantages of involvement in a long distance relationship are missing out on other activities and relationships, missing physical intimacy, spending a lot of money on phone calls/travel, and not discussing important relationship topics. Regarding the later, Stafford (2010) compared the communication topics of individuals in long distance dating relationships (LDDR) with those in geographically close distance relationships (GCDR) and found that the former avoided conflictual topics and relationship issues such as what household roles men and women should fulfill, the importance of marriage, if and how many children are desired, the importance of religion, etc. "Accentuating intimacy and positive affect in their talk, and avoiding discussion of potentially problematic or taboo topics could allow geographically separated couples to maintain a positive outlook on their relationship" (p. 292).

For couples who have the goal of maintaining their relationship and not letting the distance break them, some specific things to do include:

1. *Maintain daily contact via texting.* Texting allows individuals to stay in touch throughout the day. A husband we interviewed who must travel four to five days a week said, "Texting allows us to stay connected with what is going on in each other's lives throughout the day so when I get home on weekends we've been 'together' as much as possible during the week." Pettigrew (2009) emphasized that one of the functions of

Long-distance relationship being separated from a romantic partner by 300 or more miles which precludes regular weekly face-to-face contact.

texting is to maintain interpersonal relationships. It is viewed as more constant and private than using a cell phone.

2. *Enjoy or use the time when apart.* While separated, it is important to remain busy with study, friends, work, sports, and personal projects. Doing so will make the time pass faster.

3. *Avoid conflictual phone conversations.* Talking on the phone should involve the typical sharing of events. When the need to discuss a difficult topic arises (e.g. trust), the phone is not the best place for such a discussion. Rather, it may be wiser to wait and have the discussion face-to-face. If you decide to settle a disagreement over the phone, stick to it until you have a solution acceptable to both of you.

4. *Stay monogamous.* Agreeing not to be open to other relationships is crucial to maintaining a long-distance relationship. Individuals who say, "Let's date others to see if we are really meant to be together," often discover that they are capable of being attracted to and becoming involved with numerous "others." Such other involvements usually predict the end of an LDR. Lydon et al. (1997) studied sixty-nine undergraduates who were involved in LDRs and found that "moral commitment" predicted the survival of the relationships. Individuals committed to maintaining their relationships are often successful in doing so.

Sources

Lydon, J., T. Pierce, and S. O'Regan. 1997. Coping with moral commitment to long-distance dating relationships. *Journal of Personality and Social Psychology* 73: 104–113.

Pistole, M. C. 2010. Long distance romantic couples: An attachment theoretical perspective. *Journal of Marital and Family Therapy* 36: 115–125.

Pistole, M. C., A. Roberts, and M. L. Chapman. 2010. Attachment, relationship maintenance, and stress in long distance and geographically close romantic relationships. *Journal of Social and Personal Relationships* 27: 535–552.

Stafford, L. 2010. Geographic distance and communication during courtship. *Communication Research* 37: 275–297.

Cohabitation

Cohabitation two adults, unrelated by blood or by law, involved in an emotional and sexual relationship who sleep in the same residence at least four nights a week for three months.

Cohabitation, also known as living together, involves two adults, unrelated by blood or by law, involved in an emotional and sexual relationship who sleep in the same residence at least four nights a week for three months. Of 2,922 undergraduates, over two thirds (68%) reported that they would live with a partner they were not married to, and 16% were currently cohabitating or had already done so (Knox and Hall 2010). However, living together is typically seen, particularly by single women, as movement toward marriage. They have little interest in living together as an alternative to marriage or in serial cohabitation relationships with different partners (Lichter et al. 2010)

National **Data**

There are 6 million unmarried-couple households in the United States (*Statistical Abstract of the United States 2011*, Table 63).

Reasons for the increase in cohabitation include career or educational commitments; increased tolerance of society, parents, and peers; improved birth control technology; desire for a stable emotional and sexual relationship without legal ties; avoiding loneliness (Kasearu 2010); and greater disregard for convention.

Almost 60% of women have lived together before age 24. Most of these relationships were short-lived, with 20% resulting in marriage (Schoen et al. 2007). Whites are more likely to live together than blacks, and to be childfree (Landale et al. 2010). Johnson et al. (2011) identified parental marital status (parents were married) and conventionality as predictive of the decision to cohabit versus marry.

Same-Sex Cohabitation and Race

Of the six million unmarried partner households, less than 1% consist of two males or two females. Nevertheless, of these 565,000 same sex households we estimate that 13%, or 73,450, are black couples (*Statistical Abstract of the United States 2011*, Table 63). These couples must cope with both heterosexism and racism.

Nine Types of Cohabitation Relationships

There are various types of cohabitation:

1. *Here and now.* These new partners have an affectionate relationship and are focused on the here and now, not the future of the relationship. Rhoades et al. (2009) studied a sample of 240 cohabitating heterosexual couples and found that wanting to spend more time together was one of the top motivations for living together. Their focus was on the "here and now."

2. *Testers.* These couples are involved in a relationship and want to assess whether they have a future together. Sassler and Miller (2011a) indentified such a couple:

We're trying to see how it is going to work. So if our relationship's going to continue to grow and prosper, this would be one way for us to kind of gauge, you know, maybe we will get married. And if we can't live together, we're not going to get married. So this will help us make future decisions and stuff like that.

Rhoades et al. (2009) found that those who were motivated to live together to test their relationship (in contrast to spending more time together) reported more negative couple communication, more aggression, and lower relationship adjustment. In effect, since these cohabitants were assessing a future they had to engage more issues that might open up the potential for conflict. See Applying Social Research on page 140 which reveals taking relationship risks.

3. *Engaged.* These couples are in love and are planning to marry. Chelsea Clinton and her partner were living together and engaged as they planned their wedding. Among those who report having lived together, about two-thirds (64%) say they thought of this living arrangement as a step toward marriage (Pew Research Center 2010a). Forty percent of 1336 adults in cohabitation relationships in Australia were happy in life and in their relationship (Buchler et al. 2009).

4. *Money savers.* These couples live together primarily out of economic convenience. They are open to the possibility of a future together but regard such a possibility as unlikely. Sassler and Miller (2011b) noted that working class individuals tend to transition more quickly than middle class individuals to cohabitation out of economic necessity or to meet a housing need. Dew (2011) noted that financial disagreements was a prime factor in predicting cohabitation couples breaking up.

5. *Pension partners.* This type is a variation of the money savers category. These individuals are older, have been married before, still derive benefits from their previous relationships, and are living with someone new. Getting married would mean giving up their pension benefits from the previous marriage. An example is a widow from the war in Afghanistan who was given military benefits due to her husband's death. If she remarries, she forfeits both health and pension benefits so she now lives with a new partner and continues to get benefits from the previous marriage.

6. *Alimony maintenance.* Related to widows who cohabit are the divorced who are collecting alimony which they would forfeit should they remarry. They live with a new partner instead of marrying to maintain the benefits. The example is a

Cohabitation

Couples in love take chances- they move in together after knowing each other for a short time, they change schools to be together, and they forgo condom usage thinking "this time won't end in a pregnancy." Taking chances has been studied under the rubric of "risk taking behavior". Previous research has focused on emotional risk taking (Carter and Carter, 2010), online risk taking (Baugartner et al. 2010), and sexual risk taking behavior (Cooper, 2010). This study focused on the various risks lovers take and the consequences of their doing so.

Sample

Three hundred and eighty one undergraduates completed a 64 item Internet survey designed to assess their risk taking behavior. The majority of respondents (over 80%) were female and white (74%). Their average age was 20 years old.

Fourteen examples of chance taking behavior were presented on the questionnaire, 8 of which were identified by 25% or more of the sample as having participated in the activity (the average number of "risks" a student took was 3.44) (see Table 1).

Table 1
Most Frequent Chance Taking Behaviors
in Romantic Relationship
N = 381

Chance Taking Behavior	Percent
Unprotected sex	70%
Being involved in a "friends with benefits" relationship	63%
Broke up with a partner to explore alternative	46%
Had sex before feeling ready	41%
Disconnected w/friends because of partner	34%
Maintained long distance relationship (1 year)	32%
Cheated on partner	30%
Lied to partner about being in love	28%

Results

The major findings of the study were:

1. Love and alcohol are a context for taking risks. While 46% saw themselves as a risk taker in general, 72% reported that they were willing to take chances in their love relationship.

divorced woman receiving a hefty alimony check from her ex, a successful attorney. She was involved in a new relationship with a partner who wanted to marry her. She did not do so in order to keep the alimony flowing. In effect, her ex husband was paying the bills for his former wife and her new lover who had moved in.

7. *Security blanket cohabiters.* Some of the individuals in these cohabitation relationships are drawn to each other out of a need for security rather than mutual attraction.

8. *Rebellious cohabiters.* Some couples use cohabitation as a way of making a statement to their parents that they are independent and can make their own choices. Their cohabitation is more about rebelling from parents than being drawn to each other.

9. *Marriage never (cohabitants forever).* Nineteen percent of 1336 adults in cohabitation relationships in Australia planned never to marry (Buchler et al. 2009). These cohabitants, also known as "marriage renouncing cohabitors," are typically older, may have children, or do not want more children. Compared to couples who are married, cohabitants who plan never to marry tend to report lower life and relationship satisfaction. The "marriage never" couples are rare (celebrities Johnny Depp and Vanessa Paradis and Goldie Hawn and Kurt Russell are examples of couples who live together, have children, and have opted not to marry).

The median duration that a never-married woman cohabits with a partner in the U.S. is 14 months. In contrast, in France it is 51 months; Canada, 40 months; Germany 27 months. In European countries never-married women often have children during cohabitation, which is less common in the U.S. (Cherlin 2010).

There are various reasons and motivations for living together as a permanent alternative to marriage. Some may have been married before and don't want the entanglements of another marriage. Others feel that the real bond between two

Hence, love created a context in which individuals were willing to extend themselves into the risky waters of risk taking behavior. Alcohol had a similar effect. Almost two thirds (66.3%) reported that when they were under the influence of love, they were more vulnerable to taking chances.

2. Being in love was associated with detaching from friends. As individuals in love focus more on each other, they often leave their friends behind through spending less time with them. They know this is a risk but do it anyway (67% had misgivings of disconnecting from friends). Almost half (47%) reported that disconnecting from friends had a negative outcome.

3. Being in love was associated with higher odds of having sex before feeling ready. While one's brain may say "wait" (72% reported that they had misgivings about having sex before they were ready), love moves one forward so that the person goes ahead and has sex even though not ready (45% reported that it had a negative outcome). That individuals feel regret for having sex "too soon" has been previously examined (Merrill and Knox, 2010).

4. Males more likely to view lying about being in love as having positive outcome. One wonders what males meant by (positive outcome) (e.g. the partner had sex with them)?

5. Males more likely to report having positive outcome from being in a friends with benefits relationship.

References

Baumgartner, S. E., P. M. Valkenburg and J. Peter. 2010. Assessing causality in the relationship between adolescents' risky sexual online behavior and their perceptions of this behavior. *Journal of Youth & Adolescence*. 39: 1226–1239.

Carter, P. and D. Carter. 2010. Emotional risk-taking in marital relationships: A phenomenological approach. *Journal of Couple & Relationship Therapy*. 9: 327–343.

Cooper, M. L. 2010. Toward a person X situation model of sexual risk-taking behaviors: illuminating the conditional effects of traits across sexual situations and relationship *Journal of Personality & Social Psychology*. 98: 319–341.

Merrill, J. and D. Knox. 2010. *When I fall in love again: A new study on finding and keeping the love of your life*. Santa Barbra, California: Praeger

*Source: Abridged and adapted from Elliott, L, D. Knox, and B. Easterling. Poster, Taking Chances in Romantic Relationships, Annual Southern Sociological Society Meeting, New Orleans, New Orleans, La. March 22–24, 2012.

people is (or should be) emotional. They contend that many couples stay together because of the legal contract, even though they do not love each other any longer. "If you're staying married because of the contract," said one partner, "you're staying for the wrong reason." Some couples feel that they are "married" in their hearts and souls and don't need or want the law to interfere with what they feel is a private act of commitment. Some couples who view their living together as "permanent" seek to have it defined as a **domestic partnership**, a relationship involving two adults who have chosen to share each other's lives in an intimate and committed relationship of mutual caring (see the following Social Policy feature).

Consequences of Cohabitation

Although living together before marriage does not ensure a happy, stable marriage, it has some potential advantages.

Advantages of Cohabitation   Many unmarried couples who live together report that it is an enjoyable, maturing experience. Other potential benefits of living together include the following:

1. *Sense of well-being.* Compared to uninvolved individuals or those involved but not living together, cohabitants are likely to report a sense of well-being (particularly if the partners see a future together). They are in love, the relationship is new, and the disenchantment that frequently occurs in long-term relationships has not had time to surface. One student reported, "We have had to make some adjustments in terms of moving all our stuff into one place, but we very much enjoy our life together." Although young cohabitants report high levels of enjoyment compared to single people, cohabitants in midlife who have never married when compared to married spouses, report lower levels of rela-

Domestic partnership two adults who have chosen to share each other's lives in an intimate and committed relationship of mutual caring. These relationships are given some kind of official recognition by a city or corporation so as to receive partner benefits (for example, health insurance).

Although same-sex partners find little recognition, protection, and benefits for their relationship in terms of legal marriage, there is greater acceptance when their relationship is viewed in terms of a domestic partnership. Domestic partnerships, both heterosexual and homosexual, want their employers, whether governmental or corporate, to afford them the same rights as spouses. Specifically, employed people who pay for health insurance would like their domestic partner to be covered in the same way that one's spouse would be. Employers have been reluctant to legitimize domestic partners as qualifying for benefits because of the additional expense. One reason for the reluctance is the fear that a higher proportion of partners may be HIV-infected, which would involve considerable medical costs.

Aside from the economic issue, fundamentalist religious groups have criticized domestic partner benefits as eroding family values by giving nonmarital couples the same rights as married couples. Among the 27 states in 2010 recognizing domestic partnerships in some cities, California and New Jersey lead the way in domestic partner benefits, with the law providing rights and responsibilities in areas as varied as child custody, legal claims, housing protections, bereavement leave, and state government benefits.

To receive benefits, domestic partners must register, which involves signing an affidavit of domestic partnership verifying that they are a nonmarried, cohabitating couple 18 years of age or older and unrelated by blood close enough to bar marriage in the state of residence. Other criteria typically used to define a domestic partnership include that the individuals must be jointly responsible for debts to third parties, they must live in the same residence, they must be financially interdependent, and they must intend to remain in the intimate committed relationship indefinitely. Should they terminate their

tional and subjective well-being. However, those who have been married before and are currently cohabitating do not evidence lower levels of relational or personal well-being when compared to married people (Hansen et al. 2007).

2. *Delayed marriage.* Another advantage of living together is remaining unmarried—and the longer one waits to marry, the better. Being older at the time of marriage is predictive of marital happiness and stability, just as being young (particularly 18 years and younger) is associated with marital unhappiness and divorce. Hence, if a young couple who have known each other for a short time is faced with the choice of living together or getting married, their delaying marriage while they live together seems to be the better choice. Also, if they break up, the split will not go on their "record" as would a divorce.

When racial identity is considered, about 60% of whites who cohabit and 40% of blacks who cohabit end up getting married. For whites, cohabitation is more often viewed as a transition to marriage whereas for blacks, it is more often viewed as a permanent lifestyle (Rinelli and Brown 2010).

3. *Knowledge about self and partner.* Living with an intimate partner provides couples with an opportunity for learning more about themselves and their partner. For example, individuals in a living together relationship may find that their role expectations are more (or less) traditional than they had previously thought. Learning more about one's partner is a major advantage of living together. A person's values (calling parents daily), habits (leaving the lights on), and relationship expectations (how emotionally close or distant) are sometimes more fully revealed when living together than in a traditional dating context.

4. *Safety.* Particularly for heterosexual females, living with a partner provides a higher level of safety not enjoyed by single females who live alone—presumably the male would deter someone who broke into the apartment of the female. Of course, living with a roommate or group of friends would provide a similar margin of safety.

DIVERSITY IN OTHER COUNTRIES

Iceland is a homogeneous country of 250,000 descendants of the Vikings. Their sexual norms include early protected intercourse (at age 14), nonmarital parenthood, and living together before marriage. Indeed, a wedding photo often includes not only the couple but also the children they have already had. One American woman who was involved with an Icelander noted, "My parents were upset with me because Ollie and I were thinking about living together, but his parents were upset that we were not already living together" (personal communication).

domestic partnership, they are required to file notice of such termination.

The right to be defined as the next of kin may or may not be included in a state's domestic partnership guidelines. In Washington State, Charlene Strong and Kate Fleming (same-sex partners for ten years) enjoyed their life together but had not registered as domestic partners. So when Kate was rushed to the trauma center in Harborview Medical Center in Seattle, Charlene was stopped by a social worker from entering Kate's room because she was not "family." Kate died. Charlene, devastated, testified before the Washington State legislature and had the law changed so that same-sex relationships were included in the category of domestic partnerships.

Domestic partnerships offer a middle ground between those states that want to give full legal recognition to same-sex marriages and those that deny any legitimacy to same-sex unions. Such relationships also include long-term, committed heterosexuals who are not married. Rothblum et al. (2008) compared couples in same-sex marriages, domestic partnerships, and civil unions and found few differences.

Your Opinion?

1. To what degree do you believe benefits should be given to domestic partners?
2. What criteria should be required for a couple to be regarded as domestic partners?
3. How can abuses of those claiming to be domestic partners be eliminated?

Sources

Rothblum, E. D., K. F. Balsam, and S. E. Solomon. 2008. Comparison of same-sex couples who were married in Massachusetts, had domestic partnerships in California, or had civil unions in Vermont. 29:48–63. © 2008 by SAGE PUBLICATIONS. Reprinted by Permission of SAGE Publications.

Disadvantages of Cohabitation   There is a downside for individuals and couples who live together.

1. *More problems than marrieds.* Hsueh et al. (2009) found a number of differences in the frequency of reported relationship problems between cohabiting and married individuals. Cohabiting individuals tended to argue more, find their relationships more unstable or insecure, and had more issues with past relationships and with future goals and values. But living together also had positive aspects, particularly for women who became were pregnant while living together. Olayemi et al (2010) found that the longer the couple lived together before the woman got pregnant, the lower the blood pressure in the woman during gestation.

2. *Feeling used or tricked.* When expectations differ, the more invested partner may feel used or tricked if the relationship does not progress toward marriage. One partner said, "I always felt we would be getting married, but it turns out that he never saw a future for us."

3. *Parental problems.* Some cohabiting couples must contend with parents who disapprove of or do not fully accept their living arrangement. For example, cohabitants sometimes report that, when visiting their parents' homes, they are required to sleep in separate beds in separate rooms. Some cohabitants who have parents with traditional values respect these values, and sleeping in separate rooms is not a problem. Other cohabitants feel resentful of parents who require them to sleep separately. Some parents express their disapproval of their child's cohabiting by cutting off communication, as well as economic support, from their child. Other parents display lack of acceptance of cohabitation in more subtle ways. One woman who had lived with her partner for two years said that her partner's parents would not include her in the family's annual photo portrait. Emotionally, she felt very much a part of her partner's family and was deeply hurt that she was not included in the family portrait. Still other parents are completely supportive of their children's cohabiting and support their doing so. "I'd rather my kid live together than get married and, besides, it is safer for her and she's happier," said one father.

4. *Economic disadvantages.* Some economic liabilities exist for those who live together instead of getting married. In the Social Policy section on domestic

partnerships, we noted that cohabitants typically do not benefit from their partner's health insurance, Social Security, or retirement benefits. In most cases, only spouses qualify for such payoffs.

Given that most relationships in which people live together are not long-term and that breaking up is not uncommon, cohabitants might develop a written and signed legal agreement should they purchase a house, car, or other costly items together. The written agreement should include a description of the item, to whom it belongs, how it will be paid for, and what will happen to the item if the relationship terminates. Purchasing real estate together may require a separate agreement, which should include how the mortgage, property taxes, and repairs will be shared. The agreement should also specify who gets the house if the partners break up and how the value of the departing partner's share will be determined.

If the couple have children, another agreement may be helpful in defining custody, visitation, and support issues in the event the couple terminates the relationship. Such an arrangement may take some of the romance out of the cohabitation relationship, but it can save a great deal of frustration should the partners decide to go their separate ways.

In addition, couples who live together instead of marrying can protect themselves from some of the economic disadvantages of living together by specifying their wishes in wills; otherwise, their belongings will go to next of kin or to the state. They should also own property through joint tenancy with rights of survivorship. In this way, ownership of the entire property will revert to one partner if the other partner dies. In addition, the couple should save for retirement, because live-in companions may not access Social Security benefits, and some company pension plans bar employees from naming anyone other than a spouse as the beneficiary.

5. *Effects on children.* About 40% of children will spend some time in a home where the adults are cohabiting. In addition to being disadvantaged in terms of parental income and education, they are likely to experience more disruptions in family structure.

Schmeer (2011) studied the health differences at age 5 for children born to cohabiting versus married parents and found worse health for children born to cohabiting parents. Child health was better for those whose cohabiting parents who married than for those whose parents remained stably cohabiting.

Having Children while Cohabitating?

Sassler and Cunningham (2008) interviewed twenty-five never-married American women who were cohabiting with their heterosexual partners. Most (two-thirds) reported that they wanted to be married before having a child. Indeed,

WHAT IF?

What if Your Partner Wants to Live Together and You Do Not?

It is not unusual that partners in love have different values and goals for the relationship. Wanting to live together as well as not wanting to do so are equally valid positions.

However, the costs to the person who is asked to live together who feels this is wrong or unwise are greater than the costs to the person for not having the person live with him or her. The consequence of going against one's values is to continually feel uncomfortable while cohabiting and to risk blaming the other person if something goes wrong. Win-lose decisions and relationships are never a good idea. Manning et al. (2011) provided data on this question and found that when romantic partners disagreed about whether to cohabit, the feelings of the partner against cohabitation trumped the desires of the other—so the couple did not live together.

none of the respondents planned on having a child in the near future and none were actively trying to conceive. However, some noted that marriage made no difference. One respondent noted.

> *I don't see a reason to get married if you don't want to get married. Everyone, everyone is so traditional, which is really silly because it's not a traditional world anymore. Like when, like when my sister had her baby. I mean, she was concerned about. . . "I wonder what they are going to think if I don't have a husband?" I mean, who cares? Who cares if you don't have a husband? What does marriage have to do with anything? (p. 13)*

PERSONAL CHOICES

Will Living Together Ensure a Happy, Durable Marriage?

Couples who live together before marrying assume that doing so will increase their chances of having a happy and durable marriage relationship. But will it? The answer is, "It depends." For women who have only one cohabitation experience with the man they marry, there is no increased risk of divorce. However, if a woman is a serial cohabitant, there is an increased risk (Teachman 2003). Jose et al. (2010) also found that the negative predictive effect of cohabitation does not remain if the partners (both women and men) have only lived with each other and planned to marry. In effect, persons who cohabit only once tend to regard the experience as more permanent and have greater commitment. Never-married females are the most likely to cohabit only once and with the man they eventually marry (Lichter et al. 2010).

Whether or not the couple is engaged when they begin to cohabit is also relevant. Stanley et al. (2010) found that couples in first marriages who were engaged when they began to live together were less likely to report negative interaction, divorce proneness, or divorce than those who became engaged after they had begun living together. For those in second marriages, there was a "general risk associated with premarital cohabitation for second marriages on self-reported indices of marital quality, with or without engagement when cohabitation began".

Because people commonly have more than one cohabitation experience, the term **cohabitation effect** applies. This means that those who have multiple cohabitation experiences prior to marriage are more likely to end up in marriages characterized by violence, lower levels of happiness, lower levels of positive communication, higher levels of depression, you name it (Booth et al. 2008). Liat and Havusha-Morgenstern (2011) compared the marital adjustment among women who cohabited with their spouses before marriage versus those who did not. They found that cohabiting women reported lower levels of adjustment of spousal cohesion and display of affection than did women who did not cohabit. McKean and Dush (2010) suggested that one explanation for greater dissatisfaction among spouses who had lived together is that if they were unhappy as cohabitants they were more likely to go through with the wedding than persons who were just dating and unhappy.

In the meantime, cohabitation relationships are no match for married relationships. Hansen et al. (2007) compared Norway cohabitants (who had never been married) in midlife with spouses in midlife and found cohabitants less happy, less close, and more conflictual. When the cohabitation relationship breaks, it gets worse. Williams et al. (2008) noted the psychological distress is particularly acute for single mothers. The message to single mothers was to stay single or get married (living together had a high chance of negative consequences).

What is it about serial cohabitation relationships that predict negatively for future marital happiness and durability? One explanation is that cohabitants tend to be people who are willing to violate social norms by living together before marriage. Once they marry, they may be more willing to break another social norm and divorce if they are unhappy than are unhappily married people who tend to conform to social norms and have no history of unconventional behavior. A second explanation is that, because cohabitants are less committed to the relationship than married people, this may translate

Cohabitation effect those who have multiple cohabitation experiences prior to marriage are more likely to end up in marriages characterized by violence, lower levels of happiness, lower levels of positive communication, and depression.

into withdrawing from conflict by terminating the relationship rather than communicating about the problems and resolving them because the stakes are higher (White et al. 2004). Whatever the reason, cohabitants should not assume that cohabitation will make them happier spouses or insulate them from divorce.

Not all researchers have found negative effects of cohabitation on relationships. Reinhold (2010) found that among more recent cohabitant cohorts, the negative association between living together and marital instability is weakening. Additional research is needed to confirm the effect of cohabitation on subsequent marriage.

Sources

Booth, A., E. Rustenbach, and S. McHale. 2008. Early family transitions and depressive symptom changes from adolescence to early adulthood. *Journal of Marriage and Family* 70: 3–14.

Hansen, T., T. Moum, and A. Shapiro. 2007. Relational and individual well-being among cohabitors and married individuals in midlife: Recent trends from Norway. *Journal of Family Issues* 28: 910–933.

Jose, A., K. D. O'Leary, and A. Moyer. 2010. Does premarital cohabitation predict subsequent marital stability and marital quality? A meta-analysis. *Journal of Marriage and Family* 72: 105–116.

Liat, K. and H. Havusha-Morgenstern. 2011. Does cohabitation matter? Differences in initial marital adjustment among women who cohabited and those who did not. *Families in Society* 92: 120–127.

McKean, T., and C. K. Dush. 2010. Marriage formation among daters and cohabitors: A cohabitation effect? Paper, Annual Meeting of the National Council on Family Relations, November 3–6. Minneapolis, MN.

Lichter, D. T., R. N. Turner, and S. Sassler. 2010. National estimates of the rise in serial cohabitation *Social Science Research* 39: 754–765.

Reinhold, S. 2010. Reassessing the link between premarital cohabitation and marital instability. *Demography* 47: 719–733.

Stanley, S. M., G. K. Rhoades, P. R. Amato, H. J. Markman, and C. A. Johnson. 2010. The timing of cohabitation and engagement: Impact on first and second marriages. *Journal of Marriage and Family.* 72: 906–918.

Teachman, J. 2003. Premarital sex, premarital cohabitation, and the risk of subsequent marital disruption among women. *Journal of Marriage and Family* 65: 444–455.

White, A. M., F. S. Christopher, and T. K. Poop. 2004. Cohabitation and the early years of marriage. Poster session, National Council on Family Relations, November, Orlando, FL.

Whitty, M. T., A. J. Baker, and J. A. Inman, eds. 2007. *Online matchmaking.* New York: Palgrave Macmillan.

Williams, K., S. Sassler, and L. M. Nicholson. 2008. For better or worse? The consequences of marriage and cohabitation for single mothers. *Social Forces* 86: 1481–1512.

Legal Aspects of Living Together

In recent years, the courts and legal system have become increasingly involved in relationships in which couples live together. Some of the legal issues concerning cohabiting partners include common-law marriage, palimony, child support, and child inheritance. Lesbian and gay couples also confront legal issues when they live together.

Technically, cohabitation is against the law in some states. For example, in North Carolina, cohabitation is a misdemeanor punishable by a fine not to exceed $500, imprisonment for not more than six months, or both. Most law enforcement officials view cohabitation as a victimless crime and feel that the general public can be better served by concentrating upon the crimes that do real damage to citizens and their property.

Common-Law Marriage   The concept of common-law marriage dates to a time when couples who wanted to be married did not have easy or convenient access to legal authorities (who could formally sanction their relationship so that they would have the benefits of legal marriage). Thus, if the couple lived together, defined themselves as husband and wife, and wanted other people to view them as a married couple, they would be considered married in the eyes of the law.

Despite the assumption by some that heterosexual couples who live together a long time have a common-law marriage, only 14 states recognize such marriages (see Chapter 1 for a list). In these states a heterosexual couple may be considered married if they are legally competent to marry, if the partners agree that they are married, and if they present themselves to the public as a married couple. A ceremony or compliance with legal formalities is not required.

In common-law states, individuals who live together and who prove that they were married "by common law" may inherit from each other or receive alimony and property in the case of relationship termination. They may also receive health and Social Security benefits, as would other spouses who have a marriage license. In states not recognizing common-law marriages, the individuals who live together are not entitled to benefits traditionally afforded married individuals. More than three-quarters of the states have passed laws prohibiting the recognition of common-law marriages within their borders.

Palimony   A takeoff on the word *alimony*, **palimony** refers to the amount of money one "pal" who lives with another "pal" may have to pay if the partners end their relationship. For example, comedian Bill Maher has been the target of a $9 million palimony suit by ex-girlfriend Coco Johnsen. Similarly, Norman Greenbaum (known for his song "Spirit in the Sky") was the target of a palimony suit in 2010 by an ex-girlfriend and fan club leader. She (Tracy E. Outlaw of South Carolina) accused Greenbaum of fraud for failing to live up to the terms of an agreement in which the former Clemson University employee would quit her job and move across country in exchange for one-half interest in Greenbaum's Saddleback Court home and a third of his estate. The suit against Bill Maher has been dismissed; the Greenbaum-Outlaw case is pending.

Palimony refers to the amount of money one "pal" who lives with another "pal" may have to pay if the partners end their relationship.

Child Support   Heterosexual individuals who conceive children are responsible for those children whether they are living together or married. In most cases, the custody of young children will be given to the mother, and the father will be required to pay child support. In effect, living together is irrelevant with regard to parental obligations. However, a woman who agrees to have a child with her lesbian partner cannot be forced to pay child support if the couple breaks up. The Massachusetts Supreme Judicial Court ruled that their informal agreement to have a child together did not constitute an enforceable contract.

Couples who live together or who have children together should be aware that laws traditionally applying only to married couples are now being applied to many unwed relationships. Palimony, distribution of property, and child support payments are all possibilities once two people cohabit or parent a child.

Child Inheritance   Children born to cohabitants who view themselves as spouses and who live in common-law states are regarded as legitimate and can inherit from their parents. However, children born to cohabitants who do not present themselves as married or who do not live in common-law states are also able to inherit. A biological link between the parent and the offspring is all that need be established.

Living Apart Together

A new lifestyle and family form has emerged called living apart together (Hess 2009). The premise that is being questioned by those involved in "living apart together" is that "more is better" . . . that the more time lovers spend together, including moving in together, the better. In effect, loving and committed couples automatically assume that they will marry or live together in one residence and that to do otherwise would suggest that they do not "really" love each other and aren't "really" committed to each other.

Living apart together (LAT) committed individuals who do not live in the same residence.

The definition of **living apart together (LAT)** is a committed couple who does not live in the same home (and others such as children or elderly parents may live in those respective homes). Three criteria must be met for a couple to be defined as an LAT couple: (1) they must define themselves as a committed couple; (2) others must define the partners as a couple; and (3) they must live in separate domiciles. The lifestyle of living apart together involves partners in loving and committed relationships (married or unmarried) identifying their independent needs in terms of the degree to which they want time and space away from each other. People living apart together exist on a continuum from partners who have separate bedrooms and baths in the same house to those who live in a separate place (apartment, condo, house) in the same or different cities. LAT couples are not those couples who are forced by their career or military assignment to live separately. Rather, LAT partners choose to live in separate domiciles. Take the Self Assessment on LAT to assess the suitability of this lifestyle to your needs.

DIVERSITY IN OTHER COUNTRIES

Another version of living apart together is the "walking marriage." High in the Tibetan Himalayas, "walking marriages" occur in the Mosuo culture (Kingdom of Women) in China, which does not have traditional marriage (no "husbands" or "wives"). Rather, in this matrilineal society, women live with other women (and raise the children) and men "walk by at night and visit and leave the next morning" when there is an affectionate relationship. The women and men can have as many "walking marriages" as they want with the men never living with the woman, only visiting her (Mosuo 2010).

The living apart together lifestyle or family form is not unique to couples in the United States (e.g., the phenomenon is also evident in European countries such as France, Sweden, and Norway). Couples choose this pattern for a number of reasons, including the desire to maintain some level of independence, to enjoy their time alone, and to keep their relationship exciting.

Advantages of LAT

The benefits of LAT relationships include the following:

1. *Space and privacy.* Having two places enables each partner to have a separate space to read, watch TV, talk on the phone, or whatever. This not only provides a measure of privacy for the individuals, but also as a couple. When the couple has overnight guests, the guests can stay in one place while the partners stay in the other place. This arrangement gives guests ample space and the couple, private space and time apart from the guests.

2. *Career or work space.* Some individuals work at home and need a controlled quiet space to work on projects, talk on the phone, concentrate on their work, and so on, without the presence of someone else. The LAT arrangement is particularly appealing to musicians for practicing, artists to spread out their materials, and authors for quiet (Hemingway had his own wing of the house).

3. *Variable sleep needs.* Although some partners enjoy going to bed at the same time and sleeping in the same bed, others like to go to bed at radically

This scale will help you assess the degree to which you might benefit from living in a separate residence from your spouse or partner with whom you have a lifetime commitment. There are no right or wrong answers.

Directions

After reading each sentence carefully, circle the number that best represents the degree to which you agree or disagree with the sentence.

1	2	3	4	5
Strongly agree	Mildly agree	Undecided	Mildly disagree	Strongly disagree
SA	MA	U	MD	SD

_____	1. I prefer to have my own place (apart from my partner) to live.	1 2 3 4 5
_____	2. Living apart from my partner feels "right" to me.	1 2 3 4 5
_____	3. Too much togetherness can kill a relationship.	1 2 3 4 5
_____	4. Living apart can enhance your relationship.	1 2 3 4 5
_____	5. By living apart you can love your partner more.	1 2 3 4 5
_____	6. Living apart protects your relationship from staleness.	1 2 3 4 5
_____	7. Couples who live apart are just as happy as those who don't.	1 2 3 4 5
_____	8. Couples who LAT are just as much in love as those who live together in the same place.	1 2 3 4 5
_____	9. People who LAT probably have less relationship stress than couples who live together in the same place.	1 2 3 4 5
_____	10. LAT couples are just as committed as couples who live together in the same residence.	1 2 3 4 5

Scoring

Add the numbers you circled. The lower your total score (10 is the lowest possible score), the more suited you are to the Living Apart Together lifestyle. The higher your total score (50 is the highest possible score), the least suited you are to the Living Apart Together lifestyle. A score of 25 places you at the midpoint between being the extremes. One-hundred and thirty undergraduates completed the LAT scale with an average score of 28.92 which suggests that both sexes view themselves as less rather than more suited (30 is the midpoint between the lowest score of 10 and the highest score of 50) for a LAT arrangement with females registering greater disinterest than males.

Source

Copyright 2011 © David Knox, Ph.D. Email Dr. Knox at knoxd@ecu.edu for scale use.

different times and to sleep in separate beds or rooms. The LAT arrangement allows for partners to have their own sleep needs or schedules met without interfering with a partner. A frequent comment from LAT partners is, "My partner thrashes throughout the night and kicks me, not to speak of the wheezing and teeth grinding, so to get a good night's sleep, I need to sleep somewhere else."

4. *Allergies.* Individuals who have cat or dog allergies may need to live in a separate antiseptic environment from their partner who loves animals and won't live without them. "He likes his dog on the couch," said one woman.

5. *Variable social needs.* Partners differ in terms of their need for social contact with friends, siblings, and parents. The LAT arrangement allows for the partner who enjoys frequent time with others to satisfy that need without subjecting the other to the presence of a lot of people in one's life space. One wife from a family of seven children enjoyed both her siblings and parents being around. The LAT arrangement allowed her to continue to enjoy her family at no expense to her husband who was upstairs in another condo.

6. *Blended family needs.* A variation of the previous item is a blended family in which remarried spouses with children from previous relationships sometimes find it easier to separate out the living space of their children. An example is a married couple who bought a duplex with each spouse living with his/her own children on either side of the duplex.

Satiation a stimulus loses its value with repeated exposure (e.g., lovers tire of each other if around each other all the time); also called habituation.

7. *Keeping the relationship exciting.* Zen Buddhists remind us of the necessity to be in touch with polarities, to have a perspective where we can see and appreciate the larger picture—without the darkness, we cannot fully appreciate the light. The two are inextricably part of a whole. This is the same with relationships; time apart from our beloved can make time together feel more precious. The term *satiation* is a well-established psychological principle. Basically, **satiation** means that a stimulus loses its value with repeated exposure. Just as we tire of eating the same food, listening to the same music, or watching the same movie twice, so satiation is relevant to relationships. Indeed, couples who are in a long-distance dating relationship know the joy of "missing" each other and the excitement of being with each other again. Similarly, individuals in a LAT relationship help to ensure that they will not "satiate" on each other but maintain some of the excitement in seeing or being with each other.

8. *Self-expression and comfort.* Partners often have very different tastes in furniture, home décor, music, and temperature. With two separate places, each can arrange and furnish their respective homes according to their own individual preferences. The respective partners can also set the heat or air conditioning according to their own preferences, and play whatever music they like.

9. *Cleanliness or orderliness.* Separate residences allow each partner to maintain the desired level of cleanliness and orderliness without arguing about it. Some individuals like their living space to be as clean as a cockpit. Others simply don't care.

10. *Elder care.* One partner may be taking care of an elderly parent in the parents' house or in his or her own house. Either way, the partners may have a preference not to live in the same house as a couple with a parent. An LAT relationship allows for the partner taking care of the elderly parent to do so and a place for the couple to be alone.

11. *Maintaining one's lifetime residence.* Some retirees, widows, and widowers meet, fall in love, and want to enjoy each other's companionship. However, they don't want to move out of their own house. The LAT arrangement allows each partner to maintain a separate residence but to enjoy the new relationship.

12. *Leaving inheritances to children from previous marriages.* Having separate residences allows respective partners to leave their family home or residential property to their children from a previous relationship without displacing their surviving spouse.

Disadvantages of LAT

There are also disadvantages to the LAT lifestyle.

1. *Stigma or disapproval.* Because the norm that married couples move in together is firmly entrenched, couples who do not do so are suspect. Neighbors may feel that spouses in a living apart together relationship don't really love each other and are really "separated" and moving toward divorce. Our society is not accepting of LAT relationships.

2. *Cost.* Certainly, maintaining two separate living arrangements can be more expensive than two people living in one domicile. But there are ways LAT couples keep their housing costs low to maintain this lifestyle. One way is to live in two separate condominiums which are cheaper than two separate houses. Another way is to live out of town in less expensive neighborhoods. One partner said, "We can simply drive twenty miles out of town where the price of housing drops 50% so we can afford a duplex. We have our separation and it didn't cost us a fortune."

3. *Inconvenience.* Unless the partners live in a duplex or two units in the same condominium, going between the two places to share meals or be together can be inconvenient.

4. *Lack of shared history.* Because the adults are living in separate quarters, a lot of what goes on in each house does not become a part of the life history of the other. For example, children in one place don't benefit as much from the other adult who lives in another domicile most of the time.

5. *Waking up alone.* Although some LAT partners sleep together overnight, others say goodnight and sleep in separate beds or houses. A potential disadvantage is waking up and beginning the day alone without the early-morning connection with one's beloved.

The Future of Singlehood, Long-Term Relationships, and Cohabitation

Singlehood will (in the cultural spirit of diversity) lose some of its stigma, more young adults (particularly men) will choose this option, and those who remain single will, increasingly, find satisfaction in this lifestyle.

Individuals will continue to be in no hurry to get married. Completing their education, becoming established in their career, and enjoying "hanging out and hooking up" will continue to delay serious consideration of marriage. The median age for women getting married is 26; for men, 28. This trend will continue as individuals will keep their options open in America's individualistic society.

Cohabitation will, increasingly, become an accepted, predictable stage of courtship. The link between cohabitation and negative marital outcomes will dissolve as more individuals elect to cohabit before marriage. Previously, only risk takers and people willing to abandon traditional norms lived together before marriage. In the future, mainstream individuals will cohabit. Currently close to two thirds of individuals cohabit at least once before marriage.

Summary

What is the status of singlehood today?
Singles consist of the never married, divorced, and widowed. The never married are delaying marriage in favor of completing their education, establishing themselves in a career, and enjoying "hanging out and hooking up." The primary attraction of singlehood is the freedom to do as one chooses. Others fear getting into a bad marriage or having to go through a divorce. The Alternatives to Marriage Project (ATMP) advocates "for equality and fairness for unmarried people, including people who are single, choose not to marry, cannot marry, or live together before marriage." Successful singles view being alone positively, enjoy their work, and have adequate resources.

What are the functions of and changes in dating?
The various functions of dating include: (1) confirmation of a social self; (2) recreation; (3) companionship, intimacy and sex; (4) anticipatory socialization; (5) status achievement; (6) mate selection; and (7) health enhancement. The changes in dating in recent years include: an increase in the age at marriage, more formerly divorced in the dating pool, cohabitation becoming more normative, and women becoming more assertive in initiating relationships.

The dating pool also includes over 2 million newly divorced Americans each year, most of whom will remarry. These individuals are older, have an increased HIV risk, and most often have children. They also come with some ex-spouse baggage and are usually in a hurry as the courtship among the divorced tends to be brief.

How do individuals go about finding a partner?
Besides the traditional way of meeting people at work or school or through friends and going on a date, couples today may find each other while "hanging out" (which may lead to "hooking up"). Internet dating (e.g., Match.com) and speed-dating (including weed dating) are also new forms for finding each other. Internet dating is efficient, allows one to screen multiple partners quickly and to disappear at will. Downsides include that one cannot assess "chemistry" through a computer screen, Internet relationships allow no observation of the partner with one's friends and family, and there is considerable deception by both parties.

What is cohabitation like among today's youth?
Cohabitation, also known as living together, is becoming a "normative life experience," with almost 60% of American women reporting that they had cohabited before marriage. Reasons for an increase in living together include a delay of marriage for educational or career commitments, fear of marriage, increased tolerance of society for living together, and a desire to avoid the legal entanglements of marriage. Types of relationships in which couples live together include the here-and-now, testers (testing the relationship),

engaged couples (planning to marry), and cohabitants forever (never planning to marry). Most people who live together eventually marry but not necessarily each other.

Domestic partners are two adults who have chosen to share each other's lives in an intimate and committed relationship of mutual caring. Such cohabitants, both heterosexual and homosexual, want their employers, whether governmental or corporate, to afford them the same rights as spouses.

Although living together before marriage does not ensure a happy, stable marriage, it has some potential advantages. These include a sense of well-being, delayed marriage, learning about yourself and your partner, and being able to disengage with minimal legal hassle. Disadvantages include feeling exploited, feeling guilty about lying to parents, and not having the same economic benefits as those who are married. Social Security and retirement benefits are paid to spouses, not live-in partners.

What are the pros and cons of "living apart together"?

A new lifestyle and family form is living apart together (LAT), which means that monogamous committed partners—whether married or not—carve out varying degrees of physical space between them. People living apart together exist on a continuum from partners who have separate bedrooms and baths in the same house to those who live in separate places (apartment, condo, house) in the same or different cities. Couples choose this pattern for a number of reasons, including the desire to maintain some level of independence, to enjoy their time alone, to keep their relationship exciting, and so on.

Advantages to involvement in an LAT relationship include space and privacy, sleeping without being cramped or dealing with snoring, not living with animals if there is an allergy, having family or friends over without interfering with a partner's life space, and keeping the relationship exciting. Disadvantages include being confronted with stigma or disapproval, cost, inconvenience, and waking up alone.

What is the future of singlehood?

Singlehood will lose some of its stigma, more young adults (particularly men) will choose this option, and those who remain single will, increasingly, find satisfaction in this lifestyle. While marriage will remain the lifestyle choice, individuals will continue to be in no hurry to get married.

Key Terms

Cohabitation	Hooking up	Satiation
Cohabitation effect	Living apart together (LAT)	Singlehood
Domestic partnership	Long distance relationship	Weed dating
Hanging out	Palimony	

Web Links

Selecting a Partner

Meghan Arrowood

The minute you settle for less than you deserve, you get even less than you settled for.

Maureen Dowd, journalist

Learning Objectives

Discuss how culture restricts your choice of a marriage partner.

Identify homogamous factors operative in mate selection.

Review the psychological factors operative in selecting a mate.

Summarize the sociobiological view of mate selection.

Know what is involved in premarital counseling/premarital education.

List the factors which suggest that you might delay or call off your wedding.

Predict the future of mate selection.

1. The person who has the least interest in the relationship controls the relationship.

2. Individuals who pray keep their religion to themselves; they don't expect their romantic partner to pray.

3. Undergraduate men and women report the same interest in getting married.

4. Some partners have their DNA analyzed to see if they have compatible immune systems.

5. Daughters view their father as more concerned than their mother about the economic potential of their mate choice.

Answers: 1. T   2. F   3. F   4. T   5. T

You are reading the most important chapter in this text—selecting a partner to share your life with. In no other life decision are the consequences so important. Decisions about which college to attend, which career to pursue, and which religion/philosophy of life to follow all have implications for your life and happiness. But none of these decisions is more important than your choice of a lifetime partner.

Think of a time you have been miserable and a time you have been ecstatic. Likely, both of these times were in reference to a person in your life. The capacity of your partner/mate to create a context of utter misery or elation makes clear the importance of selecting someone to partner with whom you love, who loves you, who shares a common view of the future, and someone with whom you can communicate/negotiate issues. Since some people are more suited to you than others, this chapter is about identifying those with whom you will be more likely to share a happy and fulfilling future.

No other choice in life is as important as your choice of a marriage partner. Your day-to-day happiness, health, and economic well-being will be significantly influenced by the partner with whom you choose to share your life. Most have high hopes and even believe in the perfect mate. An anonymous comic said, "I married Miss Right. . . . I just didn't know her first name was Always." The Self-Assessment feature allows you to assess the degree to which you have already selected and are involved with a partner.

Because the heart of this text is about making wise choices in relationships, this chapter begins with an examination of the cultural, sociological, and psychological filters involved in mate selection. We end the chapter with specific questions to ask a potential partner to increase your knowledge of this partner, specific partner/relationship characteristics that predict a loving/stable future, and wrong reasons to get married.

It really doesn't matter who you marry since you are sure to find out you married someone else.

Jay Leno, Tonight Show host

Cultural Aspects of Mate Selection

Individuals are not free to marry whomever they please. Indeed, university students routinely assert, "I can marry whomever I want!" Hardly. Rather, their culture and society radically restrict and influence their choice. The best example of mate choice being culturally and socially controlled is the fact that *less than 5%* of people marry someone outside their race; less than 1% consist of a black and a white spouse (*Statistical Abstract of the United States, 2011,* Table 60). Homosexual people are also not free to marry whomever they choose. Indeed, although Massachusetts, Connecticut, Iowa, New Hampshire, and Maine recognize same-sex marriage, federal law does not, reflecting wide societal disapproval.

There are also cultural influences on the characteristics of the person one chooses to marry. For example, Kline and Zhang (2009) noted that while undergraduates in both China and the United States look for honesty, physical attractiveness, ambition, and considerateness as preferred traits, Chinese students listed filial piety (respect/care for parents) as one of their three most preferred traits in a partner. This trait had very low priority by students in the U.S. Hence, one's culture tells one what to look for in a mate.

Endogamy and exogamy are also two forms of cultural pressure operative in mate selection.

Endogamy

Endogamy is the cultural expectation to select a marriage partner within one's own social group, such as in the same race, religion, and social class. Endogamous pressures involve social approval and encouragement to select a partner within your own group (for example, someone of your own race and religion) and disapproval for selecting someone outside your own group. The pressure toward an endogamous mate choice is especially strong when race is concerned. Love may be blind, but it knows the color of one's partner as the majority of individuals end up selecting someone of the same race to marry.

Exogamy

Exogamy is the cultural pressure to marry outside the family group (e.g. you cannot marry your siblings). Woody Allen experienced enormous social disapproval because he fell in love with and married his own stepdaughter. Exogamy dictates that he marry someone outside the family.

Incest taboos are universal; as well, children are not permitted to marry the parent of the other sex in any society. In the United States, siblings and (in some states) first cousins are also prohibited from marrying each other. The reason for such restrictions is fear of genetic defects in children whose parents are too closely related.

Once cultural factors have determined the general **pool of eligibles** (the population from which a person selects a mate), individual mate choice becomes more operative. However, even when individuals feel that they are making their own choices, social influences are still operative. The Self-Assessment to follow provides a way to assess your current level of involvement.

DIVERSITY IN OTHER COUNTRIES

In most cultures, mate selection is not left to chance and parents are involved in the selection of their offspring's partner. However, who picks the best mate: parents or the individual? India is made up of twenty-nine states, reflecting a range of languages, dialects, and religions. Because Indian culture is largely collectivistic or familistic (focusing on family unity and loyalty) rather than individualistic (focusing on personal interests and freedom), most Indian youth see parents as better able to select a lifetime marital partner and defer to their judgment (Medora 2003, 227). One of our students from India noted, "My parents will know who is best for me. I trust them to make the right selection. While they take into consideration my feelings, if they do not want me to marry someone, I will not. And when people ask me, 'What about love?' I tell them that love is seen in my culture as that which follows rather than precedes marriage. A partner of similar values and good education who blends into my family is the focus. Love will follow."

Endogamy cultural expectation to select a marriage partner within one's own social group.

DIVERSITY IN THE UNITED STATES

Exogamy is operative among the Hopi Indians, about 10,000 of whom live on reservations in Arizona. The Hopi belong to different clans (Bear clan, Badger clan, Rain clan, and so on), and the youth are expected to marry someone of a different clan. By doing so, they bring resources from another clan into their own.

Exogamy the cultural pressure to marry outside the family group.

Pool of eligibles the population from which a person selects a mate.

Sociological Factors Operative in Mate Selection

Numerous sociological factors are at work in bringing two people together who eventually marry.

Homogamy

Whereas endogamy refers to cultural pressure, **homogamy** refers to the tendency for the individual to seek a mate with similar characteristics (e.g., age, race, education, etc.). Homogamy is often expressed by the sentiment "like attracts like."

Homogamy the tendency for an individual to seek a mate who has similar characteristics.

This scale is designed to assess the level of your involvement in a current relationship. Please read each statement carefully, and write the number next to the statement that reflects your level of disagreement to agreement, using the following scale.

1	2	3	4	5	6	7
Strongly Disagree						Strongly Agree

_____ 1. I have told my friends that I love my partner.

_____ 2. My partner and I have discussed our future together.

_____ 3. I have told my partner that I want to marry him/her.

_____ 4. I feel happier when I am with my partner.

_____ 5. Being together is very important to me.

_____ 6. I cannot imagine a future with anyone other than my partner.

_____ 7. I feel that no one else can meet my needs as well as my partner.

_____ 8. When talking about my partner and me, I tend to use the words "us," "we," and "our."

_____ 9. I depend on my partner to help me with many things in life.

_____ 10. I want to stay in this relationship no matter how hard times become in the future.

Scoring

Add the numbers you assigned to each item. A 1 reflects the least involvement and a 7 reflects the most involvement. The lower your total score (10 is the lowest possible score), the lower your level of involvement; the higher your total score, the greater your level of involvement. A score of 40 places you at the midpoint between a very uninvolved and very involved relationship.

Other Students Who Completed the Scale

Valdosta State University. The participants were 31 male and 86 female undergraduate psychology students haphazardly selected from Valdosta State University. They received course credit for their participation. These participants ranged in age from 18 to 59 with a mean age of 20.25 (SD = 4.52). The ethnic background of the sample included 70.9% white, 23.9% Black, 1.7% Hispanic, and 3.4% from other ethnic backgrounds. The college classification level of the sample included 46.2% freshmen, 36.8% sophomores, 14.5% juniors, and 2.6% seniors.

East Carolina University. Also included in the sample were 60 male and 129 female undergraduate students haphazardly selected from East Carolina University. These participants ranged in age from 18 to 43 with a mean age of 20.40 (SD 5 3.58). The ethnic background of the sample included 76.2% white, 14.3% Black, 0.5% Hispanic, 1.6% Asian, 2.6% American Indian, and 4.8% from other ethnic backgrounds. The college classification level of the sample included 40.7% freshmen, 19.0% sophomores, 19.6% juniors, and 20.6% seniors. All participants were treated in accordance with the ethical guidelines of the American Psychological Association.

Scores of Participants

When students from both universities were combined, the average score of the men was 50.06 (SD = 14.07) and the average score of the women was 52.93 (SD = 15.53), reflecting moderate involvement for both sexes. There was no significant difference between men and women in level of involvement. However, there was a significant difference ($p < .05$) between whites and nonwhites, with whites reporting greater relationship involvement (M = 53.37; SD = 14.97) than nonwhites (M = 48.33; SD = 15.14).

In addition, there was a significant difference between the level of relationship involvement of seniors compared with juniors ($p < .05$) and freshmen ($p < .01$). Seniors reported more relationship involvement (M = 57.74; SD = 12.70) than did juniors (M = 51.57; SD 5 = 15.37) or freshmen (M = 50.31; SD = 15.33).

Source

"The Relationship Involvement Scale" 2004 by Mark Whatley, Ph.D., Department of Psychology, Valdosta State University, Valdosta, Georgia 31698-0100. Used by permission. Other uses of this scale by written permission of Dr. Whatley only (mwhatley@valdosta.edu). Information on the reliability and validity of this scale is available from Dr. Whatley.

In general, the more couples have in common, the higher the reported relationship satisfaction and the more durable the relationship (Clarkwest 2007; Amato et al. 2007). Mcintosh et al. (2011) studied older adults who posted online ads and found that they were particularly concerned with homogamous factors such as age, race, religion, etc. and that they were willing to travel great distances to find such a partner.

National **Data**

About 13% of American married couples involve someone whose partner is of a different race/ethnicity. White-Hispanic and white-Native American pairings are the most common. African Americans are the least likely to marry whites (Burton et al. 2010).

Race   Race is a socially constructed concept and refers to physical characteristics that are given social significance. Yoo et al. (2010) noted the difference between blatant and subtle racism. Blatant racism involves outright name calling and discrimination; Tiger Woods was once denied the right to play golf on a Georgia golf course because he was black. Subtle racism involves omissions, inactions, or failure to help, rather than a conscious desire to hurt. Not sitting at a table where persons of another race are sitting or failing to stop to help a person because of his or her race is subtle racism.

Ray and Rosow (2010) documented racial differences between how black and white fraternity males treat women. Black fraternity men tended to be more romantic and respectful of women than did white fraternity men. One black undergraduate said:

> *I definitely think my black fraternity brothers do a lot of stuff that make them [women] feel appreciated like getting them flowers, writing them a poem or telling them that they look beautiful.*

White fraternity men were more likely to take the position that it was not necessary to wine and dine a woman before taking her back to the fraternity house to have sex.

As noted earlier, racial homogamy operates strongly in selecting a dating, live-in, or marital partner (with greater homogamy for marital partners). Lubin (2010) found that mixed-heritage individuals reported more positive attitudes toward interethnic relationships than minority individuals (Asian and black).

Robnett and Feliciano (2011) examined racial preferences of persons seeking a partner on the Internet and found that white men were more likely to exclude blacks as possible dates, while white women were more likely to exclude Asians. Racial homogamy also operates in hooking up. McClintock (2010) studied undergraduates at Stanford University and found racial homogamy also operative in hooking up encounters. However, Asian woman and black men were most likely to cross racial lines, since they were eroticized by white students, and therefore desired.

In general, as one moves from dating to living together to marriage, the willingness to marry interracially decreases. In a sample of 2,922 undergraduates, 35% of females and 30% of males reported that they had dated interracially

While college students are open to dating interracially, fewer than 1% of marriages consist of a black and a white spouse.

Brittany Johnson

(Knox and Hall 2010). But as noted in the national data, only about 10% actually marry interracially. And black/white marriages, like that of Robert De Niro and Grace Hightower, are less than 1%.

Although a greater number of white people are available to black people for marriage, black mothers and white fathers have different roles in the respective black and white communities, in terms of setting the norms of interracial relationships. Hence, the black mother who approves of her son's or daughter's interracial relationship may be less likely to be overruled than the white mother (the white husband may be more disapproving and force his opinion).

National **Data**

Of the almost 60 million married couples in the United States, less than 1% (0.009%) consists of a black spouse and white spouse. Those consisting of a Hispanic and a non-Hispanic spouse represent 4% (*Statistical Abstract of the United States, 2011*, Table 60).

Hohmann-Marriott and Amato (2008) found that individuals (both men and women) in interethnic (includes interracial relationships such as Hispanic-white, black-Hispanic, and black-white) marriages and cohabitation relationships have lower quality relationships than those in same-ethnic relationships. Lower relationship quality was defined in terms of reporting less satisfaction, more problems, higher conflict, and lower commitment to the relationship.

Similarly, Bratter and King (2008) analyzed national data and found higher divorce rates among interracial couples (compared to same-race couples). They also found race and gender variation. Compared to white couples, white female/black male and white female/Asian male marriages were more prone to divorce; meanwhile, those involving nonwhite females and white males and Hispanics and non-Hispanic people had similar or lower risks of divorce.

Age Most individuals select someone who is relatively close in age. Men tend to select women three to five years younger than themselves. The result is the "**marriage squeeze**," which is the imbalance of the ratio of marriageable-aged

Marriage squeeze the imbalance of the ratio of marriageable-aged men to marriageable-aged women.

White/Hispanic relationships and marriages are the most frequent among interethnic relationships.

Nicole Mallard

Chapter 6 Selecting a Partner

men to marriageable-aged women. In effect, women have fewer partners to select from because men choose from not only their same age group but also those
younger than themselves. One 40-year-old recently divorced woman said, "What
chance do I have with all these guys looking at all these younger women?"

Education   Educational homogamy (selecting a cohabitant or marital partner
with similar education) also operates strongly in selecting a live-in and marital
partner (with greater homogamy for marital partners) (Kalmijn and Flap 2001).
Not only does college provide an opportunity to meet, date, live with, and marry
another college student, but it also increases one's chance that only a college-
educated partner becomes acceptable as a potential cohabitant or spouse. The
very pursuit of education becomes a value to be shared. However, Lewis and
Oppenheimer (2000) observed that, when people of similar education are not
available, women are particularly likely to marry someone with less education.
The older the woman, the more likely she is to marry a partner with less education. In effect, the number of educated, eligible males may decrease as she ages.

Open-Mindedness   People vary in the degree to which they are **open-minded**
(an openness to understanding alternative points of view, values, and behaviors).
Homogamous pairings in regard to open mindedness are those that reflect
partners who are relatively open or closed to new points of view, behaviors, and
experiences. For example, an evangelical may not be open to people of alternative religions.

Open-minded an openness to
understanding alternative points of
view, values, and behaviors.

Social Class   A. Brown (2010) interviewed undergraduate students to assess
their awareness of social class differences. Having a large number of designer
clothes and owning one's own new car were viewed as characteristics of middle
or upper class students in contrast to students in the working or lower class. Also,
the degree to which an item was sought because it was extravagant or functional
was an indicator of class. Upper class students have the most expensive "fancy"
computers, middle class students have computers that are functional—they
work, but are not the latest model. Having discretionary income to frequently
replace items such as cars, phones, or laptops was also seen as an indicator of
one's social class.

Social class reflects your parents' occupations, incomes, and educations as
well as your residence, language, and values. If you were brought up in a home
in which both parents were physicians, you probably lived in a large house in
a nice residential area—summer vacations and a college education were givens.
Alternatively, if your parents dropped out of high school and worked "blue collar"
jobs, your home would likely be smaller and in a less expensive part of town, and
your opportunities (such as for education) would be more limited. Social class
affects one's comfort in interacting with others. We tend to select as friends and
mates those with whom we feel comfortable. One's social class is also reflected in
various relationship behaviors. For example, youth from working class families
are more likely to engage in sexual intercourse at an earlier age than middle-class
youth (17 versus 19) and to marry earlier (Chaney and Marsh 2009).

The **mating gradient** refers to the tendency for husbands to be more advanced than their wives with regard to age, education, and occupational success. Indeed, husbands are typically older than their wives, have more advanced
education, and earn higher incomes (*Statistical Abstract of the United States, 2011,*
Table 702).

Mating gradient the tendency
for husbands to be more advanced
than their wives with regard to
age, education, and occupational
success.

Physical Appearance   Homogamy is operative in regard to physical appearance in that people tend to become involved with those who are similar in degree of physical attractiveness. However, a partner's attractiveness may be a more
important consideration for men than for women. In a study of homogamous
preferences in mate selection, men and women rated physical appearance an

*Strange how much you've
got to know before you
know how little you know.*

Anonymous

*A thing of beauty
is a joy for a while.*

Hal Lee Luyah

average of 7.7 and 6.8 (out of 10) in importance, respectively (Knox et al. 1997). Jaegar (2011) also found that the very attractive are more likely to be married by age 25 (except being tall for a woman was a liability).

Marital Status   Never-married people tend to select other never-married people as marriage partners, divorced people tend to select other divorced people, and widowed people tend to select other widowed people. Similar marital status may be more important to women than to men. In the study of homogamous preferences in mate selection, women and men rated similarity of marital status an average of 7.2 and 6.3 (out of 10) in importance, respectively (Knox et al. 1997).

Religion/Spirituality   Most adults in the United States tend to be affiliated with a religion. Only 11% of two national samples reported no religious affiliation (Amato et al. 2007).

Religion may be broadly defined as a specific fundamental set of beliefs (in reference to a Supreme Being, and so on) and practices generally agreed upon by a number of people or sects. Similarly, some individuals view themselves as "not religious" but "spiritual," with spirituality defined as belief in the spirit as the seat of the moral or religious nature that guides one's decisions and behavior. Ellison et al. (2010) reconfirmed those couples' in-home family devotional activities and shared religious beliefs were positively linked with reports of relationship quality. In addition, religiosity is positively associated with delaying first intercourse, having a good relationship with one's parents, and negative attitudes toward divorce/cohabitation—all factors predictive of marital quality and stability (MacArthur 2010). The Applying Social Research feature to follow emphasizes the degree to which a person who prays may prefer a date and mate who also prays.

Religious homogamy is operative in that people of similar religion or spiritual philosophy tend to seek out each other. Over a third (33.6%) of 2,922 undergraduates agreed that "It is important that I marry someone of my same religion." Forty seven percent reported that they had dated someone of another religion (Knox and Hall 2010).

To this couple, religion and spirituality were important factors in their deciding to get engaged.

Chelsea E. Curry

The phrase "the couple that prays together, stays together" is more than just a cliché. However, it should also be pointed out that religion could serve as a divisive force. For example, when one partner becomes "born again" or "saved," the relationship can be dramatically altered and eventually terminated unless the other partner shares the experience. A former rock-and-roller, hard-drinking, drug-taking wife noted that, when her husband "got saved," it was the end of their marriage. "He gave our money to the church as the 'tithe' when we couldn't even pay the light bill," she said. "And when he told me I could no longer wear pants or lipstick or have a beer, I left." Another example of how religious disharmony has a negative effect on relationships is the marriage of Ted Turner and Jane Fonda. Soon after she became a Christian and was "saved," she and Turner split up. In an interview, Fonda noted that she feared telling her husband about her religious conversion because she knew it would be the end of their eight-year marriage.

Attachment   Individuals who report similar levels of attachment to each other report high levels of relationship satisfaction. This is the conclusion of Luo and Klohnen (2005), who studied 291 newlyweds to assess the degree to which similarity affected marital quality. Indeed, similarity of attachment level was *the* variable most predictive of relationship quality.

Personality   Helen Fisher (2009a) analyzed data from over 28,000 heterosexual members of chemistry.com. She observed a tendency for certain personality types to select other personality types. For example, "explorers" (curious, creative, adventurous, sexual) were attracted to other "explorers," and "builders" (calm, loyal, traditional) were attracted to other "builders." These pairings are an obvious example of homogamy—like seeking like.

Circadian Preference   The term refers to an individual's preference for morningness-eveningness in regard to intellectual and physical activities. In effect, some prefer morning while others prefer late afternoon or evening hours. In a study of 84 couples, Randler and Kretz (2011) found that partners in a romantic relationship tended to have similar circadian preferences. The couples also tended to have similar preferences as to when they went to bed in the evening and when they rose in the morning.

Circadian preference refers to an individual's preference for morningness-eveningness in regard to intellectual and physical activities.

Traditional Roles   Abowitz et al. (2011) found that about a third of the 692 undergraduate women at a large southeastern university reported that they wanted to marry a "traditional" man—one who viewed his primary role as that of provider and who would be supportive of his wife staying home to rear the children. In a related study of 1027 undergraduate men, 31% reported that they wanted a traditional wife—one who viewed her primary role as wife and mother, staying home and rearing the children (not having a career) (Knox and Zusman 2007).

Geographic Background   Haandrikman (2011) studied Dutch cohabitants and found that the romantic partners tended to have grown up within six kilometers of each other (**spatial homogamy**). While the university context draws people from different regions of the country, the demographics of state universities tend to reflect a preponderance of those from within the same state.

Spatial homogamy romantic partners tend to select partners who grew up/are from similar geographic areas.

Economic Values, Money Management, and Debt   Individuals vary in the degree to which they have money, spend money, and save. Some have very limited resources and are careful about all spending. Others have significant resources and buy whatever they want. Some are deeply in debt whereas others have no debt. Of 2,922 undergraduates, 9% noted that they owed more than

Now I lay me down to sleep,

I pray the Lord my soul to keep.

And if I die before I wake,

I pray the Lord my soul to take.

This prayer is familiar to those adults who (as children) got on their knees, put their hands together, and said the words before leaping into bed. It was a ritual that symbolized to the parents that their child's religious socialization was on track. Indeed 71% of a national sample reported praying or engaging in a religious activity within the last month prior to getting in bed (2010 Sleep in America Poll). This study examined the degree to which the frequency of prayer was a meaningful factor in romantic relationships, including requiring that a potential date and potential mate pray.

Sample and "Prayer Profile"

Data for this study came from a sample of 286 undergraduate volunteers at a large southeastern university who completed a 77-item questionnaire. Demographic characteristics of the sample follow:

Prayer Profile—Over three fourths (75.5%) reported that they prayed—35.3% prayed between 2 and 6 times a week;

21.7% prayed every day. About a quarter (24.5%) reported that they never prayed. Almost 60% (59.5%) believed that people who pray are happier. A prerequisite for *dating* or *marrying* someone is that they must pray "some of the time"—45.8% and 61.4% respectively. Over half (52.1%) would end a future relationship if the partner was not supportive of their praying. The stereotype of the "godless, anti-religious college student" who does not pray, or expect their romantic partner to do so, was not true of these respondents.

Sex—Almost three fourths (73.7%) of the respondents were female; 26.3% were male.

Age—The mean age of the respondents was 20.2 with a range of 17 to 54.

Race—Racial background of the respondents was 80.8% white, 11.1% black, 2.1% Hispanic, 1.0% Asian, and 2.4% biracial. Those selecting "other" as their racial identification represented 2.4%.

Findings and Discussion

Analysis of the findings (multivariate, regression) revealed that prayer was a frequent behavior in the lives of over 75% of the respondents and that the more a

a thousand dollars on a credit card (Knox and Hall 2010). Some carry significant educational debt. The median debt for those with a bachelor's degree was $19,300 (Chu 2007). Money becomes an issue in mate selection in that different economic backgrounds, values, and spending patterns are predictable conflict issues. One undergraduate noted, "There is no way I would get involved with/ marry this person as they don't know how to handle money."

WHAT IF? | What if You Love Your Partner but the Partner Has a "Serious" Flaw?

A "serious" flaw is a behavior that you find unacceptable. Examples include "does drugs," "is abusive," "requires that we live no more than two hours from their parents," or "previously married with kids." Although Ann Landers's phrase that "everyone comes with a catch" is a fact of mate selection, some factors should, indeed, eliminate a partner from consideration. Only you can decide the degree to which the behavior warrants ending the relationship. Of course, beliefs that the "partner will change because of their love for me" or "will change after marriage" are nonsense. We recommend you treat your decision to continue the relationship as seriously as you consider the "serious flaw" of the partner you love.

respondent prayed the more insistent the respondent was that a current dating partner and potential spouse also must pray. Specifically, higher prayer frequencies of the respondent were associated with being more likely to require that a person the respondent dated not only would pray themselves but also pray with the respondent. In addition, frequent prayer on the part of a respondent translated into requiring that a future mate pray. Finally, respondents who prayed more often were more likely to have prayed to find a mate and for help in resolving problems with a partner. One explanation for the insistence that one's date and mate mirror the prayer values of the respondent is the principle of homogamy.

Abundant research has confirmed that individuals are attracted to persons with similar values and that such homogamous pairings have positive future relationship outcomes (Clarkwest 2007; Amato et al. 2007). Because religious views reflect, in large part, who the person is, they have an enormous impact on one's attraction to a partner, the level of emotional engagement with that partner, and the durability of the relationship and marital happiness (Swenson et al. 2005). The current study adds to what we know about the importance of religion (with prayer as an index of devoutness) in dating and mate selection.

References

Amato, P. R., A. Booth, D. R. Johnson, and S. F. Rogers. (2007). *Alone together: How marriage in America is changing.* Cambridge, Massachusetts: Harvard University Press.

Clarkwest, A. (2007). Spousal dissimilarity, race, and marital dissolution. *Journal of Marriage and the Family* 69: 639–653.

Swenson, D., J. G. Pankhurst, and S. K. Houseknecht. (2005). Links between families and religion. In *Sourcebook of family theory & research,* ed. V. L. Bengtson, A. C. Acock, K. R. Allen, P. Dilworth-Anderson, and D. M. Klein, 530–533. Thousand Oaks, California: Sage Publications.

2010 Sleep in America Poll. http://www.sleepfoundation.org/sites/default/files/nsaw/NSF%20Sleep%20in%20%20America%20Poll%20-%20Summary%20of%20Findings%20.pdf

*Abridged from poster by Sarah Tyson, David Knox, and Beth Easterling presented at Southern Sociological Society's annual meeting, Atlanta, April 2010.

Psychological Factors Operative in Mate Selection

> *No matter how carefully one chooses a mate, there will always be qualities that the mate has that simply don't fit well.*
>
> Neil Jacobson and Andrew Christensen, psychologists

Psychologists have focused on complementary needs, exchanges, parental characteristics, and personality types with regard to mate selection.

Complementary-Needs Theory

As campaign manager for Barack Obama, David Plouffe was able to observe "up close" the relationship between Barack and his wife, Michelle. He noted the differences in the ways they prepared for a major speech. Their styles, he found, were complementary.

> *Michelle wanted a draft of her speech more than a month out so she could massage it further, get comfortable with it, and practice the delivery. Barack was always crafting his at the eleventh hour. In this regard, Michelle was a concert pianist—disciplined, regimented, methodical—and Barack was a jazz musician, riffing, improvisational, and playing by ear. Both Obamas, it turned out, were clutch performers when the curtain rose* (Plouffe 2009, 301–302).

The marriage of Academy Award winner Sandra Bullock and Jesse James is another example of complementary needs or "opposites attract." Hollywood wondered what she saw in the tattooed owner of a biker shop who was formerly married to a porn star. While his affairs were the public reason for the divorce, the effect of their seeming array of differences is an alternative explanation.

Complementary-needs theory states that we tend to select mates whose needs are opposite and complementary to our own. Partners can also be drawn to each other on the basis of nurturance versus receptivity. These complementary needs suggest that one person likes to give and take care of another, whereas the other likes to be the benefactor of such care. Other examples of complementary needs may involve responsibility versus irresponsibility, peacemaker versus troublemaker, and disorder versus order. Helen Fisher (2009a), referred to previously, also found evidence of complementary needs in her study of 28,000 participants looking for a partner on chemistry.com. In terms of personality, "directors" (analytical, decisive, focused) sought "negotiators" (introspective, verbal, intuitive) and vice versa.

The idea that mate selection is based on complementary needs was suggested by Winch (1955), who noted that needs can be complementary if they are different (for example, dominant and submissive) or if the partners have the same need at different levels of intensity.

As an example of the latter, two individuals may have a complementary relationship if they both want to pursue graduate studies but want to earn different degrees. The partners will complement each other if, for instance, one is comfortable with aspiring to a master's degree and approves of the other's commitment to earning a doctorate.

Winch's theory of complementary needs, commonly referred to as "opposites attract," is based on the observation of twenty-five undergraduate married couples at Northwestern University. Other researchers who have not been able to replicate Winch's study have criticized the findings (Saint 1994). Two researchers said, "It would now appear that Winch's findings may have been an artifact of either his methodology or his sample of married people" (Meyer and Pepper 1977). Tonight Show host, Jay Leno, revealed in his autobiography that he and his wife of almost 30 years are very different:

> We were, and are, opposites in most every way. Which I love. There's a better balance…I don't consider myself to have much of a spiritual side, but Mavis has almost a sixth sense about people and situations. She has deep focus, and I fly off in 20 directions at once. She reads 15 books a week, mostly classic literature. I collect classic cars and motorcycle books. She loves European travel; I don't want to go anywhere people won't understand my jokes…(Leno 1996, 214–215).

Three questions can be raised about the theory of complementary needs:

1. *Couldn't personality needs be met just as easily outside the couple's relationship as through mate selection?* For example, couldn't a person who has the need to be dominant find such fulfillment in a job that involved an authoritative role, such as being a supervisor?

2. *What is a complementary need as opposed to a similar value?* For example, is the desire to achieve at different levels a complementary need or a shared value?

3. *Don't people change as they age?* Could dependent people grow and develop self-confidence so that they might no longer need to be involved with a dominant person? Indeed, such a person might no longer enjoy interacting with a dominant person.

Exchange Theory

Exchange theory in mate selection is focused on finding the partner who offers the greatest rewards at the lowest cost. The following five concepts help to explain the exchange process in mate selection:

1. *Rewards.* Rewards are the behaviors (your partner looking at you with the eyes of love), words (saying "I love you"), resources (being beautiful or handsome, having a car, condo, and money), and services (cooking for you, typing for you) your partner provides that you value and that influence you to continue the relationship. Increasingly, men are interested in women who offer "financial independence." In a study of Internet ads placed by women, the woman who described herself as "financially independent . . . successful and ambitious" produced 50%

more responses than the next most popular ad, in which the woman described herself as "lovely . . . very attractive and slim" (Strassberg and Holty 2003).

2. Costs. Costs are the unpleasant aspects of a relationship. A woman identified the costs associated with being involved with her partner: "He abuses drugs, doesn't have a job, and lives nine hours away." The costs her partner associated with being involved with this woman included "she nags me," "she doesn't like sex," and "she wants her mother to live with us if we marry." Ingoldsby et al. (2003) assessed the degree to which various characteristics were associated with reducing one's attractiveness on the marriage market. The most undesirable traits were not being heterosexual, having alcohol or drug problems, having a sexually transmitted disease, and being lazy.

3. Profit. Profit occurs when the rewards exceed the costs. Individuals in a relationship who have more profit than loss are likely to continue the relationship. Those who break up and seek a new partner are probably operating at a loss in their current relationship which may make them vulnerable to look for more profit in another relationship.

4. Loss. Loss occurs when the costs exceed the rewards.

5. Alternative. Is another person currently available who offers a higher profit margin?

Most people have definite ideas about what they are looking for in a mate. For example, Xie et al. (2003) found that men with good incomes were much more likely to marry than men with no or low incomes. The currency used in the marriage market consists of the socially valued characteristics of the people involved, such as age, physical characteristics, and economic status. In our free choice system of mate selection, we typically get as much in return for our social attributes as we have to offer or trade. Indeed, individuals have a perception of what they are worth on the marriage market and adjust their expections in terms of the partner they think they can attract (Bredow et al. 2011). An unattractive, drug-abusing high school dropout with no job has little to offer an attractive, drug-free, college student. A relationship between these two is not likely since there is too little profit for the college student.

Once you identify a person who offers you a good exchange for what you have to offer, other bargains are made about the conditions of your continued relationship. Waller and Hill (1951) observed that the person who has the least interest in continuing the relationship could control the relationship. This **principle of least interest** is illustrated by the woman who said, "He wants to date me more than I want to date him, so we end up going where I want to go and doing what I want to do." In this case, the woman trades her company for the man's acquiescence to her recreational choices. In effect, the person with the least interest controls the relationship.

Role Theory

Whereas the complementary-needs and exchange theories of mate selection are relatively recent, Freud suggested that the choice of a love object in adulthood represents a shift in libidinal energy from the first love objects, the parents. **Role theory of mate selection** emphasizes that a son or daughter models after the parent of the same sex by selecting a partner similar to the one the parent selected. This means that a man looks for a wife who has similar characteristics to those of his mother and that a woman looks for a husband who is very similar to her father.

Attachment Theory

Attachment theory of mate selection has been used by family scholars to emphasize the drive toward intimacy and a social and emotional connection (Sassler 2010). One's earliest experience as a child is to be emotionally bonded to one's parents (usually the mother) in the family context. The emotional need to connect remains and expresses itself in relationships with others, most notably the romantic

The best relationships are always win-win.

Jack Turner, psychologist

Principle of least interest the person who has the least interest in a relationship controls the relationship.

Role theory of mate selection theory which focuses on the social learning of roles. A son or daughter models after the parent of the same sex by selecting a partner similar to the one the parent selected.

Attachment theory of mate selection developed early in reference to one's parents, the drive toward an intimate, social/emotional connection.

love relationship. Children diagnosed with oppositional-defiant disorder (ODD) or post-traumatic stress disorder (PTSD) have had disruptions in their early bonding and frequently display attachment problems, possibly due to early abuse, neglect, or trauma. Children reared in Russian orphanages in the fifties where one caretaker was assigned to multiple children learned "no one cares about me" and "the world is not safe." Reversing early negative or absent bonding is problematic. However, Larson et al. (2010) studied relationship satisfaction for neurotic men and found that perceptions of their partners' lack of criticism as well as high levels of kindness and empathy were associated with such satisfaction.

Positive Assortative Personality Mating

Positive assortative personality mating refers to individuals who sort each other out on the basis of similar personality characteristics. Gonzaga et al. (2010) asked both partners of 417 eHarmony couples who married to identify the degree to which various personality trait terms reflected who they, as individuals, were. Examples of the terms included warm, clever, dominant, outgoing, quarrelsome, stable, energetic, affectionate, intelligent, witty, content, generous, wise, bossy, kind, calm, outspoken, shy, and trusting. Results revealed that those who were more similar to each other in personality characteristics were also more satisfied with their relationship.

Desirable Personality Characteristics of a Potential Mate

In a study of 700 undergraduates, both men and women reported that the personality characteristics of being warm, kind, and open and having a sense of humor were very important to them in selecting a romantic or sexual partner. Indeed, these intrinsic personality characteristics were rated as more important than physical attractiveness or wealth (extrinsic characteristics) (Sprecher and Regan 2002). Similarly, adolescents wanted intrinsic qualities such as intelligence and humor in a romantic partner but looked for physical appearance and high sex drive in a casual partner (no gender-related differences in responses were found) (Regan and Joshi 2003).

Undesirable Personality Characteristics of a Potential Mate

Researchers have identified several personality factors predictive of relationships which end in divorce or are unhappy (Foster 2008; Wilson and Cousins 2005). Potential partners who are observed to consistently display these characteristics might be avoided.

   1. *Controlling.* The behavior that 60% of a national sample of adult single women reported as the most serious fault of a man was his being "too controlling" (Edwards 2000).

   2. *Narcissistic.* Foster (2008) observed that individuals high on narcissism viewed relationships in terms of what they get out of them. When satisfactions wane and alternatives are present, narcissists are the first to go. Because all relationships have difficult times, a narcissist is a high risk for a durable marriage partner.

   3. *Poor impulse control.* Lack of impulse control is problematic in relationships because such individuals are less likely to consider the consequences of their actions. For example, to some people, having an affair might seem harmless but it will have devastating consequences for the partners and their relationship in most cases. For people with poor impulse control, if they feel like having an affair, they often will—and worry about the consequences later.

   4. *Hypersensitive.* Hypersensitivity to perceived criticism involves getting hurt easily. Any negative statement or criticism is received with a greater impact than a partner intended. The disadvantage of such hypersensitivity is that a partner may learn not to give feedback for fear of hurting the hypersensitive partner. Such lack of feedback to the hypersensitive partner blocks information about what the person does that upsets the other and what could be done to make things better. Hence, the hypersensitive one has no way of learning that something is

Marriage is a gamble. That's why we speak of winning a husband or wife.

Unknown

Positive assortative personality mating individuals who sort each other out on the basis of similar personality characteristics.

Neurotics worry about things that didn't happen in the past instead of worrying like normal people about things that won't happen in the future.

Unknown

Chapter 6 Selecting a Partner

wrong, and the partner has no way of alerting the hypersensitive partner. The result is a relationship in which the partners can't talk about what is wrong, so the potential for change is limited.

5. *Inflated ego.* An exaggerated sense of oneself is another way of saying a person has a big ego and always wants things to be his or her way. A person with an inflated sense of self may be less likely to consider the other person's opinion in negotiating a conflict and prefer to dictate an outcome. Such disrespect for the partner can be damaging to the relationship.

6. *Perfectionistic.* Individuals who are perfectionists may require perfection of themselves and others. This attitude is associated with relationship problems (Haring et al. 2003).

7. *Insecure.* Feelings of insecurity also compromise marital happiness. Researchers studied the personality trait of attachment and its effect on marriage in 157 couples at two-time intervals and found that "insecure participants reported more difficulties in their relationships. . . [I]n contrast, secure participants reported greater feelings of intimacy in the relationship at both assessments" (Crowell et al. 2002).

Brumbaugh and Fraley (2010) studied unattached insecure individuals interested in attracting a partner and found that they were able to present positive aspects of themselves and hide their insecurity. Specifically, they found that avoidant individuals were able to use touch to attract a partner when in reality they felt uncomfortable with psychological closeness. The researchers noted that the insecure will often use strategies to attract a partner which may give a false impression of who the person is.

8. *Controlled.* Individuals who are controlled by their parents, grandparents, former partner, child, or whomever, compromise the marriage relationship because their allegiance is external to the couple's relationship. Unless the person is able to break free of such control, the ability to make independent decisions will be thwarted, which will both frustrate the spouse and challenge the marriage.

9. *Substance abuser.* Blair (2010) studied a national sample of adolescent females and noted a negative relationship between substance abuse (alcohol and marijuana) and lower likelihood of marriage. The researcher reasoned that being labeled as a "drunk" or "stoner" as a female may alienate her from her peers and reduce her chance of being selected as a romantic partner.

In addition to undesirable personality characteristics, Table 6.1 reflects some particularly troublesome personality types and how they may impact you negatively.

A happy wife sometimes has the best husband, but more often, makes the best of the husband she has.

Unknown

TABLE 6.1 Personality Types Problematic in a Potential Partner

Type	Characteristics	Impact on Partner
Paranoid	Suspicious, distrustful, thin-skinned, defensive	Partners may be accused of everything.
Schizoid	Cold, aloof, solitary, reclusive	Partners may feel that they can never "connect" and that the person is not capable of returning love.
Borderline	Moody, unstable, volatile, unreliable, suicidal, impulsive	Partners will never know what their Jekyll-and-Hyde partner will be like, which could be dangerous.
Antisocial	Deceptive, untrustworthy, no conscience remorseless	Such a partner could cheat on, lie, or steal from a partner and not feel guilty.
Narcissistic	Egotistical, demanding, greedy, selfish	Such a person views partners only in terms of their value. Don't expect such a partner to see anything from your point of view; expect such a person to bail in tough times.
Dependent	Helpless, weak, clingy, insecure	Such a person will demand a partner's full time and attention, and other interests will incite jealousy.
Obsessive-compulsive	Rigid, inflexible	Such a person has rigid ideas about how a partner should think and behave and may try to impose them on the partner.

Who Is the Best Person for You to Marry?

Although no perfect mate exists, some individuals are more suited for you as a marriage partner than others. As we have seen in this chapter, people who have a big ego, poor impulse control, and an oversensitivity to criticism should be considered with great caution.

Also important is selecting someone with whom you have a great deal in common. "Marry someone just like you" may be a worthy guideline in selecting a marriage partner. Homogamous matings with regard to race, education, age, values, religion, social class, and marital status (for example, never-married people marry never-married people; divorced people with children marry those with similar experience) are more likely to result in more durable, satisfying relationships. "Marry your best friend" is another worthy guideline for selecting the person you marry.

Finally, marrying someone with whom you have a relationship of equality and respect is associated with marital happiness. Relationships in which one partner is exploited or intimidated engender negative feelings of resentment and distance. One man said, "I want a co-chair, not a committee member, for a mate." He was saying that he wanted a partner to whom he related as an equal. And, keep in mind that "important as it is to choose the right partner, it's probably more important to be the right partner" (Marano 2010).

Sociobiological Factors Operative in Mate Selection

In contrast to cultural, sociological, and psychological aspects of mate selection, which reflect a social learning assumption, the sociobiological perspective suggests that biological or genetic factors may be operative in mate selection.

Definition of Sociobiology

Based on Charles Darwin's theory of natural selection, which states that the strongest of the species survive, **sociobiology** holds that men and women select each other as mates on the basis of their innate concern for producing offspring who are most capable of surviving.

According to sociobiologists, men look for a young, healthy, attractive, sexually conservative woman who will produce healthy children and invest in taking care of the children. Women, in contrast, look for an ambitious man with stable economic resources who will invest his resources in her children. That men emphasize physical attractiveness and women emphasize social/economic status is true not only in the U.S. but other countries (e.g., China) (Li et al. 2011). Earlier in this chapter, we provided data supporting the idea that men seek attractive women and women seek ambitious, financially successful men.

Criticisms of the Sociobiological Perspective

The sociobiological explanation for mate selection is controversial. Critics argue that women may show concern for the earning capacity of men because women have been systematically denied access to similar economic resources, and selecting a mate with these resources is one of their remaining options. In addition, it is argued that both women and men, when selecting a mate, think about their partners more as companions than as future parents of their offspring.

Sociobiology theory which emphasizes the biological basis for all social behavior, including mate selection.

A new form of assessing compatibility is getting DNA tested. The respective saliva of these engaged individuals will be analyzed to ascertain if their immune systems are compatible.

Biological Factors—DNA Compatibility as a Factor in Mate Selection

Aside from sociological, psychological, and sociobiological considerations in mate selection, Genepartner.com offers genetic testing to assess the degree to which two individuals have compatible DNA. Various dating sites such as Sense2Love.com include genetic matching in their profiles. Genepartner.com tests for one group of genes called HLAs (human leukocyte antigens) which play a role in one's immune system. One theory is that persons with different HLAs are more likely to have offspring who will inherit a greater variety of potential immune responses and be more resistant to disease. Another theory is that those with too similar HLAs would be too genetically close and their offspring would be compromised. Empirical findings on the validity of DNA compatibility assertions are mixed.

Figure 6.1 summarizes the various cultural, sociological, and psychological filters involved in selecting a partner.

Engagement

Engagement moves the relationship of a couple from a private love-focused experience to a public, parent-involved experience. Family and friends are invited to enjoy the happiness and commitment of the individuals to a future marriage. Unlike casual dating, **engagement** is a time in which the romantic partners are sexually monogamous, committed to marry, and focused on wedding preparations. The engagement period is the last opportunity before marriage to systematically examine the relationship, ask each other specific questions, find out about the partner's parents and family background, and participate in marriage education or counseling.

Premarital Counseling

Some clergy require one or more sessions of premarital counseling as a prerequisite to agreeing to marry a couple (other couples may seek a premarital therapist through the American Association for Marriage and Family

An engagement is a short period lacking in foresight, followed by a long period loaded with hindsight.

Unknown

Engagement time in which the romantic partners are sexually monogamous, committed to marry, and focused on wedding preparations.

Figure 6.1 Cultural, sociological, and psychological filters involved in mate selection.

Cultural Filters
For two people to consider marriage to each other,

Endogamous factors   and   Exogamous factors
(same race, age)   ↓   (not blood-related)
                           must be met.
                                 ↓

After the cultural prerequisites have been satisfied, sociological and psychological filters become operative.

Sociological Filters

Spatial homogamy = the tendency to marry someone from the same geographic area.

Homogamy = the tendency to select a mate similar to oneself with regard to the following:

Race	Physical appearance
Education	Circadian preference compatibility
Social class	Religion
Age	Marital status
Intelligence	Interpersonal values

Psychological Filters

Complementary needs
Reward-cost ratio for profit
Parental characteristics
Desired personality characteristics

Therapy—go to http://www.therapistlocator.net/DisclaimerUS.asp). Premarital counseling is a process of discovery. Because partners might hesitate to reveal information that they feel will be met with disapproval during casual dating, the sessions provide the context to be specific about the other partner's thoughts, feelings, values, goals, and expectations. The Involved Couple's Inventory (see Self-Assessment feature) is designed to help individuals in committed relationships learn more about each other by asking specific questions.

In addition, couples may take the Right Mate Checkup (see web links at end of chapter) which provides a way for partners to assess their relationship in 12 different areas. Premarital counselors will often use the Involved Couple's Inventory (see Self-Assessment) and the Right Mate Checkup (or forms like these such as RELATE—see web links at end of chapter) to elicit potential problem issues to be discussed in sessions.

In addition to seeing a counselor or completing self-help tests, many individuals get advice from friends, parents, and religious leaders. Allgood and Gordon (2010) studied 56 couples who were given advice from 412 individuals. The most frequent advice givers were friends (32%) followed by parents (30%) and religious leaders (20%). Results revealed that the closer the individual was to the "advice giver," the more likely the person was to use the advice and to regard it as useful. Both men and women regarded their close friends (who happened to be of the same sex) as sources of advice they listened to and found helpful.

Assumptions are the termites of relationships.

Henry Winkler, actor

SELF-ASSESSMENT | Involved Couple's Inventory

The following questions are designed to increase your knowledge of how you and your partner think and feel about a variety of issues. Assume that you and your partner have considered getting married. Each partner should ask the other the following questions:

Partner Feelings and Issues

1. If you could change one thing about me, what would it be?
2. On a scale of 0 to 10, how well do you feel I respond to criticism or suggestions for improvement?
3. What would you like me to say or not say that would make you happier?
4. What do you think of yourself? Describe yourself with three adjectives.
5. What do you think of me? Describe me with three adjectives.
6. What do you like best about me?
7. On a scale of 0 to 10, how jealous do you think I am? How do you feel about my level of jealousy?
8. How do you feel about me?
9. To what degree do you feel we each need to develop and maintain outside relationships so as not to focus all of our interpersonal expectations on each other? Does this include other-sex individuals?
10. Do you have any history of abuse or violence, either as an abused child or adult or as the abuser in an adult relationship?
11. If we could not get along, would you be willing to see a marriage counselor? Would you see a sex therapist if we were having sexual problems?
12. What is your feeling about prenuptial agreements?
13. Suppose I insisted on your signing a prenuptial agreement?
14. To what degree do you enjoy getting and giving a massage?
15. How important is it to you that we massage each other regularly?
16. On a scale of 0 to 10, how emotionally close do you want us to be?
17. How many intense love relationships have you had, and to what degree are these individuals still a part of your life in terms of seeing them or sending text messages?
18. Have you lived with anyone before? Are you open to our living together? What would be your understanding of the meaning of our living together—would we be "finding out more about each other" or would we be "committed to marriage"?
19. What do you want for the future of our relationship? Do you want us to marry? When?
20. On a ten-point scale (0 = very unhappy and 10 = very happy), how happy are you in general? How happy are you about us?
21. How depressed have you been? What made you feel depressed?
22. What behaviors do I engage in upset you that you want me to stop?
23. What new behaviors do you want me to develop or begin to make you happier?

Feelings about Parents and Family

1. How do you feel about your mother? Your father? Your siblings?
2. On a ten-point scale, how close are you to your mom, dad, and each of your siblings?
3. How close are your family members to one another? On a ten-point scale, what value do you place on the opinions or values of your parents?
4. How often do you have contact with your father or mother? How often do you want to visit your parents and/or siblings? How often would you want them to visit us? Do you want to spend holidays alone or with your parents or mine?
5. What do you like and dislike most about each of your parents?
6. What do you like and dislike about my parents?
7. What is your feeling about living near our parents? How would you feel about my parents living with us? How do you feel about our parents living with us when they are old and cannot take care of themselves?
8. How do your parents get along? Rate their marriage on a scale of 0 to 10 (0 = unhappy, 10 = happy).
9. To what degree do your parents take vacations alone together? What are your expectations of our taking vacations alone or with others?
10. To what degree do members of your family consult one another on their decisions? To what degree do you expect me to consult you on the decisions that I make?
11. Who is the dominant person in your family? Who has more power? Who do you regard as the dominant partner in our relationship? How do you feel about this power distribution?
12. What "problems" has your family experienced? Is there any history of mental illness, alcoholism, drug abuse, suicide, or other such problems?
13. What did your mother and father do to earn an income? How were their role responsibilities divided in terms of having income, taking care of the children, and managing the household? To what degree do you want a job and role similar to that of the same-sex parent?

Social Issues, Religion, and Children

1. How do you feel about Obama as President? How do you feel about America being in Afghanistan?
2. What are your feelings about women's rights, racial equality, and homosexuality?
3. To what degree do you regard yourself as a religious or spiritual person? What do you think about religion, a Supreme Being, prayer, and life after death?
4. Do you go to religious services? Where? How often? Do you pray? How often? How important is prayer to you? How important is it to you that we pray together? What do you pray about? When we are married, how often would you want to go to religious services? In what religion would you want our children to be reared? What responsibility would you take to ensure that our children had the religious training you wanted them to have?

(continued)

5. How do you feel about abortion? Have you had an abortion?

6. How do you feel about children? How many do you want? When do you want the first child? At what intervals would you want to have additional children? What do you see as your responsibility in caring for the children changing diapers, feeding, bathing, playing with them, and taking them to lessons and activities? To what degree do you regard these responsibilities as mine?

7. Suppose I did not want to have children or couldn't have them. How would you feel? How do you feel about artificial insemination, surrogate motherhood, in vitro fertilization, and adoption?

8. To your knowledge, can you have children? Are there any genetic problems in your family history that would prevent us from having normal children? How healthy (physically) are you? What health problems do you have? What health problems have you had? What operations have you had? How often have you seen a physician in the last three years? What medications have you taken or do you currently take? What are these medications for? Have you seen a therapist, psychologist, or psychiatrist? What for?

9. How should children be disciplined? Do you want our children to go to public or private schools?

10. How often do you think we should go out alone without our children? If we had to decide between the two of us going on a cruise to the Bahamas alone or taking the children camping for a week, what would you choose?

11. What are your expectations of me regarding religious participation with you and our children?

Sex

1. How much sexual intimacy do you feel is appropriate in casual dating, involved dating, and engagement?

2. Does "having sex" mean having sexual intercourse? If a couple has experienced oral sex only, have they "had sex"?

3. What sexual behaviors do you most and least enjoy? How often do you want to have intercourse? How do you want me to turn you down when I don't want to have sex? How do you want me to approach you for sex? How do you feel about just being physical together hugging, rubbing, holding, but not having intercourse?

4. By what method of stimulation do you experience an orgasm most easily?

5. What do you think about masturbation, oral sex, homosexuality, sadism and masochism (S & M), and anal sex?

6. What type of contraception do you suggest? Why? If that method does not prove satisfactory, what method would you suggest next?

7. What are your values regarding extramarital sex? If I had an affair, would you want me to tell you? Why? If I told you about the affair, what would you do? Why?

8. How often do you view pornographic videos or pornography on the Internet? How do you feel about my viewing porno? How do you feel about our watching porn together? How do you feel about my sending you sex texts, photos, videos?

9. How important is our using a condom to you?

10. Do you want me to be tested for human immunodeficiency virus (HIV)? Are you willing to be tested?

11. What sexually transmitted infections (STIs) have you had?

12. How much do you want to know about my sexual behavior with previous partners?

13. How many "friends with benefits" relationships have you been in? What is your interest in our having such a relationship?

14. How much do you trust me in terms of my being faithful or monogamous with you?

15. How open do you want our relationship to be in terms of having emotional or sexual involvement with others, while keeping our relationship primary?

16. What things have you done that you are ashamed of?

17. What emotional, psychological, or physical health problems do you have? What issues do you struggle with?

18. What are your feelings about your sexual adequacy? What sexual problems do you or have you had?

19. Give me an example of your favorite sexual fantasy.

Careers and Money

1. What kind of job or career will you have? What are your feelings about working in the evening versus being home with the family? Where will your work require that we live? How much education do you want? How supportive will you be for my education (e.g. move for my Ph.D.)?

2. To what degree did your parents agree on how to deal with money? Who was in charge of spending, and who was in charge of saving? Did working, or earning the bigger portion of the income, connect to control over money?

3. What are your feelings about a joint versus a separate checking account? Which of us do you want to pay the bills? How much money do you think we will have left over each month? How much of this do you think we should save?

4. When we disagree over whether to buy something, how do you suggest we resolve our conflict?

5. What jobs or work experience have you had? If we end up having careers in different cities, how do you feel about being involved in a long distance marriage?

6. What is your preference for where we live? Do you want to live in an apartment or a house? What are your needs for a car, television, cable service, phone plan, entertainment devices, and so on? What are your feelings about us living in two separate places, the "living apart together" idea whereby we can have a better relationship if we give each other some space and have plenty of room?

7. How do you feel about my having a career? Do you expect me to earn an income? If so, how much annually? To what degree do you feel it is your responsibility to cook, clean, and take care of the children? How do you feel about putting young children or infants in day-care centers? When the children are sick and one of us has to stay home, who will that be?

(continued)

8. To what degree do you want me to account to you for the money I spend? How much money, if any, do you feel each of us should have to spend each week as we wish without first checking with the other partner? What percentage of income, if any, do you think we should give to charity each year?

9. What assets or debts will you bring into the marriage? How do you feel about debt? How rich do you want to be?

10. If you have been married before, how much child support or alimony do you get or pay each month? Tell me about your divorce.

11. In your will, what percentage of your assets, holdings, and retirement will you leave to me versus anybody else (siblings, children of a previous relationship, and so on)?

Recreation and Leisure

1. What is your idea of the kinds of parties or social gatherings you would like for us to go to together?

2. What is your preference in terms of us hanging out with others in a group versus being alone?

3. What is your favorite recreational interest? How much time do you spend enjoying this interest? How important is it for you that I share this recreational interest with you?

4. What do you like to watch on television? How often and for how long do you play video games?

5. What are the amount and frequency of your current use of alcohol and other drugs (for example, marijuana, cocaine, crack, speed)? What, if any, have been your previous alcohol and other drug behaviors and frequencies? What are your expectations of me regarding the use of alcohol and other drugs?

6. Where did you vacation with your parents? Where will you want us to go? How will we travel? How much money do you feel we should spend on vacations each year?

7. What pets do you own and what pets do you want to live with us? To what degree is it a requirement that we have one or more pets? To what degree can you adapt to my pets so that they live with us?

Relationships with Friends and Coworkers

1. How do you feel about my three closest same-sex friends?

2. How do you feel about my spending time with my friends or coworkers, such as one evening a week?

3. How do you feel about my spending time with friends of the opposite sex?

4. What do you regard as appropriate and inappropriate affectional behaviors with opposite-sex friends?

Remarriage Questions

1. How and why did your first marriage end? What are your feelings about your former spouse now? What are the feelings of your former spouse toward you? How much "trouble" do you feel your former spouse will want to cause us? What relationship do you want with your former spouse? May I read your divorce settlement agreement?

2. Do you want your children from a previous marriage to live with us? What are your emotional and financial expectations of me in regard to your children? What are your feelings about my children living with us? Do you want us to have additional children? How many? When?

3. When your children are with us, who will be responsible for their food preparation, care, discipline, and driving them to activities?

4. Suppose your children do not like me and vice versa. How will you handle this? Suppose they are against our getting married?

5. Suppose our respective children do not like one another. How will you handle this?

It would be unusual if you agreed with each other on all of your answers to the previous questions. You might view the differences as challenges and then find out the degree to which the differences are important for your relationship. You might need to explore ways of minimizing the negative impact of those differences on your relationship. It is not possible to have a relationship with someone in which there is total agreement. Disagreement is inevitable; the issue becomes how you and your partner manage the differences.

Note

This self-assessment is intended to be thought-provoking and fun. It is not intended to be used as a clinical or diagnostic instrument.

Visiting Your Partner's Parents

Seize the opportunity to discover the family environment in which your partner was reared and consider the implications for your subsequent marriage. When visiting your partner's parents, observe their standard of living and the way they interact and relate (for example, level of affection, verbal and nonverbal behavior, marital roles) with one another. How does their standard of living compare with that of your own family? How does the emotional closeness (or distance) of your partner's family compare with that of your family? Such comparisons are significant because both you and your partner will reflect your respective family of origins. "This is the way we did it in my family" is a phrase you will hear your partner say from time to time. If you want to know how your partner is likely to treat you in the future, observe the way your partner's parent of the same sex treats and interacts with his or her spouse. If you want to know what your partner may be like in the future, look at your partner's parent of the same sex. There is a tendency for a man to become like his father and a woman to become like her mother. Your partner's parent of the same sex and their marital relationship is the model of a spouse and a marriage relationship your partner is likely to duplicate in the way the partner relates to you.

Premarital Education Programs

Various premarital education programs (also known as premarital prevention programs, premarital counseling, premarital therapy, and marriage preparation), both academic and religious, are formal systematized experiences designed to provide information to individuals and to couples about how to have a good relationship. About 30% of couples getting married become involved in some type of premarital education. The greatest predictor of whether a couple will become involved in a premarital education program is the desire and commitment of the female to do so. In effect, she ensures that she and her partner become involved in such a program. A second important predictor is the presence of problems in the relationship of the couple about to marry (Duncan et al. 2007).

Premarital education programs are valuable in that they not only help partners assess the degree to which they are compatible with each other but they may also provide a context to discuss some specific relationship issues. Devall et al. (2009) gave a pre- and post-test to 325 individuals who attended 12 two hour sessions provided by Family Wellness instructors covering communication, conflict resolution, parenting, and family functioning. The researchers found significant increases in relationship satisfaction and the ability to resolve conflict after sessions were held. Parents were also less likely to endorse corporal punishment.

Futris et al. (2010) assessed the value of PREPARE, a program designed for engaged couples, and found that those who participated reported gains in relationship knowledge, feeling confident about their relationship, engaging in positive conflict management behavior, and feeling more satisfied in their relationship.

Prenuptial Agreement

2008 presidential candidate John McCain has a prenuptial agreement with his wife Cindy. She is heiress to a beer fortune estimated to be $100 million. Britney Spears and Kevin Federline had a prenuptial agreement whereby he was awarded only $300,000 of over $100 million in assets. Paul McCartney and Heather Mills did not have a prenuptial agreement. She was awarded almost $50 million. Some couples, particularly those with considerable assets or those in subsequent marriages, might consider discussing and signing a prenuptial agreement. To reduce the chance that the agreement will later be challenged, each partner should hire an attorney (months before the wedding) to develop and/or review the agreement.

The primary purpose of a **prenuptial agreement** (also referred to as a premarital agreement, marriage contract, or ante nuptial contract) is to specify how property will be divided if the marriage ends in divorce or when it ends

An intelligent wife sees through her husband; an understanding wife sees him through.

Unknown

One of the best things about any prenuptial contract is that it's a reality check.

Richard Dombrow, family law attorney

Prenuptial agreement a contract between intended spouses specifying which assets will belong to whom and who will be responsible for paying what in the event of a divorce or when the marriage ends by the death of one spouse.

by the death of one partner. In effect, the value of what you take into the marriage is the amount you are allowed to take out of the marriage. For example, if you bring $250,000 into the marriage and buy the marital home with this amount, your ex-spouse is not automatically entitled to half the house at divorce. Some agreements may also contain clauses of no spousal support (alimony) if the marriage ends in divorce (but some states prohibit waiving alimony). See Appendix C for an example of a prenuptial agreement developed by a husband and wife who had both been married before and had assets and children.

Reasons for a prenuptial agreement include the following.

1. *Protecting assets for children from a prior relationship.* People who are in their middle or later years, who have considerable assets, who have been married before, and who have children are often concerned that money and property be kept separate in a second marriage so that the assets at divorce or death go to the children. Some children encourage their remarrying parent to draw up a prenuptial agreement with the new partner so that their (the offspring's) inheritance, house, or whatever will not automatically go to the new spouse upon the death of their parent.

2. *Protecting business associates.* A spouse's business associate may want a member of a firm or partnership to draw up a prenuptial agreement with a soon-to-be-spouse to protect the firm from intrusion by the spouse if the marriage does not work out.

Prenuptial agreements are not very romantic ("I love you, but sign here and see what you get if you don't please me.") and may serve as a self-fulfilling prophecy ("We were already thinking about divorce."). Indeed, 22% of 2,922 undergraduates agreed "I would not marry someone who required me to sign a prenuptial agreement" and 22% feel that couples who have a prenuptial agreement are more likely to get divorced (Knox and Hall 2010). Prenuptial contracts are almost nonexistent in first marriages and are still rare in second marriages. Whether or not signing a prenuptial agreement is a good idea depends on the circumstances. Some individuals who do sign an agreement later regret it. Sherry, then a never-married 22-year-old, signed such an agreement:

> *Paul was adamant about my signing the premarriage agreement. He said he loved me but would never consider marrying me unless I signed a prenuptial agreement stating that he would never be responsible for alimony in case of a divorce. I was so much in love, it didn't seem to matter. I didn't realize that basically he was and is a selfish person. Now, five years later after our divorce, I live in a mobile home and he lives in a big house overlooking the lake with his new wife.*

The husband viewed it differently. He was glad that she had signed the agreement and that his economic liability to his former wife was limited. He could afford the new house by the lake with his new wife because he was not sending money to Sherry. Billionaire Donald Trump attributed his economic survival of his two divorces to prenuptial agreements with his ex-wives.

Couples who decide to develop a prenuptial agreement need separate attorneys to look out for their respective interests. The laws regulating marriage and divorce vary by state, and only attorneys in those states can help ensure that the document drawn up will be honored. Individuals may not waive child support or dictate child custody. Full disclosure of assets is also important. If one partner hides assets, the prenuptial can be thrown out of court. One husband recommended that the issue of the premarital agreement should be brought up and that it be signed a minimum of six months before the wedding. "This gives the issue time to settle rather than being an explosive emotional issue if it is brought up a few weeks before the wedding." Indeed, as noted previously, if a prenuptial agreement is signed within two weeks of the wedding, that is grounds enough

Should marriage licenses be obtained so easily? Should couples be required, or at least encouraged, to participate in premarital education before saying "I do"? Given the high rate of divorce today, policy makers and family scholars are considering this issue. Data confirm the value of marriage education exposure (Devall et al. 2009). Some believe that "mandatory counseling will promote marital stability" (Licata 2002, 518).

Several states have proposed legislation requiring premarital education. For example, an Oklahoma statute provides that parties who complete a premarital education program pay a reduced fee for their marriage license. Also, in Lenawee County, Michigan, local civil servants and clergy have made a pact: they will not marry a couple unless that couple has attended marriage education classes. Other states that are considering policies to require or encourage premarital education include Arizona, Illinois, Iowa, Maryland, Minnesota, Mississippi, Missouri, Oregon, and Washington.

Proposed policies include not only mandating premarital education and lowering marriage license fees for those who attend courses but also imposing delays on issuing marriage licenses for those who refuse premarital education. However, "no state mandates premarital counseling as a prerequisite to obtaining a license" (Licata 2002, 525).

Traditionally, most Protestant pastors and Catholic priests require premarital counseling before they will perform marriage ceremonies. Couples who do not want to participate in premarital education can simply get married in secular ceremonies (justices of the peace).

Dating you is like dating a Stairmaster.

Erica Albright to Mark Zuckerberg, *The Social Network* movie

for the agreement to be thrown out of court because it is assumed that the document was executed under pressure.

Although individuals are deciding whether to have a prenuptial agreement, states are deciding whether to increase marriage license requirements, this chapter's social policy issue.

Factors which Suggest You Might Delay or Call off the Wedding

"No matter how far you have gone on the wrong road, turn back" is a Turkish proverb. Behavioral psychologist B. F. Skinner noted that one should not defend a course of action that does not feel right, but stop and reverse directions. If your engagement is characterized by the following factors, consider delaying your wedding at least until the most distressing issues have been resolved. Alternatively, break the engagement (which happens in 30% of formal engagements), which will have fewer negative consequences and is less stigmatized than ending a marriage.

Age 18 or Younger

The strongest predictor of getting divorced is getting married during the teen years. Individuals who marry at age 18 or younger have three times the risk of divorce than those who delay marriage into their late twenties or early thirties. Teenagers may be more at risk for marrying to escape an unhappy home and may be more likely to engage in impulsive decision making and behavior. Early marriage is also associated with an end to one's education, social isolation from peer networks, early pregnancy or parenting, and locking one's self into a low income. Increasingly, individuals are delaying when they marry. The median age at first marriage in the U.S. is 28 for men and 26 for women (U.S. Census 2010).

Advocates of mandatory premarital education emphasize that such courses reduce marital discord. However, questions remain about who will offer what courses and whether couples will take the content of such courses seriously. Indeed, people contemplating marriage are often narcotized with love and would doubtless not take any such instruction seriously. Love myths such as "divorce is something that happens to other people" and "our love will overcome any obstacles" work against the serious consideration of such courses.

Your Opinion?

1. To what degree do you believe premarital education should be required before the state issues a marriage license?

2. How effective do you feel such programs are for people in a hurry to marry?

3. How receptive do you feel individuals in love are to marriage education?

Sources

Devall, E., M. Montanez, and D. Vanleeuwen. 2009. Effectiveness of relationship education with single, cohabiting, and married parents. Poster, National Council on Family Relations, November.

Licata, N. 2002. Should premarital counseling be mandatory as a requisite to obtaining a marriage license? *Family Court Review* 40: 518–532.

Research by Meehan and Negy (2003) on being married while in college revealed higher marital distress among spouses who were also students. In addition, when married college students were compared with single college students, the married students reported more difficulty adjusting to the demands of higher education. The researchers concluded that "these findings suggest that individuals opting to attend college while being married are at risk for compromising their marital happiness and may be jeopardizing their education" (p. 688). Hence, waiting until one is older and through college not only may result in a less stressful marriage but may also be associated with less academic stress. Prioritizing one's education and getting a degree also has long-term positive economic consequences. We noted in Chapter 2 that each new level of education is associated with higher income.

She bid me take love easy as the leaves grow on the tree;

But I, being young and foolish, with her would not agree.

W. B. Yeats, English poet

Known Partner Less Than Two Years

Of 2,922 undergraduates, 27% agreed, "If I were really in love, I would marry someone I had known for only a short time" (Knox and Hall 2010). Impulsive marriages in which the partners have known each other for less than a month are associated with a higher-than-average divorce rate. Indeed, partners who date each other for at least two years (twenty-five months to be exact) before getting married report the highest level of marital satisfaction and are less likely to divorce (Huston et al. 2001). Nevertheless, Dr. Carl Ridley (2009) observed, "I am not sure it is about time but more about attending to self and other needs, desires, preferences and then assessing if one's potential partner can meet these needs or be willing to negotiate so that mutual needs are met. For some, this takes lots of time and for others, not very long."

A short courtship does not allow partners to learn about each other's background, values, and goals and does not permit time to observe and scrutinize each other's behavior in a variety of settings (for example, with one's close friends, parents, or siblings). Indeed, some individuals may be more prone to fall in love at first sight and to want to hurry the partner into a committed love

Marry'd in haste, we may repent at leisure.

William Congreve, playwright

Being involved for a couple of years before getting married predicts well for a couple's future marriage. Getting married a short time after meeting is putting one's future marriage at risk.

Chelsea E. Curry

relationship *before* the partner can find out about who they really are ("Let the buyer beware!"). If your partner is pressuring you to marry after dating for a short time and your senses tell you that this is too fast, tell you partner you need to slow the relationship down and don't want to get married now.

To increase the knowledge you and your partner have about each other, find out the answers from each other identified in the Involved Couple's Inventory, take a five-day "primitive" camping trip, take a fifteen-mile hike together, wallpaper a small room together, or spend several days together when one partner is sick. If the couple plans to have children, they may want to take care of a 6-month-old together for a weekend. Time should also be spent with each other's friends.

Abusive Relationship

As we will discuss in Chapter 13: Violence and Abuse in Relationships, partners who emotionally and/or physically abuse their partners while dating and living together continue these behaviors in marriage. Abusive lovers become abusive spouses, with predictable negative outcomes. Though extricating oneself from an abusive relationship is difficult before the wedding, it becomes even more difficult after marriage and children.

Before you run in double harness, look well to the other horse.

Ovid, poet

One characteristic of an abusive partner is their attempt to systematically detach their intended spouse from all other relationships ("I don't want you spending time with your family and friends; you should be here with me."). This is a serious flag of impending relationship doom, should not be overlooked, and one should seek the exit ramp as soon as possible.

High Frequency of Negative Comments/Low Frequency of Positive Comments

Markman et al. (2010) studied couples across the first five years of marriage and found that more negative and less positive communication before marriage tended to be associated with subsequent divorce. In addition, they emphasized that "negatives tend to erode positives over time." Individuals who criticize each other end up damaging their relationship in a way which does not make it easy for their relationship to recover.

178   **Chapter 6** Selecting a Partner

Numerous Significant Differences

Relentless conflict often arises from numerous significant differences. Though all spouses are different from each other in some ways, those who have numerous differences in key areas such as race, religion, social class, education, values, and goals are less likely to report being happy or to have durable relationships. Amato et al. (2007) found that the less couples had in common, the more their marital distress. People who report the greatest degree of satisfaction in durable relationships have a great deal in common (Wilson and Cousins 2005).

On-and-Off Relationship

A roller-coaster premarital relationship is predictive of a marital relationship that will follow the same pattern. Partners who break up and get back together several times have developed a pattern in which the dissatisfactions in the relationship become so frustrating that separation becomes the antidote for relief. In courtship, separations are of less social significance than marital separations. "Breaking up" in courtship is called "divorce" in marriage. Couples who routinely break up and get back together should examine the issues that continue to recur in their relationship and attempt to resolve them.

Dramatic Parental Disapproval

Parents usually have an opinion of their son or daughters' mate choice. Dubbs and Buunk (2010) found that daughters view their mother as more disapproving of their mate choice if the mother sees him as not having the potential to be a good father. In contrast, fathers are viewed as being more disapproving if the mate choice has traits indicating low social status (e.g. will provide lower income/style of life for the daughter). Perilloux et al. (2011) compared student and parent rating of mate selection traits and found that parents ranked religion higher than offspring, whereas offspring ranked physical attractiveness higher than parents. As noted above, parents preferred high earning capacity and being a college graduate more in daughters' mates than sons' mates.

A parent recalled, "I knew when I met the guy it wouldn't work out. I told my daughter and pleaded that she not marry him. She did, and they divorced." Although parental and in-law dissatisfaction is rare (Amato et al. [2007] found that only 13% of parents disapprove of the partner their son or daughter plans to marry), such parental predictions (whether positive or negative) often come true. If the predictions are negative, they sometimes contribute to stress and conflict once the couple marries.

Even though parents who reject the commitment choice of their offspring are often regarded as uninformed and unfair, their opinions should not be taken lightly. The parents' own experience in marriage and their intimate knowledge of their offspring combine to put them in a unique position to assess how their child might get along with a particular mate. If the parents of either partner disapprove of the marital choice, the partners should try to evaluate these concerns objectively. The insights might prove valuable. The value of parental approval is illustrated in a study of Chinese marriages. Pimentel (2000) found that higher marital quality was associated with parents' approving of the mate choice of their offspring.

Oh, what a tangled web we weave when first we practice to conceive.

Don Herold, humorist

Low Sexual Satisfaction

Sexual satisfaction is linked to relationship satisfaction, love, and commitment. Sprecher (2002) followed 101 dating couples across time and found that low sexual satisfaction (for both women and men) was related to reporting low relationship quality, less love, lower commitment, and breaking up. Hence, couples who are dissatisfied with their sexual relationship might explore ways of improving it (alone or through counseling) or consider the impact of such dissatisfaction on the future of their relationship.

Marrying for the Wrong Reason

Some reasons for getting married are more questionable than others. These reasons include the following.

1. *Rebound*. A rebound marriage results when you marry someone immediately after another person has ended a relationship with you. It is a frantic attempt on your part to reestablish your desirability in your own eyes and in the eyes of the partner who dropped you. One man said, "After she told me she wouldn't marry me, I became desperate. I called up an old girlfriend to see if I could get the relationship going again. We were married within a month. I know it was foolish, but I was very hurt and couldn't stop myself." To marry on the rebound is questionable because the marriage is made in reference to the previous partner and not to the partner being married. In reality, you are using the person you intend to marry to establish yourself as the "winner" in the previous relationship.

Analysis of survey data from 1002 undergraduates (Knox and Zusman 2009) at a large southeastern university revealed differences between the 535 or 53.4% who had become involved (while on the rebound from a previous love relationship) in a new relationship compared to 316 or 31.5% who had not become involved in a new relationship while on the rebound. A profile of the rebounder emerged as being impatient for a new love, deceptive/unfaithful, hedonistic, and a person who does not practice safe sex. Caution about becoming involved with someone on the rebound may be warranted. One answer to the question, "How fast should you run?" may be "as fast as you can."

To avoid the negative consequences of marrying on the rebound, wait until the negative memories of your past relationship have been replaced by positive aspects of your current relationship. In other words, marry when the satisfactions of being with your current partner outweigh any feelings of revenge. This normally takes between 12 and 18 months. In addition, be careful about getting involved in a relationship with someone who is recently divorced or just emerged from a painful ending of a previous relationship.

2. *Escape*. A person might marry to escape an unhappy home situation in which the parents are oppressive, overbearing, conflictual, alcoholic and/or abusive. One woman said, "I couldn't wait to get away from home. Ever since my parents divorced, my mother has been drinking and watching me like a hawk. 'Be home early, don't drink, and watch out for those horrible men,' she would always say. I admit it. I married the first guy that would have me. Marriage was my ticket out of there." Marriage for escape is a bad idea. It is far better to continue the relationship with the partner until mutual love and respect become the dominant forces propelling you toward marriage, rather than the desire to escape an unhappy situation. In this way you can evaluate the marital relationship in terms of its own potential and not solely as an alternative to an unhappy situation.

3. *Unplanned pregnancy*. Getting married just because a partner becomes pregnant is usually a bad idea. Indeed, the decision of whether to marry should be kept separate from the fact that there is now a pregnancy. Adoption, abortion, single parenthood, and unmarried parenthood (the couple can remain together as an unmarried couple and have the baby) are all alternatives to simply deciding to marry if a partner becomes pregnant. Avoiding feelings of being trapped or "You married me just because of the pregnancy" is one of the reasons for not rushing into marriage because of pregnancy. Couples who marry when the woman becomes pregnant have an increased chance of divorce.

4. *Psychological blackmail*. Some individuals get married because their partner takes the position that "I can't live without you" or "I will commit suicide if you leave me." Because the person fears that the partner may commit suicide, he or she agrees to the wedding. The problem with such a marriage is that one partner has learned to manipulate the relationship to get what he or she wants.

Use of such power often creates resentment in the other partner, who feels trapped in the marriage. Escaping from the marriage becomes even more difficult. One way of coping with a psychological blackmail situation is to encourage the person to go with you to a therapist to "discuss the relationship." Once inside the therapy room, you can tell the counselor that you feel pressured to get married because of the suicide threat. Counselors are trained to respond to such a situation.

5. Insurance Benefits. In a poll conducted by the Kaiser Family Foundation, a health policy research group, 7% of adults said someone in their household had married in the past year to gain access to insurance. In effect, marital decisions are being made to gain access to health benefits. "For today's couples, 'in sickness and in health' may seem less a lover's troth than an actuarial contract. They marry for better or worse, for richer or poorer, for co-pays and deductibles" (Sack 2008). While selecting a partner who has resources (which may include health insurance) is not unusual, to select a partner solely because he or she has health benefits is yet another matter. Both parties might be cautious if the alliance is more about "benefits" than the relationship.

6. Pity. Some partners marry because they feel guilty about terminating a relationship with someone whom they pity. The fiancé of one woman got drunk one Halloween evening and began to light fireworks on the roof of his fraternity house. As he was running away from a Roman candle he had just ignited, he tripped and fell off the roof. He landed on his head and was in a coma for three weeks. A year after the accident his speech and muscle coordination were still adversely affected. The woman said she did not love him anymore but felt guilty about terminating the relationship now that he had become physically afflicted.

She was ambivalent. She felt it was her duty to marry her fiancé, but her feelings were no longer love feelings. Pity may also have a social basis. For example, a partner may fail to achieve a lifetime career goal (for example, he or she may flunk out of medical school). Regardless of the reason, if one partner loses a limb, becomes brain-damaged, or fails in the pursuit of a major goal, it is important to keep the issue of pity separate from the advisability of the marriage. The decision to marry should be based on factors other than pity for the partner.

7. Filling a void. A former student in the authors' classes noted that her father died of cancer. She acknowledged that his death created a vacuum, which she felt driven to fill immediately by getting married so that she would have a man in her life. Because she was focused on filling the void, she had paid little attention to the personality characteristics of or her relationship with the man who had asked to marry her.

She reported that she discovered on her wedding night that her new husband had several other girlfriends whom he had no intention of giving up. The marriage was annulled.

In deciding whether to continue or terminate a relationship, listen to what your senses tell you ("Does it feel right?"), listen to your heart ("Do you love this person or do you question whether you love this person?"), and evaluate your similarities ("Are we similar in terms of core values, goals, view of life?"). Also, be realistic. It would be unusual if none of the factors listed applied to you. Indeed, most people have some negative and some positive indicators before they marry. Table 6.2 summarizes some of the wrong reasons for getting married.

Whether to Continue or End a Relationship

Rhoades et al. (2010) identified four factors involved in whether a person continues or ends a relationship.

1. **Dedication**—motivation to build/maintain a high quality relationship, have a long-term relationship.

2. **Perceived constraints**—internal feelings over the possibility of ending a relationship. For example: social pressure to stay together from parents/friends/

TABLE 6.2	Wrong Reasons for Getting Married	
Type	**Characteristics**	**Potential Consequences**
Rebound	Still recovering from a recently ended relationship	Transfer negative effect to new partner
Escape	Leaving a negative context	May be focused on old context rather than new relationship
Pregnancy	Speeds up relationship	May feel marriage is not by choice
Blackmail	Marriage by force	Anger over being forced to marry
Insurance	Goal is to obtain benefits	Relationship takes back seat to money
Pity	Marriage out of obligation	Partners don't feel they have choice
Rebellion	Go against parents	Focus not on new partner

church group, intangible investments (e.g. years together) lost if relationship ends, belief that one's quality of life would deteriorate should the relationship end, concern for the welfare of one's partner, fear that taking the steps (e.g., talking to partner) to end the relationship would be difficult, or fear of finding a suitable replacement if the relationship ends.

3. Material constraints—observable changes which would occur if the relationship ends. For example: dealing with shared debt, apartment, furniture, pets, etc. Having to move/set up new residence.

4. Feeling Trapped—individuals rate the degree to which they feel trapped by the investments in the relationship. The more trapped, the less likely they are to stay in the relationship.

The researchers analyzed data on 1184 unmarried adults in a romantic relationship and found that each of the four factors predicted the relationship continuing or ending. Specifically, being less dedicated, having fewer constraints, and feeling trapped predicted the break-up.

Basically it is time to end a relationship when the gain or advantages of staying together no longer outweigh the pain and disadvantages of leaving—the pain of leaving is less than the pain of staying. Of course, all relationships go through periods of time when the disadvantages outweigh the benefits, so one should not bail out without careful consideration.

The Future of Selecting a Partner

The future of selecting a lifetime partner will involve the increased use of Internet dating sites. Use of technology permeates the lives of today's youth. Just as cell phone calls and text messaging are commonplace, the use of Match.com and other such sites will become normative. While Internet dating sites will not replace traditional methods of meeting (at parties, in class, through friends, hanging out, etc.), Internet use will lose its stigma and will be an addendum to one's partnering. Hence, in addition to meeting the "old-fashioned way," individuals will be logging on and fishing in the online pool.

DIVERSITY IN OTHER COUNTRIES

Matchmakers and astrologers were the precursors to Match.com and eHarmony. Prior to 1950, Chinese parents arranged for the marriages of their children with the help of a professional matchmaker. The matchmaker was "usually an elderly woman who knew the birthday, temperament, and appearance of every unmarried man and woman in her community" (Xia and Zhou 2003, 231). This woman would visit parents with female children who were ready for marriage and propose specific individuals, usually of similar social and economic status. If the parents liked a man the matchmaker proposed for their daughter, the matchmaker would meet with the man's parents and alert them of the family's interest in their son marrying their daughter. If the parents agreed, a Chinese astrologer would be consulted to see if their signs of the zodiac were compatible. If the signs were off, the marriage would be, too, and the families would have no further contact.

Summary

What are the cultural factors that affect your selection of a mate?

Two types of cultural influences in mate selection are endogamy (to marry someone inside one's own social group such as race, religion, social class) and exogamy (to marry someone outside one's own family).

What are the sociological factors that influence mate selection?

Sociological aspects of mate selection involve homogamy—"like attracts like"—or the tendency to be attracted to people similar to oneself. Variables include race, age, religion, education, social class, personal appearance, attachment, personality, and open-mindedness. Undergraduates who pray also have a preference to date and marry partners who pray. Couples who have a lot in common are more likely to have a happy and durable relationship.

What are the psychological factors operative in mate selection?

Psychological aspects of mate selection include complementary needs, exchange theory, and parental characteristics. Complementary-needs theory suggests that people select others who have characteristics opposite to their own. For example, a highly disciplined, well-organized individual might select a free-and-easy, worry-about-nothing mate. Most researchers find little evidence for complementary-needs theory.

Exchange theory suggests that one individual selects another on the basis of rewards and costs. As long as an individual derives more profit from a relationship with one partner than with another, the relationship will continue. Exchange concepts influence who dates whom, the conditions of the dating relationship, and the decision to marry. Parental characteristics theory suggests that individuals select a partner similar to the opposite-sex parent.

Desirable personality characteristics of a potential mate desired by both men and women include being warm, kind, and open and having a sense of humor.

Undesirable personality characteristics of a potential mate include being too controlling, narcissistic, having poor impulse control, being hypersensitive to criticism, having an inflated ego, etc. Paranoid, schizoid, and borderline personalities also require one to be cautious.

What are the sociobiological factors operative in mate selection?

The sociobiological view of mate selection suggests that men and women select each other on the basis of their biological capacity to produce and support healthy offspring. Men seek young women with healthy bodies, and women seek ambitious men who will provide economic support for their offspring. There is considerable controversy about the validity of this theory. Critics argue that women may show concern for the earning capacity of men because women have been systematically denied access to similar economic resources, and selecting a mate with these resources is one of their remaining options. Meanwhile, one's DNA may be obtained and compatibility assessed.

What factors should be considered when becoming engaged?

The engagement period is the time to ask specific questions about the partner's values, goals, and marital agenda, to visit each other's parents to assess parental models, and to consider involvement in premarital educational programs and/or counseling.

Some couples (particularly those with children from previous marriages) decide to write a prenuptial agreement to specify who gets what and the extent of spousal support in the event of a divorce. To be valid, the document should be developed by an attorney in accordance with the laws of the state in which the partners reside. Last-minute prenuptial agreements put enormous emotional strain on the couple and are often considered invalid by the courts. Discussing a prenuptial agreement six months in advance is recommended.

What factors suggest you might consider calling off the wedding?

Factors suggesting that a couple may not be ready for marriage include being in their teens, having known each other less than two years, and having a relationship characterized by significant differences and/or dramatic parental disapproval. Some research suggests that partners with the greatest number of similarities in values, goals, and common interests are most likely to have happy and durable marriages. Negative reasons for getting married include being on the rebound, escaping from an unhappy home life, psychological blackmail, and pity.

What is the future of selecting a partner?

The future of mate selection will involve the increased use of dating sites to find a marriage partner. Use of technology permeates the lives of today's youth, so use of Match.com and other such sites will become normative and lose their stigma.

Key Terms

Attachment theory of mate selection
Circadian preference
Complementary-needs theory
Endogamy
Engagement
Exchange theory

Exogamy
Homogamy
Marriage squeeze
Mating gradient
Open-minded
Pool of eligibles

Positive assortative personality mating
Prenuptial agreement
Principle of least interest
Role theory of mate selection
Spatial homogamy
Sociobiology

Web Links

PAIR Project
    www.utexas.edu/research/pair
RELATE Institute
    https://www.relate-institute.org/

Right Mate Checkup
    http://www.heartchoice.com/hc/rightmate/rmsurvey.php

> *I have great hopes that we shall love each other all our lives as much as if we had never married at all.*
>
> Lord Byron, poet

David Knox

Learning Objectives

Discuss the individual motivations for and societal functions of marriage.

Review how marriage is a commitment to one's partner, family, and state.

Identify how weddings and honeymoons are rites of passage.

Summarize the various changes from being a lover to being a spouse.

Review Hispanic, Canadian, Muslim-American, and military marriages.

List the characteristics of successful marriages.

Predict the future of marriage relationships.

1. Military wives whose husbands are deployed report that talking with other military wives is the most effective mechanism for coping with their husbands' absence.

2. In regard to their sex lives, married people report greater emotional and physical satisfaction than single people.

3. The stereotypes about in-law relationships are true—most spouses report that they do not get along with their in-laws.

4. Grooms report more delight on their wedding night than brides.

5. Spouses who are close in age are happier than spouses who have a greater age difference between them.

Answers: 1. T   2. T   3. F   4. T   5. F

Since marriage is primarily conducted outside the public view, outsiders never know what goes on in the private lives of the individuals and their relationship. The marriage of Tiger Woods and Elin Nordegren became a public focus in 2010. Previously, Woods had said of his marriage:

> I have a balance in my life. Ever since Elin came into my life, things just became a lot better. Someone you can bounce things off, somebody who is a great friend. We do just about everything together. It's nice having that type of person around you. She's so much like me. She's very competitive, very feisty, just like I am.

Following Woods' admission of infidelity, the couple divorced. All marriages begin at a time of great promise, but things change across time. Only the couple knows the details, and the details are as different as the couples.

The title of this chapter, with plural "relationships" confirms that all marriages are different. *Diversity* is the term that best describes relationships, marriages, and families today. No longer is there a one-size-fits-all cultural norm of what a relationship, marriage, or family should be. Rather, individuals, couples, and families select their own path. In this chapter, we review the diversity of relationships. We begin with looking at some of the different reasons people marry.

Motivations for and Functions of Marriage

In this section, we discuss both why people marry and the functions that getting married serve for society.

Individual Motivations for Marriage

We have defined marriage in the U.S. as a legal contract between two heterosexual adults that regulates their economic and sexual interaction. However, individuals in the United States tend to think of marriage in more personal than legal terms. The following are some of the reasons people give for getting married.

Love Many couples view marriage as the ultimate expression of their love for each other—the desire to spend their lives together in a secure, legal, committed relationship. In U.S. society, love is expected to precede marriage—thus, only couples in love consider marriage. Those not in love would be ashamed to admit it.

*Some pray to marry the
man they love,
My prayer will
somewhat vary:
I humbly pray
to Heaven above,
That I love the man
I marry.*

Rose Pastor Stokes, *My Prayer*

Personal Fulfillment   We marry because we feel a sense of personal fulfillment in doing so. We were born into a family (family of origin) and want to create a family of our own (family of procreation). We remain optimistic that our marriage will be a good one. Even if our parents divorced or we have friends who have done so, we feel that our relationship will be different.

Companionship   Talk show host Oprah Winfrey once said that lots of people want to ride in her limo, but what she wants is someone who will take the bus when the limo breaks down. One of the motivations for marriage is to enter a structured relationship with a genuine companion, a person who will take the bus with you when the limo breaks down.

Although marriage does not ensure it, companionship is the greatest expected benefit of marriage in the United States. Coontz (2000) noted that it has become "the legitimate goal of marriage" (p. 11). Johnson et al. (2010) noted that people often marry when they "find the right person" rather than marry to avoid living together since they may not be morally comfortable with this choice.

Parenthood   Most people want to have children. In response to the statement, "Someday, I want to have children," 84% of 2,922 undergraduates (90% of females; 77% of males) responded, "yes" (Knox and Hall 2010). The amount of time parents spend in rearing their children has increased. Sayer et al. (2004) documented that, contrary to conventional wisdom, both mothers and fathers report spending greater amounts of time in child-care activities in the late 1990s than in the "family-oriented" 1960s.

Although some people are willing to have children outside marriage (in a cohabitating relationship or in no relationship at all), most Americans prefer to have children in a marital context. Previously, a strong norm existed in our society (particularly among whites) that individuals should be married before they have children. This norm has relaxed, with more individuals willing to have children without being married.

Economic Security   Married people report higher household incomes than do unmarried people. Indeed, national data from the Health and Retirement Survey revealed that individuals who were not continuously married had significantly lower wealth than those who remained married throughout the life course. Remarriage offsets the negative economic effect of marital dissolution (Wilmoth and Koso 2002).

Although individuals may be drawn to marriage for the preceding reasons on a conscious level, unconscious motivations may also be operative. Individuals reared in a happy family of origin may seek to duplicate this perceived state of warmth, affection, and sharing. Alternatively, individuals reared in unhappy, abusive, drug-dependent families may inadvertently seek to recreate a similar family because that is what they are familiar with. In addition, individuals are motivated to marry because of the fear of being alone, to better themselves economically, to avoid birth out of wedlock, and to prove that someone wants them.

Just as most individuals want to marry (regardless of the motivation), most parents want their children to marry. If their children do not marry too young and if they marry someone they approve of, parents feel some relief from the economic responsibility of parenting, anticipate that marriage will have a positive, settling effect on their offspring, and look forward to the possibility of grandchildren.

Societal Functions of Marriage

As noted in Chapter 1, important societal functions of marriage are to bind a male and female together who will reproduce, provide physical care for their dependent young, and socialize them to be productive members of society who will replace those who die (Murdock 1949). Marriage helps protect children by giving the state

TABLE 7.1 Traditional versus Egalitarian Marriages	
Traditional Marriage	**Egalitarian Marriage**
There is limited expectation of husband to meet emotional needs of wife and children.	Husband is expected to meet emotional needs of his wife and children.
Wife is not expected to earn income.	Wife is expected to earn income.
Emphasis is on ritual and roles.	Emphasis is on companionship.
Couples do not live together before marriage.	Couples often lives together before marriage.
Wife takes husband's last name.	Wife may keep her maiden name. In some cases, he will take her last name.
Husband is dominant; wife is submissive.	Neither spouse is dominant.
Roles for husband and wife are rigid.	Roles for spouses are flexible.
Husband initiates sex; wife complies.	Either spouse initiates sex.
Wife takes care of children.	Fathers and mothers share parenting equally
Education is important for husband, not for wife.	Education is important for both spouses.
Husband's career decides family residence.	Career of either spouse may determine family residence.

A great marriage is not when the 'perfect couple' come together. It is when an imperfect couple learns to enjoy their differences.

Dave Meurer, author/humorist

legal leverage to force parents to be responsible to their offspring whether or not they stay married. If couples did not have children, the state would have no interest in regulating marriage.

Additional functions include regulating sexual behavior (spouses are expected to be faithful, which results in less exposure to sexually transmitted infections [STIs] than singles) and stabilizing adult personalities by providing a companion and "in-house" counselor. In the past, marriage and family have served protective, educational, recreational, economic, and religious functions.

However, as these functions have gradually been taken over by police or legal systems, schools, the entertainment industry, workplace, and church or synagogue, only the companionship-intimacy function has remained virtually unchanged.

The emotional support each spouse derives from the other in the marital relationship remains one of the strongest and most basic functions of marriage (Coontz 2000). In today's social world, which consists mainly of impersonal, secondary relationships, living in a context of mutual emotional support may be particularly important. Indeed, the companionship and intimacy needs of contemporary U.S. marriage have become so strong that many couples consider divorce when they no longer feel "in love" with their partner. Of 2,922 undergraduates, 40% reported that they would divorce their spouse if they no longer loved the spouse (Knox and Hall 2010).

The very nature of the marriage relationship has also changed from being very traditional or male-dominated to being very modern or egalitarian. A summary of these differences is presented in Table 7.1. Keep in mind that these are stereotypical marriages and that only a small percentage of today's modern marriages have all the traditional or egalitarian characteristics that are listed.

Unless commitment is made, there are only promises and hopes; but no plans.

Peter F. Drucker, author

Commitment the intent to maintain a relationship.

Marriage as a Commitment

Marriage represents a multilevel commitment—person-to-person, family-to-family, and couple-to-state.

Person-to-Person Commitment

Commitment is the intent to maintain a relationship. Behavioral indexes of commitment (928 of them) were identified by 248 people who were committed to someone. These behaviors were then coded into ten major categories and included

Chapter 7 Marriage Relationships

What if My Partner is Not as Interested in the Relationship as I Am?

Both partners in a relationship rarely have the same level of interest, involvement, and desire for the future. These differences are due to personality, previous relationships, and the rebound effect. In regard to personality, some individuals forge ahead and want to escalate new relationships whereas others are very cautious. As for previous relationships, people who have been betrayed in a former relationship are slow to reengage. The rebound effect is that people who have been recently dumped are often quick to reengage. Unless the different levels of interest are dramatic such that one is unsure whether to remain in the relationship at all, one option is to continue the relationship to give time for one partner to catch up while the other partner slows down so that the partners are closer in their walk together.

providing affection, providing support, providing companionship, making an effort to communicate, showing respect, creating a relational future, working on relationship problems together, and expressing commitment (Weigel and Ballard-Reisch 2002). Weigel (2010) also found that greater mutual commitment was associated with higher relationship quality and that females were more likely to perceive that commitment levels were higher. Bartle-Haring (2010) studied 112 couples in marriage counseling and found that a commitment change in one partner was associated with commitment change in the other.

Family-to-Family Commitment

Whereas love is private, marriage is public. Marriage is the second of three times that one's name can be expected to appear in the local newspaper. When individuals marry, the parents and extended kin also become involved. In many societies (for example, Kenya), the families arrange for the marriage of their offspring, and the groom is expected to pay for his new bride. How much is a bride worth? In some parts of rural Kenya, premarital negotiations include the determination of **bride wealth**—also known as bride price or bride payment, the amount of money or goods paid by the groom or his family to the wife's family for giving her up. Such a payment is not seen as "buying the woman" but compensating the parents for the loss of labor from their daughter. Forms of payment include livestock ("I am worth many cows," said one Kenyan woman), food, and/or money. The man who raises the bride wealth also demonstrates not only that he is ready to care for a wife and children but also that he has the resources to do so (Wilson et al. 2003).

Bride wealth also known as bride price or bride payment, the amount of money or goods paid by the groom or his family to the bride's family for giving her up.

Marriage also involves commitments by each of the marriage partners to the family members of the spouse. Married couples are often expected to divide their holiday visits between both sets of parents.

Couple-to-State Commitment

In addition to making person-to-person and family-to-family commitments, spouses become legally committed to each other according to the laws of the state in which they reside. This means they cannot arbitrarily decide to terminate their own marital agreement.

Just as the state says who can marry (not close relatives, the insane, or the mentally deficient) and when (usually at age 18 or older), legal procedures must be instituted if the spouses want to divorce. The state's interest is that a couple stays married, have children, and take care of them. Should they divorce, the state will dictate how the parenting is to continue, both physically and economically.

Social policies designed to strengthen marriage through divorce law reform reflect the value the state places on stable, committed relationships.

Marriage as a Commitment

Covenant Marriage: A Stronger Commitment?

Covenant marriage is an agreement between persons getting married that reflects a serious regard for their marriage and their future (Cade 2010). In a covenant marriage, the spouses sign a Declaration of Intent specifying that they have had premarital education/counseling, that they will seek marriage counseling before seeking a divorce, and that they will have a "cooling off" period of two years if they have children before they seek divorce. Furthermore, they agree that divorce is not acceptable just because they are "unhappy;" rather, adultery, abandonment, or abuse are the only reasons regarded as legitimate.

Louisiana (1997), Arizona (1998), and Arkansas (2001) are the only three states to recognize covenant marriage. Couples opting for covenant marriage are rare; only 2% of individuals getting married in Louisiana choose this option.

While the divorce rate among those electing covenant marriage is lower, data on Louisiana couples reveal that only 71% of them reported getting the required marital counseling and many (29%) said they did it only to meet the legal requirement. Seemingly, when the American cultural value of individualism (Cherlin 2009) kicked in, the strategies designed as safeguards to maintain the marriage evaporated for about 30%. Only 25% of couples getting divorced who had a covenant marriage reported thinking more seriously about divorce as a result of their covenant marriage (Nock et al. 2008).

Data on 600 Louisiana marriages allowed for a comparison of covenant and "regular" couples. The covenant couples were more religious and traditional, reported happier marriages, fewer marital problems, and lower divorce rates. Their primary reason for selecting the covenant option was to show their sacred attitude toward marriage. A unique finding between the two groups was that covenant wives reported an increase in marital quality following the birth of a child compared to wives in standard marriages who reported decreased marital quality (including increased thoughts about divorce) (Nock et al. 2008). Since many of those who chose covenant marriage were religious and traditional, it is not known if the positive effects were due to the covenant marriage itself or to the fact that the participants were religious or conventional.

Perhaps the greatest impact of covenant marriage is that it has brought a "national conversation about marriage. These conversations are rare. . . . For getting that conversation started, covenant marriage deserves an 'A'" (Nock et al. 2008, 151–152).

Cold Feet?

Having cold feet about getting married is not unusual. Crystal Harris called off the wedding to Hugh Hefner four days before the event. Other real life examples of cold feet follow. The following are examples of persons who "knew" they were doing the wrong thing the closer they got to the wedding:

> *"I knew the day of the wedding that I did not want to marry. I told my dad, and he said, 'Be a man.' I went through with the marriage and regretted it ever since."* (This person divorced after twenty-five years of marriage and remarried his "soul mate.")
>
> *"I said 'Holy Jesus' just before I walked down the aisle with my dad. He said, 'What's the matter, honey?' I couldn't tell him, went through with the wedding and later divorced."* (This person divorced after twelve years.)
>
> *"I never really believed I was getting married till I saw my name in the paper that I was soon to be married. It scared me. I called it off after the announcements had been sent out and we had been to see the preacher. It was a real mess. She kept the ring."*

Cold feet may be a signal to the brain to jump off the train speeding toward the altar. Listening to one's senses may be well advised.

Covenant marriage type of marriage whereby the spouses agree to have marriage counseling before getting married, to have a cooling off period of two years if they have children before they seek divorce, and to divorce only for serious faults (such as abuse, adultery, or imprisonment for a felony).

Marriage—a word which should be pronounced "mirage."

Herbert Spencer, philosopher

Marriage as a Rite of Passage

A **rite of passage** is an event that marks the transition from one social status to another. Starting school, getting a driver's license, and graduating from high school or college are events that mark major transitions in status (to student, to driver, and to graduate). The wedding itself is another rite of passage that marks the transition from fiancé to spouse. Preceding the wedding is the traditional bachelor party for the soon-to-be groom. What is new on the cultural landscape is the bachelorette party (sometimes more wild than the bachelor party), which conveys the message of equality and that great changes are ahead (Montemurro 2006).

Weddings

While weddings are often a source of stress, they are a rite of passage that is both religious and civil. To the Catholic Church, marriage is a sacrament that implies that the union is both sacred and indissoluble. According to Jewish and most Protestant faiths, marriage is a special bond between the husband and wife sanctified by God, but divorce and remarriage are permitted. Wedding ceremonies still reflect traditional cultural definitions of women as property. For example, the father of the bride usually walks the bride down the aisle and "hands her over to the new husband." In some cultures, the bride is not even present at the time of the actual marriage. For example, in the upper-middle-class Muslim Egyptian wedding, the actual marriage contract signing occurs when the bride is in another room with her mother and sisters. The father of the bride and the new husband sign the actual marriage contract (identifying who is marrying whom, the families they come from, and the names of the two witnesses). The father will then place his hand on the hand of the groom, and the maa'zun, the official presiding, will declare that the marriage has occurred (Sherif-Trask 2003).

Blakely (2008) noted that weddings are increasingly becoming outsourced, commercialized events. The latest wedding expense is to have the wedding webcast so that family and friends afar can actually see and hear the wedding vows in real time. One such website is http://webcastmywedding.net/, where the would-be bride and groom can arrange the details. That marriage is a public experience is emphasized by weddings in which the couple invites family and friends of both parties to participate. The wedding is a time for the respective families to learn how to cooperate with each other for the benefit of the couple. Conflicts over number of bridesmaids and ushers, number of guests to invite, and place of the wedding are not uncommon. Though some families harmoniously negotiate all differences, others become so adamant about their preferences that the prospective bride and groom elope to escape or avoid the conflict. However, most families recognize the importance of the event in the life of their daughter or son and try to be helpful and nonconflictual. To obtain a marriage license, some states require the partners to have blood tests to certify that neither has an STI. The document is then taken to the county courthouse, where the couple applies for a marriage license. Two-thirds of states require a waiting period between the issuance of the license and the wedding. A member of the clergy marries 80% of couples; the other 20% (primarily in remarriages) go to a justice of the peace, judge, or magistrate.

Rite of passage an event that marks the transition from one social status to another.

Don't marry the person you think you can live with; marry only the individual you think you can't live without.

James Dobson, psychologist

Love is the fever which marriage puts to bed and cures.

Richard J. Needham, Canadian humorist

Colleen Breen

"A kiss for luck and we're on our way."

Some family scholars and policy makers advocate strengthening marriage by reforming divorce laws to make divorce harder to obtain. Because California became the first state to implement "no-fault" divorce laws in 1969, every state has passed similar laws allowing couples to divorce without proving in court that one spouse was at fault for the marital breakup. The intent of no-fault divorce legislation was to minimize the acrimony and legal costs involved in divorce, making it easier for unhappy spouses to get out of a marriage. Under the system of no-fault divorce, a partner who wanted a divorce could get one, usually by citing irreconcilable differences, even if their spouse did not want a divorce.

Other states believe the no-fault system has gone too far and have taken measures designed to make breaking up harder to do by requiring proof of fault (such as infidelity, physical or mental abuse, drug or alcohol abuse, and desertion) or extending the waiting period required before granting a divorce. In most divorce law reform proposals, no-fault divorces would still be available to couples who mutually agree to end their marriages.

Opponents argue that divorce law reform measures would increase acrimony between divorcing spouses (which harms the children as well as the adults involved), increase the legal costs of getting a divorce (which leaves less money to

---

Artifact concrete symbol that reflects the existence of a cultural belief or activity (e.g. wedding ring).

Brides often wear traditional **artifacts** (concrete symbols that reflect the existence of a cultural belief or activity) something old, new, borrowed, and blue. The "old" wedding artifact is something that represents the durability of the impending marriage (for example, an heirloom gold locket). The "new" wedding artifact, perhaps in the form of new, unlaundered undergarments, emphasizes the new life to begin. The "borrowed" wedding artifact is something that has already been worn by a currently happy bride (for example, a wedding veil). The "blue" wedding artifact represents fidelity (for example, those dressed in blue or in blue ribbons have lovers true). When the bride throws her floral bouquet, it signifies the end of girlhood; the rice thrown by the guests at the newly married couple signifies fertility.

Couples now commonly have weddings that are neither religious nor traditional. In the exchange of vows, neither partner may promise to obey the other, and the couple's relationship may be spelled out by the partners rather than by tradition. Vows often include the couple's feelings about equality, individualism, humanism, and openness to change. In 2011, the average wedding for a couple getting married for the first time was estimated to be $27,800 (www.theknot.com).

*I dreamed of a wedding of
elaborate elegance,
A church filled with
family and friends.
I asked him what kind of
a wedding he wished for,
He said one that would
make me his wife.*

Unknown

Ways in which couples lower the cost of their wedding include marrying any day but Saturday, or marrying off-season (not June), off-locale (in Mexico or a Caribbean Island where fewer guests will attend), and, as mentioned previously, broadcasting their wedding ceremony live over the Internet. The latter means that the couple can get married in Hawaii and have their ceremony beamed back to the states where well-wishers can see the wedding without leaving home.

Honeymoons

Traditionally, another rite of passage follows immediately after the wedding—the **honeymoon** (the time following the wedding whereby the couple becomes isolated to recover from the wedding and to solidify their new status change from lovers to spouses). The functions of the honeymoon are both personal and social.

Honeymoon the time following the wedding whereby the couple becomes isolated to recover from the wedding and to solidify their new status change from lovers to spouses.

The personal function is to provide a period of recuperation from the usually exhausting demands of preparing for and being involved in a wedding ceremony and reception. The social function is to provide a time for the couple to be alone to solidify the change in their identity from that of an unmarried to a married couple. Now that they are married, their sexual expression and childbearing with each other achieves full social approval and legitimacy.

support any children), and delay court decisions on child support, custody, and distribution of assets. In addition, critics point out that ending no-fault divorce would add countless court cases to the dockets of an already over-loaded court system. Efforts to repeal no-fault divorce laws in many state legislatures have largely failed.

Your Opinion?

1. To what degree do you believe the government can legislate "successful" marriage relationships?
2. How has no-fault divorce gone too far?

3. Will there be a return to laws which make it more difficult to divorce?

Sources

Brown, M. 2008. The state of our unions. *Redbook*, June 38.

Licata, N. 2002. Should premarital counseling be mandatory as a requisite to obtaining a marriage license? *Family Court Review* 40:518–532.

Sears, L. W. 2007. The "marriage gap": A case for strengthening marriage in the 21st century. *New York University Law Review* 82:1243–1253.

Changes after Marriage

After the wedding and honeymoon, the new spouses begin to experience changes in their legal, personal, and marital relationship.

Legal Changes

Unless the partners have signed a prenuptial agreement specifying that their earnings and property will remain separate, after the wedding, each spouse becomes part owner of what the other earns in income and accumulates in property. Although the laws on domestic relations differ from state to state, courts typically award to each spouse half of the assets accumulated during the marriage (even though one of the partners may have contributed a smaller proportion).

For example, if a couple buys a house together, even though one spouse invested more money in the initial purchase, the other will likely be awarded half of the value of the house if they divorce. (Having children complicates the distribution of assets because the house is often awarded to the custodial parent.) In the case of death of the spouse, the remaining spouse is legally entitled to inherit between one-third and one-half of the partner's estate, unless a will specifies otherwise.

Personal Changes

New spouses experience an array of personal changes in their lives. One initial consequence of getting married may be an enhanced self-concept. Parents and close friends usually arrange their schedules to participate in your wedding and give gifts to express their approval of you and your marriage. In addition, the strong evidence that your spouse approves of you and is willing to spend a lifetime with you also tells you that you are a desirable person.

Married people also begin adopting new values and behaviors consistent with the married role. Although new spouses often vow that "marriage won't change me," it does. For example, rather than stay out all night at a party, which is not uncommon for single people who may be looking for a partner, spouses (who are already paired off) tend to go home early. Their roles of spouse, employee, and parent result in their adopting more regular, alcohol- and drug-free hours.

Marriage is like a deck of cards
In the beginning all you need is two hearts and a diamond
By the end you'll wish you had a freaking club and a spade

Unknown

Wedding traditions vary widely throughout the world. In this photo essay, we look at Jewish weddings, Hindu Indian weddings, and Islamic Turkish weddings.

JEWISH WEDDINGS

In a traditional Jewish wedding, the bride and groom do not see each other for a week before the wedding, and they both fast on the day of the wedding until after the ceremony. Before the ceremony, while the bride is being veiled, the groom signs the *ketubah*, which is a Jewish premarital contract stating that the husband commits to provide food, clothing, and sexual relations to his wife, and that he will pay a specified sum of money if he divorces her. The groom and bride may not engage in marital relations unless the groom and two witnesses have signed the *ketubah*.

Israelimages / Israel Talby / Israel images / Alamy

The wedding ceremony takes place underneath a *chuppah*—a cloth canopy supported by four poles, symbolizing the home the couple will build. The groom enters the chuppah first, then the bride enters, approaches the groom, and circles him seven times, symbolizing the seven days of creation.

Tomi Junger Photography/ PhotoStock-Israel / Alamy

The ceremony ends with the groom smashing a glass (or a small symbolic piece of glass) with his right foot, to symbolize the destruction of the Temple. At this point, the guests shout "mazel tov!" which literally means "good fortune" and also means "congratulations."

Buccina Studios/Stockbyte (RF)/Jupiter Images

HINDU INDIAN WEDDINGS

About 13% of the world's population is Hindu. In India, where most Hindus live, parents or a matchmaker arrange most marriages. Weddings vary according to the wealth ▶

Panorama Productions Inc. / Alamy

and status of the couple. Traditionally, some of the prayers of a Hindu wedding are recited at least partially in Sanskrit—the classical language of India. Hindu weddings are very festive and can last several days. Prior to the wedding ceremony, there is music and dancing and other pre-ceremony activities such as fireworks and throwing flowers. The groom arrives at the wedding riding a decorated horse or elephant. The bride wears a colorful sari and has elaborate henna applied to her feet, hands, and perhaps face.

Plush Studios/Jupiter Images

The Hindu Indian wedding ceremony takes place under a decorated canopy called a *mandap*. The ceremony lasts for about two hours, during which the Brahmin priests intone Sanskrit chants to various Hindu deities and throw grains of rice on the couple and into a small sacred fire that is supposed to be sustained and nurtured through a couple's entire marriage, with oblations offered into the fire each day.

ISLAMIC TURKISH WEDDING

About one in five people in the world is Muslim; they follow the religion of Islam. Prior to the traditional Islamic Turkish wedding, a bride attends a henna ceremony at the home of the bride's in-laws. The groom sends the henna paste, which is placed on the bride's palms and on the palms of other women from both families who attend.

Joan gravell / Alamy

Prior to the wedding, the bride's father or brother puts a red belt around the bride's waist to symbolize the bride's virginity.

After a fifteen-minute civil marriage ceremony at the municipality's wedding hall, the couple and parents receive presents and congratulations from guests outside the wedding hall entrance. A wedding reception with food and drink may be held on the evening of the marriage ceremony. If a reception is held, it may be a small dinner party at a restaurant or at the home of the groom's parents or a big event with a couple hundred guests at a hotel. If the families are conservative, male and female guests may not sit together, but are seated in different parts of the same hall.

Images & Stories / Alamy

There is considerable interest about a couple's wedding night. Stereotypes are that they are either wonderful or disastrous, but what do the data tell us?

Sample

This sample consisted of 95 spouses, 75% women and 25% men. Over three-fourths had been married once, with 20% in their second marriages. The respondents were asked to report on their "most recent wedding night." Over 80% (81.1%) were virgins on their wedding night.

Findings

The questionnaire was designed to assess the quality of the wedding night experience, the best and worst experiences, and some recommendations.

1. *Rating.* When asked the question, "On a scale of one to ten with zero being 'awful' and ten being 'wonderful', what number would you select to describe your wedding night experience?," the average was 6.81 when we looked at all marriages. Grooms reported more positive experiences than brides, 7.19 to 6.68, respectively. When second marriages were the focus, the average was 8.4.

2. *Summaries.* When asked to "summarize your wedding night experience" some of the comments were:

   a. "We only went away for 1 night as the next day was Christmas. I was 17, it was lots of fun, we ate out, and had a great night."

   b. "It was a fulfilling, happy occasion, knowing that I was starting a new chapter in life."

   c. "It was a disaster. He was thinking about his old girlfriend and wishing he had married her."

   d. "We had an amazing night. It was so hard to keep our hands off each other."

   e. "It was so wonderful but we were tired. It could have been better."

Friendship Changes

Marriage also affects relationships with friends of the same and other sex. Less time will be spent with friends because of the new role demands as a spouse. More time will be spent with other married couples who will become powerful influences on the new couple's relationship. Indeed, to the degree that married couples have the *same* friends is the degree the marital quality of the couple will be high.

What spouses give up in friendships, they gain in developing an intimate relationship with each other. However, abandoning one's friends after marriage may be problematic because one's spouse cannot be expected to satisfy all of one's social needs. Because many marriages end in divorce, friendships that have been maintained throughout the marriage can become a vital source of support for a person adjusting to a divorce.

PERSONAL CHOICES

Is "Partner's Night Out" a Good Idea?

Although spouses may want to spend time together, they may also want to spend time with their friends—shopping, having a drink, fishing, golfing, seeing a movie, or whatever. Some spouses have a flexible policy based on trust with each other. Other spouses are very suspicious of each other. One husband said, "I didn't want her going out to bars with her girlfriends after we were married. You never know what someone will do when they get three drinks in them." For "partner's night out" to have a positive impact on the couple's relationship, it is important that the partners maintain emotional and sexual fidelity to each other, that each partner have a night out, and that the partners spend some nights alone with each other. Friendships can enhance a marriage relationship by making the individual partners happier, but friendships cannot replace the marriage relationship. Spouses must spend time alone to nurture their relationship.

3. *Best part.* When asked to identify "the best part of the wedding night," 29.7% listed "just being with my new partner"; 17.8% listed "sex"; 10.5% said "nothing"; and 9.4% listed the "reception."

4. *Worst part.* When asked to identify the "worst part of the wedding night," 23.1% listed "accommodations and the partner's demeanor"; 21% said "being so tired"; 9.4% said the "end of celebration"; 8.4% said "nothing"; 6.3% listed "sex"; and 3.1% listed "pain."

5. *Change.* To the question, "If you could replay your wedding night, what would you change?" 34.7% of respondents answered, "nothing"; 18.9% responded "not be tired;" 14.7% listed "different time/place;" and 9.5% listed "different person."

The data suggest that the wedding nights for these respondents were mostly positive experiences. Recommendations were to plan weddings early in the day so that the reception is over early. Also, the couple might plan to spend the first night a short drive from the reception. Avoid leaving a reception at 11:00 p.m. and driving four hours. Many newlyweds suggested staying in a hotel close to the reception and flying to a honeymoon destination late the following afternoon. Also, avoid an early-morning flight the day after your wedding.

Source

This research is based on unpublished data collected for this text. Appreciation is expressed to Kelly Woody for distribution of the questionnaires (that were mailed to us) and to Emily Richey who tabulated the data.

Marital Changes

A happily married couple of forty-five years spoke to our class and began their presentation with, "Marriage is one of life's biggest disappointments." They spoke of the difference between all the hype and cultural ideal of what marriage is supposed to be . . . and the reality. One effect of getting married is **disenchantment**—the transition from a state of newness and high expectation to a state of mundaneness tempered by reality. It may not happen in the first few weeks or months of marriage, but it is almost inevitable. Whereas courtship is the anticipation of a life together, marriage is the day-to-day reality of that life together—and reality does not always fit the dream. "Moonlight and roses become daylight and dishes" is an old adage reflecting the realities of marriage. Disenchantment after marriage is also related to partners shifting their focus away from each other to work or children; each partner usually gives and gets less attention in marriage than in courtship. College students are not oblivious to the change after marriage. Of 2,922 respondents, 25% of the males and 17% of the females agreed that "most couples become disenchanted with marriage within five years" (Knox and Hall 2010).

A couple will experience other changes when they marry, as the following details:

1. *Experiencing loss of freedom.* Single people do as they please. They make up their own rules and answer to no one. Marriage changes that as the expectations of the spouse impact the freedom of the individual. In a study of 1,001 married adults, although 41% said that they missed "nothing" about the single life, 26% reported that they most missed not being able to live by their own rules (Cadden and Merrill 2007).

2. *Feeling more responsibility.* Single people are responsible for themselves. Spouses are responsible for the needs of each other and sometimes resent it. One wife said she loved when her husband went out of town on business because she then did not feel the responsibility to cook for him. One husband said that

Marriage is a covered dish.

Swiss Proverb

Disenchantment the transition from a state of newness and high expectation to a state of mundaneness tempered by reality.

Love is an obsessive delusion that is cured by marriage.

Dr. Karl Boman, psychiatrist

he felt burdened by having to help his wife care for her aging parents. In the study of 1,001 married adults referred to previously, 25% reported that having less responsibility was what they missed most about being single.

3. *Missing alone time.* Aside from the few spouses who live apart, most live together. They wake up together, eat their evening meals together, and go to bed together. Each may feel too much togetherness. "This altogether, togetherness thing is something I don't like," said one spouse. In the study referred to previously, 24% reported that "having time alone for myself" was what they missed most about being single (Cadden and Merrill 2007). One wife said, "My best time of the day is at night when everybody else is asleep."

4. *Change in how money is spent.* Entertainment expenses in courtship become allocated to living expenses and setting up a household together. In the same study, 17% reported that they missed managing their own money most (Cadden and Merrill 2007).

5. *Sexual changes.* The sexual relationship of the couple also undergoes changes with marriage. First, because spouses are more sexually faithful to each other than are dating partners or cohabitants (Treas and Giesen 2000), their number of sexual partners will decline dramatically. Second, the frequency with which they have sex with each other decreases. One wife said the following:

> *The urgency to have sex disappears after you're married. After a while you discover that your husband isn't going to vanish back to his apartment at midnight. He's going to be with you all night, every night. You don't have to have sex every minute because you know you've got plenty of time. Also, you've got work and children and other responsibilities, so sex takes a lower priority than before you were married.*

Although married couples may have intercourse less frequently than they did before marriage, marital sex is still the most satisfying of all sexual contexts. Of married people in a national sample, 85% reported that they experienced extreme physical pleasure and extreme emotional satisfaction with their spouses. In contrast, 54% of individuals who were not married or not living with anyone said that they experienced extreme physical pleasure with their partners, and 30% said that they were extremely emotionally satisfied (Michael et al. 1994). Fisher and McNulty (2008) studied seventy-two couples just after their weddings and found that high sexual satisfaction was associated with high marital satisfaction one year later.

6. *Power changes.* The distribution of power changes after marriage and across time. The way wives and husbands perceive and interact with each other continues to change throughout the course of the marriage. Two researchers studied 238 spouses who had been married more than thirty years and observed that (across time) men changed from being patriarchal to collaborating with their wives and that women changed from deferring to their husbands' authority to challenging that authority (Huyck and Gutmann 1992). In effect, men tend to lose power and women gain power. However, such power changes may not always occur. In abusive relationships, abusive partners may increase the display of power because they fear the partner will try to escape from being controlled.

7. *Discovering that one's mate is different from one's date.* Courtship is a context of deception. Marriage is one of reality. Spouses sometimes say, "He (she) is not the person I married." Jay Leno once quipped, "It doesn't matter who you marry since you will wake up to find that you have married someone else."

Parents and In-Law Changes

Marriage affects relationships with parents. Time spent with parents and extended kin radically increases when a couple has children. Indeed, a major difference between couples with and without children is the amount of time they

Should a Married Couple Have Their Parents Live with Them?

This question is more often asked by individualized Westernized couples who live in isolated nuclear units. Asian couples reared in extended-family contexts expect to take care of their parents and consider it an honor to do so. American spouses with aging parents must make a decision about whether to have the parents live with them. Usually the parent in need of care is the mother of either spouse because the father is more likely to die first. One wife said, "We didn't have a choice. His mother is 82 and has Alzheimer's disease. We couldn't afford to put her in a nursing home at $5,200 a month, and she couldn't stay by herself. So we took her in. It's been a real strain on our marriage, since I end up taking care of her all day. I can't even leave her alone to go to the grocery store."

Some elderly people have resources for nursing home care, or their married children can afford such care. However, even in these circumstances, some spouses decide to have their parents live with them. "I couldn't live with myself if I knew my mother was propped up in a wheelchair eating Cheerios when I could be taking care of her," said one daughter.

When spouses disagree about parents in the home, the result can be devastating. According to one wife, "I told my husband that Mother was going to live with us. He told me she wasn't and that he would leave if she did. She moved in, and he moved out (we were divorced). Five months later, my mother died."

spend with relatives. Parents and kin rally to help with the newborn and are typically there for birthdays and family celebrations.

Emotional separation from one's parents is an important developmental task in building a successful marriage. When choices must be made between one's parents and one's spouse, more long-term positive consequences for the married couple are associated with choosing the spouse over the parents. However, such choices become more complicated and difficult when one's parents are old, ill, or widowed.

Only a minority of spouses (3% to 4%) report that they do not get along with their in-laws (Amato et al. 2007). Nuner (2004) interviewed twenty-three daughters-in-law married between five and ten years (with no previous marriages and at least one child from the marriage) and found that most reported positive relationships with their mothers-in-law. The evaluation by the daughters depended on the role of the mothers-in-law within the family (for example, mothers-in-law as grandmothers were perceived more positively).

The behavior in-laws engage in affects how their children and their spouses like and perceive them. Morr Serewicz and Canary (2008) investigated newlyweds' perceptions of private disclosures received from their in-laws and found that, when these were positive (for example, the in-law talked positively about family members), the spouse perceived the in-law positively. Conversely, when the in-law talked negatively about family members or told negative stories, the spouse viewed the in-law negatively.

Financial Changes

An old joke about money in marriage says that "two can live as cheaply as one as long as one doesn't eat." The reality behind the joke is that marriage involves the need for spouses to discuss and negotiate how they are going to get and spend money in their relationship. Some spouses bring considerable debt into the marriage. Dew (2008) observed that debt was associated with marital dissatisfaction. Spouses who were in debt reported spending less leisure time together and arguing more about money. We will discuss more about debt in Chapter 12: Work and Family Life.

Diversity in Marriage

Any study of marriage relationships emphasizes the need to understand the diversity of marriage/family life. In this section, we review Hispanic, Canadian, Muslim-American, and military families. We also look at other examples of family diversity: interracial, interreligious, cross-national, and age-discrepant.

Hispanic Families

The panethnic term *Hispanic* refers to both immigrants and U.S. natives with an ancestry to one of twenty Spanish-speaking countries in Latin America and the Caribbean (Landale and Oropesa 2007). There are about 50 million Hispanics in the United States who represent 15% of the population (*Statistical Abstract of the United States, 2011,* Table 6). Hispanic families vary not only by where they are from but by whether they were born in the United States. About 40% of U.S. Hispanics are foreign-born and immigrated here, 32% have parents who were born in the United States, and 28% were born here of parents who were foreign-born.

Great variability exists among Hispanic families. Although it is sometimes assumed that immigrant Hispanic families come from rural impoverished Mexico where family patterns are traditional and unchanging, immigrants may also come from economically developed urbanized areas in Latin America (Argentina, Uruguay, and Chile), where family patterns include later family formation, low fertility, and nuclear family forms.

Nevertheless, Hispanics tend to have higher rates of marriage, early marriage, higher fertility, nonmarital child rearing, and prevalence of female householder. They also have two micro family factors: male power and strong familistic values.

1. *Male power.* The husband and father is the head of the family in most Hispanic families. Rodriguez et al. (2010) also found that Mexican husbands reported more marital satisfaction. The children and wife respect him as the source of authority in the family. The wife assumes the complementary role where her focus is taking care of the home and children. Sayer and Fine (2011) found that Hispanic women do more cooking and cleaning compared with white and black women. Rodriguez et al. (2010) also found that husbands had more prestigious jobs, worked more hours, and earned more money (by $9000) than their wives.

2. *Strong familistic values.* The family is the most valued social unit in the society—not only the parents and children but also the extended family. Hispanic families have a moral responsibility to help family members with money, health, or transportation needs. Children are also taught to respect their parents as well as the elderly. Indeed, elderly parents may live with the Hispanic family where children may address their grandparents in a formal way. Spanish remains the language spoken in the home as a way of preserving family bonds.

Family patterns may be changing with assimilation. For example, foreign-born individuals tend to bring their traditional values with them. However, these may erode over time. "With respect to family patterns, a key question is whether some of the strengths of Hispanic families are eroded as they spend more time in the United States. Several patterns suggest that this is indeed the case. For example, levels of divorce, nonmarital childbearing, and female family headship increase across generations, while the prevalence of family extension declines" (Landale and Oropesa 2007, 400).

Canadian Families

Although Canada stretches from coast to coast, its population of 33 million is only about 10% of the U.S. population of over 315 million. Although much of marriage and family life in Canada is similar to that in the United States, some of the differences include the following:

1. *Language.* Canada is officially bilingual with both English and French being spoken there.

2. *Definition of family.* Although only thirteen states in America recognize common-law marriages, common-law couples with or without children are officially included in the definition of "family" (along with married couples with and without children and single parents).

3. *Same-sex relationships.* Although three of the ten Canadian provinces (Quebec, British Columbia, and Ontario) have legalized same-sex marriage, national court rulings provide greater protection for same-sex relationships than in the United States. In Canada, individuals in same-sex couple's relationships have the right to inherit from each other, the right to each other's pension benefits, and are protected under the law from discrimination based on sexual orientation.

4. *Children.* "Not so soon, not so many" is the norm for having children in Canada. Families are slightly smaller, and children are born later. A stronger patrilineal norm is operative—women usually take the husband's last name, live where the husband works, and the children are given the last name of the father.

5. *Government programs for families.* Quebec offers universal access to childcare centers for a low fee, medical costs are covered by the state, and parental leave for up to a year is paid for at the rate of employment insurance.

6. *Divorce.* Canada has half the divorce rate of the United States. A strong Catholic religious presence is the most likely explanation for a lower divorce rate. One also gets the sense that Canadian spouses have a rural mentality and take their relationships more seriously (Harvey 2005).

Muslim-American Families

Although Islam (the religious foundation for Muslim families in sixty nations) is the second largest religion (next to Christianity) in North America, 9/11 resulted in an increased awareness that Muslim families are part of American demographics. These Muslim-American families hardly represent the extremists responsible for terrorism, but more than six million adults in the United States and 1.3 billion worldwide self-identify with the Islamic religion (there are now more Muslims than Christians in the world). The three largest American Muslim groups in the United States are African Americans, Arabs, and South Asians (for example, from Pakistan, Bangladesh, Afghanistan, and India).

Islamic tradition emphasizes close family ties with the nuclear and extended family, social activities with family members, and respect for the authority of the elderly and parents. Religion and family are strong sources of a Muslim's personal identity. Muslim families provide a strong sense of emotional and social support. Breaking from one's religion and family comes at a great cost because alternatives are perceived as limited. Parents of Muslim children who are reared in America struggle to maintain traditional values while allowing their children (particularly sons) to pursue higher education and professional training.

One of the striking features of Muslim American families is the strong influence parents have over the behavior of their children. Because the families control the property and economic resources and generally provide total financial support to the children, and because the offspring may not be able to find adequate work outside the family system, the children generally acquiesce to parents' wishes. Such acquiescence does not imply the nonexistence of genuine love and affection children may have for their parents, however.

Military Families

The war in the Middle East continues to be a focus of the news media. Behind the news reports are individuals and families who are coping with danger and the sacrifices of deployments to Afghanistan and other peacekeeping and humanitarian operations. Approximately 1.4 million are active-duty military personnel. Another 1.1 million are in the military reserve and 358,000 in the National Guard (*Statistical Abstract of the United States, 2011,* Tables 507, 511, 513). About 60% of military personnel are married and/or have children (NCFR Policy Brief 2004).

There are three main types of military marriages. One, those in which the soldier falls in love with a high school sweetheart, marries the person, and subsequently joins the military. A second type of military marriage consists of those who meet and marry after one of them has signed up for the military. This is a typical marriage where the partners fall in love on the job and one or both of them happens to be the military. The final and least common military marriage is known as a contract marriage in which a person will marry a civilian to get more money and benefits from the government. For example, a soldier might decide to marry a platonic friend and split the money from the additional housing allowance (which is sometimes a relatively small amount of money and varies depending on geographical location and rank). Other times, the military member keeps the extra money and the civilian will take the benefit of health insurance. Often, in these types of military marriages, the couple does not reside together. There is no emotional connection because the marriage is mercenary. Contract military marriages are not common but they do exist.

Some ways in which military families are unique include:

1. *Traditional sex roles.* Although both men and women are members of the military service, the military has considerably more men than women. In the typical military family, the husband is deployed (sent away) and the wife is expected to "understand" his military obligations and to take care of the family in his absence. Her duties include paying the bills, keeping up the family home, and taking care of the children; a military wife must often play the role of both spouses due to the demands of her husband's military career and obligations. The wife often has to sacrifice her career to follow (or stay behind in the case of deployment) and support her husband in his fulfillment of military duties (Easterling 2005).

In the case of wives or mothers who are deployed, the rare husband is able to switch roles and become Mr. Mom. One military career wife said of her husband,

Chelsea Spalding

Deployment is a major source of stress in relationships. This couple is parting again.

whom she left behind when she was deployed, "What a joke. He found out what taking care of kids and running a family was really like and he was awful. He fed the kids Spaghetti Os for the entire time I was deployed."

There are also circumstances in which both parents are military members, and this can blur traditional sex roles because the woman has already deviated from a traditional "woman's job." Military families in which both spouses are military personnel are rare.

2. *Loss of control—deployment.* Military families have little control over their lives as the specter of deployment is ever-present. Where one of the spouses will be next week and for how long are beyond the control of the spouses and parents. Easterling and Knox (2010) surveyed 259 military wives (whose husbands had been deployed) who reported feelings of loneliness, fear, and sadness. Some women had gone extended periods of time without communicating with their husbands and in constant worry over their well-being. Talking with other military wives who 'understood' was the primary mechanism for coping with the husband's deployment. Getting a job, participating in military-sponsored events, and living with a family were also helpful. On the positive side, wives of deployed husbands reported feelings of independence and strength. They were the sole family member available to take care of the house and children, and they rose to the challenge. Adjusting to the return of the deployed spouse has its own challenges. A team of researchers observed an increased incidence of spousal violence related to PTSD as a result of having been deployed (Teten et al. 2010).

3. *Infidelity.* Although most spouses are faithful to each other, the context of separation from each other for months (sometimes years) at a time increases the vulnerability of both spouses to infidelity. The double standard may also be operative, whereby "men are expected to have other women when they are away" and "women are expected to remain faithful and be understanding." Separated spouses try to bridge the time they are apart with e-mails and phone calls (when possible), but sometimes the loneliness becomes more difficult than anticipated. One enlisted husband said that he returned home after a year-and-a-half deployment to be confronted with the fact that his wife had become involved with someone else. "I absolutely couldn't believe it," he noted. "In retrospect, I think the separation was more difficult for her than it was for me."

4. *Frequent moves and separation from extended family or close friends.* Because military couples are often required to move to a new town, parents no longer have doting grandparents available to help them rear their children. As well, although other military families become a community of support for each other, the consistency of such support may be lacking. "We moved seven states away from my parents to a town in North Dakota," said one wife. "It was very difficult for me to take care of our three young children with my husband deployed."

Similar to being separated from parents and siblings is the separation from one's lifelong friends. Although new friendships and new supportive relationships develop within the military community to which the family moves, the relationships are sometimes tenuous and temporary as the new families move on. The result is the absence of a stable, predictable social structure of support, which may result in a feeling of alienation and not belonging in either the military or the civilian community. The more frequent the moves, the more difficult the transition and the more likely the alienation of new military spouses.

5. *Lower marital satisfaction and higher divorce rates among military families.* Solomon et al. (2011) compared 264 veterans who experienced combat stress reaction (CSR) with 209 veterans who did not experience such stress. Results show that traumatized veterans reported lower levels of marital adjustment and more problems in parental functioning. Similarly, Orthner and Roderick (2009) examined the marriages of 8,056 wives whose husbands were deployed and found that both spouse psychological well-being and marital satisfaction were at risk. Another researcher (Lundquist 2007) found higher divorce rates in military than civilian marriages.

6. *Employment of spouses.* Military spouses are at a disadvantage when it comes to finding and maintaining careers or even finding a job they can enjoy. Employers in military communities are often hesitant to hire military spouses because they know the demands that are placed on them in the absence of the deployed military member can be enormous. They are also aware of frequent moves that military families make and may be reluctant to hire employees for what may be a relatively short amount of time. The result is a disadvantaged wife who has no job and must put her career on hold. Military spouses, when they do find employment, are often underemployed, which can lead to low levels of job satisfaction. They also make less, on average, than their civilian counterparts with similar characteristics. All of these factors can contribute to distress among military spouses (Easterling 2005).

7. *Resilient military families.* In spite of these challenges, there are also enormous benefits to being involved in the military, such as having a stable job (one may get demoted but it is much less frequent that one is "fired") and having one's medical bills paid for. In addition, most military families are amazingly resilient. Not only do they anticipate and expect mobilization and deployment as part of their military obligation, they respond with pride. Indeed, some reenlist eagerly and volunteer to return to military life even when retired. One military captain stationed at Fort Bragg, in Fayetteville, North Carolina, noted, "It is part of being an American to defend your country. Somebody's got to do it and I've always been willing to do my part." He and his wife made a presentation in our classes. She said, "I'm proud that he cares for our country and I support his decision to return to Afghanistan to help as needed. And most military wives that I know feel the same way."

Although military families face great challenges and obstacles, many adopt the philosophy of "whatever doesn't kill us makes us stronger." Facing deployments and frequent moves often forces a military couple to learn to rely on themselves as well as each other. They make it through difficult life events, unique to their lifestyle, which can make day-to-day challenges seem trivial. The strength that is developed within a military marriage through all the challenges they face has the potential to build a strong, resilient marriage.

Interracial Marriages

National Data

About 13% of American married couples involve someone whose partner is of a different race. White-Hispanic and White-Native American pairings are the most common. African Americans are the least likely to marry whites (Burton et al. 2010).

Interracial marriages, of which there are about 300,000 annually, involve many combinations, including American white, American black, Indian, Chinese, Japanese, Korean, Mexican, Malaysian, and Iraqi mates. White-Hispanic couples account for 41% of new interracial marriages; white-Asian, 15%; white-black couples are the least likely (Passel et al. 2010).

In discussing interracial marriages, a complicating factor is that one's racial identity may be mixed. Tiger Woods refers to his race as "Cablinasian," which combines Caucasian, Black, Native American (Indian), and Asian origins—he is one quarter Chinese, one quarter Thai, one quarter Black, one eighth Native American, and one eighth Dutch.

Interracial marriages are becoming more frequent. About 15% of all marriages in the U.S. are mixed racially, with Hispanic–Non-Hispanic being the most frequent. Nine percent of whites, 16% of blacks, 26% of Hispanics, and 31% of Asians marry someone whose race or ethnicity is different from their

The goal of marriage is not to think alike but to think together.

Robert C. Dodds, counselor

own (Passel et al. 2010). Rates vary by gender and region. Black males and those from the Western states are more likely to be involved in black-white marriage than females and those from the South.

Black-white marriages are the most infrequent. Fewer than 1% of all marriages are between a black person and a white person (*Statistical Abstract of the United States, 2011*, Table 60). Segregation in religion (the races worship in separate churches), housing (white and black neighborhoods), and education (white and black colleges), not to speak of parental and peer endogamous pressure to marry within one's own race, are factors that help to explain the low percentage of interracial black and white marriages. The spouses in black and white couples are more likely to have been married before, to be age-discrepant, to live far away from their families of orientation, to have been reared in racially tolerant homes, and to have educations beyond high school. Some may also belong to religions that encourage interracial unions. The Baha'i religion, which has more than 6 million members in 200 countries worldwide (85,000 members in the United States), teaches that God is particularly pleased with interracial unions. Finally, interracial spouses may tend to seek contexts of diversity. "I have been reared in a military family, been everywhere and met people of different races and nationalities throughout my life. I seek diversity," noted one student.

Kennedy (2003) identified three reactions to a black-white couple who cross racial lines to marry: (1) approval (increases racial open-mindedness, decreases social segregation), (2) indifference (interracial marriage is seen as a private choice), and (3) disapproval (reflects racial disloyalty, impedes perpetuation of black culture). As Kennedy notes, "The argument that intermarriage is destructive of racial solidarity has been the principal basis of black opposition" (p. 115). There is also the concern for the biracial identity of offspring of mixed-race parents.

Fusco et al. (2010) studied biracial children in the child welfare system and found that, compared to White or African-American children, biracial children were more likely to be rated as high risk and investigated. In addition, their mothers were younger, and they were more often assessed as having physical, intellectual, or emotional problems.

Interracial partners sometimes experience negative reactions to their relationship. Black people partnered with white people have their blackness and racial identity challenged by other black people. White people partnered with black people may lose their white status and have their awareness of whiteness heightened more than ever before. At the same time, one partner is not given full status as a member of the other partner's race (Hill and Thomas 2000). Gaines and Leaver (2002) also note that the pairing of a black male and a white female is regarded as "less appropriate" than that of a white male and a black female. In the former, the black male "often is perceived as attaining higher social status (i.e., the white woman is viewed as the black man's 'prize,' stolen from the more deserving white man)" (p. 68). In the latter, when a white male pairs with a black female, "no fundamental change in power within the American social structure is perceived as taking place" (p. 68). Interracial marriages are also more likely to dissolve than same-race marriages (Fu 2006).

Black-white interracial marriages are likely to increase–slowly. Not only has white prejudice against African Americans in general declined, but also segregation in school, at work, and in housing has decreased, permitting greater contact between the races. One third of 2,922 undergraduates reported that they have dated someone of another race (20% reported that they would marry across racial lines) (Knox and Hall 2010). Most Americans say they approve of racial or ethnic intermarriage—not just in the abstract, but in their own families. More than six-in-ten say it "would be fine" with them if a family member told them they were going to marry someone from any of three major race/ethnic groups other than their own (Passel et al. 2010).

The Self-Assessment feature allows you to assess your openness to involvement in an interracial relationship.

To feel real in-your-face racism- to be called a nigger and be treated bad, bad, bad,- we have to go where white people live and vacation. Like the Hamptons. The only people around are those working for white people. Walk into a store and if you're black, no one says hello. They say, "What do you want?" as in "What the hell are you doing here and how soon are you leaving?"

Chris Rock, comedian

SELF-ASSESSMENT | Attitudes toward Interracial Dating Scale

Interracial dating or marrying is the dating or marrying of two people from different races. The purpose of this survey is to gain a better understanding of what people think and feel about interracial relationships. Please read each item carefully, and in each space, score your response using the following scale. There are no right or wrong answers to any of these statements.

1	2	3	4	5	6	7
Strongly Disagree						Strongly Agree

_____ 1. I believe that interracial couples date outside their race to get attention.

_____ 2. I feel that interracial couples have little in common.

_____ 3. When I see an interracial couple, I find myself evaluating them negatively.

_____ 4. People date outside their own race because they feel inferior.

_____ 5. Dating interracially shows a lack of respect for one's own race.

_____ 6. I would be upset with a family member who dated outside our race.

_____ 7. I would be upset with a close friend who dated outside our race.

_____ 8. I feel uneasy around an interracial couple.

_____ 9. People of different races should associate only in non-dating settings.

_____ 10. I am offended when I see an interracial couple.

_____ 11. Interracial couples are more likely to have low self-esteem.

_____ 12. Interracial dating interferes with my fundamental beliefs.

_____ 13. People should date only within their race.

_____ 14. I dislike seeing interracial couples together.

_____ 15. I would not pursue a relationship with someone of a different race, regardless of my feelings for that person.

_____ 16. Interracial dating interferes with my concept of cultural identity.

_____ 17. I support dating between people with the same skin color, but not with a different skin color.

_____ 18. I can imagine myself in a long-term relationship with someone of another race.

_____ 19. As long as the people involved love each other, I do not have a problem with interracial dating.

_____ 20. I think interracial dating is a good thing.

Scoring

First, reverse the scores for items 18, 19, and 20 by switching them to the opposite side of the spectrum. For example, if you selected 7 for item 18, replace it with a 1; if you selected 3, replace it with a 5, and so on. Next, add your scores and divide by 20. Possible final scores range from 1 to 7, with 1 representing the most positive attitudes toward interracial dating and 7 representing the most negative attitudes toward interracial dating.

Norms

The norming sample was based upon 113 male and 200 female students attending Valdosta State University. The participants completing the Attitudes toward Interracial Dating Scale (IRDS) received no compensation for their participation. All participants were U.S. citizens. The average age was 23.02 years (standard deviation = 5.09), and participants ranged in age from 18 to 50 years. The ethnic composition of the sample was 62.9% white, 32.6% black, 1% Asian, 0.6% Hispanic, and 2.2% other. The classification of the sample was 9.3% freshmen, 16.3% sophomores, 29.1% juniors, 37.1% seniors, and 2.9% graduate students. The average score on the IRDS was 2.88 (SD = 1.48), and scores ranged from 1.00 to 6.60, suggesting very positive views of interracial dating. Men scored an average of 2.97 (SD = 1.58), and women, 2.84 (SD = 1.42). There were no significant differences between the responses of women and men.

Source

"Attitudes Toward Interracial Dating Scale," 2004 by Mark Whatley, Ph.D., Department of Psychology, Valdosta State University, Valdosta, Georgia 31698-0100. Used by permission. Other uses of this scale by written permission of Dr. Whatley only (mwhatley@valdosta.edu). Information on the reliability and validity of this scale is available from Dr. Whatley.

Interreligious Marriages

National Data

Of married couples in the United States, 37% have an interreligious marriage (Pew Research 2008).

Although religion may be a central focus of some individuals and their marriage, Americans in general have become more secular, and religion has become less influential as a criterion for selecting a partner as a result. In a survey of 2,922 undergraduates, slightly over a third (33.6%) reported that marrying someone of the same religion was important for them (Knox and Hall 2010). Chelsea Clinton, a Methodist, married Marc Mezvinsky, who is Jewish. They had an interfaith ceremony.

Are people in interreligious marriages less satisfied with their marriages than those who marry someone of the same faith? The answer depends on a number of factors. First, people in marriages in which one or both spouses profess "no religion" tend to report lower levels of marital satisfaction than those in which at least one spouse has a religious tie. People with no religion are often more liberal and less bound by traditional societal norms and values; they feel less constrained to stay married for reasons of social propriety.

The impact of a mixed religious marriage may also depend more on the devoutness of the partners than on the fact that the partners are of different religions. If both spouses are devout in their respective religious beliefs, they may expect some problems in the relationship. Less problematic is the relationship in which one spouse is devout but the partner is not. If neither spouse in an interfaith marriage is devout, problems regarding religious differences may be minimal or nonexistent. In their marriage vows, one interfaith couple who married (he was Christian, she was Jewish) said that they viewed their different religions as an opportunity to strengthen their connections to their respective faiths and to each other. "Our marriage ceremony seeks to celebrate both the Jewish and Christian traditions, just as we plan to in our life together."

Cross-National Marriages

Of 2,922 undergraduates, 77.3% of the men and 67.4% of the women reported that they would be willing to marry someone from another country (Knox and Hall 2010). The opportunity to meet someone from another country is increasing as more than 700,000 foreign students are studying at American colleges and universities. Because not enough Americans are going into math and engineering, these foreign students are wanted because they may find the cure for cancer or invent a vaccine for HIV (Marklein 2008).

This couple met when the female was on a study abroad program in Australia. They married and have traveled with their children throughout Australia.

Because American students take classes with foreign students, there is the opportunity for dating and romance between the two groups, which may lead to marriage. Some people from foreign countries marry an American citizen to gain citizenship in the United States, but immigration laws now require the marriage to last two years before citizenship is granted. If the marriage ends before two years, the foreigner must prove good faith (that the marriage was not just to gain entry into the country) or will be asked to leave the country.

When the international student is male, more likely than not his cultural mores will prevail and will clash strongly with his American bride's expectations, especially if the couple should return to his country. One female American student described her experience of marriage to a Pakistani, who violated his parents' wishes by not marrying the bride they had chosen for him in childhood. The marriage produced two children before the four of them returned to Pakistan.

The woman felt that her in-laws did not accept her and were hostile toward her. The in-laws also imposed their religious beliefs on her children and took control of their upbringing. When this situation became intolerable, the woman wanted to return to the United States. Because the children were viewed as being "owned" by their father, she was not allowed to take them with her and was banned from even seeing them. Like many international students, the husband was from a wealthy, high-status family, and the woman was powerless to fight the family. The woman has not seen her children in six years.

Cultural differences do not necessarily cause stress in cross-national marriage; the degree of cultural difference is not necessarily related to degree of stress. Much of the stress is related to society's intolerance of cross-national marriages, as manifested in attitudes of friends and family. Japan and Korea place an extraordinarily high value on racial purity. At the other extreme is the racial tolerance evident in Hawaii, where a high level of out-group marriage is normative.

Age-Discrepant Relationships and Marriages

Although people in most pairings are of similar age, sometimes the partners are considerably different in age. In marriage, these are referred to as ADMs (age-dissimilar marriages) and are in contrast to ASMs (age-similar marriages). ADMs are also known as **May-December marriages**. Typically, the woman is in the spring of her youth (May) whereas the man is in the later years of his life (December). There have been a number of May-December celebrity marriages, including that of Celine Dion, who is twenty-six years younger than René Angelil (in 2011, aged 42 and 68). Larry King is also twenty-six years older than his seventh wife, Shawn (in 2011 he was 76). Michael Douglas is twenty-five years older than his wife, Catherine Zeta Jones, and Ellen DeGeneres is fifteen years older than Portia de Rossi, her partner (they "married" in 2008). At age 84, Hugh Hefner became engaged to Crystal Harris, age 24. "She deserves to be my widow" he said. (She subsequently called all the wedding)

One might assume that these marriages are less happy because the spouses were born into such different age contexts. Research shows otherwise. Barnes and Patrick (2004) compared thirty-five ADMs (in which spouses were fourteen or more years apart) and thirty-five ASMs (in which spouses were less than five years apart) and found no difference in reported marital satisfaction between the two groups. As is true in other research, wives reported lower marital satisfaction and more household responsibilities in both groups.

May-December marriage age dissimilar marriage (ADM) in which the woman is typically in the spring of her life (May) and her husband is in the later years (December).

Thirty two years separate this couple, but their spirits and hearts are very much together.

Perhaps the greatest example of a May-December marriage that "worked" is of Oona Chaplin, wife of Charles Chaplin. She married him when she was age 18 (he was age 54). Their May-December alliance was expected to last the requisite six months, but they remained together for 34 years (until his death at 88) and raised eight children, the last of which was born when Chaplin was 73.

There are definite benefits for the older man (but not the younger woman) in the age-discrepant relationship. Drefahl (2010) studied the relationship between age gap and longevity and found that having a younger spouse is beneficial for men but detrimental for women. One explanation is health selection—healthier males are able to attract younger partners. Hence, an older male married to a younger woman may have lower mortality since he is healthier. He may also benefit since the younger woman provides health care and social support. Women tend not to benefit from having younger partners since social norms are less open for women marrying younger men and women have more social relationships/contacts/supports than men (so they have less need for men).

Although less common, some age-discrepant relationships are those in which the woman is older than her partner. Mary Tyler Moore (married twenty-five years) is eighteen years older than her husband, Robert Levine. Demi Moore is sixteen years older than husband Ashton Kutcher (she was 40 and he was 27 at the time of the wedding). Valerie Gibson (2002) is the author of *Cougar: A Guide for Older Women Dating Younger Men*. She noted that the current use of the term **cougars** refers to "women, usually in their thirties and forties, who are financially stable and mentally independent and looking for a younger man to have fun with." Gibson noted that one-third of women between the ages of 40 and 60 are dating younger men. Financially independent women need not select a man in reference to his breadwinning capabilities. Instead, these "cougars" are looking for men, not to marry but to enjoy. The downside of such relationships comes if the man gets serious and wants to have children, which may spell the end of the relationship.

Marriage Quality

A successful marriage is the goal of most couples. But how is the term successful marriage defined and what are its characteristics?

Cougar a woman, usually in her thirties or forties, who is financially stable and mentally independent and looking for a younger man with whom to have fun.

A successful marriage requires falling in love many times, always with the same person.

Mignon McLaughlin,
American journalist

Marriage Quality

209

Definition and Characteristics of Successful Marriages

Sometimes love will bloom in the springtime

Then like flowers in summer it will grow

Just to fade away in the winter

When the cold winds begin to blow

But when it's evergreen, evergreen

It will last through the summer and winter too

When love is evergreen, evergreen

Like my love for you

Roy Orbison, "Evergreen"

Marital success refers to the quality of the marriage relationship measured in terms of marital stability and marital happiness. Stability refers to how long the spouses have been married and how permanent they view their relationship, whereas marital happiness refers to more subjective/emotional aspects of the relationship. In describing marital success, researchers have used the terms *satisfaction, quality, adjustment, lack of distress,* and *integration.* Marital success is often measured by asking spouses how happy they are, how often they spend their free time together, how often they agree about various issues, how easily they resolve conflict, how sexually satisfied they are, how equitable they feel their relationship is, and how often they have considered separation or divorce. Researchers Van Laningham and Johnson (2009) found that a one item question such as "How happy are you in your marriage?" with options "very happy," "pretty happy," and "not very happy' is just as effective in assessing one's marital happiness as asking a series of questions. Lavner and Bradbury (2010) studied marital satisfaction in newlywed couples over a four year period. Marital satisfaction declined across the four years with steeper declines associated with couples where one or both partners evidenced relatively high levels of negative personality, chronic stress, and aggression. Similarly, higher levels of relationship satisfaction were associated with the absence of these characteristics and high levels of positive affect.

Researchers have also identified characteristics associated with enduring happy marriages (DeMaris 2010; Amato et al. 2007). Their findings and those of other researchers include the following:

1. *Intimacy.* Patrick et al. (2010) found that intimacy (feeling close to spouse, showing physical affection, sharing ideas/events, sharing hobbies/leisure activities) was related to marital satisfaction. They studied both spouses in 124 marriages of employees at two major state universities in the Midwest. Most were in the mid-to-late forties, married an average of 20 years, and 80% in their first marriages.

2. *Communication.* Gottman and Carrere (2000) studied the communication patterns of couples over an eleven-year period and emphasized that those spouses who stay together are five times more likely to lace their arguments with positives ("I'm sorry I hurt your feelings.") and to consciously choose to say things to each other that nurture the relationship rather than destroy it. Successful spouses also feel comfortable telling each other what they want and not being defensive at feedback from the partner.

3. *Common interests.* Spouses in happy stable marriages talked of sharing interests, values, goals, children, and the desire to be together. They "respect" and "cherish" each other and the time they spend together.

DIVERSITY IN OTHER COUNTRIES

Though marital happiness is an important concept in American marriages, this is "not necessarily a salient concept in Chinese society" (Pimentel 2000, 32). What is important to the Chinese is the allegiance of the husband to his family of origin (his parents). Also, the five basic relationships in order of importance in Confucian philosophy are ruler-minister, father-son, elder brother-younger brother, husband-wife, and friend-friend.

4. *Not materialistic.* Being nonmaterialistic was a characteristic of these happily married couples. Although the couples may have lived in nice houses and had expensive possessions (for example, a boat and camper), they were tied to nothing. "You can have my things, but don't take away my people," is a phrase from one husband reflecting his feelings about his family. Being nonmaterialistic also implies not focusing on work exclusively but making time for one's relationships.

5. *Role models.* The couples spoke of positive role models in their parents. Good marriages beget good marriages; i.e., good marriages run in families. It is said that the best gift you can give your children is a good marriage.

6. *Religiosity.* Feeling close to God and frequent church attendance increase one's commitment to the marriage (Dowd 2009). One Mennonite man said, "The way to get close to your spouse is to get close to God." People with no religious affiliation report more marital problems and are more likely to divorce

(Amato et al. 2007). Marks et al. (2010) also noted the impact of religion on couples. They looked at gift giving on the part of religious couples, how doing so is viewed as a gift rather than a sacrifice, and what this says about the solidarity of the couple. A quote reflects the thinking:

> I sacrifice nothing because I don't look at it as a sacrifice. I look at it as: God has given me a gift to do what He would want me to do, so I can't say that I sacrifice anything. That's the same way that I look at [what] my wife [gives and does] . . . [I]f we have a relationship together, we both do know God . . . I don't look at it as a sacrifice. I think it's a gift . . . [We] work together, understanding that God is in charge of everything that we have to do in a household. (p. 442)

7. _Trust._ Trust in the partner provided a stable floor of security for the respective partners and their relationship. Neither partner feared that the other partner would leave or become involved in another relationship. "She can't take him anywhere he doesn't want to go" is a phrase from a country-and-western song that reflects the trust that one's partner will be faithful.

8. _Personal and emotional commitment to stay married._ Divorce was not considered an option. The spouses were committed to each other for personal reasons rather than societal pressure. In addition, the spouses were committed to maintain the marriage out of emotional rather than economic need (DeOllos 2005).

9. _Sexual desire._ Wilson and Cousins (2005) confirmed that partners' similar rankings of sexual desire is important in predicting long-term relationship success. Earlier, we noted the superiority of marital sex over sex in other relationship contexts in terms of both emotion and physical pleasure.

10. _Equitable relationships._ Amato et al. (2007) observed that the decline in traditional gender attitudes and the increase in egalitarian decision making were related to increased happiness in today's couples. DeMaris (2010) also found that spouses who regarded their relationships as those with equal contribution reported higher quality marriages.

11. _Marriage/connection rituals._ **Marriage rituals** are deliberate repeated social interactions that reflect emotional meaning to the couple. **Connection rituals** are those which occur daily in which the couple share time and attention. Allgood and Bakker (2009) surveyed eighty couples (married from 1 to 50 years) who identified a total of 2,812 connection rituals in their relationships. The most frequent categories were physical (kissing, hugging, massaging, sexual activity), communication (face-to-face, phone calls, texts), religiosity (prayer, church attendance, scripture reading), and idioms (nicknames, pet names, "secret" language). These rituals were associated with marital satisfaction.

12. _Absence of negative attributions._ Spouses who do not attribute negative motives to their partner's behavior report higher levels of marital satisfaction than spouses who ruminate about negative motives. Dowd et al. (2005) studied 127 husbands and 132 wives and found that the absence of negative attributions was associated with higher marital quality.

13. _Forgiveness._ At some time in all marriages, each spouse engages in behavior that may hurt the partner. Forgiveness rather than harboring and nurturing resentment allows spouses to move forward. Spouses who do not "drop the lowest test score" (an academic metaphor) find that they inadvertently create a failing marriage in which they then must live. McNulty (2008) noted the value of forgiveness particularly when the married partners rarely behaved negatively. Couple recovery from a transgression not only involves forgiveness but reconciliation—the mutual commitment to an enhanced positive future (Eckstein et al. 2009).

14. _Economic security._ Although money does not buy happiness, having a stable, secure economic floor is associated with marital quality (Amato et al. 2007) and marital happiness (Mitchell 2010). North et al. (2008) examined the role of income and social support in predicting concurrent happiness and change in happiness among 274 married adults across a ten-year period. They

Marriage rituals deliberate repeated social interactions that reflect emotional meaning to the couple.

Connection rituals rituals which occur daily in which the couple share time and attention.

A married man should forget his mistakes; no use two people remembering the same thing.

Duane Dewel, journalist and state legislator

found that income had a small, positive impact on happiness, which diminished as income increased. In contrast, family social support, as reflected in cohesion, expressiveness, and low conflict showed a substantial, positive association with concurrent happiness, particularly when income was low.

15. *Health.* Increasingly, research emphasizes that the quality of family relationships affects family member health and that the health of family members influences the quality of family relationships and family functioning (Proulx and Snyder 2009). Indeed, such an association begins early as the stress of a dysfunctional family environment can activate the physiological responses to stress, change the brain structurally, and leave children more vulnerable to negative health outcomes. High conflict spouses also create chronic stress, high blood pressure, and depression.

The family is also the primary socialization unit for physical health in reference to eating nutritious food, avoidance of smoking, getting regular exercise, etc. The first author was reared in a home where a high fat diet was routine and the father was a chronic smoker who never exercised. He died of a coronary at age 46, illustrating the need for attending to one's health.

Corra et al. (2009) analyzed data collected over a thirty-year period, from the 1972 to 2002 General Social Surveys, to discover the influence of sex (male or female) and race (white or black) on the level of reported marital happiness. Findings indicated greater levels of marital happiness among males and white people than among females and black people. The researchers suggested that males make fewer accommodations in marriage and that white people are not burdened with racism and have less economic stress. Amato et al. (2007) also reported less marital happiness and more marital problems among wives than husbands.

Theoretical Views of Marital Happiness and Success

Interactionists, developmentalists, exchange theorists, and functionalists view marital happiness and success differently. Symbolic interactionists emphasize the subjective nature of marital happiness and point out that the definition of the situation is critical.

A happy marriage exists only when spouses define the verbal and nonverbal behavior of their partner as positive, and only when they label themselves as being

SELF-ASSESSMENT | Satisfaction with Married Life Scale

The Satisfaction with Married Life Scale (SMLS) consists of five items which measure marital satisfaction.

The following are five statements with which you may agree or disagree. Using the 1–7 scale, indicate your agreement with each item by writing the appropriate number on the line in front of that item. Please be open and honest in responding to each item; there are no right or wrong answers.

Strongly disagree (1), Disagree (2), Slightly disagree (3), Neither agree nor disagree (4), Slightly agree (5), Agree (6), Strongly agree (7).

_____ 1. In most ways my married life is close to ideal.
_____ 2. The conditions of my married life are excellent.
_____ 3. I am satisfied with my married life.
_____ 4. So far I have gotten the important things I want in my married life.
_____ 5. If I could live my married life over, I would change almost nothing.

Scoring

Add the numbers you wrote. The marital satisfaction score will range from 5 to 35. For purposes of this study a couple's combined marital satisfaction score was calculated by summing both partners' scores, resulting in a possible score range of 10 to 70, with higher scores indicating greater marital satisfaction for the couple. The internal consistency of the SWML has been reported with a Cronbach's alpha of .92 along with some evidence of construct validity (Johnson et al. 2006).

Source

Johnson, H. A., Zabriskie, R. B., and Hill, B. (2006) The contribution of couple leisure involvement, leisure time, and leisure satisfaction to marital satisfaction. *Marriage and Family Review* 40: 69–91.

Ward, Peter J., Lundberg, Neil R., Zabriskie, Ramon B. and Berrett, Kristen (2009) "Measuring Marital Satisfaction: A Comparison of the Revised Dyadic Adjustment Scale and the Satisfaction with Married Life Scale," *Marriage & Family Review*, 45:4, 412–429. Used with permission.

in love. Hence, marital happiness is not defined by the existence of eight or more specific criteria but is subjectively defined by the respective partners.

Family developmental theorists emphasize the developmental tasks that must be accomplished to enable a couple to have a happy marriage. Wallerstein and Blakeslee (1995) identified several of these tasks, including separating emotionally from one's parents, building a sense of "we-ness," establishing an imaginative and pleasurable sex life, and making the relationship safe for expressing differences.

Exchange theorists focus on the exchange of behavior of a kind and at a rate that is mutually satisfactory to both spouses. When spouses exchange positive behaviors at a high rate, they are more likely to feel marital happiness than when the exchange is characterized by high-frequency negative behavior.

Structural functionalists see marital happiness as contributing to marital stability, which is functional for society. When two parents are in love and happy, the likelihood that they will stay together to provide physical care and emotional nurturing for their offspring is increased. Furthermore, when spouses take care of their own children, society is not burdened with having to pay for the children's care through welfare payments, paying foster parents, or paying for institutional management (group homes) when all else fails. Happy marriages also involve limiting sex to each other. In their national sex survey, Michael et al. (1994) reported, "[H]appiness is clearly linked to having just one partner— which may not be too surprising since that is the situation that society smiles upon" (p. 130). Fewer cases of HIV also mean lower medical bills for society. Similarly, marriage is associated with improved health (Stack and Eshleman 1998) because spouses monitor each other's health and encourage or facilitate medical treatment as indicated.

Couple Identification of the Conditions of Marital Happiness

South et al. (2010) confirmed that marital satisfaction is associated with how much spouses accept the frequency of both positive and negative behaviors in their relationship. In general, spouses look for a high frequency of positive behavior and a low frequency of negative behavior.

Mitchell (2010) interviewed 390 married couples who were British, Chinese, Indo/East Indian, and Southern-European and asked them why they were happy or unhappy. The happy couples reported that they had adequate "quality time" to spend together and were compatible with each other due to common interests. Those who were unhappy said that they were too busy with competing demands of caregiving and work, that they had no common interests, that they had different goals/values, and that there was lack of trust, love, and passion.

Phillips and Wilmoth (2010) reported survey data on 71 African-American married couples who were identified as "long term couples" by their clergy. The spouses were in their mid-fifties and had been married an average of 32 years. "God/Jesus" and "love" were identified by 51% and 31% of the spouses as the top reason their marriage had lasted so long. Trust/infidelity (25%) was the biggest challenge/obstacle to overcome in their marriage and money (24%) and decision making/communication (20%) were the top issues over which they disagreed.

Marital Happiness across Time

Anderson et al. (2010) analyzed longitudinal data of 706 individuals over a 20 year period. Over 90% were in their first marriage; most had two children and 14 years of education. Reported marital happiness, marriage problems, time spent together and economic hardship were assessed. Five patterns emerged:

1. High stable 2 (started out happy and remained so across time) = 21.5%
2. High stable 1 (started out slightly less happy and remained so across time) = 46.1%
3. Curvilinear (started out happy, slowly declined, followed by recovery) = 10.6%

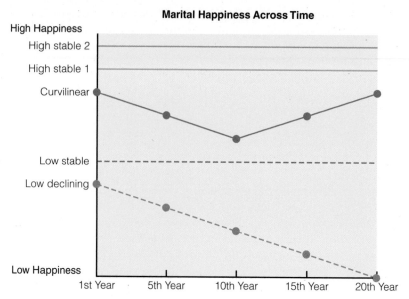

Figure 7.1 Trajectories of Marital Happiness.

4. Low stable (started out not too happy and remained so across time) = 18.3%

5. Low declining (started out not too happy and declined across time) = 3.6%

The researchers found that, for couples who start out with a high level of happiness, they are capable of rebounding if there is a decline. But for those who start out at a low level, the capacity to improve is more limited. These data are in contrast to previous research by Valliant and Vaillant (1993) who suggested that all couples show a gradual decline and a bounce when the children leave home. This new research shows the complexity of marital happiness patterns.

Plagnol and Easterlin (2008) conceptualized happiness as the ratio of aspirations and attainments in the areas of family life and material goods. They studied 47,000 women and men and found that up to age 48, women are happier than men in both domains. After age 48, a shift causes women to become less satisfied with family life (their children are gone) and material goods (some are divorced and have fewer economic resources). In contrast, men become more satisfied with family life (the empty nest is more a time of joy) and finances (men typically have more economic resources than women, whether married or divorced). Individuals who are black and/or with lower education report less happiness.

Healthy Marriage Initiative

The Healthy Marriage Initiative (http://www.acf.hhs.gov/healthymarriage/) was set up in the Department of Health and Human Services to provide education and support for relationships, marriages, and families. Research has confirmed the positive effect of such programs not only on the adult couple relationship but also on the parent-child relationship with positive implications for child well being (Calligas et al. 2010). Tompkins et al. (2010) studied the success of one such program (Family Success in Adams County) which used a holistic approach whereby relationship education, parenting skills, and financial management were provided across 8 hours of class. Results revealed significant positive changes in parenting alliance, psychological well-being, and stress levels. Of note is that over half of the participants were minorities below the poverty line, confirming success of the program across income and ethnicity lines.

The Future of Marriage Relationships

Diversity will continue to characterize marriage relationships of the future. The traditional model of the husband provider, stay-at-home mom, and two children will continue to transition to other forms including more women in the work force, single parent families, and smaller families. What will remain is the intimacy/companionship focus that spouses expect from their marriages.

Openness to interracial, interreligious, cross-national, age-discrepant relationships will increase. The driving force behind this change will be individualism which discounts parental familial concerns. An increased global awareness, international travel, and international students will facilitate a mindset of stretched norms in terms of one's selection of a partner.

Summary

What are the individual motivations for and societal functions of marriage?

Individual motives for marriage include personal fulfillment, companionship, legitimacy of parenthood, and emotional and financial security. Societal functions include continuing to provide society with socialized members, regulating sexual behavior, and stabilizing adult personalities.

What are three levels of commitment in marriage?

Marriage involves a commitment—person-to-person, family-to-family, and couple-to-state. Covenant marriages, available in Louisiana, Arizona, and Arkansas, are selected by couples who are more religious/traditional, report happier marriages, fewer marital problems, and lower divorce. However, only 2% of those getting married in Louisiana elected to have a covenant marriage.

What are two rites of passage associated with marriage?

While weddings are often a source of stress, they are a rite of passage that is both religious and civil. To the Catholic Church, marriage is a sacrament that implies that the union is both sacred and indissoluble. According to Jewish and most Protestant faiths, marriage is a special bond between the husband and wife sanctified by God, but divorce and remarriage are permitted. Wedding ceremonies still reflect traditional cultural definitions of women as property.

The wedding is a rite of passage signifying the change from the role of fiancé to the role of spouse. Women, more than men, are more invested in preparation for the wedding, the wedding is more for the bride's family, and women prefer a traditional wedding. Most spouses report a positive wedding night experience with exhaustion from the wedding or reception being a problem. The honeymoon is a time of personal recuperation and making the transition to the new role of spouse.

What changes might a person anticipate after marriage?

Changes after the wedding are legal (each becomes part owner of all income and property accumulated during the marriage), personal (enhanced self-concept), social (less time with friends), economic (money spent on entertainment in courtship is diverted to living expenses and setting up a household), sexual (less frequency), and parental (improved relationship with parents). The greatest change is disenchantment—moving from a state of exhilaration to a state of mundaneness. New spouses also report experiencing a loss of freedom, increased responsibility, and changes to how money is spent.

What are examples of diversity in marriage relationships?

Hispanic families tend to marry earlier, and have higher rates of marriage and higher fertility. Male power and strong familistic values also characterize Hispanic families. Canadian families are unique in that they exist in a bilingual (English and French) country, may be common-law with full legal recognition, and have half the divorce rate of marriages in the United States. Same-sex relationships in Canada also have greater approval and government protection, including legal marriage in three provinces. In Muslim-American families, norms involve no premarital sex, close monitoring of children, and intense nurturing of the parental-child bond. Military families cope with deployment and the double standard.

Mixed marriages include interracial, interreligious, and age-discrepant. About 15% of all marriages in the U.S. are interracial with Hispanic–non-Hispanic the most frequent pairing. Interracial marriages are also more likely to dissolve than same-race marriages. The success of interreligious marriage is related to the degree of devoutness of the partners in the respective religions. If both are very devout, the conflict is more likely to surface. In regard to age-discrepant relationships, when age-discrepant and age-similar marriages are compared, there are no differences in regard to marital happiness. Men who marry younger women seem to benefit in terms of living longer.

What are the characteristics associated with successful marriages?

Marital success is defined in terms of both quality and durability. Characteristics associated with marital success include intimacy, commitment, common interests, communication, religiosity, trust, and nonmaterialism, and having positive role models, low stress levels, and sexual desire.

Marital couples reflect various happiness patterns across time. While some start out happy and remain happy, others start out low and go lower. Still others start out moderately happy and continue this pattern.

What is the future of marriage relationships?

Marriages and families will continue to be characterized by diversity. The traditional model of the family will continue to transition to other forms. What will remain is the intimacy/companionship focus that spouses expect from their marriages. Openness to interracial, interreligious, cross-national, age-discrepant relationships will increase. An increased global awareness, international travel, and international students will facilitate a mindset of stretched norms in terms of one's selection of a partner.

Key Terms

Artifact

Bride wealth

Commitment

Connection rituals

Cougar

Covenant marriage

Disenchantment

Honeymoon

Marital success

Marriage rituals

May-December marriage

Rite of passage

Web Links

African American Marriage
http://www.healthymarriageinfo.org/marriage-and-culture/african-americans-and-the-black-community

Asian and Pacific Islander Marriages
http://www.healthymarriageinfo.org/marriage-and-culture/hispanics-and-latinos

Bridal Registry
http://www.theknot.com

Brides and Grooms
http://www.bridesandgrooms.com/

Facts about Marriage
http://www.cdc.gov/nchs/fastats/divorce.htm

Hispanic and Latino Marriage
http://www.healthymarriageinfo.org/marriage-and-culture/hispanics-and-latinos

Marriage Relationships
http://www.heartchoice.com/hc/marriage/marriage.php

National Healthy Marriage Resource Center
http://www.healthymarriageinfo.org/

Marriage Builders
http://marriagebuilders.com/

Marriage Success Training
http://www.stayhitched.com/

Military Marriages
http://www.nmfa.org/ (The National Military Family Association)

Native American Marriage
http://www.healthymarriageinfo.org/marriage-and-culture/native-americans

Smart Marriages
http://www.smartmarriages.com/

Traditional Korean Marriage
http://www.lifeinkorea.com/culture/marriage/marriage.cfm

Wedding Webcasts
http://www.webcastmywedding.net/

CHAPTER 8

Same-Sex Couples and Families

I'm a supporter of gay rights. And not a closet supporter either. From the time I was a kid, I have never been able to understand attacks upon the gay community. There are so many qualities that make up a human being. . . by the time I get through with all the things that I really admire about people, what they do with their private parts is probably so low on the list that it is irrelevant.

The late Paul Newman

Olivia Holloway

Learning Objectives

Specify the prevalence of homosexuality, bisexuality, and same-sex couples.

Discuss the origins of homosexuality and the degree to which gays can "change."

Review homonegativity, homophobia, and biphobia.

Summarize gay, lesbian, bisexual, and mixed-orientation relationships.

Identify the process of coming out and the issues of same-sex marriage.

Review GLBT Parenting Issues.

Predict the future of gay issues in the United States.

1. Homosexuals are more likely to keep a secret from a romantic partner than heterosexuals.

2. Heterosexual females who have kissed a girl have more favorable attitudes toward homosexuals than those who have not done so.

3. Gay individuals are most likely to come out to a friend, then their mother, then their father.

4. Children reared in lesbian families are more likely to feel that they are gay than children reared in heterosexual families.

5. Homosexuals are far more likely than any other minority group in the United States to be victimized by violent hate crime.

Answers: 1. T   2. T   3. T   4. F   5. T

Aware of prejudice and discrimination against LGBT (lesbian, gay, bisexual, transgender) individuals on campus, a number of colleges and universities have implemented educational interventions with names such as Safe Zone, Safe Space, Safe Harbor, and Safe on Campus. After the training sessions each participant (usually faculty/staff) receives a Safe Zone sticker to put on their door which indicates they have been through the training and can be counted on to be understanding, supportive, and available for help/advice for those with concerns of sexual orientation and gender identity. The symbol also means that homophobic and heterosexist comments and actions will not be tolerated, but will be addressed in an educational and informative manner. The mere existence of such "safe zone" programs on campus reveals that prejudice and discrimination against LGBT individuals is alive and well. This chapter embraces persons whose sexual orientation is not heterosexual.

Some same-sex identified individuals do not like the term homosexual. It originated in the late 19th- and early 20th-century when there was a search for the "cure" for homosexuality (Adams et al. 2007). In the United States, homosexuality remains a subject over which there continues to be wide differences of approval. Although some states grant marriage licenses to gay couples (see Figure 8.2 later in the chapter), others have defined marriage as the exclusive union between a woman and a man. Worldwide, approval differences are considerable, with the Netherlands, Spain, Belgium, Norway, and South Africa granting equal marriage rights to same-sex couples, whereas intense discrimination is the norm in other countries (for example, Pakistan and Kenya). Hate crimes against gays are not uncommon, even in America. Matthew Shepard was a 21-year-old student at the University of Wyoming who was tortured and murdered because he was gay.

In this chapter, we discuss same-sex couples' and families' relationships that are, in many ways, similar to heterosexual ones. A major difference, however, is that gay and lesbian couples and families are subjected to prejudice and discrimination. Although other minority groups (e.g., racial) also experience prejudice and discrimination, sexual identity

Chelsea E. Curry

Many universities are now encouraging faculty to put "safe zone" stickers on their office doors to alert gay/transgendered individuals that their office is an approving "safe" context.

minorities are denied federal legal marital status and the benefits and responsibilities that go along with marriage (which we discuss later in this chapter). Also, gay couples are sometimes rejected by their own parents, siblings, and other family members. One father told his son, "I'd rather have a dead son than a gay son."

Same-sex behavior has existed throughout human history and in most (perhaps all) human societies (Kirkpatrick 2000). In this chapter, we focus on Western views of sexual diversity that define **sexual orientation** (also known as **sexual identity**) as a classification of individuals as heterosexual, bisexual, or gay, based on their emotional, cognitive, and sexual attractions and self-identity. **Heterosexuality** refers to the predominance of cognitive, emotional, and sexual attraction to individuals of the other sex. **Homosexuality** refers to the predominance of cognitive, emotional, and sexual attraction to individuals of the same sex, and **bisexuality** is cognitive, emotional, and sexual attraction to members of both sexes. The term **lesbian** refers to women who prefer same-sex partners; **gay** can refer to either women or men who prefer same-sex partners.

The term queer is typically used by males, but can also be used by females, as a self-identifier that the person has a sexual orientation other than heterosexual. Traditionally, the term queer was used to denote a gay person and the connotation was negative. More recently individuals use the term queer with pride much the same way African-Americans called themselves black during the 1960s Civil Rights Era as a matter of building ethnic pride and identity. It also has shock value, which some people seem to savor when they introduce themselves as being queer.

Alissa R. King (2010) uses queer:

> as an 'umbrella' term that does not necessarily designate the user's sexual identity. It's a more inclusive term; thus someone who labels him/herself as "queer" could be gay, lesbian, bisexual, pansexual, trans, intersexed, non-conforming heterosexual,

Sexual orientation classification of individuals as heterosexual, bisexual, or homosexual, based on their emotional, cognitive, sexual attractions and self-identity.

Sexual identity term used synonymously with sexual orientation.

Heterosexuality emotional and sexual attraction to individuals of the other sex.

Homosexuality refers to the predominance of cognitive, emotional, and sexual attraction to individuals of the same sex.

Bisexuality cognitive, emotional, and sexual attraction to members of both sexes.

Lesbian a woman who prefers same-sex partners.

Gay term which refers to women or men who prefer same-sex partners.

If advances in the understanding of sexual orientations are to be made, it is critical that definitions and measures of sexual orientation be standardized.

Randall Sell, researcher

A person may use the term "queer" with pride, which suggests an ambiguous sexual identity.

etc. "Queer" is a tricky term to explain because of its ugly history and it is helpful to be aware of all the "possibilities" under that umbrella—i.e., pansexual, intersexed, trans, genderqueer, etc. The clear distinguisher between gay and queer is that if someone identifies as bisexual or pansexual, he or she may not feel as at home using the label "gay" because that implies sexual attraction/behavior with one other person of a specific gender identity, whether it's used to describe same-sex identified men or women. However, "queer" provides a bit more maneuvering room (fluidity) to encompass deviations from same- or other-sex identifications (gay/lesbian or heterosexual).

Queer theory refers to a movement/theory dating from the early 1990s. Queer theorists want there to be less labeling of sexual orientation and a stronger "anyone can be anything they want" attitude. The theory is that society should support a more fluid range of sexual orientations and that individuals can move through the range as they become more self-aware. Queer theory may be seen as a very specific subset of gender/human sexuality studies (Jones 2010).

The terms LGBT or **GLBT** are often used to refer collectively to lesbians, gays, bisexuals, and transgendered individuals. A more inclusive term today is LGBITQ (lesbian, gay, bisexual, intersexed, transgender, queer) or GLBITQ (gay, lesbian, bisexual, intersexed, transgender, queer). Some researchers regard the Q as meaning "questioning."

GLBT general term which refers to gay, lesbian, bisexual, and transgender individuals.

Identifying and Classifying Sexual Orientation

The classification of individuals into sexual orientation categories (for example, heterosexual, homosexual, bisexual) is problematic for a number of reasons (Savin-Williams 2006). First, because of the social stigma associated with nonheterosexual identities, many individuals conceal or falsely portray their sexual-orientation identities to avoid prejudice and discrimination. In a study of 420 GLBT respondents at a large southeastern university, almost three fourths (72%) reported that they "sometimes or often avoided disclosing their GLBT status due to fear of negative consequences, harassment, or discrimination;" 67% sometimes or often concealed their GLBT status to avoid intimidation. The most common sources of negative treatment due to GLBT status were students (48%), followed by fraternity and sorority members (30%), and colleagues/co-workers (26%) (Mooney 2009).

Second, not all people who are sexually attracted to or have had sexual relations with individuals of the same-sex view themselves as homosexual or bisexual. A final difficulty in labeling a person's sexual orientation is that an individual's sexual attractions, behavior, and identity may change across time. One's sexual orientation is, indeed, fluid. For example, in a longitudinal study of 156 lesbian, gay, and bisexual youth, 57% consistently identified as gay or lesbian and 15% consistently identified as bisexual over a one-year period, but 18% transitioned from bisexual to lesbian or gay (Rosario et al. 2006).

Early research on sexual behavior by Kinsey and his colleagues (1948; 1953) found that, although 37% of men and 13% of women had had at least one same-sex sexual experience since adolescence, few of the individuals reported exclusive homosexual behavior. In a study of 243 undergraduates, Vrangalova and Savin-Williams (2010) found that 84% of the heterosexual women and 51% of the heterosexual men reported the presence of at least one same-sex quality—sexual attraction, fantasy, or behavior. These data emphasize that people are not exclusively heterosexual or homosexual. Rather, Kinsey suggested an individual's sexual orientation may have both heterosexual and homosexual elements. In other words, Kinsey suggested that heterosexuality and homosexuality represent two ends of a sexual-orientation continuum and that most individuals are neither

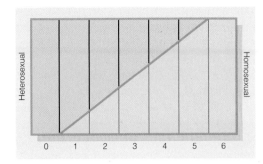

Based on both psychologic reactions and overt experience, individuals rate as follows:

0. Exclusively heterosexual with no homosexual
1. Predominantly heterosexual, only incidentallly homosexual
2. Predominantly heterosexual, but more than incidentally homosexual
3. Equally heterosexual and homosexual
4. Predominantly homosexual, but more than incidentally heterosexual
5. Predominantly homosexual, but incidentally heterosexual
6. Exclusively homosexual

W.B. Saunders, *Sexual Behavior in the Human Male*, The Kinsey Institute. Reprinted by permission of the Kinsey Institute for Research in Sex, Gender, and Reproduction, Inc.

Figure 8.1 The Heterosexual-Homosexual Rating Scale.

entirely homosexual nor entirely heterosexual, but fall somewhere along this continuum (Drucker 2010).

The Heterosexual-Homosexual Rating Scale that Kinsey et al. (1953) developed allows individuals to identify their sexual orientation on a continuum. Individuals with ratings of 0 or 1 are entirely or largely heterosexual; 2, 3, or 4 are more bisexual; and 5 or 6 are largely or entirely homosexual (see Figure 8.1). Very few individuals are exclusively a 0 or 6, prompting Kinsey to believe that most individuals are bisexual.

Sexual-orientation classification is also complicated by the fact that sexual behavior, attraction, love, desire, and sexual-orientation identity do not always match. For example, "research conducted across different cultures and historical periods (including present-day Western culture) has found that many individuals develop passionate infatuations with same-gender partners in the absence of same-gender sexual desires . . . whereas others experience same-gender sexual desires that never manifest themselves in romantic passion or attachment" (Diamond 2003, 173).

Prevalence of Homosexuality, Heterosexuality, and Bisexuality

Despite the difficulties inherent in categorizing individuals' sexual orientation, considerable data reveal the prevalence of individuals in the United States who identify as lesbian, gay, or bisexual. In a national survey by Michael et al. (1994), fewer than 2% of women and 3% of the men identified themselves as homosexual or bisexual. Berg and Lein (2006) estimated that 7% of males and 4% of

What if I Am Attracted to Someone of the Same Sex?

Morgan et al. (2010) observed that questioning one's sexual identity is not unusual. In a study of 184 male college students, even among those who reported that they were "exclusively straight," over half (53%) reported having questioned their sexual orientation. The researchers noted that conventional notions of a "standardized" heterosexual identity "appear simplistic."

Although most people tend to be attracted to other-sex people most of the time, it is not uncommon to find one's self attracted to someone of the same sex. Such attractions are usually suppressed because of the cultural heterosexual bias. Where there are physical attractions as well, the person has a harder time diverting them. Gay people typically say that sex with another-sex partner was "OK but nothing like the explosive feelings I felt with a same-sex partner." What one does with these feelings depends on the strength of the feelings, the imperative to "be true to one's self," and the social context. Although some deny or suppress the feelings or attractions, others explore them.

females were not heterosexual. A 2004 national poll showed that about 5% of U.S. high school students identified themselves as lesbian or gay (Curtis 2004). Tao (2008) analyzed U.S. women ages 15 to 44 and found that 1.6% and 4%, respectively, self-identified as being lesbian and bisexual. In a sample of 2,922 undergraduates, 1.01%, 1.6%, and 3.7% reported that they were lesbian, gay male, or bisexual, respectively (Knox and Hall 2010).

National **Data**

According to the previously cited research, estimates of the U.S. GLBT population range from about 2% to 7% of the U.S. adult population. An easy-to-remember percentage is 5%, which translates into 16 million gay individuals in the United States (population of approximately 312 million) and 875,000 gay college students (of 17.5 million total college students in the United States).

Prevalence of Same-Sex Couples/Households

Although U.S. Census surveys do not ask about sexual orientation or gender identity, same-sex cohabiting couples may identify themselves as "unmarried partners." Those couples in which both partners are men or both are women are considered to be same-sex couples or households for purposes of research. In 2009, there were 565,000 unmarried same-sex couple households (*Statistical Abstract of the United States, 2011,* Table 63).

Carpenter and Gates (2008) analyzed data in California and noted that, although 62% of heterosexual couples cohabit, about 40% of gay males and about 60% of lesbians cohabit. Same-sex couples are more likely to live in metropolitan areas than in rural areas. However, the largest proportional increases in the number of same-sex couples self-reporting in 2000 versus 1990 came in rural, sparsely populated states.

National **Data**

The number of formally/legally married same-sex couples in the U.S. is 35,000; there are 86,000 legally recognized civil unions and domestic partnerships (Gates 2009).

Why are data on the numbers of GLBT individuals and couples in the United States relevant? The primary reason is that census numbers on the prevalence of GLBT individuals and couples can influence laws and policies that affect gay individuals and their families. The slogan "The more we are counted, the more we count" (Bradford et al. 2002, 3) speaks to the importance of being visible. "The fact that the Census documents the actual presence of same-sex couples in nearly every state legislative and U.S. Congressional district means anti-gay legislators can no longer assert that they have no gay and lesbian constituents" (p. 8).

Gay Male Singlehood

Approximately 50% of gay men describe themselves as single (i.e., without a committed relationship). Hostetler (2009) interviewed 94 single middle class gay men (over 30% were men of color) between the ages of 35 and 70. Just over a fifth (20.2%) viewed their singlehood as a potentially permanent choice. Various reasons for remaining unattached included:

1. *Personal Past.* One attributes one's single status to events or circumstances in one's childhood, adolescence, and/or early-adult years. For example, growing up in a home where one's parents were viewed as having a stale marriage and not wanting to model this experience.

2. *Collective Past.* One attributes one's single status to membership in a particular historical cohort. An example might be belonging to a group who served in the military in Iraq or Afghanistan, being traumatized by war and feeling "unfit to marry anybody."

3. *Previous Relationship.* One attributes one's single status to past relationship experiences. For example, being burned in a previous relationship and wanting to steer clear of new romances/commitments.

4. *Particular Tastes.* One attributes one's single status to sexual tastes that are seen as incompatible with the establishment and/or maintenance of a long-term relationship. For example, feeling that one could never be faithful due to the need for constant sexual variety or being gay and not having the option to marry.

5. *Still Searching.* One maintains the conviction that one has not yet met the right person.

6. *Loner.* One attributes one's single status to seeing oneself as a loner; attributing one's single status to preferences and/or behavioral patterns acquired as a result of long-term singlehood. For example, feeling that one is set in one's ways and not willing to adapt to a relationship.

The majority (80%) of Hostetler's respondents reported that they were not happy being single and that it was not their preference. A common revelation was that the "decision" to be single was made in retrospect—after a while, they just noticed that they were not partnered. Over 60% (61.7%) felt that they would be happier if they had a relationship.

Origins of Sexual-Orientation Diversity

Much of the biomedical and psychological research on sexual orientation attempts to identify one or more "causes" of sexual-orientation diversity. The driving question behind this research is, "Is sexual orientation inborn or is it learned or acquired from environmental influences?" Although a number of factors have been correlated with sexual orientation, including genetics, gender role behavior in childhood, and fraternal birth order, no single theory can explain diversity in sexual orientation.

Beliefs about What "Causes" Homosexuality

Aside from what "causes" homosexuality, social scientists are interested in what people believe about the "causes" of homosexuality. Most gay people believe that homosexuality is an inherited, inborn trait. In a national study of homosexual men, 90% reported that they believed that they were born with their homosexual orientation; only 4% believed that environmental factors were the sole cause (Lever 1994).

Individuals who believe that homosexuality is genetically determined tend to be more accepting of homosexuality and are more likely to be in favor of equal rights for lesbians and gays (Tyagart 2002). In contrast, "those who believe homosexuals choose their sexual orientation are far less tolerant of gays and lesbians and more likely to feel that homosexuality should be illegal than those who think sexual orientation is not a matter of personal choice" (Rosin and Morin 1999, 8). Although the terms *sexual preference* and *sexual orientation* are often used interchangeably, the term *sexual orientation* avoids the implication that homosexuality, heterosexuality, and bisexuality are determined. Hence, those who believe that sexual orientation is inborn more often use the term *sexual orientation*, and those who think that individuals choose their sexual orientation use *sexual preference* more often. The term sexual identity is, increasingly, being used since it connotes more about the whole person than sexuality.

Can Homosexuals Change Their Sexual Orientation?

Individuals who believe that homosexual people choose their sexual orientation tend to think that homosexuals can and should change their sexual orientation. Various forms of reparative therapy or **conversion therapy** are focused on changing

I am a heterosexual. I don't know why I'm like this. I was born this way.

Rosanne Barr, comedian

If homosexuality is a disease, let's all call in queer to work: "Hello. Can't work today, still queer."

Robin Tyler, author

Conversion therapy also called reparative therapy, focuses on changing the sexual orientation of homosexuals.

homosexuals' sexual orientation. Some religious organizations sponsor "ex-gay ministries," which claim to "cure" homosexuals and transform them into heterosexuals by encouraging them to ask for "forgiveness for their sinful lifestyle," through prayer and other forms of "therapy." Serovich et al. (2008) reviewed twenty-eight empirically based, peer-reviewed articles and found them methodologically problematic, which threatens the validity of interpreting available data on this topic.

Parelli (2007) attended private as well as group therapy sessions to change his sexual orientation from homosexuality to heterosexuality. It did not work. In retrospect, he identified seven reasons for the failure of reparative therapy. In effect, he noted that these therapies focus on outward behavioral change and do not acknowledge the inner yearnings:

> *By my mid-forties, I was experiencing a chronic need for appropriately affectionate male touch. It was so acute I could think of nothing else. Every cell of my body seemed relationally isolated and emotionally starved. Life was so completely and fatally ebbing out of my being that my internal life-saving system kicked in and put out a high-alert call for help. I desperately needed to be held by loving, human, male arms. (p. 32)*

Critics of reparative therapy and ex-gay ministries take a different approach—gay people are not the problem; social disapproval is the problem. The National Association for the Research and Therapy of Homosexuality (NARTH) has been influential in moving public opinion from "gays are sick" to "society is judgmental." The American Psychiatric Association, the American Psychological Association, the American Academy of Pediatrics, the American Counseling Association, the National Association of School Psychologists, the National Association of Social Workers, and the American Medical Association agree that homosexuality is not a mental disorder and needs no cure—that efforts to change sexual orientation do not work and may, in fact, be harmful (Human Rights Campaign 2000). An extensive review of the ex-gay movement concludes, "There is a growing body of evidence that conversion therapy not only does not work, but also can be extremely harmful, resulting in depression, social isolation from family and friends, low self-esteem, internalized homophobia, and even attempted suicide" (Cianciotto and Cahill 2006, 77). According to the American Psychiatric Association, "clinical experience suggests that any person who seeks conversion therapy may be doing so because of social bias that has resulted in internalized homophobia, and that gay men and lesbians who have accepted their sexual orientation are better adjusted than those who have not done so" (quoted by Holthouse 2005, 14).

New research brings into the question that reparative therapies "never work." Karten and Wade (2010) reported that the majority of 117 men who were dissatisfied with their sexual orientation and who sought sexual orientation change efforts (SOCE) were able to reduce their homosexual feelings and behaviors and increase their heterosexual feelings and behaviors. The primary motivations for their seeking change was religion ("homosexuality is wrong") and emotional dissatisfaction with the homosexual lifestyle. Being married, feeling disconnected with other men prior to seeking help, and feeling able to express nonsexual affection toward other men were the factors predictive of greatest change. Better understanding the causes of homosexuality and one's homosexuality and one's emotional needs and issues as well as developing nonsexual relationships with same-sex peers were also identified as most helpful in change.

Heterosexism, Homonegativity, Homophobia, and Biphobia

Heterosexism the institutional and societal reinforcement of heterosexuality as the privileged and powerful norm. Assumes that homosexuality is "bad."

The United States, along with many other countries throughout the world, is predominantly heterosexist. **Heterosexism** refers to "the institutional and societal reinforcement of heterosexuality as the privileged and powerful norm." For example, abstinence-only-until-marriage educational programs are silent about gay love/sexual relationships—no information is provided and gay individuals are left to feel "less than human" (Fisher 2009b). Heterosexism is based on the belief that heterosexuality is superior to homosexuality. Of 2,922 undergraduates, 28% agreed, "It is better to be heterosexual than homosexual" (Knox and Hall 2010). Heterosexism results in prejudice and discrimination against homosexual and bisexual people. Prejudice refers to negative attitudes, whereas discrimination refers to behavior that denies equality of treatment for individuals or groups. Gilla et al. (2010) found evidence that gays are subjected to negative comments and name calling in contexts such as physical education classes or physical activities where they are perceived to be small or gay. That the "don't ask, don't tell" policy of the military changed and then changed again in 2010 implies that being gay is "bad and we don't want to hear about it." (Both the President and military brass supported the repeal.) Before reading further, you may wish to complete the Self-Assessment on page 226 which assesses behaviors toward individuals perceived to be homosexual.

"Don't Ask Don't Tell" Military Policy

"Don't Ask, Don't Tell" has been the military policy and guideline concerning gays and lesbians in the military since 1993. It was a compromise that helped Bill Clinton make good on his promise to allow gays and lesbians to serve in the military. The policy states that gays and lesbians may serve in the military, but will be discharged if certain conditions are not met. The conditions are that gays and lesbians serving in the military may not engage in homosexual conduct, say that they are homosexual or bisexual, and they cannot marry or already be married to a person of the same sex (Congressional Digest 2010a and b). Since its inception, Don't Ask, Don't Tell has resulted in 12,000 men and women being discharged from the military.

The University of California Blue Ribbon Commission estimated that the cost of replacing service men and women who have been discharged including recruiting and training new personnel to be $363 million. Defense Secretary Robert Gates, siding with the Pentagon Report (which showed no significant upheaval with gays being open), supported the repeal of Don't Ask, Don't Tell. In December of 2010, Congress passed and the President signed the repeal of the Don't Ask, Don't Tell policy.

Attitudes toward Homosexuality, Homonegativity, and Homophobia

Adolfsen et al. (2010) noted that there are multiple dimensions of attitudes about homosexuality and identified five:

1. *General attitude.* Is homosexuality considered to be normal or abnormal? Do people think that homosexuals should be allowed to live their lives just as freely as heterosexuals?

2. *Equal rights.* Should homosexuals be granted the same rights as heterosexuals in regard to marriage and adoption?

3. *Close quarters.* Feelings in regard to the same a gay neighbor or a lesbian colleague.

4. *Public display.* Reactions to a gay couple kissing in public.

5 *Modern homonegativity.* Feeling that homosexuality is accepted in society and that all kinds of special attention are unnecessary.

The term **homophobia** is commonly used to refer to negative attitudes and emotions toward homosexuality and those who engage in homosexual behavior. Homophobia is not necessarily a clinical phobia (that is, one involving a compelling desire to avoid the feared object despite recognizing that the fear is unreasonable). Other terms that refer to negative attitudes and emotions toward homosexuality include **homonegativity** (attaching negative connotations to being gay) and antigay bias. Dysart-Gale (2010) noted that these prejudicial negative attitudes exert a negative impact.

No matter how I look at the issue, I cannot escape being troubled by the fact that we have in place a policy which forces young men and women to lie about who they are in order to defend their fellow citizens.

Mike Mullen, chairman of the Joint Chiefs of Staff

Personally, I think they are just afraid of a thousand guys with M16s going, "Who'd you call a faggot?"

Jon Stewart, comedian, in reference to the issue of gays in the military

You could move.

Abigail Van Buren, "Dear Abby," in response to a reader who complained that a gay couple was moving in across the street and wanted to know what he could do to improve the quality of the neighborhood

Homophobia negative attitudes and emotions toward homosexuality and those who engage in homosexual behavior.

Homonegativity attaching negative connotations to homosexuality.

This questionnaire is designed to examine which of the following statements most closely describes your behavior during past encounters with people you thought were homosexuals. Rate each of the following self-statements as honestly as possible by choosing the frequency that best describes your behavior: Never = 1; Rarely = 2; Occasionally = 3; Frequently = 4; Always = 5.

_____ 1. I have spread negative talk about someone because I suspected that the person was gay.

_____ 2. I have participated in playing jokes on someone because I suspected that the person was gay.

_____ 3. I have changed roommates and/or rooms because I suspected my roommate was gay.

_____ 4. I have warned people who I thought were gay and who were a little too friendly with me to keep away from me.

_____ 5. I have attended antigay protests.

_____ 6. I have been rude to someone because I thought that the person was gay.

_____ 7. I have changed seat locations because I suspected the person sitting next to me was gay.

_____ 8. I have had to force myself to keep from hitting someone because the person was gay and very near me.

_____ 9. When someone I thought to be gay has walked toward me as if to start a conversation, I have deliberately changed directions and walked away to avoid the person.

_____ 10. I have stared at a gay person in such a manner as to convey my disapproval of the person being too close to me.

_____ 11. I have been with a group in which one (or more) person(s) yelled insulting comments to a gay person or group of gay people.

_____ 12. I have changed my normal behavior in a restroom because a person I believed to be gay was in there at the same time.

_____ 13. When a gay person has checked me out, I have verbally threatened the person.

_____ 14. I have participated in damaging someone's property because the person was gay.

_____ 15. I have physically hit or pushed someone I thought was gay because the person brushed against me when passing by.

_____ 16. Within the past few months, I have told a joke that made fun of gay people.

_____ 17. I have gotten into a physical fight with a gay person because I thought the person had been making moves on me.

_____ 18. I have refused to work on school and/or work projects with a partner I thought was gay.

_____ 19. I have written graffiti about gay people or homosexuality.

_____ 20. When a gay person has been near me, I have moved away to put more distance between us.

Scoring

The Self-Report of Behavior Scale (SBS-R) is scored by totaling the number of points endorsed on all items, yielding a range from 20 to 100 total points. The higher the score, the more negative the attitudes toward homosexuals.

Comparison Data

Sunita Patel (1989) originally developed the Self-Report of Behavior Scale in her thesis research in her clinical psychology master's program at East Carolina University. College men (from a university campus and from a military base) were the original participants (Patel et al. 1995). The scale was revised by Shartra Sylivant (1992), who used it with a coed high school student population, and by Tristan Roderick (1994), who involved college students to assess its psychometric properties. The scale was found to have high internal consistency. Two factors were identified: a passive avoidance of homosexuals and active or aggressive reactions.

In a study by Roderick et al. (1998), the mean score for 182 college women was 24.76. The mean score for 84 men was significantly higher, at 31.60. A similar-sex difference, although with higher (more negative) scores, was found in Sylivant's high school sample (with a mean of 33.74 for the young women, and 44.40 for the young men).

The following table provides detail for the scores of the college students in Roderick's sample (from a mid-sized state university in the southeast):

	N	Mean	Standard Deviation
Women	182	24.76	7.68
Men	84	31.60	10.36
Total	266	26.91	9.16

Sources

Patel, S. 1989. Homophobia: Personality, emotional, and behavioral correlates. Master's thesis, East Carolina University.

Patel, S., T. E. Long, S. L. McCammon, and K. L. Wuensch. 1995. Personality and emotional correlates of self reported antigay behaviors. *Journal of Interpersonal Violence* 10: 354–366.

Roderick, T. 1994. Homonegativity: An analysis of the SBS-R. Master's thesis, East Carolina University.

Roderick, T., S. L. McCammon, T. E. Long, and L. J. Allred. 1998. Behavioral aspects of homonegativity. *Journal of Homosexuality* 36: 79–88.

Sylivant, S. 1992. The cognitive, affective, and behavioral components of adolescent homonegativity. Master's thesis, East Carolina University. The SBS-R is reprinted by the permission of the students and faculty who participated in its development: S. Patel, S. L. McCammon, T. E. Long, L. J. Allred, K. Wuensch, T. Roderick, and S. Sylivant.

The Sex Information and Education Council of the United States notes that "individuals have the right to accept, acknowledge, and live in accordance with their sexual orientation, be they bisexual, heterosexual, gay or lesbian. The legal system should guarantee the civil rights and protection of all people, regardless of sexual orientation" (SIECUS 2009, retrieved March 23). Nevertheless, negative attitudes toward homosexuality continue.

Characteristics associated with positive attitudes toward homosexuals, gay rights, etc. are younger age, advanced education, no religious affiliation, liberal political party affiliation, and personal contact with homosexual individuals (the more the better). Jenkins et al. (2009) found no significant difference between black and white students in reported levels of homophobia in a sample of 551 Midwestern college students. Whitley et. al. (2011) studied racial differences in attitudes toward lesbians and gay men (ATLGM) and found that black students held generally neutral ATLGM whereas white students' attitudes were slightly positive. Having had contact with someone who is gay is particularly associated with lower homophobia (Walch et al. 2010). Also, see Applying Social Research feature.

Negative social meanings associated with homosexuality can affect the self-concepts of LGBT individuals. **Internalized homophobia** is a sense of personal failure and self-hatred among lesbians and gay men resulting from social rejection and stigmatization of being gay and has been linked to increased risk for depression, substance abuse and addiction, anxiety, and suicidal thoughts (Rubinstein 2010; Bobbe 2002; Gilman et al. 2001). Lalicha and McLarena (2010) noted how being homosexual and growing up in a religious context that is antigay is particularly difficult. A gay Jehovah's Witness recalled:

> *All my life, for as long as I can remember, I struggled with who I was and my attraction to the same sex. I was taught that this was wrong. So, I grow up not having anyone to talk to about my thoughts and my feelings. Knowing that if I said anything that God would hate me, and that I would be turned on by the people that loved and accepted me for who they thought I was.*

Homophobia also results in gays staying "in the closet" (hiding their sexual orientation for fear of negative reactions/consequences). Such hiding may also have implications for gay interpersonal relationships. Brackett et al. (2010) found that gays were significantly more likely to keep a secret from their romantic partner than straights. The researchers hypothesized that gays learn early to keep secrets and that this skill slides over into their romantic relationships.

Biphobia

Just as the term *homophobia* is used to refer to negative attitudes toward homosexuality, gay men, and lesbians, **biphobia** (also referred to as binegativity) refers to a parallel set of negative attitudes toward bisexuality and those identified as bisexual. Although heterosexuals often reject both homosexual- and bisexual-identified individuals, bisexual-identified women and men also face rejection from many homosexual individuals. Thus, bisexuals experience "double discrimination."

Lesbians are a major source of negative views towards bisexuals. In a study of 346 self-identified lesbians, Rust (1993) found that lesbians view bisexuals as less committed to other women than are lesbians, perceive bisexuals as disloyal to

Internalized homophobia a sense of personal failure and self-hatred among lesbians and gay men resulting from social rejection and stigmatization of being gay.

Biphobia parallel set of negative attitudes toward bisexuality and those identified as bisexual.

This is the Gay World Series, not the Bisexual World Series.

North American Gay Amateur Athletic Association (NAGAAA) officials

Heterosexism, Homonegativity, Homophobia, and Biphobia

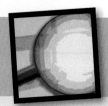

After a few drinks, it is not unusual that heterosexual females at a bar will kiss each other. Their motives are typically geared towards "entertainment for the guys," experimentation, or curiosity. The goal of this research was to identify correlations of such heterosexual female same-sex kissing on attitudes toward same-sex relationships. Specifically, are heterosexual females who kiss other females more likely to approve of same-sex marriage, approve of same-sex couples having their own children, and approve of same-sex couples adopting children?

Methodology

Data for this study came from a sample of 436 undergraduate female student volunteers at two large southeastern universities who completed a 50-item questionnaire (approved by the Institutional Review Board) on "Attitudes toward Homosexuality of College Students." Respondents completed the questionnaire over the Internet.

The mean age of the respondents was 19.83. Over 40% of the respondents were freshmen (44.3%) with sophomores representing 33%; juniors, 13.2% and seniors, 9.5%. In comparing the responses of those heterosexual women who reported having kissed another woman (out of sexual experimentation/curiosity) with women who had not had this experience, cross-classification was conducted to determine any relationships by using Chi Square to assess statistical significance.

Findings—Sam-Sex Kissing

Almost half (47.9%) of the self-identified heterosexual women reported that they had kissed another woman out of sexual experimentation/curiosity. When the heterosexual women who reported same-sex kissing were compared with those who reported no such kissing, several statistically significant findings emerged in regard to pro gay attitudes.

*Abridged from a poster by Tiffany Beaver, David Knox and Vaiva Kiskute, Southern Sociological Society Annual Meeting, Atlanta, April 2010.

lesbians and resent bisexual women who have close relationships with the "enemy" of lesbians—men. Some negative attitudes toward bisexual individuals "are based on the belief that bisexual individuals are really lesbian or gay individuals who are in transition or in denial about their true sexual orientation" (Israel and Mohr 2004, 121). Rust (1993) found that lesbians were likely to get critical of bisexuals if the transition period from identifying as bisexual to identifying as a lesbian took too long. According to this view, bisexual people fear coming out as lesbian or gay, or they are trying to maintain heterosexual privilege. Rust (1993) found lesbians perceived it easier for bisexuals to pass as heterosexual than for lesbians to do so. Given the cultural bias against being gay, gay individuals must be careful about "coming out." (This view also minimizes the reality of bisexuality as an identity.)

In addition to double discrimination (from both gays and straights), bisexuals also have a lack of resources and support which adds to the stress of coming out (Rust 2003).

Effects of Antigay Bias and Discrimination on Heterosexuals

As a junior and senior at Homewood-Flossmoor High School in the suburbs of Chicago, Myka Held played a key role in leading a campaign to promote tolerance of gay and lesbian students. The campaign involved selling gay-friendly T-shirts to students and teachers and having as many people as possible wear the T-shirts to school on a designated day. The T-shirts, made by Duke University, say, "gay? fine by me." "I think it's really important for gay people out there to know that there are straight people who support them," Ms. Held said (quoted in Puccinelli 2005, 20). "I have always supported equal rights for every person and have been disgusted by discrimination and prejudice. As a young Jewish woman, I believe it is my duty to stand up and support minority groups. . . . In my mind, fighting for gay rights is a proxy for fighting for every

1. HSSK (heterosexual same-sex kissers) were less likely to think that homosexuality is immoral (p < .000).

2. HSSK were less likely to report getting sick if they watched two gay people being intimate (p < .000).

3. HSSK were less likely to report hating gay people (p < .000).

4. HSSK were more likely to be in favor of same-sex marriage (p < .000).

5. HSSK were more likely to be in favor of gay people having their own children (p < .000).

Discussion

Psychologist Gordon Allport (1954) asserted that contact between groups is necessary for the reduction of prejudice—an idea known as the contact hypothesis. In general, heterosexuals have more favorable attitudes toward gay men and lesbians if they have had prior contact with or know someone who is gay or lesbian (Mohipp & Morry 2004). Contact with openly gay individuals reduces negative stereotypes and ignorance and increases support for gay and lesbian equality (Wilcox and Wolpert 2000).

References

Allport, G. W. (1954). *The nature of prejudice.* Cambridge, MA: Addison-Wesley.

Mohipp, C., and Morry, M. M. (2004). "Relationship of symbolic beliefs and prior contact to heterosexuals' attitudes toward gay men and lesbian women." *Canadian Journal of Behavioral Science* 36(1): 36–44.

Schiappa, E., Gregg, P. B., and Hewes, D. E. (2005). The parasocial contact hypothesis. *Communication Monographs* 72(1): 92–115.

Wilcox, C., and Wolpert, R. (2000). "Gay rights in the public sphere: Public opinion on gay and lesbian equality." In C. A. Rimmerman, Wald, K. D., & Wilcox, C. (Eds.), *The politics of gay rights* (p. 409–432). Chicago: University of Chicago Press.

person's rights" (Held 2005). Myka Held's T-shirt campaign illustrates that fighting prejudice and discrimination against sexual-orientation minorities is an issue not just for lesbians, gays, and bisexuals but also for all those who value fairness and respect for human beings in all their diversity.

The antigay and heterosexist social climate of our society is often viewed in terms of how it victimizes the gay population. However, heterosexuals are also victimized by heterosexism and antigay prejudice and discrimination. Some of these effects follow:

1. *Heterosexual victims of hate crimes.* As discussed earlier in this chapter, extreme homophobia contributes to instances of violence against homosexual—acts known as hate crimes. Hate crimes are crimes of perception, meaning that victims of antigay hate crimes may not be homosexual; they may just be perceived as being homosexual. The National Coalition of Anti-Violence Programs (2005) reported that, in 2004, 192 heterosexual individuals in the United States were victims of antigay hate crimes, representing 9% of all antigay hate crime victims.

2. *Concern, fear, and grief over well-being of gay or lesbian family members and friends.* Many heterosexual family members and friends of homosexual people experience concern, fear, and grief over the mistreatment of their gay or lesbian friends and/or family members. For example, heterosexual parents who have a gay or lesbian teenager often worry about how the harassment, ridicule, rejection, and violence experienced at school might affect their gay or lesbian child. Will their child drop out of school, as one-fourth of gay youth do (Chase 2000), to escape the harassment, violence, and alienation they endure there? Will the gay or lesbian child respond to the antigay victimization by turning to drugs or alcohol or by committing suicide, as there is an increased risk in this population (Haas et al., 2011)? Such fears are not unfounded: lesbian, gay, and bisexual youth who report high levels of victimization at school also have higher levels of substance use and suicidal thoughts than heterosexual peers who report high levels of at-school victimization (Bontempo and D'Augelli 2002). A survey of youths' risk behavior conducted by

the Massachusetts Department of Education in 1999 revealed that 30% of gay teens had attempted suicide in the previous year, compared with 7% of their straight peers (Platt 2001). Indeed, four gay teens (Billy Lucas, Tyler Clementi, Asher Brown, and Seth Walsh) committed suicide in 2010 in response to being bullied about their sexuality. The suicides generated media attention and started an online video campaign called the "It Gets Better Project" (http://www.itgetsbetter.org/).

Meyer et al. (2008) studied the lifetime prevalence of mental disorders and suicide attempts of a diverse group of lesbian, gay, and bisexual individuals and found higher rates of substance abuse among bisexual people than lesbians and gay men. Also, Latino respondents attempted suicide more often than white respondents.

Heterosexual individuals also worry about the ways in which their gay, lesbian, and bisexual family members and friends could be discriminated against in the workplace.

To heterosexuals who have lesbian and gay family members and friends, lack of family protections such as health insurance and rights of survivorship for same-sex couples can also be cause for concern. Finally, heterosexuals live with the painful awareness that their gay or lesbian family member or friend is a potential victim of antigay hate crime. Imagine the lifelong grief experienced by heterosexual family members and friends of hate crime murder victims, such as Matthew Shepard, a 21-year-old college student who was brutally beaten to death in 1998, for no apparent reason other than he was gay.

3. *Restriction of intimacy and self-expression.* Because of the antigay social climate, heterosexual individuals, especially males, are hindered in their own self-expression and intimacy in same-sex relationships. "The threat of victimization

National **Data**

Only twenty states have laws banning discrimination based on sexual orientation (National Gay and Lesbian Task Force 2010—see http://www.thetaskforce.org/).

(i.e., antigay violence) . . . causes many heterosexuals to conform to gender roles and to restrict their expressions of (nonsexual) physical affection for members of their own sex" (Garnets et al. 1990, 380). Homophobic epithets frighten youth who do not conform to gender role expectations, leading some youth to avoid activities such as arts for boys, athletics for girls that they might otherwise enjoy and benefit from (Gay, Lesbian, and Straight Education Network 2000). A male student in our class revealed that he always wanted to work with young children and had majored in early childhood education. His peers teased him relentlessly about his choice of majors, questioning both his masculinity and his heterosexuality. Eventually, this student changed his major to psychology, which his peers viewed as an acceptable major for a heterosexual male.

4. *Dysfunctional sexual behavior.* Some cases of rape and sexual assault are related to homophobia and compulsory heterosexuality. For example, college men who participate in gang rape, also known as "pulling train," entice each other into the act "by implying that those who do not participate are unmanly or homosexual" (Sanday 1995, 399). Homonegativity also encourages early sexual activity among adolescent men. Adolescent male virgins are often teased by their male peers, who say things like "You mean you don't do it with girls yet? What are you, a fag or something?" Not wanting to be labeled and stigmatized as a "fag," some adolescent boys "prove" their heterosexuality by having sex with girls.

5. *School shootings.* Antigay harassment has also been a factor in many of the school shootings in recent years. In March 2001, 15-year-old Charles Andrew Williams fired more than thirty rounds in a San Diego suburban high school, killing two and injuring thirteen others. A woman who knew Williams reported that the students had teased him and called him gay (Dozetos 2001). According to the Gay, Lesbian, and Straight Education Network (GLSEN), Williams's story is not unusual. Referring to a study of harassment of U.S. students that was commissioned by the American Association of University Women, a GLSEN report concluded, "For boys, no other type of harassment provoked as strong a reaction on average; boys in this study would be less upset about physical abuse than they would be if someone called them gay" (Dozetos 2001).

6. *Loss of rights for individuals in unmarried relationships.* The passage of state constitutional amendments that prohibit same-sex marriage can also result in denial of rights and protections to other-sex unmarried couples. For example, in 2005, Judge Stuart Friedman of Cuyahoga County, Ohio agreed that a man who was charged with assaulting his girlfriend could not be charged with a domestic violence felony because the Ohio state constitutional amendment granted no such protections to unmarried couples (Human Rights Campaign 2005). As discussed earlier, some antigay marriage measures also threaten the provision of domestic partnership benefits to unmarried heterosexual couples.

Gay, Lesbian, Bisexual, and Mixed-Orientation Relationships

Research suggests that gay and lesbian couples tend to be more similar than different from heterosexual couples (Kurdek 2005; 2004). However, there are some unique aspects of intimate relationships involving gay, lesbian, and bisexual individuals. In this section, we note the similarities as well as differences between heterosexual, gay male, and lesbian relationships in regard to relationship satisfaction, conflict and conflict resolution, and monogamy and sexuality. We also look at relationship issues involving bisexual individuals and mixed-orientation couples.

Relationship Satisfaction

Ackbar and Senn (2010) studied 77 women in long-term, lesbian relationships and found that women who reported showing greater closeness toward their partners were more satisfied in their relationships. The authors suggested that marriage and family therapists might avoid pathologizing high levels of closeness in lesbian relationships simply because they may deviate from a heterosexual norm.

However, for both heterosexual and GLBT partners, relationship satisfaction tends to be high in the beginning of the relationship and decreases over

This couple is in a stable monogamous relationship.

time. In a review of literature on lesbian and gay couples, Kurdek (1994b) concluded, "The most striking finding regarding the factors linked to relationship satisfaction is that they seem to be the same for lesbian couples, gay couples, and heterosexual couples" (p. 251). These factors include having equal power and control, being emotionally expressive, perceiving many attractions and few alternatives to the relationship, placing a high value on attachment, and sharing decision making. Kurdek (2008) compared relationship quality of cohabitants over a ten-year period of both partners from 95 lesbian, 92 gay male, and 226 heterosexual couples living without children, and both partners from 312 heterosexual couples living with children. Lesbian couples showed the highest levels of relationship quality averaged over all assessments.

Researchers who studied relationship quality among same-sex couples noted that, "in trying to create satisfying and long-lasting intimate relationships, LGBT individuals face all of the same challenges faced by heterosexual couples, as well as a number of distinctive concerns" (Otis et al. 2006, 86). These concerns include if, when, and how to disclose their relationships to others and how to develop healthy intimate relationships in the absence of same-sex relationship models.

In one review of research on gay and lesbian relationships, we concluded that the main difference between heterosexual and nonheterosexual relationships is that, "Whereas heterosexuals enjoy many social and institutional supports for their relationships, gay and lesbian couples are the object of prejudice and discrimination" (Peplau et al. 1996, 268). Both gay male and lesbian couples must cope with the stress created by antigay prejudice and discrimination and by "internalized homophobia" or negative self-image and low self-esteem due to being a member of a stigmatized group. Not surprisingly, higher levels of such stress are associated with lower reported levels of relationship quality among LGBT couples (Otis et al. 2006).

Despite the stresses and lack of social and institutional support LGBT individuals experience, gay men and lesbians experience relationship satisfaction at a level that is at least equal to that reported by married heterosexual spouses (Kurdek 2005). Partners of the same sex enjoy the comfort of having a shared gender perspective, which is often accompanied by a sense of equality in the relationship. For example, contrary to stereotypical beliefs, same-sex couples (male or female) typically do not assign "husband" and "wife" roles in the division of household labor; as well, they are more likely than heterosexual couples to achieve a fair distribution of household labor and at the same time accommodate the different interests, abilities, and work schedules of each partner (Kurdek 2005). In contrast, division of household labor among heterosexual couples tends to be unequal, with wives doing the majority of such tasks. Rock et al. (2010) studied 190 students from accredited couple and family therapy programs in regard to their training in working with GLBT clients. While most held positive attitudes toward GLBT clients, less than half reported any training in affirmative therapy (e.g. exploring homophobic and heterosexist beliefs, being comfortable with GLBT individuals, etc.).

Monogamy and Sexuality

Like many heterosexual women, most gay women value stable, monogamous relationships that are emotionally as well as sexually satisfying. Gay and heterosexual women in U.S. society are taught that sexual expression should occur in the context of emotional or romantic involvement.

A common stereotype of gay men is that they prefer casual sexual relationships with multiple partners versus monogamous long-term relationships. However, although most gay men report having more casual sex than heterosexual men (Mathy 2007), most gay men prefer long-term relationships, and sex outside of the primary relationship is usually infrequent and not emotionally involving (Green et al. 1996).

The degree to which gay males engage in casual sexual relationships is better explained by the fact that they are male than by the fact that they are gay. In this regard, gay and straight men have a lot in common: they both tend to have

When I work with gay couples who want to explore the option of nonmonogamy, I have to monitor my own inner voices—the one that celebrates the gay culture's permissiveness and sexual freedom, as well as the one that wants to protect the dyad from possible emotional disintegration.

Michael Shernoff, counselor of gay individuals

fewer barriers to engaging in casual sex than do women (heterosexual or lesbian). One way that gay men meet partners is through the Internet. Ogilvie et al. (2008) surveyed men who have sex with men (MSM) who found partners using the Internet; he noted that they were more likely to have had ten sexual partners in the last year and to agree with the statement, "I think most guys in relationships have condom-free sex."

Such nonuse of condoms results in the high rate of human immunodeficiency virus (HIV) infection and acquired immunodeficiency syndrome (AIDS). Although most worldwide HIV infections occur through heterosexual transmission, male-to-male sexual contact is the most common mode of HIV transmission in the United States. Women who have sex exclusively with other women have a much lower rate of HIV infection than do men (both gay and straight) and women who have sex with men. Many gay men have lost a love partner to HIV infection or AIDS; some have experienced multiple losses. Those still in relationships with partners who are HIV-positive experience profound changes, such as developing a sense of urgency to "speed up" their relationship because they may not have much time left together (Palmer and Bor 2001).

Monogamous Gay Male Sex (with casual sex on the side)

Bonello and Cross (2010) interviewed 8 gay men in monogamous relationships who also had extradyadic (nonmonogamous) sex to discover the mechanisms that were operative in this disconnect. The men noted that they started out being faithful but as differences in sex drive emerged, level of sexual satisfaction declined, or they began to perceive the partner as unfaithful, casual sex with new partners was initiated. Also, the respondents talked of "emotional monogamy" whereby they felt as long as the extradyadic sex was not emotional, it was not really wrong or a threat. One respondent said, "It's about the emotion. I don't feel that I've given any of my emotion away . . . if I was meeting someone from the Internet and seeing them on a weekly basis and falling in love with them and planning for the future then I would feel that I was cheating because I should be wanting to do that with my partner." Other individuals included their partner in the venture for casual sex with others, such as having threesomes or an open relationship.

The outcomes of casual sex in these gay male monogamous relationships varied—some respondents said it enhanced the primary relationship by meeting the need for variety the partner could not meet. Others felt guilty and ashamed. Some disclosed their extradyadic sex and the relationship was over. Others disclosed and the couple went to counseling.

Hoff and Beougher (2010) conducted interviews with 39 gay male couples about their sexual agreements and found a range of agreements existed. While some agreed on monogamy, others had open agreements which included rules or conditions limiting when (e.g., when the partner was not available), where (e.g., not in the couple's home), how often (e.g., infrequent), and with whom outside sex was permitted (e.g., the partner had veto power over a particular individual). Partners also varied in how they handled breaks in their agreements, which depended on what condition was broken, whether it was disclosed, and the partner's reaction. Overall, the various agreements had a positive consequence for the couples by providing boundaries for the relationship, supporting a nonheteronormative identity, and fulfilling the sexual needs of the couple.

Bisexuality immediately doubles your chances for a date on Saturday night.

Woody Allen, director

Relationships of Bisexuals

Individuals who identify as bisexual have the ability to form intimate relationships with both sexes. However, research has found that the majority of bisexual women and men tend toward primary relationships with the other sex (McLean 2004). Contrary to the common myth that bisexuals are, by definition, nonmonogamous, some bisexuals prefer monogamous relationships (especially in light of the widespread concern about HIV). In another study of sixty bisexual women and men, 25% of the men and 35% of the women were in exclusive relationships; 60% of the men and 53% of the women were in "open" relationships in which both partners agreed to allow each other to have sexual and or emotional relationships with others, often under specific conditions or rules about how this would occur (McLean 2004). In these "open" relationships, nonmonogamy was not the same as infidelity, and the former did not imply dishonesty. The researcher concluded:

> *Despite the stereotypes that claim that bisexuals are deceitful, unfaithful, and untrustworthy in relationships, most of the bisexual men and women I interviewed demonstrated a significant commitment to the principles of trust, honesty, and communication in their intimate relationships and made considerable effort to ensure both theirs and their partner's needs and desires were catered for within the relationship. (McLean 2004, 96)*

Monogamous bisexual women and men find that their erotic attractions can be satisfied through fantasy and their affectional needs through nonsexual friendships (Paul 1996). Edser and Shea (2002), in a study of bisexual married men, observed that they had been married over 20 years (and thus able to deal with same-sex attractions). However, de Vries (2007) found that bisexual men and women who were seeking monogamy in a relationship had difficulty finding partners who would believe them. Even in a monogamous relationship, "the partner of a bisexual person may feel that a bisexual person's decision to continue to identify as bisexual . . . is somehow a withholding of full commitment to the relationship. The bisexual person may be perceived as holding onto the possibility of other relationships by maintaining a bisexual identity and, therefore, not fully committed to the relationship" (Ochs 1996, 234). However, this perception overlooks the fact that one's identity is separate from one's choices about relationship involvement or monogamy. Miller et al. (2002) pointed out that bisexuals maintain a bisexual identity regardless of the sex of their partners. Ochs notes that "a heterosexual's ability to establish and maintain a committed relationship with one person is not assumed to falter, even though the person retains a sexual identity as 'heterosexual' and may even admit to feeling attractions to other people despite her or his committed status" (p. 234).

Bisexual couples experiencing relationship problems might benefit from cognitive behavioral marital therapy. This approach focuses on cognitions (perceptions, beliefs, assumptions, attributions, expectancies), behaviors and affect (positive/negative emotional experience), and how these factors are expressed in the couple's relationship (Deacon et al. 1996). Deacon et al. (1996) stressed the importance of finding a therapist that is knowledgeable of the needs of bisexuals and their relationships.

Division of Labor

Sutphina (2010) studied the division of labor patterns reported by 165 respondents in same-sex relationships. Partners with greater resources (e.g., income, education) performed fewer household tasks. Satisfaction with division of labor and sense of being appreciated for one's contributions to household tasks were positively correlated with overall relationship satisfaction.

Esmaila (2010) noted that lesbians struggle to maintain a sense of fairness in regard to the division of labor. This includes invoking how heterosexual couples often slip into an unequal division of labor and how gay couples try to raise the bar on this issue.

Mixed-Orientation Relationships

Mixed-orientation couples are those in which one partner is heterosexual and the other partner is gay, lesbian, or bisexual. Up to 2 million gay, lesbian, or bisexual people in the United States have been in heterosexual marriages at some point (Buxton 2004). Some lesbians and gays do not develop same-sex attractions and feelings until after they have been married. Others deny, hide, or repress their same-sex desires.

In a study of twenty gay or bisexual men who had disclosed their sexual orientation to their wives, most of the men did not intentionally mislead or deceive their future wives with regard to their sexuality. Rather, they did not fully grasp their feelings toward men, although they had a vague sense of their same-sex attraction prior to the marriage (Pearcey 2004). The majority of the men in this study (14 of 20) attempted to stay married after disclosure of their sexual orientation to their wives, and nearly half (9 of 20) stayed married for at least three years.

A team of researchers (Yarhouse et al. 2009) provided information about 13 heterosexually married couples in which one partner expressed same-sex attraction and both partners reported satisfaction with their marriage. The average age of the spouses was 46; the average length of the marriage was 17 years. The level of same-sex orientation (self-report) on a ten point scale prior to the marriage was 6.6; currently 3.3. In regard to level of marital happiness from 0 (extremely unhappy) to 6 (perfect), "strugglers" averaged 4.07; spouses, 3.71.

The top reasons given by those with same-sex attraction for maintaining the marriage were commitment to spouse, commitment to children, covenant, obedience to God, and love of spouse. The top reasons given by their spouses for staying in the marriage with the partner who was attracted to others of the same sex were covenant before God, love for spouse, and commitment to spouse. Worship and prayer were identified as the top coping mechanisms for both partners in the marriage (hence, these were very religious couples).

The most difficult aspects of the marriage from the viewpoint of the spouse with same-sex attraction were communication, lack of time, intimacy, finances, and sex. From the viewpoint of the straight spouse, lack of time, communication, sex and sexuality, and life stresses were the most difficult. Most of the spouses continued to have sex with each other. None of the spouses who had same-sex attractions reported sex outside the marriage in the last four years.

What emerges from studying these couples is their religious faith in which they viewed their marriage as a covenant with each other and God took priority over the issue of one having same-sex attraction. The spouses also shared values, love, and affection. The Straight Spouse Network (www.straightspouse.org) provides support to heterosexual spouses or partners, current or former, of GLBT mates.

Lifetime Sexual Abuse

Gay, lesbian, and bisexual individuals are not immune to coping with sexual abuse in their previous relationships (as adults or children). Rothman et al. (2011) reviewed studies which involved a total sample of 139,635 gay, lesbian, and bisexual women and men and found that the highest estimates reported for lifetime sexual assault for lesbian and bisexual women was 85%, 76% for childhood sexual assault of lesbian and bisexual women, and 59% for childhood sexual assault of gay and bisexual men. Peterson et al. (2011) reviewed the literature on male victims of adult sexual assault (ASA) and found notable adverse physical and psychological consequences for some men.

Coming Out to a Partner and Same-Sex Marriage

Coming out is a major decision gay individuals struggle with. The personal choices feature details what is involved in this decision.

It's not cheating because she knows what I'm doing, but something about it is not right for me, and then the experience doesn't seem worth it. I feel like I need to find the balance somehow of living both sides.

Jim Will, husband in a mixed-orientation marriage

That word "lesbian" sounds like a disease. And straight men know because they're sure that they're the cure.

Denise McCanles, author

Are the Benefits of "Coming Out" Worth the Risks?

Coming out being open about one's sexual orientation and identity.

In a society where heterosexuality is expected and considered the norm, heterosexuals do not have to choose whether or not to tell others that they are heterosexual. However, decisions about **"coming out,"** or being open and honest about one's sexual orientation and identity (particularly to one's parents; Heatherington and Lavner (2008)), are some of the most difficult and important choices that gay, lesbian, and bisexual individuals face. Choices about coming out include whether to come out to others, who to come out to, and when and how to come out. Rossi (2010) studied the coming out experiences of 53 young adults and noted that most came out to a friend first, then their mothers, then their fathers.

Risks of Coming Out

Whether GLBT individuals come out is influenced by the degree to which they are tired of hiding their sexual orientation, the degree to which they feel more "honest" about being open, their assessment of the risks of coming out, and their prediction of how others will respond. Some of the risks involved in coming out include disapproval and rejection by parents and other family members; harassment and discrimination at school; discrimination and harassment in the workplace; and hate crime victimization.

1. *Parental and family members' reactions.* Padilla et al. (2010) found that parental reaction to a son or daughter coming out had a major effect on the development of their child. Acceptance had an enormous positive effect. When GLBT individuals come out to their parents, parental reactions range from "I already knew you were gay and I'm glad that you feel ready to be open with me about it" to "get out of this house, you are no longer welcome here." Mary Cheney reported that, when she told her father, former Vice President Dick Cheney, that she is a lesbian, his response was, "You're my daughter and I love you and I just want you to be happy" (quoted in Walsh 2006, 27). When Reverend Mel White, a closeted gay Christian man who was nearly driven to suicide after two decades of struggling to save his marriage and his soul with "reparative therapies," finally came out to his mother, her response was, "I'd rather see you at the bottom of that swimming pool, drowned, than to hear this" (White 2005, 28).

According to *The Resource Guide to Coming Out* (Human Rights Campaign 2004), "many parents are shocked when their children say they are gay, lesbian, or bisexual. Some parents react in ways that hurt. Some cry. Some get angry. Some ask where they went wrong as a parent. Some call it a sin. Some insist it's a phase. Others try to send their child to counselors or therapists who attempt to change gay people into heterosexuals . . ." (p. 24).

Because black individuals are more likely than white individuals to view homosexual relations as "always wrong," African Americans who are gay or lesbian are more likely to face disapproval from their families (and straight friends) than are white lesbians and gays (Lewis 2003). The result is that African Americans are more likely to stay closeted and not let their parents know of their homosexuality (Grov et al. 2006). One effect of deciding not to come out to one's family is to replace one's disapproving family with one's supportive friends. This phenomenon has been particularly observed in Finland (Dewaele et al. 2011).

The Resource Guide notes, however, that "for many parents it's very hard to completely reject their children" and that it takes time for parents to adjust—sometimes months, sometimes years. In some families with a GLBT member, the "gay issue" is not openly discussed, even though family members may know or suspect that a loved one is gay, lesbian, or bisexual. One gay male student explained, "My parents know I live with my 'friend' Christopher, and they have invited Christopher to our family holiday gatherings and family vacations. I am sure my parents know that I'm gay. . . . We just don't talk about it" (personal communication). Scherrer (2010) noted that grandparents may be an overlooked resource for GLBT individuals struggling with acceptance issues.

Parents and other family members can learn more about homosexuality from the local chapter of Parents, Families, and Friends of Lesbians and Gays (PFLAG) and from books and online resources, such as those found at Human Rights Campaign's National Coming Out Project (http://www.hrc.org/).

2. *Harassment and discrimination at school.* In a national survey of students aged 13 to 18 years, 65% of LGBT students reported that they had been verbally harassed, 16% physically harassed, and 8% physically assaulted because of their sexual orientation (Harris Interactive and GLSEN 2005). "Students who openly identify as lesbian, gay, bisexual, and transgender (LGBT) have a more acute problem with being harassed at school" (p. 4). This survey found that LGBT students are over three times more likely than non-LGBT students to report that they feel unsafe at school (20% versus 6%). Only eight states and the District of Columbia have statewide policies that prohibit antigay harassment in public schools (Snorton 2005).

A study of GLBT college students, faculty, and staff or administrators found that 51% had concealed their sexual orientation or gender identity to avoid intimidation (Rankin 2003). The same study found that in the previous year 36% of undergraduates experienced harassment in the form of derogatory remarks, verbal threats, antigay graffiti, threats of physical violence, denial of services, and physical assault.

3. *Discrimination and harassment at the workplace.* The 2005 Workplace Fairness Survey found that 39% of lesbian and gay employees reported experiencing some form of discrimination or harassment in the workplace (Lambda Legal and Deloitte Financial Advisory Services, LLP 2006). Although 88% of U.S. adults support equal employment rights for gays and lesbians, firing, declining to hire or promote, or otherwise discriminating against an employee because of sexual orientation was legal in thirty-three states as of March 2006 (National Gay and Lesbian Task Force 2006; Saad 2005).

4. *Hate crime victimization.* Another risk of coming out is that of being victimized by antigay hate crimes against individuals or their property that are based on bias against the victim because of their perceived sexual orientation. Such crimes include verbal threats and intimidation, vandalism, sexual assault and rape, physical assault, and murder. "Homosexuals are far more likely than any other minority group in the United States to be victimized by violent hate crime" (Potok 2010, p. 29).

Benefits of Coming Out

Given the risks of coming out, some GBLT individuals (not surprisingly) never come out, live a life of repressed feelings, and deny who they are to others and (sometimes) to themselves. However, according to the Human Rights Campaign, "most people come out because, sooner or later, they can't stand hiding who they are any more. Once they've come out, most people acknowledge that it feels much better to be open and honest than to conceal such an integral part of themselves" (2004, 16). Political gay activist Harvey Milk emphasized the importance of coming out.

In addition to the benefits for individuals who choose to come out, there are also benefits for the entire GLBT population. Research has found that, in general, heterosexuals have more positive attitudes toward gays and lesbians if they have had prior contact with or know someone who is gay (Mohipp and Morry 2004). One woman describes how her brother's coming out changed her views on homosexuality (Yvonne 2004):

I was raised in a devout born-again Christian family . . . to believe that homosexuality was evil and a perversion. When I was growing up, I used to wonder if any of the kids at my school could be gay. I couldn't imagine it could be so. As it happens, there was a gay individual even closer than I imagined. My brother Tommy came out of the closet in the early 1990s. After he came out, I had to confront my own denial about the fact that, in my heart of hearts, I had always known that Tommy was gay.

Over the years, his partner Rod has come to be a loved and cherished member of our family, and we have all had to confront the prejudices and stereotypes we have held onto for so long about sexual orientation. It seems to me that coming out of the closet is the greatest weapon that gays and lesbians have. If my own brother had never come out, my family would never have been forced to confront the deep-seated prejudices we were raised with. . . . I am still a Christian, but my husband (who also has a gay brother) and I attend a church that truly puts the teachings of Christ into practice teaching about love, tolerance and inclusivity. As a Christian who grew up with a gay family member, I know that the propaganda put forth by the religious right on this issue is founded in fear, hatred, and preju-dice. None of these are values taught by Jesus!

Cheryl Jacques, a Massachusetts state senator who came out publicly in the *Boston Globe*, recognizes that coming out is a risk. But she suggests the following:

Coming out is a risk worth taking because it is one of the most powerful things any of us can do. I've yet to meet anyone who regretted the decision to live life truthfully. . . . That's why while coming out may be just one step in the life of a gay, lesbian, bisexual, or transgender person, it contributes to a giant leap for all GLBT people. (Human Rights Campaign 2004, 4)

Coming out to a Partner

Being true to oneself often includes coming out not only to oneself but also to one's partner. Spouses who are shocked by the revelations of their gay partner discover that they are not alone through such websites as www.straightspouse-connection.com, which are focused on helping the heterosexual spouse as he or she absorbs and reacts to the knowledge that his or her partner is gay.

Another website, www.straightspouse.org, features the Straight Spouse Network (SSN), an international organization that provides personal, confidential support and information to heterosexual spouses/partners, current or former, of gay, lesbian, bisexual, or transgender mates and mixed-orientation couples for constructively resolving coming-out problems. SSN also offers research-based information about spouse, couple, and family issues and resources to other family members, professionals, community organizations, and the public. SSN is the only support network of its kind in the world.

The basic trajectory of reacting to a partner's homosexuality is not unlike learning of the death of one's beloved. Life as one knew it is altered. The stages of this transition are shock, disbelief, numbness, and mourning for the partner/ life that was, readjustment, and moving on. The last two stages may involve staying in the relationship with the gay partner (the choice of about 15% of straight spouses) or divorcing/ending the relationship (the choice of 85%).

Same-Sex Marriage

As of 2011, six states and the District of Columbia now offer civil marriage licenses to same-sex couples—Connecticut, Iowa, Massachusetts, New Hampshire, Vermont, and Maine. However, unlike marriages between a man and a woman, the same-sex marriages in these states are not recognized in other states, nor does the federal government recognize them. Same-sex marriage is still hotly contested in many states. California overturned Proposition 8, which banned same-sex marriage. Maine voted to ban same-sex marriage right after the law passed granting same-sex marriage.

Antigay Marriage Legislation   In a national survey, 32% of U.S. adults support gay marriage, whereas 59% are opposed (Pew Research Center 2008a). In a national study of first-year freshmen in colleges and universities throughout the United States, 72.4% were in favor of same-sex marriage (Pryor et al. 2008). Where disapproval exists, the primary reason is morality. Gay marriage is viewed as "immoral, a sin, against the Bible." However, support for gay marriage varies by age; about half of young adults (18 to 29) in the same survey support gay marriage. Moskowitz et al. (2010) also found that men are more supportive of lesbian marriage than gay male marriage. In 1996, Congress passed and former President Clinton signed the **Defense of Marriage Act** (DOMA), which states that marriage is a "legal union between one man and one woman" and denies federal recognition of same-sex marriage. In effect, this law allows states to either recognize or not recognize same-sex marriages performed in other states. As of 2011, thirty-six states have banned gay marriage either through statute or a state constitutional amendment, and seventeen states have passed broader

---

I condone what she has done 100 percent, and it's her business to talk about it, not mine. It doesn't change anything. I'm proud of who she is.

Stephen Moyer, actor, when his then fiancé (now wife) came out as bisexual

---

Gay people got a right to be as miserable as everybody else.

Chris Rock, comedian

Defense of Marriage Act legislation which says that marriage is a "legal union between one man and one woman" and denies federal recognition of same-sex marriage.

Chapter 8 Same-Sex Couples and Families

antigay family measures that ban other forms of partner recognition in addition to marriage, such as domestic partnerships and civil unions. These broader measures, known as "Super DOMAs," potentially endanger employer-provided domestic partner benefits, joint and second-parent adoptions, health care decision-making proxies, or any policy or document that recognizes the existence of a same-sex partnership (Cahill and Slater 2004). Some of these "Super DOMAs" ban partner recognition for unmarried heterosexual couples as well.

Arguments in Favor of Same-Sex Marriage   Advocates of same-sex marriage argue that banning or refusing to recognize same-sex marriages granted in other states is a violation of civil rights that denies same-sex couples the many legal and financial benefits that are granted to heterosexual married couples. Rights and benefits that married spouses have include the following:

- The right to inherit from a spouse who dies without a will;
- No inheritance taxes between spouses;
- The right to make crucial medical decisions for a partner and to take care of a seriously ill partner or parent of a partner under current provisions in the federal Family and Medical Leave Act;
- Social Security survivor benefits; and
- Health insurance coverage under a spouse's insurance plan.

Other rights bestowed on married (or once-married) partners include assumption of a spouse's pension, bereavement leave, burial determination, domestic violence protection, reduced-rate memberships, divorce protections (such as equitable division of assets and visitation of partner's children), automatic housing lease transfer, and immunity from testifying against a spouse. As noted earlier, same-sex couples are taxed on employer-provided insurance benefits for domestic partners, whereas married spouses receive those benefits tax-free. Finally, unlike seventeen other countries that recognize same-sex couples for immigration purposes, the United States does not recognize same-sex couples in granting immigration status because such couples are not considered "spouses." Another argument for same-sex marriage is that it would promote relationship stability among gay and lesbian couples. "To the extent that marriage provides status, institutional support, and legitimacy, gay and lesbian couples, if allowed to marry, would likely experience greater relationship stability" (Amato 2004, 963). Indeed, same-sex relationships, like cohabitation relationships, end at a higher rate than marriage relationships (Wagner 2006).

Recognized marriage, argues Amato, would be beneficial to the children of same-sex parents. Without legal recognition of same-sex families, children living in gay- and lesbian-headed households are denied a range of securities that protect children of heterosexual married couples. These include the right to get health insurance coverage and Social Security survivor benefits from a nonbiological parent. In some cases, children in same-sex households lack the automatic right to continue living with their nonbiological parent should their biological mother or father die (Tobias and Cahill 2003). It is ironic that the same pro-marriage groups that stress that children are better off in married-couple families disregard the benefits of same-sex marriage to children.

Opponents of gay marriage sometimes suggest that gay marriage leads to declining marriage rates, increased divorce rates, and increased nonmarital births. However, data in Scandinavia reflects that these trends were in place ten years before Scandinavian adopted registered partnership laws, liberalized alternatives to marriage (such as cohabitation), and expanded exit options (such as no-fault divorce) (Pinello 2008).

Finally, there are religious-based arguments in support of same-sex marriage. Although many religious leaders teach that homosexuality is sinful and prohibited by God, some religious groups, such as the Quakers and the United Church of Christ (UCC), accept homosexuality, and other groups have made reforms toward increased acceptance of lesbians and gays. In 2005, the UCC became the largest Christian denomination to endorse same-sex marriages. In a sermon titled

"The Christian Case for Gay Marriage," Jack McKinney (2004) interprets Luke 4: "Jesus is saying that one of the most fundamental religious tasks is to stand with those who have been excluded and marginalized. . . . [Jesus] is determined to stand with them, to name them beloved of God, and to dedicate his life to seeing them empowered." McKinney goes on to ask, "Since when has it been immoral for two people to commit themselves to a relationship of mutual love and caring? No, the true immorality around gay marriage rests with the heterosexual majority that denies gays and lesbians more than 1,000 federal rights that come with marriage."

Arguments Against Same-Sex Marriage Whereas advocates of same-sex marriage argue that they will not be regarded as legitimate families by the larger society so long as same-sex couples cannot be legally married, opponents do not want to legitimize same-sex couples and families. Opponents of same-sex marriage who view homosexuality as unnatural, sick, and/or immoral do not want their children to view homosexuality as socially acceptable.

Opponents of same-sex marriage commonly argue that such marriages would subvert the stability and integrity of the heterosexual family. However, Sullivan (1997) suggests that homosexuals are already part of heterosexual families:

> [Homosexuals] are sons and daughters, brothers and sisters, even mothers and fathers, of heterosexuals. The distinction between "families" and "homosexuals" is, to begin with, empirically false; and the stability of existing families is closely linked to how homosexuals are treated within them. (p. 147)

Many opponents of same-sex marriage base their opposition on their religious views. In a Pew Research Center national poll, the majority of Catholics and Protestants opposed legalizing same-sex marriage, whereas the majority of secular respondents favored it (Green 2004). However, churches have the right to deny marriage for gay people in their congregations. Legal marriage is a contract between the spouses and the state; marriage is a civil option that does not require religious sanctioning.

In previous years, opponents of gay marriage have pointed to public opinion polls that suggested that the majority of Americans are against same-sex marriage. However, public opposition to same-sex marriage is decreasing. We previously noted that a 2008 Pew Research Center national poll found that 59% of U.S. adults oppose legalizing gay marriage, down from 63% in 2004 (Pew Research Center 2006). We also noted that support for gay marriage is higher among young adults.

State Legal Recognition of Same-Sex Couples There is no federal recognition of same-sex couples in the United States. However, a number of U.S. states allow same-sex couples legal status that entitles them to many of the same rights and responsibilities as married couples. See Figure 8.2 which identifies the laws in each state.

International **Data**

In 2001, the Netherlands became the first country in the world to offer full legal marriage to same-sex couples. Same-sex married couples and other-sex married couples in the Netherlands are treated identically, with two exceptions. Unlike other-sex marriages, same-sex couples married in the Netherlands are unlikely to have their marriages recognized as fully legal abroad. Regarding children, parental rights will not automatically be granted to the nonbiological spouse in gay couples. To become a fully legal parent, the spouse of the biological parent must adopt the child. In 2003, Belgium passed a law allowing same-sex marriages but disallowing any adoptions. In June 2005, Spain became the third country to legalize same-sex marriage; shortly thereafter, Canada became the fourth, and South Africa became the fifth. Mexico City now gives legal recognition to gay couples who want to marry and to adopt children.

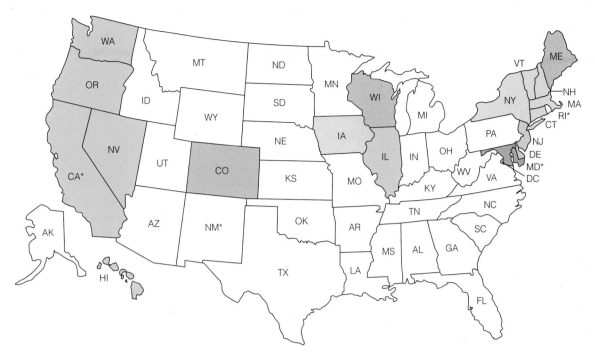

- ☐ State issues marriage licenses to same-sex couples
- ☐ State recognizes marriages by same-sex couples legally entered into in another jurisdiction
- ☐ Statewide law providing the equivalent of state-level spousal rights to same-sex couples within the state
- ☐ Statewide law providing some statewide spousal rights to same-sex couples within the state

Map accurate as of June 25, 2011

Figure 8.2 States Recognizing Same Sex Marriage.

GLBT Parenting Issues

National **Data**

Nearly one-quarter of all same-gender couples are raising children; 34.3% of lesbian couples are raising children, and 22.3% of gay male couples are raising children (compared with 45.6% of married heterosexual and 43.1% of unmarried heterosexual couples raising children) (Pawelski et al. 2006). Over half (54%) of U.S. adults feel that same-sex parents can be good parents (Pew Research Center 2008a).

Gay and lesbian people fall in love. We settle down. We commit our lives to one another. We raise our children. We protect them. We try to be good citizens.

Sheila Kuehl, California senator

Of the more than 600,000 same-sex-couple households identified in the 2000 census, 162,000 had one or more children living in the household. This is a low estimate of children who have gay or lesbian parents, as it does not count children in same-sex households who did not identify their relationship in the census, those headed by gay or lesbian single parents, or those whose gay parent does not have physical custody but is still actively involved in the child's life.

National **Data**

Estimates of the number of U.S. children with gay or lesbian parents range from 1 to 14 million (Howard 2006).

Gay Families—Lesbian Mothers and Gay Fathers

Many gay and lesbian individuals and couples have children from prior heterosexual relationships or marriages. Up to 2 million gay, lesbian, or bisexual people in the United States have been in heterosexual marriages at some point (Buxton 2005). Some of these individuals married as a "cover" for their homosexuality; others discovered their interest in same-sex relationships after they married. Children with mixed-orientation parents may be raised by a gay or lesbian parent, a gay or lesbian stepparent, a heterosexual parent, and a heterosexual stepparent.

A gay or lesbian individual or couple may have children through the use of assisted reproductive technology, including in vitro fertilization, surrogate mothers, and donor insemination. *The Kids are All Right* is a 2010 movie which featured some of the issues of a lesbian couple having children via donor insemination. Others gay couples adopt or become foster parents. Less commonly, some gay fathers are part of an emergent family form known as the hetero-gay family. In a hetero-gay family, a heterosexual mother and gay father conceive and raise a child together but reside separately.

While the struggle for acceptance of gay parenting has been difficult (Rivers 2010), data on lesbian mothers has revealed that they tend to have high levels of shared decision making, parenting, and family work, all reflecting an egalitarian ideology. When compared to heterosexual mothers, gay mothers tended to spend more time with their children and expressed more warmth and affection (Biblarz et al. 2010).

Data on gay fathers has also revealed that they are more likely to co-parent equally and compatibly than fathers in heterosexual relationships. Gay fathers are also less likely to spank their children as a method of discipline (Biblarz et al. 2010). Bergman et al. (2010) also found that gay men who became parents had heightened self-esteem as a result of becoming parents and raising children. So just the act of becoming a father had a very positive outcome on gay men's sense of self-worth. Part of this effect may be due to the fact that some gay men think that gay fatherhood is an unattainable role. While both gay females and gay males report increases in individual happiness during the first year of having a baby/adopting a child, relationship happiness decreases (Goldberg et al. 2010). Such a decrease in relationship happiness following the birth of a baby is similar to what happens with heterosexual couples.

Antigay views concerning gay parenting include the belief that homosexual individuals are unfit to be parents and that children of lesbians and gays will not develop normally and/or that they will become homosexual. As the following section suggests, research findings paint a more positive picture of the development and well-being of children with gay or lesbian parents.

Development and Well-Being of Children with Gay or Lesbian Parents

A growing body of research on gay and lesbian parenting supports the conclusion that children of gay and lesbian parents are just as likely to flourish as are children of heterosexual parents.

There is the belief that children need both a mother and father together, which are critical to the total development of a child. Biblarz and Stacey (2010) compared children from two-parent families with same or different sex co-parents and single-mother with single-father families and found that the strengths typically associated with married mother-father families appeared to the same extent in families with two mothers and potentially in those with two fathers. Hence, children seem to benefit when there are two parents in the household (rather than a single mom or dad) but that the gender of these parents is irrelevant—the structure can be a woman and a man, two women, or two men. Crowl et al. (2008) also reviewed nineteen studies on the developmental outcomes and quality of parent-child relationships among children raised by gay and lesbian parents. Results confirmed previous studies that children raised by same-sex parents fare equally well to children raised

They say gay marriage ruins families and hurts kids. Well, I've had the privilege of seeing my gay friends being parents and watching their kids grow up in a loving environment.

Brad Pitt, actor

by heterosexual parents. Regardless of family type, adolescents were more likely to show positive adjustment when they perceived more caring from adults and when parents described having close relationships with them. Thus, the qualities of adolescent-parent relationships rather than the sexual orientation of the parents were significantly associated with adolescent adjustment.

Bos and Gartrell (2010) assessed the presence of homophobic stigmatization on the well-being of 39 female and 39 male 17-year-old adolescents whose mother was lesbian and who were conceived through donor insemination. Forty-one percent reported experiencing stigmatization based on homophobia, and the greater the stigmatization was, the more problem behavior was evident in these adolescents. However, close, positive relationships with their lesbian mothers muted the effect of stigmatization and allowed the adolescents to be resilient.

DIVERSITY IN THE UNITED STATES

Female same-sex couples in which both partners are Hispanic are raising children at over twice the rate of white, non-Hispanic female same-sex couples (66% versus 32%). Male same-sex couples in which both partners are Hispanic are raising children at more than three times the rate of white, non-Hispanic male couples (58% versus 19%) (Cianciotto 2005).

Bos and Sandfort (2010) studied children in lesbian and heterosexual families and found that children in lesbian families felt less parental pressure to conform to gender stereotypes, were less likely to experience their own gender as superior, and were more likely to be uncertain about future heterosexual romantic involvement. The last finding does not suggest that the children were more likely to feel that they were "gay," but that they were living in a social context which allowed for more than the heterosexual adult model.

In another study, researchers examined the quality of parent-child relationships and the socioemotional and gender development of a sample of 7-year-old children with lesbian parents, compared with 7-year-olds from two-parent heterosexual families and with single heterosexual mothers (Golombok et al. 2003). No significant differences between lesbian mothers and heterosexual mothers were found for most of the parenting variables assessed, although lesbian mothers reported smacking their children less and playing more frequently with their children than did heterosexual mothers. No significant differences were found in psychiatric disorders or gender development of the children in lesbian families versus heterosexual families. The findings also suggest that having two parents is associated with more positive outcomes for children's psychological well-being, but the gender of the parents is not relevant. One difference for adolescents who have two gay mothers is that over 40% report that they are teased and ridiculed about the sexual orientation of their parents. But they are not subjected to serious bullying over this or other issues (Biblarz et al. 2010). The American Psychological Association (2004) noted that "results of research suggest that lesbian and gay parents are as likely as heterosexual parents to provide supportive and healthy environments for their children" and that "the adjustment, development, and psychological well-being of children [are] unrelated to parental sexual orientation and that the children of lesbian and gay parents are as likely as those of heterosexual parents to flourish." Indeed, Pro-Family Pediatricians cheered when a proposal to ban gay marriage was defeated. "Our duty as pediatricians is to see that all children have the same security and protection regardless of the sexual orientation of their parents. Denying legal rights to same-sex couples injures their children," noted Asch-Goodkin (2006, 2).

Discrimination in Child Custody, Visitation, Adoption, and Foster Care

A student in one of our classes reported that, after she divorced her husband, she became involved in a lesbian relationship. She explained that she would like to be open about her relationship to her family and friends, but she was afraid that if her ex-husband found out that she was in a lesbian relationship, he might take her to court and try to get custody of their children. Although several respected national organizations including the American Academy of

Should GLBT Individuals/Couples be Prohibited from Adopting Children?

According to the U.S. Children's Bureau, 119,000 children in the U.S. child welfare system were waiting to be adopted in 2003, only 20,000 of whom were in pre-adoptive homes (Howard 2006). Thousands of these children will never be adopted and will never have stable, permanent homes and families. Most adoptive parents want infants or young children, yet two-thirds of children waiting to be adopted are over 5 years old. A report by the Evan B. Donaldson Adoption Institute concludes that "laws and policies that preclude adoption by gay or lesbian parents disadvantage the tens of thousands of children mired in the foster care system who need permanent, loving homes" and that "adoption by gays and lesbians holds promise as an avenue for achieving permanency for many of the waiting children in foster care" (pp. 2, 3).

As noted earlier in this chapter, Crowl et al. (2008) reviewed nineteen studies on the quality of parent-child relationships among children raised by gay and lesbian parents and found that children raised by same-sex parents fare equally well as children raised by heterosexual parents. A study of adoptive parents showed no significant differences between gay and lesbian adoptive parents and heterosexual adoptive parents on measures of family functioning and child behavior problems (Erich et al. 2005). Despite this and other research that finds positive outcomes for children raised by gay or lesbian parents, and despite the support for gay adoption by child advocacy organizations, placing children for adoption with gay or lesbian parents remains controversial. A 2006 poll reveals that 46% of U.S. adults support gay adoption, up from 38% in 1999 (Pew Research

Pediatrics, the Child Welfare League of America, the American Bar Association, the American Medical Association, the American Psychological Association, the American Psychiatric Association, and the National Association of Social Workers have gone on record in support of treating gays and lesbians without prejudice in parenting and adoption decisions (Howard 2006; Landis 1999), lesbian and gay parents are often discriminated against in child custody, visitation, adoption, and foster care.

Lehman (2010) identified the three issues the courts have used to deny homosexuals custody of their children—per se, presumption, and nexus. Per se takes the position that if the parent is homosexual, he or she is unfit to parent. The presumption position is similar to the per se position except the parent has the right to prove otherwise. The nexus approach focuses on the connection between being homosexual and there being negative outcomes for the child. These outcomes include that the child will be ridiculed, the parent will turn the child gay, and that homosexuality is immoral. Each of these positions is rarely taken seriously by the courts.

Gay and lesbian individuals and couples who want to adopt children can do so through adoption agencies or through the foster care system in at least twenty-two states and the District of Columbia. Florida most recently allowed gays and lesbians to adopt (Mississippi forbids such adoptions). Utah forbids adoption by any unmarried couple (which includes all same-sex couples), and Arkansas prohibits lesbians and gay men from serving as foster parents (National Gay and Lesbian Task Force 2004). Goldberg et al. (2011a) studied 30 gay male, 45 lesbian, and 51 heterosexual couples who had recently adopted a child. The researchers found that gay women perceived higher levels of stigma than gay men and that higher levels of stigma were associated with depression. Gay men may find it particularly difficult to adopt. Wells (2011) studied gay men who decide to become fathers and how they navigate pervasive heterosexism and antigay prejudice.

This chapter's Social Policy section asks whether gay and lesbian individuals and couples should be prohibited from adopting.

Most adoptions by gay people are second-parent adoptions. A **second-parent adoption** (also called co-parent adoption) is a legal procedure that allows

Second-parent adoption legal procedure that allows individuals to adopt their partner's biological or adoptive child without terminating the first parent's legal status as parent.

Center 2006). Although public support for gay adoption has increased in recent years, fewer than half of U.S. adults reported support for gay adoption in 2006. Social agencies are mixed in their policies toward gay adoption. Hall (2010b) surveyed Northern California adoption agency caseworkers and found that more than 95% of questionnaire respondents stated that GLB people should be allowed to adopt.

Social policies that prohibit LGBT individuals and couples from adopting children result in fewer children being adopted. What happens to children who are not adopted? The thousands of children who "age out" of the foster care system annually experience high rates of homelessness, incarceration, early pregnancy, failure to graduate from high school, unemployment, and poverty (Howard 2006). Essentially, social policies that prohibit gay adoption are

policies that deny thousands of children the opportunity to have a nurturing family.

Your Opinion?

1. Do you believe children who grow up with same-sex parents are disadvantaged?

2. Do you believe that children without homes should be prohibited by law from being adopted by lesbian or gay individuals or couples?

3. If you were a 6-year-old child with no family, would you rather remain in an institutional setting or be adopted by a gay or lesbian couple?

individuals to adopt their partner's biological or adoptive child without terminating the first parent's legal status as parent. Second-parent adoption gives children in same-sex families the security of having two legal parents. Second-parent adoption potentially benefits a child by:

- Placing legal responsibility on the parent to support the child;
- Allowing the child to live with the legal parent in the event that the biological (or original adoptive) parent dies or becomes incapacitated;
- Enabling the child to inherit and receive Social Security benefits from the legal parent;
- Enabling the child to receive health insurance benefits from the parent's employer; and
- Giving the legal parent standing to petition for custody or visitation in the event that the parents break up. (Clunis and Green 2003)

However, in four states (Colorado, Nebraska, Ohio, and Wisconsin), court rulings have decided that the state adoption law does not allow for second-parent adoption by members of same-sex couples, and it is unclear whether the state adoption laws in twenty-two states allows second-parent adoption (National Gay and Lesbian Task Force 2005). Second-parent adoption is not possible when a parent in a same-sex relationship has a child from a previous heterosexual marriage or relationship, unless the former spouse or partner is willing to give up parental rights. Although "third-parent" adoptions have been granted in a small number of jurisdictions, this option is not widely available (National Center for Lesbian Rights 2003).

The Future of Same-Sex Relationships

Moral acceptance and social tolerance/acceptance of gays and lesbians as individuals, spouses, and parents will come slowly (Kozloskia 2010). Homonegativity is entrenched in American society. However, as more states recognize the legitimacy of same-sex marriage, acceptance will increase. Similarly as more individuals "come out," their social networks will become more tolerant and supportive of same-sex individuals and relationships.

The future will also include GLBT suicide prevention efforts such as identifying at-risk individuals early and intervening through national and state programs (Haas et al. 2011). Such programs will include "educating community gatekeepers including teachers and staff in youth programs, senior centers, aging services agencies and others who come in contact with at-risk individuals." Finally, emphasis will be on "training in LGBT suicide risk for staff and volunteers of suicide crisis lines, law enforcement, emergency care professionals, and others who work with suicidal individuals."

Summary

How prevalent are homosexuality/bisexuality and same-sex households and families?

Estimates of the GLBT population in the United States range from about 2% to 5% of the total U.S. population. There are also about a half million same-sex couples in the United States with about a quarter self-identifying as married and three fourths identifying as together but not married. About one-fifth (22.3%) of gay male couples and one-third (34.3%) of lesbian couples have children in the home.

What are the origins of sexual-orientation diversity?

Although a number of factors have been correlated with sexual orientation, including genetics, gender role behavior in childhood, and fraternal birth order, no single theory can explain diversity in sexual orientation. Most gay people believe that homosexuality is an inherited, inborn trait. In a national study of homosexual men, 90% reported that they believed that they were born with their homosexual orientation; only 4% believed that environmental factors were the sole cause.

Some individuals who believe that homosexuals choose their sexual orientation think that homosexual people can and should change their sexual orientation. Various forms of reparative therapy or conversion therapy are dedicated to changing homosexuals' sexual orientation. However, various organizations agree that homosexuality is not a mental disorder and needs no cure. Furthermore, efforts to change sexual orientation typically do not work and may, in fact, be harmful. These organizations also support the notion that societal reaction to homosexuals rather than "curing homosexuality" is a more appropriate focus.

What are heterosexism, homonegativity, homophobia, and biphobia?

Heterosexism refers to "the institutional and societal reinforcement of heterosexuality as the privileged and powerful norm." Heterosexism is based on the belief that heterosexuality is superior to homosexuality. The term homophobia is commonly used to refer to negative attitudes and emotions toward homosexuality and those who engage in it. Homophobia is not necessarily a clinical phobia (that is, one involving a compelling desire to avoid the feared object despite recognizing that the fear is unreasonable). Other terms that refer to negative attitudes and emotions toward homosexuality include homonegativity and antigay bias.

Homophobia also results in gays staying "in the closet" (hiding their sexual orientation for fear of negative reactions/consequences). Such hiding may also have implications for gay interpersonal relationships. Gays are more likely to keep a secret from their romantic partner than straights. One hypothesis is that gays learn early to keep secrets and that this skill slides over into their romantic relationships.

How are gay and lesbian couples different from heterosexual couples?

Research suggests that gay and lesbian couples tend to be more similar to than different from heterosexual couples. Gay and lesbian couples also tend to disagree about the same issues that heterosexual couples argue about. Same-sex couples' relationships are different from heterosexual relationships in that same-sex couples have more concern about when and how to disclose their relationships to others. Same-sex couples are also more likely than heterosexual couples to achieve a fair distribution of household labor and to argue more effectively and resolve conflict in a positive way. Most significantly, whereas heterosexuals enjoy many social and institutional supports for their relationships, gay and lesbian couples face prejudice and discrimination.

What is involved in coming out to oneself or one's partner and same-sex marriage?

The risks of coming out include potential rejection from family, being bullied at school, being harassed at work, and being the victim of a hate crime. The benefits of coming out include personal satisfaction in having done so and benefits to the entire GLBT population in terms of greater visibility.

The argument for same-sex marriage is that banning such unions is a violation of civil rights that denies same-sex couples the many legal and financial benefits that are granted to heterosexual married couples. Children of gay/lesbian individuals would also benefit by being brought up in a more stable, socially approved relationship. Arguments against same-sex marriage focus on the "pathology" of homosexuality, the "immorality," and the subversion of traditional marriage.

What does research on gay and lesbian parenting conclude?

A growing body of credible, scientific research on gay and lesbian parenting concludes that children raised by gay and lesbian parents adjust positively and their families function well. Lesbian and gay parents are as likely as heterosexual parents to provide supportive and healthy environments for their children, and the children of lesbian and gay parents are as likely as those of heterosexual parents to flourish.

What is the future of same-sex relationships?

Societal acceptance of gays and lesbians as individuals, spouses, and parents will come slowly. Homonegativity is entrenched in American culture. Hate crimes will continue but their frequency will drop.

Key Terms

Biphobia

Bisexuality

Coming out

Conversion therapy

Defense of Marriage Act

Gay

GLBT

Heterosexism

Heterosexuality

Homosexuality

Homophobia

Internalized homophobia

Lesbian

Second-parent adoption

Sexual identity

Sexual orientation

Web Links

Advocate (Online Newspaper for LGBT News)
http://www.advocate.com/

Bisexual Resource Center
http://www.biresource.org

Children of Lesbians & Gays Everywhere (COLAGE)
http://www.pflag.org/fileadmin/user_upload/
Publications/Be_Yourself.pdf

Gay Dating Site
http://www.compatiblepartners.net/

Gay and Lesbian Support Groups for Parents
http://www.gayparentmag.com/

Human Rights Campaign
http://www.hrc.org

Mixed sexual orientation marriages
http://www.straightspouseconnection.com/
http://www.straightspouse.org/home.php

National Gay and Lesbian Task Force
http://www.thetaskforce.org/

Other Sheep (Christian organization for empowering sexual minorities)
http://www.othersheepexecsite.com/
http://www.othersheep.org/

PFLAG (Parents, Families, and Friends of Lesbians and Gays)
http://community.pflag.org/Page.aspx?pid=194&srcid=-2

PlanetOut (Online News Source)
http://www.planetout.com

Same-sex Relationships
http://www.alternet.org/sex/

CHAPTER 9

Sexuality in Relationships

WIFE: *I say foreplay begins in the morning.*

HUSBAND: *It seems to me being sexual would make us closer, but she says it works the other way—if she felt closer, there'd be more sex.*

Lillian Rubin, *Intimate Strangers*

Chelsea E. Curry

Learning Objectives

Be aware of various sexual values which implicate subsequent sexual behaviors.

Identify the factors one is to consider in deciding to have sex with a new person.

Review the sources of sexual values.

Summarize the frequency and outcome of various sexual behaviors.

Review the expression of sexuality in the context of various relationships.

Know how there is an illusion of monogamy in sexual relationships.

List the factors involved in a sexually fulfilling relationship.

Predict the future of sexual relationships.

"Like a match, sex ignites and brings life, light, and warmth; it also burns" (Merrill and Knox 2010). These words are in reference to a discussion of office romances—once sex is introduced into a relationship, the relationship is changed forever. These changes are often positive—sex is one of the delights of being involved in a relationship. Indeed, sexuality is a major relationship aspect and the subject of this chapter. We begin by discussing the definition of sex and sexual values lovers bring to the sexual encounter.

Sexual Definitions and Sexual Values

Like the definition of love, there are different understandings of what sex means. Bill Clinton's statement (at a White House press conference) "I did not have sexual relations with that woman, Miss Lewinsky" revealed the different conceptions individuals have of sex. Schwarz et al. (2010) asked 122 students, "What do you associate with the word sex?" Their top ten connotations (in order) were love, fun, lust, trust, erotic, contraception, orgasm, tenderness, closeness, and passion. As might be expected, women were more likely to provide referents on relationships and love; men focused more on body parts and sexual acts.

Regardless of how sex is conceptualized, there are sexual values. The following are some examples of choices with which individuals in a new relationship are confronted. Their decision will be made on the basis of their sexual values.

How much sex/how soon in a relationship is appropriate?

Do you tell your partner the *actual* number of previous sexual partners?

Do you tell your partner your sexual fantasies?

Do you require a condom for oral, vaginal, or anal sex?

Do you require a condom and/or dental dam (vaginal barrier) for oral sex?

Before having sex with someone, do you require that they have been recently tested for sexually transmitted infections (STIs) and the human immunodeficiency virus (HIV)?

Do you reveal to your partner previous or current same-sex behavior or interests?

Sexual values are moral guidelines for making sexual choices in nonmarital, marital, heterosexual, and homosexual relationships. Attitudes and values sometimes predict sexual behavior. One's sexual values may be identical to one's sexual choices. For example, a person who values abstinence until marriage may choose to remain a virgin until marriage. One's behavior does not always correspond with one's values. Some who express a value of waiting until marriage have intercourse

Between two evils, I always pick the one I haven't tried before.

Mae West, actress

Sexual values moral guidelines for making sexual choices in nonmarital, marital, heterosexual, and homosexual relationships.

TABLE 9.1	Sexual Value by Sex of Undergraduate Respondent		
Respondents	Absolutism	Relativism	Hedonism
Male students	13.9%	51.1%	27.0%
Female students	19.8%	64.3%	15.6%

Source: Knox, D., and S. Hall. 2010. Relationship and sexual behaviors of a sample of 2,922 university students. Unpublished data collected for this text. Department of Sociology, East Carolina University, and Department of Family and Consumer Sciences, Ball State University.

Absolutism a sexual value system which is based on unconditional allegiance to tradition or religion (e.g., waiting until marriage to have sexual intercourse).

before marriage. One explanation for the discrepancy between values and behavior is that a person may engage in a sexual behavior, then decide the behavior was wrong, and adopt a sexual value against it.

There are at least three sexual value perspectives that guide choices in sexual behavior: absolutism, relativism, and hedonism. See Table 9.1 for the respective sexual values of almost 3000 undergraduates. Individuals often have different sexual values at different stages of the family life cycle. For example, elderly individuals are more likely to be absolutist, whereas those in the middle years are more likely to be relativistic. Young unmarried adults are more likely than the elderly to be hedonistic.

Absolutism

Fourteen percent of males and 20% of females of 2,922 undergraduates identified absolutism as their sexual value (Knox and Hall 2010). **Absolutism** is a sexual value system which is based on unconditional allegiance to tradition or religion (i.e. waiting until marriage to have sexual intercourse). People who are guided by absolutism in their sexual choices have a clear notion of what is right and wrong.

The official creeds of fundamentalist Christian and Islamic religions encourage absolutist sexual values. Intercourse is solely for procreation, and any sexual acts that do not lead to procreation (masturbation, oral sex, homosexuality) are immoral and regarded as sins against God, Allah, self, and community. Waiting until marriage to have intercourse is also an absolutist sexual value. This value is often promoted in the public schools (see Social Policy feature).

This couple has decided that they will wait until they are married to have sexual intercourse.

Chelsea E. Curry

Chapter 9 Sexuality in Relationships

Virginity loss for heterosexuals typically refers to vaginal sex (though some would say they are no longer virgins if there has been oral or anal sex). Lesbian and gay males typically refer to virginity loss if there has been oral or anal sex. Regardless of the definition, typical meanings associated with virginity loss are in reference to it being a gift, stigma, or rite of passage (Carpenter 2009). Virginity loss as a gift suggests that giving oneself sexually the first time is a special gift which may or may not require reciprocation. Two virgins who "save" themselves until their wedding night may view their first intercourse as a gift to each other. Boislard and Poulin (2011) were able to predict early virginity loss. In a sample of 402 youth, those engaging in more antisocial behaviors, having a lot of other sex friends and being from a non-intact family predicted first intercourse at age 13 or less when compared to those who reported few antisocial behaviors, having few other sex friends, and being from an intact family.

Some may view being a virgin as a stigma which they need to get rid of through vaginal sex. Virginity may also be seen as shameful since virgins have never had sex and are thought of as being very sexually inept. Using this perspective, virginity loss would be removing the stigma of a sexually inept shameful state.

Virginity may also be seen as a rite of passage so that the individual who has had sex has moved from childhood to adulthood and who has personal knowledge of sexual behavior, gaining new knowledge about sex. "True Love Waits" is an international campaign designed to challenge teenagers and college students to remain sexually abstinent until marriage. Under this program, created and sponsored by the Baptist Sunday School Board, young people are asked to agree to the absolutist position and sign a commitment to the following: "Believing that true love waits, I make a commitment to God, myself, my family, my friends, my future mate, and my future children to be sexually abstinent from this day until the day I enter a biblical marriage relationship" (True Love Waits 2011, http://www.lifeway.com/tlw/students/join.asp).

How effective are these "True Love Waits" and "virginity pledge" programs in delaying sexual behavior until marriage? Data from the National Longitudinal Study of Adolescent Health revealed that, although youth who took the pledge were more likely than other youth to experience a later "sexual debut," have fewer partners, and marry earlier, most eventually engaged in premarital sex, were less likely to use a condom when they first had intercourse, and were more likely to substitute oral and/or anal sex in the place of vaginal sex. There was no significant difference in the occurrence of STIs between "pledgers" and "nonpledgers" (Brucker and Bearman 2005). The researchers speculated that the emphasis on virginity may have encouraged the pledgers to engage in non-coital (nonintercourse) sexual activities (for example, oral sex), which still exposed them to STIs and to be less likely to seek testing and treatment for STIs. Similarly, Hollander (2006) collected national data on two waves of adolescents. Half of those who had taken the virginity pledge reported no such commitment a year later. Males and black individuals were particularly likely to retract their pledge. Landor and Simons (2010) studied 1215 undergraduates and also found that male pledgers (in contrast to female pledgers) were less likely to remain virgins.

A similar cultural ritual to encourage and celebrate virginity until one's wedding day is the "Father-Daughter Purity Ball." Over 4,000 such events occur annually that involve fathers and daughters as young as 4 years old taking mutual pledges to be pure. Fathers promise to protect their daughters and as well to be faithful and shun pornography themselves. Daughters promise to be pure, which implies the value of absolutism (Gibbs 2008).

Some individuals still define themselves as virgins even though they have engaged in oral sex. Of 2,922 university students, 60% agreed that, "If you have oral sex, you are still a virgin." Hence, according to these undergraduates, having

Sexuality education was introduced in the American public school system in the late nineteenth century with the goal of combating STIs (sexually transmitted infections) and instilling sexual morality (typically understood as abstinence until marriage). Over time, the abstinence agenda became more evident. In the Bush administration, only sex education programs that emphasized or promoted abstinence were eligible to qualify for federal funding. Programs that also discussed contraception and other means of pregnancy protection, referred to as **comprehensive sex education programs**, were not eligible. The Obama administration has been in favor of the latter. While these philosophical differences have been pervasive, a trend has emerged whereby schools and communities provide both abstinence education and contraception information (more comprehensive sex education).

Chen et al. (2011) evaluated the cost effectiveness of such health education intervention programs which improve preadolescents' attitudes toward abstinence and pregnancy avoidance through contraceptive use. For each $1000 spent in these programs 13.67 unintended pregnancies among preadolescents were avoided.

Abstinence-only sex education programs are associated with delay of first intercourse. Jemmott et al. (2010) studied 662 African American students in grades 6 and 7 (mean age of 12.2 years), some of whom were exposed to abstinence-only sex education and some not. Follow up at 24 months resulted in model estimated chance of first intercourse at 33.5% in the abstinence-only intervention and 48.5% in the control group. Abstinence-only intervention did not affect condom use. No other differences between interventions and controls were significant.

Comprehensive sex education program learning experience which recommends abstinence but also discusses contraception and other means of pregnancy protection.

Secondary virginity a sexually initiated person's deliberate decision to refrain from intimate encounters for a set period of time and to refer to that decision as a kind of virginity (rather than "mere" abstinence).

Asceticism sexual belief system which emphasizes that giving in to carnal lust is unnecessary and one should attempt to rise above the pursuit of sensual pleasure into a life of self-discipline and self-denial.

Relativism value system emphasizing that sexual decisions should be made in the context of a particular relationship.

oral sex with someone is not really having sex (Knox and Hall 2010). Individuals may engage in oral sex rather than sexual intercourse to avoid getting pregnant, to avoid getting a STI, to keep their partner interested, to avoid a bad reputation, and to avoid feeling guilty over having sexual intercourse (Vazonyi and Jenkins 2010).

Carpenter (2010) discussed the concept of **secondary virginity**—a sexually-initiated person's deliberate decision to refrain from intimate encounters for a set period of time and to refer to that decision as a kind of virginity (rather than "mere" abstinence). Secondary virginity may be a result of physically painful, emotionally distressing, or romantically disappointing sexual encounters. Of sixty-one young adults interviewed (most of whom were white, conservative, religious women), more than half believed that a person could, under some circumstances, be a virgin more than once. Fifteen people contended that people could resume their virginity in an emotional, psychological, or spiritual sense. Terence Duluca, a 27-year-old, heterosexual, white, Roman Catholic, explained:

> There is a different feeling when you love somebody and when you just care about somebody. So I would have to say if you feel that way, then I guess you could be a virgin again. Christians get born all the time again, so. . . . When there's true love involved, yes, I believe that.

A subcategory of absolutism is **asceticism**. Ascetics believe that giving in to carnal lust is unnecessary and attempt to rise above the pursuit of sensual pleasure into a life of self-discipline and self-denial. Accordingly, spiritual life is viewed as the highest good, and self-denial helps one to achieve it. Catholic priests, monks, nuns, and some other celibate people have adopted the sexual value of asceticism.

Relativism

Fifty-one percent of males and 64% of females of 2,922 undergraduates identified relativism as their sexual value (Knox and Hall 2010). **Relativism** is a

Landry et al. (2011) reviewed how faith based organizations are providing comprehensive sexuality education in some contexts. Lamb (2010) emphasized that the school system should not focus on just abstinence or condom use but teach ethical reasoning. Indeed, sex education should be a "form of citizenship education, focusing on justice, equity, and caring for the other person as well as the self." Such a perspective would imply the use of a condom as an expression of concern about self and others.

Your Opinion?

1. To what degree do you support abstinence education in public schools?
2. Should condoms be made available for students already having sex?
3. Should parents control the content of sex education in public schools?

Sources

Chen, C., T. Yamada and E. M. Walker. (2011) Estimating the cost-effectiveness of a classroom-based abstinence and pregnancy avoidance program targeting preadolescent sexual risk behaviors. *Journal of Children & Poverty* 17: 87–109.

Jemmott III, J. B., L. S. Jemmott, and G. T. Fong. 2010. Efficacy of a theory-based abstinence only intervention over 24 months: A randomized controlled trial with young adolescents. *Archives of Pediatric and Adolescent Medicine* 164: 152–159.

Lamb, S. 2010. Toward a sexual ethics curriculum: Bringing philosophy and society to bear on individual development. *Harvard Educational Review* 80: 81–105.

Landry, D. J., L. D. Lindberg, A. Gemmill, H. Boonstra, and L. B. Finer. 2011. Review of the role of faith- and community-based organizations in providing comprehensive sexuality. *Education American Journal of Sexuality Education* 6: 75–103.

value system emphasizing that sexual decisions should be made in the context of a particular relationship. Whereas absolutists might feel that having intercourse is wrong for unmarried people, relativists might feel that the moral correctness of sex outside marriage depends on the particular situation. For example, a relativist might feel that in some situations, sex between casual dating partners is wrong (such as when one individual pressures the other into having sex or lies to persuade the other to have sex). However, in other cases—when there is no deception or coercion and the dating partners are practicing "safer sex"—intercourse between casual dating partners may be viewed as acceptable.

Sexual values and choices that are based on relativism often consider the degree of love, commitment, and relationship involvement as important factors. In a study designed to assess "turn-ons" and "turn-offs" in sexual arousal, women spoke of "feeling desired versus feeling used" by the partner. "Many women talked about how their arousal was increased with partners who seemed particularly interested in them as individual women, rather than someone that they just wanted to have sex with" (Graham et al. 2004).

A disadvantage of relativism as a sexual value is the difficulty of making sexual decisions on a relativistic case-by-case basis. The statement "I don't know what's right anymore" reflects the uncertainty of a relativistic view. Once a person decides that mutual love is the context justifying intercourse, how often and how soon is it appropriate for the person to fall in love? Can love develop after some alcohol and two hours of conversation? How does one know that love feelings are genuine? The freedom that relativism brings to sexual decision making requires responsibility, maturity, and judgment. In some cases, individuals may convince themselves that they are in love so that they will not feel guilty about having intercourse. Though one may feel "in love," "secure," and "committed" at the time first intercourse occurs, only 17% of all first intercourse experiences that women reported are with the person they eventually marry (Raley 2000). See the self-assessment to evaluate your attitudes toward premarital sex.

Among men, sex sometimes results in intimacy; among women, intimacy sometimes results in sex.

Barbara Cartland, English romance novelist

Attitudes toward Premarital Sex Scale

Premarital sex is defined as engaging in sexual intercourse prior to marriage. The purpose of this survey is to assess your thoughts and feelings about intercourse before marriage. Read each item carefully and consider how you feel about each statement. There are no right or wrong answers to any of these statements, so please give your honest reactions and opinions. Please respond by using the following scale:

1   2   3   4   5   6   7
Strongly   Strongly
Disagree   Agree

_____ 1. I believe that premarital sex is healthy.
_____ 2. There is nothing wrong with premarital sex.
_____ 3. People who have premarital sex develop happier marriages.
_____ 4. Premarital sex is acceptable in a long-term relationship.
_____ 5. Having sexual partners before marriage is natural.
_____ 6. Premarital sex can serve as a stress reliever.
_____ 7. Premarital sex has nothing to do with morals.
_____ 8. Premarital sex is acceptable if you are engaged to the person.
_____ 9. Premarital sex is a problem among young adults.
_____ 10. Premarital sex puts unnecessary stress on relationships.

Scoring

Selecting a 1 reflects the most negative attitude toward premarital sex; selecting a 7 reflects the most positive attitude toward premarital sex. Before adding the numbers you assigned to each item, change the scores for items #9 and #10 as follows: replace a score of 1 with a 7; 2 with a 6; 3 with a 5; 4 with a 4; 5 with a 3; 6 with a 2; and 7 with a 1. After changing these numbers, add your ten scores. The lower your total score (10 is the lowest possible score), the less accepting you are of premarital sex; the higher your total score (70 is the highest possible score), the greater your acceptance of premarital sex. A score of 40 places you at the midpoint between being very disapproving of premarital sex and very accepting of premarital sex.

Scores of Other Students Who Completed the Scale

The scale was completed by 252 student volunteers at Valdosta State University. The mean score of the students was 40.81 (standard deviation [SD] = 13.20), reflecting that the students were virtually at the midpoint between a very negative and a very positive attitude toward premarital sex. For the 124 males and 128 females in the total sample, the mean scores were 42.06 (SD = 12.93) and 39.60 (SD = 13.39), respectively (not statistically significant). In regard to race, 59.5% of the sample was white and 40.5% was nonwhite (35.3% black, 2.4% Hispanic, 1.6% Asian, 0.4% American Indian, and 0.8% other). The mean scores of whites, blacks, and nonwhites were 41.64 (SD = 13.38), 38.46 (SD = 13.19), and 39.59 (SD = 12.90) (not statistically significant). Finally, regarding year in college, 8.3% were freshmen, 17.1% sophomores, 28.6% juniors, 43.3% seniors, and 2.8% graduate students. Freshman and sophomores reported more positive attitudes toward premarital sex (mean = 44.81; SD = 13.39) than did juniors (mean = 40.32; SD = 12.58) or seniors and graduate students (mean = 38.91; SD = 13.10) (p = .05).

Source

"Attitudes toward Premarital Sex Scale," 2006 by Mark Whatley, Ph.D., Department of Psychology, Valdosta State University, Valdosta, Georgia 31698-0100. Used by permission. Other uses of this scale by written permission of Dr. Whatley only (mwhatley@valdosta.edu). Information on the reliability and validity of this scale is available from Dr. Whatley.

Absolutists and relativists have different views on whether or not two unmarried people should have intercourse. Whereas an absolutist would say that having intercourse is wrong for unmarried people and right for married people, a relativist would say, "It depends on the situation." Suppose, for example, that a married couple do not love each other and intercourse is an abusive, exploitative act. Suppose also that an unmarried couple love each other and their intercourse experience is an expression of mutual affection and respect. A relativist might conclude that, in this particular situation, having intercourse is "more right" for the unmarried couple than the married couple. Students who become involved in a "friends with benefits" relationship reflect a specific expression of relativism.

Women want the "friends" while men want the "benefits."

Result of research study on FWB

Friends with Benefits

Friends with benefits is becoming part of the relational sexual landscape of youth. **Friends with benefits** (FWB) is a relationship of nonromantic friends who also have a sexual relationship. Owen and Fincham (2011) studied a sample of 889 individuals and focused on those who had had at least one FWB relationship in the last year. More men than women (54% versus 43%) reported having had

Friends with benefits a relationship of nonromantic friends who also have a sexual relationship.

such an experience and more men than women reported a positive emotional reaction to the experience. Greater alcohol use was associated with being more likely to report having been in a **FWB** relationship.

Puentes et al. (2008) analyzed data from 1,013 undergraduates (over half of which reported experience with a "friends with benefits" relationship), and compared the background characteristics of participants with nonparticipants in such a relationship. Findings revealed that participants were significantly more likely to be males, casual daters, hedonists, nonromantics, jealous, black, juniors or seniors, those who have had sex without love, and those who regard "financial security" as their top value. The friends with benefits relationship is primarily sexual and engaged in by nonromantic hedonists who have a pragmatic view of relationships. However there are those in friends with benefits relationships who use the label to hide behind the fact that they are actually more involved emotionally but do not acknowledge these feelings.

These two friends are often seen together and thought of as being a couple—they are simply friends with benefits.

In a smaller sample of 170 undergraduates at the same university, 57.3% of these undergraduates reported that they were or had been involved in an FWB relationship. There were no significant differences between the percentages of women and men reporting involvement in an FWB relationship. This is one of the few studies finding no difference in sexual behavior between women and men (for example, one would expect men to have more FWB relationships than women). However, the percentages of women and men were very similar in their reported rates of FWB involvement—57.1% and 57.9%, respectively. Is a new sexual equality operative in FWB relationships? (McGinty et al. 2007). Analysis of the data revealed other significant differences between female and male college students in regard to various aspects of the FWB relationship.

1. *Women were more emotionally involved.* Women were significantly more likely than men (62.5% versus 38.1%) to view their current FWB relationship as an emotional relationship. In addition, women were significantly more likely than men to be perceived as being more emotionally involved in the FWB relationship. Of the men, 43.5%, compared with 13.6% of the women, reported "my partner is more emotionally involved than I am." This finding was NOT reflected in the film *Up in the Air* (2009) which depicted George Clooney as Ryan Bingham and Vera Farmiga as Alex in a friends with benefits relationship. Unlike the traditional FWB relationship, it was the character of Ryan Bingham who ended up falling in love and wanting more than just the sexual benefits of the relationship.

2. *Men were more sexually focused.* As might be expected from the first finding, men were significantly more likely than women to agree with the statement, "I wish we had sex more often than we do" (43.5% versus 13.6%).

3. *Men were more polyamorous.* With polyamory defined as desiring to be involved in more than one emotional or sexual relationship at the same time, men were significantly more likely than women to agree that "I would like to have more than one FWB relationship going on at the same time" (34.8% versus 4.5%). Serial FWB relationships may already be occurring. Of the men, 52.2%, compared with 24.6% of women, reported that they had been involved in more than one FWB relationship. Hughes et al. (2005) studied 143 undergraduates in FWB relationships and noted that a ludic, playful, noncommittal love characterized them.

"Sex is the icing on the relationship cake" is an accepted truth. Few believe that individuals can "screw themselves into a good relationship." Rather, it is the relationship which is more likely to have a positive influence on the couple's sex life. Nevertheless, there may be certain sexual factors associated with relationship satisfaction/happiness—the focus of this study.

Sample and Methods

The data for analysis were taken from a nonrandom sample of 1319 undergraduate volunteers at a large southeastern university who responded to an online 100-item questionnaire. Since there were no identifying codes associated with the submitted questionnaires, the identity of the respondents was anonymous.

Almost three-fourths (73.1%) of the respondents were female; over a fourth (26.7%) were male. The average age of the respondents was 20.1 with a median age of 19 years. Regarding race, 74% self identified as white; 17.1% self identified as black (African-American, African Black, and Caribbean Black were combined); 3% Asian; 2.1% Hispanic; 1.7% biracial, and 1.7% other. Regarding relationship status of the respondents, almost half (49%) were emotionally involved with one person, 29.3% were not dating, and 21.7% were casually dating.

The dependent variable, relationship happiness, was measured by asking respondents to respond to the following statement: "In regard to my current relationship, on a scale of 1 to 10, with 1 being extremely unhappy and 10 being extremely happy, I rate my current level of happiness as:" by circling the number that described their level of relationship happiness. Various independent variables which were found to be statistically significant are included in the results section. Ordinal Least Squares (OLS) regression analysis was used to determine sexual predictors of relationship happiness among the respondents.

Results

The results are summarized into those variables associated with relationship happiness (see Table 1).

*Abridged and adapted from an original paper, 2011, by Dantzler, T., D. Knox, V. Kriskute and K. Vail-Smith, East Carolina University.

Concurrent sexual partnership relationship in which the partners have sex with several individuals concurrently.

When I'm good, I'm very good. When I'm bad, I'm better.

Mae West, actress

Hedonism the belief that the ultimate value and motivation for human actions lies in the pursuit of pleasure and the avoidance of pain.

Concurrent Sexual Partnerships

Concurrent sexual partnerships are those in which the partners have sex with several individuals concurrently. Paik (2010) analyzed data on 783 adults aged 18–59 and found that 10% reported that both they and their partners had had other sexual partners during their own relationship. Men were more likely than women to have been nonmonogamous (17% vs. 5%). A serious relationship was the context where nonrelationship sex was least likely. However, if the partners had a casual or friends with benefits relationship the chance of nonmonogamy rose 30% and 44% respectively.

Hedonism

Twenty-seven percent of males and 16% of females of 2,922 undergraduates identified hedonism as their primary sexual value (Knox and Hall 2010). **Hedonism** is the belief that the ultimate value and motivation for human actions lie in the pursuit of pleasure and the avoidance of pain. The hedonistic value is reflected in the statement, "If it feels good, do it." Hedonism assumes that sexual desire, like hunger and thirst, is an appropriate appetite and its expression is legitimate. Of 2,922 undergraduates, 20% reported that they had hooked up (had oral or sexual intercourse) the first time they met someone (Knox and Hall 2010).

Hedonism is sometimes viewed as sexual addiction. Levine (2010) noted that the term sexual addition is often misused and applied to those who watch

Sexual Correlates of Relationship Happiness

Emotional involvement—being involved with one person, being engaged, being married

Absolutist sexual value—waiting until marriage to have sexual intercourse

Sex with love—history of having had sex in the context of a love relationship

Open sexual communication—expressing preferences to one's sexual partner

Use of birth control—not allowing sexual intercourse to occur without it

No forced sex—absence of having been forced to have sex by a partner

No regrets—about timing of first sexual experience

Many of the findings of this study were consistent with previous research. Being in love (Impett et al. 2005), delaying intercourse in a new relationship (Merrill and Knox 2010), having open communication (MacNeil and Byers 1997), and not being forced to have sex (Oswalt 2005) are standard findings in the sexuality/relationship literature. Some unique findings of this study included that relationship happiness was *unrelated* to one's sexual orientation, having had sexual intercourse, and having engaged in various sexual behaviors—masturbation, oral sex, and anal sex.

References

Impett, E. A., L. A. Peplau, and S. L. Gable. 2005. Approach and avoidance sexual motives: Implications for personal and interpersonal well-being. *Personal Relationships* 12:465–482.

Merrill, J., and D. Knox. 2010. *When I fall in love again: A new study on finding and keeping the love of your life*. Santa Barbra, California: Praeger.

MacNeil, S., and Byers, E. S. 1997. The relationships between sexual problems, communication, and sexual satisfaction. *The Canadian Journal of Human Sexuality* 6: 277–284.

Oswalt, S. B., K. A. Cameron, and J. J. Koob. 2005. Sexual regret in college students. *Archives of Sexual Behavior* 34:663–669.

pornography, have commercial sex, and engage in cybersex. More accurately, sex addition applies to those who have lost control over their sexual behavior which is often accompanied by spiraling psychological deterioration—i.e., depression.

Sexual Double Standard

The **sexual double standard** is the view that encourages and accepts sexual expression of men more than women. Men are almost two times more hedonistic than women (Knox and Hall 2010). Acceptance of the double standard is evident in that hedonistic men are thought of as "studs" but hedonistic women as "sluts." The sexual double standard is also evident in Canada. Lai and Hynie (2011) sampled 305 Canadians and found women more interested in traditional sex whereas men were more interested in experimental sex (extramarital sex, viewing of erotic material).

The sexual double standard is also evident in that there is lower disapproval of men having higher numbers of sexual partners but high disapproval of women for having the same number of sexual partners as men. In a recent study, England and Thomas (2006) noted that the double standard was operative in hooking up. Women who hooked up too often with too many men and had sex too easily were vulnerable to getting bad reputations. Men who did the same thing got a bad reputation among women, but with fewer stigmas. In addition, men gained status among other men for their exploits; women were quieter. Similarly, women who looked at pornography (in contrast to men who viewed pornography) were viewed as "loose" (19.4% versus 4.5%) (O'Reilly et al. 2007). (see What If? Box which focuses on pornography.)

What men desire is a virgin who is a whore.

Edward Dahlberg, *Reasons of the Heart*

Sexual double standard the view that encourages and accepts the sexual expression of men more than women.

What if My Partner Wants me to Watch Pornography with Him?

Maddox et al. (2011) studied the associations between viewing sexually-explicit material (SEM) and relationship functioning in a random sample of 1291 un-married individuals in romantic relationships. As expected, more men (76.8%) than women (31.6%) reported having viewed SEM on their own—but almost half (45%) reported sometimes viewing SEM with their partner (44.8%). The researchers looked at communication, relationship adjustment, commitment, sexual satisfaction, and infidelity and found that those who never viewed SEM reported higher relationship quality on all indices than those who viewed SEM alone. In addition, those who viewed SEM only with their partners reported more dedication and higher sexual satisfaction than those who viewed SEM alone. Hence, viewing pornography is a behavior individuals may feel very dif-ferent about; should they feel comfortable viewing SEM with their partner, there appear to be benefits.

Kim et al. (2007) found evidence for the double standard in their review of twenty-five prime-time television programs. In addition, Greene and Faulkner (2005) studied 689 heterosexual couples and found that women were disadvan-taged in negotiating sexual issues with their partners, particularly when their traditional gender roles were operative.

PERSONAL CHOICES

Deciding to Have Intercourse with a New Partner

The following are issues you might consider in making the decision to have sexual in-tercourse with a new partner:

1. Personal consequences. How do you predict you will feel about yourself after you have had intercourse with a new partner? An increasing percentage of college students are relativists and feel that the outcome will be positive if they are in love. The following quote is from a student in our classes:

> I believe intercourse before marriage is OK under certain circumstances. I believe that when a person falls in love with another and the relationship is stable, it is then appropriate. This should be thought about very carefully for a long time, so as not to regret engaging in intercourse.

Those who are not in love and have sex in a casual context often report sexual regret about their decision.

> When I have the chance to experience sex with someone new, my first thought is to jump right in, ask questions later. However, my very second thought is, "If we do this, will he think I'm slutty?" followed by "Is it too early to be doing this? What if I regret it?" Looking back at my dating experiences over the past several years, it seems that earlier on, I let a guy talk me into doing something sexual, even though those questions were running circles in my head. However, in the most recent instances, I remained firm that sex wasn't going to happen so soon into dating someone, and I do not regret my decision. In the future, I don't plan on being intimate with anyone until we have gone from "going out on dates" to boyfriend/girlfriend status. (Merrill and Knox 2010, 3)

The effect intercourse will have on you personally will be influenced by your personal values, your religious values, and your emotional involvement with your partner. Some people prefer to wait until they are married to have intercourse and feel that this is the best course for future marital stability and happiness. There is often, but not necessarily, a religious basis for this value.

Strong personal and religious values against nonmarital intercourse may result in guilt and regret following an intercourse experience. In a sample of 270 unmarried undergraduates who had had intercourse, 71.9% regretted their decision to do so at least once (for example, they may have had intercourse more than once or with multiple partners and reported regret at least once). Higgins et al. (2010a) analyzed data on first intercourse behavior from a cross-sectional survey of 1,986 non-Hispanic white and black 18- to 25-year-old respondents from four university campuses. Respondents were asked to rate the degree to which their first vaginal intercourse was physiologically and psychologically satisfying. Women were less likely than men to report a positive first vaginal intercourse experience, particularly physiological satisfaction. Being in a committed relationship was the condition associated with the most positive experience.

2. *Timing.* Delaying intercourse with a new partner is important for achieving a positive outcome and avoiding regret. The table below reveals that a third of 429 undergraduates regretted having sexual intercourse "too soon" in a relationship (Merrill and Knox 2010, p. 2).

Regret for Engaging in Behavior "Too Soon" or "Too Late"

(N = 429)

Behaviors	"Too Soon"	"Too Late"	"Perfect Timing"	"Did not do"
Sexual Intercourse	33.3%	3.3%	48.4%	15.0%
Spent the night	26.6%	3.7%	58.8%	11.4%
Saying "I love you"	26.1%	3.9%	50.8%	19.2%
Kissing	11.1%	3.1%	82.8%	3.1%

3. *Partner consequences.* Because a basic moral principle is to do no harm to others, it is important to consider the effect of intercourse on your partner. Whereas intercourse may be a pleasurable experience with positive consequences for you, your partner may react differently. What is your partner's religious background, and what are your partner's sexual values? A highly religious person with absolutist sexual values will typically have a very different reaction to sexual intercourse than a person with low religiosity and relativistic or hedonistic sexual values. In the study of 270 undergraduates previously referred to, 23% (more women than men) reported regret due to pressure from the partner (Oswalt et al. 2005).

4. *Relationship consequences.* What is the effect of intercourse on a couple's relationship? One's personal reaction to having intercourse may spill over into the relationship. Individuals might predict how they feel having intercourse will affect their relationship before including it in their relationship.

5. *Contraception.* Another potential consequence of intercourse is pregnancy. Once a couple decides to have intercourse, a separate decision must be made as to whether intercourse should result in pregnancy. Most sexually active undergraduates do not want children. People who want to avoid pregnancy must choose and plan to use a contraceptive method.

6. *HIV and other sexually transmissible infections.* Engaging in casual sex has potentially fatal consequences. Avoiding HIV infection and other STIs is an important consideration in deciding whether to have intercourse in a new relationship. The increase in the number of people having more partners results in the rapid spread of the bacteria and viruses responsible for numerous varieties of STIs. However, in a sample of 2,922 undergraduates, only a quarter reported consistent condom use (Knox and Hall 2010).

Serovich and Mosack (2003) reported that only 37% of men with HIV infection told a casual partner of their positive HIV status. Those who did so reported, "I thought [my partner] had a right to know." Although depending on a person's integrity is laudable, it may not be wise. Indeed, a condom should be used routinely to help protect one from contracting a STI.

7. *Influence of alcohol and other drugs.* A final consideration with regard to the decision to have intercourse in a new relationship is to be aware of the influence of alcohol and other drugs on such a decision. Of students who reported regretting having had intercourse, 31.9% noted that alcohol was involved in their decision to have intercourse (Oswalt et al. 2005). A term has emerged to describe these pregnancies: **alcohol-exposed pregnancy** (AEP) (Ingersoll et al. 2005).

8. *OK to change decision about including intercourse in a relationship.* Although most couples who include intercourse in their relationship continue the pattern, some decide to omit it from their sexual agenda. One female student said, "Since I did not want to go on the pill and would be frantic if I got pregnant, we decided the stress was not worth having intercourse. While we do have oral sex, we don't even think about having intercourse any more."

Alcohol-exposed pregnancy sexual intercourse that occurs in the context of alcohol which involves little consideration for using contraception.

Sources

Ingersoll, K. S., S. D. Ceperich, M. D. Nettleman, K. Karanda, S. Brocksen, and B. A. Johnson. 2005. Reducing alcohol-exposed pregnancy risk in college women: Initial outcomes of a clinical trial of a motivational intervention. *Journal of Substance Abuse Treatment* 29: 173–180.

Knox, D., and S. Hall. 2010 Relationship and sexual behaviors of a sample of 2,922 University students. Unpublished data collected for this text. Department of Sociology, East Carolina University, and Department of Family and Consumer Sciences, Ball State University.

Merrill, J. and D. Knox. 2010. *When I fall in love again: A new study on finding and keeping the love of your life.* Santa Barbra, California: Praeger.

Oswalt, S. B., K. A. Cameron, and J. J. Koob. 2005. Sexual regret in college students. *Archives of Sexual Behavior* 34: 663–669.

Serovich, J. M., and K. E. Mosack. 2003. Reasons for HIV disclosure or nondisclosure to casual sexual partners. *AIDS Education and Prevention* 15: 70–81.

Sources of Sexual Values

I used to be Snow White . . . but I drifted.

Mae West, actress

The sources of one's sexual values are numerous and include one's school, religion, and family, as well as technology, television, social movements, and the Internet. Previously we noted that public schools in the United States promote absolutist sexual values through abstinence education and that the effectiveness of these programs has been questioned.

Religion is also an important influence. Miller et. al. (2011) sampled 1289 adults at a large southeastern university between the ages of 18 and 30 and found that 23% identified themselves as very religious/spiritual, 50% somewhat religious/spiritual, 18% not very religious/spiritual, and 9% not at all religious/spiritual. In regard to sexual behavior the researchers found that "religious activity and levels of religiosity impacted young adults' attitudes about acceptable behaviors prior to initiating a dating relationship. For example, religiously active young adults were more likely to indicate that sharing intimate details, kissing/hand holding/other physical expression, sexting, sexual intercourse and non-intercourse sexual activities were inappropriate activities to engage in prior to a committed dating relationship."

Family is also influential. Kaye et al. (2009) found that adolescents who had strong positive relationships with their parents were more likely to delay first intercourse. This association holds true whether the child is in an intact family or a stepfamily. Wetherill et al. (2010) also found that individuals with high levels of both parental and peer PAC (perceived awareness and caring) engaged in less frequent sexual behaviors.

Siblings are also influential. Kornreich et al. (2003) found that girls who had older brothers held more conservative sexual values. "Those with older brothers in the home may be socialized more strongly to adhere to these traditional standards in line with power dynamics believed to shape and reinforce more submissive gender roles for girls and women" (p. 197).

Reproductive technologies such as birth control pills, the morning-after pill, and condoms influence sexual values by affecting the consequences of behavior. Being able to reduce the risk of pregnancy and HIV infection with the pill and condoms allows one to consider a different value system than if these methods of protection did not exist.

The media is also a source of sexual values. A television advertisement shows an affectionate couple with minimal clothes on in a context where sex could occur. "Be ready for the moment" is the phrase of the announcer, and Levitra, the new quick-start Viagra, is the product for sale. The advertiser uses sex to get the attention of the viewer and punches in the product.

However, as a source of sexual values and responsible treatments of contraception, condom usage, abstinence, and consequences of sexual behavior, television is woefully inadequate. Indeed, viewers learn that sex is romantic and exciting but learn nothing about discussing the need for contraception or HIV and STI protection. With few exceptions, viewers are inundated with role models who engage in casual sex without protection.

In regard to magazines as media, exposure outcome is more positive. Walsh and Ward (2010) assessed sexual health behaviors and magazine reading among 579 undergraduate students. They found that more frequent reading of mainstream magazines was associated with greater sexual health knowledge, safe-sex self-efficacy, and consistency of using contraception (although results varied across sex and magazine genre). Social movements such as the women's movement affect sexual values by empowering women with an egalitarian view of sexuality. This translates into encouraging women to be more assertive about their own sexual needs and giving them the option to experience sex in a variety of contexts (for example, without love or commitment) without self-deprecation. The net effect is a potential increase in the frequency of recreational, hedonistic sex. The gay liberation movement has also been influential in encouraging values that are accepting of sexual diversity.

Another influence on sexual values is the Internet; its sexual content is extensive. The Internet features erotic photos, videos, and "live" sex acts/stripping by webcam sex artists. Individuals can exchange nude photos, have explicit sex dialogue, arrange to have "phone sex" or meet in person, or find a prostitute. Indeed, the adult section of Craig's List was shut down because it featured blatant prostitution.

Sexual Behaviors

We have been discussing the various sources of sexual values. We now focus on what people report that they do sexually. Some individuals are **asexual**—absence of sexual behavior with a partner and oneself (masturbation). About 4% of females and 11% of males reported being asexual in the last twelve months (DeLamater and Hasday 2007). In contrast, most individuals report engaging in various sexual behaviors. Penhollow et al. (2010) found that, for both male and female students, participation in recreational sexual behaviors (with or without a partner) enhanced their overall sexual satisfaction. The following discussion includes kissing, masturbation, oral sex, vaginal intercourse, and anal sex.

Kissing
Kissing has been the subject of literature and science. The meanings of a kiss are variable—love, approval, hello, goodbye, or as a remedy for a child's hurt knee. There is also a kiss for luck, a stolen kiss, and a kiss to seal one's marriage vows. Kisses have been used to denote hierarchy. In the Middle Ages, only peers

Asexual the absence of sexual behavior with a partner as well as oneself (masturbation).

Kissing—and I mean like, yummy, smacking kissing—is the most delicious, most beautiful and passionate thing that two people can do, bar none. Better than sex, hands down.

Drew Barrymore, actress

Kissing is an outward physical behavior that often reflects intimacy (and heat) between the couple.

Frank Walsh

Masturbation stimulating one's own body with the goal of experiencing pleasurable sexual sensations.

kissed on the lips, a person of lower status kissed someone of higher status on the hand, and a person of low status showed great differential of status by kissing on the foot. Kissing has a negative connotation in the "kiss of death," which reflects the kiss Judas gave Jesus as he was about to betray him. Kissing may be an aggressive act as some do not want to be kissed. Kissing, particularly French kissing, may be very dangerous. Individuals who engage in French kissing with multiple partners face a fourfold risk of contracting meningococcal disease, a bacterial infection that may lead to meningitis, swelling of the brain and spinal cord (Gross 2006).

Although the origin of kissing is unknown, one theory posits that kissing is associated with parents putting food into their offspring's mouth . . . the bird pushes food down the throat of a chick in the nest. Some adult birds also exchange food by mouth during courtship. Anthropologists note that some cultures (for example, Eskimos, Polynesians) promote meeting someone by rubbing noses. Kissing is not too far a leap from rubbing noses.

The way a person kisses reflects the person's country, culture, and society. The French kiss each other once on each cheek or three times in the same region. Greeks tend to kiss on the mouth, regardless of the sex of the person. Eskimos, we have noted, rub noses. The Chinese rarely kiss in public.

Masturbation

Chelsea E. Curry

Sex toys are used to enhance one's sexual experience, with or without a partner. Sex toy parties are regularly held on many college campuses where women are exposed to an array of vibratory aids.

Masturbation involves stimulating one's own body with the goal of experiencing pleasurable sexual sensations. Westbrooke (2011) analyzed data from 158 undergraduates and found that men were much more likely to masturbate once a week or more (82%) while only 45.7% of women reported this frequency. In addition, 74% of the men answered that they began masturbating between the ages of 11 and 14, while only 36.7% of women answered did so. Woman were more likely to start at different times, with 14.1% starting at age 10 or younger and 21.2% at age 17 or older.

Of 2,922 undergraduates, 76.2% reported having masturbated (83.4% men, 69.0% women (Knox and Hall 2010)). Previous research on masturbation among college students revealed that being male, hedonistic, nonreligious, and cohabitants were characteristics of people who have masturbated (O'Reilly et al. 2006). Alternative terms for masturbation include *autoeroticism, self-pleasuring, solo sex,* and *sex without a partner.* An appreciation of the benefits of masturbation has now replaced various myths about it (for example, it causes blindness). Most health care providers and therapists today regard masturbation as a normal and healthy sexual behavior. Furthermore, masturbation has become known as a form of safe sex in that it involves no risk of transmitting diseases (such as HIV) or producing unintended pregnancy. Masturbation is also associated with orgasm during intercourse. In a study by Thomsen and Chang (2000), 292 university undergraduates reported whether they had ever masturbated and whether they had an orgasm during their first intercourse experience. The researchers found that the strongest single predictor of orgasm and emotional satisfaction with first intercourse was previous masturbation.

Masturbation—it's sex with someone I love.

Woody Allen, director

Oral Sex

National **Data**

Of 5,865 men and women ages 18 to 49, in the past year, more than half of women and men reported engaging in oral sex (Herbenick et al. 2010).

In a sample of 2,922 undergraduates, 69% reported that they had given oral sex (men 62.7%; women 76.2%); 73% report having received oral sex (men 69%; women 77.4%) (Knox and Hall 2010). Fellatio is oral stimulation of the man's genitals by his partner. In many states, legal statutes regard fellatio as a "crime against nature." "Nature" in this case refers to reproduction, and the "crime" is sex that does not produce babies. Nevertheless, most men have experienced fellatio. Cunnilingus is oral stimulation of the woman's genitals by her partner.

Malacad and Hess (2010) observed among Canadian adolescent females that oral sex has become an increasingly common and casual activity. In a study of 181 women aged 18–25 years, approximately three quarters reported having engaged in oral sex, a prevalence rate almost identical to that for vaginal intercourse. The respondents were an average age of 17 (for both oral sex and intercourse) at their first experience. Most reported that their most recent oral sex experience was positive and in a committed relationship. Adolescent females who reported oral sex with a partner who they were not in love with reported the most negative experience.

We noted earlier that, increasingly, youth who have oral sex regard themselves as virgins, believing that only sexual intercourse constitutes "having sex." As noted previously, of 2,922 university students, 60% agreed that, "If you have oral sex, you are still a virgin." This perspective allows the person to avoid the social stigma of having had sexual intercourse, which puts them in a more positive light for subsequent partners ("I'm a virgin."). Dotson-Blake et al. (2009) developed a profile of those most likely to believe that "oral sex is not sex": they tended to be first-year students, white,

DIVERSITY IN OTHER COUNTRIES

More than a thousand high school students (1,368) in the United Kingdom revealed their expectations regarding oral sex. Of the sexually experienced males, 48.9% expected oral sex during a sexual experience; 46.5% of the sexually experienced females expected oral sex. However, only 14% of the males and 29% of the females reported that they believed it was important to use a condom during fellatio (Stone et al. 2006).

nonreligious, and hedonistic in sexual values with experience in hooking up, oral sex, and cohabitation.

There is the mistaken belief that only intercourse carries the risk of contracting a STI. However, STIs as well as HIV can be contracted orally. A team of researchers confirmed that a person may contract not only the herpes virus from oral sex but also HIV (Mbopi-Keou et al. 2005). Use of a condom or dental dam, a flat latex device that is held over the vaginal area, is recommended.

Vaginal Intercourse

Coitus the sexual union of a man and woman by insertion of the penis into the vagina.

Vaginal intercourse, or **coitus**, refers to the sexual union of a man and woman by insertion of the penis into the vagina. In a study of 2,922 undergraduates, 67% reported that they had had sexual intercourse (men 61.3%; women 72.5%) (Knox and Hall 2010). In regard to university students, each academic year, there are fewer virgins.

Meston and Buss (2007) surveyed 1,549 university students and identified 237 reasons for having sexual intercourse. The top reasons included: (1) pure attraction to the other person in general; (2) physical pleasure; (3) expression of love; (4) feelings of being desired by the other; (5) escalation of the depth of the relationship; (6) curiosity for new experiences; (7) special occasion for celebration; (8) mere opportunity; and (9) seemingly uncontrollable circumstances. Indeed, twenty of the twenty-five top reasons were remarkably similar for women and men.

In a study of 92 heterosexual couples (Simrns and Byers 2009), the men's ideal frequency for engaging in sexual behaviors was greater than the women's, both partners perceived each other's ideal frequencies to be more dissimilar than they actually were, and women perceived greater dissimilarity between their own and their partner's ideal frequencies than did the men. Also, the more frequently partners engaged in sexual activities and the more similar men perceived their own and their partner's ideal sexual frequency to be, the higher the sexual satisfaction of both partners.

First Intercourse

Cavazos-Rehg et al. (2010) found that having first sexual intercourse at or before age 16 was associated with having an alcoholic parent, being black, and being born to a teenage mother. Intimacy was an important motivation for women to engage in first intercourse (Patrick and Lee 2010).

A team of researchers (Else-Quest et al. 2005) examined data from the National Health and Social Life Survey on the outcome of the respondent's first vaginal intercourse. The first experience was premarital for 82.9% of the respondents, at an average age of 17.7 years. The relationship status of the respondent was not associated with later psychological or physical health outcomes. However, if the first was "prepubertal, forced, with a blood relative or stranger, or the result of peer pressure, drugs, or alcohol, poorer psychological and physical outcomes were consistently reported in later life" (p. 102). The takeaway message of this research is to delay first intercourse, which might best occur with a partner in an emotional relationship where no pressure, alcohol, or drugs are involved.

Sexual readiness factors such as autonomy of decision (not influenced by alcohol or peers), consensuality (both partners equally willing), and absence of regret "the right time for me" are more important than age as meaningful criteria for determining when a person is ready for first intercourse.

Hawes et al. (2010) studied first intercourse experiences in the United Kingdom and emphasized that **sexual readiness**, not age, is the more meaningful criteria. Such readiness can be determined in reference to contraception, autonomy of decision (not influenced by alcohol or peer pressure), consensuality (both partners equally willing), and the absence of regret (the right time for me). Using these criteria, the negative consequences of first intercourse are minimized.

Chapter 9 Sexuality in Relationships

What if My Partner Does Not Like Sex or is Hypersexual?

Sex is an important part of a couple's relationship, particularly in new relationships. However, individuals vary in their interest in, capacity for, and preference for different sexual behaviors. Although some need sex daily, are orgasmic, and enjoy a range of sexual behaviors, others never think of sex, have never had an orgasm, and want to get any sexual behavior over with as soon as possible. When two people of widely divergent views on sex end up in the same relationship, clear communication and choices are necessary. The person who has no interest must decide if developing an interest is a goal and be open to learning about masturbation and sexual fantasy. Where becoming interested in sex is not a goal, the partner must decide the degree to which this is an issue. Some will be pleased the partner has no interest because this means no sexual demands, whereas others will bolt. Openness about one's sexual feelings and hard choices will help resolve the dilemma.

While hypersexuality (also known as hyperphilia, hypersexual disorder, and compulsive sexual disorder) has no agreed upon definition in terms of how much sex is excessive, it may become a problem for the partner through its expression of being driven to and feeling a loss of control over spending hours viewing pornography over the Internet, having cybersex, visiting strip clubs, having affairs, etc. If there is a feeling of not having control over one's sexual appetite, behavioral sex therapy may be helpful (Kaplan and Krueger 2010). Alternatively, sex addition is treated in the same way as any other addition (Niven 2010).

Anal Sex

National **Data**

Of 5,865 men and women, more than 20% of men ages 25–49 and women ages 20–39 reported anal sex in the past year (Herbenick et al. 2010). Lifetime experience of heterosexuals with anal sex ranges from 6% to 40% (McBride and Fortenberry 2010).

In a study of 2,922 undergraduates (primarily heterosexuals), 24% reported that they had had anal sex (men 23.8%; women 24%) (Knox and Hall 2010). Younger individuals, those with higher numbers of partners, and those with STIs are more likely to report having participated in anal sex. Motivations associated with anal sex include intimacy (anal sex is more intimate than regular sex), enjoyment in variety, domination by male, breaking taboos, and pain-pleasure enjoyment (McBride and Fortenberry 2010).

The greatest danger of anal sex is that the rectum might tear, in which case blood contact can occur; STIs (including HIV infection) may then be transmitted. Partners who use a condom during anal intercourse reduce their risk of not only HIV infection but also other STIs. Pain (physical as well as psychological) may occur for anyone of any sexual identity involved in receiving anal sex. Such pain is called **anodyspareunia**—frequent, severe pain during receptive anal sex.

A team of researchers (Maynard et al. 2009) interviewed 28 women who reported having had unprotected anal intercourse in the last year with a man who was HIV-positive or whose status was unknown. Experiencing physical pleasure, enhancing emotional intimacy, pleasing their male partners, and avoiding

Anodyspareunia frequent, severe pain during receptive anal sex.

violence were their primary motives. Male partners usually initiated anal sex, which occurred in the context of vaginal and oral sex. Reasons for not using condoms were familiarity with their partner and feeling that condoms made anal sex less pleasurable.

Cybersex

Cybersex is "any consensual, computer mediated, participatory sexual experience involving two or more individuals" (Hamman 2007, 34). In effect, this means that the individuals are typing or sending text about what they are doing to the other sexually (or doing to themselves—for example, masturbation). Or, they are writing sex stories with the goal of arousal. Hamman noted that, "increasing numbers of teenagers and young people will have their first sexual experiences, the ones that will shape their sex lives forever, online" (p. 38).

Internet use also varies by whether the user is male or female. Males are more likely than females to meet a partner for sex online. Similarly, Internet use varies by sexual orientation with gay and bisexual males more likely than heterosexual males to meet a partner for sex online. Bisexual females are more likely than gay females and heterosexual females to meet a partner for sex online (Mathy 2007).

Sexuality in Relationships

Sexuality occurs in a social context that influences its frequency and perceived quality.

Sexual Relationships among Never-Married Individuals

Never-married individuals and those not living together report more sexual partners than those who are married or living together. In one study, 9% of never-married individuals and those not living together reported having had five or more sexual partners in the previous twelve months; 1% of married people and 5% of cohabitants reported the same. However, unmarried individuals, when compared with married individuals and cohabitants, reported the lowest level of sexual satisfaction. One-third of a national sample of people who were not married and not living with anyone reported that they were emotionally satisfied with their sexual relationships. In contrast, 85% of the married and pair-bonded individuals reported emotional satisfaction in their sexual relationships. Hence, although never-married individuals have more sexual partners, they are less emotionally satisfied (Michael et al. 1994).

Sexual Relationships among Married Individuals

Marital sex is distinctive for its social legitimacy, declining frequency, and satisfaction (both physical and emotional).

1. *Social legitimacy.* In our society, marital intercourse is the most legitimate form of sexual behavior. Those who engage in homosexual, premarital, and extramarital intercourse do not experience as high a level of social approval as do those who engage in marital sex. It is not only okay to have intercourse when married, it is expected. People assume that married couples make love and that something is wrong if they do not.

2. *Declining frequency.* Sexual intercourse between spouses occurs about six times a month, which declines in frequency as spouses age. Pregnancy also decreases the frequency of sexual intercourse (Lee et al. 2010b). In addition to biological changes due to aging and pregnancy, satiation also contributes to the declining frequency of intercourse between spouses and partners in long-term relationships. Psychologists use the term **satiation** to mean that repeated exposure

---

Cybersex any consensual, computer mediated, participatory sexual experience involving two or more individuals.

It doesn't matter where you get your appetite as long as you eat at home.

Unknown

Satiation repeated exposure to a stimulus that results in the loss of its ability to reinforce.

Chapter 9 Sexuality in Relationships

This couple have been married over TWENTY years and keep the affection/sex alive in their relationship.

Chelsea E. Curry

to a stimulus results in the loss of its ability to reinforce. For example, the first time you listen to a new CD, you derive considerable enjoyment and satisfaction from it. You may play it over and over during the first few days. After a week or so, listening to the same music is no longer new and does not give you the same level of enjoyment that it first did. So it is with sex with the same person. The thousandth time that a person has sex with the same partner is not as new and exciting as the first few times.

Polyamorists use the term **new relationship energy (NRE)** to refer to the euphoria of a new emotional/sexual relationship which dissipates over time. Polyamorists often talk with a long-term partner about the NRE they are feeling and both watch its eventual decline. Hence, polyamorists don't get upset when they see their partner experiencing NRE with a new partner since they view it as having a cycle that will not last forever (Wilson and Rodrigous 2010).

Satiation in combination with age can include long periods of no sexual activity. When Nevada governor Jim Gibbons was accused of sexual harassment, he testified in a deposition that "I'm living proof that you can survive without sex for that long." In effect, he testified that he had not been intimate with any woman, including his wife, for ten years.

3. *Satisfaction (emotional and physical).* Despite declining frequency and less satisfaction over time (Liu 2003), marital sex remains a richly satisfying experience. Contrary to the popular belief that unattached singles have the best sex, the married and pair-bonded adults enjoy the most satisfying sexual relationships. In a national sample, 88% of married people said they received great physical pleasure from their sexual lives, and almost 85% said they received great emotional satisfaction (Michael et al. 1994). Individuals least likely to report being physically and emotionally pleased in their sexual relationships are those who are not married, not living with anyone, or not in a stable relationship with one person (ibid.).

Sexual Relationships among Divorced Individuals

Of the almost 2 million people getting divorced, most will have intercourse within one year of being separated from their spouses. The meanings of intercourse for separated or divorced individuals vary. For many, intercourse is a way to reestablish—indeed, repair—their crippled self-esteem. Questions like, "What did I do wrong?" "Am I a failure?" and "Is there anybody out there

There are a number of mechanical devices which increase sexual arousal, particularly in women. Chief among these is the Mercedes-Benz 380SL convertible.

P.J. O'Rourke, American political satirist

New relationship energy (NRE) the euphoria of a new emotional/sexual relationship which dissipates over time.

who will love me again?" loom in the minds of divorced people. One way to feel loved, at least temporarily, is through sex. Being held by another and being told that it feels good gives people some evidence that they are desirable. Because divorced people may be particularly vulnerable, they may reach for sexual encounters as if for a lifeboat. "I felt that, as long as someone was having sex with me, I wasn't dead and I did matter," said one recently divorced person.

Because divorced individuals are usually in their thirties or older, they may not be as sensitized to the danger of contracting HIV as people in their twenties. Divorced individuals should always use a condom to lessen the risk of STI, including HIV infection and AIDS.

National **Data**

When asked whether intercourse was occurring less frequently than desired, 38% of cohabiting, 49% of married, 60% of widowed, 65% of single, and 74% of divorced people answered "yes." Hence, those most dissatisfied with frequency of intercourse are the divorced, and those most satisfied are cohabitants (Dunn et al. 2000, 145).

Effect of Work on Sexual Frequency

Gager and Yabiku (2010) studied the work patterns of spouses to determine the effect on frequency of sexual behavior. They found that both women and men who "work hard" also "play hard" in regard to their sex lives. Indeed, wives and husbands who spent more hours in housework and paid work also reported more frequent sex.

Maintaining Safe Sex in One's Relationship

One of the negative consequences of sexual behavior is the risk of contracting a sexually transmitted infection (STI). Also known as sexually transmitted disease, or *STD*, **STI** refers to the general category of sexually transmitted infections such as chlamydia, genital herpes, gonorrhea, and syphilis. The most lethal of all STIs is that due to human immunodeficiency virus (HIV), which attacks the immune system and can lead to autoimmune deficiency syndrome (AIDS). Here we note the importance of lowering one's risk by assessing one's risk (see Self-Assessment feature).

The Illusion of Safety in a "Monogamous" Relationship

Most individuals in a serious "monogamous" relationship assume that they are at zero risk for contracting an STI from their partner. Vail-Smith et al. (2010) analyzed data from 1,341 undergraduates at a large southeastern university and found that 27.2% of the males and 19.8% (almost one in five) females reported having oral, vaginal, or anal sex outside of a relationship that their partner considered monogamous. People most likely to cheat were men over the age of 20, those who were binge drinkers, members of a fraternity, male college athletes, or nonreligious people. Paik (2010) also found in a sample of 783 adults ages 18 to 59 that 17% of the males and 5% of the females reported sexual involvement in multiple relationships. These data suggest the need for educational efforts to encourage individuals in committed relationships to reconsider their STI risk and to protect themselves via condom usage.

The false face must hide what the false heart doeth know.

Shakespeare, *Macbeth*

STI (sexually transmitted infection) refers to the general category of sexually transmitted infections such as chlamydia, genital herpes, gonorrhea, and syphilis.

DIVERSITY IN OTHER COUNTRIES

How to put on a condom, sex positions, and same-sex relationships are all part of the curriculum for 14-year-old students in Swedish high schools. But Muslim immigrants with extremely conservative values who require their daughters to wear head scarves to ensure modesty, prevent their children from attending these classes. Under a new Swedish law, these parents would no longer be able to opt their children out of such classes. "All students have the right to take part in the compulsory school education, regardless of whether their parents approve or disapprove," said Sweden's education secretary, Jan Bjorklund (Oscarsson 2009).

Safer sex means making choices that reduce the risk of transmitting STIs, including the AIDS virus. Using condoms is an example of safer sex. Unsafe, risky, or unprotected sex refers to sex without a condom, or to other sexual activity that might increase the risk of transmitting the AIDS virus. For each of the following items, check the response that best characterizes your option.

A = Agree

U = Undecided

D = Disagree

_____ 1. If my partner wanted me to have unprotected sex, I would probably give in.

_____ 2. The proper use of a condom could enhance sexual pleasure.

_____ 3. I may have had sex with someone who was at risk for HIV/AIDS.

_____ 4. If I were going to have sex, I would take precautions to reduce my risk for HIV/AIDS.

_____ 5. Condoms ruin the natural sex act.

_____ 6. When I think that one of my friends might have sex on a date, I remind my friend to take a condom.

_____ 7. I am at risk for HIV/AIDS.

_____ 8. I would try to use a condom when I had sex.

_____ 9. Condoms interfere with romance.

_____ 10. My friends talk a lot about safer sex.

_____ 11. If my partner wanted me to participate in risky sex and I said that we needed to be safer, we would still probably end up having unsafe sex.

_____ 12. Generally, I am in favor of using condoms.

_____ 13. I would avoid using condoms if at all possible.

_____ 14. If a friend knew that I might have sex on a date, the friend would ask me whether I was carrying a condom.

_____ 15. There is a possibility that I have HIV/AIDS.

_____ 16. If I had a date, I would probably not drink alcohol or use drugs.

_____ 17. Safer sex reduces the mental pleasure of sex.

_____ 18. If I thought that one of my friends had sex on a date, I would ask the friend if he or she used a condom.

_____ 19. The idea of using a condom doesn't appeal to me.

_____ 20. Safer sex is a habit for me.

_____ 21. If a friend knew that I had sex on a date, the friend wouldn't care whether I had used a condom or not.

_____ 22. If my partner wanted me to participate in risky sex and I suggested a lower-risk alternative, we would have the safer sex instead.

_____ 23. The sensory aspects of condoms (smell, touch, and so on) make them unpleasant.

_____ 24. I intend to follow "safer sex" guidelines within the next year.

_____ 25. With condoms, you can't really give yourself over to your partner.

_____ 26. I am determined to practice safer sex.

_____ 27. If my partner wanted me to have unprotected sex and I made some excuse to use a condom, we would still end up having unprotected sex.

_____ 28. If I had sex and I told my friends that I did not use condoms, they would be angry or disappointed.

_____ 29. I think safer sex would get boring fast.

_____ 30. My sexual experiences do not put me at risk for HIV/AIDS.

_____ 31. Condoms are irritating.

_____ 32. My friends and I encourage each other before dates to practice safer sex.

_____ 33. When I socialize, I usually drink alcohol or use drugs.

_____ 34. If I were going to have sex in the next year, I would use condoms.

_____ 35. If a sexual partner didn't want to use condoms, we would have sex without using condoms.

_____ 36. People can get the same pleasure from safer sex as from unprotected sex.

_____ 37. Using condoms interrupts sex play.

_____ 38. Using condoms is a hassle.

Scoring (to be read after completing the previous scale)

Begin by giving yourself 80 points. Subtract one point for every undecided response. Subtract two points every time that you disagreed with odd-numbered items or with item number 38. Subtract two points every time you agreed with even-numbered items 2 through 36.

Interpreting Your Score

Research shows that students who make higher scores on the SSRS are more likely to engage in risky sexual activities, such as having multiple sex partners and failing to consistently use condoms during sex. In contrast, students who practice safer sex tend to endorse more positive attitudes toward safer sex, and tend to have peer networks that encourage safer sexual practices. These students usually plan on making sexual activity safer, and they feel confident in their ability to negotiate safer sex, even when a dating partner may press for riskier sex. Students who practice safer sex often refrain from using alcohol or drugs, which may impede negotiation of safer sex, and often report having engaged in lower-risk activities in the past. How do you measure up?

(Below 15) Lower Risk

Congratulations! Your score on the SSRS indicates that, relative to other students, your thoughts and behaviors are more supportive of safer sex. Is there any room for improvement in your score? If so, you may want to examine items for which you lost points and try to build safer sexual strengths in those areas. You can help protect others from HIV by educating your peers about making sexual activity safer. (Of 200 students surveyed by DeHart and Berkimer, 16% were in this category.)

(15 to 37) Average Risk

Your score on the SSRS is about average in comparison with those of other college students. Though it is good that you don't fall

into the higher-risk category, be aware that "average" people can get HIV, too. In fact, a recent study indicated that the rate of HIV among college students is ten times that in the general heterosexual population. Thus, you may want to enhance your sexual safety by figuring out where you lost points and work toward safer sexual strengths in those areas. (Of 200 students surveyed by DeHart and Berkimer, 68% were in this category.)

(38 and Above) Higher Risk
Relative to other students, your score on the SSRS indicates that your thoughts and behaviors are less supportive of safer sex. Such high scores tend to be associated with greater HIV-risk behavior. Rather than simply giving in to riskier attitudes and behaviors, you may want to empower yourself and reduce your risk by critically examining areas for improvement. On which items did you lose

points? Think about how you can strengthen your sexual safety in these areas. Reading more about safer sex can help, and sometimes colleges and health clinics offer courses or workshops on safer sex. You can get more information about resources in your area by contacting the Center for Disease Control's HIV/AIDS Information Line at 1-800-342-2437. (Of 200 students surveyed by DeHart and Birkimer, 16% were in this category.)

Source

DeHart, D. D. and J. C. Birkimer. 1997. The Student Sexual Risks Scale (modification of SRS for popular use; facilitates student self-administration, scoring, and normative interpretation). Developed specifically for this text by Dana D. DeHart, College of Social work at the University of South Carolina; John C. Birkimer, University of Louisville. Used by permission of Dana DeHart.

Sexual Fulfillment: Some Prerequisites

> *Know thyself.*
>
> Socrates, philosopher

There are several prerequisites for having a good sexual relationship.

Self-Knowledge, Self-Esteem, and Health

Sexual fulfillment involves knowledge about yourself and your body. Such information not only makes it easier for you to experience pleasure but also allows you to give accurate information to a partner about pleasing you. It is not possible to teach a partner what you don't know about yourself.

Sexual fulfillment also implies having a positive self-concept. To the degree that you have positive feelings about yourself and your body, you will regard yourself as a person someone else would enjoy touching, being close to, and making love with. If you do not like yourself or your body, you might wonder why anyone else would. Lemer et al. (2010) studied 502 women ages 18 to 47 and found that a positive body image was related to sexual frequency. Meltzer and McNulty (2010) studied 53 married couples and found that wives who viewed themselves as sexually attractive reported higher levels of marital satisfaction.

Effective sexual functioning also requires good physical and mental health. This means regular exercise, good nutrition, lack of disease, and lack of fatigue. Performance in all areas of life does not have to diminish with age—particularly if people take care of themselves physically (see Chapter 16 on The Later Years).

> *A good relationship is an exchange of behavior of a kind (positive) and at a rate (high frequency) that is mutual.*
>
> Jack Turner, psychologist

Good health also implies being aware that some drugs may interfere with sexual performance. Alcohol is the drug most frequently used by American adults. Although a moderate amount of alcohol can help a person become aroused through a lowering of inhibitions, too much alcohol can slow the physiological processes and deaden the senses. Shakespeare may have said it best: "It [alcohol] provokes the desire, but it takes away the performance" (*Macbeth*, act 2, scene 3). The result of an excessive intake of alcohol for women is a reduced chance of orgasm; for men, overindulgence results in a reduced chance of attaining an erection.

The reactions to marijuana are less predictable than the reactions to alcohol. Though some individuals report a short-term enhancement effect, others say that marijuana just makes them sleepy. In men, chronic use may decrease sex drive because marijuana may lower testosterone levels.

A Good Relationship, Positive Motives

A guideline among therapists who work with couples who have sexual problems is to treat the relationship before focusing on the sexual issue. The sexual relationship is part of the larger relationship between the partners, and what happens outside the bedroom in day-to-day interaction has a tremendous influence on what happens inside the bedroom. The statement, "I can't fight with you all day and want to have sex with you at night" illustrates the social context of the sexual experience.

Women most valued a partner who was open to discussing sex, who was knowledgeable about sex, who clearly communicated his desires, who was physically attractive, and who paid her compliments during sex. Being easily sexually aroused and being uninhibited were also important.

Sexual interaction communicates how the partners are feeling and acts as a barometer for the relationship. Each partner brings to a sexual encounter, sometimes unconsciously, a motive (pleasure, reconciliation, procreation, duty), a psychological state (love, hostility, boredom, excitement), and a physical state (tense, exhausted, relaxed, turned on). The combination of these factors will change from one encounter to another. Tonight a wife may feel aroused and loving and seek pleasure, but her husband may feel exhausted and hostile and have sex only out of a sense of duty. Tomorrow night, both partners may feel relaxed and have sex as a means of expressing their love for each other.

One's motives for a sexual encounter are related to the outcome. Impett et al. (2005) found that, when individuals have intercourse out of the desire to enhance personal and interpersonal or relationship pleasure, the personal and interpersonal effect on well-being is very positive. However, when sexual motives were to avoid conflict, the personal and interpersonal effects did not result in similar positive outcomes. In a study of 1,002 French adults, sexuality was more synonymous with pleasure (44.0%) and love (42.1%) than with procreation, children, or motherhood (7.8%) (Colson et al. 2006).

An Equal Relationship

Laumann et al. (2006) surveyed 27,500 individuals in twenty-nine countries and found that reported sexual satisfaction was higher where men and women were considered equal. Austria topped the list, with 71% reporting sexual satisfaction; only 25.7% of those surveyed in Japan reported sexual satisfaction. The United States was among those countries in which a high percentage of the respondents reported sexual satisfaction.

Open Sexual Communication and Feedback

Sexually fulfilled partners are comfortable expressing what they enjoy and do not enjoy in the sexual experience. Unless both partners communicate their needs, preferences, and expectations to each other, neither is ever sure what the other wants. In essence, the Golden Rule ("Do unto others as you would have them do unto you") is *not* helpful, because what you like may not be the same as what your partner wants. A classic example of the uncertain lover is the man who picks up a copy of *The Erotic Lover* in a bookstore and leafs through the pages until the topic on how to please a woman catches his eye. He reads that women enjoy having their breasts stimulated by their partner's tongue and teeth. Later that night in bed, he rolls over and begins to nibble on his partner's breasts. Meanwhile, she wonders what has possessed him and is unsure what to make of this new (possibly unpleasant) behavior.

Women need a reason to have sex. Men just need a place.

Billy Crystal, comedian

One of the things about equality is not just that you be treated equally to a man, but that you treat yourself equally to the way you treat a man.

Marlo Thomas, actress

Sexually fulfilled partners take the guesswork out of their relationship by communicating preferences and giving feedback. This means using what some therapists call the touch-and-ask rule. Each touch and caress may include the question, "How does that feel?" It is then the partner's responsibility to give feedback. If the caress does not feel good, the partner should say what does feel good. Guiding and moving the partner's hand or body are also ways of giving feedback.

Having Realistic Expectations

To achieve sexual fulfillment, expectations must be realistic. A couple's sexual needs, preferences, and expectations may not coincide. It is unrealistic to assume that your partner will want to have sex with the same frequency and in the same way that you do on all occasions. It may also be unrealistic to expect the level of sexual interest and frequency of sexual interaction in long-term relationships to remain consistently high.

Sexual fulfillment means not asking things of the sexual relationship that it cannot deliver. Failure to develop realistic expectations will result in frustration and resentment. One's health, feelings about the partner, age, and previous sexual experiences (including child sexual abuse, rape, and so on) will have an effect on one's sexuality and one's sexual relationship.

Sexual Compliance

Sexual compliance an individual willingly agrees to participate in sexual behavior without having the desire to do so.

Given that partners may differ in sexual interest and desire, Vannier and O'Sullivan (2010) identified the concept of **sexual compliance** whereby an individual willingly agrees to participate in sexual behavior without having the desire to do so. The researchers studied 164 heterosexual young (18-24) adult couples in committed relationships to assess the level of sexual compliance. Almost half (46%) of the respondents reported at least one occasion of sexual compliance with sexual compliance comprising 17% of all sexual activity recorded over a three week period. Indeed, sexual compliance was a mechanism these individuals used in their committed relationships to resolve the issue of different levels of sexual desire that is likely to happen over time in a stable couple's relationship. Others felt guilty they did not desire sex and still others did it because their partner provided sex when the partner was not in the mood.

There were no gender differences in sexual desire and no gender difference in providing sexual compliant behavior. And the majority of participants reported enjoying the sexual activity despite not wanting to engage in it at first.

Avoiding Spectatoring

Spectatoring involves mentally observing your sexual performance and that of your partner.

One of the obstacles to sexual functioning is **spectatoring**, which involves mentally observing your sexual performance and that of your partner. When the researchers in one extensive study observed how individuals actually behave during sexual intercourse, they reported a tendency for sexually dysfunctional partners to act as spectators by mentally observing their own and their partners' sexual performance. For example, the man would focus on whether he was having an erection, how complete it was, and whether it would last. He might also watch to see whether his partner was having an orgasm (Masters and Johnson 1970).

Spectatoring, as Masters and Johnson conceived it, interferes with each partner's sexual enjoyment because it creates anxiety about performance, and anxiety blocks performance. A man who worries about getting an erection reduces his chance of doing so. A woman who is anxious about achieving an orgasm probably will not. The desirable alternative to spectatoring is to relax, focus on and enjoy your own pleasure, and permit yourself to be sexually responsive.

Female Vibrator Use, Orgasm, and Partner Comfort

It is commonly known that vibrators (also known as sex toys and novelties) are beneficial for increasing the probability of orgasmic behavior in women. During intercourse women typically report experiencing a climax 30% of the time; vibrator use increases orgasmic reports to over 90%. Herbenick et al (2010) studied women's use of vibrators within sexual partnerships. They analyzed data from 2056 women aged 18–60 years in the United States. Partnered vibrator use was common among heterosexual-, lesbian-, and bisexual-identified women. Most vibrator users indicated comfort using them with a partner and vibrator use was related to positive sexual function. In addition, partner knowledge and perceived liking of vibrator use was a significant predictor of sexual satisfaction for heterosexual women.

That men are accepting of using a vibrator with their female partners was confirmed by Reece et al. (2010) who surveyed a nationally representative sample of heterosexual men in the U.S. Forty-three percent reported having used a vibrator. Of those who had done so, most vibrator use had occurred within the context of sexual interactions with a female partner. Indeed, 94% of male vibrator users reported that they had used a vibrator during sexual play with a partner, and 82% reported that they had used a vibrator during sexual intercourse. These data support recommendations of therapists and educators who often suggest the incorporation of vibrators into partnered relationships.

Sexual Insights of an Undergraduate Female

One of our students, a senior, wrote a paper on her sexual experiences and insights. These reveal that college is a context in which one discovers one's sexuality in reference to others, a context in which choices about what characteristics of a lover are desired, and a context to make sexual choices which have positive outcomes for the individual and for the couple. Reflecting on one's experiences allows one to enhance one's relationships and sexuality in those relationships.

What are the behaviors of an outstanding lover?

- *passionate, enthusiastic*
- *wants to make sure to please you—asks you verbally*
- *stamina*
- *interest in variety and trying new things to please each other—toys, different positions*

What are the behaviors of a terrible lover?

- *trouble with keeping a hard-on*
- *jack hammer style loving*
- *doesn't ask about your pleasure*

Do guys ask for an evaluation..."How was I?"

Some do, but typically they don't—I think most guys either assume you are happy or they are afraid of the answer. The more common question is "Did you come?" I used to always lie about this because I have rarely had an orgasm during intercourse but I made a decision about a year ago and have told the truth since. Guys want to work to make you [orgasm] but need you to be honest about it, if you actually want a satisfying sexual experience.

If a guy really loves having sex with you, does this move him toward an emotional relationship with you or do you just remain the "sex girl" he loves having sex with, but never escalates the relationship emotionally?

I have had experience with both of these types. Most of the time I just have remained the "sex girl" but a couple times the guys have fallen in love because of the sexual aspect. One of my exes that I had a long off and on serious relationship with started out as a completely casual sexual relationship.

To what degree is emotional involvement or the potential for a future an important variable in terms of your enjoyment of the sexual experience?

Emotional involvement has always been important to me—sex is always so much better if you are emotionally connected. I went through a period where I worked hard at keeping my heart in check and just keeping an attitude of conquering and using boys like they use girls. During that time I could forget about the potential (or lack of) future with whomever I was with and just enjoy myself. But sleeping with someone always leads me to be emotionally attached to them so it was always hard to keep my emotions in check.

In regard to the guy, I think if a guy already has feelings for you then it only gets better but if he doesn't have feelings developed before you have sex then it usually doesn't develop on his part.

The Future of Sexual Relationships

The future of sexual relationships will involve continued individualism as the driving force in selecting sexual partners. Numerous casual partners ("hooking up") with predictable outcomes—higher frequencies of STI's, unexpected pregnancies, sexual regret, and relationships going nowhere—will continue to characterize individuals in late adolescence and early twenties. As these persons reach their late twenties, the goal of sexuality begins to transition to seeking a partner not just to hook up and have fun with but to mate with. This new goal is accompanied by new sexual behaviors such as delayed first intercourse in the new relationship, exclusivity, and movement toward marriage. The monogamous toward marriage context creates a transitioning of sexual values from hedonism to relativism to absolutism where strict morality rules become operative in the relationship (expected fidelity).

Summary

What are three main sexual values and the sources of these values?

Sexual values are moral guidelines for making sexual choices in nonmarital, marital, heterosexual, and homosexual relationships.

Three sexual values are absolutism (rightness is defined by official code of morality), relativism (rightness depends on the situation—who does what, with whom, in what context), and hedonism ("if it feels good, do it"). Relativism is the sexual value most college students hold, with women being more relativistic than men and men being more hedonistic than women. Blacks report having more absolutist values.

School polices include absolutist/abstinence teaching. There is no evidence that abstinence-based sex education programs are effective in stopping unmarried youth from having sex. Half of those who take the "virginity pledge" withdraw the pledge within one year (male and black individuals most likely). About three-fourths of college students believe that if they have oral sex, they are still virgins. About half of undergraduates reported involvement in a "friends with benefits" relationship. Women are more likely to focus on the "friendship" aspect, men on the "benefits" (sex) aspect of the "friends with benefits" relationship.

The sexual double standard is the view that encourages and accepts sexual expression of men more than women. For example, men may have more sexual partners than women without being stigmatized. The double standard is also reflected in movies, TV, and other popular media.

The sources of sexual values include one's school, family, and religion as well as technology and television. Public schools in the United States promote absolutist sexual values through abstinence education—the effectiveness of these programs has been questioned. Religious activity and religiosity are associated with lower levels of sexual behavior and sexting outside of a committed relationship. Families are influential in that close relationships with children are associated with a delay in first intercourse. Media affects sexual values in that models are often about casual sex.

What are various sexual behaviors?

Some individuals are asexual—having an absence of sexual behavior with a partner and one's self (masturbation). About 4% of females and 11% of males report being asexual in the last twelve months. Kissing involves various meanings including love, approval, hello, goodbye, or as an ointment such as when a parent kisses the hurt knee of a child, and so on. French kissing may be dangerous by increasing risk of a bacterial infection that may lead to meningitis.

Masturbation involves stimulating one's own body with the goal of experiencing pleasurable sexual sensations. Fellatio is oral stimulation of the man's genitals by his partner. In many states, legal statutes regard fellatio as a "crime against nature," in that the sex does not produce babies. Cunnilingus is oral stimulation of the woman's genitals by her partner. Increasingly, youth who have oral sex regard themselves as virgins, believing that only sexual intercourse constitutes "having sex."

Anal (not vaginal) intercourse is the sexual behavior associated with the highest risk of HIV infection. The potential for the rectum to tear and blood contact to occur presents the greatest danger. AIDS is lethal. Partners who use a condom during anal intercourse reduce their risk of not only HIV infection but also other STIs.

Sexual behavior with a partner is associated with feelings of well-being. Indeed, researchers have found that experiencing affection and sex with a partner on one day predicted lower negative mood, higher positive mood, and lower stress the following day.

Cybersex is "consensual, computer mediated, participatory sexual experience involving two or more individuals." Men are more likely to engage in cybersex behavior.

What are the sexual relationships of the never-married, married, and divorced?

Never-married and noncohabiting individuals report more sexual partners than those who are married or living with a partner. Marital sex is distinctive for its social legitimacy, declining frequency, and satisfaction (both physical and emotional). Divorced individuals have a lot of sexual partners but are the least sexually fulfilled.

How might it be an illusion that monogamy means safe sex?

The person most likely to get an STI has sexual relations with a number of partners or with a partner who has a variety of partners. Even if you are in a mutually monogamous relationship, you may be at risk for acquiring an STI, as 30% of male undergraduate students and 20% of female undergraduate students in "monogamous" relationships reported having oral, vaginal, or anal sex with another partner outside of the monogamous relationship.

What are the prerequisites of sexual fulfillment?

Fulfilling sexual relationships involve self-knowledge, self-esteem, health, a good nonsexual relationship, open sexual communication, safer sex practices, and making love with, not to, one's partner. Other variables include realistic expectations ("my partner will not always want what I want"), and avoiding spectatoring (not being self conscious and observing one's "performance").

Key Terms

Absolutism	Concurrent sexual partnership	Secondary virginity
Alcohol-exposed pregnancy	Cybersex	Sexual compliance
Anodyspareunia	Friends with benefits	Sexual double standard
Asceticism	Hedonism	Sexual readiness
Asexual	Masturbation	STI (Sexually transmitted infection)
Coitus	New relationship energy (NRE)	Sexual values
Comprehensive sex education program	Relativism	Spectatoring
	Satiation	

Web Links

Body Health: A Multimedia AIDS and HIV Information Resource
http://www.thebody.com

Centers for Disease Control and Prevention (CDC)
http://www.cdc.gov

Father Daughter Purity Ball
http://www.lifeway.com/tlw/students/join.asp

Go Ask Alice: Sexuality
http://www.goaskalice.columbia.edu/

National Center for Health Statistics
http://www.cdc.gov/nchs/

Kinsey Confidential
http://kinseyconfidential.org/

Sex Education Library
http://www.sexedlibrary.org/

Kinsey Institute
http://www.kinseyinstitute.org/

Sexual Intimacy
http://www.heartchoice.com/hc/intimacy/index.php

Sexual Health Network
http://www.sexualhealth.com/

Sexuality Information and Education Council of the United States (SIECUS)
http://www.siecus.org

Sex Survey
http://sex.healthguru.com/applications/surveys?RedElephant37

CHAPTER 10

Planning for Children

Love is a fourteen letter word—family planning.

Planned Parenthood

David Knox

Learning Objectives

Discuss the social influences motivating an individual to have a child.

Review the individual motivations for having a child.

Identify the various categories/motivations of individuals who decide to remain childfree.

Know the causes of infertility and technology available to help induce a pregnancy.

Be aware of the motives for adoption and the demographics of those who adopt.

Discuss the types of abortion and the outcome of having an abortion.

Predict the future of planning for children.

1. About a third of births in the United States are unintended.

2. Most women obtaining abortions are able to rely on their male partners for social support.

3. Women who have had a tubal ligation are less likely to experience extremely high levels of sexual satisfaction, relationship satisfaction, and sexual pleasure.

4. With the availability of emergency contraception pills, unintended pregnancy rates have decreased.

5. American couples who adopt tend to prefer out of country adoption over domestic adoption.

Answers: 1. T   2. T   3. F   4. F   5. T

Before I got married I had six theories about bringing up children; now I have six children, and no theories.

John Wilmot, 2nd Earl of Rochester

Pregnancy coercion coercion by a male partner for the woman to become pregnant.

Birth control sabotage partner interference with contraception.

While over four million babies are born in the U.S. annually, the cultural message on having children is mixed. Although young married individuals are encouraged to "have fun and travel before they begin their family" and to "strap on their seat belts when their children become teenagers," spouses are also socialized to believe that "marriage has no real meaning without children" and "aren't they precious?" In spite of these conflicting messages, having children continues to be a major goal of young adults. Among youth between the ages of 18 and 29, almost three fourths (74%) in a Pew Research Center report noted that they wanted to have children. And most said that "being a good parent" was more important than "having a successful marriage" (52% to 30%) (Wang and Tayloer 2011).

Planning children, or failing to do so, is a major societal issue. Planning when to become pregnant has benefits for both the mother and the child. Having several children at short intervals increases the chances of premature birth, infectious disease, and death of the mother or the baby. Would-be parents can minimize such risks by planning fewer children with longer intervals in between. Women who plan their pregnancies can also modify their behaviors and seek preconception care from a health care practitioner to maximize their chances of having healthy pregnancies and babies. For example, women planning pregnancies can make sure they eat properly and avoid alcohol and other substances (such as cigarettes) that could harm developing fetuses. Partners who plan their children also benefit from family planning by pacing the financial demands of their offspring. Having children four years apart helps to avoid having more than one child in college at the same time. Conscientious family planning will also help to reduce the number of unwanted pregnancies.

One third (34%) of the births in the United States are unintended (Wildsmith et al. 2010). In addition to "forgetting" to use contraception and assuming one will not become pregnant with just "one" exposure, Miller et al. (2010) noted physical or sexual partner force/violence including **pregnancy coercion** (coercion by a male partner for the woman to become pregnant) and **birth control sabotage** (partner interference with contraception—e.g., the woman will stop taking the pill) as reasons for unintended pregnancies. The researchers studied 1278 females ages 16–29 years and found that 53% reported physical or sexual partner violence, 19% reported pregnancy coercion, and 15% reported birth control sabotage.

Your choices in regard to whether you want to have children and the use of contraception have important effects on your happiness, lifestyle, and resources. These choices, in large part, are influenced by social and cultural factors that may operate without your awareness. We now discuss these influences.

Do You Want to Have Children?

Beyond a biological drive to reproduce (which not all adults experience), societies socialize their members to have children. This section examines the social influences that motivate individuals to have children, the lifestyle changes that result from such a choice, and the costs of rearing children.

Social Influences Motivating Individuals to Have Children

Our society tends to encourage childbearing, an attitude known as **pronatalism**. Our family, friends, religion, and government help to develop positive attitudes toward parenthood. Cultural observances also function to reinforce these attitudes.

Family   Our experience of being reared in families encourages us to have families of our own. Our parents are our models. They married; we marry. They had children; we have children. We also expect to have a "happy family." Malinen et al. (2010) examined the satisfaction with family life among a sample of Finnish and Dutch dual earners and found that almost three-fourths (73%) reported satisfactory family relationships. In contrast, 13% regarded their family life in terms of parent-child relationships as unsatisfying. Families with either dissatisfied men or dissatisfied women were 6 and 7%, respectively.

Friends   Our friends who have children influence us to do likewise. After sharing an enjoyable weekend with friends who had a little girl, one husband wrote to the host and hostess, "Lucy and I are always affected by Karen—she is such a good child to have around. We haven't made up our minds yet, but our desire to have a child of our own always increases after we leave your home." This couple became parents sixteen months later.

Religion   Religion is a strong influence on an individual's decision to have children—lots of them. Twenty percent of Mormons and 15% of Muslims have at least three children (Pew Research 2008b). Catholics are taught that having children is the basic purpose of marriage and gives meaning to the union. Mormonism and Judaism also have a strong family orientation. Petroni (2011) noted that the Religious Right has had an impact on the role of U.S. policy makers in developing countries in regard to family planning—e.g., they are against abortion.

Race   Hispanics have the highest fertility rate of any racial/ethnic category.

Government   The tax structures that our federal and state governments impose support parenthood. Married couples without children pay higher taxes than couples with children, although the reduction in taxes is not sufficient to offset the cost of rearing a child and is not large enough to be a primary inducement to have children. Individuals have children for emotional, not financial, reasons.

Economy   Times of affluence are associated with a high birth rate. The postwar expansion of the 1950s resulted in the oft-noted "baby boom" generation. Similarly, couples are less likely to decide to have a child during economically depressed times. In addition, the necessity of two wage earners in our postindustrial economy is associated with a reduction in the number of children. The result is a birth rate in the U.S. that has been steadily declining since 1990.

Kids spell love T-I-M-E.

John Crudele, advice columnist

Pronatalism cultural attitude which encourages having children.

Rajo Devi was age 70 when she gave birth to a baby girl in Calcutta in 2008. Her pregnancy was the result of in vitro fertilization. Jaci Dalenberg, of Wooster, Ohio, was 56 when she gave birth to triplets in 2008. She carried the babies as a surrogate for her daughter, Kim Coseno. Hence, Jaci gave birth to her own grandchildren. The two identical twins and their sister were born October 11, at Cleveland Clinic's Hillcrest Hospital. Jaci is thought to be the oldest woman in America to give birth to twins. Bretherick et al. (2010) noted that undergraduate women are not aware of the steep rate of fertility decline with age. Popular media revealing women having children in their seventies encourages them to feel that they have forever to get pregnant. Talk show host Larry King and rock star Rod Stewart each fathered a child at the age of 65. Births to older parents are becoming more common, and questions are now being asked about the appropriateness of elderly individuals becoming parents. Should social policies on this issue be developed?

There are advantages and disadvantages of having a child as an elderly parent. The primary developmental advantage for the child of retirement-aged parents is the attention the parents can devote to their offspring. Not distracted by their careers, these parents have more time and interest to nurture, play with, and teach their children. Although they may have less energy, their experience and knowledge are doubtless better.

The primary disadvantage of having a child in the later years is that the parents are likely to die before, or early in, the child's adult life. Larry King and Rod Stewart will need to live until their mid-eighties to experience the high school graduation and marriage of their infant children.

There are also medical concerns for both the mother and the baby during pregnancy in later life. They include an increased risk of morbidity (chronic illness and disease) and mortality (death) for the mother. These risks are typically a function of chronic disorders that go along with aging, such as diabetes, hypertension, and cardiac disease. Stillbirths, miscarriages, ectopic pregnancies, multiple births, and congenital malformations are also more

Children love for parents to play with them. . . and parents enjoy it too.

David Knox

Cultural Observances Our society reaffirms its approval of parents every year by identifying special days for Mom and Dad. Each year on Mother's Day and Father's Day (and now Grandparents' Day), parenthood is celebrated across the nation with cards, gifts, and embraces. People choosing not to have children have no cultural counterpart (for example, Childfree Day). In addition to influencing individuals to have children, society and culture also influence feelings about the age parents should be when they have children. Recently, couples have been having children at later ages. Is this a good idea? The Social Policy feature discusses this issue.

Individual Motivations for Having Children

Individual motivations, as well as social influences, play an important role in the decision to have children. Some of these are conscious, as in the desire to love and to be loved by one's own child, companionship, and the desire to be personally fulfilled as an adult by having a child. Some also want to recapture their own childhood and youth by having a child. Unconscious motivations for parenthood may also be operative. Examples include wanting a child to avoid career tracking and to gain the acceptance and approval of one's parents and peers. Teenagers sometimes want to have a child to have someone to love them. Later in the chapter we detail teenage motherhood as a major social issue.

frequent for women with advancing age. However, prenatal testing can identify some potential problems such as the risk of Down syndrome and any chromosome abnormality—negative neonatal outcomes are not inevitable. Because an older woman can usually have a healthy baby, government regulations on the age at which a woman can become pregnant are not likely.

Age of the father may also be an issue in older parenting. Krishnaswamy et al. (2011) found that Malaysian children born to fathers aged 50 or above had an increased risk of having CMD (common mental disorders) compared to children who were fathered by young men. However, Romkens et al. (2005) reviewed both the medical and psychological literature of older men (age 50 years and up) having children and concluded that there are no medical or psychosocial data-based justifications to support an age limit for men having children. Given the lack of scientific support and the cultural norm that older men may fertilize younger women, governmental regulations on age limits for parenting are unlikely.

Your Opinion?

1. How old do you think is "too old" to begin being a parent?
2. Do you think the government should attempt to restrict people from having a biological child in their fifties?
3. Who do you feel benefits most and least from having a child in later life?

Sources

Bretherick, K., N. Fairbrother, L. Avila, S. Harbord, and W. Robinson. 2010. Fertility and aging: do reproductive-aged Canadian women know what they need to know? *Fertility & Sterility* 93: 2162–2168.

Krishnaswamy, S., K. Subramaniam, P. Ramachandran, T. Indran and J. Abdul Aziz. 2011. Delayed fathering and risk of mental disorders in adult offspring. *Early Human Development* 87: 171–175.

Romkens, M., B. Gordijn, C. M. Verhaak, E. J. H. Meuleman, and D. D. M. Braat. 2005. No arguments to support an age limit for men entering an in vitro fertilization or intracytoplasmic sperm injection programme. *Nederlands Tijdschrift Voor Geneeskunde* 149: 992–995.

Lifestyle Changes and Economic Costs of Parenthood

Although becoming a parent has numerous potential positive outcomes, parenting also has drawbacks. Every parent knows that parenthood involves difficulties as well as joys. Some of the difficulties associated with parenthood are discussed next.

Lifestyle Changes   Becoming a parent often involves changes in lifestyle. Daily living routines become focused around the needs of the children. Living arrangements change to provide space for another person in the household. Some parents change their work schedule to allow them to be home more. Food shopping and menus change to accommodate the appetites of children. A major lifestyle change is the loss of freedom of activity and flexibility in one's personal schedule. Lifestyle changes are particularly dramatic for women. The time and effort required to be pregnant and rear children often compete with the time and energy needed to finish one's education. Building a career is also negatively impacted by the birth of children. Parents learn quickly that being both involved, on-the-spot parents and climbing the career ladder are difficult. The careers of women may suffer most.

To give our students a simulated exposure to the effect of a baby on their lifestyle we asked them to take care of a "fake baby" for a week. Baby Think It Over (BTIO) is a life-sized computerized infant simulation doll that has been used in pregnancy prevention programs with adolescents (de Anda 2006). The following is part of the write-up of two of our students:

> *The whole idea of the electronic baby was to see if I was ready to be a mother. I am sad to say that I failed the test. I am not ready to be a mother. This whole experience was extremely difficult for me because I am a full-time student and I work. It was really hard to get the things that I needed to get done. Suddenly, I couldn't just think about myself but I also had a little one to think about.*

Now the thing about having a baby—and I can't be the first person to have noticed this—is that thereafter you have it.

Jean Kerr, author and playwright

I love my husband and my children but I want something more—like a life.

Roseanne Barr, comedian

Although undergraduate life includes parties, alcohol, and sex, the larger personal agenda of pair-bonding, procreation, and socialization of one's offspring remains a major life goal. Indeed, 73% of 201,818 freshmen in 279 colleges and universities identified "raising a family" as the most important "objective considered 'essential' or 'very important'" (from a list of 20 items (Pryor et al. 2011).

Sample

Data for the study involved a sample of 293 undergraduate student volunteers at a large southeastern university who completed a fifty-item questionnaire. Of the respondents, 73.4% were female, and 26.6% were male. The median age of the respondents was 20 with a range of 17 to 46. Of the respondents, 40.8% were freshmen, 20.5% sophomores, 17.5% juniors, and 21.2% seniors. Racial background of the respondents was 82% white people and 17.9% black people (respondent self-identified as African American Black, African Black, or Caribbean Black). In comparing the responses of women and men, cross-classification was conducted to determine any relationships with chi-square utilized to assess statistical significance.

Findings

Analysis of the data revealed the following ten statistically significant gender differences in regard to attitudes toward children:

1. *Females were more likely to feel having children is important* (94.3% versus 90.5%) ($p < .007$). This finding is consistent with previous research. Indeed, De Marneffe (2004) emphasized that women "naturally" have the desire to care for children and that this is one of life's great pleasures. She suggested that the desire to nurture is a biological imperative and crucial to the survival of the species.

2. *Females were more likely to view children as providing a reason to live* (40.1% versus 23.2%) ($p < .008$).

3. *Females were more likely to enjoy being around toddlers or young children* (93.2% versus 74.1%) ($p < .001$).

4. *Females were more likely to have taken care of an infant.* The pro-child value females have is related to the fact that they have almost three times the experience of taking care of an infant than a male. Almost 60% of the female respondents, compared to less than a fourth of male respondents (59.7% versus 23.1%) ($p < .000$), reported that they had taken care of an infant.

5. *Females were less likely to be annoyed by crying babies.* In response to the statement, "crying babies drive me crazy," 32.4% of the female respondents compared to 56.2% ($p < .004$) of the male respondents agreed.

6. *Females were more likely to view "wanting children" as an important criteria for a mate* (91.7% versus 70.5%) ($p < .001$).

7. *Females were less likely to marry a man who did not want children.* Consistent with the previous finding, females were more likely to eliminate from consideration marrying a person who could not or would not have children (44.3% versus 21.5%) ($p < .007$).

8. *Females were more likely to divorce a spouse if the spouse turned against children.* Were a marriage to occur with the woman assuming that her husband wanted children and she were to find out that he had changed his mind, she would divorce him. Almost three times as many female as male (29.5% versus 10.8%) respondents agreed that they would divorce their spouse if "I was married and my spouse turned against having children."

9. *Females were more likely to consider adoption if the spouse were sterile* (89.3% versus 76.2%) ($p < .001$).

10. *Females were more likely to be open to having a child of either sex.* Although spouses tend to prefer having two children (one of each sex) (Overington 2006), if they could have only one child, 46.2% of the female respondents compared to 33.8% of the male respondents reported that they "couldn't care less" whether they had a male or female child.

The baby seemed to cry a lot, even if I had just changed her, she still cried. The experience really hit me when I had to wake up four and five times in the night to feed and to change the baby. I learned that when the baby sleeps, I need to sleep as well. I also learned that if I had to take care of the baby by myself, I just couldn't do it. What is sad is that single moms do it every day. If I had a supportive boyfriend or husband to help, the whole idea of having a baby wouldn't be so bad. But my boyfriend told me to call him when the project was over and I had given the baby back.

Theoretical Explanation for the Findings

Both sociobiology and biosocial theoretical perspectives are helpful in explaining the findings of this study. The data emphasized that women evidenced significantly more interest in having children, in selecting a mate who wanted children, and in divorcing a husband who turned against children. Sociobiology (social behavior can be explained on the basis of biology) emphasizes the fact that women carrying their babies to term and providing milk for their survival are a reflection of a biological genetic wiring that predisposes women to greater interest in having a baby and in bonding with them. Indeed the species demands that at least one parent take responsibility for ensuring the survival of the species. That men do not have such a biological link but more often derive their reward from social approval for economic productivity in the workplace may help to explain the discrepancy in female and male attitudes and behaviors.

The biosocial theoretical framework emphasizes the interaction of one's biological or genetic inheritance with one's social environment to explain and predict human behavior (Ingoldsby et al. 2004). Borgerhoff Mulder and McCabe (2006) noted that, although sociobiology is sometimes dismissed as purely genetic determinism, the biosocial perspective is not merely genetics but also interested in the environmental and social context. Hence, although human behavior can be explained as having an evolutionary function, it operates in a social context. In effect, women may not only have the genetic wiring for motherhood but are more likely to experience the social rewards from parents and peers for acknowledging and acting on this biological imperative.

Implications

One, traditional gender roles with the female more focused on children may be a genetically wired and culturally supported norm. Women who move into demanding career roles may continue to be challenged by the need to balance career and family. The reluctance some women feel when they leave their baby at a day-care center and the tug they feel at work to return to pick up their baby may have its origin in evolutionary biology. This dilemma was a consistent theme in Pamela Stone's book on *Opting Out* (2007).

Two, egalitarian gender role relationships may be more of an illusion than a reality. As long as women prioritize children, their nurturing and child-care behavior will follow. Because males are less likely to prioritize children, their lack of commitment in this area often translates into less involvement.

Three, the finding that females are more likely to "care less" whether their baby is a boy or a girl reflects the value that children are valued per se. In contrast, men may be more likely to value a child if it is a male because this may be connected to issues of masculinity.

Sources

Borgerhoff Mulder, M., and C. McCabe. 2006. Whatever happened to human sociobiology? *Anthropology* 22: 21–22.

De Marneffe, D. 2004. *Maternal desire: On children, love, and the inner life.* New York: Little, Brown, and Company.

Ingoldsby, B. B., S. R. Smith, and J. E. Miller. 2004. *Exploring family theories.* Los Angeles, CA: Roxbury Publishing Co.

Overington, C. 2006. Desire for children of each sex grows families. *Australasian Business Intelligence*, May 28.

Pryor, J. H., S. Hurtado, L. DeAngelo, L. P. Blake and S. Tran. 2011. *The American freshman: National norms fall 2010.* Los Angeles: Higher Education Research Institute. U.C. LA.

Stone, P. 2007. *Opting out?: Why women really quit careers and head home.* Berkeley: University of California Press.

*Note: Abridged from B. Bragg, D. Knox, and M. Zusman. 2008. The Little Ones: Gender differences in attitudes toward children among university students. Poster, Southern Sociological Society, April, Richmond, VA.

Other students had a very positive experience with the baby and did not want to part with it at the end of the week. Indeed, in one case the "fake baby" became a part of the family. One student said that she awakened early one morning to find her father rocking the baby.

Financial Costs Meeting the financial obligations of parenthood is difficult for many parents. The costs begin with prenatal care and continue at childbirth.

Being responsible for an electronic baby that cries and needs "feeding" and "diapering" can give childfree, unmarried undergraduates some idea of the responsibility of taking care of an infant. This student is already frustrated with taking care of the "baby" and her boyfriend wants her to call him when the project is over.

Chelsea E. Curry

For an uncomplicated vaginal delivery, with a two-day hospital stay, the cost may total $10,000, whereas a cesarean section birth may cost $14,000. The annual cost of a child less than 2 years old for middle-income parents ($56,670 to $98,120)—which includes housing ($3,890), food ($1,340), transportation ($1,420), clothing ($750), health care ($790), child care ($2,630), and miscellaneous ($880)—is $11,700. For a 15- to 17-year-old, the cost is $13,530 (*Statistical Abstract of the United States, 2011*, Table 688). These costs do not include the wages lost when a parent drops out of the workforce to provide child care.

Most parents value their children attending college. See the website at the end of this chapter in regard to the College Cost Calculator. The price varies depending on whether a child attends a public or private college. The annual cost for a child attending a four-year public college in the state of residence is around $15,000 (including tuition, board, dorm); for a private college the cost is around $40,000 annually. Collegeboard.com will identify the cost of a specific college.

My mom used to say it doesn't matter how many kids you have . . . because one kid'll take up 100% of your time so more kids can't possibly take up more than 100% of your time.

Karen Brown

How Many Children Do You Want?

National Data

Livingston and Chon (2010) analyzed national data, which reflects that almost 20% of Americans never bear a child. Of those who do not have children, 44% are voluntarily childfree; 40% involuntarily childfree; 16% plan a child in the future (Smock and Greenland 2010). The White House Study on Women in America confirmed that about twenty percent (18%) of women ages 40 to 44 have never had a child—this percent has almost doubled since 1976 (Department of Commerce et al. 2011).

Procreative liberty the freedom to decide to have children or not.

Procreative liberty is the freedom to decide whether or not to have children. More women are deciding not to have children or to have fewer children (Department of Commerce et al. 2011).

Childfree Marriage?

Just as cultural forces help to ensure that a person will marry (over 90% do) and select someone of the same race (95% do), there is similar pressure to have children. The intentionally childfree may be viewed with suspicion ("they are selfish"), avoidance ("since they don't have children they won't like us or support our family values"), discomfort ("what would I have in common with these people?"), rejection ("they are wrong not to want children/I don't want to spend time with these people"), and pity ("they don't know what they are missing") (Scott 2009). Indeed "the mere existence of a growing childless by choice population is a challenge to people who believe procreation is instinctive, intrinsic, biological, or obligatory" (p. 191). Stereotypes about couples who deliberately elect not to have children include that they don't like kids, are immature, and are not fulfilled since they don't have a child to make their lives "complete." The reality is that they may enjoy children and some deliberately choose careers to work with them (e.g., elementary school teacher). But they don't want the full-time emotional and economic responsibility of having their own children. Pelton and Hertlein (2011) emphasized that voluntary childfree couples do not progress through the traditional family life cycle. Indeed they pass through a different set of stages specific to these couples.

Koropeckyj-Cox and Pendell (2007a) examined attitudes about childlessness in the United States. They used a national sample and found that college-educated, white females had the most favorable attitudes toward childlessness. Those who were least likely to regard childlessness as a desirable lifestyle were not college-educated, were black, male, and held conservative religious beliefs. In general, there seems to be an acceptance of childlessness, not an endorsement of the lifestyle. The data reflecting adults in a national study revealed that 55.9% of females compared to 48.1% of males disagreed that "People who have never had children have empty lives." Hence, a stigma is still associated with not having children, and men buy into this more than women (Koropeckyj-Cox and Pendell 2007b).

Some people simply do not like children. Aspects of our society reflect **anti-natalism** (a perspective against children). Indeed, there is a continuous fight for corporations to implement or enforce any family policies (from family leaves to flex time to on-site day care). Profit and money—not children—are priorities. In addition, although people are generally tolerant of their own children, they often exhibit antinatalistic behavior in reference to the children of others. Notice the unwillingness of some individuals to sit next to a child on an airplane.

Laura Scott (2009), a childfree wife, set up the Childless by Choice Project and surveyed 171 childfree adults (ages 22 to 66, 71% female, 29% male) to identify their motivations not to have children. The categories of her respondents and the percentage of each follow:

Early articulators (66%)—these adults knew early that they did not want children.

Postponers (22%)—adults who kept delaying when they would have children and remained childless.

Acquiescers (8%)—those who made the decision to remain childless because their partner did not want children.

Undecided (4%)—those who are childless but still in the decision making process.

The top five reasons individuals gave for wanting to remain childfree were (Scott 2009):

1. Life/relationship satisfaction was great and they feared that parenthood would not enhance it.

2. Being free and independent were strong values that they feared would be affected by children.

3. They wanted to avoid the responsibility of rearing a child.

I never missed having children.

Gloria Steinem, feminist

Antinatalism opposition to children.

The purpose of this scale is to assess your attitudes toward having a childfree lifestyle. After reading each statement, select the number that best reflects your answer, using the following scale:

1	2	3	4	5	6	7
Strongly Disagree						Strongly Agree

_____ 1. I do not like children.

_____ 2. I would resent having to spend all my money on kids.

_____ 3. I would rather enjoy my personal freedom than have it taken away by having children.

_____ 4. I would rather focus on my career than have children.

_____ 5. Children are a burden.

_____ 6. I have no desire to be a parent.

_____ 7. I am too "into me" to become a parent.

_____ 8. I lack the nurturing skills to be a parent.

_____ 9. I have no patience for children.

_____ 10. Raising a child is too much work.

_____ 11. A marriage without children is empty.

_____ 12. Children are vital to a good marriage.

_____ 13. You can't really be fulfilled as a couple unless you have children.

_____ 14. Having children gives meaning to a couple's marriage.

_____ 15. The happiest couples that I know have children.

_____ 16. The biggest mistake couples make is deciding not to have children.

_____ 17. Childfree couples are sad couples.

_____ 18. Becoming a parent enhances the intimacy between spouses.

_____ 19. A house without the "pitter patter" of little feet is not a home.

_____ 20. Having a child means your marriage is successful.

Scoring

Reverse score items 11 through 20. For example if you wrote a 1 for item 20, change this to a 7. If you wrote a 2, change it to a 6, etc. Add the numbers. The higher the score (140 is the highest possible score), the greater the value for a childfree lifestyle. The lower the score (20 is the lowest possible score), the less the desire to have a childfree lifestyle. The midpoint between the extremes is 80: Scores below 80 suggest less preference for a childfree lifestyle and scores above 80 suggest a desire for a childfree lifestyle. The average score of 52 male and 138 female undergraduates at Valdosta State University was below the midpoint (M = 68.78, SD = 17.06), suggesting a tendency toward a lifestyle that included children. A significant difference was found between males, who scored 72.94 (SD = 16.82), and females, who scored 67.21 (SD = 16.95), suggesting that males are more approving of a childfree lifestyle. There were no significant differences between whites and blacks or between students in different ranks (freshmen, sophomore, junior, or senior).

Source

"The Childfree Lifestyle Scale" 2010 by Mark A. Whatley, Ph.D., Department of Psychology, Valdosta State University, Valdosta, Georgia 31698-0100. Used by permission. Other uses of this scale by written permission of Dr. Whatley only (mwhatley@valdosta.edu). Information on the reliability and validity of this scale is available from Dr. Whatley.

I founded the organization "No Kidding" because I got tired of losing a lot of my close buddies to their children. Once they had kids, they were no longer available for friendship. Our organization allows adults to continue adult friendships unencumbered by the demands of rearing children.

Jerry Steinberg, founder of No Kidding

4. They experienced an absence of maternal/paternal instinct.

5. Their desire to accomplish and experience things in life would be difficult as a parent.

Some of Scott's interviewees also had an aversion to children, having had a bad childhood or concerns about childbirth.

One Child?

Citing a Pew Research study, Sandler (2010) emphasized that only 3% of adults view one child as the ideal family size. Those that have only children may do so because they want the experience of parenthood without children markedly interfering with their lifestyle and careers. Still others have an only child because of the difficulty in pregnancy or birthing the child. One mother said, "I threw up every day for nine months including on the delivery table." Another said, "I was torn up giving birth to my child." Still another mother said, "It took two years for my body to recover. Once is enough for me." There are also those who have only one child because they can't get pregnant a second time.

Two Children?

The most preferred family size in the United States (for non-Hispanic white women) is the two-child family (1.9 to be exact!). Reasons for this preference include feeling that a family is "not complete" without two children, having a companion for the first child, having a child of each sex, and repeating the positive experience of parenthood enjoyed with their first child. Some couples may not want to "put all their eggs in one basket." They may fear that, if they have only one child and that child dies or turns out to be disappointing, they will not have another opportunity to enjoy parenting.

Meanwhile, when a sibling joins an only child, new feelings become possible as a result of the interaction. Yoshimura (2010) noted that siblings tend to be envious of each other and that sibling rivalry is real.

Three Children?

Couples are more likely to have a third child, and to do so quickly, if they already have two girls rather than two boys. They are least likely to bear a third child if they already have a boy and a girl. Some individuals may want three children because they enjoy children and feel that "three is better than two." In some instances, a couple that has two children may simply want another child because they enjoy parenting and have the resources to do so.

Having a third child creates a "middle child." This child is sometimes neglected because parents of three children may focus more on the "baby" and the firstborn than on the child in between. However, an advantage to being a middle child is the chance to experience both a younger and an older sibling. Each additional child also has a negative effect on the existing children by reducing the amount of parental time available to existing children. The economic resources for each child are also affected for each subsequent child.

DIVERSITY IN OTHER COUNTRIES

Traditionally, couples in China typically have had one child due to China's One Child Policy, which has led to forced abortions, sterilizations, and economic penalties for having more than one child. But a need for more young workers has caused China to reexamine the policy. Beginning in 2011 in some provinces, couples may be allowed to have a second child (if at least one of the spouses is an only child) (MacLeod 2010b). Meanwhile, in India, the Indian Medical Association has recommended that the country implement a one-child policy (similar to China) to slow down population growth. This policy is to take effect in 2015.

Whenever I held my newborn baby in my arms, I used to think that what I said and did to him could have an influence not only on him but on all whom he met, not only for a day or a month or a year, but for all eternity—a very challenging and exciting thought for a mother.

Rose Kennedy,
matriarch of Kennedy family

Michelle Six

This woman met her husband on a train in Paris...they now have three children.

Hispanics are more likely to want larger families than are white or African American people. Larger families have complex interactional patterns and different values. The addition of each subsequent child dramatically increases the possible relationships in the family. For example, in a one-child family, four interpersonal relationships are possible: mother-father, mother-child, father-child, and father-mother-child. In a family of four, eleven relationships are possible; in a family of five, 26; and in a family of six, 57.

I have 8 children—I'm a busy dad.

Eddie Murphy, celebrity

Competitive birthing having the same number (or more) of children in reference to one's peers.

The great advantage of living in a large family is that early lesson of life's essential unfairness.

Nancy Mitford

Four Children—New Standard for the Affluent?

Smith (2007) noted that, among affluent couples, four children may be the new norm. Fueled by competitive career moms who have opted out of the workforce and who find themselves in suburbia surrounded by other moms with resources and time on their hands, having a large family is being reconsidered. A pattern has begun called **competitive birthing**, where individuals have the same number of children (or more) as their peers. Subsequent research will need to confirm if the pattern is widespread.

PERSONAL CHOICES

Is Genetic Testing for You?

Because each of us may have flawed genes that carry increased risk for diseases such as cancer and Alzheimer's, the question of whether to have a genetic test before becoming pregnant becomes relevant. The test involves giving a blood sample. About 900 genetic tests are available. The advantage is the knowledge of what defective genes you may have and what diseases you may pass to your children. McKee and Blow (2010) emphasized that genetic testing is a collective decision since it impacts one's partner and parents and that others might be consulted.

The disadvantage is stress or anxiety (for example, what do you do with the information?) as well as discrimination from certain health insurance companies (who may deny coverage). The validity of the test is also problematic. Hunter et al. (2008) warned, ". . . even the ardent proponents of genomic susceptibility testing would agree that for most diseases, we are still at the early stages of identifying the full list of susceptibility-associated variants" (p. 106). Because no treatment may be available if the test results are positive, the knowledge that they might pass diseases to their children can be devastating to a couple. Finally, genetic testing is expensive, ranging from $200 to $2,400. The National Society of Genetic Counselors (http://www.nsgc.org) offers information about genetic testing.

Most of us become parents long before we have stopped being children.

Mignon McLaughlin, author

Sex Selection?

Some couples use various sex selection technologies to help ensure a boy or a girl. MicroSort is the new preconception sperm sorting technology, which allows parents to increase the chance of having a girl or boy baby. The process is successful 75 to 90% of the time depending on whether a girl or boy baby is desired. No longer are girl babies being preferred by couples who use sex selection technology.

Puri and Nachtigall (2010) examined ethical considerations in regard to sex selection. One perspective held by sex-selection technology providers argues that sex selection is an expression of reproductive rights and a sign of female empowerment to prevent unwanted pregnancies and abortions. A contrasting view, more likely to be held by primary care physicians, is that sex selection contributes to gender stereotypes that could result in neglect of children of the lesser-desired sex.

Teenage Motherhood

Teenage motherhood is typically described as a social problem. The various conceptualizations include it as an issue for "Public Health" where early motherhood is viewed as an issue requiring surveillance and a public health response, as an "Economic" drain where teenage mothers are seen as a financial drain on society, as an "Ethnicity" issue that classifies young mothers into ethnic groups, and as a "Eugenics" issue which highlights the "unsuitability" of young mothers as parents (Breheny and Stephens 2010).

Double Dutch strategy of using both the pill and condom by sexually active youth in the Netherlands.

National **Data**

Forty-one percent of all births are to unmarried women, most of whom are teenagers. Seventy-two percent of these births are to black women; 53% to Hispanic women (Pew Research Center 2010a).

Reasons for teenagers having a child include not being socialized as to the importance of contraception, having limited parental supervision, and perceiving few alternatives to parenthood. Indeed, motherhood may be one of the only remaining meaningful roles available to young teen mothers. In addition, some teenagers feel lonely and unloved and have a baby to create a sense of being needed and wanted. In contrast, in Sweden, eligibility requirements for welfare payments make it almost necessary to complete an education and get a job before becoming a parent.

Problems Associated with Teenage Motherhood

Teenage parenthood is associated with various negative consequences, including the following:

1. *Stigmatization and marginalization.* Wilson and Huntington (2006) noted that, because teen mothers resist the typical life trajectory of their middle-class peers, they are stigmatized and marginalized. In effect, they are a threat to societal goals of economic growth through higher education and increased female workforce participation. In spite of such stigmatization and marginalization, Rolfe (2008) interviewed thirty-three young women who were mothers before the age of 21 and discovered three themes of their experience of teenage motherhood—as "hardship and reward," "growing up and responsibility," and "doing things differently." The researcher noted that the respondents were "active in negotiating and constructing their own identities as mothers, careers and women" (p. 299).

2. *Poverty among single teen mothers and their children.* Many teen mothers are unwed. Livermore and Powers (2006) studied a sample of 336 unwed mothers and found them plagued with financial stress; almost 20% had difficulty providing food for themselves and their children (18.5%), had their electricity cut off for nonpayment (19.7%), and had no medical care for their children (18.2%). Almost half (47%) reported experiencing "one or more financial stressors" (p. 6).

3. *Poor health habits.* Teenage unmarried mothers are less likely to seek prenatal care and are more likely than older and married women to smoke, drink alcohol, and take other drugs. These factors have an adverse effect on the health of the baby. Indeed, babies born to unmarried teenage mothers are more likely to have low birth weights (less than five pounds, five ounces) and to be born prematurely. Children of teenage unmarried mothers are also more likely to be developmentally delayed. These outcomes are largely a result of the association between teenage unmarried childbearing and persistent poverty.

4. *Lower academic achievement* Lipman et al. (2011) analyzed data from the Ontario Child Health Study (OCHS) and found that being born to a teen mother is associated with poorer educational achievement, personal income, and life satisfaction.

5. *Personal health and psychosocial adjustment.* Amato and Kane (2011) examined the general health and psychosocial adjustment among 2,290 women and found that young women who became teenage mothers had poorer general health, were more depressed, and had lower self esteem than women who were not teen mothers (the latter also attended college and were full time employed). However, those women who had become teenage mothers drank significantly less than women who were not teen mothers. The latter emphasizes a positive outcome of teenage motherhood often ignored.

Infertility

There is an epidemic of infertility in this country. There are more women who have put off child bearing in favor of their professional lives. For them, the only way they are going to have a family is adopt from China.

Iris Chang,
journalist

Infertility the inability to achieve a pregnancy after at least one year of regular sexual relations without birth control, or the inability to carry a pregnancy to a live birth.

A woman's chance of getting pregnant in any one month is about 20% for women under 30, but only 5% for women over 40.

Diane Aronson, Executive Director of Resolve (support group for individuals coping with infertility and loss)

Conception refers to the fusion of the egg and sperm. Also known as fertilization.

Pregnancy when the fertilized egg is implanted (typically in the uterine wall).

Infertility is defined as the inability to achieve a pregnancy after at least one year of regular sexual relations without birth control, or the inability to carry a pregnancy to a live birth. Different types of infertility include the following:

1. *Primary infertility.* The woman has never conceived even though she wants to and has had regular sexual relations for the past twelve months.

2. *Secondary infertility.* The woman has previously conceived but is currently unable to do so even though she wants to and has had regular sexual relations for the past twelve months.

3. *Pregnancy wastage.* The woman has been able to conceive but has been unable to produce a live birth.

Approximately 12% of all 15- to 44-year-old women have what is termed "impaired fecundity." These women report (a) that it is physically impossible for them or their husbands/partners to have a baby (excluding those who have been sterilized), (b) that it is physically difficult or dangerous to carry a baby to term, or (3) that they had been continuously married or cohabiting for 3 years, not used birth control, and not become pregnant (Smock and Greenland 2010). Infertility risks increase with age. The take-home message for women is that to delay getting pregnant is to delay the chance of getting pregnant.

Causes of Infertility

Although popular usage does not differentiate between the terms *fertilization* and the *beginning of pregnancy*, fertilization or **conception** refers to the fusion of the egg and sperm, whereas **pregnancy** is not considered to begin until five to seven days later, when the fertilized egg is implanted (typically in the uterine wall). Hence, not all fertilizations result in a pregnancy. An estimated 30% to 40% of conceptions are lost prior to or during implantation. Forty percent of infertility problems are attributed to the woman, 40% to the man, and 20% to both of them. Some of the more common causes of infertility in men include low sperm production, poor semen motility, effects of STIs (such as chlamydia, gonorrhea, and syphilis), and interference with passage of sperm through the genital ducts due to an enlarged prostate. Additionally, men who are infertile may have an increased risk of having prostate cancer (Walsh et al. 2010).

The causes of infertility in women include blocked fallopian tubes, endocrine imbalance that prevents ovulation, dysfunctional ovaries, chemically hostile cervical mucus that may kill sperm, and effects of STIs. Brandes et al. (2011) noted that unexplained infertility is one of the most common diagnoses in fertility care and associated with a high probability of achieving a pregnancy—most

What if Your Partner is Infertile—Would you Marry the Partner?

Being infertile (for the woman) may have a negative lifetime effect. Wirtberg et al. (2007) interviewed fourteen Swedish women twenty years after their infertility treatment and found that childlessness had had a major impact on all the women's lives and remained a major life theme. The effects were both personal (sad) and interpersonal (half were separated and all reported negative effects on their sex lives). The effects of childlessness were especially increased at the time the study was conducted, as the women's peer group was entering the "grandparent phase." The researchers noted that infertility has lifetime consequences for the individual woman and her relationships. One can feel the emotional pain of Celine Dion, who at age 41, revealed her 4th failed IVF attempt to get pregnant. "I'm going to try till it works." It worked on the sixth try—with twins.

spontaneously. An at-home fertility kit, Fertell, allows women to measure the level of their follicle-stimulating hormone on the third day of their menstrual cycles. An abnormally high level means that egg quality is low. The test takes thirty minutes and involves a urine stick. The same kit allows men to measure the concentration of motile sperm. Men provide a sample of sperm (for example, via masturbation) that swim through a solution similar to cervical mucus. This procedure takes about eighty minutes. Fertell has been approved by the Food and Drug Administration (FDA), no prescription is necessary, and costs around $100.

Assisted Reproductive Technology (ART)

A number of technological innovations are available to assist women and couples in becoming pregnant. These include hormonal therapy, artificial insemination, ovum transfer, in vitro fertilization, gamete intrafallopian transfer, and zygote intrafallopian transfer.

Hormone Therapy   Drug therapies are often used to treat hormonal imbalances, induce ovulation, and correct problems in the luteal phase of the menstrual cycle. Frequently used drugs include Clomid, Pergonal, and human chorionic gonadotropin (HCG), a hormone extracted from human placenta. These drugs stimulate the ovary to ripen and release an egg. Although they are fairly effective in stimulating ovulation, hyperstimulation can occur, which may result in permanent damage to the ovaries.

Hormone therapy also increases the likelihood that multiple eggs will be released, resulting in multiple births. The increase in triplets and higher order multiple births over the past decade in the United States is largely attributed to the increased use of ovulation-inducing drugs for treating infertility. Infants of higher order multiple births are at greater risk of having low birth weight and their mortality rates are higher. Mortality rates have improved for these babies, but these low birth-weight survivors may need extensive neonatal medical and social services.

Artificial Insemination   When the sperm of the male partner are low in count or motility, sperm from several ejaculations may be pooled and placed directly into the cervix. This procedure is known as *artificial insemination by husband* (AIH). When sperm from someone other than the woman's partner are used to fertilize a woman, the technique is referred to as *artificial insemination by donor* (AID).

This gay couple are in a stable monogamous relationship. Their baby is the product of artificial insemination from a donor friend.

Heather Barnes

Women who want to become pregnant obtain sperm from a friend or from a sperm bank (e.g. http://www.cryolab.com/). Regardless of the source of the sperm, it should be screened for genetic abnormalities and STIs, quarantined for 180 days, and retested for human immunodeficiency virus (HIV); also, the donor should be younger than 50 to diminish hazards related to aging. These precautions are not routinely taken—let the buyer beware.

How do children from donor sperm feel about their fathers? A team of researchers (Scheib et al. 2005) studied twenty-nine individuals (41% from lesbian couples, 38% from single women, and 21% from heterosexual couples) and found that most (75%) always knew about their origin and were comfortable with it. All but one reported a neutral to positive impact with the birth mother. Most (80%) indicated a moderate interest in learning more about the donor. No youths reported wanting money, and only 7% reported wanting a father-child relationship. Berger and Paul (2008) studied the effects of disclosing or not disclosing to the child that he or she is from a donor sperm. The results were inconclusive but favored disclosure.

Artificial Insemination of a Surrogate Mother   In some instances, artificial insemination does not help a woman get pregnant. (Her fallopian tubes may be blocked, or her cervical mucus may be hostile to sperm.) The couple that still wants a child and has decided against adoption may consider parenthood through a surrogate mother. There are two types of surrogate mothers. One is the contracted surrogate mother who supplies the egg, is impregnated with the male partner's sperm, carries the child to term, and gives the baby to the man and his partner. A second type is the surrogate mother who carries to term a baby to whom she is not genetically related (a fertilized egg from the "infertile couple" who can't carry a baby to term is implanted in her uterus). As with AID, the motivation of the prospective parents is to have a child that is genetically related to at least one of them. For the surrogate mother, the primary motivation is to help childless couples achieve their aspirations of parenthood and to make money. Although some American women are willing to "rent their wombs," women in India have also begun to provide this service. For $5,000, an Indian wife who already has a child will carry a baby to term for an infertile couple (for a fraction of the cost of an American surrogate).

California is one of twelve states in which entering into an arrangement with a surrogate mother is legal. The fee to the surrogate mother is $20,000 to $25,000. Other fees (travel, hospital, lawyers, and so on) can run the figure to as high as $125,000. Surrogate mothers typically have their own children, making giving up a child that they carried easier. For information about the legality of surrogacy in your state, see http://www.surrogacy.com/legals/map.html.

Rosenberg (2010) emphasized the need to have every aspect of involving a surrogate detailed in a formal agreement including:

> who the parties are, where the genetic material is coming from, the medical and psychological screening required, the duties and expectations for each party, who will choose the OB and hospital for delivery and how the choice should be made, how decisions will be made during the course of the arrangement, particularly decisions about termination of pregnancy or reduction, addressing the issue of a woman's right to choose, medical testing that might be required and when, any payments for expenses covered, and payments focused on time, effort, and inconvenience, whether the funds will be escrowed, any limitations on the surrogates travel or diet, time limitations of the agreement, choice of law (not all states are surrogacy friendly), and all other relevant matters. The agreement should detail who is the legal parent or parents and how that will be recognized and the appropriate birth certificate will be issued. Such an agreement mandates that both sides be represented by independent counsel to ensure proper understanding of its meaning and of all of its implications. The legal agreement is very specific to surrogacy, so will require the advice and assistance of an ART attorney knowledgeable in the field (sounds a bit like a broken record, but I cannot stress the importance of this enough) (p. 98).

In Vitro Fertilization   About 2 million couples cannot have a baby because the woman's fallopian tubes are blocked or damaged, preventing the passage of eggs to the uterus. In some cases, blocked tubes can be opened via laser surgery or by inflating a tiny balloon within the clogged passage. When these procedures are not successful (or when the woman decides to avoid invasive tests and exploratory surgery), *in vitro* (meaning "in glass") *fertilization* (IVF), also known as test-tube fertilization, is an alternative.

Using a laparoscope (a narrow, telescope-like instrument inserted through an incision just below the woman's naval to view tubes and ovaries), the physician is able to see a mature egg as it is released from the woman's ovary. The time of release can be predicted accurately within two hours. When the egg emerges, the physician uses an aspirator to remove the egg, placing it in a small tube containing stabilizing fluid. The egg is taken to the laboratory, put in a culture petri dish, kept at a certain temperature-acidity level, and surrounded by sperm from the woman's partner (or donor). After one of these sperm fertilizes the egg, the egg divides and is implanted by the physician in the wall of the woman's uterus. Usually, several eggs are implanted in the hope one will survive. This was the case of Nadya Suleman, who ended up giving birth to eight babies. Eight embryos were transferred in 2008, at Duke University's in vitro fertilization program, into her body with the thought that some would not survive . . . all did (Rochman 2009).

Some couples want to ensure the sex of their baby. In a procedure called "family balancing" because couples that already have several children of one sex often use it, the eggs of a woman are fertilized and the sex of the embryos three and eight days old is identified. Only those of the desired sex are then implanted in the woman's uterus.

Alternatively, the Y chromosome of the male sperm can be identified and implanted. The procedure is accurate 75% of the time for producing a boy baby and 90% of the time for a girl baby. On page 288 we discussed sex selection by means of MicroSort.

Occasionally, some fertilized eggs are frozen and implanted at a later time, if necessary. This procedure is known as **cryopreservation**. Separated or divorced couples may disagree over who owns the frozen embryos, and the legal system is still wrestling with the fate of their unused embryos, sperm, or ova after a divorce or death.

Cryopreservation the freezing of fertilized eggs for implantation at a later stage.

Ovum transfer a fertilized egg is implanted in the uterine wall.

Ovum Transfer   In conjunction with in vitro fertilization is **ovum transfer**, also referred to as embryo transfer. In this procedure, an egg is donated, fertilized in vitro with the husband's sperm, and then transferred to his wife. Alternatively, a physician places the sperm of the male partner in a surrogate woman. After about five days, her uterus is flushed out (endometrial lavage), and the contents are analyzed under a microscope to identify the presence of a fertilized ovum.

The fertilized ovum is then inserted into the uterus of the otherwise infertile partner. Although the embryo can also be frozen and implanted at another time, fresh embryos are more likely to result in successful implantation. Infertile couples that opt for ovum transfer do so because the baby will be biologically related to at least one of them (the father) and the partner will have the experience of pregnancy and childbirth. As noted earlier, the surrogate woman participates out of her desire to help an infertile couple or to make money.

Other Reproductive Technologies   A major problem with in vitro fertilization is that only about 15% to 20% of the fertilized eggs will implant on the uterine wall. To improve this implant percentage (to between 40% and 50%), physicians place the egg and the sperm directly into the fallopian tube, where they meet and fertilize. Then the fertilized egg travels down into the uterus and implants.

Because the term for sperm and egg together is *gamete,* this procedure is called *gamete intrafallopian transfer,* or GIFT. This procedure, as well as in vitro fertilization, is not without psychological costs to the couple.

Gestational surrogacy, another technique, involves fertilization in vitro of a woman's ovum and transfer to a surrogate. Trigametic IVF also involves the use of sperm in which the genetic material of another person has been inserted. This technique allows lesbian couples to have a child genetically related to both women. Infertile couples hoping to get pregnant through one of the more than 400 in vitro fertilization clinics should make informed choices by asking questions such as, "What is the center's pregnancy rate for women with a similar diagnosis?"

What percentage of these women has a live birth? According to the Centers for Disease Control and Prevention, the typical success rate (live birth) for infertile couples who seek help in one of the 400 fertility clinics is 28% (Lee 2006). Beginning assisted-reproductive technology as early after infertility is suspected is important. Wang et al. (2008) analyzed data on 36,412 patients to assess success of actual births for infertile women using assisted-reproductive technology and found that, for women age 30 and above, each additional year in age was associated with an 11% reduction in the chance of achieving pregnancy and a 13% reduction in the chance of a live delivery. If women aged 35 years or older would have had their first treatment one year earlier, 15% more live deliveries would be expected.

Finally, Hammarberg et al. (2008b) studied 166 women who had conceived through assisted-reproductive technology to identify any differences in birthing. They did find that ART participants were more likely to have a cesarean birth (51% versus 25%) and to report disappointment with the birth event when compared with those who had a vaginal birth.

The cost of treating infertility is enormous. Katz et al. (2011) examined the costs for 398 women in 8 infertility practices over an 18 month period. For the half who pursued IVF, the median per-person medication costs ranged from $1,182 for medications only to $24,373 and $38,015 for IVF and IVF–donor egg groups, respectively. In regard to the costs of successful outcomes (delivery or ongoing pregnancy by 18 months) for IVF—$61,377. Within the time frame of the study, costs were not significantly different for women whose outcomes were successful and women whose outcomes were not.

Adoption

National **Data**

Adoption is rare. Just over 1% of 18- to 44-year-old women reported having adopted a child (Smock and Greenland 2010).

Angelina Jolie and Brad Pitt are celebrities who have given national visibility to adopting children. They are not alone in their desire to adopt children. The various routes to adoption are public (children from the child welfare system), private agency (children placed with nonrelatives through agencies), independent adoption (children placed directly by birth parents or through an intermediary such as a physician or attorney), kinship (children placed in a family member's home), and stepparent (children adopted by a spouse). Motives for adopting a child include wanting a child because of an inability to have a biological child (infertility), a desire to give an otherwise unwanted child a permanent loving

Digga and Teresa Schacht

An adoption is a win-win for everyone—the adoptive parents are joyful to have their new baby, the infant is fortunate to be in the home of a loving couple, and the biological parents feel contentment in knowing that their baby will be reared in a loving context.

home, or a desire to avoid contributing to overpopulation by having more biological children. Some couples may seek adoption for all of these motives. Adoption is actually quite rare, with less than 5% of couples adopting; 15% of these adoptions will be children from other countries.

Demographic Characteristics of People Seeking to Adopt a Child

Whereas demographic characteristics of those who typically adopt are white, educated, and high-income, adoptees are being increasingly placed in nontraditional families, including with older, gay, and single individuals. Sixteen states have taken steps to ban adoption by gay couples on the grounds that, because "marriage" is "heterosexual marriage," children do not belong in homosexual relationships (Stone 2006). Leung et al. (2005) compared children adopted or reared by gay or lesbian and heterosexual parents. They found no negative effects when the adoptive parents were gay or lesbian.

Characteristics of Children Available for Adoption

Adoptees in the highest demand are healthy, white infants. Those who are older, of a racial or ethnic group different from that of the adoptive parents, of a sibling group, or with physical or developmental disabilities have been difficult to place. Baden and Wiley (2007) reviewed the literature on adoptees as adults and found that the mental health of most was on par with those who were not adopted. However, a small subset of the population showed concerns that may warrant therapeutic intervention.

Costs of Adoption

Adopting from the U.S. foster care system is generally the least expensive type of adoption, usually involving little or no cost, and states often provide subsidies to adoptive parents. However, a couple can become foster care parents to a child and become emotionally bonded with the child, and the birth parents can reappear and request their child back.

Stepparent and kinship adoptions are also inexpensive and have less risk of the child being withdrawn. Agency and private adoptions can range from $5,000 to $40,000 or more, depending on travel expenses, birth mother expenses, and requirements in the state. International adoptions can range from $7,000 to $30,000 (see http://costs.adoption.com/).

Transracial adoption adopting children of a race different from that of the parents.

Transracial Adoption   Sandra Bullock is one of the latest Hollywood celebrities to adopt a child of another race. Her son Louis is from an adoption service in New Orleans, which called in early 2010 to alert her that her newborn was ready for her. **Transracial adoption** is defined as the practice of adopting children of a race different from that of the parents—for example, a white couple adopting a Korean or African American child. Perry (2010) assessed attitudes towards transracial adoption in a sample of 1721 undergraduates and found no racial differences. You can assess your attitudes toward interracial adoption using the scale on page 297. Most college students who took the scale reported favorable attitudes.

Transracial adoptions are controversial. The motivations of persons wanting to adopt cross racially is sometimes questioned . . . are they just making a political statement? Another controversy is whether it is beneficial for children to be adopted by parents of the same racial background. In regard to the adoption of African American children by same-race parents, the National Association of Black Social Workers (NABSW) passed a resolution against transracial adoptions, citing that such adoptions prevented black children from developing a positive sense of themselves "that would be necessary to cope with racism and prejudice that would eventually occur" (Hollingsworth 1997, 44).

SELF-ASSESSMENT | Attitudes toward Transracial Adoption Scale

Transracial adoption is the adoption of children of a race other than that of the adoptive parents. Please read each item carefully and consider what you believe about each statement. There are no right or wrong answers to any of these statements, so please give your honest reaction and opinion. After reading each statement, select the number that best reflects your answer, using the following scale:

1	2	3	4	5	6	7
Strongly Disagree						Strongly Agree

_____ 1. Transracial adoption can interfere with a child's well being.

_____ 2. Transracial adoption should not be allowed.

_____ 3. I would never adopt a child of another race.

_____ 4. I think that transracial adoption is unfair to the children.

_____ 5. I believe that adopting parents should adopt a child within their own race.

_____ 6. Only same-race couples should be allowed to adopt.

_____ 7. Biracial couples are not well prepared to raise children.

_____ 8. Transracially adopted children need to choose one culture over another.

_____ 9. Transracially adopted children feel as though they are not part of the family they live in.

_____ 10. Transracial adoption should occur only between certain races.

_____ 11. I am against transracial adoption.

_____ 12. A person has to be desperate to adopt a child of another race.

_____ 13. Children adopted by parents of a different race have more difficulty developing socially than children adopted by foster parents of the same race.

_____ 14. Members of multiracial families do not get along well.

_____ 15. Transracial adoption results in "cultural genocide."

Scoring

After assigning a number to each item, add the numbers and divide by 15. The lower the score (1 is the lowest possible), the more positive one's view of transracial adoptions. The higher the score (7 is the highest possible), the more negative one's view of transracial adoptions. The norming sample was based upon thirty-four male and sixty-nine female students attending Valdosta State University. The average score was 2.27 (SD = 1.15), suggesting a generally positive view of transracial adoption by the respondents, and scores ranged from 1.00 to 6.60.

The average age of participants completing the scale was 22.22 years (SD = 4.23), and ages ranged from 18 to 48. The ethnic composition of the sample was 74.8% white, 20.4% black, 1.9% Asian, 1.0% Hispanic, 1.0% American Indian, and one person of nonindicated ethnicity. The classification of the sample was 15.5% freshmen, 6.8% sophomores, 32.0% juniors, 42.7% seniors, and 2.9% graduate students.

Source

"Attitudes Toward Transracial Adoption Scale," 2004 by Mark Whatley, Ph.D., Department of Psychology, Valdosta State University, Valdosta, Georgia 31698-0100. Used by permission. Other uses of this scale by written permission of Dr. Whatley only (mwhatley@valdosta.edu). Information on the reliability and validity of this scale is available from Dr. Whatley.

The counterargument is that healthy self-concepts, an appreciation for one's racial heritage, and coping with racism or prejudice can be learned in a variety of contexts. Legal restrictions on transracial adoptions have disappeared, and social approval for transracial adoptions is increasing. However, a substantial number of studies conclude that "same-race placements are preferable and that special measures should be taken to facilitate such placements, even if it means delaying some adoptions" (Kennedy 2003, 469).

Samuels (2010) noted the issues relevant to 25 adult multiracial adoptees, including: (1) claiming whiteness culturally but not racially, (2) learning to "be black"—peers as agents of enculturation, (3) bicultural kinship beyond black and white. One 26-year-old black female was asked how she felt about being reared by white parents and replied, "Again, they are my family and I love them, but I am black. I have to deal with my reality as a black woman" (Simon and Roorda 2000, 41). A black man reared in a white home advised white parents considering a transracial adoption, "Make sure they have the influence of blacks in their lives; even if they have to go out and make friends with black families—it's a must" (p. 25). However, Huh and Reid (2000) found that positive adjustment by adoptees was associated with participation in the cultural activities of the race of the parents who adopted them.

Open versus Closed Adoptions

Another controversy is whether adopted children should be allowed to obtain information about their biological parents. In general, there are considerable benefits for having an open adoption—the biological parent has the opportunity to stay involved in the child's life. Adoptees learn early that they are adopted and who their biological parents are. Birth parents are more likely to avoid regret and to be able to stay in contact with their child. Adoptive parents have information about the genetic background of their adopted child. Ge et al. (2008) studied birth mothers and adoptive parents and found that increased openness between the two sets of parents was positively associated with greater satisfaction for both birth mothers and adoptive parents. Goldberg et al. (2011b) studied lesbian, gay, and heterosexual couples who were involved in an open adoption. While there were some tensions with the birth parents over time, most of the 45 adoptive couples reported satisfying relationships.

Intercountry Adoptions

Zhang and Lee (2011) emphasized that the United States is one of the major "baby-receiving countries in the world" with a high demand for intercountry adoption. They interviewed American adoptive parents and noted a preference to adopt foreign-born children instead of adopting minority children domestically. In addition to the uneven domestic supply and demand of "desirable" children, the authors noted a perception that American children available for adoption presented difficult problems whereas foreign children presented interesting challenges. The study revealed why so many black children are left behind in foster care, which provides an insight into current race relations in the U.S.

Ethiopia has become a unique country from which to adopt a child. Not only is the adoption time shorter (four months) and less expensive, but the children there are also psychologically very healthy. "You don't hear crying babies [in the orphanages]. . . they are picked up immediately" (Gross and Conners 2007, A16). In addition, "adoption families are encouraged to meet birth families and visit the villages where the children are raised . . ." (ibid.). Ethiopian adoptions have received considerable visibility in the United States due to the involvement of celebrity Angelina Jolie, who adopted an Ethiopian child.

Internet Adoption

Roby and White (2010) noted that some couples use the Internet to adopt a baby. However, the researchers warned that such use can pose serious problems of potential fraud, exploitation, and, most important, lack of professional consideration of the child's best interest. Couples should proceed with great caution.

Children who are Adopted

Children who are adopted have an enormous advantage over those who are not adopted. Juffer et al. (2011) examined 270 research articles including more than 230,000 children to compare the physical growth, attachment, cognitive development, school achievement, self-esteem and behavioral problems of adopted and non-adopted children. Results revealed that adopted children out-performed their non-adopted peers who remained in institutions and they showed a dramatic recovery in practically all areas of development.

"Who are your real parents?", "Why did your mother give you up?", and "Are those your real parents?" are questions children who are adopted must sometimes cope with. W.I.S.E. UP is a tool provided to adopted children to help them cope with these intrusive uncomfortable questions (Singer 2010). W.I.S.E. is an acronym for: Walk away Ignore or change the subject, Share what you are comfortable sharing, and Educate about adoption in general. The tool emphasizes that adopted children are wiser about adoption than their peers and can educate them or remove themselves from the situation. Adopted children vary in the degree to which being adopted is a problem for them. While some define being adopted as being abandoned by a birth mother, others view their birth mother putting them up for adoption as an act of love and view themselves as "special" in that they were selected by their adoptive parents.

Foster Parenting

Some individuals seek the role of parent via foster parenting. A **foster parent**, also known as a family caregiver, is neither a biological nor an adoptive parent but is a person who takes care of and fosters a child taken into custody. A foster parent has made a contract with the state for the service, has judicial status, and is reimbursed by the state. Foster parents are screened for previous arrest records and child abuse and neglect. Foster parents are licensed by the state; some states require a "foster parent orientation" program. Rhode Island, for example, provides a twenty-seven-hour course. Brown (2008) asked sixty-three foster parents what they needed to allow them to have a successful foster parenting experience. They reported that they needed the right personality (for example, patience and nurturance), information about the foster child, a good relationship with the fostering agency, linkages to other foster families, and supportive immediate and extended families. Other research has found the need for formal foster parent organizations.

Children placed in foster care have typically been removed from parents who are abusive, who are substance abusers, and/or who are mentally incompetent. Although foster parents are paid for taking care of children in their home, they are also motivated by love of children. The goal of placing children in foster care is to remove them from a negative family context, improve that context, and return them, or find a more permanent home than foster care. Some couples become foster parents in hopes of being able to adopt a child that is placed in their custody. Meyer et al. (2010) studied the conditions under which parental rights of children in foster care are terminated and found that parents who were incarcerated and who had mental health problems were the most vulnerable.

Due to tighter restrictions on foreign adoptions (for example, it typically takes three years to complete a foreign adoption; China excludes people seeking to adopt who are unmarried, obese, or over age 50) and due to the limited number of domestic infants, more couples are considering adoption of a foster child. Tax credits are available for up to $11,650 for adopting a special needs child (Block 2008).

Foster parent neither a biological nor an adoptive parent but a person who takes care of and fosters a child taken into custody by social services.

Sterilization

National **Data**

Worldwide over 220 million couples have used sterilization as their method of birth control—nearly 43 million men and 180 million women. The lifetime chance of a pregnancy after sterilization is one in 200 (Beerthuizen 2010).

Sterilization is a permanent surgical procedure that prevents reproduction. Sterilization may be a contraceptive method of choice when the woman should not have more children for health reasons or when individuals are certain about their desire to have no more children or to remain childfree. Most couples complete their intended childbearing in their late twenties or early thirties, leaving more than fifteen years of continued risk of unwanted pregnancy. Because of the risk of pill use at older ages and the lower reliability of alternative birth control

Sterilization a permanent surgical procedure that prevents reproduction.

methods, sterilization has become the most popular method of contraception among married women who have completed their families.

Slightly more than half of all sterilizations are performed on women. Although male sterilization is easier and safer than female sterilization, women feel more certain they will not get pregnant if they are sterilized. "I'm the one that ends up being pregnant and having the baby," said one woman. "So I want to make sure that I never get pregnant again."

Female Sterilization

Oophorectomy form of female sterilization whereby the woman's ovaries are removed.

Hysterectomy form of female sterilization whereby the woman's uterus is removed.

Salpingectomy type of female sterilization whereby the fallopian tubes are cut and the ends are tied.

Although a woman may be sterilized by removal of her ovaries (**oophorectomy**) or uterus (**hysterectomy**), these operations are not normally undertaken for the sole purpose of sterilization because the ovaries produce important hormones (as well as eggs) and because both procedures carry the risks of major surgery. Sometimes, however, another medical problem requires hysterectomy.

The usual procedures of female sterilization are the salpingectomy and a variant of it, the laparoscopy. **Salpingectomy**, also known as tubal ligation or tying the tubes (see Figure 10.1), is often performed under a general anesthetic while the woman is in the hospital just after she has delivered a baby. An incision is made in the lower abdomen, just above the pubic line, and the fallopian tubes are brought into view one at a time. A part of each tube is cut out, and the ends are tied, clamped, or cauterized (burned). The operation takes about thirty minutes. Smith et al. (2010a) studied 3448 women (ages 16 to 64) to assess if having a tubal ligation was related the frequency of sexual problems and ratings of sexual satisfaction, relationship satisfaction, and sexual pleasure. Results revealed that having had a tubal ligation was not associated with any specific sexual problem, such as physical pain during sex or an inability to reach orgasm. In fact, sterilized women were more likely to experience extremely high levels of sexual satisfaction, relationship satisfaction, and sexual pleasure.

Laparoscopy a form of tubal ligation that involves a small incision through the woman's abdominal wall just below the navel.

A less expensive and quicker (about fifteen minutes) form of salpingectomy, which is performed on an outpatient basis, is the **laparoscopy**. Often using local anesthesia, the surgeon inserts a small, lighted viewing instrument (laparoscope) through the woman's abdominal wall just below the navel, through which the uterus and the fallopian tubes can be seen. The surgeon then makes another small incision in the lower abdomen and inserts a special pair of forceps that carry electricity to cauterize the tubes. The laparoscope and the forceps are then withdrawn, the small wounds are closed with a single stitch, and small bandages are placed over the closed incisions. (Laparoscopy is also known as "the Band-Aid operation.") As an alternative to reaching the fallopian tubes through an opening below the navel, the surgeon may make a small incision in the back of the vaginal barrel (vaginal tubal ligation).

Essure is a permanent sterilization procedure that requires no cutting and only a local anesthetic in a half-hour procedure that blocks the fallopian tubes. Women typically may return home within forty-five minutes (and to work the next day). These procedures for female sterilization are greater than 95% effective, but sometimes they have complications. In rare cases, a blood vessel in the abdomen is torn open during the sterilization and bleeds into the abdominal cavity. When this happens, another operation is necessary to tie the bleeding vessel closed. Occasionally, injury occurs to the small or large intestine, which may cause nausea, vomiting, and loss of appetite. The fact that death may result, if only rarely, is a reminder that female sterilization is surgery and, like all surgery, involves some risks. In addition, although some

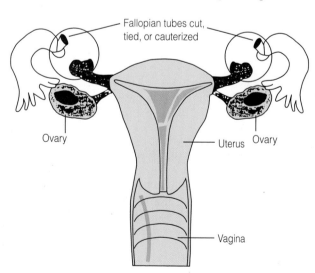

Fallopian tubes cut, tied, or cauterized

Ovary

Uterus   Ovary

Vagina

Figure 10.1
Female Sterilization: Tubal Sterilization.

Chapter 10 Planning for Children

female sterilizations may be reversed, a woman should become sterilized only if she does not want to have a biological child.

Beerthuizen (2010) compared female sterilization via the transcervical route with abdominal procedures and emphasized that the complication rate of the former is low and should be the preferred method provided the equipment and the experience required are available. The ten-year cumulative pregnancy rate of sterilization techniques ranges from 0.1 to 3.6 per 1000 procedures.

Male Sterilization

Vasectomies are the most frequent form of male sterilization. They are usually performed in the physician's office under a local anesthetic. Michielsen and Beerthuizen (2010) reviewed the data on record and note that it convincingly demonstrates that **vasectomy** is a safe and cost-effective intervention for permanent male contraception. They recommend the no-scalpel vasectomy under local anesthesia. Sperm are still produced in the testicles, but because there is no tube to the penis, they remain in the epididymis and eventually dissolve.

The procedure takes about fifteen minutes. The man can leave the physician's office within a short time. Because sperm do not disappear from the ejaculate immediately after a vasectomy (some remain in the vas deferens above the severed portion), a couple should use another method of contraception until the man has had about twenty ejaculations. In about 1% of the cases, the vas deferens grows back and the man becomes fertile again. A vasectomy does not affect the man's desire for sex, ability to have an erection or an orgasm, amount of ejaculate (sperm account for only a minute portion of the seminal fluid), health, or chance of prostate cancer. Although a vasectomy may be reversed (with a 30 to 60% success rate), a man should get a vasectomy only if he does not want to have a biological child.

Vasectomy form of male sterilization whereby the vas deferens is cut so that sperm cannot continue to travel outside the body via the penis.

Talking with a Partner about Contraception

Brown and Guthrie (2010) interviewed 24 women ages 16 to 20 who were due to have, or had recently had, a surgical abortion about their views on contraception. The most common reasons for not using contraception were alcohol, being 'in the moment' (i.e., being 'in the mood', not wishing to 'break the spell'), and pressure from young men not to use condoms. Lack of knowledge was rarely cited as a reason.

A good beginning is to be sober and to plan to have sex/use contraception. Having a conversation about birth control is a good way to begin sharing responsibility for it; one can learn of the partner's interest in participating in

WHAT IF?

What if the Condom Breaks—Do you Seek Emergency Contraception?

For couples using a condom for the first time, the condom will break about 7% of the time. Responses vary from nothing ("I'm not going to worry about it") to worrying but doing nothing ("I hope I don't get pregnant") to worrying and doing something—seeking **emergency contraception** (EC). "Better safe than sorry" requires immediate action because the sooner the EC pills are taken, the lower the risk of pregnancy—twelve hours is best, and seventy-two is the latest. The medication is available over the counter—no prescription is necessary (and no pregnancy test is required). Although the side effects (nausea, vomiting, and so on) may occur, they will be over in a couple of days and the risk of being pregnant is minimal.

Emergency contraception refers to various types of morning-after pills.

the choice and use of a contraceptive method. Men can also share responsibility by purchasing and using condoms, paying for medical visits and the pharmacy bill, reminding his partner to use the method, assisting with insertion of barrier methods, checking contraceptive supplies, and having a vasectomy if that is an appropriate option. However, in addition, women need to take steps to protect themselves from unwanted pregnancy and from exposure to STIs.

Abortion

An abortion may be either an **induced abortion**, which is the deliberate termination of a pregnancy through chemical or surgical means, or a **spontaneous abortion (miscarriage)**, which is the unintended termination of a pregnancy. Geller et al. (2010) emphasized how miscarriage is often a significant loss provoking both depression and anxiety and that health care professionals are often oblivious to its treatment.

In this text we will use the term *abortion* to refer to induced abortion. In general, abortion is legal in the United States but had been challenged under the Bush administration. Specifically, federal funding was withheld if an aid group offered abortion or abortion advice. However, the election of Obama restored approval from the administration for abortion. Obama said that denying such aid undermined "safe and effective voluntary family planning in developing countries."

Incidence of Abortion

From 1990 to 2005, the incidence of abortion declined nearly every year. Acceptability of having a child without a partner, increased use of contraception, and lack of access to abortion clinics were the suspected causes. More recently, the declining abortion trend has stalled and there have been slight increases in abortion. In 2008, an estimated 1.21 million abortions were performed in the United States. The **abortion rate** (the number of abortions per thousand women aged 15 to 44) increased 1% between 2005 and 2008, from 19.4 to 19.6 abortions per 1,000 women aged 15-44; the total number of abortion providers was virtually unchanged (Jones and Kooistra 2011).

Parental consent means that a woman needs permission from a parent to get an abortion if under a certain age, usually 18. **Parental notification** means that a woman has to tell a parent she is getting an abortion if she is under a certain age, usually 18, but she doesn't need parental permission. Laws vary by states. Call the National Abortion Federation Hotline at 1-800-772-9100 for the laws in your state.

The Self-Assessment provides a way for you to assess your abortion views.

Reasons for an Abortion

In a survey of 1,209 women who reported having had an abortion, the most frequently cited reasons were that having a child would interfere with a woman's education, work, or ability to care for dependents (74%); that she could not afford a baby now (73%); and that she did not want to be a single mother or was having relationship problems (48%). Nearly four in ten women said they had completed their childbearing, and almost one-third of the women were not ready to have a child. Fewer than 1% said their parents' or partner's desire

DIVERSITY IN OTHER COUNTRIES

There are wide variations in the range of cultural responses to the abortion issue. On one end of the continuum is the Kafir tribe in Central Asia, where an abortion is strictly the choice of the woman. In this preliterate society, there is no taboo or restriction with regard to abortion, and the woman is free to exercise her decision to terminate her pregnancy. One reason for the Kafirs' approval of abortion is that childbirth in the tribe is associated with high rates of maternal mortality. Because birthing children may threaten the life of significant numbers of adult women in the community, women may be encouraged to abort. Such encouragement is particularly strong in the case of women who are viewed as too young, old, sick, or small to bear children.

A tribe or society may also encourage abortion for a number of other reasons, including practicality, economics, lineage, and honor. Abortion is practical for women in migratory societies. Such women must control their pregnancies, because they are limited in the number of children they can nurse and transport. Economic motivations become apparent when resources are scarce—the number of children born to a group must be controlled. Abortion for reasons of lineage or honor involves encouragement of an abortion in those cases in which a woman becomes impregnated in an adulterous relationship. To protect the lineage and honor of her family, the woman may have an abortion.

This is not a test. There are no wrong or right answers to any of the statements, so just answer as honestly as you can. The statements ask your feelings about legal abortion (the voluntary removal of a human fetus from the mother during the first three months of pregnancy by a qualified medical person). Tell how you feel about each statement by giving only one response. Use the following scale for your answers:

Strongly Agree	Slightly Agree	Slightly Disagree	Strongly Agree	Disagree
5	4	3	2	1

_____ 1. The Supreme Court should strike down legal abortions in the United States.

_____ 2. Abortion is a good way of solving an unwanted pregnancy.

_____ 3. A mother should feel obligated to bear a child she has conceived.

_____ 4. Abortion is wrong no matter what the circumstances are.

_____ 5. A fetus is not a person until it can live outside its mother's body.

_____ 6. The decision to have an abortion should be the pregnant mother's.

_____ 7. Every conceived child has the right to be born.

_____ 8. A pregnant female not wanting to have a child should be encouraged to have an abortion.

_____ 9. Abortion should be considered killing a person.

_____ 10. People should not look down on those who choose to have abortions.

_____ 11. Abortion should be an available alternative for unmarried pregnant teenagers.

_____ 12. People should not have the power over the life or death of a fetus.

_____ 13. Unwanted children should not be brought into the world.

_____ 14. A fetus should be considered a person at the moment of conception.

Scoring and Interpretation

As its name indicates, this scale was developed to measure attitudes toward abortion. Sloan (1983) developed the scale for use with high school and college students. To compute your score, first reverse the point scale for items 1, 3, 4, 7, 9, 12, and 14. For example, if you selected a 5 for item one, this becomes a 0; if you selected a 1, this becomes a 4, etc. After reversing the scores on the seven items specified, add the numbers you circled for all the items. Sloan provided the following categories for interpreting the results:

70–56	Strong pro-abortion
54–44	Moderate pro-abortion
43–27	Unsure
26–16	Moderate pro-life
15–0	Strong pro-life

Reliability and Validity

The Abortion Attitude Scale was administered to high school and college students, Right to Life group members, and abortion service personnel. Sloan (1983) reported a high total test estimate of reliability (0.92). Construct validity was supported in that the mean score for Right to Life members was 16.2; the mean score for abortion service personnel was 55.6; and other groups' scores fell between these values.

Source

"Abortion Attitude Scale" by L. A Sloan. *Journal of Health Education* Vol. 14, No. 3, May/June 1983. *The Journal of Health Education* is a publication of the American Allegiance for Health, Physical Education, Recreation and Dance, 1900 Association Drive, Reston, VA 20191. Reprinted by permission.

for them to have an abortion was the most important reason (Finer et al. 2005). Falcon et al. (2010) confirmed that the use of drugs was related to unintended pregnancy and the request for an abortion. Clearly, some women get pregnant when high on alcohol or other substances, regret the pregnancy, and want to reverse it.

Abortions performed to protect the life or health of the woman are called **therapeutic abortions**. However, there is disagreement over this definition. Garrett et al. (2001) noted, "Some physicians argue that an abortion is therapeutic if it prevents or alleviates a serious physical or mental illness, or even if it alleviates temporary emotional upsets. In short, the health of the pregnant woman is given such a broad definition that a very large number of abortions can be classified as therapeutic" (p. 218).

Some women with multifetal pregnancies (a common outcome of the use of fertility drugs) may have a procedure called *transabdominal first-trimester selective termination.* In this procedure, the lives of some fetuses are terminated to increase the chance of survival for the others or to minimize the health risks

Induced abortion the deliberate termination of a pregnancy through chemical or surgical means.

Spontaneous abortion (miscarriage) the unintended termination of a pregnancy.

Abortion rate the number of abortions per thousand women aged 15 to 44.

Parental consent a woman needs permission from a parent to get an abortion if under a certain age, usually 18.

associated with multifetal pregnancy for the woman. For example, a woman carrying five fetuses may elect to abort three of them to minimize the health risks of the other two.

Pro-Life and Pro-Choice Abortion Positions

A dichotomy of attitudes toward abortion is reflected in two opposing groups of abortion activists. Individuals and groups who oppose abortion are commonly referred to as "pro-life" or "antiabortion."

Of 657 undergraduates in a random sample at a large southeastern university, 40% reported that they were pro-life in regard to their feelings about abortion (Bristol and Farmer 2005). Pro-life groups favor abortion policies or a complete ban on abortion. They essentially believe the following:

1. The unborn fetus has a right to live and that right should be protected.

2. Abortion is a violent and immoral solution to unintended pregnancy.

3. The life of an unborn fetus is sacred and should be protected, even at the cost of individual difficulties for the pregnant woman.

Individuals who are over the age of 44, male, mothers of three or more children, married to white-collar workers, affiliated with a religion, and Catholic are most likely to be pro-life (Begue 2001). Pro-life individuals emphasize the sanctity of human life and the moral obligation to protect it. The unborn fetus cannot protect itself so is literally dependent on others for life. Naomi Judd noted that if she had had an abortion she would have deprived the world of one of its greatest singers—Wynonna Judd.

In a large, nonrandom sample, 52% of 2,922 undergraduates reported that, "abortion is acceptable under certain conditions" (Knox and Hall 2010). Pro-choice advocates support the legal availability of abortion for all women. They essentially believe the following:

1. Freedom of choice is a central value—the woman has a right to determine what happens to her own body.

2. Those who must personally bear the burden of their moral choices ought to have the right to make these choices.

3. Procreation choices must be free of governmental control.

People most likely to be pro-choice are female, are mothers of one or two children, have some college education, are employed, and have annual income of more than $50,000. Although many self-proclaimed feminists and women's organizations, such as the National Organization for Women (NOW), have been active in promoting abortion rights, not all feminists are pro-choice.

Physical Effects of Abortion

Part of the debate over abortion is related to the presumed effects of abortion. In regard to the physical effects, legal abortions, performed under safe medical conditions, are safer than continuing the pregnancy. The earlier in the pregnancy the abortion is performed, the safer it is. Vacuum aspiration, a frequently used method in early pregnancy, does not increase the risks to future childbearing. However, late-term abortions do increase the risks of subsequent miscarriages, premature deliveries, and babies of low birth weight.

Post-abortion complications include the possibility of incomplete abortion, which occurs when the initial procedure misses the fetus and the procedure must be repeated. Other possible complications include uterine infection; excessive bleeding; perforation or laceration of the uterus, bowel, or adjacent organs; and an adverse reaction to a medication or anesthetic. After having an abortion, women are advised to expect bleeding (usually not heavy) for up to two weeks and to return to their health care provider thirty days after the abortion to check that all is well.

I've noticed that everyone who is for abortion is already born.

Ronald Reagan, former U.S. President

Parental notification a woman has to tell a parent she is getting an abortion if she is under a certain age, usually 18, but she doesn't need parental permission.

Therapeutic abortion abortions performed to protect the life or health of the woman.

Psychological Effects of Abortion

Of equal concern are the psychological effects of abortion. The American Psychological Association reviewed all outcome studies on the mental health effects of abortion and concluded, "Based on our comprehensive review and evaluation of the empirical literature published in peer-reviewed journals since 1989, this Task Force on Mental Health and Abortion concludes that the most methodologically sound research indicates that among women who have a single, legal, first-trimester abortion of an unplanned pregnancy for nontherapeutic reasons, the relative risks of mental health problems are no greater than the risks among women who deliver an unplanned pregnancy" (Major et al. 2008, 71). Steinberg and Russo (2008) also looked at national data and did not find a significant relationship between first pregnancy abortion and subsequent rates of generalized anxiety disorder, social anxiety, or post-traumatic stress disorder.

Knowledge and Support of Partners of Women who Have Abortion

Jones et al. (2011) examined data from 9,493 women who had obtained an abortion to find out the degree to which their male partners knew of the abortion and their feelings about the abortion. The overwhelming majority of women reported that the men with whom they got pregnant knew about the abortion, and most perceived these men to be supportive. Cohabiting men were particularly supportive. The researchers concluded that most women obtaining abortions are able to rely on male partners for social support.

In a previous study, Kero and Lalos (2004) conducted interviews with men four and twelve months after their partners had had an abortion. Overwhelmingly, the men (at both time periods) were happy with the decision of their partners to have an abortion. More than half accompanied their partner to the abortion clinic (which they found less than welcoming).

PERSONAL CHOICES

Should You Have an Abortion?

The decision to have an abortion continues to be a complex one. Women who are faced with the issue may benefit by considering the following guidelines:

1. *Consider all the alternatives available to you, realizing that no alternative will have only positive consequences and no negative consequences.* As you consider each alternative, think about both the short-term and the long-term outcomes of each course of action, what you want, and what you can live with.

2. *Obtain information about each alternative course of action.* Inform yourself about the medical, financial, and legal aspects of abortion, childbearing, parenting, and placing the baby up for adoption.

3. *Talk with trusted family members, friends, or unbiased counselors.* Consider talking with the man who participated in the pregnancy. If possible, also talk with women who have had abortions as well as with women who have kept and reared a baby or placed a baby for adoption. If you feel that someone is pressuring you in your decision making, look for help elsewhere.

4. *Consider your own personal and moral commitments in life.* Understand your own feelings, values, and beliefs concerning the fetus and weigh those against the circumstances surrounding your pregnancy.

The Future of Planning for Children

While having children will continue to be the option selected by most couples, fewer will select this option and being childfree will lose some of its stigma. Individualism and economics are the primary factors responsible for reducing the obsession to have children. To quote Laura Scott, "having children will change from an assumption to a decision." Once the personal, social, and economic consequences of having children comes under close scrutiny the automatic response to have children will be tempered.

Summary

What are the social influences and individual motivations for having children?

Having children continues to be a major goal of most individuals (women more than men). Social influences to have a child include family, friends, religion, government, favorable economic conditions, and cultural observances. The reasons people give for having children include love and companionship with one's own offspring, the desire to be personally fulfilled as an adult by having a child, and the desire to recapture one's youth. Having a child (particularly for women) reduces one's educational and career advancement. The cost for housing, food, transportation, clothing, health care, and child care for a child up to age 2 is over $10,000 annually.

What are the categories/motivations of individuals who decide to remain childfree?

About 20% of women aged 40 to 44 do not have children. Whether these women will remain childfree or eventually have children is unknown. Categories of those who do not have children include early articulators who knew early that they did not want children, postponers who kept delaying when they would have children and remained childfree, acquiescers who made the decision to remain childless because their partner did not want children, and undecided, who are childfree but still in the decision-making process.

The top five reasons individuals give for wanting to remain childfree are a high level of current life satisfaction, being free/independent, avoiding the responsibility of rearing a child, the absence of maternal/paternal instinct, and a desire to accomplish/experience things in life which would be difficult as a parent.

What are the causes of infertility and the technology available to help?

Infertility is defined as the inability to achieve a pregnancy after at least one year of regular sexual relations without birth control, or the inability to carry a pregnancy to a live birth. Forty percent of infertility problems are attributed to the woman, 40% to the man, and 20% to both of them. The causes of infertility in women include blocked fallopian tubes, endocrine imbalance that prevents ovulation, dysfunctional ovaries, chemically hostile cervical mucus that may kill sperm, and effects of STIs.

A number of technological innovations are available to assist women and couples in becoming pregnant. These include hormonal therapy, artificial insemination, ovum transfer, in vitro fertilization, gamete intrafallopian transfer, and zygote intrafallopian transfer. Being infertile (for the woman) may have a negative lifetime effect, both personal and interpersonal (half the women in one study were separated or reported a negative effect on their sex lives).

What are the motives for adoption and the demographics of those who adopt?

Motives for adoption include a couple's inability to have a biological child (infertility), their desire to give an otherwise unwanted child a permanent loving home, or their desire to avoid contributing to overpopulation by having more biological children. Adoption is actually quite rare, with less than 5% of couple adopting; 15% of these adoptions will be children from other countries.

Whereas demographic characteristics of those who typically adopt are white, educated, and of high income, adoptees are increasingly being placed in nontraditional families, including with older, gay, and single individuals; it is recognized that these individuals may also be white, educated, and of high income. Most college students are open to transracial adoption.

What are the motives and outcome for an abortion?

The most frequently cited reasons for induced abortion were that having a child would interfere with a woman's education, work, or ability to care for dependents (74%); that she could not afford a baby now (73%);

and that she did not want to be a single mother or was having relationship problems (48%). Less than 1% said their parents' or partner's desire for them to have an abortion was the most important reason. In regard to the psychological effects of abortion, the American Psychological Association reviewed the literature and concluded that "among women who have a single, legal, first-trimester abortion of an unplanned pregnancy for nontherapeutic reasons, the relative risks of mental health problems are no greater than the risks among women who deliver an unplanned pregnancy."

Key Terms

Abortion rate

Antinatalism

Birth control sabotage

Competitive birthing

Conception

Cryopreservation

Double Dutch

Emergency contraception

Foster parent

Hysterectomy

Induced abortion

Infertility

Laparoscopy

Oophorectomy

Ovum transfer

Parental consent

Parental notification

Pregnancy

Pregnancy coercion

Procreative liberty

Pronatalism

Salpingectomy

Spontaneous abortion (miscarriage)

Sterilization

Therapeutic abortion

Transracial adoption

Vasectomy

Web Links

Alan Guttmacher Institute
www.guttmacher.org/

Childfree by Choice
http://www.childfreebychoice.com/
http://www.childlessbychoiceproject.com/

College Cost
http://apps.collegeboard.com/fincalc/college_cost.jsp

Engenderhealth
http://www.engenderhealth.org/

The Evan B. Donaldson Adoption Institute
http://www.adoptioninstitute.org/

Fetal Fotos (bonding with your fetus)
http://www.fetalfotosusa.com/

Georgia Reproduction Specialists (male infertility)
http://www.ivf.com/male.html

National Right to Life
http://www.nrlc.org/

No Kidding!
http://www.nokidding.net/

MicroSort
http://www.microsort.net/

Planned Parenthood Federation of America, Inc.
http://www.plannedparenthood.org

NARAL Pro-Choice America (reproductive freedom and choice)
http://www.naral.org/

Selecting a Method of Contraception
http://www.mayoclinic.com/health/birth-control/BI99999

CHAPTER 11

Parenting

> *Nothing you do for children is ever wasted. They seem not to notice us, hovering, averting our eyes, and they seldom offer thanks, but what we do for them is never wasted.*
>
> Garrison Keillor, American author

David Knox

Learning Objectives

Discuss the various roles involved in parenting children.

Review the nature of parenting choices.

Identify how women and men make the transition to parenthood.

Summarize the facts of parenthood.

Review the various principles of effective parenting.

Know the unique challenges of single parents.

Predict the future of parenting.

1. Students who feel accepted by their parents report less casual sex and less hard drug use.

2. Close parent-child relationships are associated with adolescents/young adults living at home and not enrolling in 4-year colleges.

3. Fathers and mothers of today spend an equal amount of "emotion" work with their children.

4. Parents have similar relationships with all of their children and do not report being closer to one child over another or having less conflict with one child over another.

5. Mothers who adopt babies show increases in depression just as do mothers who have infants "naturally."

Answers: 1. T   2. T   3. F   4. F   5. T

❝I didn't have any talent to be a father" said the late Paul Newman of his relationship with his six children (three each by two wives) (Levy 2009, p. 256). "The process of really connecting is very long and painful for me...I sometimes have a hard time talking because I have a hard time talking to anybody" (p. 256). **Parenting** was also difficult for his Academy Award winning wife Joanne Woodward. "I don't like children....I like my own children; I occasionally like other people's children. But I don't like babies per se....My career suffered because of children and my children suffered because of my career" (p. 257). While other parents find the meaning of life in having and rearing children, enjoy it and are good at it, the majority of parents fall somewhere in between.

Whether or not one likes children, there are guidelines for effective parenting. In this chapter we review them with the goal of facilitating happy, economically independent, socially contributing adult offspring. We begin by looking at the various roles of parenting in a photo essay on page 310.

Parenting defined in terms of roles including caregiver, emotional resource, teacher, and economic resource.

The Choices Perspective of Parenting

Although both genetic and environmental factors are at work, the choices parents make have a dramatic impact on their children. In this section, we review the nature of parental choices and some of the basic choices parents make.

Nature of Parenting Choices

Parents might keep the following points in mind when they make choices about how to rear their children:

1. *Not to make a parental decision is to make a decision.* Parents are constantly making choices even when they think they are not doing so. When a child is impolite and the parent does not provide feedback and encourage polite behavior, the parent has chosen to teach the child that being impolite is acceptable. When a child makes a promise ("I'll call you when I get to my friend's house") and does not do as promised, the parent has chosen to allow the child to not take commitments seriously. Hence, parents cannot choose not to make choices in their parenting, because their inactivity is a choice that has as much impact as a deliberate decision to reinforce politeness and responsibility.

2. *All parental choices involve trade-offs.* Parents are also continually making trade-offs in the parenting choices they make. The decision to take on a second job or to work overtime to afford the larger house will come at the price of

Although finding one definition of **parenting** is difficult, there is general agreement about the various roles parents play in the lives of their children. New parents assume at least seven roles:

CAREGIVER

A major role of parents is the physical care of their children. From the moment of birth, when infants draw their first breath, parents stand ready to provide nourishment (milk), cleanliness (diapers), and temperature control (warm blanket). The need for such sustained care continues and becomes an accepted and anticipated role of parents. Parents who excuse themselves early from a party because they "need to check on the baby" are alerting the hostess of their commitment to the role of caregiver.

Chelsea E. Curry

Chelsea E. Curry

EMOTIONAL RESOURCE

Beyond providing physical care, parents are sensitive to the emotional needs of children in terms of their need to belong, to be loved, and to develop positive self-concepts. In hugging, holding, and kissing an infant, parents not only express their love for the infant but also reflect awareness that such displays of emotion are good for the child's sense of self-worth. Willer and Soliz (2010) found that needing positive feedback/approval for the maintenance of one's self esteem is true not just for children, but continues into adulthood.

Parents also provide "emotion work" for children—listening to their issues, helping them figure out various relationships they are struggling with, etc. Minnottea et al. (2010) studied the emotion work of parents with their children in a sample

of 96 couples and found that women did more of it. Indeed, the greater the number of labor hours by men, the fewer the number of emotion work hours for their children and the higher the number of hours for women.

TEACHER

All parents think they have a philosophy of life or set of principles their children will benefit from. Parents soon discover that their children may not be interested in their religion or philosophy and, indeed, may rebel against it. This possibility does not deter parents from their role as teacher. Most parents also hope to have a direct effect on their children going to college. López Turley et al. (2010) discovered an unanticipated connection between positive parent-child relations and academic achievement. In a study of over ten thousand adolescents, although positive parent-child relations were associated with better academic achievement in high school, they were also associated with an increased desire to live at home during college, which in turn decreased the chance that the student would enroll in a 4-year college.

An array of self-help parenting books provide parents with ideas about the essentials children need that parents must teach. Galinsky (2010) identified seven skills all parents should be responsible for teaching their children: focus and self control, seeing some else's point of view, communicating, making connections, critical thinking, taking on challenges, and self-directed engaged learning.

Chelsea E. Curry

Chelsea E. Curry

Parents may also teach their children unconsciously. Cui et al. (2008) noted that parents (whether together or divorced) who reported high conflict in their marriage tended to have young adults who also reported high conflict and low quality in their own romantic relationships.

ECONOMIC RESOURCE
New parents are also acutely aware of the costs for medical care, food, and clothes for infants, and seek ways to ensure that such resources are available to their children. Working longer hours, taking second jobs, and cutting back on leisure expenditures are attempts to ensure that money is available to meet the needs of the children.

Sometimes the pursuit of money for the family has a negative consequence for children. Rapoport and Le Bourdais (2008) investigated the effects of parents' working schedules on the time they devoted to their children and confirmed that the more parents worked, the less time they spent with their children. In view of extensive work schedules, parents are under pressure to spend "quality time" with their children, and it is implied that putting children in day care robs children of this time. However, Booth et al. (2002) compared children in day care with those in home care in terms of time the mother and child spent together per week. Although the mothers of children in day care spent less time with their children than the mothers who cared for their children at home, the researchers concluded that the "groups did not differ in the quality of mother-infant interaction" and that the difference in the "quality of the mother-infant interaction may be smaller than anticipated" (p. 16).

PROTECTOR
Parents also feel the need to protect their children from harm. This role may begin in pregnancy. Castrucci et al. (2006) interviewed 1,451 women about their smoking behavior during their pregnancy. Although 89% reduced their smoking during pregnancy, 24.9% stopped smoking during pregnancy.

Chelsea E. Curry

Other expressions of the protective role include insisting that children wear seat belts in cars and jackets in cold weather, protecting them from violence or nudity in the media, and protecting them from strangers. Taubman-Ben-Ari and Noy (2011) found that while parents of young children may initially become safer drivers, over time, most revert back to their previous driving behaviors prior to becoming a parent.

Diamond et al. (2006) studied forty middle-class mothers of young children and identified fifteen strategies they used to protect their children. Their three principal strategies were to educate, control, and remove risk. The strategy used depended on the age and temperament of the child. For example, some parents felt protecting their children from certain television content was important. TV monitoring ranges from families that do not allow a television in their home, to monitoring everything their children watch on television.

Some parents feel that protecting their children from harm implies appropriate discipline for inappropriate behavior. Galambos et al. (2003) noted "parents' firm behavioral control seemed to halt the upward trajectory in externalizing problems among adolescents with deviant peers." In other words, parents who intervened when they saw a negative context developing were able to help their children avoid negative peer influences. Kolko et al. (2008) compared clinically referred boys and girls (ages 6 to 11) diagnosed with **oppositional defiant disorder** (children do not comply with requests of authority figures) to a matched sample of healthy control children and found that the former had greater exposure to delinquent peers. Hence, parents who monitor their children's peer relationships minimize children's exposure to delinquent models, making it a wise time investment.

Increasingly, parents are joining the technological age and learning how to text message. In their role as protector, this allows parents to text-message their child to tell them to come home, phone home, or to work out a logistical problem—"meet me at the food court in the mall." Children can also use text-messaging to their parents to let them know that they arrived safely at a destination, when they need to be picked up, or when they will be home.

Protection also becomes relevant to teens who may be asked to sign a parent-teen driving contract which specifies details about use of the family car—no texting, drinking, or drugs while driving; keep a quarter of a tank of gas in the car; meet curfew; etc. Such contracts may also include the child paying for car insurance with penalties for late payments (Copeland 2010).

HEALTH PROMOTER

The family is a major agent for health promotion. Children learn from the family context about healthy food. Indeed, one-third of U.S. children are overweight. A major cause of overweight children is parents who do not teach healthy food choices, let their children watch TV all day, and are bad models (eat junk food and don't exercise). Health promotion also involves sunburn protection, responsible use of alcohol, and safe driving skills.

Chelsea E. Curry

RITUAL BEARER

To build a sense of family cohesiveness, parents often foster rituals to bind members together in emotion and in memory. Prayer at meals and before bedtime, birthday and holiday celebrations, and vacationing at the same place (beach, mountains, and so on) provide predictable times of togetherness and sharing.

David Knox

having less time to spend with one's children and being more exhausted when such time is available. The choice to enroll one's child in the highest-quality day care (which may also be the most expensive) will mean less money for family vacations. The choice to have an additional child will provide siblings for the existing children but will mean less time and fewer resources for those children. Parents should increase their awareness that no choice is without a tradeoff and should evaluate the costs and benefits in making such decisions.

Parents also recognize that there is a balance they must strike between helping their children and impeding their own growth and development. One example is making decisions on how long to provide free room and board for an adult child. See the following Personal Choices section, which details this issue.

PERSONAL CHOICES

Should Parents Provide Housing for Their Adult Children?

In the respective roles of Tripp and Paula, Matthew McConaughey and Sarah Jessica Parker are the stars of the movie *Failure to Launch,* a romantic comedy. Tripp is 35 years old and still living with his parents. Paula has been hired by Tripp's parents to be his girlfriend and get him to move out of the house. The film reflected reality in that about 20% of males and 12% of females 25 years old live with their parents (Seiffge-Krenke 2010). About 40% of those adults who have lived on their own for at least four months will return to the parental home to live at least four months (Sassler et al. 2008). Fifteen percent of a sample of 2,922 undergraduates said that they could see themselves living with their parents after college; 57% said that the primary reason for doing so would be economic (Knox and Hall 2010).

Oppositional defiant disorder disorder in which children fail to comply with requests of authority figures.

Although some offspring continue to live with their parents during and after college, others have moved away but move back home because of a return to school, loss of a job, or a divorce. In the case of the latter, they may bring young children back into the parental home with them. As noted, saving money is the primary reason adult children reside with parents.

Parents vary widely in how they view, adapt to, and carry out adult children living with them. Some parents prefer that their children live with them, enjoy their company, and hope the arrangement continues indefinitely (the norm in Italy). They have no rules about their children living with them and expect nothing from them. The adult children can come and go as they please, pay for nothing, and have no responsibilities or chores. Mothers typically report more stress when their children leave than fathers (Seiffge-Krenke 2010).

Other parents develop what is essentially a rental agreement, whereby their children are expected to pay rent, cook and clean the dishes, mow the lawn, and service the cars. No overnight guests are allowed, and a time limit is specified as to when the adult child is expected to move out.

A central issue from the point of view of young adults who live with their parents is whether their parents perceive them as adults or children. This perception has implications of whether they are free to come and go as well as to make their own decisions (Sassler et al. 2008). Males are generally left alone whereas females are often under more scrutiny.

Adult children also vary in terms of how they view the arrangement. Some enjoy living with their parents, volunteer to pay rent (though most do not), take care of their own laundry, and participate in cooking and housekeeping. Others are depressed that they are economically forced to live with their parents, embarrassed that they do so, pay nothing, and do nothing to contribute to the maintenance of the household. In a study of thirty young adults who had returned to the parental home, two-thirds paid nothing to live there and wanted to keep it that way. Those who did pay paid considerably less than what rent would cost on the open market. Most paid for their cell phones, long-distance charges, and personal effects like clothing and toiletries (Sassler et al. 2008).

Whether parents and adult children discuss the issues of their living together will depend on the respective parents and adult children. Although there is no best way, clarifying expectations might prevent some misunderstandings. For example, what is the norm about bringing new pets into the home? Take, for example, a divorced son who moved back in with his parents along with his 6-year-old son *and* dog. His parents enjoyed being with them but were annoyed that the dog chewed on the furniture. Parental feelings eventually erupted that dismayed their adult child. He moved out, and the relationship with his parents became very strained. However, about three-fourths of the respondents in the Sassler et al. (2008) study reported generally satisfactory feelings about returning to their parental home. Most looked forward to moving out again but, in the short run, were content to stay to save money.

Sources

Knox, D., and S. Hall. 2010. Relationship and sexual behaviors of a sample of 2,922 university students. Unpublished data collected for this text. Department of Sociology, East Carolina University, and Department of Family and Consumer Sciences, Ball State University.

Sassler, S., D. Ciambrone, and G. Benway. 2008. Are they really mamma's boys/daddy's girls? The negotiation of adulthood upon returning home to the parental home. *Sociological Forum* 23:670–698.

Seiffge-Krenke, I. 2010. Predicting the timing of leaving home and related developmental tasks: Parents' and children's perspectives. *Journal of Social and Personal Relationships* 27: 495–518.

The important thing is not that every child should be taught, but that every child should be given the wish to learn.

John Lubbock, biologist and politician

3. *Reframe "regretful" parental decisions.* All parents regret a previous parental decision (for example, they should have held their child back a year in school or not done so; they should have intervened in a bad peer relationship; they should have handled their child's drug use differently). Whatever the issue, parents chide themselves for their mistakes. Rather than berate themselves as parents, they might emphasize the positive outcome of their choices: not holding the child back made the child the "first" to experience some things among his or her peers; they made the best decision they could at the time; and so on. Children might also be encouraged to view their own decisions positively.

4. *Parental choices are influenced by society and culture.* Parents are continually assaulted by commercial interests to get them to buy products for their children. Corporations regularly market to young parents to get them to buy the latest "learning aid" for their child which promises a genius by age 5. See Applying Social Research feature "Buy Buy Baby" on page 316.

Six Basic Parenting Choices

The six basic choices parents make include deciding (1) whether or not to have a child, (2) the number of children to have, (3) the interval between children, (4) one's method of discipline and guidance, (5) the degree to which one will be invested in the role of parent and (6) whether or not to co-parent. Morrill and Morrill (2010) emphasized that co-parenting has benefits to both the children and the quality of the relationship between the spouses.

> *To picture this, consider a family with both parents and their two children together in the living room. The wife wants to help one child with her homework, but the husband does not wish to give the other child a bath, and they argue. This ineffective co-parenting decreases the quality of each child's experience with his or her parent; the father may angrily leave the room feeling guilty, the mother may make a spiteful comment to the daughter about her dad. At the same time, neither parent is feeling positive about or intimately connected to their spouse, an emotional state that will likely persist through the evening. On the other hand, if these parents had started off collaborating and supporting each other throughout the evening with their children, it is easy to imagine a much different experience with each child, as well as greater feelings of satisfaction in their marital relationship (p, 69).*

Though all of the above decisions are important, the relative importance one places on parenting as opposed to one's career will have implications for the parents, their children, and their children's children.

Transition to Parenthood

The **transition to parenthood** refers to that period from the beginning of pregnancy through the first few months after the birth of a baby. The mother, father, and couple all undergo changes and adaptations during this period.

Transition to parenthood
period from the beginning of pregnancy through the first few months after the birth of a baby during which the mother and father undergo changes.

Vicki Oliver

This grown man has turned into a playful father in reference to his son. Children are the cause of a major role transition from spouses to parents.

Janis Kennedy

Janis Kennedy

The transition to motherhood begins with pregnancy…and so does the increase in appetite.

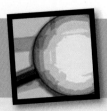
Susan G. Thomas is an investigative journalist (and mother of two toddler daughters when she wrote her book) who examined how corporations use and fund child development research to sell directly to infants and toddlers. She published her findings in *Buy, Buy Baby: How Consumer Culture Manipulates Parents and Harms Young Minds*. The exposure of infants from 6 to 23 months old to television (including videos and DVDs) is extensive: 61% watch television. By age 3, 88% of children watch just under two hours of TV a day. Parents are led to believe by "experts" (sometimes anxious to have their research funded by toy manufacturers who may turn them into celebrities) that a toy is of enormous educational value. Such value is expressed in terms of improving the cognitive development or performance of the young mind. In fact, "there is little evidence that any of these products was any more stimulating than

shaking a rattle, playing with blocks, mucking around in the backyard or just hanging out, or playing with a beloved caregiver or parent" (2007, 18).

However, what advertisers found worked for baby boomer mothers (mothers of babies born between 1946 and 1964) would not work for Generation X mothers. The former were socialized to "fast-track" their kids; the latter were more focused on always being there emotionally for their children. Thus, "mapping the mind of the Generation X mom" became a focus of corporate America to get her to spend money on her infants. For example, Generation X moms would not hear of sticking their infant alone in a room with a gadget, toy, or video because this would be considered abandonment. Therefore, products were disguised as beneficial to the child because they taught the child to "self-soothe." As an example, Fisher-Price marketed the Slumbertime Soother

Transition to Motherhood

Having a baby changes a woman's life forever. Whatever her previous focus, it is now the well being of her infant. One woman expressed: "Our focus shifts from our husbands to our children, who then take up all our energy. We become angry and lonely and burdened... and that's if we LIKE it." The Self-Assessment feature on page 318 examines one's view of traditional motherhood. Although childbirth is sometimes thought of as a painful ordeal, some women describe the experience as fantastic, joyful, and unsurpassed. A strong emotional bond between the mother and her baby usually develops early, and both the mother and infant resist separation.

Sociobiologists suggest that the attachment between a mother and her offspring has a biological basis (one of survival). The mother alone carries the fetus in her body for nine months, lactates to provide milk, and produces **oxytocin**, a hormone from the pituitary gland during the expulsive stage of labor that has been associated with the onset of maternal behavior in lower animals.

Not all mothers feel joyous after childbirth. Emotional bonding may be temporarily impeded by feeling overworked, exhaustion, mild depression, irritability, crying, loss of appetite, and difficulty in sleeping. Many new mothers experience **baby blues**, transitory symptoms of depression twenty-four to forty-eight hours after the baby is born. A few, about 10%, experience postpartum depression, a more severe reaction than baby blues.

Postpartum depression may occur in reference to a very complicated delivery (Filippi et al. 2010) as well as the numerous physiological and psychological changes occurring during pregnancy, labor, and delivery. Although the woman may become depressed during pregnancy or in the hospital, she more often experiences these feelings within the first month after returning home with her baby (sometimes the woman does not experience postpartum depression until a couple of years later). Chich-Hsiu et al. (2011) studied 859 women in southern Taiwan who recently gave birth and identified the factors predictive of postpartum stress. These factors included women with minor psychiatric morbidity, already having one or two children, formula feeding for their infants, preference

Oxytocin hormone from the pituitary gland during the expulsive stage of labor that has been associated with the onset of maternal behavior in lower animals.

Baby blues transitory symptoms of depression in a mother twenty-four to forty-eight hours after her baby is born.

Postpartum depression a more severe reaction following the birth of a baby which occurs in reference to a complicated delivery as well as numerous physiological and psychological changes occurring during pregnancy, labor, and delivery; usually in the first month after birth but can be experienced after a couple of years have passed.

with the package showing the mother "standing in the doorway holding the remote control, while her baby gazed contentedly at the mobile apparently feeling the mother's presence but not interrupted by it" (Thomas 2007, 69).

Thomas concludes that Generation X parents should recognize that keeping children continually stimulated through structured activities, DVDs, or gadgets "actually starves children's imagination, curiosity, and ability to relax. . . . All the childhood experts I spoke with said that spending time hanging out together is the best possible thing parents can do for their young children's development" (2007, 228). Indeed, she recommends that for Generation X parents to "fight for the right to do nothing with their children may be the perfect cause to celebrate" (p. 229).

More ominous is *BabyFirstTV*, available via satellite to infants as young as 6 months, which is seeking cable companies to carry its programming. The American Academy of Pediatrics recommends no television for children under 2 years. The Campaign for a Commercial-Free Childhood (CCFC) cites no evidence that television is beneficial to the cognitive development of children and has filed suit against *BabyFirstTV* for making false and deceptive marketing claims that its programs are educational for babies and toddlers.

Sources

S. G. Thomas. 2007 *Buy, Buy Baby: How Consumer Culture Manipulates Parents and Harms Young Minds.* Boston: Houghton Mifflin. http://www.commercialfreechildhood.org/pressreleases/bftcable.pdf

for an infant boy, and a low level of social support. To minimize baby blues and postpartum depression, antidepressants such as Zoloft and Prozac have been used. Most women recover within a short time; some (between 5% and 10%) become suicidal (Pinheiro et al. 2008).

Fathers too may also experience depression following the birth of a baby. Quing et. al. (2011) examined mothers' and fathers' postnatal (378 pairs) reactions and found that some parents of both genders reported postnatal depression (15% for mothers and 13% for fathers). The preference for a male baby was associated with fathers' depression.

Postpartum psychosis, a reaction in which a woman wants to harm her baby, is experienced by only one or two women per 1,000 births (British Columbia Reproductive Mental Health Program 2005). One must recognize that having misgivings about a new infant is normal. In addition, a woman who has negative feelings about her new role as mother should elicit help with the baby from her family or other support network so that she can continue to keep up her social contacts with friends and to spend time by herself and with her partner. Regardless of the cause, a team of researchers noted that maternal depression is associated with subsequent antisocial behavior in the child (Kim-Cohen et al. 2005).

To what degree is getting depressed after having an infant related to whether the mother goes back to work soon after the baby is born, whether the child is adopted, or the child is a product of assisted reproductive technology? In regard to work, Marshall and Tracy (2008) studied 700 working mothers of infants and found that working mothers in poorer quality jobs, as well as working mothers who were single or whose infant's health was poorer than that of other infants, reported greater depressive symptomatology. In addition, working a greater number of hours increased work/family conflict.

In regard to whether the infant was adopted, Payne et al. (2010) studied 112 adoptive mothers of infants under 12 months of age and found that almost 28% and 26% reported depressive symptoms at 0–4 and 5–12 weeks, respectively. At 13–52 weeks the rates of depression had dropped to 13%. Hence, significant

Parenthood remains the greatest single preserve of the amateur.

Alvin Toffler, author

Postpartum psychosis
a reaction in which a woman wants to harm her baby.

The purpose of this survey is to assess the degree to which students possess a traditional view of motherhood. Read each item carefully and consider what you believe. There are no right or wrong answers, so please give your honest reaction and opinion. After reading each statement, select the number that best reflects your level of agreement, using the following scale:

1	2	3	4	5	6	7
Strongly Disagree						Strongly Agree

_____ 1. A mother has a better relationship with her children than a father does.

_____ 2. A mother knows more about her child than a father, thereby being the better parent.

_____ 3. Motherhood is what brings women to their fullest potential.

_____ 4. A good mother should stay at home with her children for the first year.

_____ 5. Mothers should stay at home with the children.

_____ 6. Motherhood brings much joy and contentment to a woman.

_____ 7. A mother is needed in a child's life for nurturance and growth.

_____ 8. Motherhood is an essential part of a female's life.

_____ 9. I feel that all women should experience motherhood in some way.

_____ 10. Mothers are more nurturing than fathers.

_____ 11. Mothers have a stronger emotional bond with their children than do fathers.

_____ 12. Mothers are more sympathetic to children who have hurt themselves than are fathers.

_____ 13. Mothers spend more time with their children than do fathers.

_____ 14. Mothers are more lenient toward their children than are fathers.

_____ 15. Mothers are more affectionate toward their children than are fathers.

_____ 16. The presence of the mother is vital to the child during the formative years.

_____ 17. Mothers play a larger role than fathers in raising children.

_____ 18. Women instinctively know what a baby needs.

Scoring

After assigning a number from 1 (strongly disagree) to 7 (strongly agree), add the numbers and divide by 18. The higher your score (7 is the highest possible score), the stronger the traditional view of motherhood. The lower your score (1 is the lowest possible score), the less traditional the view of motherhood.

Norms

The norming sample of this self-assessment was based upon 20 male and 86 female students attending Valdosta State University. The average age of participants completing the scale was 21.72 years (SD = 2.98), and ages ranged from 18 to 34. The ethnic composition of the sample was 80.2% white, 15.1% black, 1.9% Asian, 0.9% American Indian, and 1.9% other. The classification of the sample was 16.0% freshmen, 15.1% sophomores, 27.4% juniors, 39.6% seniors, and 1.9% graduate students.

Participants responded to each of the eighteen items according to the 7-point scale. The most traditional score was 6.33; the score reflecting the least support for traditional motherhood was 1.78. The midpoint (average score) between the top and bottom score was 4.28 (SD = 1.04); thus, people scoring above this number tended to have a more traditional view of motherhood and people scoring below this number have a less traditional view of motherhood.

There was a significant difference (p .05) between female participants' scores (mean = 4.19; SD = 1.08) and male participants' scores (mean = 4.68; SD = 0.73), suggesting that males had more traditional views of motherhood than females.

Source

"Attitudes Toward Motherhood Scale," 2004 by Mark Whatley, Ph.D., Department of Psychology, Valdosta State University, Valdosta, Georgia 31698-0100. Used by permission. Other uses of this scale by written permission of Dr. Whatley only (mwhatley@valdosta.edu). Information on the reliability and validity of this scale is available from Dr. Whatley.

depressive symptoms are not uncommon in the early months whether the woman has the baby "naturally" or by adoption.

As for assisted reproductive technology (ART), Hammarberg et al. (2008a) assessed the different parenting experiences of those who had difficulty getting pregnant or who used ART compared to those who did not use ART. The researchers concluded that, although the evidence is inconclusive, those couples who become pregnant via ART may possibly idealize parenthood and this might then hinder adjustment and the development of a confident parental identity.

Finally, is transition to motherhood similar for lesbian and heterosexual mothers? Not according to Cornelius-Cozzi (2002), who interviewed lesbian mothers and found that the egalitarian norm of the lesbian relationship had been altered; for example, the biological mother became the primary caregiver, and the co-parent, who often heard the biological mother refer to the child as "her child," suffered a lack of validation.

While most mothers do *not* get depressed, they often pride themselves on "not picking favorites." Pillemer et al. (2010) studied mothers in their mid sixties to mid seventies (and their children) to identify differentiation patterns on the part of the mothers—the degree to which they favored one child over another. Seventy percent named a child to whom they felt closest, 79% named a child as the most likely caregiver, and 73% specified a child with whom she had the most arguments and disagreements.

Transition to Fatherhood

Brizendine (2010) found that testosterone levels in fathers drop about three weeks before a baby is born, which may be functional for keeping dad at home rather than out looking for a new sex partner. Schindler (2010) studied fathers in two-parent families and found that their engagement in parenting and financial contributions to the family predicted improvements in fathers' psychological well-being. Hence, fathers benefit from active involvement with their children. Schoppe-Sullivan et al. (2008) emphasized that mothers are the "gatekeepers" of the father's involvement with his children. A father may be involved or not involved with his children to the degree that a mother encourages or discourages a father's involvement. The **gatekeeper role** is particularly pronounced in divorce where the mother ends up with custody of the children (the role of father may be severely limited). When a mother is not present, the role of the father may be enormous. Such was the case of Joe Biden's involvement with his three children. His first wife and young child were killed in an automobile accident that resulted in him becoming the sole parent until his remarriage. One of Biden's sons noted that, "Our dad had his job in Washington but his family in his heart."

The Self-Assessment on traditional fatherhood examines one's view of traditional fatherhood. The importance of the father in the lives of his children is enormous and goes beyond his economic contribution (Bronte-Tinkew et al. 2008; Flouri and Buchanan 2003; Knox with Leggett, 2000). Children from

Gatekeeper role term used to refer to the influence of the mother on the father's involvement with his children.

This father says his life began with the birth of his daughter.

David Knox

The purpose of this survey is to assess the degree to which students have a traditional view of fatherhood. Read each item carefully and consider what you believe. There are no right or wrong answers, so please give your honest reaction and opinion. After reading each statement, select the number that best reflects your level of agreement, using the following scale:

1	2	3	4	5	6	7
Strongly Disagree						Strongly Agree

_____ 1. Fathers do not spend much time with their children.

_____ 2. Fathers should be the disciplinarians in the family.

_____ 3. Fathers should never stay at home with the children while the mother works.

_____ 4. The father's main contribution to his family is giving financially.

_____ 5. Fathers are less nurturing than mothers.

_____ 6. Fathers expect more from children than their mothers do.

_____ 7. Most men make horrible fathers.

_____ 8. Fathers punish children more than mothers do.

_____ 9. Fathers do not take a highly active role in their children's lives.

_____ 10. Fathers are very controlling.

Scoring

After assigning a number from 1 (strongly disagree) to 7 (strongly agree), add the numbers and divide by 10. The higher your score (7 is the highest possible score), the stronger the traditional view of fatherhood. The lower your score (1 is the lowest possible score), the less traditional the view of fatherhood.

Norms

The norming sample was based upon 24 male and 69 female students attending Valdosta State University. The average age of participants completing the Traditional Fatherhood Scale was 22.15 years (SD = 4.23), and ages ranged from 18 to 47. The ethnic composition of the sample was 77.4% white, 19.4% black, 1.1% Hispanic, and 2.2% other. The classification of the sample was 16.1% freshmen, 11.8% sophomores, 23.7% juniors, 46.2% seniors, and 2.2% graduate students.

Participants responded to each of the ten items on the 7-point scale. The most traditional score was 5.50; the score representing the least support for traditional fatherhood was 1.00. The average score was 3.33 (SD = 1.03), suggesting a less-than-traditional view.

There was a significant difference (p .05) between female participants' attitudes (mean-3.20; SD = 1.01) and male participants' attitudes toward fatherhood (mean-3.69; SD = 1.01), suggesting that males had more traditional views of fatherhood than females. There were no significant differences between ethnicities.

Source

"Traditional Fatherhood Scale," 2004 by Mark Whatley, Ph.D., Department of Psychology, Valdosta State University, Valdosta, Georgia 31698-0100. Used by permission. Other uses of this scale by written permission of Dr. Whatley only (mwhatley@valdosta.edu). Information on the reliability and validity of this scale is available from Dr. Whatley.

---

homes in which their fathers maintained an active involvement in their lives tend to:

Make good grades	Have higher incomes as adults
Be less involved in crime	Have higher education levels
Have good health/self-concept	Have higher cognitive functioning
Have a strong work ethic	Have stable jobs
Have durable marriages	Have fewer premarital births
Have a strong moral conscience	Have lower incidences of child sex abuse
Have higher life satisfaction	Exhibit fewer anorectic symptoms

Gavin et al. (2002) noted that parental involvement was predicted most strongly by the quality of the parents' romantic relationship. If the father was emotionally and physically involved with the mother, he was more likely to take an active role in the child's life. Fathers whose wives worked more hours than the fathers worked also reported more involvement with their children (McBride et al. 2002). Eggebeen et. al. (2010) found that a transforming effect of fatherhood was to increase the male's altruism and involvement in service organizations, particularly if he became a father in middle age.

Thomas et al. (2008) noted the inadequacy of the stereotype of the "uninvolved" African American father due to data that they often do not live in the household with the mother. The researchers sought to redefine father presence in the context of feelings of closeness to the father as well as frequency of father visitation. Their findings confirmed that a considerable portion of African American nonresident fathers visit their children on a daily or weekly basis. In addition, African American adult children with nonresident fathers often feel significantly closer to their fathers than do their white peers. Finally, African American adult children were more likely than their white peers to believe that their mothers supported their relationship with their father and to have positive perceptions of their parents' relationship.

Transition from a Couple to a Family

Gameiro et al. (2010) noted that following the birth of their first child, parents increase contact with family members and diminish contact with friends. While the couple may spend more time at home, their marital satisfaction is likely to decline rather than increase. A decrease in relationship quality across the first year of parenthood occurs regardless of whether the child is a biological birth or adopted and regardless of whether the parents are heterosexual or homosexual. Women experience the steeper decline (Goldberg et al. 2010). Twenge et al. (2003) reviewed 148 samples representing 47,692 individuals in regard to the effect children have on marital satisfaction. They found that (1) parents (both women and men) reported lower marital satisfaction than nonparents; (2) mothers of infants reported the most significant drop in marital satisfaction; (3) the higher the number of children, the lower the marital satisfaction; and (4) the factors in depressed marital satisfaction were conflict and loss of freedom. Claxton and Perry-Jenkins (2008) confirmed a decrease in marital leisure time after the birth of a baby but an increase when the wife went back to work. In addition, they noted that wives who reported high levels of prenatal joint leisure reported greater marital love and less conflict the first year after the baby's birth.

Having children makes you no more a parent than having a piano makes you a pianist.

Michael Levine, pianist

For parents who experience a pattern of decreased happiness, it bottoms out during the teen years. Facer and Day (2004) found that adolescent problem behavior, particularly that of a daughter, is associated with increases in marital conflict. Of even greater impact was their perception of the child's emotional state. Parents who viewed their children as "happy" were less maritally affected by their adolescent's negative behavior.

Regardless of how children affect the feelings that spouses have about their marriage, spouses report more commitment to their relationship once they have children (Stanley and Markman 1992). Figure 11.1 illustrates that the more children a couple has, the more likely the couple will stay married. A primary reason for this increased commitment is the desire on the part of both parents to provide a stable family context for their children. In addition, parents of dependent children may keep their marriage together to maintain continued access to and a higher standard of living for their children. Finally, people (especially mothers) with small children feel more pressure to stay married (if the partner provides sufficient economic resources) regardless of how unhappy they may be. Hence, though children may decrease happiness, they increase stability, because pressure exists to stay together (the larger the family is, the more difficult the economic post-divorce survival is).

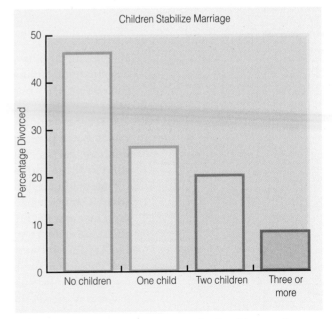

Figure 11.1 Percentage of Couples Getting Divorces by Number of Children.

Parenthood: Some Facts

Parenting is only one stage in an individual's or couple's life (children typically live with an individual 30% of that person's life and with a couple 40% of their marriage). Parenting involves responding to the varying needs of children as they grow up, and parents require help from family and friends in rearing their children.

Some additional facts of parenthood follow.

When I was a boy of fourteen, my father was so ignorant I could hardly stand to have the old man around. But when I got to be twenty-one, I was astonished at how much the old man had learned in seven years.

Mark Twain, Attributed in Error

Views of Children Differ Historically

Whereas children of today are thought of as dependent, playful, and adventurous, they have been historically viewed quite differently (Beekman 1977). Indeed, the concept of childhood, like gender, has been socially constructed rather than a fixed life stage. From the thirteenth through the sixteenth centuries, children were viewed as innocent, sweet, and a source of amusement for adults. From the sixteenth through the eighteenth centuries, they were viewed as in need of discipline and moral training to educate them into adult salvation. In the nineteenth century, whippings were routine as a means of breaking children's spirits and bringing them to submission. Although remnants of both the innocent and moralistic views of children exist today, the lives of children are greatly improved. Child labor laws protect children from early forced labor, education laws ensure a basic education, and modern medicine has been able to increase the life span of children.

DIVERSITY IN THE UNITED STATES

Parents are diverse. Amish parents rear their children in homes without electricity (no television, phones, or MP3 players). Charismatics and evangelicals (members of the conservative religious denominations) often educate their children in Christian schools, emphasizing a Biblical and spiritual view of life and the world. More secular parents tend to bring up their children amid individualistic liberal views.

Parents Are Only One Influence in a Child's Development

Although parents often take the credit and the blame for the way their children turn out, they are only one among many influences on child development. Although parents are the first significant influence, peer influence becomes increasingly important during adolescence. For example, Ali and Dwyer (2010) studied a nationally representative sample of adolescents and found that having peers who drank alcohol was related to the adolescent drinking alcohol.

Siblings also have an important and sometimes lasting effect on each other's development. Siblings are social mirrors and models (depending on the age) for each other. They may also be sources of competition and can be jealous of each other.

Teachers are also significant influences in the development of a child's values. Some parents send their children to religious schools to ensure that they will have teachers with conservative religious values. This may continue into the child's college and university education.

Media in the form of television replete with MTV and "parental discretion advised" movies are a major source of language, values, and lifestyles for children that may be different from those of the parents. Parents are also concerned about the violence to which television and movies expose their children.

Another influence of concern to parents is the Internet. Though parents may encourage their children to conduct research and write term papers using the Internet, they may fear their children are accessing pornography and related sex sites. Parental supervision of teens on the Internet and the right of the teen for privacy remain potential conflict issues (see the Social Policy feature). Parents are also concerned about **sextortion** (online sexual extortion). Sextortion often takes the form of teenage girls with a webcam at a party who will visit an Internet chat room and flash their breasts. A week later they will get an email from

Sextortion online sexual extortion.

a stranger who informs them that he has captured their video from the Internet and will post their photos on Facebook and MySpace if they do not send him explicit photos (Wilson et al. 2010).

Each Child Is Unique

Children differ in their genetic makeup, physiological wiring, intelligence, tolerance for stress, capacity to learn, comfort in social situations, and interests. Parents soon become aware of the uniqueness of each child—of her or his difference from every other child they know. Parents of two or more children are often amazed at how children who have the same parents can be so different. Children also differ in their mental and physical health. Mental and physical disabilities of children present emotional and financial challenges to their parents.

National **Data**

In the United States, there are more than 6 million children with disabilities—almost half (43%) of these children have learning disabilities, 8% have mental retardation, and 7% have a serious emotional disturbance (*Statistical Abstract of the United States 2011*, Table 185).

Although parents often contend "we treat our children equally," Tucker et al. (2003) found that parents treat children differently, with firstborns usually receiving more privileges than children born later. Suitor and Pillemer (2007) found that elderly mothers reported that they tended to establish a closer relationship with the firstborn child, on whom they are more likely to call in later life when there is a crisis. Birth order is discussed in more detail in the following section.

Birth Order Effects

Psychologist Frank Sulloway (1996) examined how a child's birth order influenced the development of various personality characteristics. His thesis is that children with siblings develop different strategies to maximize parental investment in them. For example, children can promote parental favor directly by "helping and obeying parents" (p. 67). Sulloway identified some personality characteristics that have their basis in a child's position in the family.

1. *Conforming/traditional*. First borns are the first on the scene with parents and always have the "inside track." They want to stay that way so they are traditional and conforming to their parent's expectations.

2. *Experimental/adventurous*. Later borns learn quickly that they enter an existing family constellation where everyone is bigger and stronger. They cannot depend on having established territory so must excel in ways different from the first born. They are open to experience, adventurousness, and trying new things since their status is not already assured.

3. *Neurotic/emotionally unstable*. Since first borns are "dethroned" by younger children to whom parents had to divert their attention, they tend to be more jealous, anxious, and fearful. Later borns were never number one to begin with so did not experience this trauma.

As support for his ideas, Sulloway (1996) cited 196 controlled birth order studies and contended that, while there are exceptions, one's position in the family is a factor influencing personality outcomes. Nevertheless, he acknowledged that researchers disagree on the effects of birth order on the personalities of children and that birth order research is incomplete in that it does not consider the position of each child in families that vary in size, gender, and number. Indeed, Sulloway (2007) has continued to conduct research, the findings of which do not always support his prediction. For example, he examined

SOCIAL POLICY | Government or Parental Control of Internet Content for Children?

Over 40% (42%) of 10- to 17-year-olds have been exposed to Internet pornography (Preston 2007). Governmental censoring of content on the Internet is the focus of an ongoing public debate. At issue is what level of sexual content children should be exposed to and should parents or the government make the decision? This issue has already been decided in China, where the government ensures that all major search engines filter what is made available to Internet users. In the United States, Congress passed the Communications Decency Act in 1996, which prohibited sending "indecent" messages over the Internet to people under age 18. However, the Supreme Court struck down the law in 1997, holding that it was too broadly worded and violated free speech rights by restricting too much material that adults might want to access. Other laws to restrict content access on the Internet have been passed but struck down on the basis that they violate First Amendment rights. However, The Protection of Children from Sexual Predators Act of 1998 required Internet Service Providers (ISPs) to report any knowledge of child victimization or child pornography to law enforcement.

Many object to government control of sexual content on the Internet on the grounds of First Amendment rights. Many others believe government restrictions are necessary to protect children from inappropriate sexual content.

An overriding question is, "Should parents or the government be in control of what children are exposed to?" In the United States, parents, not the government, prefer to be responsible for regulating their children's use of the Internet. Beyond exposure to Internet pornography, children sometimes post information about themselves on the Internet that encourages pedophiles to e-mail them. MySpace has more than 100 million profiles, with 230,000 new members signing up every day (Andrews 2006). Some of these profiles may be more revealing than intended. Software products, such as Net Nanny, Surfwatch, CYBERsitter, CyberPatrol, and Time's Up, are being marketed to help parents control what their children view on the Internet. These software programs allow parents to block unapproved websites and categories (such as pornography), block transmission of personal data (such as address and

---

intelligence and birth order among 241,310 Norwegian 18- and 19-year-olds and found no relationship. He recommended that researchers look to how a child was reared rather than birth order for understanding a child's IQ in the family.

Parenting Styles Differ

Diana Baumrind (1966) developed a typology of parenting styles that has become classic in the study of parenting. She noted that parenting behavior has two dimensions: responsiveness and demandingness. **Responsiveness** refers to the extent to which parents respond to and meet the needs of their children. In other words, how supportive are the parents? Warmth, reciprocity, person-centered communication, and attachment are all aspects of responsiveness. **Demandingness**, on the other hand, is the manner in which parents place demands on children in regard to expectations and discipline. How much control do they exert over their children? Monitoring and confrontation are also aspects of demandingness. Categorizing parents in terms of their responsiveness and their demandingness creates four categories of parenting styles: permissive (also known as indulgent), authoritarian, authoritative, and uninvolved.

1. *Permissive parents are high on responsiveness and low on demandingness.* They are very lenient and allow their children to largely regulate their own behavior. Walcheski and Bredhoft (2010) found that the permissive parenting style is associated with overindulgence. For example, these parents state punishments but do not follow through, give in to their children, and allow their children to interrupt others. These parents act out of fear that disciplining the child will cause the child to dislike his or her parents.

> *The most important thing that parents can teach their children is how to get along without them.*
>
> Frank A. Clark, politician

Responsiveness refers to the extent to which parents respond to and meet the needs of their children.

Demandingness the manner in which parents place demands on children in regard to expectations and discipline.

telephone numbers), scan pages for sexual material before they are viewed, and track Internet usage.

Another alternative is for parents to use the Internet with their children both to monitor what their children are viewing and to teach their children values about what they believe is right and wrong on the Internet. Some parents believe that children must learn how to safely surf the Internet. Internet sites helpful in this regard are BlogSafety.com, forums for parents to discuss blogging and other aspects of social networking; NetSmartz.org service, which teaches kids 5 to 17 how to be safe on the Internet; and WiredSafety.org, which posts information on Internet safety.

One parent reported that the Internet is like a busy street, and just as you must teach your children how to safely cross in traffic, you must teach them how to avoid giving information to strangers on the Internet. As noted in the text, young people are vulnerable to "sextortion" (Wilson 2010). By sending a nude photo to a boyfriend the image can be captured and put online on the Internet and the blackmail may begin. The parents will be unaware while their offspring spirals into increasing demands from the extortionist.

Your Opinion?

1. To what degree do you believe the government should control Internet content?

2. To what degree do you believe that parents should monitor what their children are exposed to on the Internet?

3. What can parents do to teach their children responsible use of the Internet?

Sources

Andrews, M. 2006. Decoding MySpace. *U.S. News and World Report* Sept 18, 46–47.

Preston, C. B. 2007. Making family-friendly Internet a reality: The Internet Community Ports Act. *Brigham Young University Law Review* 2007:1471–1534.

Wilson, C. 2010. Feds: Online 'sextortion' of teens on the rise. Associated Press. August 14 http://news.yahoo.com/s/ap/us_sextortion_teens

2. *Authoritarian parents are high on demandingness and low in responsiveness.* They feel that children should obey their parents no matter what, and they provide a great deal of structure in the child's world.

3. *Authoritative parents are both demanding and responsive.* They impose appropriate limits on their children's behavior but emphasize reasoning and communication. This style offers a balance of warmth and control and is associated with the most positive outcome for children. Walcheski and Bredehoft (2010) emphasized the value of the authoritative parenting style and gave examples—parents telling the child their expectations of the child's behavior before the child engages in the activity, giving the child reasons why rules should be obeyed, talking with the child when he or she has misbehaved, and explaining consequences.

4. *Uninvolved parents are low in responsiveness and demandingness.* These parents are not invested in their children's lives.

McKinney and Renk (2008) identified the differences between maternal and paternal parenting styles, with mothers tending to be authoritative and fathers tending to be authoritarian. Mothers and fathers also use different parenting styles for their sons and daughters, with fathers being more permissive with their sons than daughters. Overall, this study emphasized the importance of examining the different parenting styles of parents on adolescent outcome and suggested that having at least one authoritative parent may be a protective factor for late adolescents. Recall that the authoritative parenting style is the combination of warmth, guidelines, and discipline—"I love you but you need to be in by midnight or you'll lose the privileges of having a cell phone and a car."

What if You and Your Partner Disagree over Parenting Styles?

Because spouses grow up in families where different parenting philosophies are operative, it is not unusual for them to disagree over how to rear or discipline their own children. The respective typologies previously identified, responsiveness and demandingness, which may be regarded as "loose-goosey, anything goes" and "iron discipline," will cause obvious conflicts with no easy solutions. Some parents divorce because they argue constantly over how to rear their children. One alternative is for parents to agree that the most involved parent's child-rearing style will prevail. Such an agreement is obviously a concession on the part of one parent.

Another alternative is to use the parenting philosophy that research suggests has the best outcome. As previously noted, authoritative parenting, which is a blend of compassion and discipline, seems to have the best outcome in terms of fewer problems with children and their development of high levels of social competence.

In summary, parenting does not come naturally to individuals and couples. After decades of exploring how to provide an optimal environment for the development of children, social scientists have identified various essential elements which are detailed in the section to follow.

The childhood shows the man,
As morning shows the day.

John Milton, poet

Principles of Effective Parenting

Numerous principles are involved in being effective parents. Dumka et al. (2010) noted that **parenting self-efficiency** (feeling competent as a parent) was associated with parental control. In effect, parents who viewed themselves as good teachers felt confident in their role which, in turn, translated into their ability to manage their adolescents effectively. We begin the discussion of effective parenting with the most important of these principles, which involves giving time/love to your children as well as praising and encouraging them.

Give Time, Love, Praise, Encouragement, and Acceptance

Children most need to feel that they are worth spending time with and that someone loves them. Because children depend first on their parents for the development of their sense of emotional security, it is critical that parents provide a warm emotional context in which the children can develop. Feeling loved as an infant also affects one's capacity to become involved in adult love relationships. Abundant evidence from children reared in institutions where nurses attended only to their physical needs testify to the negative consequences of early emotional neglect. **Reactive attachment disorder** is common among children who were taught as infants that no one cared about them. Such children have no capacity to bond emotionally with others since they have no learning history of the experience and do not trust adults/caretakers/parents.

Acceptance is also important and has been associated with positive outcomes. Schwartz et al. (2009) studied the association between 1,728 undergraduates feeling acceptance from their parents and engaging in high-risk behaviors.

Parenting self-efficiency feeling competent as a parent.

Reactive attachment disorder common among children who were taught as infants that no one cared about them; these children have no capacity to bond emotionally with others since they have no learning history of the experience and do not trust adults, caretakers, or parents.

Students who perceived acceptance from their parents (as opposed to rejection) were less likely to engage in various health-risk behaviors such as hard drug use, inhalant use, prescription drug misuse, casual sex, driving while intoxicated, and riding with an impaired driver.

Even in the midst of a heated quarrel, some words should never be spoken by parents. To do so is to destroy the relationship with their children forever. William Berle, adopted son of comedian Milton Berle, recalled being in an argument with his dad who lashed out at him, "Oh yeah? Well, I did make one mistake and that was twenty-seven years ago when we adopted you.... how do you like that you little prick!" (Berle and Lewis 1999, 188). The son recalled being devastated and walking into the next room to take out a pistol to kill himself. Luckily, a knock on the door interrupted his plan.

Avoid Overindulgence

Overindulgence may be defined as more than just giving children too much, but includes overnurturing and providing too little structure. Using this definition, a study of 466 participants revealed that those who are overindulged tend to hold materialist values for success, are not able to delay gratification, and are less grateful for things and to others. Indeed, not being overindulged promotes the ability to delay gratification, be grateful, and experience a view (nonmaterialism) that is associated with happiness (Slinger and Bredehoft 2010). The book *How Much is Enough?* emphasizes attentiveness to avoiding overindulgence (Dawson et al. 2003).

Parents typically overindulge because they feel guilty or because they did not have certain material goods in their own youth. In a study designed to identify who overindulged, Clarke (2004) found that mothers were four times more likely to overindulge as fathers. Another result of overindulgence is that children grow up without consequences, and they avoid real jobs where employers expect them to show up at 8:00 A.M.

Monitor Child's Activities/Drug Use

Abundant research suggests that parents who monitor their children and teens, know where their children are, who they are with, what they are doing, and so on are less likely to report that their adolescents receive low grades, or are engaged in early sexual activity, delinquent behavior, and drug use (Crosnoe and Cavanagh 2010).

Parents who used marijuana or other drugs themselves wonder how to go about encouraging their own children to be drug-free. Drugfree.org has some recommendations for parents, including being honest about previous drug use, making clear that you do not want your children to use drugs, and explaining that although not all drug use leads to negative consequences, staying clear of such possibilities is the best course of action.

Monitor Television and Pornography Exposure

Parents might also monitor the amount and content of media exposure their children's experience. A Kaiser Foundation (2010) study of media use of 8- to 18-year-olds found that they were watching television an average of 4.5 hours per day. Nonschool computer use was an hour and a half a day—much of which was unsupervised (two thirds said there were no rules about Internet use).

Looking at pornography by children and youth is not unusual. Flood (2009) noted the negative effects of pornography exposure among children including their developing sexist and unhealthy notions of sex and relationships. In addition, among boys and young men, frequent consumption of pornography is associated with supportive attitudes of sexual coercion and increases their likelihood of perpetrating assault. Pornography is a poor, and indeed dangerous, sex educator.

Overindulgence defined as more than just giving children too much, includes overnurturing and providing too little structure.

The surest way to make it hard for children is to make it easy for them.

Eleanor Roosevelt, First Lady of the U.S. 1933 to 1945

Monitor Cell Phone/Text Messaging Use

Some parents are concerned about their child's use of a cell phone and texting. Not only may predators contact children/teenagers without the parents' notice, 39% of teens report having sent sexually suggestive text messages and 20% have sent nude or semi-nude photos or videos of themselves. New technology called "mobile protector" permits parents to effectively take over their child's mobile phone. A parent can screen a child's incoming and outgoing calls and messages, block particular numbers, and even listen in on a conversation. A dashboard on a parent's phone or a personal computer shows the cell phones being monitored and the permitted callers such as friends and family. Another protective device is MyMobileWatchdog (known as MMWD) which monitors a child's cell phone use and instantly alerts the parents online if their son or daughter receives unapproved e-mail, text messages, or phone calls.

Time-out a noncorporal form of punishment that involves removing the child from a context of reinforcement to a place of isolation.

Set Limits and Discipline Children for Inappropriate Behavior

The goal of guidance is self-control. Parents want their children to be able to control their own behavior and to make good decisions without their parents.

Guidance may involve reinforcing desired behavior or providing limits to children's behavior. This sometimes involves disciplining children for negative behavior. Unless parents provide negative consequences for lying, stealing, and hitting, children can grow up to be dishonest, to steal, and to be inappropriately aggressive. **Time-out** (a noncorporal form of punishment that involves removing the child from a context of reinforcement to a place of isolation for one minute for each year of the child's age) has been shown to be an effective consequence for inappropriate behavior (Morawska and Sanders 2011). Withdrawal of privileges (watching television, playing with friends), pointing out the logical consequences of the misbehavior ("you were late; we won't go"), and positive language ("I know you meant well but . . .") are also effective methods of guiding children's behavior.

Physical punishment is less effective in reducing negative behavior (see the Personal Choices section); it teaches the child to be aggressive and encourages negative emotional feelings toward the parents. When using time-out or the withdrawal of privileges, parents should make clear to the child that they disapprove of the child's behavior, not the child. Some evidence suggests that consistent discipline has positive outcomes for children. Lengua et al. (2000) studied 231 mothers of 9- to 12-year-olds and found that inconsistent discipline was related to adjustment problems, particularly for children high in impulsivity.

PERSONAL CHOICES

Should Parents Use Corporal Punishment?

Parents differ in the type of punishment they feel is appropriate for children. Some parents use corporal punishment as a means of disciplining their children. These parents tend to be Protestant rather than Catholic, have younger rather than older children, and are black rather than white or Latino (Grogan-Kaylor and Otis 2007).

The decision to choose a corporal or noncorporal method of punishment should be based on the consequences of use. In general, the use of time-out and withholding of privileges seems to be more effective than corporal punishment in stopping undesirable behavior (Paintal 2007). Though beatings and whippings will temporarily decrease negative verbal and nonverbal behaviors, they have major side effects. First, punishing children by inflicting violence teaches them that it is acceptable to physically hurt someone you love. Hence, parents may be inadvertently teaching their children to use violence in the family. Children who are controlled by corporal punishment grow up to be violent toward their own children, spouses, and friends (Paintal 2007). Second, parents who beat their children should be aware that they are teaching their children to fear and avoid them. The result is that children who grow up in homes where corporal punishment is used have more distant relationships with their parents (Paintal 2007). Third, children who grow up in homes in which corporal punishment is used are more likely to feel helpless, to have low self-esteem, and to withdraw (Paintal 2007). In recognition of the negative consequences of corporal punishment, the law in Sweden forbids parents to spank their children. In the United States, Zolotor et al. (2011) examined data on corporal punishment and physical abuse of three-to 11-year-old children from 1975 and 2002. They found that there was an overall decline in the spanking and slapping of children in the time period that they investigated.

So, what kind of discipline is best? Parents who reason with their children and back up their reasoning with consequences report the greatest consistent behavior change. Consequences that are effective include withdrawing privileges such as use of cell phone, computer, television, and friends (for example, grounding so that the child or teen is not allowed to see friends on the weekends) (Crisp and Knox 2009). We earlier noted that the authoritative parenting style (which emphasizes a balance of structure, control, warmth, and consequences) seems to have a good outcome for children and parents.

A review of some of the alternatives to corporal punishment include the following:

1. *Be a positive role model.* Children learn behaviors by observing their parents' actions, so parents must model the ways in which they want their children to behave. If a parent yells or hits, the child is likely to do the same.

2. *Set rules and consequences.* Make rules that are fair, realistic, and appropriate to a child's level of development. Explain the rules and the consequences of not following them. If children are old enough, they can be included in establishing the rules and the consequences for breaking them.

3. *Encourage and reward good behavior.* When children are behaving appropriately, give them verbal praise and occasionally reward them with tangible objects, privileges, or increased responsibility.

4. *Use charts.* Charts to monitor and reward behavior can help children learn appropriate behavior. Charts should be simple and focus on one behavior at a time, for a certain length of time.

5. *Use time-out.* "Time-out" involves removing children from a situation following a negative behavior. This can help children calm down, end an inappropriate behavior, and reenter the situation in a positive way. Explain what the inappropriate behavior is, why the time-out is needed, when it will begin, and how long it will last. Set an appropriate length of time for the time-out based on age and level of development, usually one minute for each year of the child's age (see the following Self-Assessment on Spanking versus Time-Out).

Remember, when they have a tantrum, don't have a tantrum of your own.

Judith Kuriansky, psychiatrist

Sources

Crisp, B., and D. Knox. 2009. *Behavioral family therapy: An evidence based approach.* Durham, NC: Carolina Academic Press.

Grogan-Kaylor, A., and M. D. Otis. 2007. The predictors of parental use of corporal punishment *Family Relations* 56: 80–91.

Paintal, S. 2007. Banning corporal punishment of children. *Childhood Education* 83:410–421.

Zolotor, A., A. Theodore, D. Runyan, K. Desmond, J. J. Chang, A. Laskey, L. Antoinette and L. Izvor. 2011. Corporal punishment and physical abuse: population-based trends for three-to-11-year-old children in the United States. *Child Abuse Review* 20: 57-66.

SELF-ASSESSMENT | Spanking versus Time-Out Scale

Parents discipline their children to help them develop self-control and correct misbehavior. Some parents spank their children; others use time-out. Spanking is a disciplinary technique whereby a mild slap (that is, a "spank") is applied to the buttocks of a disobedient child. Time-out is a disciplinary technique whereby, when a child misbehaves, the child is removed from the situation. The purpose of this survey is to assess the degree to which you prefer spanking versus time-out as a method of discipline. Please read each item carefully and select a number from 1 to 7, which represents your belief. There are no right or wrong answers; please give your honest opinion.

1	2	3	4	5	6	7
Strongly Disagree						Strongly Agree

_____ 1. Spanking is a better form of discipline than time-out.

_____ 2. Time-out does not have any effect on children.

_____ 3. When I have children, I will more likely spank them than use a time-out.

_____ 4. A threat of a time-out does not stop a child from misbehaving.

_____ 5. Lessons are learned better with spanking.

_____ 6. Time-out does not give a child an understanding of what the child has done wrong.

_____ 7. Spanking teaches a child to respect authority.

_____ 8. Giving children time-outs is a waste of time.

_____ 9. Spanking has more of an impact on changing the behavior of children than time-out.

_____ 10. I do not believe "time-out" is a form of punishment.

_____ 11. Getting spanked as a child helps you become a responsible citizen.

_____ 12. Time-out is only used because parents are afraid to spank their kids.

_____ 13. Spanking can be an effective tool in disciplining a child.

_____ 14. Time-out is watered-down discipline.

Scoring

If you want to know the degree to which you approve of spanking, reverse the number you selected for all odd-numbered items (1, 3, 5, 7, 9, 11, and 13) after you have selected a number from 1 to 7 for each of the fourteen items. For example, if you selected a 1 for item 1, change this number to a 7 (1 = 7; 2 = 6; 3 = 5; 4 = 4; 5 = 3; 6 = 2; 7 = 1). Now add these seven numbers. The lower your score (7 is the lowest possible score), the lower your approval of spanking; the higher your score (49 is the highest possible score), the greater your approval of spanking. A score of 21 places you at the midpoint between being very disapproving of or very accepting of spanking as a discipline strategy.

If you want to know the degree to which you approve of using time-out as a method of discipline, reverse the number you selected for all even-numbered items (2, 4, 6, 8, 10, 12, and 14).

For example, if you selected a 1 for item 2, change this number to a seven (that is, 1 = 7; 2 = 6; 3 = 5; 4 = 4; 5 = 3; 6 = 2; 7 = 1). Now add these seven numbers. The lower your score (7 is the lowest possible score), the lower your approval of time-out; the higher your score (49 is the highest possible score), the greater your approval of time-out. A score of 21 places you at the midpoint between being very disapproving of or very accepting of time-out as a discipline strategy.

Scores of Other Students Who Completed the Scale

The scale was completed by 48 male and 168 female student volunteers at East Carolina University. Their ages ranged from 18 to 34, with a mean age of 19.65 (SD = 2.06). The ethnic background of the sample included 73.1% white, 17.1% African American, 2.8% Hispanic, 0.9% Asian, 3.7% from other ethnic backgrounds; 2.3% did not indicate ethnicity. The college classification level of the sample included 52.8% freshman, 24.5% sophomore, 13.9% junior, and 8.8% senior. The average score on the spanking dimension was 29.73 (SD = 10.97), and the time-out dimension was 22.93 (SD = 8.86), suggesting greater acceptance of spanking than time-out.

Time-out differences. In regard to sex of the participants, female participants were more positive about using time-out as a discipline strategy (M = 33.72, SD = 8.76) than were male participants (M = 30.81, SD = 8.97; p = .05). In regard to ethnicity of the participants, white participants were more positive about using time-out as a discipline strategy (M = 34.63, SD = 8.54) than were nonwhite participants (M = 28.45, SD = 8.55; p = .05). In regard to year in school, freshmen were more positive about using spanking as a discipline strategy (M = 34.34, SD = 9.23) than were sophomores, juniors, and seniors (M = 31.66, SD = 8.25; p = .05).

Spanking differences. In regard to ethnicity of the participants, nonwhite participants were more positive about using spanking as a discipline strategy (M = 35.09, SD = 10.02) than were white participants (M = 27.87, SD = 10.72; p = .05). In regard to year in school, freshmen were less positive about using spanking as a discipline strategy (M = 28.28, SD = 11.42) than were sophomores, juniors, and seniors (M = 31.34, SD = 10.26; p = .05). There were no significant differences in regard to sex of the participants (p = .05) in the opinion of spanking.

Overall differences. There were no significant differences in overall attitudes to discipline in regards to sex of the participants, ethnicity, or year in school.

Source

"The Spanking vs. Time-Out Scale," 2004 by Mark Whatley, Ph.D., Department of Psychology, Valdosta State University, Valdosta, Georgia 31698-0100. Used by permission. Other uses of this scale by written permission of Dr. Whatley only (mwhatley@valdosta.edu). Information on the reliability and validity of this scale is available from Dr. Whatley.

Have Family Meals

Parents who stay connected with their children build strong relationships with them and report fewer problems. Bisakha (2010) found that families who have regular meals with their adolescents report fewer behavioral problems such as less substance use and running away (for females) and less drinking, physical violence, property-destruction, stealing, and running away (for males). The researcher recommends that family meals be made a regular family ritual.

Encourage Responsibility

Giving children increased responsibility encourages the autonomy and independence they need to be assertive and independent. Giving children more responsibility as they grow older can take the form of encouraging them to choose healthy snacks and letting them decide what to wear and when to return from playing with a friend (of course, the parents should praise appropriate choices).

Children who are not given any control and responsibility for their own lives remain dependent on others. Successful parents can be defined in terms of their ability to rear children who can function as independent adults. A dependent child is a vulnerable child.

Some American children remain in their parents' house into their twenties and thirties. Their doing so is primarily because it is cheaper to do so. Ward and Spitze (2007) analyzed data from the National Survey of Families and Households in regard to children aged 18 and older who lived with their parents. Findings revealed that, although disagreements between parents and children increased, the quality of parent-child or husband-wife relations did not change.

Provide Sex Education

Wilson et al. (2010) found that while parents value talking with their children about sex, they often do not do so since they feel their children are "too young" or they don't know how to talk with them about sex. Those who found it easiest to talk about sex with their children reported very positive parent-child relationships and talked about sex when the children were young. There are definite benefits when parents talk with their children about sex. Usher-Seriki et al. (2008) studied mother-daughter communication about sex and sexual intercourse in a sample of 274 middle- to upper-income African American adolescent girls. They found that daughters were more likely to delay sexual intercourse when they had close relationships with their mother, when they perceived that their mother disapproved of premarital sex for moral reasons, and when their mothers emphasized the negative consequences of premarital sex.

Express Confidence

"One of the greatest mistakes a parent can make," confided one mother, "is to be anxious all the time about your child, because the child interprets this as your lack of confidence in his or her ability to function independently." Rather, this mother noted that it is best to convey to a child that you know that he or she will be fine because you have confidence in him/her. "The effect on the child," said this mother, "is a heightened sense of self-confidence." Another way to conceptualize this parental principle is to think of the self-fulfilling prophecy as a mechanism that facilitates self-confidence. If parents show a child that they have confidence in him or her, the child begins to accept these social definitions as real and becomes more self-confident.

Nelson (2010) noted that some parents have become "helicopter parents" (also referred to as hovercrafts and PFH—Parents from Hell) in that they are constantly hovering at school and in the workplace to ensure their child's "success." The workplace has become the new field where parents negotiate the benefits and salaries of their children with employers they feel are out to take advantage of inexperienced workers. However, employers may not appreciate the tampering, and the parents risk hampering their child's "ability to develop self-confidence" (p. D1). Other employers are engaging parents and know that, unless the parents are convinced, the offspring won't sign on (Hira 2007).

Respond to the Teen Years Creatively

Parenting teenage children presents challenges that differ from those in parenting infants and young children. Becker (2010) informs parents that adolescents tend to engage in a high rate of risky behavior—smoking, alcohol consumption, hazardous driving, drug use, delinquency, dares, sporting risks, rebellious behavior, and sexual intercourse. Seeking novelty, peer influences, genetic factors, and brain maturation (or lack of maturation) are among the factors accounting for the vulnerability of adolescents.

National **Data**

Sometimes strain between a teen and parent(s) ends in the teen running away. The National Runaway Switchboard receives more than 100,000 calls annually, 72% from females. The most frequent reason (29%) given for running away is "family dynamics" (National Runaway Switchboard 2010).

Conflicts between parents and teenagers often revolve around money and independence. The desire for a car, cell phone, DVD player, and high-definition TV can outstrip the budget of many parents. Teens also increasingly want more freedom. However, neither of these issues needs to result in conflicts. When they do, the effect on the parent-child relationship may be inconsequential. One parent tells his children, "I'm just being the parent, and you're just being who you are; it is OK for us to disagree but you can't go." The following suggestions can help to keep conflicts with teenagers at a low level:

1. *Catch them doing what you like rather than criticizing them for what you don't like.* Adolescents are like everyone else—they don't like to be criticized, but they do like to be noticed for what they do that is good.

2. *Be direct when necessary.* Though parents may want to ignore some behaviors of their children, addressing some issues directly may also be effective. Regarding the avoidance of STI or HIV infections and pregnancy, Dr. Louise Sammons (2008) tells her teenagers, "It is utterly imperative to require that any potential sex partner produce a certificate indicating no STIs or HIV infection and to require that a condom or dental dam be used before intercourse or oral sex."

3. *Provide information rather than answers.* When teens are confronted with a problem, try to avoid making a decision for them. Rather, providing information on which they may base a decision is helpful. What courses to take in high school and what college to apply for are decisions that might be made primarily by the adolescent. The role of the parent might best be that of providing information or helping the teenager to obtain information.

4. *Be tolerant of high activity levels.* Some teenagers are constantly listening to loud music, going to each other's homes, and talking on cell phones for long periods of time. Parents often want to sit in their easy chairs and be quiet. Recognizing that it is not realistic to expect teenagers to be quiet and sedentary may be helpful in tolerating their disruptions.

5. *Engage in some activity with your teenagers.* Whether renting a DVD, eating a pizza, or taking a camping trip, structuring some activities with your teenagers is important. Such activities permit a context in which to communicate with them. Fulkerson et. al. (2010) studied 4,750 adolescents and their relationships with their parents and noted the benefits to the parent-teenager relationship and communication with parents when parents and teens regularly eat dinner together. Sometimes teenagers present challenges with which the parents feel unable to cope. Aside from monitoring their behavior closely, behavioral family

You have a wonderful child. Then when he's thirteen, gremlins carry him away and leave in his place a stranger who gives you not a moment's peace.

Jill Eikenberry, actress

While we try to teach our children all about life,

Our children teach us what life is all about.

Angela Schwindt, home schooling mom

therapy may be helpful. A major focus of such therapy is to increase the emotional bond between the parents and the teenagers and to encourage positive consequences for desirable behavior (for example, concert tickets for good grades) and negative consequences for undesirable behavior (for example, loss of car privileges for getting a speeding ticket).

Child Rearing Theories

Parenting professionals offer parents advice that has changed over time. For example, in 1914, parents who wanted to know what to do about their child's thumb sucking were told to try to control such a bad impulse by pinning the sleeves of the child to the bed if necessary. Today, parents are told that thumb sucking meets an important psychological need for security and they should not try to prevent it. If a child's teeth become crooked as a result, an orthodontist should be consulted.

Advice may also be profit driven. As noted in the Applying Social Research section earlier in this chapter, there is alarming information that some of what passes for "research-based advice" to provide parents with information about specific toys or learning programs turns out to be child development research funded by corporations who are intent on selling parents merchandise for their children. In her book *Buy, Buy Baby: How Consumer Culture Manipulates Parents and Harms Young Minds,* Thomas (2007) identified the dangerous economic and cultural shift that encourages parents to buy things for their toddlers and infants that may actually harm them.

There are several theories about rearing children. These are reviewed in the following and summarized in Table 11.1. In examining these approaches, it is important to keep in mind that no single approach is superior to another. What works for one child may not work for another. Any given approach may not even work with the same child at two different times. When child development theories are viewed as socialization theories, Grusec and Davidov (2010) emphasized that "there is no general, all-purpose principle or mechanism of socialization, but rather each form of relationship between the object and the agent of socialization serves a different function, involves different rules and mechanisms for effecting behavior change, and facilitates different outcomes" (p. 687).

It is important for new parents to use a cafeteria approach when examining parenting advice from health care providers, family members, friends, parenting educators, and other well-meaning individuals. Take what makes sense, works, and feels right as a parent and leave the rest behind. Parents know their own child better than anyone else and should be encouraged to combine different approaches to find what works best for them and their unique child.

Developmental-Maturational Approach

For the past 60 years, Arnold Gesell and his colleagues at the Yale Clinic of Child Development have been known for their ages-and-stages approach to childrearing (Gesell et al. 1995). The developmental-maturational approach has been widely used in the United States. The basic perspective, some considerations for childrearing, and some criticisms of the approach follow.

Basic Perspective Gesell views what children do, think, and feel as being influenced by their genetic inheritance. Although genes dictate the gradual unfolding of a unique person, every individual passes through the same basic pattern of growth. This pattern includes four aspects of development: motor behavior (sitting, crawling, walking), adaptive behavior (picking up objects and walking around objects), language behavior (words and gestures), and personal-social behavior (cooperativeness and helpfulness). Through the observation of

Parents ought, through their own behavior and values by which they live, to provide direction for their children. But they need to rid themselves of the idea that there are surefire methods which, when well applied, will produce certain predictable results.

Bruno Bettelheim,
child development specialist

hundreds of normal infants and children, Gesell and his coworkers have identified norms of development. Although there may be large variations, these norms suggest the ages at which an average child displays various behaviors. For example, on the average, children begin to walk alone (although awkwardly) at age 13 months and use simple sentences between the ages of 2 and 3.

Considerations for Childrearing   Gesell suggested that if parents are aware of their children's developmental clock, they will avoid unreasonable expectations. For example, a child cannot walk or talk until the neurological structures necessary for those behaviors have matured. Also, the hunger of a 4-week-old must be immediately appeased by food, but at 16 to 28 weeks, the child has some capacity to wait because the hunger pains are less intense. In view of this and other developmental patterns, Gesell suggested that infants need to be cared for on a demand schedule; instead of having to submit to a schedule imposed by parents, infants are fed, changed, put to bed, and allowed to play when they want. Children are likely to be resistant to a hard-and-fast schedule because they may be developmentally unable to cope with it.

In addition, Gesell emphasized that parents should be aware of the importance of the first years of a child's life. In Gesell's view, these early years assume the greatest significance because the child's first learning experiences occur during this period.

Criticisms of the Developmental-Maturational Approach   Gesell's work has been criticized because of (1) its overemphasis on the idea of a biological clock, (2) the deficiencies of the sample he used to develop maturational norms, (3) his insistence on the merits of a demand schedule, and (4) the idea that environmental influences are weak.

Most of the children who were studied to establish the developmental norms were from the upper middle class. Children in other social classes are exposed to different environments, which influence their development. So norms established on upper-middle-class children may not adequately reflect the norms of children from other social classes.

Gesell's suggestion that parents do everything for the infant when the infant wants has also been criticized. Rearing an infant on the demand schedule can drastically interfere with the parents' personal and marital interests. In the United States, with its emphasis on individualism, many parents feed their infants on a demand schedule but put them to bed to accommodate the parents' schedule.

Behavioral Approach

The behavioral approach to childrearing, also known as the social learning approach, is based on the work of B. F. Skinner and the theme of Behavioral Family Therapy (Crisp and Knox 2009). Behavioral approaches to child behavior have received the most empirical study. Public health officials and policymakers are encouraged to promote programs with empirically supported treatments. We now review the basic perspective, considerations, and criticisms of this approach to childrearing.

Basic Perspective   Behavior is learned through classical and operant conditioning. Classical conditioning involves presenting a stimulus with a reward. For example, infants learn to associate the faces of their parents with food, warmth, and comfort. Although initially only the food and feeling warm will satisfy the infant, later just the approach of the parent will soothe the infant. This may be observed when a father hands his infant to a stranger. The infant may cry because the stranger is not associated with pleasant events. But when the stranger hands the infant back to the parent, the crying may subside because the parent

Find out what a child will work for and what a child will work to avoid, systematically apply these consequences and you can change behavior.

Jack Turner, psychologist

represents positive events and the stimulus of the parent's face is associated with pleasurable feelings and emotional safety.

Other behaviors are learned through operant conditioning, which focuses on the consequences of behavior. Two principles of learning are basic to the operant explanation of behavior—reward and punishment. According to the reward principle, behaviors that are followed by a positive consequence will increase. If the goal is to teach the child to say "please," doing something the child likes after he or she says "please" will increase the use of "please" by the child. Rewards may be in the form of attention, praise, desired activities, or privileges. Whatever consequence increases the frequency of an occurrence is, by definition, a reward. If a particular consequence doesn't change the behavior in the desired way, a different reward needs to be tried.

The punishment principle is the opposite of the reward principle. A negative consequence following a behavior will decrease the frequency of that behavior; for example, the child could be isolated for 5 or 10 minutes following an undesirable behavior. The most effective way to change behavior is to use the reward and punishment principles together to influence a specific behavior. Praise children for what you want them to do and provide negative consequences for what they do that you do not like.

Considerations for Childrearing   Parents often ask, "Why does my child act this way, and what can I do to change it?" The behavioral approach to childrearing suggests the answer to both questions. The child's behavior has been learned through being rewarded for the behavior, and it can be changed by eliminating the reward for or punishing the undesirable behavior and rewarding the desirable behavior.

The child who cries when the parents are about to leave home to go to dinner or see a movie is often reinforced for crying by the parents' staying home longer. To teach the child not to cry when the parents leave, the parents should reward the child for not crying when they are gone for progressively longer periods of time. For example, they might initially tell the child they are going outside to walk around the house and they will give the child a treat when they get back if he or she plays until they return. The parents might then walk around the house and reward the child for not crying. If the child cries, they should be out of sight for only a few seconds and gradually increase the amount of time they are away. The essential point is that children learn to cry or not to cry depending on the consequences of crying. Because children learn what they are taught, parents might systematically structure learning experiences to achieve specific behavioral goals.

Criticisms of the Behavioral Approach   Professionals and parents have attacked the behavioral approach to childrearing on the basis that it is deceptively simple and does not take cognitive issues into account. Although the behavioral approach is often presented as an easy-to-use set of procedures for child management, many parents do not have the background or skill to implement the procedures effectively. What constitutes an effective reward or punishment, how it should be presented, in what situation and with what child, to influence what behavior are all decisions that need to be made before attempting to increase or decrease the frequency of a behavior. Parents often do not know the questions to ask or lack the training to make appropriate decisions in the use of behavioral procedures. One parent locked her son in the closet for an hour to punish him for lying to her a week earlier—a gross misuse of learning principles.

Behavioral childrearing has also been said to be manipulative and controlling, thereby devaluing human dignity and individuality. Some professionals feel that humans should not be manipulated to behave in certain ways through the use of rewards and punishments. Behaviorists counter that rewards and punishments are always involved in any behavior the child engages in and parents are simply encouraged to arrange them in a way that benefits the child in terms of learning appropriate behavior.

Finally, the behavioral approach has been criticized because it de-emphasizes the influence of thought processes on behavior. Too much attention, say the critics, has been given to rewarding and punishing behavior and not enough attention has been given to how the child perceives a situation. For example, parents might think they are rewarding a child by giving her or him a bicycle for good behavior. But the child may prefer to upset the parents by rejecting the bicycle and may be more rewarded by their anger than by the gift. Behaviorists counter by emphasizing they are very much focused on cognitions and always include the perception of the child in developing a treatment program (Crisp and Knox 2009).

Parent Effectiveness Training Approach

Thomas Gordon (2000) developed a model of childrearing based on parent effectiveness training (PET).

Basic Perspective   Parent effectiveness training focuses on what children feel and experience in the here and now—how they see the world. The method of trying to understand what the child is experiencing is active listening, in which the parent reflects the child's feelings. For example, the parent who is told by the child, "I want to quit taking piano lessons because I don't like to practice" would reflect, "You're really bored with practicing the piano and would rather have fun doing something else." PET also focuses on the development of the child's positive self-concept.

To foster a positive self-concept in their child, parents should reflect positive images to the child—letting the child know he or she is loved, admired, and approved of.

Considerations for Childrearing   To assist in the development of a child's positive self-concept and in the self-actualization of both children and parents, Gordon (2000) recommended managing the environment rather than the child, engaging in active listening, using "I" messages, and resolving conflicts through mutual negotiation.

An example of environmental management is putting breakables out of reach of young children. Rather than worry about how to teach children not to touch breakable knickknacks, it may be easier to simply move the items out of the children's reach.

The use of active listening becomes increasingly important as the child gets older. When Joanna is upset with her teacher, it is better for the parent to reflect the child's thoughts than to take sides with the child. Saying "You're angry that Mrs. Jones made the whole class miss play period because Becky was chewing gum" rather than saying "Mrs. Jones was unfair and should not have made the whole class miss play period" shows empathy with the child without blaming the teacher.

Gordon also suggested using "I" rather than "you" messages. Parents are encouraged to say "I get upset when you're late and don't call" rather than "You're an insensitive, irresponsible kid for not calling me when you said you would." The former avoids damaging the child's self-concept but still expresses the parent's feelings and encourages the desired behavior.

Gordon's fourth suggestion for parenting is the no-lose method of resolving conflicts. Gordon rejects the use of power by parent or child. In the authoritarian home, the parent dictates what the child is to do and the child is expected to obey. In such a system, the parent wins and the child loses. At the other extreme is the permissive home, in which the child wins and the parent loses. The alternative, recommended by Gordon, is for the parent and the child to seek a solution that is acceptable to both and to keep trying until they find one. In this way, neither parent nor child loses and both win.

Criticisms of the Parent Effectiveness Training Approach   Although much is commendable about PET, parents may have problems with two of Gordon's suggestions.

First, he recommends that because older children have a right to their own values, parents should not interfere with their dress, career plans, and sexual behavior. Some parents may feel they do have a right (and an obligation) to "interfere." Second, the no-lose method of resolving conflict is sometimes unrealistic.

Suppose a 16-year-old wants to spend the weekend at the beach with her boyfriend and her parents do not want her to. Gordon advises negotiating until a decision is reached that is acceptable to both. But what if neither the daughter nor the parents can suggest a compromise or shift their position? The specifics of how to resolve a particular situation are not always clear.

Socioteleological Approach

Alfred Adler, a physician and former student of Sigmund Freud, saw a parallel between psychological and physiological development. When a person loses her or his sight, the other senses (hearing, touch, taste) become more sensitive—they compensate for the loss. According to Adler, the same phenomenon occurs in the psychological realm. When individuals feel inferior in one area, they will strive to compensate and become superior in another. Rudolph Dreikurs, a student of Adler, developed an approach to childrearing that alerts parents as to how their children might be trying to compensate for feelings of inferiority (Soltz and Dreikurs 1991). Dreikurs's socioteleological approach is based on Adler's theory.

Basic Perspective   According to Adler, it is understandable that most children feel they are inferior and weak. From the child's point of view, the world is filled with strong giants who tower above him or her. Because children feel powerless in the face of adult superiority, they try to compensate by gaining attention (making noise, becoming disruptive), exerting power (becoming aggressive, hostile), seeking revenge (becoming violent, hurting others), and acting inadequate (giving up, not trying). Adler suggested that such misbehavior is evidence that the child is discouraged or feels insecure about her or his place in the family. The term socioteleological refers to social striving or seeking a social goal. In the child's case, the goal is to find a secure place within the family—the first "society" the child experiences.

Considerations for Childrearing   When parents observe misbehavior in their children, they should recognize it as an attempt to find security. According to Dreikurs, parents should not fall into playing the child's game by, say, responding to a child's disruptiveness with anger but should encourage the child, hold regular family meetings, and let natural consequences occur. To encourage the child, the parents should be willing to let the child make mistakes. If John wants to help Dad carry logs to the fireplace, rather than saying, "You're too small to carry the logs," Dad should allow John to try and should encourage him to carry the size limb or stick that he can manage. Furthermore, Dad should praise John for his helpfulness.

As well as being constantly encouraged, the child should be included in a weekly family meeting. During this meeting, such family issues as bedtimes, the appropriateness of between-meal snacks, assignment of chores, and family fun are discussed. The meeting is democratic; each family member has a vote. Participation in family decision making is designed to enhance the self-concept of each child. By allowing each child to vote on family decisions, parents respect the child as a person as well as the child's needs and feelings.

Resolutions to conflicts with the child might also be framed in terms of choices the child can make. "You can go outside and play only in the backyard, or you can play in the house" gives the child a choice. If the child strays from the backyard, he or she can be brought in and told, "You can go out again later." Such a framework teaches responsibility for and consequences of one's choices.

Finally, Driekurs suggested that the parents let natural consequences occur for their child's behavior. If a daughter misses the school bus, she walks or is charged taxi fare out of her allowance. If she won't wear a coat and boots in bad weather, she gets cold and wet. Of course, parents are to arrange logical consequences when natural consequences will not occur or would be dangerous if they did. For example, if a child leaves the television on overnight, access might be taken away for the next day or so.

Criticisms of the Socioteleological Approach   The socioteleological approach is sometimes regarded as impractical, since it teaches the importance of letting children take the natural consequences for their actions. Such a principle may be interpreted to let the child develop a sore throat if he or she wishes to go out in the rain without a raincoat. In reality, advocates of the method would not let the child make a dangerous decision. Rather, they would give the child a choice with a logical consequence such as "You can go outside wearing a raincoat, or you can stay inside—it is your choice."

Attachment Parenting

Dr. William Sears, along with his wife, Martha Sears, developed an approach to parenting called attachment parenting (Sears and Sears 1993). This "commonsense parenting" approach focuses on parents connecting with their baby.

Basic Perspective   The emotional attachment process between mother and child is thought to begin prior to birth and continues to be established during the next 3 years. Sears identified three parenting goals: to know your child, to help your child feel right, and to enjoy parenting. He also suggested five concepts or tools (identified in the following section) that comprise attachment parenting that will help parents to achieve these goals. Overall, the ultimate goal is for parents to get connected with their baby. Once parents are connected, it is easy for parents to figure out what works for them and to develop a parenting style that fits them and their baby. Meeting a child's needs early in life will help him or her form a secure attachment with parents. This secure attachment will help the child to gain confidence and independence as he or she grows up. Attachment Parenting International is a nonprofit organization committed to educating society and parents about the critical emotional and psychological needs of infants and children.

Considerations for Childrearing   The first attachment tool is for parents to connect with their baby early. The initial months of parenthood are a sensitive time for bonding with your baby and starting the process of attachment.

The second tool is to read and respond to the baby's cues. Parents should spend time getting to know their baby and learn to recognize his or her unique cues. Once a parent gets in tune with the baby's cues, it is easy to respond to the child's needs. Sears encourages parents to be open and responsive and to pick their baby up if he or she cries. Responding to a baby's cries helps the baby to develop trust and encourages good communication between child and parent. Eventually, babies who are responded to will internalize their security and will not be as demanding.

The third attachment tool is for mothers to breast-feed their babies and to do this on demand rather than trying to follow a schedule. He emphasizes the important role that fathers also play in successful breast-feeding by helping to create a supportive environment.

The fourth concept of attachment parenting is for parents to wear their baby by using a baby sling or carrier. The closeness is good for the baby and it makes life easier for the parent. Wearing your baby in a sling or carrier allows parents to engage in regular day-to-day activities and makes it easier to leave the house.

Finally, Sears advocates that parents let the child sleep in their bed with them since it allows parents to stay connected with their child throughout the night. However, some parents and babies often sleep better if the baby is in a separate crib, and Sears recognizes that wherever parents and their baby sleep best is the best policy. Wherever you choose to have your baby sleep, Sears is clear on one thing—it is never acceptable to let your baby cry when he or she is going to sleep! Parents need to parent their child to sleep rather than leaving him or her to cry.

TABLE 11.1 Theories of Child Rearing				
Theory	**Major Contributor**	**Basic Perspective**	**Focal Concerns**	**Criticisms**
Developmental-Maturational	Arnold Gesell	Genetic basis for child passing through predictable stages	Motor behavior Adaptive behavior Language behavior Social behavior	Overemphasis on biological clock Inadequate sample to develop norms Demanding schedule questionable Upper-middle class bias
Behavioral	B. F. Skinner B. Crisp	Behavior is learned through operant and classical conditioning	Positive reinforcement Negative reinforcement Punishment Extinction Stimulus response	De-emphasis on cognitions of child Theory too complex for parent to accurately or appropriately apply Too manipulative or controlling Difficult to know reinforcers and punishers in advance
Parent Effectiveness Training	Thomas Gordon	The child's world view is the key to understanding the child	Change the environment before attempting to change the child's behavior Avoid hurting the child's self-esteem Avoid win-lose solutions	Parents must sometimes impose their will on the child's How to achieve win-win solutions is not specified
Socioteleological	Alfred Adler	Behavior is seen as attempt of child to secure a place in the family	Insecurity Compensation Power Revenge Social striving Natural consequences	Limited empirical support Child may be harmed taking "natural consequences"
Attachment	William Sears	Goal is to establish a firm emotional attachment with child	Connecting with baby Responding to cues Breast-feeding Wearing the baby Sharing sleep	May result in spoiled, overly dependent child Exhausting for parents

Criticisms of Attachment Parenting Some parents feel that responding to their baby's cries, carrying or wearing their baby, and sharing sleep with their baby will lead to a spoiled baby who is overly dependent. Some parents may feel more tied down using this parenting approach and may find it difficult to get their child on a schedule. Many women return to work after the baby is born and find some of the concepts difficult to follow. Some women choose not to breast-feed their children for a variety of reasons. Finally, the idea of sharing sleep has resulted in a lot of criticism. Some parents might be nervous that they might roll over on the child, that the child might disturb their sleep or intimacy with their partner, or that sharing a bed will mean that they will never get their child to sleep on his or her own. However, children who grow up in an emotionally secure environment and have strong attachment report less behavioral and substance abuse problems as adolescents (Elgar et al. 2003).

Each of the approaches to child rearing has some value. How a child misbehaves can be viewed as responding to previous learning (behavioral approach), trying to enhance connectedness (attachment approach) or establish one's place in the family (socioteleological approach). Approaches which result in behavior change might be favored. Children benefit from an environment conducive to encouraging desirable behavior (e.g. being polite, respspectful, helpful). Such behaviors engnder the positive reaction of others (which have implications for the child's positive self-concept).

Single-Parenting Issues

Kate Gosselin is one of the most notable single moms. She was seen on the reality show *Jon and Kate Plus 8* (later *Kate Plus 8*), a real life story with her former husband and eight young children. Now divorced from husband Jon, she has learned how to be a single parent. At least half of all children will spend one-fourth of their lives in a female-headed household (Webb 2005). The stereotype of the single parent is the unmarried black mother. In reality, 40% of single mothers are white and only 33% are black (Sugarman 2003).

Distinguishing between a single-parent "family" and a single-parent "household" is important. A **single-parent family** is one in which there is only one parent; the other parent is completely out of the child's life through death, sperm donation, or complete abandonment, and no contact is ever made with the other parent. In contrast, a **single-parent household** is one in which one parent typically has primary custody of the child or children but the parent living out of the house is still a part of the child's family. This is also referred to as a binuclear family. In most divorce cases where the mother has primary physical custody of the child, the child lives in a single-parent household because the child is still connected to the father, who remains part of the child's family. In cases in which one parent has died, the child or children live with the surviving parent in a single-parent family because there is only one parent.

Single Mothers by Choice

Single parents enter their role though divorce or separation, widowhood, adoption, or deliberate choice to rear a child or children alone. Jodie Foster, Academy Award winning actress, has elected to have children without a husband. She now has two children and smiles when asked, "Who's the father?" The implication is that she has a right to her private life and that choosing to have a single-parent family is a viable option. Jennifer Aniston was the actress in *Switch*, a movie about a single woman who turns to a sperm donor to have a baby. In her private life, Jennifer Aniston is in her early forties and single. She is on record as saying,

Single-parent family family in which there is only one parent and the other parent is completely out of the child's life through death, sperm donation, or abandonment and no contact is made with the other parent.

Single-parent household one parent has primary custody of the child/children with the other parent living outside of the house but still being a part of the child's family; also called binuclear family.

"single woman don't have to settle with a man just to have a baby." An organization for women who want children and who may or may not marry is Single Mothers by Choice.

Bock (2000) noted that single mothers by choice are, for the most part, in the middle- to upper-class, mature, well-employed, politically aware, and dedicated to motherhood. Interviews with twenty-six single mothers by choice revealed their struggle to avoid stigmatization and to seek legitimization for their choice. Most felt that their age (older), sense of responsibility, maturity, and fiscal capability justified their choice. Their self-concepts were those of competent, ethical, mainstream mothers.

Challenges Faced by Single Parents

The single-parent lifestyle involves numerous challenges, including some of the following issues:

1. *Responding to the demands of parenting with limited help.* Perhaps the greatest challenge for single parents is taking care of the physical, emotional, and disciplinary needs of their children alone. Solem et al. (2011) noted that single parents are more often less educated, less likely to be employed, and have less social support, and that these factors weigh against a protective frame around a child resulting in more behavioral problems in the children.

2. *Coping with adult psychological/assaultive issues alone.* Samuels-Dennis et al. (2011) noted that single mothers are not immune to psychological and assaultive traumas. They studied a sample of 247 single mothers and found that 31% met the criteria for a probable PTSD diagnosis—between 78% and 80% reported 1 or more instances of lifetime adversities, psychological traumas or assaultive traumas. In the absence of a partner, single parents must find other ways of coping with these events since children alone do not have the capacity to be in that role. One single mother said, "I'm working two jobs (and being sexually harassed at one of them), taking care of my kids, and trying to go to school. Plus my mother has cancer. Who am I going to talk to to help me get through all of this?" Many single women solve the dilemma with a network of friends.

3. *Resolving the issue of adult sexual needs.* Some single parents regard their parental role as interfering with their sexual relationships. They may be concerned that their children will find out if they have a sexual encounter at home or be frustrated if they have to go away from home to enjoy a sexual relationship. Some choices with which they are confronted include, "Do I wait until my children are asleep and then ask my lover to leave before morning?" or "Do I openly acknowledge my lover's presence in my life to my children and ask them not to tell anybody?" and "Suppose my kids get attached to my lover, who may not be a permanent part of our lives?"

4. *Lack of money.* Single-parent families, particularly those headed by women, report that money is always lacking.

National **Data**

The median income of a single-woman householder is $25,014, much lower than that of a single-man householder ($36,006) or a married couple ($73,010) (*Statistical Abstract of the United States 2011*, Table 691).

5. *Guardianship.* If the other parent is completely out of the child's life, the single parent needs to appoint a guardian to take care of the child in the event of the parent's death or disability.

6. *Prenatal care.* Single women who decide to have a child have poorer pregnancy outcomes than married women. Their children are likely to be born prematurely and to have low birth weight (Mashoa et al. 2010) The reason for such

an association may be the lack of economic funds (no male partner with economic resources available) as well as the lack of social support for the pregnancy or the working conditions of the mother, all of which result in less prenatal care for their babies.

7. *Absence of a father.* Another consequence for children of single-parent mothers is that they often do not have the opportunity to develop an emotionally supportive relationship with their father. Barack Obama noted in one of his campaign speeches, "I know what it is like to grow up without a father." The late Rodney Dangerfield said he spent an average of two hours a year with his father in his entire life and that he missed such love and nurturing. In contrast, Ansel Adams, the late photographer, attributed his personal and life success to his father, who was steadfastly involved in his life and guided his development. Shook et al. (2010) noted that when single mothers receive support from the father who is active in the role of co-parent, child competence resulted. When neither of these occurs, child maladjustment was more likely.

8. *Negative life outcomes for the child in a single-parent family.* Researcher Sara McLanahan, herself a single mother, set out to prove that children reared by single parents were just as well off as those reared by two parents. McLanahan's data on 35,000 children of single parents led her to a different conclusion: children of only one parent were twice as likely as those reared by two married parents to drop out of high school, get pregnant before marriage, have drinking problems, and experience a host of other difficulties, including getting divorced themselves (McLanahan and Booth 1989; McLanahan 1991). In addition, Freeman and Temple (2010) found that adolescents from single parent homes were more likely to be raped than those from two parent homes. Lack of supervision, fewer economic resources, and less extended family support were among the culprits.

9. *Perpetuation of single family structure.* Growing up in a single family home increases the likelihood that the adult child will have a first child while unmarried, thus perpetuating the single family structure. Hofferth and Goldscheider (2010) found this outcome particularly true for males. If they are born into a single parent family they are more likely to have children outside of marriage.

Though the risk of negative outcomes is higher for children in single-parent homes, most are happy and well-adjusted. Benefits to single parents themselves include a sense of pride and self-esteem that results from being independent.

The Future of Parenting

The future of parenting will involve new contexts for children and new behaviors that children learn and parents tolerate. While parents will continue to be the primary context in which their children are reared, because the financial need for both parents to earn an income will increase, children will, increasingly, end up in day care, afterschool programs, and day camps during the summer. These changed contexts will result in new parental norms where a wider range of behaviors on the part of their children are accepted. Hence, since parents will be increasingly preoccupied with their job/careers, the norms their children are learning in day care and other contexts will be more readily accepted since parents will have less time and energy trying to reverse them. Hence, children may be less polite, less obedient, and less compliant to authority resulting more often in parental acceptance than addressing or correcting each behavior. We are not suggesting that children will become wild hellions, only that the behaviors they learn in contexts other than home will be less restrictive.

Summary

What are the basic roles of parents?

Parenting includes providing physical care for children, loving them, being an economic resource, providing guidance as a teacher or model, protecting them from harm, promoting their health, and providing meaningful family rituals. Seven skills all children must have (that parents are responsible for teaching) are: focus and self control, seeing someone else's point of view, communicating, making connections, critical thinking, taking on challenges, and self-directed engaged learning. One of the biggest problems confronting parents today is the societal influence on their children. These include drugs and alcohol; peer pressure; TV, Internet, and movies; and crime or gangs.

What is a choices perspective of parenting?

Although both genetic and environmental factors are at work, the choices parents make have a dramatic impact on their children. Parents who don't make a choice about parenting have already made one. The five basic choices parents make include deciding (1) whether to have a child, (2) the number of children, (3) the interval between children, (4) one's method of discipline and guidance, and (5) the degree to which one will be invested in the role of parent.

What is the transition to parenthood like for women, men, and couples?

Transition to parenthood refers to that period of time from the beginning of pregnancy through the first few months after the birth of a baby. The mother, father, and couple all undergo changes and adaptations during this period. Most mothers relish their new role; some may experience the transitory feelings of baby blues; a few report postpartum depression.

The father's involvement with his children is sometimes predicted by the quality of the parents' romantic relationship. If the father is emotionally and physically involved with the mother, he is more likely to take an active role in the child's life. In recent years, there has been a renewed cultural awareness of fatherhood.

A summary of almost 150 studies involving almost 50,000 respondents on the question of how children affect marital satisfaction revealed that parents (both women and men) reported lower marital satisfaction than nonparents. In addition, the higher the number of children, the lower the marital satisfaction; the factors that depressed marital satisfaction were conflict and loss of freedom. The new couple also gradually withdraws from friends and spends increasing amounts of time at home.

What are several facts about parenthood?

The concept of childhood, like gender, has been socially constructed rather than being a fixed life stage. From the thirteenth through the sixteenth centuries, children were viewed as innocent, sweet, and a source of amusement for adults. From the sixteenth through the eighteenth centuries, they were viewed as in need of discipline and moral training (to educate them into adult salvation). In the nineteenth century, whippings were routine as a means of breaking children's spirits and bringing them to submission. Although remnants of both the innocent and moralistic views of children exist today, the lives of children are greatly improved. Child labor laws protect children from early forced labor and education laws ensure a basic education.

Parenthood will involve about 40% of the time a couple live together, parents are only one influence on their children, each child is unique, and parenting styles differ. Research suggests that an authoritative parenting style, characterized as both demanding and warm, is associated with positive outcomes. In addition, being emotionally connected to a child, respecting the child's individuality, and monitoring the child's behavior to encourage positive contexts have positive outcomes.

What are some of the principles of effective parenting?

Giving time, love, praise, and encouragement; monitoring the activities of one's child; setting limits; encouraging responsibility; and providing sexuality education are aspects of effective parenting. Acceptance is associated with positive outcomes for the child. Undergraduates who perceived acceptance from their parents (as opposed to rejection) are less likely to engage in various health risk behaviors such as hard drug use, inhalant use, prescription drug misuse, casual sex, driving while intoxicated, and riding with an impaired driver.

What are the issues of single parenting?

About 40% of all children will spend one-fourth of their lives in a female-headed household. The challenges of single parenthood for the parent include taking care of the emotional and physical needs of a child alone, meeting one's own adult emotional and sexual needs, money, and rearing a child without a father (the influence of whom can be positive and beneficial).

What is the future of parenting?

Focused on income getting, parents will depend more on secondary/nonfamily resources to take care of and rear their children. The result will be a wider set of norms that children learn and behaviors they engage in with less parental correction. Parents will remain in control of their children but less so than in the past.

Key Terms

Baby blues

Demandingness

Gatekeeper role

Oppositional defiant disorder

Overindulgence

Oxytocin

Parenting

Parenting self-efficiency

Postpartum depression

Postpartum psychosis

Reactive attachment disorder

Responsiveness

Sextortion

Single-parent family

Single-parent household

Time-out

Transition to parenthood

Web Links

Attachment Parenting International
http://www.attachmentparenting.org/

Family Wellness Workshops
http://www.familywellness.com/

The Partnership for a Drug-Free America
http://www.drugfree.org/

The Children's Partnership Online
http://www.childrenspartnership.org

National Fatherhood Initiative
http://www.fatherhood.org/

CHAPTER 12

Work and Family Life

Coming together is a beginning.
Keeping together is progress.
Working together is success.

Henry Ford, founder of Ford Motor

Chelsea E. Curry

Learning Objectives

Discuss how money, and the lack of it—poverty—affects relationships.

Review money as power, office romance, and working wives.

Identify the effect of a mother's employment on her children.

Review the various strategies of balancing work and family life.

Predict the future of the effect of work on family life.

1. Undergraduates fantasize about having an emotional and sexual relationship with someone at work about as often as such emotional and sexual relationships actually occur.

2. "Supermoms" with competent, helping husbands may have a lower sense of self-competence since they "should be able to take care of everything."

3. Economists agree that the way to evaluate the worth of a wife is to calculate what it would cost to replace her labor.

4. One's emotional well being tops out at an income of about $50,000; over that amount there are few emotional gains.

5. A woman who is unhappy in her marriage and takes a well-paying job is vulnerable to divorcing her husband.

Answers: 1. F   2. T   3. F   4. F   5. T

The devastating oil spill off the coast of Louisiana in 2010 made clear how money affects the family. With oil destroying the wetlands, fish, shrimp, and other seafood, the way of life for many of the local population halted. Life as they knew it had come to an end. Although the oil leak has been plugged, the spouses, marriages, and families continue to recover.

The larger economy (e.g., unemployment, etc.) has also had an effect on marriages and families throughout the United States A deep recession beginning in mid-2008 and a slow economic recovery has been reflected in job layoffs, housing foreclosures, and couples fearful of their economic future. The consequences can be catastrophic. Diem and Pizarro (2010) found a link between homicides in the family—intimate partner (murder of a spouse), filicide (murder of an offspring by a parent), parricide (murder of a parent by an offspring), and siblicide (murder of a sibling)—and economic deprivation. Impoverished families, including those riddled with unemployment, were contexts for homicide. Policy implications include the need for society to help those in need before financial strain reaches a point of violence and murder.

Compounding the effect of the environment and the economy on marriages and families is the entrenched value of **consumerism**—to buy everything and to have everything now. Hence, the real stresses that money inflicts on relationships are the result of internalizing the societal expectations of who one is, and should be, in regard to the pursuit of money. Taking a promotion, getting a new car, and buying the bigger house come with economic, emotional, and relationship stress. A wife whose husband had just bought a second McDonald's franchise said, "It's not fun anymore. We have money but I never see him. It isn't worth it." (The husband subsequently sold both stores. The couple had less money, but their marriage recovered.)

This chapter focuses on how money or the lack of it affects the health and well-being of spouses, parents, and children. We also discuss the effect of work on marriage and family life. Because the American economy is in the process of enormous change, we begin with the effect on American workers.

Consumerism economic societal value that one must buy everything and have everything now.

We worry too much about something to live on—and too little about something to live for.

Jimmy Carter, former President

Money and Relationships

While individuals may marry for love, it is money which pays the rent, utilities, and food bills. After looking at how money is exchanged between adult children and their parents, we look at the effects of the recession on spouses, and debt/poverty as they affect individuals, marriages, and families.

Reciprocity—Time/Care for Money between Children and Parents

There is reciprocity in most relationships. Once adult children leave the family home, their continued involvement with children may include reciprocity from parents in the form of money. Leopold and Raab (2011) found that parents give money to their children who reciprocate by spending time with the parents. In those cases where there is intense investment and care given by children to their elderly parents, there was considerable financial reciprocation of money by parents.

Effects of the Recession

Rick Newman (2010) identified six economic changes that affect American workers in their relationships, marriages, and families.

1. *Tax increases.* With the government's debt at 14 trillion dollars (which Obama and the Republications are in conflict over how to reduce) and more than two thirds of the states facing budget shortfalls, individuals will find that they pay more taxes.

2. *Government services decreases.* Examples are education, Medicaid, and other services for the poor. The postal service will reduce hours and service, which is only the beginning. Medicare and Social Security are not immune from being cut back.

3. *Retirement age increases.* The official retirement age—the point at which you can claim full Social Security benefits—is rising gradually from 65 to 67 by the year 2027. The result is that people will end up working longer.

4. *Decreased income.* With an excess supply of workers, employers will be able to pay less. U.S. companies are also substituting technology or cheap foreign labor for American workers, further depressing hiring and incomes.

5. *Uncertainty.* As foreign competition gets tougher, technology affects the way companies operate and their needs.

6. *Big institutions will no longer take care of their workers.* The days of a secure pension a worker can count on are over.

It isn't so much that hard times are coming; the change observed is mostly soft times going.

Grocho Marx, comedian

Effects of Poverty on Marriages and Families

Poverty is devastating to spouses, parents, and children (Edin and Kissane 2010). Those living in poverty have poorer physical and mental health, report lower personal and marital satisfaction, and die sooner. The anxiety over lack of money may result in relationship conflict. Hardie and Lucas (2010) analyzed data from over 4,000 respondents in both married and cohabitation contexts and

WHAT IF?

What if You and Your Partner Feel Differently about Being in Debt?

Some individuals have always been in debt, regard it as a given in life, and have no anxiety about debt. Others have never been in debt, view it as something to avoid, and become suicidal when thinking about it (Meltzer et al. 2011). These different philosophies may create conflict. Resolution of this dilemma usually comes through deferring to the preferences of the partner with the most anxiety. For example, the partner who can't sleep if in debt will be in greater distress and will create more stress in the relationship than the partner who can tolerate a great deal of debt. Hence, the least negative consequences for the couple are in favor of having minimal debt.

Beware of the little expenses; a small leak will sink a great ship.

Benjamin Franklin, Founding Father of the United States

When money is tight, yard sales are beneficial to both seller and buyer—the seller generates money by selling old belongings and the buyer saves money by purchasing used items cheaply.

David Knox

found that economic hardship was associated with more conflict in both sets of relationships. Predictably, money (the lack of it, disagreement over how it is spent) is a frequent problem that couples report in marriage counseling. Cokes and Kornblum (2010) noted the negative mental health outcome of a severe economic recession on individuals' reported mental health and emphasized that the effects cut across gender and race. Baek and DeVaney (2010) confirmed that most families experiencing economic hardship responded by increasing their debt and reducing their savings to make up the difference between income and spending.

Kahneman and Deaton (2010) defined emotional well-being as the quality of an individual's everyday experience—the frequency and intensity of experiences of joy, stress, sadness, anger, and affection that make one's life pleasant or unpleasant. They found that reporting emotional well being rises with income and tops out at $75,000 a year. Increases in income above this figure are not associated with increases in satisfaction. Stanley and Einhorn (2007) suggested that the reason money is such a profound issue in marriage is its symbolic significance (for example, power and control) as well as the fact that individuals do something daily in reference to money—spend, save, or worry about it.

The stresses associated with low income also contribute to substance abuse, domestic violence, child abuse and neglect, divorce, and questionable parenting practices. For example, economic stress is associated with greater marital discord, and couples with incomes less than $25,000 are 30% more likely to divorce than couples with incomes greater than $50,000 (Whitehead and Popenoe 2004). Child neglect is more likely to be found with poor parents who are unable to afford child care or medical expenses and leave children at home without adult supervision or fail to provide needed medical care. Indeed, Rafferty and Wiggan (2011) studied mothers in the United Kingdom on welfare and found that a substantial number do not want a job and that their main reported reason is that they are looking after their children. Those poor parents who are with their children are more likely than other parents to use harsh physical disciplinary techniques.

Another family problem associated with poverty is teenage pregnancy. Poor adolescent girls are more likely to have babies as teenagers or to become young

Get as much pleasure from your saving as from your spending.

Suze Ormond, investment guru

Chapter 12 Work and Family Life

single mothers. Early childbearing is associated with numerous problems, such as increased risk of having premature babies or babies of low birth weight, dropping out of school, and earning less money as a result of lack of academic achievement.

One of the quickest ways to poverty is to become overwhelmed with medical debt. Indeed, most individuals filing for bankruptcy do so because of medical bills. Below are basic provisions of the "Health Care Bill" (officially called the Patient Protection and Affordable Healthcare Act) passed in 2010. These provisions include:

1. *Coverage till age 26.* After graduating from college, some students can't find a job immediately and have no employer to provide health care. The bill allows them to be covered by their parents' policy up until age 26. Previously many insurance policies would not allow children who reached the age of 19 to be covered.

2. *Policy can't be canceled.* Previously, insurance companies could drop a person's policy if they got sick, but this will be no more.

3. *Insurance cannot be denied.* Previously, if you or your children had preexisting conditions or were a bad health risk, you could not get insurance. Thirty-six percent of Americans could not get coverage (Tumulty 2010). Though this won't go into full effect until 2014, this will also be no more.

4. *No cap on the coverage you need.* Previously, an insurance company would only pay a limited amount for an illness. Now there is no limit.

5. *Coverage available now.* Previously, insurance companies could force you to wait six months before coverage could begin. You can now buy insurance with a cap on what it will cost ($6000 per individual; $12,000 per family).

6. *You must be insured.* Starting in 2014, the lowest fine for not buying insurance will be $95 or 1% of a person's income (whichever is greater) and then increase to a high of $695 or 2.5% of an individual's taxable income by 2016.

7. *Persons with higher incomes will pay more taxes.* Starting in 2018, if your combined family income exceeds $250,000 you are going to be taking less money home each pay period. This is because you will have more money deducted from your paycheck to go toward increased Medicare payroll taxes. In addition to higher payroll taxes you will also have to pay 3.8% tax on any unearned income, which is currently tax-exempt.

Financial Behaviors of Undergraduates and Young Adults

Gutter et al. (2010) examined the financial behaviors of 15,797 undergraduate students. In general, more than half (51.7%) reported that they do not use a budget and a majority (52.2%) reported that they are saving money. Students who budget and save reported more discussion of finances with parents and friends and observed more parents' and friends' financial behaviors. Being white, pair-bonded, and older was associated with budgeting and saving.

Rutherford and Fox (2010) provided data on the "financial wellness" of a sample of 458 young adults between the ages of 18 and 30 who were the head of household. Such wellness was defined as whether the individual had enough cash stashed away to cover bills for 2.5 months if there was a sudden loss of income. Less than half (47%) met the criterion. Asset allocation was another index of financial wellness. Such wellness was defined as having at least 15% of all household investments in liquid assets; only 35% met the criterion. Only 28% of young households met the guidelines for both the liquidity and asset allocation ratios. "This finding is critical to financial planners, financial educators, and legislators in realizing the magnitude of the problem and taking action to improve financial literacy and financial behavior among young adults." One answer is communication with parents. Serido et al. (2010) noted that parents who have open and supportive discussions with their first year college students about financial topics increase the chance that their children will acquire financial competence and contribute to self-sufficiency in adulthood.

If you want to feel rich, just count the things you have that money can't buy.

Chinese Proverb

We don't need to increase our goods nearly as much as scaling down our wants. Not wanting something is as good as possessing it.

Donald Horban, pastor

Palmer et al. (2010a) surveyed 170 undergraduates and found that they were stressed and depressed about their economic situation; they reported spending too much money eating out, buying alcohol, etc. The students participated in an online money management program from the Financial Wellness Group (see Weblinks at end of chapter) and reported significant improvement in their spending behavior, reduction of stress, etc.

Work and Marriage: Effects on Spouses

A couple's marriage is organized around the work of each spouse. Where the couple lives is determined by where the spouses can get jobs. Jobs influence what time spouses eat, which family members eat with whom, when they go to bed, and when, where, and for how long they vacation. In this section, we examine some of the various influences of work on a couple's relationship. We begin by looking at money as power.

Money as Power in a Couple's Relationship

Money is a central issue in relationships because of its association with power, control, and dominance. Generally, the more money a partner makes, the more power that person has in the relationship. Males make considerably more money than females and generally have more power in relationships. However, Morin and Cohn (2008) reported on a national sample and found that, although two-thirds of all husbands in dual-income families say they make more money than their wives, women are still more likely to make the decisions in more areas (42% versus 30%). In addition, Dema-Moreno and Díaz-Martínez (2010) studied Spanish dual income couples and found that a higher income on the part of the wife does not necessarily translate into a gender balance in the relationship since more traditional gender roles and attitudes may persist. Other studies (Murphy-Graham 2010) reveal that higher education increases a woman's power in the relationship, which empowers her to negotiate for more equitable roles.

When we try in good faith to believe in materialism . . . we are disavowing the very realm where we exist and where all things precious are kept—the realm of emotion and conscience; of memory and intention and sensation.

John Updike, novelist

When a wife earns an income, her power then increases in the relationship. We know of a married couple in which the wife recently began to earn an income. Before doing so, her husband's fishing boat was in the protected carport. With her new job and increased power in the relationship, she began to park her car in the carport and her husband put his fishing boat underneath the pine trees to the side of the house. Money also provides an employed woman the power to be independent and to leave an unhappy marriage. Indeed, the higher a wife's income, the more likely she is to leave an unhappy relationship

(Schoen et al. 2002). Similarly, because adults are generally the only source of money in a family, they have considerable power over children, who have no money.

To some individuals, money also means love. While admiring the engagement ring of her friend, a woman said, "What a big diamond! He must really love you." A cultural assumption is that a big diamond equals an expensive diamond and a lot of sacrifice and love. Similar assumptions are often made when gifts are given or received. People tend to spend more money on presents for the people they love, believing that the value of the gift symbolizes the depth of their love. People receiving gifts may make the same assumption. "She must love me more than I thought," mused one man. "I gave her a Blu-Ray movie for Christmas, but she gave me a Blu-Ray player. I felt embarrassed."

PERSONAL CHOICES

Work or Relationships?

People who consistently choose their work over their relationships either have partners who have also made such choices or partners who may be disenchanted. Traditionally, men have chosen their work over relationships; women have chosen their relationships over work. Most choose to balance the two to afford a lifestyle they enjoy. Professions or careers that inherently provide the opportunity for balance include elementary school teaching, where one is home by 4:00 with a couple of months off in the summer. The role of college or university teacher is even better, with greater flexibility during the day, week, and year.

Alternatively, a **momprenuer** is a woman who has a successful at-home business. She conducts her business from home, which provides her with maximum flexibility for her family. Barbie Tew is an example of a momprenuer who operates a successful "passion party" business out of her home (http://barbiespassion.com).

Some individuals require very little income and have no interest in material wealth. We have a friend who has little to no regard for material wealth. He has been homeless and now runs a homeless shelter where he daily feeds two meals a day to over 100 individuals. At night, he goes back to his $200 a month loft where he reads and paints until the next day when he feeds the homeless though private funding and donations. We asked him how his lack of concern for money affects the interest women have in establishing a long-term relationship with him. "It kills it," he noted. "They simply have no interest in being pair-bonded to someone living this vagabond existence."

Momprenuer a woman who has a successful at-home business.

Chase your passion, not your pension.

Denis Waitley, writer

Working Wives

Driven primarily by the need to provide income for the family, 70% of all U.S. wives with children are in the labor force. The time wives are most likely to be in the labor force is when their children are teenagers (between the ages of 14 and 17), the time when food and clothing expenses are the highest (*Statistical Abstract of the United States 2011,* Tables 598 and 599). The stereotypical family consisting of a husband who earns the income and a wife who stays at home with two or more children is no longer the norm. Only 13% are "traditional" in the sense of a breadwinning husband, a stay-at-home wife, and their children (Stone 2007). In contrast, most marriages may be characterized as dual-earner couples (Amato et al. 2007).

Because women still take on more child care and household responsibilities than men, women in dual-earner marriages (marriages in which both spouses provide significant income to the family unit) are more likely than men to want to be employed part-time rather than full-time. If this is not possible, many women prefer to work only a portion of the year (the teaching profession allows employees to work about ten months and to have two months in the summer free). Although many low-wage earners need two incomes to afford basic housing and a minimal standard of living, others have two incomes to afford expensive homes, cars, vacations, and educational opportunities for their children. Whether it makes economic sense is another issue.

Some parents wonder if the money a wife earns by working outside the home is worth the sacrifices to earn it. Not only is the mother away from their children but she must also pay for strangers to care for their children. Sefton (1998) calculated that the value of a stay-at-home mother is $36,000 per year in terms of what a dual-income family spends to pay for all services that she provides (domestic cleaning, laundry, meal planning and preparation, shopping, providing transportation to activities, taking the children to the doctor, and running errands). Adjusting for changes in the consumer price index, this figure was around $50,000 in 2012. The value of a househusband would be the same. However, because males typically earn higher incomes than females, the loss of income would be greater than for a female.

Economist Allen (2010) observed that this economic value of the wife is off the mark—that spouses select each other on the basis of their perceived willingness to contribute equally to the relationship. He further states that the wife gives birth to and raises children and that this value cannot be calculated; economists who think otherwise (quoting from Oscar Wilde) "know the price of everything but the value of nothing." Putting a price on the value the wife contributes to the marriage is to attempt to "measure the unmeasurable." Those who use the dollar figure of the wife's value may conclude that working outside the home may not be as economically advantageous as one might think. Of course there are other reasons for employment than money, such as psychological (enhanced self-concept) and social (enlarged social network) benefits.

The **mommy track** (stopping paid employment to spend time with young children) is another potential cost to those women who want to build a career. Taking time out to rear children in their formative years can derail a career. Noonan and Corcoran (2004) found that female lawyers who took time out for child responsibilities were less likely to make partner and more likely to earn less money if they did make partner. Aware that executive women have found it difficult to reenter the workforce after being on the mommy track, the Harvard Business School created an executive training program for mothers to improve their technical skills and to help them return to the workforce (Rosen 2006).

Wives Who "Opt Out"

Pamela Stone (2007) published *Opting Out*, which revealed content from fifty-four interviews with women representing a broad spectrum of professions: doctors, lawyers, scientists, bankers, management consultants, editors, and teachers. **Opting out** involved women leaving their careers and returning home to take care of their children for a variety of reasons. However, two reasons stood out: (1) husbands who were unavailable or unable to "shoulder significant portions of caregiving and family responsibilities" (p. 68); and (2) employers who had a lot of policies on the books to encourage and support women parental leave "but not much in the way of making it possible for them to return or stay once they had babies" (p. 119). Part-time work didn't work out for these women because the work was not really "part-time"; the employer kept wanting more.

Keller (2008b) revealed a similar story, that career women who left work for the home found a lower salary and a demotion, and were sidelined on their return. They were, indeed, punished at work for prioritizing family. Nevertheless, women who are flexible, creative, and determined can find work that meets their needs and can be a good fit for their employer.

There is disagreement about the wisdom of "opting out." Bennetts (2007) argued that educated women with careers are foolish to leave their lucrative careers and put on their aprons. She emphasized that men cannot

Mommy track stopping paid employment to spend time with young children.

An individual is rich in proportion to the number of things he or she can live without.

Henry David Thoreau, American author

Opting out when a woman leaves a career and returns home to take care of her children.

This couple worked together in the same office for five years before they began to see each other outside of work. They are now engaged to be married.

be counted on—that they leave the marriage for younger women or die. The wife is then left to support herself and her children and will need a good income to do so.

Office Romance

The office or workplace is where people earn money to pay bills and pay down their debt. It is also a place where they meet and establish relationships, including love and sexual relationships. In a survey conducted by Vault.com of 1050 employees in the workplace, almost 60% (59%) said that they had been involved in an office romance (Stott 2010). CareerBuilder.com (2010) also surveyed 5231 full-time, not self-employed, nongovernment employees and found that almost 40% (37%) of their respondents reported an office romance. About a third (32%) of those in the CareerBuilder.com survey who had had an office romance reported that it ended in marriage. Merrill interviewed 70 adults who had been involved in an office romance. Of these interviews she reported that ". . . the office is still a scene of seduction and amorality. The hallmark of office romances is secrecy—either hiding a steamy short or long love relationship between two single people or where one or both is married or in a serious relationship" (Merrill and Knox 2010). Undergraduates have also experienced romance on the job (see Applying Social Research feature).

Types of Dual-Career Marriages

A **dual-career marriage** is defined as one in which both spouses pursue careers and maintain a life together that may or may not include dependents. A career is different from a job in that the former usually involves advanced education or training, full-time commitment, working nights and weekends "off the clock," and a willingness to relocate. Dual-career couples operate without a person who stays home to manage the home and care for dependents.

Nevertheless, four types of dual-career marriages are those in which the husband's career takes precedence (**HIS/her career**), the wife's career takes precedence (**HER/his career**), both careers are regarded equally (**HIS/HER career marriage**), or both spouses share a career or work together (**THEIR career marriage**).

When couples hold traditional gender role attitudes, the husband's career is likely to take precedence (HIS/her career). This situation translates into the wife being willing to relocate and to disrupt her career for the advancement of

Dual-career marriage a marriage in which both spouses pursue careers.

HIS/her career dual career marriage in which husband's career takes precedence.

HER/his career dual career marriage in which wife's career takes precedence.

HIS/HER career marriage dual career marriage in which both careers are viewed as equal.

THEIR career marriage dual career marriage in which spouses share a career or work together.

Seven-hundred seventy-four undergraduates in North Carolina, Florida, and California completed an Internet survey on office romance. The goal was to identify the percent of the sample reporting that they fantasized about love and sex on the job and the percent actually experiencing these phenomena. Other goals included the percent reporting having kissed a coworker, whether the involvement was with someone of equal or higher status, disclosure to others, sexual harassment, etc.

Over three-fourths (78%) of the respondents were female; 22% were male. Forty-seven percent were employed in a service job such as sales, fast food, or retail; 9% worked in an academic context, 12% in an "office," and 4% in a medical context. Most (73%) regarded where they worked as a "job," not a "career," and 13% saw it as a place to meet a future spouse.

Over 40% of the student employees reported fantasies of both love and sex with a coworker in contrast to a quarter who *actually* experienced love and sex with a coworker (see Table 1).

TABLE 1 Fantasies and Realities On The Job

(N = 774)

	Fantasized About	Actually Experienced
Love Relationship at the Office	43%	25%
Sexual Relationship at the Office	41%	24%

Other findings included:

Kissing

Almost 30% (28.6%) reported that they had kissed a fellow employee at work.

Rank of Person Undergraduate Had Sex With

Almost 80% (79%) reported having had sex with a peer/co-worker, 11% had sex with a boss, supervisor, or someone above them.

her husband's career. In this arrangement, the wife may also have children early, which has an effect on the development of her career. Gordon and Whelan-Berry (2005) interviewed thirty-six professional women and found that, in 22% of the marriages, the husband's career took precedence. The primary reasons for this arrangement were that the husband earned a higher salary, going where the husband could earn the highest income was easier because the wife could more easily find a job wherever he went (than vice versa), and ego needs (the husband needed to have the dominant career). This arrangement is sometimes at the expense of the wife's career. Mason and Goulden (2004) studied women in academia and found that those who had a child within five years of earning their PhDs were less likely to achieve tenure.

For couples who do not have traditional gender role attitudes, the wife's career may take precedence (HER/his career). Of the marriages in the Gordon and Whelan-Berry study (2005) mentioned previously, 19% could be categorized as giving precedence to a wife's career. In such marriages, the husband is willing to relocate and to disrupt his career for his wife's. Such a pattern is also likely to occur when a wife earns considerably more money than her husband. In some cases, the husband, who is downsized or who prefers the role of full-time parent, becomes "Mr. Mom."

Over 100,000 husbands (and parents of at least one child under the age of 6) are married to wives who work full-time in the labor force (Tucker 2005). The incidence of Mr. Moms has increased almost 30% in the past ten years (Society for the Advancement of Education 2005); almost 80% of men in Europe report that they would be happy to stay at home with the kids (Pepper 2006). A major advantage of men assuming this role is a more lasting emotional bond with their children (Tucker 2005).

When the careers of both the wife and husband are given equal status in the relationship (HIS/HER career), they may have a commuter marriage in which they follow their respective careers wherever they lead. Alternatively, Deutsch et al. (2007) surveyed 236 undergraduate senior women to assess their views on

Telling Others about the Office Romance

About 30% (28.1%) reported that they told someone else about their office romance.

Sexual Harassment

About 20% (19%) reported that they had been physically touched at work in a way that made them uncomfortable. A higher percentage (30%) said that a person at work said something sexual to them that made them uncomfortable (only 3% filed a formal complaint of sexual harassment).

Duration of Office Romance

Most (18%) of the office romances lasted less than six months, 5% lasted six months to a year, 5% lasted between one and two years, and 2% lasted five years and continued. The remaining respondents said they had not been involved in an office romance so the question did not apply.

After the Office Romance Ends

Of the workplace romances that had ended, almost three quarters (73%) ended positively with over half (54%) remaining friends. Thirteen percent still see each other, 5% are married, and 1% live together.

Losing One's Job

Less than 2% (1.7%) ended up losing their job at the place where they had the office romance.

Source

Merrill, J. and D. Knox. 2010. *Finding Love from 9 to 5: Trade Secrets of an Office Love.* Santa Barbara, CA: Praeger.

Finding Love from 9 to 5: Trade Secrets of an Office Love by Jane Merrill and David Knox. Copyright © 2010 by Jane Merrill and David Knox. Reproduced with permission of ABC-CLIO, LLC.

husbands, managing children, and work. Most envisioned two egalitarian scenarios in which both spouses would cut back on their careers and/or both would arrange their schedules to devote time to child care. Still other couples may hire domestic child care help so that neither spouse functions in the role of housekeeper. In reality, equal status to HIS and HER careers is not a dominant pattern (Stone 2007). In regard to leisure, Roeters and Treas (2011) studied 898 Dutch couples and found that when both partners are engaged in full-time work they participate more in leisure time activities as a couple than do couples where only one of the partners is full time employed.

Finally, some couples have the same career and may work together (THEIR career). Some news organizations hire both spouses to travel abroad to cover the same story. These careers are rare. In the following sections, we look at the effects on women, men, their marriage, and their children when a wife is employed outside the home.

Money often costs too much.

Ralph Waldo Emerson,
American essayist

Effects of the Wife's Employment on the Wife

Whether a wife is satisfied with her job is related to the degree to which the job takes a toll on her family life. Grandey et al. (2005) found that, when the wife's employment interfered with her family life, the wife reported less job satisfaction. Such a relationship with husbands was not found. Jobs that interfered with family life did not result in decreased job satisfaction for men. Consistent with this finding, Kiecolt (2003) noted that most women with young children much prefer to be at home and view home, not work, as their primary haven of satisfaction. The following is the experience of a woman going back to work after the birth of her first child.

Going Back to Work—A New Mother Speaks
You probably get the most advice of your life during pregnancy, especially if it's your first. Some advice is very welcomed and some, not so much. Actually, I wanted as much advice as possible about anything and everything baby. I figured it would

better prepare me . . . HA! "Be prepared" is something I heard repeatedly after I answered yes to "Are you going back to work?" I knew I would be very sad, but I also knew there was no other choice. I was too busy worrying about making sure all the cute pink things were washed and folded (and refolded) in her drawer as I anticipated her arrival. Thinking about having her and then having to leave her were distant thoughts. I mean, 8–10 weeks of time off is a really long time, isn't it? More time off than a summer between grade school. It seemed like an eternity. But there is no way to prepare yourself for the reality. She is born, your life is forever different. Your time in the hospital is a whirlwind. You blow in, and blow out.

Then you are home, with your precious baby and you have 10 weeks . . . 10 long weeks.

But no matter how much time you take with your newborn, it flies by. It's never enough time. You are trying to heal from delivery and learn about being a new mother at the same time. You're losing sleep and using more energy. A spinning tornado of being tired and falling in love. . . . and POOF . . . it's gone. You go through periods of guilt for leaving her, sadness, missing her, worrying to death and even a slight bit of anger at your spouse. Hey, "if you made more money, we could afford for me to stay home."

Your first day back at work will be very hard. If you work somewhere with many employees like I did, I had people coming to my office every few minutes "It's nice to have you back!" What can you possibly say to that? It's good to be back . . . hell no. But you are not going to tell them how much you would rather be home with your baby. If you are lucky, you have an amazing mother who always says the right thing at the right time. When I was pouring my heart out to her over the phone my last day at the house with my daughter, she said, "Yes, I couldn't imagine leaving my babies with anyone else. It's going to be really hard when she does everything first with them and not with you. I sure hope she says mama before her caregiver's name." Talk about a freak out.

The best thing you can do if going back to work is hard for you is to talk to other mothers that went through it. One of my friends stayed home for a year before she put her son in daycare. She says now that she would feel like a bad mother if she took him out of his daycare because he learns so much and really enjoys it. When I'm feeling really sad about being at work I try to think of all the things we will be able to afford for her because I work. Like sports teams, music lessons, dance, vacations, etc. that are also important for her growth and development. Also remember that no matter what, you will always be mommy, and no one can take that away.

Amanda Kinsch

Role overload not having the time or energy to meet the demands or responsibilities in the roles of wife, parent, and worker.

Second shift the housework and child care that employed women engage when they return home from their jobs.

Indeed, family and work become spheres to manage, and sometimes women experience **role overload**—not having the time or energy to meet the demands of their responsibilities in the roles of wife, parent, and worker. Because women have traditionally been responsible for most of the housework and child care, employed women come home from work to what Hochschild (1989) calls the second shift: the housework and child care that employed women do when they return home from their jobs. According to Hochschild, the **second shift** has the following result:

". . . women tend to talk more intently about being overtired, sick, and 'emotionally drained.' Many women could not tear away from the topic of sleep. They talked about how much they could 'get by on' . . . six and a half, seven, seven and a half, less, more. . . . Some apologized for how much sleep they needed. . . . They talked about how to avoid fully waking up when a child called them at night, and how to get back to sleep. These women talked about sleep the way a hungry person talks about food" (p. 9).

Part of this role overload may be that women have more favorable attitudes toward housework and childcare than men. Two researchers assessed the attitudes of 732 spouses and found that women had more favorable attitudes toward cleaning, cooking, and child care than did men—women enjoyed it more, set higher standards for it, and felt more responsible for it. The researchers also noted that

356   **Chapter 12** Work and Family Life

these favorable and men's unfavorable attitudes were associated with women's greater contribution to household labor (Poortman and Van der Lippe 2009).

Another stressful aspect of employment for employed mothers in dual-earner marriages is role conflict being confronted with incompatible role obligations. For example, the role of an employed mother is to stay late and prepare a report for the following day. However, the role of a mother is to pick up her child from day care at 5 P.M. When these roles collide, there is role conflict. Although most women resolve their role conflicts by giving preference to the mother role, some give priority to the career role and feel guilty about it.

Role strain, the anxiety that results from being able to fulfill only a limited number of role obligations, occurs for both women and men in dual-earner marriages. No one is at home to take care of housework and children while they are working, and they feel strained at not being able to do everything.

Effects of the Wife's Employment on Her Husband

Husbands also report benefits from their wives' employment. These include being relieved of the sole responsibility for the financial support of the family and having more freedom to quit jobs, change jobs, or go to school.

Because men traditionally had no options but to work full-time, men now benefit by having a spouse with whom to share the daily rewards and stresses of employment. To the degree that women find satisfaction in their work role, men benefit by having a happier partner. Finally, men benefit from a dual-earner marriage by increasing the potential to form a closer bond with their children through active child care. Some prefer the role of househusband and stay-at-home dad. Patrick (2005) confirmed that stay-at-home dads have a stronger emotional bond with their children than traditional working dads. Though such dads are clearly the minority, their visibility is increasing.

Effects of the Wife's Employment on the Marriage

Helms et al. (2010) analyzed the relationship between employment patterns and marital satisfaction of 272 dual earner couples and found that co-provider couples (in contrast to those where there were distinct primary and secondary providers) reported the highest marital satisfaction. In addition, these couples reported the most equitable division of housework.

Are marriages in which the wife has her own income more vulnerable to divorce? Not if the wife is happy; but if she is unhappy, her income will provide her a way to take care of herself when she leaves. Schoen et al. (2002) wrote, "Our results provide clear evidence that, at the individual level, women's employment does not destabilize happy marriages but increases the risk of disruption in unhappy marriages" (p. 643). Hence, employment won't affect a happy marriage but it can in an unhappy one. Further research by Schoen et al. (2006) revealed that full-time employment of the wife is actually associated with marital stability.

Couples may be particularly vulnerable when a wife earns more money than her husband. Thirty percent of working wives earn more than their husbands (Tyre and McGinn 2003). Meisenbach (2010) interviewed 15 U.S. female breadwinners (FBWs) and identified their issues, concerns, and worries. They enjoyed having control but were not sure they should, they valued independence, they felt pressure for being responsible for the money and needing to keep their jobs, they valued and needed to be attentive to their partners' contributions to the family, they felt guilty for not being the involved mother, and they were proud of and valued their career progress.

The wife earning more than her husband may affect the couple's marital happiness in several ways. Marriages in which the spouses view the provider role as the man's responsibility, in which the husband cannot find employment, and in which the husband is jealous of his wife's employment are vulnerable to dissatisfaction. Cultural norms typically dictate that the man is supposed to earn more money

Role strain the anxiety that results from being able to fulfill only a limited number of role obligations.

Money is the most egalitarian force in society. It confers power on whoever holds it.

Roger Starr, columnist, and editor

than his partner and that something is wrong with him if he doesn't. Jalovaara (2003) found that the divorce rate was higher among couples where the wife's income exceeded her husband's. On the other hand, some men want an ambitious, economically independent woman who makes a high income. Over a third of men in a national *Newsweek* poll said that they'd consider quitting their job or reducing their hours if their wife earned more money (Tyre and McGinn 2003).

Do the hours worked (in terms of night or weekend) make any difference? Yes. Strazdins et al. (2006) compared over 4,000 families where parents worked standard weekday times against those parents who worked nonstandard schedules and found that couples working nonstandard schedules reported worse family functioning, more depressive symptoms, and less effective parenting. Their children were also more likely to have social and emotional difficulties. Olsen and Dahl (2010) confirmed that having control over the hours one works is related to positive family functioning.

Work and Family: Effects on Children

Independent of the effect on the wife, husband, and marriage, what is the effect of the wife earning an income outside of the home on the children? Individuals disagree on the effects of maternal employment on children. The Self-Assessment provides a way to assess your beliefs in this regard.

Effects of the Wife's Employment on the Children

Mothers with young children are the least likely to be in the labor force. Mothers most likely to be in the work force are those with children between the ages of 14 and 17—the teen years. The children are no longer dependent on the physical care of their mother and teenagers cost more money, often requiring a second income.

National **Data**

Data from the National Survey of America's Families indicate that most children under the age of 5 whose parents are in the workforce are in a nonparental care arrangement. For white, black, and Hispanic children, the proportions receiving such care are 87%, 81%, and 80%, respectively (Capizzano et al. 2006).

Dual-earner parents want to know how children are affected by maternal employment. An abundance of research has resulted in the finding of few negative effects (Perry-Jenkins et al. 2001). Hence, children do not appear to suffer cognitively or emotionally as long as positive, consistent child care alternatives are in place. Children may even benefit in terms of exercise and leisure activities. Sener et al. (2008) studied leisure activities of children as influenced by the employment patterns of parents and found that children in households with parents who are employed, and with higher income or higher education, participate in structured outdoor activities at higher rates.

However, children of two-earner parents also receive less supervision. Leaving children to come home to an empty house is particularly problematic. In addition, in a study of 2,246 adults who had lived with two biological parents until age 16, those whose mothers had been employed during most of their childhood reported less support and less discipline from both parents than those who had stay-at-home mothers (Nomaaguchi and Milkie 2006).

Self-Care/Latchkey Children   Long work hours, not being able to control their work schedule, and leaving their children unsupervised result in consid-

Directions

Using the following scale, please mark a number on the blank next to each statement to indicate how strongly you agree or disagree.

1	2	3	4	5	6
Disagree Very Strongly	Disagree Strongly	Disagree Slightly	Agree Slighty	Agree Strongly	Agree Very Strongly

_____ 1. Children are less likely to form a warm and secure relationship with a mother who works full-time outside the home.

_____ 2. Children whose mothers work are more independent and able to do things for themselves.

_____ 3. Working mothers are more likely to have children with psychological problems than mothers who do not work outside the home.

_____ 4. Teenagers get into less trouble with the law if their mothers do not work full-time outside the home.

_____ 5. For young children, working mothers are good role models for leading busy and productive lives.

_____ 6. Boys whose mothers work are more likely to develop respect for women.

_____ 7. Young children learn more if their mothers stay at home with them.

_____ 8. Children whose mothers work learn valuable lessons about other people they can rely on.

_____ 9. Girls whose mothers work full-time outside the home develop stronger motivation to do well in school.

_____ 10. Daughters of working mothers are better prepared to combine work and motherhood if they choose to do both.

_____ 11. Children whose mothers work are more likely to be left alone and exposed to dangerous situations.

_____ 12. Children whose mothers work are more likely to pitch in and do tasks around the house.

_____ 13. Children do better in school if their mothers are not working full-time outside the home.

_____ 14. Children whose mothers work full-time outside the home develop more regard for women's intelligence and competence.

_____ 15. Children of working mothers are less well-nourished and don't eat the way they should.

_____ 16. Children whose mothers work are more likely to understand and appreciate the value of a dollar.

_____ 17. Children whose mothers work suffer because their mothers are not there when they need them.

_____ 18. Children of working mothers grow up to be less competent parents than other children because they have not had adequate parental role models.

_____ 19. Sons of working mothers are better prepared to cooperate with a wife who wants both to work and have children.

_____ 20. Children of mothers who work develop lower self-esteem because they think they are not worth devoting attention to.

_____ 21. Children whose mothers work are more likely to learn the importance of teamwork and cooperation among family members.

_____ 22. Children of working mothers are more likely than other children to experiment with alcohol, other drugs, and sex at an early age.

_____ 23. Children whose mothers work develop less stereotyped views about men's and women's roles.

_____ 24. Children whose mothers work full-time outside the home are more adaptable; they cope better with the unexpected and with changes in plans.

Scoring

Items 1, 3, 4, 7, 11, 13, 15, 17, 18, 20, and 22 refer to "costs" of maternal employment for children and yield a Costs Subscale score. High scores on the Costs Subscale reflect strong beliefs that maternal employment is costly to children. Items 2, 5, 6, 8, 9, 10, 12, 14, 16, 19, 21, 23, and 24 refer to "benefits" of maternal employment for children and yield a Benefits Subscale score. To obtain a Total Score, reverse the score of all items in the Benefits Subscale so that 1 = 6, 2 = 5, 3 = 4, 4 = 3, 5 = 2, and 6 = 1. The higher one's Total Score, the more one believes that maternal employment has negative consequences for children.

Source

E. Greenberger, W. A. Goldberg, T. J. Crawford, and J. Granger. *Psychology of Women Quarterly*, 12. pp. 35–59, copyright © 1988 by *Sage* Publications. Reprinted by Permission of *Sage* Publications.

erable concern for parents (Barnett et al. 2010b). Although these self-care, or latchkey, children often fend for themselves very well, some are at risk. Over 230,000 are between the ages of 5 and 7 and are vulnerable to a lack of care in case of an accident or emergency. Children who must spend time alone at home should know the following:

1. How to reach their parents at work (the phone number, extension number, and name of the person to talk to if the parent is not there)

2. Their home address and phone number in case information must be given to the fire department or an ambulance service

3. How to call emergency services, such as the police and fire departments

4. The name and number of a relative or neighbor to call if the parent is unavailable

5. To keep the door locked and not let anyone in

6. How to avoid telling callers their parents are not at home; instead, they should tell them their parents are busy or can't come to the phone

7. How to avoid playing with appliances, matches, or the fireplace.

Parents should also consider the relationship of the children they leave alone. If the older one terrorizes the younger one, the children should not be left alone. Also, if the younger one is out of control, it is inadvisable to put the older one in the role of being responsible for the child. If something goes wrong (such as a serious accident), the older child may be unnecessarily burdened with guilt.

Well, it's 1 A.M. Better go home and spend some quality time with the kids.

Homer Simpson, *The Simpons*

Quality Time

Dual-income parents struggle with having "quality time" with their children. The term *quality time* has become synonymous with good parenting. Snyder (2007) studied 220 parents from 110 dual-parent families and found that "quality time" is defined in different ways. Some parents (structured-planning parents) saw quality time as planning and executing family activities. Many Mormons set aside "Monday home evenings" as a time to bond, pray, and sing together (thus, quality time). Other parents (child-centered parents) noted that "quality time" occurred when they were having heart-to-heart talks with their children. Still other parents believed that all the time they were with their children was quality time. Whether they were having dinner together or riding to the post office, quality time was occurring if they were together. As might be expected, mothers assumed greater responsibility for "quality time."

Day Care Considerations

Parents going into or returning to the workforce are intent on finding high-quality day care. Rose and Elicker (2008) surveyed the various characteristics of day care that are important to 355 employed mothers of children under 6 years of age and found that warmth of caregivers, a play-based curriculum, and educational level of caregivers emerge as the first-, second-, and third-most important factors in selecting a day-care center.

Quality of Day Care   More than half of U.S. children are in center-based child-care programs. Most mothers prefer relatives for the day-care arrangement for

Working parents depend on day care for their young children.

Chelsea E. Curry

their children. Researchers Gordon and Högnäs (2006) analyzed data from the National Institute of Child Health and Human Development Study of Early Child Care and found that nearly two-thirds of the mothers reported such a preference for relatives, including 15% who preferred their spouse or partner, 28% who preferred another relative in their home, and 22% who preferred another relative in another home. An additional 16% preferred care by a nonrelative in their own home and 11% preferred care by a nonrelative in another private home. Just 9% expressed a preference for center-based care. However, most children end up in center-based child-care programs.

National **Data**

Forty-two percent of 3-year-olds and 69% of 5-year-old children are in center-based day-care programs (*Statistical Abstract of the United States 2011*, Table 576).

Employed parents are concerned that their children get good-quality care. Their concern is warranted. Cortisol is a steroid hormone that plays critical roles in adaptation to stress. Higher levels of cortisol reflect higher levels of stress experienced by the individual. Gunnar et al. (2010) assessed the cortisol stress levels of 151 children in full-time, home-based day-care centers and compared these levels to those of children the same age who were at home. Increases were noted in the majority of children (63%) at day care, with 40% classified as a stress response. These increased cortisol levels began in the morning and continued throughout the afternoon.

Vandell et al. (2010) examined the effects of early child care 15 years later and found that higher-quality care predicted higher-cognitive–academic achievement at age 15, with escalating positive effects at higher levels of quality. These findings do not mean that lower quality day care has negative outcomes for cognitive development but that parents might be discriminating in their day care selection.

Warash et al. (2005) reviewed the literature on day-care centers and found that the average quality of such centers is mediocre "unsafe, unsanitary, non-educational, and inadequate in regard to the teacher-child ratio for a classroom." Care for infants was particularly lacking. Of 225 infant or toddler rooms observed, 40% were rated "less than minimal" with regard to hygiene and safety. Because of the low pay and stress of the occupation, the rate of turnover for family child-care providers is very high (estimated at between 33 and 50%) (Walker 2000). However, De Schipper et al. (2008) noted that day-care workers who engage in high-frequency positive behavior engender secure attachments with the children they work with. Hence, children of such workers don't feel they are on an assembly line but bond with their caretakers. Parents concerned about the quality of day care their children receive might inquire about the availability of webcams. Some day-care centers offer full-time webcam access so that parents or grandparents can log onto their computers and see the interaction of the day-care worker with their children.

Ahnert and Lamb (2003) emphasized that attentive, sensitive, loving parents can mitigate any potential negative outcomes in day care. "Home remains the center of children's lives even when children spend considerable amounts of time in child care. . . . [A]lthough it might be desirable to limit the amount of time children spend in child care, it is much more important for children to spend as much time as possible with supportive parents" (pp. 1047–48).

Cost of Day Care   Day-care costs are a factor in whether a low-income mother seeks employment, because the cost can absorb her paycheck. Even for dual-earner families, cost is a factor in choosing a day-care center. Day care costs vary widely from nothing, where friends trade off taking care of the children, to very expensive institutionalized day care in large cities.

The cost of high quality infant care can be as high as $1300 per month; when the child turns 2 the cost can drop to $1000. These costs are for one child; most spouses have two children. Do the math.

We e-mailed a dual-earner metropolitan couple (who use day care for their two children and are planning for them to enter school) to inquire about the institutional costs of day care in the Baltimore area. The father's response follows:

> There is a sliding scale of costs based on quality of provider, age of child, full-time or part-time. Full-time infant care can be hard to find in this area. Many people put their names on a waiting list as soon as they know they are pregnant. We had our first child on a waiting list for about nine months before we got him into our first choice of providers.
>
> Infant care runs $1,250 per month for high-quality day care. That works out to $15,000 per year and you thought college was expensive.
>
> The cost goes down when the child turns 2 to around $900–$1,000 per month. In Maryland, this is because the required ratio of teachers to children gets bigger at age 2. This cost break doesn't last long. Preschool programs (more academic in structure than day care) begin at age 3 or 4. When our second child turned 4, he started an academic preschool in September. The school year lasts nine and a half months, and it costs about $14,000. Summer camps or summer day-care costs must be added to this amount to get the true annual cost for the child. My wife and I have budgeted about $30,000 total for our two children to attend private school and summer camps this year.
>
> The news only gets worse when the children get older. A good private school for grades K–5 runs $15,000 per school year. Junior high (6–8) is about $20,000, and private high school here goes for $20,000 to $35,000 per nine-month school year.
>
> Religious private schools run about half the costs above. However, they require that you be a member in good standing in their congregation and, of course, your child undergoes religious indoctrination.
>
> There is always the argument of attending a good public school and thus not having to pay for private school. However, we have found that home prices in the "good school neighborhoods" were out of our price range. Some of the better-performing public schools also now have waiting lists even if you move into that school's district.
>
> In most urban and suburban areas, cost is secondary to admissions. Getting into any good private school is difficult. In order to get into the good high schools, it's best to be in one of the private elementary or middle schools that serves as a "feeder" school. Of course, to get into the "right" elementary school, you must be in a good feeder kindergarten. And of course to get into the right kindergarten, you have to get into the right feeder preschool.
>
> Parents have A LOT of anxiety about getting into the right preschool because this can put your child on the path to one of the better private high schools. Getting into a good private preschool is not just about paying your money and filling out applications. Yes, there are entrance exams. Both of our children underwent the following process to get into preschool: First you must fill out an application. Second, your child's day-care records/transcripts are forwarded for review. If your child gets through this screening, you and your child are called in for a visit. This visit with the child lasts a few hours and your child goes through evaluation for physical, emotional, and academic development. Then you wait several agonizing months to see if your child has been accepted.
>
> Though quality day care is expensive, parents delight in the satisfaction that they are doing what they feel is best for their child.

Balancing Work and Family Life

Work is definitely stressful on relationships. Lavee and Ben-Ari (2007) noted that one of the effects of work stress is that spouses have greater emotional distance between them. The researchers suggested that such stress may signal a deterioration of a relationship or a way that couples minimize interaction so as to protect the relationship.

One of the major concerns of employed parents and spouses is how to juggle the demands of work and family simultaneously and achieve a sense of accomplishment and satisfaction in each area. Cinamon (2006) noted that women experience work interfering with family at higher levels. Women are more likely to resolve the conflict by giving precedence to family. Kiecolt (2003) examined national data and concluded that employed women with young children are "more likely to find home a haven, rather than finding work a haven" (p. 33). Recall the research by Stone (2007) in regard to women opting out of high-income or high-status work in favor of taking care of their children.

Nevertheless, the conflict between work and family is substantial, and various strategies are employed to cope with the stress of role overload and role conflict, including (1) the superperson strategy, (2) cognitive restructuring, (3) delegation of responsibility, (4) planning and time management, and (5) role compartmentalization (Stanfield 1998).

Superperson Strategy

The superperson strategy involves working as hard and as efficiently as possible to meet the demands of work and family. The person who uses the superperson strategy often skips lunch and cuts back on sleep and leisure to have more time available for work. Women are particularly vulnerable because they feel that if they give too much attention to child-care concerns, they will be sidelined into lower-paying jobs with no opportunities.

Hochschild (1989) noted that the terms **superwoman** or **supermom** are cultural labels that allow a woman to regard herself as very efficient, bright, and confident. However, Hochschild noted that this is a "cultural cover-up" for an overworked and frustrated woman. Not only does the woman have a job in the workplace (first shift), she comes home to another set of work demands in the form of house care and child care (second shift). Finally, she has a "third shift" (Hochschild 1997).

The **third shift** is the expense of emotional energy by a spouse or parent in dealing with various issues in family living. An example is the emotional energy needed for children who feel neglected by the absence of quality time. Although young children need time and attention, responding to conflicts and problems with teenagers also involves a great deal of emotional energy. Minnottea et al. (2010) studied 96 couples and found that women perform more "emotion work."

Mothers who try to escape the supermom trap through the competent help of their husbands may experience a downside. Sasaki et al (2010) studied 78 dual career couples with an 8-month-old infant and found that when wives felt that their husbands were skillful caregivers, greater husbands' contribution to caregiving was associated with the wives' lower self-competence. The authors concluded that "despite increasingly egalitarian sex roles, employed mothers seem to be trapped between their desire for help with childrearing and the threat to their personal competence posed by failure to meet socially constructed ideals of motherhood."

Cognitive Restructuring

Another strategy used by some women and men experiencing role overload and role conflict is cognitive restructuring, which involves viewing a situation in positive terms. Exhausted dual-career earners often justify their time away from their children by focusing on the benefits of their labor: their children live in a nice house in a safe neighborhood and attend the best schools. Whether these outcomes offset the lack of "quality time" may be irrelevant; the beliefs serve simply to justify the two-earner lifestyle.

If evolution really works, how come mothers only have two hands?

Milton Berle, actor and comedian

Superwoman/supermom a cultural label that allows a woman to regard herself as very efficient, bright, and confident; usually a cultural cover-up for an overworked and frustrated woman.

Third shift the expenditure of emotional energy by a spouse or parent in dealing with various emotional issues in family living.

Under the Family and Medical Leave Act, all companies with fifty or more employees are required to provide each worker with up to twelve weeks of unpaid leave for reasons of family illness, birth, or adoption of a child. In a subsequent amendment, the Family Leave Act permits states to provide unemployment pay to workers who take unpaid time off to care for a newborn child or a sick relative.

Under the Obama administration, the benefits would be available in businesses with twenty-five or more employees and workers could take leave for elder care needs. All fifty states would also be encouraged to adopt paid leave and would provide a $1.5 billion fund to assist states with start-up costs. Currently, the United States is the only industrialized country that does not provide paid child leave. Over 160 countries provide paid leave for mothers to birth their babies; forty-five countries provide the same benefits for fathers. Australia guarantees a year of leave to all new mothers (Thomas 2007). Germany provides fourteen weeks off with 100% salary.

Aside from government-mandated work-family policies, corporations and employers have begun to initiate policies and programs that address the family concerns of their employees. Employer-provided assistance with child care, assistance with elderly parent care, options in work schedules, and job relocation assistance are becoming more common. Widener (2008) noted that some U.S. companies have advanced strategies to balance work and life and gender equity, and that these family-friendly policies "can garner a competitive edge, attracting and retaining young, early career professionals." Hill et al. (2010) studied the value of working at home and having flexible work hours in a sample of 24,436 on work-life conflict and found the expected—the greater the flexibility of work hours was, the greater the life satisfaction.

While the various family friendly options are commendable, Stone (2007) noted that one of the reasons women opt out of their career paths is the inflexibility of corporations. Her interviewees noted that, when a woman drops out to have or to take care of her children, her clients are assigned to someone else and, when she returns, she never seems able to get back on the same standing as those who did not drop out. She suggested family-friendly policies are

Delegation of Responsibility and Limiting Commitments

A third way couples manage the demands of work and family is to delegate responsibility to others for performing certain tasks. Because women tend to bear most of the responsibility for child care and housework, they may choose to ask their partner to contribute more or to take responsibility for these tasks. Although some husbands are involved, cooperative, and contributing, Stone (2007) revealed that fifty-three respondents in her study opted out of their careers to return home because they had a lack of available help either from husbands who were not at home or who did not do much when they were home.

Another form of delegating responsibility involves the decision to reduce one's current responsibilities and not take on additional ones. For example, women and men may give up or limit agreeing to volunteer responsibilities or commitments. One woman noted that her life was being consumed by the responsibilities of her church; she had to change churches because the demands were relentless. In the realm of paid work, women and men can choose not to become involved in professional activities beyond those that are required.

Time Management

While two-thirds of women prefer to work part-time as opposed to full-time, Vanderkam (2010) argues that women who work part-time end up spending just 41 more minutes daily on child care and 10 minutes more per day playing with the child. By working full-time she says the woman affords high-quality day care and can focus on the child when she is not at work. The full-time worker is not exhausted from being with the children all day but can look forward to spending dinner, bath, and reading time with their children at night.

Other women use time management by prioritizing and making lists of what needs to be done each day. This involves trying to anticipate stressful periods,

simply window dressings and that, in fact, corporations view families as interfering with production.

The Obama administration emphasized informing businesses about the benefits of flexible work schedules, helping businesses create flexible work opportunities, and increasing federal incentives for telecommuting. Moen (2011) emphasized the need for corporations to "break open the time clocks around paid work—the tacit, taken-for-granted beliefs, rules, and regulations about the time and timing of work days, work weeks, work years, and work lives." Similarly, Soo Jung and Zippay (2011) emphasized the tensions associated with the demands of employment and home life, which may have negative effects on mental and physical health, and called for employer-based policies to facilitate work-life balance and well-being.

Your Opinion?

1. Argue for and against the fact that businesses benefit from having family-friendly policies.

2. Argue for and against the theory that childfree workers should work later and on holidays so that parents can be with their children.

3. Why do you think that the United States lags behind other industrialized nations in terms of paid leave for parents?

Sources

Hill, E. J., J. J. Erickson, E. K. Holmes, and M. Ferris. 2010. Workplace flexibility, work hours, and work-life conflict: Finding an extra day or two. *Journal of Family Psychology* 24: 349–358.

Moen, P. 2011. From 'work-family' to the 'gendered life course' and 'fit': five challenges to the field. *Community, Work & Family* 14: 81–96.

Soo Jung, J. and A. Zippay. 2011. Juggling act: Managing work-life conflict and work-life balance. *Families in Society* 92: 84–90.

Stone, P. 2007. *Opting out?* Berkley: University of California Press.

Thomas, S. G. 2007. *Buy, Buy Baby: How consumer culture manipulates parents and harms young minds.* Houghton Mifflin Publisher.

Widener, A. J. 2008. Family-friendly policy: Lessons from Europe-Part II *Public Manager* Winter 2007/2008. 36:44–50.

planning ahead for them, and dividing responsibilities with the spouse. Such division of labor allows each spouse to focus on an activity that needs to be done (grocery shopping, picking up children at day care) and results in a smoothly functioning unit.

Having flexible jobs and/or careers is particularly beneficial for two-earner couples. Being self-employed, telecommuting, or working in academia permits flexibility of schedule so that individuals can cooperate on what needs to be done. Alternatively, some dual-earner couples attempt to solve the problem of child care by having one parent work during the day and the other parent work at night so that one parent can always be with the children. Shift workers often experience sleep deprivation and fatigue, which may make fulfilling domestic roles as a parent or spouse difficult for them. Similarly, shift work may have a negative effect on a couple's relationship because of their limited time together.

Presser (2000) studied the work schedules of 3,476 married couples and found that recent husbands (married less than five years) who had children and who worked at night were six times more likely to divorce than husbands or parents who worked days.

Role Compartmentalization

Some spouses use **role compartmentalization,** separating the roles of work and home so that they do not think about or dwell on the problems of one when they are at the physical place of the other. Spouses

Role compartmentalization strategy used to separate the roles of work and home so that an individual does not think about or dwell on the problems of one when they are at the physical place of the other.

DIVERSITY IN OTHER COUNTRIES

France, Denmark, and Germany have experimented with fewer required work hours per week. France tried the 35-hour workweek with eight weeks of vacation. The goal was to increase employment. However, over the ten years during which the 35-hour workweek was in place, the desired gains did not occur. In 2008, France moved to increase its workweek, with the result that the French now work an average of 41 hours a week; in America, the average is 41.7 hours (Keller 2008a). In his article on "Vacation, All I Never Wanted," Walter Kirn (2007) suggested that Americans are satisfied with their workweek. "Strip-mall day spas. Corner yoga studios. Suburban mega gyms. . . . This is the land of leisure. And you don't have to leave town to get it" (p. 12). Time away from work seems to have benefits for the well-being of individuals (Verbakel and DiPrete 2008).

unable to compartmentalize their work and home feel role strain, role conflict, and role overload, with the result that their efficiency drops in both spheres. Some families look to the government and their employers for help in balancing the demands of family and work.

Balancing Work and Leisure Time

The ant is knowing and wise, but he doesn't know enough to take a vacation.

Clarence Day, author

The workplace has become the home place. Because of the technology of the workplace, such as laptops, cell phones, Blackberries, iPods, iPads, and so on, some spouses work all the time, wherever they are. One couple noted that they are working on their laptops and talking on their cell phones when they are at home so that they rarely have any time when they are not working and communicating directly with each other.

Even on vacation, spouses are at work. An Ipsos poll of 1,000 randomly chosen adults revealed that about 40% of adult vacationers checked their office e-mail and messages, and 50% checked their voice mail via laptops and cell phones, with 20% taking the former and 80% the latter. Reasons for staying tethered to the job included worry about missing important information, feeling an expectation from bosses or coworkers that they stay connected, and enjoyment of staying involved. People most likely to use technology on vacation were under 40, white, and male (Fram 2007). Although the family is supposed to be a place of relaxation and recovery, it has become another place where spouses work. Balance is needed. Some data suggest that balance may be occurring. In a study of adults who were asked about their need for the Internet on vacation, 77% noted that they wanted access during their vacation. But their use of the Internet on vacation, in descending order, was "to check personal e-mail," "find trip information," "online banking," "read the news," "social media profiles," and "work e-mail" (Carey and Trap 2010b).

If all the year were playing holidays,

To sport would be as tedious as to work.

Shakespeare, *Henry IV*

Leisure the use of time to engage in freely chosen activities perceived as enjoyable and satisfying.

Definition and Importance of Leisure
Leisure refers to the use of time to engage in freely chosen activities perceived as enjoyable and satisfying, including exercise. Tucker et al. (2008) identified three conditions of leisure—quiet leisure at home (for example, reading a book), active

Technology is such that a person may never detach from work and stay connected to the office even on vacation.

David Knox

leisure (for example, playing basketball), and more work. Ratings of rest, recuperation, and satisfaction were lowest in the additional work condition.

The value of family leisure was confirmed by West and Merriam (2009). They interviewed 306 families who camped at the St. Croix State Park, Pine County, Minnesota and found that cohesiveness as was measured by the amount of intimate communication of troubles, secrets, and mood among family members was created and maintained by families who engaged in outdoor recreation. In addition, analysis of data on 898 families from throughout the United States revealed a positive relationship between all family leisure satisfaction variables and satisfaction with family life (Agate et al. 2009). The take home message of these two studies is that parents might recognize the benefits of family leisure and make it happen as often as possible. Leisure also has benefits for the individual. Losada et al. (2010) studied 134 caregivers of the elderly who reported significant reductions in feeling burdened and having mental health problems (e.g., depression) as a result of taking off time for leisure.

View of Work/Leisure by Millenials

Indeed, corporations are learning that individuals are no longer interested in being consumed by their work. This new work ethic is particularly operative among **millenials**, the 80 million workers born between 1980 and 1995. Having been told that they are special, that performance is not required for praise, and that fun, leisure, and lifestyle come before giving one's life to a job or career, these millenials are socializing their bosses to be more flexible or to find someone else to work for them. Indeed, corporations today are hiring consultants to socialize management how to cope with a workforce that wants a job on their own terms. They are told that they must be not just a boss but also a life coach and a shrink, and that they must motivate rather than demand. These millenials don't need the money, as one-half of college seniors will live with their parents after graduation (Textor 2007).

Functions of Leisure

Leisure fulfills important functions in our individual and interpersonal lives. Leisure activities may relieve work-related stress and pressure; facilitate social interaction and family togetherness; foster self-expression, personal growth, and skill development; and enhance overall social, physical, and emotional well-being.

He does not seem to me to be a free man who does not sometimes do nothing.

Cicero, Roman philosopher

Millenials workers born between 1980 and 1995.

Leisure provides excitement and a break from the routine of work. This man landed a four pound puppy drum.

David Knox

Video game playing is part of the collegiate landscape. In a national study of 1,162 undergraduates in 27 colleges and universities, 65% reported playing video games regularly or occasionally (Jones et al. 2003). However, individual's relationships may pay a price—the focus of this study.

Sample and Methods

The data consisted of a nonrandom sample of 148 undergraduate volunteers at a large southeastern university who responded to a 45-item questionnaire on "Video Game Survey" (approved by the Institutional Review Board of the university). The objective of the study was to identify current uses and consequences of playing video games.

General Findings

Of the respondents, 97.6% of the men and 92.6% of the women reported having played a video game. *Madden Football* was the most commonly played video. Although previous researchers (Rau et al. 2006; Wan and Chiou 2006) reported the capacity of video games to become addictive, these respondents were *not* obsessed by or addicted to playing video games; they rarely skipped class nor played relentlessly for days. Indeed only two males and one female (less than 3% of the respondents) used the term *obsessive* to describe their video game playing. The average GPA of these students was 2.96, suggesting that their academic performance was standard for their year in school.

Significant Gender Differences

In addition to the previous general findings, three significant sex differences emerged.

1. *Women viewed video games as more experimental than recreational.* Women and men differed in why they played video games. Female respondents were significantly more likely than male respondents (33.7% versus 7.5%) to report that

playing video games was experimental in the sense of curiosity, just seeing what it was like, and so on. In contrast, male respondents were significantly more likely to report that playing video games was recreational (65% versus 41.3%; p ⩽ .000).

Previous researchers have examined sex differences in playing video games. Hayes (2005) found that women prefer video games that involve cooperation and interaction. They may play combat games, but their focus is on learning to protect themselves and to defend themselves. Similarly, Carr (2005) studied gaming at an all-girls state school in the United Kingdom and found that patterns of access and peer culture determined which games girls played. What emerges from the literature and this study is that playing video games for females is less about having fun or recreation and more about experimenting, connecting, interacting, and protecting.

2. *Male partners of females were more likely to play video games.* The data reflected that females were much more likely to have a partner who played video games. Of the undergraduate females in this study, 75.6% reported having been involved in a relationship with a partner who played video games (in contrast to 42.4% of male undergraduates involved with a female who played video games) (p ⩽ .001). In effect, the undergraduate males played video games and the undergraduate females coped with their doing so. Previous researchers (Jones et al. 2003) have found that males report higher use of playing video games than females (57% versus 17%).

3. *Females were more likely to be angry over a partner playing video games.* Of the female respondents, 32.9% reported that they had become upset with their partners' relentless video game playing. Of the males, only 4.5% reported having told their partner that the partner spent too much time playing video games. The difference was statistically significant (p ⩽ .005). We asked an undergraduate female who reported that her partner's game playing was a

Though leisure represents a means of family togetherness and enjoyment, it may also represent an area of stress and conflict. Some couples function best when they are busy with work and child care so that they have limited time to interact or for the relationship. Indeed, some prefer not to have a lot of time alone together.

Individual and Relationship Problems Related to Leisure

Individual or relationship problems may be involved in leisure. Excessive drinking may occur in leisure contexts with friends. Grekin et al. (2007) analyzed the alcohol consumption of 3,720 undergraduates (mean age 17.96) who spent spring break with either friends or parents and found that students who vacationed with friends during spring break dramatically increased their alcohol use. In contrast, students who stayed home or vacationed with parents during spring break were at low risk for excessive alcohol use.

problem in their relationship to explain the source of her frustration:

Not only are video games expensive (not to speak of the equipment) but he hogs the TV for an entire evening and acts like I don't exist. Quality time—what is that?

Previous researchers have confirmed the deleterious effects of video game playing on relationships. Lo et al. (2005) surveyed 174 Taiwanese college students and found that the quality of their interpersonal relationships decreased as the amount of time spent playing online games increased.

Theoretical Framework

Social exchange theory is one of the more common theories for explaining human interaction (Taylor and Bagd 2005). The social exchange framework operates from a premise of utilitarianism- that individuals rationally weigh the rewards and costs associated with their involvement in relationships. Each interaction between two actors can be understood in terms of each individual seeking the most benefits at the least cost so as to have the highest "profit" and to avoid a "loss" (White and Klein 2002). The framework provides a clear understanding for the frustration of the female who is emotionally invested in her partner and wants him to reciprocate her interest. His choosing to play video games over her translates into a "cost" and a "loss" in terms of "less profit" in the relationship. Telling her partner that he spends too much time playing video games is her way of trying to increase the benefits in the relationship to experience a "profit."

Implications

Three implications emerged from the data. First, in spite of the stereotype that video game players are geeks who are focused on machines and have no relationships, 74.5% of the sample reported that they were dating or involved with someone. Indeed, over half were emotionally involved, engaged, or married. Second, these respondents were neither addicted to nor obsessed with playing video games. Missing class and being ensconced in their room playing video games for days on end was virtually nonexistent. Third, video game playing was a problem in relationships for about a third of the female respondents. Feeling neglected, disrespected ("doesn't care what I want to watch on TV"), and wasting money were among the dissatisfactions.

Sources

Carr, D. 2005. Contexts, gaming pleasures, and gendered preferences. *Simulation & Gaming* 36: 464–482.

Hayes, E. 2005. Women, video gaming & learning: Beyond the stereotypes. *Tech Trends* 49: 23–28.

Jones, S., L. N. Clarke, S. Cornish, M. Gonzales, C. Johnson, J. N. Lawson, S. Smith, et al. 2003. Let the games begin: Gaming technology and entertainment among college students. Pew Internet and American Life. July 6. www.pewinternet.org.

Lo, S. K., C. C. Wang, and W. Fang. 2005. Physical interpersonal relationships and social anxiety among online game players. *CyberPsychology & Behavior* 8: 15–20.

Rau, P. L. P., S. Y. Peng, and C. C. Yang. 2006. Time distortion for expert and novice online game players. *CyberPsychology & Behavior* 9: 396–403.

Taylor, A. C., and A. Bagd. 2005. The lack of explicit theory in family research: The case analysis of the *Journal of Marriage and the Family 1990–1999*. In *Sourcebook of family theory & research*, ed. Vern L. Bengtson, Alan C. Acock, Katherine R. Allen, Peggye Dilworth-Anderson, and David M. Klein, 22–25. Thousand Oaks, CA: Sage Publications.

Wan, C. S., and W. B. Chiou. 2006. Psychological motives and online games addiction: A test of flow theory and humanistic needs theory of Taiwanese adolescents. *CyberPsychology & Behavior* 9: 317–324.

White, J. M., and D. M. Klein. 2002. *Family theories,* 2d ed. Thousand Oaks, CA: Sage Publications.

*Abridged from D. Knox, M. Zusman, A. White, and G. Haskins. 2009. "Coed anger over romantic partner's video game playing." *Psychology Journal* 6: 10–16.

Relationship problems may also occur in reference to leisure. In a study of 102 older couples (average age of spouses was 69), "leisure activities" was the most frequently cited problem area (accounting for 23% of problems) (Henry et al. 2005). The ways in which this problem expressed itself included the following:

TV watching: He wanted to watch football and she wanted to watch *American Idol.*

Travel: She wanted to stay at home and he wanted to travel, or they differed in travel preferences.

Time: Both were very busy with their lives and did not take time to spend together.

Video games are also a source of stress in some relationships (see the "Applying Social Research" feature beginning on page 368).

What if Your Partner Sends Text Messages, Uses a Cell Phone, and E-Mails throughout Your Vacation?

Doing "business" via texting, cell phones, and e-mail while on vacation is not unusual. Where both partners do so or the other partner is indifferent, there is no problem. However, some partners are angered that their vacation partner never detaches and is always texting and "on." Of course, a partner legitimizes texting, the cell phone and e-mail as necessary "business" for income. One solution is a discussion before leaving town about the definition of the "vacation"—whether it is a "working" vacation, a pure vacation, or what? Although couples will vary in their agreements, the issue is that they have an agreement. Otherwise, their vacation might become not only "all about business," but also about a great deal of arguing.

Vacation Stress

Vacations can be stressful, beginning with the preparation. A national sample identified some of the problems as taking care of one's work schedule (28%), arranging for pet care (23%), making travel plans (11%), and finding someone to look after one's home (11%) (Umminger and Parker 2005). Travel through airports has also become time-consuming and stressful.

The Future of Work and Family Life

Families will continue to be stressed by work. Employers will, increasingly, ask employees to work longer and do more without the commensurate increases in salary or benefits. Businesses are struggling to stay solvent and the workers will take the brunt of the instability.

Wives who work outside the home will increase—the economic needs of the family will demand it. Husbands will adapt, reluctantly. Children will become aware that budgets are tight, tempers will be strained, and leisure in the summer is something to look forward to.

Summary

What is the effect of money on relationships?

Changes ahead for American workers and their families include an increase in taxes and the retirement age, a decrease in government services and company pensions, and a more precarious work situation with fewer infrequent or non existent pay increases.

Poverty is devastating to couples and families. Those living in poverty have poorer physical and mental health, report lower personal and marital satisfaction, and die sooner.

The anxiety over lack of money may result in relationship conflict. Money (the lack of it, disagreement over how it is spent) is a frequent problem that couples report in marriage counseling.

Emotional well being rises with income and tops out at $75,000 a year. Increases in income above this figure are not associated with increases in satisfaction. The reason money is such a profound issue in marriage is its symbolic significance as well as the fact that individuals do something daily in reference to money—spend, save, or worry about it.

How do work and money affect relationships?

Money is a central issue in relationships because of its association with power, control, and dominance. Generally, the more money a partner makes, the more power that person has in the relationship. Males make considerably more money than females and generally have more power in relationships. When a wife earns an income, her power increases in the relationship. Money also provides an employed woman the power to be independent and to leave an unhappy marriage.

Driven primarily by the need to provide income for the family, about 70% of all U.S. wives are in the labor force. The time wives are most likely to be in the labor force is when their children are teenagers (between the ages of 14 and 17), the time when food and clothing expenses are the highest.

The office or workplace is often a place where people meet and establish relationships—including love and sexual relationships. Between 40% and 60% of workers report having been involved in an office romance. Whether a wife is satisfied with her job is related to the degree to which the job takes a toll on her family life. When the wife's employment interferes with her family life, she reports less job satisfaction. Husbands also report benefits from their wives' employment. These include being relieved of the sole responsibility for the financial support of the family and having more freedom to quit jobs, change jobs, or go to school.

In regard to marriage, women's employment does not destabilize happy marriages but increases the risk of disruption in unhappy marriages. Couples may be particularly vulnerable when a wife earns more money than her husband. Thirty percent of working wives earn more than their husbands. The wife earning more than her husband may affect the couple's marital happiness in several ways. Marriages in which the spouses view the provider role as the man's responsibility, in which the husband cannot find employment, and in which the husband is jealous of his wife's employment are vulnerable to dissatisfaction. Cultural norms typically dictate that the man is supposed to earn more money than his partner and that something is wrong with him if he doesn't.

What is the effect of parents' work decisions on the children?

Children do not appear to suffer cognitively or emotionally from their parents working as long as positive, consistent child-care alternatives are in place. However, less supervision of children by parents is an outcome of having two-earner parents. Leaving children to come home to an empty house is particularly problematic.

Parents view quality time as structured, planned activities, talking with their children, or just hanging out with them. Day care is typically mediocre but any negatives are offset by consistent parental attention.

What are the various strategies for balancing the demands of work and family?

Strategies used for balancing the demands of work and family include the superperson strategy, cognitive restructuring, delegation of responsibility, planning and time management, and role compartmentalization. Government and corporations have begun to respond to the family concerns of employees by implementing work-family policies and programs. These policies are typically inadequate and cosmetic.

What is the importance of leisure and what are its functions?

Corporations are learning that the new millenials value leisure and will not let their work life end it. Leisure helps to relieve stress, facilitate social interaction and family togetherness, and foster personal growth and skill development. However, leisure time may also create conflict over how to use it. Females sometimes become angry at their romantic partners for playing video games.

What is the future of work and the family?

Families will continue to be stressed by work. Employers will, increasingly, ask employees to work longer and do more. Businesses are struggling to stay solvent and the workers will take the brunt of the instability. Additionally, more and more women and wives will work outside the home, leading to more children in day-care centers.

Key Terms

Consumerism	Leisure	Role overload
Dual-career marriage	Millenials	Role strain
HER/his career	Mommy track	Second shift
HIS/her career	Momprenuer	Superwoman/supermom
HIS/HER career marriage	Opting out	Third shift
THEIR career marriage	Role compartmentalization	

Web Links

At Home Dads
http://athomedad.org/

Dual Academic Couples
http://www.stanford.edu/group/gender/

Identity Theft: Federal Trade Commission
http://www.consumer.gov/idtheft/

Identity Theft: Prevention and Survival
http://www.identitytheft.org/

Money Planner Online
http://www.thefinancialwellnessgroup.com/individual-online.php

Ms. Money (budgeting documents)
http://www.msmoney.com

Network on the Family and the Economy
http://www.olin.wustl.edu/macarthur/

Quality Day Care
http://www.familylifecenters.org/

Violence and Abuse in Relationships

> *He shook Love and her head repeatedly hit the wall.*
>
> **Police affidavit of George Huguely**

Photographer/Photo Agency

Learning Objectives

Identify the various types of abuse.

Discuss the four explanations for abuse in relationships.

Review abuse in undergraduate relationships and in marriage.

Summarize the effects of abuse and the cycle of abuse.

Review general and sexual child abuse.

Be aware of parent, sibling, and elder abuse.

Predict the future of abuse in relationships.

1. Emotional attachment is the primary reason a woman does not leave an abusive relationship.

2. Women are more likely to snoop/read partner's text messages.

3. Some victims of sexual abuse view their lives as being changed for the better.

4. Yelling and screaming is the most frequent emotionally abusive behavior partners engage in.

5. Forcible rape is more likely to be associated with PTSD and major depression than alcohol-induced rape.

Answers: 1. T   2. T   3. T   4. F   5. T

The photo that opens this chapter is of Yeardley Love, a 22-year-old senior and lacrosse player at the University of Virginia who was beaten to death in 2010 by her estranged former boyfriend George Huguely. He was charged with first degree murder. Television media (e.g., *Dateline, American Justice*) regularly feature horror stories of women who are beaten up or killed by their partners. Whether the motive is revenge, to be free for another partner, financial gain, or jealousy, the result is the same. What began as an intimate love relationship ends in the death of their partner or spouse. In this chapter, we examine the other side of intimacy in one's relationships. We begin by defining some terms.

Love does not dominate,
it cultivates.

Goethe, German philosopher

Nature of Relationship Abuse

Abuse is not uncommon. The Self-Assessment Abusive Behavior Inventory allows you to assess the degree of abuse in your current or most recent relationship. There are several types of abuse in relationships.

Violence and Homicide

Also referred to as physical abuse, **violence** may be defined as the intentional infliction of physical harm by either partner on the other. Examples of physical violence include pushing, throwing something at the partner, slapping, hitting, and forcing sex on the partner. **Intimate-partner violence** (IPV) is an all-inclusive term that refers to crimes committed against current or former spouses, boyfriends, or girlfriends. There are two types of violence (Brownridge 2010). One type is **situational couple violence** (SCV) where conflict escalates over an issue and one or both partners lose control. The person feels threatened and seeks to defend himself or herself. Control is lost and the partner strikes out. Both partners may lose control at the same time so it is symmetrical. A second type of violence, referred to as **intimate terrorism** (IT) is designed to control the partner. Witte and Kendra (2010) found that individuals who had been victims of interpersonal violence were less likely to discern when a social encounter was becoming dangerous and moving toward violence. The authors emphasize that learning to recognize cues (e.g., being overly aggressively, encouraging alcohol consumption, etc.) is essential to avoiding such experiences. A syndrome related to violence is **battered-woman syndrome**, which refers to the general pattern of battering that a woman is subjected to and is defined in terms of the frequency, severity, and injury she experiences. Battering is severe if the person's injuries require medical treatment or the perpetrator could be prosecuted.

Violence intentional infliction of physical harm by either partner on the other.

Intimate partner violence (IPV) an all-inclusive term that refers to crimes committed against current or former spouses, boyfriends, or girlfriends.

Situational couple violence (SCV) conflict escalates over an issue and one or both partners lose control.

Intimate terrorism (IT) behavior designed to control the partner.

Battered-woman syndrome the general pattern of battering that a woman is subjected to, defined in terms of the frequency, severity, and injury she experiences.

SELF-ASSESSMENT | Abusive Behavior Inventory

Write the number that best represents your closest estimate of how often each of the behaviors have happened in the relationship with your current or former partner during the previous six months.

1	2	3	4	5
Never	Rarely	Occasionally	Frequently	Very Frequently

_____ 1. Called you a name and/or criticized you

_____ 2. Tried to keep you from doing something you wanted to do (for example, going out with friends or going to meetings)

_____ 3. Gave you angry stares or looks

_____ 4. Prevented you from having money for your own use

_____ 5. Ended a discussion and made a decision without you

_____ 6. Threatened to hit or throw something at you

_____ 7. Pushed, grabbed, or shoved you

_____ 8. Put down your family and friends

_____ 9. Accused you of paying too much attention to someone or something else

_____ 10. Put you on an allowance

_____ 11. Used your children to threaten you (for example, told you that you would lose custody or threatened to leave town with the children)

_____ 12. Became very upset with you because dinner, housework, or laundry was not done when or how it was wanted

_____ 13. Said things to scare you (for example, told you something "bad" would happen or threatened to commit suicide)

_____ 14. Slapped, hit, or punched you

_____ 15. Made you do something humiliating or degrading (for example, begging for forgiveness or having to ask permission to use the car or do something)

_____ 16. Checked up on you (for example, listened to your phone calls, checked the mileage on your car, or called you repeatedly at work)

_____ 17. Drove recklessly when you were in the car

_____ 18. Pressured you to have sex in a way you didn't like or want

_____ 19. Refused to do housework or child care

_____ 20. Threatened you with a knife, gun, or other weapon

_____ 21. Spanked you

_____ 22. Told you that you were a bad parent

_____ 23. Stopped you or tried to stop you from going to work or school

_____ 24. Threw, hit, kicked, or smashed something

_____ 25. Kicked you

_____ 26. Physically forced you to have sex

_____ 27. Threw you around

_____ 28. Physically attacked the sexual parts of your body

_____ 29. Choked or strangled you

_____ 30. Used a knife, gun, or other weapon against you

Scoring

Add the numbers you wrote down and divide the total by 30 to determine your score. The higher your score (five is the highest score), the more abusive your relationship.

The inventory was given to 100 men and 78 women equally divided into groups of abusers or abused and nonabusers or nonabused. The men were members of a chemical dependency treatment program in a veterans' hospital and the women were partners of these men. Abusing or abused men earned an average score of 1.8; abusing or abused women earned an average score of 2.3. Nonabusing, abused men and women earned scores of 1.3 and 1.6, respectively.

Source

M. F. Shepard and J. A. Campbell. The Abusive Behavior Inventory: A measure of psychological and physical abuse. *Journal of Interpersonal Violence* 7(3): 291 – 305. ©1992 by Sage Publications Inc. Journals. Reproduced with permission of Sage Publications Inc. Journals in the format Other book via Copyright Clearance Center.

Battering may lead to murder. **Uxoricide** is the murder of a woman by a romantic partner. The murder of Yeardley Love by George Huguely was uxoricide. Thirty percent of homicides of women 18–24 are by an intimate partner (Teten et al. 2009). Most perpetrators are a male partner. Blacks and Hispanic women have higher rates of being killed by their intimate partners than non-Hispanic women (Azziz-Baumgartner et al. 2011). **Intimate partner homicide** is the murder of a spouse.

Other forms of murder in the family are **filicide** (murder of an offspring by a parent—the Casey Anthony trial captured nationwide attention), **parricide** (murder of a parent by an offspring), and **siblicide** (murder of a sibling). Family homicide is rare—less than one in 100,000 (Diem and Pizarro 2010). Whether it is abuse by the parent that influences a child to kill for protection of themselves or other family members, or recurring abuse that eventually ends in murder of the abuser, abuse is a common background factor in murder with the family.

Uxoricide the murder of a woman by a romantic partner.

Intimate partner homicide murder of a spouse.

Filicide murder of an offspring by a parent.

Parricide murder of a parent by an offspring.

Siblicide murder of a sibling.

National **Data**

In the U.S. adult population, over 4 million women are physically harmed by their husband, boyfriend, or other intimate partner annually (Torpy et al. 2008). In a national sample of Canadians age 15 and over, about three quarters of cohabiting and married female victims reported having experienced SCV, and about one quarter of cohabiting and married female victims reported having experienced IT in the 5 years preceding the survey. The percents were 80% (SCV) and 20% (IT) in the 5 years before the survey for cohabiting/married male victims (Brownridge 2010).

A great deal more abuse occurs than is reported. In a national sample of women who had been physically assaulted, almost three-fourths (73%) did not report the incidence to the police. The primary reason for no report was the belief that the police could not help (Dugan et al. 2003). Yet the damage is enormous. Johnson et al. (2008) emphasized that IPV is associated with post-traumatic stress disorder (PTSD) and results in social maladjustment and personal or social resource loss.

Regardless of the type, violence is not unique to heterosexual couples. Physical violence is also reported to occur in 11% to 12% of same-sex couples. Differences between violence in opposite-sex couples versus same-sex couples include that the latter is more mild, the threat of "outing" is present, and the violence is more isolated because it often occurs in a context of "being in the closet." Finally, victims of same-sex relationship abuse lack legal protections and services that are available to abuse victims in heterosexual couples. For example, most battered women's shelters do not serve lesbian women (or gay men); in nine states, legal protections for victims of domestic violence are granted only when the relationship is between a man and a woman or between spouses or former spouses (Rohrbaugh 2006).

Emotional Abuse

Emotional abuse abuse designed to make the partner feel bad, to denigrate the partner, to reduce the partner's status, and to make the partner vulnerable to being controlled by the abuser.

In addition to being physically violent, partners may also engage in **emotional abuse** (also known as psychological abuse, verbal abuse, or symbolic aggression). Whereas 8% of 2,922 undergraduates reported being involved in a physically abusive relationship, 30% reported that they "had been involved in an *emotionally* abusive relationship with a partner" (Knox and Hall 2010). Although emotional abuse does not involve physical harm, it is designed to make the partner feel bad—to denigrate the partner, reduce the partner's status, and make the partner vulnerable to being controlled by the abuser. Abowitz et al. (2010) revealed from a sample of 1,027 that undergraduates most likely to report emotional abuse were women, seniors (by class rank), participants in a "friends-with-benefits" relationship, students who had sought a partner via the Internet, students who believed in love at first sight ("romantics"), and students who reported other intimate partner violence (sexual coercion or physical abuse) or prior victimization (child physical abuse).

Follingstad and Edmundson (2010) identified the top emotionally abusive behaviors of one's partner in a national sample of adults in the United States: refusal to talk to the partner as a way of punishing the partner, making personal decisions for the partner (e.g., what to wear, what to eat, whether to smoke), throwing a temper tantrum and breaking things to frighten the partner, criticizing/belittling the partner to make him or her feel bad, and acting jealous when the partner is observed talking to a potential romantic partner. Other emotionally abusive behaviors include:

Yelling and screaming as a way of intimidating
Staying angry/pouting until the partner gives in

Requiring an accounting of the partner's time
Treating the partner with contempt
Making the partner feel stupid
Withholding emotional and physical contact
Isolation—prohibiting the partner from spending time with friends, siblings, and parents
Threats—threatening the partner with abandonment or threats of harm to oneself, family, or pets
Public demeaning behavior—insulting the partner in front of others
Restricting behavior—restricting the partner's mobility (for example, use of the car) or telling the partner what to do
Demanding behavior—requiring the partner to do as the abuser wishes (for example, have sex).

According to Follingstad and Edmundson (2010), both partners in a relationship report that they engage in emotionally abusive behavior. But each partner reports that they engage in less frequent and less severe emotionally abusive behavior. In effect, the respondents "created a picture of their own use of psychological abuse as limited in scope and not harmful in nature" (p. 506).

Sometimes abuse is both emotional and physical. An experience of a student follows. Readers who have been emotionally or physically abused may want to skip this section.

Going with Mr. Bad Boy

I met this boy at the end of my sophomore year. The first month was unbelievably perfect. We started fighting when I heard he had been cheating on me. He continually threatened me and told everyone we had been sleeping together, which was not true.

The fights got progressively worse but I couldn't let him go. I thought the first month would come back to us. Shortly he started pushing me during our fights, which evolved to hitting. I'd come home with bruises and have to make up stories to my parents on how I got them. I was scared and didn't know what to do so I just stayed. If I had known what I know now I would have immediately left. We went out one night and he called me a slut and I walked away and I said I was done so he grabbed me by my wrists and slammed me against a wall, pulled me up by my neck and yelled at me until I was blue in the face and his friend pulled him away from me. Things got worse because I told him that was the end. Yet he stalked and verbally attacked me over MySpace and text messages. I was a mess. I lost it and went into really deep depression. I started burning myself with curling irons.

About a week later I went out to a party with some friends where he showed up wanting to "straighten things out." He pulled me aside trying to talk gently to me and so I allowed the conversation. We all were crashing there and he ended up raping me. He stole something from me that I couldn't get back and now uses it against me. I have a lot of scars not to mention what he has done to my head since. He has ruined me. I'm now free of him but I can't stay in any relationship no matter how good because I'm so scared of it happening again.

Female Abuse of Partner

While it is assumed that men are more often the perpetrators of abuse, Brownridge (2010) noted that both women and men report abuse with equal frequency. Some researchers focus on women as abusive partners. Swan et al. (2008) noted that women's physical violence may be just as prevalent as men's violence but is more likely to be motivated by self-defense and fear, whereas men's physical violence is more likely than women's to be driven by control motives. Hence, women who are violent toward their partners tend to be striking back rather than throwing initial blows. An abusive female wrote the following:

> When people think of physical abuse in relationships, they get the picture of a man hitting or pushing the woman. Very few people, including myself, think about the "poor man" who is abused both physically and emotionally by his partner—but this is what I do to my boyfriend. We have been together for over a year now and things have gone from really good to terrible.
>
> The problem is that he lets me get away with taking out my anger on him. No matter what happens during the day, it's almost always his fault. It started out as kind of a joke. He would laugh and say "oh, how did I know that this was going to somehow be my fault." But then it turned into me screaming at him, and not letting him go out and be with his friends as his "punishment."
>
> Other times, I'll be talking to him or trying to understand his feelings about something that we're arguing about and he won't talk. He just shuts off and refuses to say anything but "alright." That makes me furious, so I start verbally and physically attacking him just to get a response.
>
> Or, an argument can start from just the smallest thing, like what we're going to watch on TV, and before you know it, I'm pushing him off the bed and stepping on his stomach as hard as I can and throwing the remote into the toilet. That really gets to him.

When women are abusive, they are rarely arrested by the police unless there is injury to the partner.

Spiritual Abuse

Spiritual abuse any attempt to impair the woman's spiritual life, spiritual self, or spiritual well-being.

In addition to physical and psychological abuse, Dehan and Levi (2009) identified spiritual abuse in a study of Haredi (ultraorthodox) Jewish wives. **Spiritual abuse** is defined as any attempt to impair the woman's spiritual life, spiritual self, or spiritual well-being. Ways in which men spiritually abuse their partners include: (a) belittling their spiritual worth, beliefs, or deeds; (b) preventing them from performing spiritual acts; and (c) causing them to transgress spiritual obligations or prohibitions. Examples of spiritual abuse include a husband belittling his wife for praying by telling her "your praying has no worth" and not allowing his wife to buy yeast and flour which are necessary for her to bake bread for the Shabbat (Saturday). Again, he would belittle her by telling her that this ritual is a waste of time.

An example of causing her to transgress spiritual prohibitions is the husband who rapes his ultraorthodox Jewish wife during or right after her menstrual period. According to Jewish law, during her menstrual period, a woman has the status of niddah, which requires that she not have intercourse until she has had seven clean days with no blood. An early death to oneself or to one's children awaits the woman who does not follow this prohibition. Hence, the husband who rapes his wife during or shortly after her period subjects her psychologically to devastating consequences.

Stalking in Person

Stalking unwanted following or harassment that induces fear in a target person.

Abuse may take the form of stalking. **Stalking** is defined as unwanted following or harassment that induces fear in a target person. Stalking is a crime. A stalker is one who has been rejected by a previous lover or is obsessed with a stranger or acquaintance who fails to return the stalker's romantic overtures. In about 80% of cases, the stalker is a heterosexual male who follows his previous lover. Women who stalk are more likely to target a married male. Stalking is a pathological emotional or motivational state (Meloy and Fisher 2005). Exactly one fourth of 2,922 university students reported that they had been stalked (followed and harassed) (Knox and Hall 2010). Such stalking is usually designed either to seek revenge or to win a partner back.

What if I'm Not Sure I am being Abused?

Sometimes it is not easy to know what constitutes being abused. Is a wife's PMS irritability and criticism considered abuse? Is calling a partner stupid considered abuse? Is shoving a partner an abusive act? In general, any verbal or physical behavior that is designed to denigrate and belittle the other (or take away the freedom of the other) is abuse.

How much should a person tolerate? The answer is very little, because the partner will be reinforced for being abusive and will continue to do so. In only a short time, being demeaning and pushing or shoving a partner can become routine acts considered normative in a couple's relationship. If you don't like the way your partner is treating you, it is important to register your disapproval quickly so that being abused is not an option. In Chapter 1, we discussed, "Not to decide is to decide." If you don't make a decision that you will not tolerate being abused, you have made a decision whereby you will be abused . . . and the abuse will escalate.

There's a fine line between support and stalking and let's all stay on the right side of that.

Joss Whedon, screenwriter

Who are the stalkers? Most often (in 85% of cases), men are stalkers and women are their victims. Two primary reasons for stalking are rejection by a sexual intimate (e.g., the male has been in a previous emotional or sexual relationship with the woman and obsessively tries to win her back) or rejection by a stranger with whom the stalker is infatuated and who fails to return romantic overtures. The stalking of celebrity females (for example, Jodie Foster) sometimes becomes visible in the media. The two most common emotions of the stalker are anger (over being rejected) and jealousy (at being replaced) (Meloy and Fisher 2005). Exercising a great deal of control in an existing relationship is predictive that the controlling partner will become a stalker when the other partner ends the relationship (King 2003). For the 15% of stalkers who are women, the most common victim is another woman. These may be partners in lesbian relationships that have ended where the stalker feels rejected and wants to renew the relationship.

Stalkers are obsessional and very controlling. Obsessional thinking is their most common cognitive trait (Meloy and Fisher 2005). They are typically mentally ill and have one or more personality disorders involving paranoid, antisocial, or obsessive-compulsive behaviors. There are three primary brain systems involved in stalking: sex drive, whereby the individual is motivated to achieve sexual gratification; attraction, whereby the individual is driven to emotionally connect with a specific mating partner; and attachment, whereby the individual is motivated to experience a secure relationship with a long-term partner. In effect, the stalker feels a barrier in access to the beloved (physically and emotionally), which intensifies the drive to be with the rejecting partner (referred to as abandonment rage). Indeed, brain activity can be observed with magnetic resonance imaging (MRI) that reveals differences between a person who is happily in love and a person who is the "spurned or unrequited stalker" (Meloy and Fisher 2005).

Although various coping strategies have been identified, additional research is needed on how to manage unwanted attention. A survey of young adults suggested the following general coping categories (Spitzberg and Cupach 1998):

1. Make a direct statement to the person ("I am not interested in dating you, my feelings about you will not change, and I know that you will respect my decision and direct your attention elsewhere") (Regan 2000, 266).

Nature of Relationship Abuse   379

2. Seek protection through formal channels (for example, police or court restraining order).

3. Avoid the perpetrator (ignore, don't walk with or talk to, hang up if the person calls).

4. Use informal coping methods (seek advice from others in regard to how they have coped with stalking).

Direct statements and actions that unequivocally communicate lack of interest are probably the most effective types of intervention.

National **Data**

Between 8% and 15% of women and between 2% and 4% of men in the United States will be stalked in their lifetime (Meloy and Fisher 2005).

Stalking Online—Cybervictimization

Cybervictimization includes being sent threatening e-mail, unsolicited obscene e-mail, computer viruses, or junk mail (spamming). It may also include flaming (online verbal abuse), and leaving improper messages on message boards. Cybervictimization in reference to Internet dating is "minimal." This is the conclusion of Jerin and Dolinsky (2007) and is based on a study of 134 people who completed profiles on a variety of Internet dating sites. Cyberangels.org is a website which promotes online safety.

Prior to stalking is **obsessive relational intrusion (ORI)**, the relentless pursuit of intimacy with someone who does not want it. The person becomes a nuisance but does not have the goal of harm as does the stalker. When ORI and stalking are combined, as many as 25% of women and 10% of men can expect to be pursued in unwanted ways (Spitzberg and Cupach 2007).

People who cross the line in terms of pursuing an ORI relationship or responding to being rejected (stalking) engage in a continuum of eight forms of behavior. Spitzberg and Cupach (2007) have identified these:

1. *Hyperintimacy*—telling a person that they are beautiful or desirable to the point of making them uncomfortable.

2. *Relentless electronic contacts*—flooding the person with e-mail messages, cell phone calls, text messages, or faxes.

3. *Interactional contacts*—showing up at the person's work or gym. The intrusion may also include joining the same volunteer groups as the pursued.

4. *Surveillance*—monitoring the movements of the pursued such as following the person or driving by the person's house.

5. *Invasion*—breaking into the person's house and stealing objects that belong to the person; identity theft; or putting Trojan horses (viruses) in the person's computer. One woman downloaded child pornography on her boyfriend's computer and called the authorities to arrest him. He is now in prison.

6. *Harassment or intimidation*—leaving unwanted notes on one's desk or a dead animal on one's doorstep.

7. *Threat or coercion*—threatening physical violence or harm to the person or one's family or friends.

8. *Aggression or violence*—carrying out a threat by becoming violent (for example, kidnapping or rape).

Cybercontrol

Short of cybervictimization is **cybercontol** whereby individuals use communication technology, such as cell phones, e-mail, and social networking sites, to monitor or control partners in intimate relationships. In one study of 745 undergraduates, half of both the men and the women reported this behavior either as the initiator or victim. Close to 25% of female college students monitored

Cybervictimization harassing behavior which includes being sent threatening e-mail, unsolicited obscene e-mail, computer viruses, or junk mail (spamming); can also include flaming (online verbal abuse) and leaving improper messages on message boards.

Obsessive relational intrusion (ORI) the relentless pursuit of intimacy with someone who does not want it.

Cybercontrol use of communication technology, such as cell phones, e-mail, and social networking sites, to monitor or control partners in intimate relationships.

their partner's behavior by checking e-mails, even password protected accounts, versus only 6% of males. In addition, more females than males thought it was appropriate to check e-mail and cell phone call histories of their partners. A sociobiological explanation can be used to explain that more females are involved in monitoring their partner's behavior than vice versa—males are more likely to cheat and women are protecting their relationship (Burke et al. 2011).

Explanations for Violence and Abuse in Relationships

Teenage girls can't tell their parents that their boyfriend beat them up. You don't dare let your neighbor know that you fight. It's one of the things we [women] will hide, because it's embarrassing.

Rihanna, singer

Research suggests that numerous factors contribute to violence and abuse in intimate relationships. These factors include those that occur at the cultural, community, and individual and family levels.

Cultural Factors

In many ways, American culture tolerates and even promotes violence. Violence in the family stems from the acceptance of violence in our society as a legitimate means of enforcing compliance and solving conflicts at interpersonal, familial, national, and international levels. Violence and abuse in the family may be linked to cultural factors, such as violence in the media, acceptance of corporal punishment, gender inequality, and the view of women and children as property. The context of stress is also conducive to violence.

Violence in the Media One need only watch the evening news to see the violence in Afghanistan, Egypt, and sub-Saharan Africa. Feature films and TV movies regularly reflect themes of violence. Football, which dominates Saturday, Sunday, and Monday television in the fall is a very violent sport with some players who become injured for life. Boxing matches continue to draw large paid TV audiences.

DIVERSITY IN OTHER COUNTRIES

On 1979, Sweden passed a law that effectively abolished corporal punishment as a legitimate child-rearing practice. Fifteen other countries, including Italy, Germany, and Ukraine, have banned all corporal punishment in all settings, including the home.

Corporal Punishment of Children The use of physical force with the intention of causing a child to experience pain, but not injury, for the purpose of correction or control of the child's behavior is **corporal punishment**. In the United States, it is legal in all fifty states for a parent to spank, hit, belt, paddle, whip, or otherwise inflict punitive pain on a child so long as the corporal punishment does not meet the individual state's definition of child abuse. Violence has become a part of our cultural heritage through the corporal punishment of children. In a review of the literature, 94% of parents of toddlers reported using corporal punishment; children who are victims of corporal punishment display more antisocial behavior, are more violent, and have an increased incidence of depression as adults (Straus 2000). Child development specialists recommend an end to corporal punishment to reduce the risk of physical abuse, harm to other children, and to break the cycle of abuse.

Corporal punishment the use of physical force with the intention of causing a child to experience pain, but not injury, for the purpose of correction or control of the child's behavior.

Gender Inequality Domestic violence and abuse may also stem from traditional gender roles. Traditionally, men have also been taught that they are superior to women and that they may use their aggression toward women, believing that women need to be "put in their place." The greater the inequality and dependence of the woman on the man is, the more likely it is that he will abuse her. Conversely, women with higher income and education than their partner report more frequent abuse since these achievements may be a threat to the male's masculinity (Anderson 2010b).

Honor crime/Honor killing refers to unmarried women who are killed because they have sexual intercourse which brings shame shame on their parents and siblings; occurs in Middle Eastern countries such as Jordan.

Some occupations, such as police officers and military personnel, lend themselves to contexts of gender inequality. In military contexts, men notoriously devalue, denigrate, and sexually harass women. In spite of the rhetoric about gender equality in the military, women in the Army, Navy, and Air Force academies continue to be sexually harassed. These male perpetrators may not separate their work roles from their domestic roles. One student in our classes noted that she was the ex-wife of a Navy Seal and that "he knew how to torment someone, and I was his victim."

View of Women and Children as Property   Prior to the late nineteenth century, a married woman was considered the property of her husband. A husband had a legal right and marital obligation to discipline and control his wife through the use of physical force.

Stress   Our culture is also a context of stress. The stress associated with getting and holding a job, rearing children, staying out of debt, and paying bills may predispose one to lash out at one's partner and/or children. Gormley and Lopez (2010) found that undergraduate males who reported high stress levels also reported higher rates of psychological abuse of their partners.

Community Factors

Community factors that contribute to violence and abuse in the family include social isolation, poverty, and inaccessible or unaffordable health care, day care, eldercare, and respite care services and facilities.

Social Isolation   Living in social isolation from extended family and community members increases the risk of being abused. Spouses whose parents live nearby are least vulnerable.

Poverty   Abuse in adult relationships occurs among all socioeconomic groups. However, poverty and low socioeconomic development are associated with crime and higher incidences of violence. This violence may spill over into interpersonal relationships as well as the frustration of living in poverty.

Inaccessible or Unaffordable Community Services   Failure to provide medical care to children and elderly family members sometimes results from the lack of accessible or affordable health care services in the community. Failure to provide supervision for children and adults may result from inaccessible day-care and eldercare services. Without eldercare and respite care facilities, families living in social isolation may not have any help with the stresses of caring for elderly family members and children.

Individual Factors

Individual factors associated with domestic violence and abuse include psychopathology, personality characteristics, and alcohol or substance abuse. A number of personality characteristics have been associated with people who are abusive in their intimate relationships. Some of these characteristics follow:

   1. *Dependency.* Therapists who work with batterers have observed that they are extremely dependent on their partners. Because the thought of being left by their partners induces panic and abandonment anxiety, batterers use physical aggression and threats of suicide to keep their partners with them.

2. *Jealousy.* Along with dependence, batterers exhibit jealousy, possessiveness, and suspicion. An abusive husband may express his possessiveness by isolating his wife from others; he may insist she stay at home, not work, and not socialize with others. His extreme, irrational jealousy may lead him to accuse his wife of infidelity and to beat her for her presumed affair.

3. *Need to control.* Abusive partners have an excessive need to exercise power over their partners and to control them. The abusers do not let their partners make independent decisions, and they want to know where they are, whom they are with, and what they are doing. They like to be in charge of all aspects of family life, including finances and recreation.

4. *Unhappiness and dissatisfaction.* Abusive partners often report being unhappy and dissatisfied with their lives, both at home and at work. Many abusers have low self-esteem and high levels of anxiety, depression, and hostility. They may expect their partner to make them happy.

5. *Anger and aggressiveness.* Abusers tend to have a history of interpersonal aggressive behavior. They have poor impulse control and can become instantly enraged and lash out at the partner. Battered women report that episodes of violence are often triggered by minor events, such as a late meal or a shirt that has not been ironed.

6. *Quick involvement.* Because of feelings of insecurity, the potential batterer will move his partner quickly into a committed relationship. If the woman tries to break off the relationship, the man will often try to make her feel guilty for not giving him and the relationship a chance.

7. *Blaming others for problems.* Abusers take little responsibility for their problems and blame everyone else. For example, when they make mistakes, they will blame their partner for upsetting them and keeping them from concentrating on their work. A man may become upset because of what his partner said, hit her because she smirked at him, and kick her in the stomach because she poured him too much alcohol.

8. *Jekyll-and-Hyde personality.* Abusers have sudden mood changes so that a partner is continually confused. One minute an abuser is nice, and the next minute angry and accusatory. Explosiveness and moodiness are typical.

9. *Isolation.* An abusive person will try to cut off a partner from all family, friends, and activities. Ties with anyone are prohibited. Isolation may reach the point at which an abuser tries to stop the victim from going to school, church, or work.

10. *Alcohol and other drug use.* Whether alcohol reduces one's inhibitions to display violence, allows one to avoid responsibility for being violent, or increases one's aggression, alcohol and substance abuse are associated with violence and abuse.

11. *Emotional deficit.* Some abusing spouses and parents may have been reared in contexts that did not provide them with the capacity to love, nurture, or be emotionally engaged.

12. *Criminal/Psychiatric background.* Eke et al. (2011) examined the characteristics of 146 men who committed an actual or attempted act of murdering their intimate partner. Of these, 42% had prior criminal charges, 15% had a psychiatric history, and 18% had both. Hence, individuals with a criminal record of abuse or psychiatric problems should be regarded with caution.

Family Factors

Family factors associated with domestic violence and abuse include being abused as a child, having parents who abused each other, and not having a father in the home.

Child Abuse in Family of Origin   Individuals who were abused as children were more likely to be abusive toward their partners as adults.

Family Conflict Children learn abuse from their family context. Fathers who were not affectionate were also more vulnerable to being abusive.

Parents Who Abused Each Other Busby et al. (2008) reconfirmed in a study of 30,600 individuals that the family of orientation is the context where individuals who observe their parents being violent with each other and who perpetuate the violence as children end up being more likely to be violent in their adult relationships. However, a majority of children who witness abuse do not continue the pattern. A family history of violence is only one factor out of many that may be associated with a greater probability of adult violence.

Sexual Abuse in Undergraduate Relationships

Palmer et al. (2010b) surveyed 370 college students regarding their past year experiences and found that 34% of women reported unwanted sexual contact. Flack et al. (2008) found no evidence of the **red zone**, the first month of the first year of college when women are most likely to be victims of sexual abuse. In a sample of 2,922 undergraduates at a large southeastern university, 33% reported being pressured to have sex by a partner they were dating (Knox and Hall 2010). Katz et al. (2008) noted that some women experience sexual abuse in addition to a larger pattern of physical abuse and that the combination is associated with less general satisfaction, less sexual satisfaction, more conflict, and more psychological abuse from the partner. In effect, women in these co-victimization relationships are miserable.

Acquaintance and Date Rape

The word *rape* often evokes images of a stranger jumping out of the bushes or a dark alley to attack an unsuspecting victim. However, most rapes are perpetrated not by strangers but by people who have a relationship with the victim. About 85% of rapes are perpetrated by someone the woman knows. This type of rape is known as **acquaintance rape**, which is defined as nonconsensual sex between adults (of the same or other sex) who know each other. The behaviors of sexual coercion occur on a continuum from verbal pressure and threats to the use of physical force to obtain sexual acts, such as kissing, oral sex or intercourse.

The perpetrator of a rape is likely to believe in various rape myths. **Rape myths** are beliefs that deny victim injury or cast blame on the woman for her own rape. These beliefs are false, widely held, and justify male aggression. Examples include: women deserve to be raped, particularly when they drink too much and are provocatively dressed; rape isn't rape but consensual sex; women fantasize about and secretly want to be raped; women who really don't want to be raped resist more—they could stop a guy if they really wanted to; and a woman cannot be raped by her husband. McMahon (2010) found that rape myths were more likely to be accepted by males, those pledging a fraternity/sorority, athletes, those without previous rape education, and those who did not know someone who had been sexually assaulted.

Men are also raped. In a sample of 1,400 men ages 18–24, 6% reported that they were coerced to have vaginal sex with a female; 1% by a male to have oral or anal sex (Smith et al. 2010b). Palmer et al. (2010b) surveyed 370 college students regarding their past year experiences and almost a third of the men (31%) reported unwanted sexual contact.

Rape myths abound as it is assumed that "men can't be raped," "men who are raped are gay," and that "men always want sex" (Chapleau et al. 2008). Not only is there a double standard of perceptions that only women can be raped

Rape's not something where you just go, "Well, get over it" or "Believe in love and peace, my child, and it'll all be over."

Anonymous post to dancinginthedarkness.com

Red zone the first month of the first year of college when women are most likely to be victims of sexual abuse.

Acquaintance rape nonconsensual sex between adults (of same or other sex) who know each other.

Rape myths beliefs that deny victim injury or cast blame on the woman for her own rape.

but a double standard is operative in the perception of the gender of the person engaging in sexual coercion. Men who rape are aggressive; women who rape are promiscuous (Oswald and Russell 2006).

One type of acquaintance rape is **date rape**, which refers to nonconsensual sex between people who are dating or on a date. Women who dress seductively, even wives, are viewed by both undergraduate men and women as partly responsible for being raped by their partners (Whatley 2005).

Women are also vulnerable to repeated sexual force. Daigle et al. (2008) noted that 14% to 25% of college women experience repeat sexual victimization during the same academic year—they are victims of rape or other unwanted sexual force more than once, often in the same month. The primary reason is that they do not change the context; they may stay in the relationship with the same person and continue to use alcohol or drugs in that context.

Both women and men may pressure a partner to have sex. Weiss (2010) focused on male rape.

Date rape one type of acquaintance rape which refers to nonconsensual sex between people who are dating or are on a date.

Alcohol and Rape

Alcohol is the most common "rape drug." A person under the influence cannot give consent. Hence, a person who has sex with someone who is impaired is engaging in rape. In a study of 340 college rape victims, 41% reported being impaired and 21% reported being incapacitated—hence over 60% were in an altered state with 38% reporting no impairment (Littleton et al. 2009). About two thirds of all three groups—impaired, incapacitated, and not impaired—reported moderate physical force by the rapist. The nonimpaired reported the highest percent of verbal threats (27%) and severe physical force (9%). The greatest resistance (56% nonverbal; 53% verbal; 45% physical) came from the nonimpaired. In regard to the nature of the relationship, in over half of the cases in which the girl was raped, the assailant was an acquaintance. Farris et al. (2010) also found that alcohol dose was related to men interpreting a woman's friendliness as sexual interest and being more coercive. Swartout and White (2010) also found that drug use (other than alcohol) in general was associated with sexual aggression. The take away message is that alcohol and drug use or being impaired increases the risk of being raped and perpetrating sexual coercion (Palmer et al. 2010b). This finding is further supported by Krebs et al. (2009) who found that getting drunk and using marijuana were "strongly associated with experiencing incapacitated sexual assault."

Chelsea E. Curry

A fun night like these two friends are having can quickly become dangerous. The woman who trusts this guy can easily drink until she loses her judgment as they go back to his apartment for one more drink where he rapes her. Eighty-five percent of rapes are not by a stranger, but by someone the partner knows . . . and trusts.

Rophypnol—The Date Rape Drug

While 85% of those occasions where the woman is incapacitated due to alcohol or drugs is voluntary (the woman is willingly drinking or doing drugs), in 15% of the cases, drugs are used against her will (Lawyer et al. 2010). **Rophypnol**—also known as the date rape drug, rope, roofies, Mexican Valium, or the "forget (me) pill"—causes profound, prolonged sedation and short-term memory loss. Similar to Valium but ten times as strong, Rophypnol is a prescription drug in Europe and used as a potent sedative. It is sold in the United States for about $5, is dropped in a drink (where it is tasteless and odorless), and causes victims to lose their memory for eight to ten hours. During this time, victims may be raped yet be unaware until they notice signs of it. A former student in our classes reported being drugged ("he put something in my drink") by a "family friend" when she was 16. She noticed blood in her panties the next morning but had no memory of the previous evening. The "friend" is currently being prosecuted.

The Drug-Induced Rape Prevention and Punishment Act of 1996 makes it a crime to give a controlled substance to anyone without their knowledge and with the intent of committing a violent crime (such as rape). Violation of this law is punishable by up to twenty years in prison and a fine of $250,000.

Women are defenseless when drugged. When an assault begins when the woman is not raped, her responses may vary from pleading to resistance including "turning cold" and "running away." Gidycz et al. (2008) discussed the various responses in terms of background. Women who had been victimized as children were more likely to "freeze and turn cold."

The effect of rape is negative to devastating, including loss of self-esteem, loss of trust, and the inability to be sexual. Zinzow et al. (2010) studied a national sample of women and found that forcible rape was more likely to be associated with PTSD and with major depression than drug/alcohol/incapacitated rape.

Acknowledging that one is a rape victim is associated with fewer negative psychological symptoms (e.g., depression) and increased coping (e.g., moving beyond the event) (Clements and Ogle 2009). Denial, not discussing it with others, and avoiding the reality of having been victimized prolongs one's recovery.

Rape Prevention

In regard to preventing rape, Brecklin and Ullman (2005) analyzed data on 1,623 women and noted that those who had had self-defense or assertiveness training reported that their resistance stopped the offender or made him less aggressive than victims without such training. Women with the training also noted that they were less scared during the attack. Cowburn (2010) emphasized a broader strategy—that sexual coercion by men can best be understood in the context of masculine socialization and that attention to this issue must be part of the prevention strategies for developing community safety. In this regard, Foubert et al. (2010) reported positive attitude and behavior change on the part of undergraduate sophomore males who were involved in an all-male sexual assault peer education/rape-prevention program (The Men's Program) at two year follow-up. Not only were these men less likely to be perpetrators, they were more likely to intervene when they observed male peers about to hook up with an intoxicated female.

Abuse in Marriage Relationships

The chance of abuse in a relationship increases with marriage. Indeed, the longer individuals know each other and the more intimate the relationship, the more likely the abuse.

Rophypnol causes profound, prolonged sedation and short-term memory loss; also known as the date rape drug, roofies, Mexican Valium, or the "forget (me) pill."

By not coming forward [about rape], you make yourself a victim forever.

Kelly McGillis, American actress

General Abuse in Marriage

Abuse in marriage may differ from unmarried abuse in that the husband may feel "ownership" of the wife and feel the need to "control her." But the behaviors of abuse are the same—belittling the spouse, controlling the spouse, hitting the spouse, etc. Deciding to end a marriage in which there is abuse is considered in the Personal Choices section that follows.

PERSONAL CHOICES

Would You End an Abusive Marital Relationship?

Students in our classes were asked whether they would end a marriage if the spouse hit or kicked them. Some of their comments follow.

Many felt that marriage was too strong a commitment to end if the abuse could be stopped.

I would not divorce my spouse if she hit or kicked me. I'm sure that there's always room for improvement in my behavior, although I don't think it's necessary to assault me. I recognize that under certain circumstances, it's the quickest way to draw my attention to the problems at hand. I would try to work through our difficulties with my spouse.

***

The physical contact would lead to a separation. During that time, I would expect him to feel sorry for what he had done and to seek counseling. My anger would be so great, it's quite hard to know exactly what I would do.

***

I wouldn't leave him right off. I would try to get him to a therapist. If we could not work through the problem, I would leave him. If there was no way we could live together, I guess divorce would be the answer.

***

I would not divorce my husband, because I don't believe in breaking the sacred vows of marriage. But I would separate from him and let him suffer!

***

I would tell her I was leaving but that she could keep me if she would agree for us to see a counselor to ensure that the abuse never happened again.

Some said, "Seek a divorce." Those opting for divorce (a minority) felt they couldn't live with someone who had abused or might abuse them again.

I abhor violence of any kind, and since a marriage should be based on love, kicking is certainly unacceptable. I would lose all respect for my husband and I could never trust him again. It would be over.

Many therapists emphasize that a pattern of abuse develops and continues if such behavior is not addressed immediately. The first time abuse occurs, the couple should seek therapy. The second time it occurs, they should separate. The third time, they might consider a divorce. Later in the chapter we discussed disengaging from an abusive relationship.

Men who Abuse

Henning and Connor-Smith (2010) studied a large sample of men who were recently convicted of violence toward a female intimate partner (N = 1,130). More than half of the men (59%) reported that they were continuing or planning to continue their relationship. Reasons included older age, being married to the victim, having children together, attributing less blame to the victim for the recent offense, and having a childhood history of family violence.

Marital rape forcible rape by one's spouse—a crime in all states.

Rape in Marriage

Marital rape, now recognized in all states as a crime, is forcible rape by one's spouse. The forced sex may take the form of sexual intercourse, fellatio, or anal intercourse. Sexual violence against women in an intimate relationship is often repeated.

Effects of Abuse

Abuse affects the physical and psychological well-being of victims. Abuse between parents also affects the children.

Effects of Partner Abuse on Victims

Effects are always in reference to perception. Becker et al. (2010) confirmed that being a victim of intimate partner violence is associated with symptoms of PTSD—loss of interest in activities/life in general, feeling detached from others, inability to sleep, irritability, etc. Rhatigan and Nathanson (2010) found in a study of 293 college women that women took into account their own behavior in evaluating their boyfriends' abusive behavior. If they felt they had "set him up" by their own behavior, they were more understanding of his aggressiveness. Additionally, those with low self-esteem were more likely to take some responsibility for their boyfriends' abusive behavior. While some partners will rationalize or overlook their partners, abusive behavior, others are hurt by it and the effect is negative.

Sarkar (2008) identified the negative impact of intimate partner violence. IPV affected the woman's physical and mental health, and increased the risk for unintended pregnancy and multiple abortions. These women also reported high levels of anxiety and depression that often led to alcohol and drug abuse. Violence on pregnant women significantly increased the risk for infants of low birth weight, preterm delivery, and neonatal death. Katz and Myhr (2008) noted that 21% of 193 female undergraduates were experiencing verbal sexual coercion in their current relationships. The effects included feeling psychologically abused, arguing, and decreased relationship satisfaction and sexual functioning.

Effects of Partner Abuse on Children

In the most dramatic effect, some women are abused during their pregnancy, resulting in a high rate of miscarriage and birth defects, affecting the child directly. Negative effects may also accrue to children who witness domestic abuse. Twenty percent of undergraduates report having observed their parents engage in abusive behavior toward each other (DiLillo et al. 2010). Russella et al. (2010) found that children who observed parental domestic violence were more likely to be depressed as adults. Hence, a child need not be a direct target to abuse, but merely a witness to incur negative effects. However, Kulkarni et. al. (2011) compared the effects of witnessing domestic violence and actually experiencing it on PTSD in later life and found that witnessing alone was not associated with PTSD.

The Cycle of Abuse

The following reflects the cycle of abuse.

"I Got Flowers Today"

I got flowers today. It wasn't my birthday or any other special day. We had our first argument last night, and he said a lot of cruel things that really hurt me. I know he is sorry and didn't mean the things he said, because he sent me flowers today.

What If My Close Friend is Being Abused and Keeps Going Back?

As noted, an abused wife will leave and return to her husband an average of seven times before she finally makes the break. So an abused individual must reach a point where staying is more aversive than leaving (or leaving more attractive than staying). Lacey et al. (2011) studied women who left an intimate relationship and found that those who had children and who had been together less than five years were more likely to leave. The event that triggered the leaving was threat with a weapon or psychological abuse. Most who left returned within a month.

Until your friend makes a decision to leave, you can do little but make it clear that you are available and have a place for your friend to stay when ready to make the break. Without a place to go, many will remain in an abusive relationship until they are, literally, beaten to death. Your promise of a safe place to stay is a stimulus to action . . . . you just won't know when your friend will take you up on your offer. In addition, you might consider suggesting reasons to leave "to save your children" and to empower them with your confidence that they are "strong enough to leave . . . actually go through with it." Such a strong motivation (children's safety) and a confident peer have been associated with leaving an abusive relationship (Baly 2010).

I got flowers today. It wasn't our anniversary or any other special day. Last night, he threw me into a wall and started to choke me. It seemed like a nightmare. I couldn't believe it was real. I woke up this morning sore and bruised all over. I know he must be sorry, because he sent me flowers today.

Last night, he beat me up again. And it was much worse than all the other times. If I leave him, what will I do? How will I take care of my kids? What about money? I'm afraid of him and scared to leave. But I know he must be sorry, because he sent me flowers today.

I got flowers today. Today was a very special day. It was the day of my funeral. Last night, he finally killed me. He beat me to death.

If only I had gathered enough courage and strength to leave him, I would not have gotten flowers today.

Author unknown

The cycle of abuse begins when a person is abused and the perpetrator feels regret, asks for forgiveness, and starts acting nice (for example, gives flowers). The victim, who perceives few options and feels guilty terminating the relationship with the partner who asks for forgiveness, feels hope for the relationship at the contriteness of the abuser and does not call the police or file charges. Forgiving the partner and taking him back usually occurs seven times before the partner leaves for good. Shakespeare (in *As You Like It*) said of such forgiveness, "Thou prun'st a rotten tree."

After the forgiveness, couples usually experience a period of making up or honeymooning, during which the victim feels good again about the partner and is hopeful for a nonabusive future. However, stress, anxiety, and tension mount again in the relationship, which is relieved by violence toward the victim. Such violence is followed by the familiar sense of regret and pleadings for forgiveness, accompanied by being nice (a new bouquet of flowers, and so on).

As the cycle of abuse reveals, some victims do not prosecute their partners who abuse them. To deal with this problem, Los Angeles has adopted a "zero

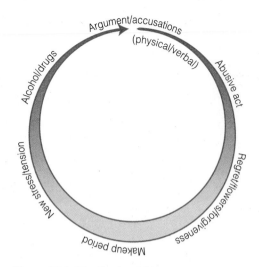

Figure 13.1 The Cycle of Abuse

Guess I'd rather hurt than feel nothing at all.

"Need You Now," Lady Antebellum

Entrapped stuck in an abusive relationship and unable to extricate oneself from the abusive partner.

tolerance" policy toward domestic violence. Under the law, an arrested person is required to stand trial and his victim required to testify against the perpetrator. The sentence in Los Angeles County for partner abuse is up to six months in jail and a fine of $1,000.

Figure 13.1 illustrates this cycle, which occurs in clockwise fashion. In the rest of this section, we discuss reasons why people stay in an abusive relationships and how to get out of such relationships.

Why Victims Stay in Abusive Relationships

One of the most frequently asked questions of people who remain in abusive relationships is, "Why do you stay?" Alexander et al. (2009) noted that the primary reason a woman returns again and again to an abusive relationship is the emotional attachment to her partner. While someone who criticizes (for example, "you're ugly/stupid/pitiful"), is dishonest (is sexually unfaithful, for instance), or physically harms a partner will create a context of interpersonal misery, the victim might still love and feel emotionally drawn to that person.

Most marriage therapists suggest examining why you love and continue to stay with such a person. Do you feel that you deserve this treatment because you are "no good" or that you would not be able to find a better alternative (low self-concept)? Do you feel you would rather be with a person who treats you badly than be alone (fear of the unknown)? Or do you hang on because you look forward to a better tomorrow (hope)?

Another explanation for why some people with abusive partners continue to be in love is that the abuse is only one part of the relationship. When such partners are not being abusive, they may be kind, loving, and passionate. Shackelford et al. (2005) analyzed data on 1,461 men and found that they often used "mate-retention behaviors," such as giving flowers or gifts to entice a partner to stay in the relationship, which can become even more abusive.

The presence of these mate-retention behaviors, which happen every now and then (a periodic reinforcement), keeps the love feelings alive. Love stops when insufficient positive behaviors counteract the extent of the abusive behaviors. One abused partner said, "when he started abusing my kids, that was it."

The presence of children also has an influence, which works in conflicting ways (Rhodes et al. 2010b). Some mothers will leave their abusive partners so as not to subject their children to the abuse. But others will stay in an abusive marriage since the mothers want to keep the family together.

In earlier research, Few and Rosen (2005) interviewed twenty-five women who had been involved in abusive dating relationships from three months to nine years (average = 2.4 years) to find out why they stayed. The researchers conceptualized the women as **entrapped**—stuck in an abusive relationship and unable to extricate oneself from the abusive partner. Indeed, these women escalated their commitment to stay in hopes that doing so would eventually pay off. Among the factors of their perceived investment were the time they had already spent in the relationship, the sharing of their emotional self with their partner, and the relationships to which they were connected because of the partner. In effect, they had invested time with a partner they were in love with and wanted to turn the relationship around into a safe, nonabusive one. The following are some of the factors explaining how abused women become entrapped:

- Fear of loneliness ("I'd rather be with someone who abuses me than alone.")
- Love ("I love him.")
- Emotional dependency ("I need him.")
- Commitment to the relationship ("I took a vow 'for better or for worse.'")
- Hope ("He will stop.")
- A view of violence as legitimate ("All relationships include some abuse.")

- Guilt ("I can't leave a sick man.")
- Fear ("He'll kill me if I leave him.")
- Economic dependence ("I have no place to go.")
- Isolation ("I don't know anyone who can help me.")

Battered women also stay in abusive relationships because they rarely have escape routes related to educational or employment opportunities, their relatives are critical of plans to leave the partner, they do not want to disrupt the lives of their children, and they may be so emotionally devastated by the abuse (anxious, depressed, or suffering from low self-esteem) that they feel incapable of planning and executing their departure.

Disengaging is a process that takes time. One should not be discouraged when they see a loved one return to an abusive context but recognize that progress in disengagement is occurring (she did it once) and that positive movement is predictive of eventually getting out.

*You may trod me
in the very dirt
But still, like dust,
I'll rise.*

Maya Angelou, poet

Strategies Abused Women Use in Coping with the Abuse

Smith et al. (2010b) identified several coping strategies of women in abusive relationships.

These include:
- Self-Talk—talking to oneself about the necessity to end the relationship with the abuser while not actually taking action (Saying "I really need to end this," but doing nothing)
- Keeping It at Bay—making light of the partner's abuse, laughing at how silly the abuse is (e.g., "He broke my frigging jaw." [laughing])
- Taking Blame—excusing the behavior of the partner ("I set him off when I asked if he had had enough alcohol.")
- Spirituality—praying for things to get better
- Release (Intrapersonal)—exercise/journaling
- Drugs and or alcohol—medicating the problem to a zone where it no longer hurts
- Thoughts of Death—thinking of death as an escape

How One Leaves an Abusive Relationship

Leaving an abusive partner begins with the decision to do so (and includes involving family and friends to provide a safe supportive context). Such a decision often follows the acknowledgement that one has enough and the belief that one must withdraw and move on as the relationship will only deteriorate. A plan comes into being and the person acts (for example, moves in with parents, a sister, friend, or goes to a homeless shelter). If the new context is better than being in the abusive context, the person will stay away. Otherwise, the person may go back and start the cycle all over. As noted previously, this leaving and returning typically happens seven times.

Sometimes the woman does not just disappear while the abuser is away but calls the police and has the man arrested for violence and abuse. While the abuser is in jail, she may move out and leave town. In either case, disengagement from the abusive relationship takes a great deal of courage. Calling the National Domestic Violence Hotline (800-799-7233 [SAFE]), available twenty-four hours, is a point of beginning. Involvement with an intimate partner who is violent may be life-threatening. Particularly if the individual decides to leave the violent partner, the abuser may react with more violence and murder the person who has left. Indeed, a third of murders that occur in domestic violence cases occur shortly after a breakup. Specific signs that could be precursors to someone about to murder an intimate partner are stalking, strangulation, forced sex, physical abuse, gun ownership, and drug or alcohol use on the part of the violent partner (Kress et al. 2008).

Individuals in such relationships should be cautious about how they react and develop a safe plan of withdrawal. Safety plans will vary but include the following:

- Identifying a safe place an individual can go the next time she needs to leave the house. Staying with one's parents, friends, or in a women's shelter must be set up in advance. The victim needs to stay in a protected context until the abuser learns that he is being left and the relationship is over. His first response will often be to get to the victim and hurt her or try to talk her out of it.
- Telling friends or neighbors about the violence and requesting that they call the police if they hear suspicious noises or witness suspicious events.
- Storing an escape kit (for example, keys, money, checks, important phone numbers, medications, Social Security cards, bank documents, birth certificates, change of clothes, and so on) somewhere safe (and usually not in the house) (Kress et al. 2008).

Above all, individuals should trust their instincts and do what they can to de-escalate the situation.

Healing from being in an abusive relationship takes time. Allen and Wozniak (2011) discouraged repetitive disclosure of one's abuse and history. Instead they focused on "holistic, integrative, and alternative healing approaches such as prayer, meditation, yoga, creative visualization, and art therapy". They provided positive quantitative and qualitative results of this approach.

Treatment of Partner Abusers

LeCouteur and Oxlad (2011) interviewed men who talked of why they abused their partners and found that they constructed a justification that their partner had "breached the normative moral order." Successful therapy required the men to acknowledge the wrongness of their violent/abusive actions and to move forward. Shamai and Buchbinder (2010) surveyed men who had participated in a treatment program for partner-violent men. Most experienced therapy as positive and meaningful and underwent personal changes, especially the acquisition of self-control. Taking a time out, counting to ten, and reassessing the situation before reacting were particularly helpful techniques. While gains were made, the men still tended to create dominant relationships with women, which left open the potential to explode again. Silvergleid and Mankowski (2006) identified learning a new way of masculinity and making a personal commitment to change as crucial to the rehabilitation of the male batterer. In addition, because alcohol or drug abuse and violence toward a partner are often related, addressing one's alcohol or substance abuse problem is often a prerequisite for treating partner abuse.

The involvement of the partner is also important. Some men stop abusing their partners only when their partners no longer put up with it. One abusive male said that his wife had to leave him before he learned not to be abusive toward women. "I've never touched my second wife," he said.

This concludes our discussion of abuse in adult relationships. In the following pages, we discuss other forms of abuse, including child abuse and parent, sibling, and elder abuse.

Child abuse casts a shadow the length of a lifetime.

Herbert Ward, English footballer

General Child Abuse

Child abuse may take many forms, including physical abuse, neglect, and sexual abuse.

Physical Abuse and Neglect

DiLillo et al. (2010) surveyed a geographically diverse sample of college students (N = 1398) and found that 31% reported being physically abused as a child. Most (67%) reported 2 adults as perpetrators and 98% of these were family members. Such abuse occurred numerous times (39% reported 11 or more times) and lasted more than 2 years. The evaluation of the abuse was that it was severe (52%) with bruises, cuts, scratches the most frequent type of physical abuse (26%).

Child abuse can be defined as any interaction or lack of interaction between children and their parents or caregivers that results in nonaccidental harm to the children's physical or psychological well-being. Child abuse includes physical abuse, such as beating and burning; verbal abuse, such as insulting or demeaning children; and neglect, such as failing to provide adequate food, hygiene, medical care, or adult supervision for children. Children can also experience emotional neglect by their parents. Children from unintended pregnancies or with poor health and developmental problems are more likely to be abused.

The percentages of various types of child abuse in substantiated victim cases are illustrated in Figure 13.2. Notice that "neglect" is the largest category of abuse.

Although our discussion will focus on physical abuse, it is important to keep in mind that children are often not fed, are not given medical treatment, and are left to fend for themselves.

Child abuse any interaction or lack of interaction between children and their parents or caregivers that results in nonaccidental harm to the children's physical or psychological well-being.

Factors Contributing to General Child Abuse

A variety of the following factors contribute to child abuse:

1. *Parental psychopathology.* Symptoms of parental psychopathology that may predispose a parent to abuse or neglect children include low frustration tolerance, inappropriate expression of anger, and alcohol or substance abuse.

2. *Unrealistic expectations.* Abusive parents often have unrealistic expectations of their children's behavior. For example, a parent might view the crying of

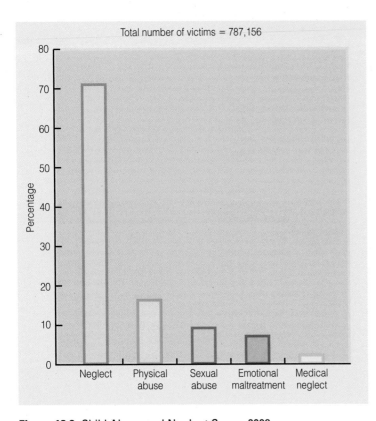

Total number of victims = 787,156

Figure 13.2 Child Abuse and Neglect Cases: 2008
Source: *Statistical Abstract of the United States 2011* 130th ed. Washington, DC. U.S. Bureau of the Census, Table 339

DIVERSITY IN OTHER COUNTRIES

College students in Turkey reported on their abuse as children. In a study of 988 students, over half (53.3%) reported a history of childhood physical abuse. Men were 1.5 times more likely to report abuse. The most frequent reasons for the physical violence were "loss of perpetrator's self-control" and "establishment of discipline at home." Humiliation after subjection to physical violence was the most frequent outcome reported by these students (Turla et al. 2010).

a baby as a deliberate attempt on the part of the child to irritate the parent. Shaken baby syndrome—whereby the caretaker, most often the father, shakes the baby to the point of causing the child to experience brain or retinal hemorrhage—most often occurs in response to a baby who won't stop crying. Abusive head trauma (AHT) refers to nonaccidental head injury in infants and toddlers. Head injury is the leading cause of death among abused children (Rubin et al. 2003). Managing parents' unrealistic expectations and teaching them to cope with frustration are important parts of an overall plan to reduce physical child abuse.

3. *History of abuse.* Individuals who were abused as children are more likely to report abusing their own children (Dixon et al. 2005). The researchers observed that the absence of positive parenting carries over into the next generation. Although parents who were physically or verbally abused or neglected as children are somewhat more likely to repeat that behavior than parents who were not abused, the majority of parents who were abused do *not* abuse their own children. Indeed, many parents who were abused as children are dedicated to ensuring nonviolent parenting of their own children precisely because of their own experience of abuse.

4. *Displacement of aggression.* One cartoon shows several panels consisting of a boss yelling at his employee, the employee yelling at his wife, the wife yelling at their child, and the child kicking the dog, who chases the cat up a tree. Indeed, frustration may spill over from the adults to the children, the latter being less able to defend themselves.

5. *Social isolation.* The saying, "it takes a village to raise a child," is relevant to child abuse. Unlike inhabitants of most societies of the world, many Americans rear their children in closed and isolated nuclear units. In extended kinship societies, other relatives are always present to help with the task of child rearing. Isolation means parents have no relief from the parenting role as well as no supervision by others who might interrupt observed child abuse.

6. *Disability of a child.* Jaudesa and Mackey-Bilaverb (2008) found that children who were developmentally delayed were twice as likely to be victims of abuse and neglect.

7. *Other factors.* In addition to the factors just mentioned, the following factors are also associated with child abuse and neglect:
    a. The pregnancy is premarital or unplanned, and the father or mother does not want the child.
    b. Mother-infant attachment is lacking.
    c. Child-rearing techniques are harsh, with little positive reinforcement.
    d. The parents are unemployed.
    e. Abuse between the husband and wife is present.
    f. The children are adopted or are foster children.

Effects of General Child Abuse

With more than 1,500 deaths resulting from child abuse in the United States annually, death is the most devastating effect for child abuse (Paluscia et al. 2010). There are profound effects in later life for those who are not killed.

In general, the effects are negative and vary according to the intensity and frequency of the abuse. In a study of 816 married women in Bangkok, Kerley et al. (2010) confirmed what has been found by U.S. researchers: that children who are victims of violence or witnessed violence in the family of origin are more likely to offend and be victims in their family of procreation. Researchers have also found that children who have been abused are more likely to display the following (Reyome, 2010):

1. Few close social relationships and an inability to love or trust
2. Communication problems and learning disabilities, particularly in children who have been neglected. These children lag in language development (Sylvestrea and Mérettec 2010).
3. Aggression, low self-esteem, depression, and low academic achievement
4. Increased risk of alcohol or substance abuse and suicidal tendencies as adults

In addition, students in a sample of 1,728 who reported maltreatment as a child also reported higher levels of interpersonal problems. Females who had been emotionally neglected were nonassertive and distant. Males who were physically abused were more likely to be domineering and distant (Paradis and Boucher 2010). Greenfield and Mark (2010) confirmed poorer psychological functioning for adults who had experienced physical and psychological violence in childhood. Berzenski and Yates (2010) found that childhood emotional abuse was particularly related to later relationship violence.

Parent Training Programs

Through pre- and post-tests and preventative and control groups, Letartea et al. (2010) evaluated the effectiveness of a parent training program designed to improve parenting practices, which was offered through child protection services. Called "The Incredible Years," the program lasted 16 weeks. Results revealed that the program had a positive impact on parenting practices (praise/incentive, appropriate discipline, and positive verbal discipline) and parents' perception of their child's behavior (frequency of behavioral problems and number of problematic behaviors).

I would certainly choose my jobs depending on the actions of the character. I won't do anything that has to do with child abuse or women's abuse.

Kathleen Turner, actress

Child Sexual Abuse

National **Data**

In the United States, of 3.5 million children who were investigated annually for abuse, 7.6% of the cases are for sexual abuse (Fallona et al. 2010).

DiLillo et al. (2010) surveyed a geographically diverse sample of college students (N = 1398) and found that 8% reported sex abuse as a child. Most reported one perpetrator who was a nonfamily member (77%), which aggressed one to two times (50%), and that these events lasted less than a year (69%). Most (58%) of the sexual acts reported by the respondents did not involve penetration. Verbal tactics (45%) were used most often to manipulate the child; in 39% of the cases, the child was held down.

Child sexual abuse is one of two types—**extrafamilial** and **intrafamilial**.

Extrafamilial child sexual abuse in which the perpetrator is someone outside the family who is not related to the child.

Intrafamilial child sexual abuse referring to exploitive sexual contact or attempted sexual contact between relatives before the victim is 18.

Extrafamilial Child Sexual Abuse

In extrafamilial child sexual abuse, a perpetrator is someone outside the family who is not related to the child. Extrafamilial child sexual abuse received national attention with the accusation that priests in the Catholic Church had had sex with young children. By 2010, the Catholic Church had paid over two billion dollars in settlement claims to tens of thousands of victims. The extent of abuse by other trusted adults—Scoutmasters and public school teachers—is unknown.

Extrafamilial sex offenders are not just men. Strickland (2008) studied sixty incarcerated female sex offenders who had had sex with a child under the age of 18. When compared to seventy incarcerated females who were not sex offenders, the sex offending females were more likely to report their own childhood abuse, specifically sexual abuse, and to be inadequate in forming adult relationships.

Intrafamilial Child Sexual Abuse

A more frequent type of child sexual abuse is intrafamilial (formerly referred to in professional literature as incest). This refers to exploitive sexual contact or attempted sexual contact between relatives before the victim is 18. Sexual contact or attempted sexual contact includes intercourse, fondling of the breasts and genitals, and oral sex. Relatives include biologically-related individuals but may also include stepparents and stepsiblings.

Maikovich-Fonga et al. (2010) surveyed a national sample of youth and found that females were more likely than males (48% versus 31%) to report penetrative sex abuse. About two thirds of both sexes reported that the perpetrator was a family member. Both sexes experienced emotional and behavioral problems related to the abuse—depression, suicidal thoughts/attempts, anxiety, self-blame, shame, guilt, negative self-concept, impaired relationships, sexual dysfunction, loss of trust, and insomnia.

Intrafamilial child sexual abuse involves an abuse of power and authority, particularly when the perpetrator is a parent. The following was written by an undergraduate in our classes who was raped and impregnated by her stepfather at age 12. Readers who have been raped or sexually abused may wish to skip this section.

Until that night

She nervously bites her nails as she awaits her results
In just seconds she would receive news that was sure to forever change her life
Pacing the floor of the waiting room, she begins to recall the events that have led her
  here
Just a few months prior her life had seemed so perfect
She had been a happy, carefree twelve year old girl full of hopes and dreams
Nothing seemed to damper her zest for life or her energetic personality it seemed

That is, until that night

Her room was completely dark except for the occasional bolts of lightning that gave it
a florescent glow
She tries hard to find sleep but her mind just won't let her go
She sits up when he enters her room, knowing he must act fast, he coaxes her from her
bed into the den of sin, his bedroom, she is lead
Once inside the spacious bedroom she is startled by a loud clap of thunder
"My mom isn't home so what does he want with me"? she begins to wonder
He pulls her closer to him and starts caressing her undeveloped body
Trying to find new ways to please himself as if it's his new hobby
Taken aback and now frightened, she backs away from him but he runs and locks the
  door
Unsure of what to do next, she watches helplessly as her nightgown falls to the floor
Exposing her small and fragile frame, truly terrified now she begins to tremble

He pulls her back towards him and she can feel his fully erect, throbbing member
His fingers palm her small breast as his hand slide down to her small mound
As his fingers enter her, he covers her mouth so she can't make a sound
He forces himself deep inside of her with one mighty thrust
Moaning and groaning he's filled with so much lust
She begs and pleads for him to stop but her cries are unheard
Picking up his pace he concentrates on pleasing himself never saying a word
As he begins to climax, he pushes his body down on her throat and she begins to choke
Once he is satisfied, he climbs off of her and begins to clean himself
He tries to bring her into his arms but she'd rather run into the arms of death
Wanting to never again feel that type of pain
She locks herself in her bathroom where she is determined to remain
Never in a million years would she have envisioned this type of thing happening to her, especially by someone she loved, respected and trusted

That is, until that night

For weeks following the incident she feels completely disgusted with herself
Why her? What did she do to deserve it? Why not someone else?
Things quickly turned ugly with her mother, who believes her daughter made up the entire story to cause drama between her and her significant other
Now the only emotion she feels towards her parents is hate
And just when she thinks things can't get any worse, her period is two weeks late
Ignoring all the obvious pregnancy symptoms, she is in complete denial
How would she ever be able to explain that she is pregnant with her stepfather's child?
She would never have thought that at the age of twelve she would be faced with the hardest decision of her life

That is, until that night

As soon as the nurse enters the room, the look on her face says it's not good news
It is quickly confirmed that she is, indeed, five weeks pregnant
Unfairly given this responsibility, she has nowhere to turn and is unsure of what to do
She bows her head and prays that God forgive her for the sin that she is about to commit
Only once did she hesitate
For she knew her pregnancy she would have to terminate
She never would have dreamed that by the age of twelve she would be having an abortion to kill her stepfather's child

That is, until that night

She has tried and tried since then to move on but it's still something that hurts her to this day.
Every single man that has come into her personal life has lied to or hurt her in some way.
She believes the damage is too deep to ever repair.
RLB

Effects of Child Sexual Abuse

Child sexual abuse is associated with negative long-term consequences. Chartiera et al. (2010) investigated the relative contribution of childhood physical/sexual abuse and other adverse childhood experiences (e.g., physical/emotional neglect, poverty, parental psychopathology) to adult health in a sample of 9,953 respondents aged 15 years and older. They found that almost three fourths (72%) reported at least one adverse childhood experience with childhood physical and sexual abuse having a stronger influence on negative health outcomes than other types of adverse childhood experiences. These negative health outcomes included multiple health problems, rating one's health as poor, experiencing pain, and using emergency rooms, general practitioners, and health professionals more often.

There are other emotional, social, and psychological consequences. Adult-child sexual contact is, in most cases, a child's first introduction to adult sexuality. The sexual script acquired during such relationships forms the basis on which other sexual experiences are assimilated. The negative effects include the following:

1. PTSD (Posttraumatic stress disorder). Cantón-Cortés and Cantóna (2010) studied 138 victims of childhood sexual abuse and found PTSD a likely outcome, particularly if the victimization was continuous rather than an isolated case and where the perpetrator was inside rather than outside the family. Seifert and Murdoch (2011) studied a sample of U.S. Army soldiers and found that those who had been both victims of childhood physical and sexual abuse were more likely to develop posttraumatic stress disorder symptoms.

2. Early forced sex is associated with being withdrawn, anxious, depressed, and a substance abuser. Women who have experienced childhood sexual abuse are 7.75 times more likely to attempt suicide (Anderson et al. 2003).

3. Spouses who were physically and sexually abused as children report lower marital satisfaction, higher individual stress, and lower family cohesion than do couples with no abuse history (Nelson and Wampler 2000).

4. Adult males who were sexually abused as children are more likely to become child molesters themselves (Lussier et al. 2005).

Successful movement beyond negative consequences of being sexually molested as a child involves accepting that the adult molester (not the child) was responsible for the abuse (some children feel that they "led the adult on" and are responsible for the abuse). In addition, the child might be physically removed from the abusive context and placed in foster care. McWey et al. (2010) found that such a change is associated with positive outcomes in regard to a decrease in externalization (aggression and delinquent behaviors) and internalization (withdrawn, anxious/depressed). Nolla (2008) also noted that a substantial portion of sex abuse victims is remarkably resilient. Factors associated with such resilience include positive self-esteem, ego control, emotional support, therapy, stable home environments, and having an ability to recall and integrate past adversity into current life circumstances. Victims of sexual abuse do not always view it negatively. Cleverly and Boyle (2010) identified 12 respondents who experienced sexual abuse (including rape, molestation, sexual assault) before the age of 22 and came to view it as changing their life for the "better." Descriptions include "I am a stronger person now and wise in life decisions," "[it] enabled me to take ownership of only those responsibilities one can control," and "[it] made me stronger."

The sexual abuse and exploitation of children is one of the most vicious crimes conceivable, a violation of mankind's most basic duty to protect the innocent.

James T. Walsh, U.S. Congressman

Strategies to Reduce Child Sexual Abuse

Strategies to reduce child sexual abuse include regendering cultural roles, providing specific information and training children in public schools to be alert to inappropriate touching, improving the safety of neighborhoods, providing healthy sexuality information for both teachers and children in public schools at regular intervals, and promoting public awareness campaigns. Finally, helping to ensure that children live in neighborhoods safe from convicted child sex offenders may reduce child sexual abuse. Protecting children from former convicted child molesters is the basis of Megan's Law (see the Social Policy feature).

Parent, Sibling, and Elder Abuse

As we have seen, intimate partners and children may be victims of relationship violence and abuse. Parents, siblings, and the elderly may also be abused by family members.

Parent Abuse

Some people assume that, because parents are typically physically and socially more powerful than their children, they are immune from being abused by their children. However, parents are often targets of their children's anger, hostility, and frustration. Routt and Anderson (2011) emphasized the prevalence of adolescent to parent violence and provided details from interviews with these youth and their parents. Teenage and even younger children may physically and verbally lash out at their parents. Children have been known to push parents down stairs, set the house on fire while their parents are in it, and use weapons such as guns or knives to inflict serious injuries or even to kill a parent. Background characteristics of children who abuse their parents include observing parents' violence toward each other and having parents who used corporal punishment on their children as a method of discipline (Ulman and Straus 2003).

Sibling Abuse

Observe a family with two or more children and you will likely observe some amount of sibling abuse. Even in "well-adjusted" families, some degree of fighting among children is expected. Most incidents of sibling violence consist of slaps, pushes, kicks, bites, and punches. What passes for "normal," "acceptable," or "typical" behavior between siblings would often be regarded as violent and abusive behavior outside the family context.

Sibling abuse is said to be the most prevalent form of abuse. Ninety-eight percent of the females and 89% of the males in one study reported having received at least one type of emotionally aggressive behavior from a sibling; 88% of the females and 71% of the males reported having received at least one type of physically aggressive behavior from a sibling (Simonelli et al. 2002).

Sibling abuse may include sexual exploitation, whereby an older brother will coerce younger female siblings into nudity or sex. Though some sex between siblings is consensual, often it is not. "He did me *and* all my sisters before he was done," reported one woman. Another woman reported that, as a child and young adolescent, she performed oral sex on her brother three times a week for years because he told her that ingesting a man's semen was the only way a woman would be able to have babies as an adult.

The most prevalent abuse in families is sibling abuse.

n 1994, Jesse Timmendequas lured 7-year-old Megan Kanka into his Hamilton Township house in New Jersey to see a puppy. He then raped and strangled her and left her body in a nearby park. Prior to his rape of Megan, Timmendequas had two prior convictions for sexually assaulting girls. Megan's mother, Maureen Kanka, argued that she would have kept her daughter away from her neighbor if she had known about his past sex offenses. She campaigned for a law, known as **Megan's Law**, requiring that communities be notified of a neighbor's previous sex convictions. New Jersey and forty-five other states have enacted similar laws.

The 1994 Jacob Wetterling Act requires states to register individuals convicted of sex crimes against children. The law requires that convicted sexual offenders register with local police in the communities in which they live. It also requires the police to go out and notify residents and certain institutions (such as schools) that a previously convicted sex offender has moved into the area. This provision of the law has been challenged on the belief that individuals should not be punished forever for past deeds. Critics of the law argue that convicted child molesters who have been in prison have paid for their crime. To stigmatize them in communities as sex offenders may further alienate them from mainstream society and increase their vulnerability for repeat offenses.

In many states, Megan's Law is not operative because it is on appeal. Parents ask, "Would you want a convicted sex offender, even one who has completed his prison sentence, living next door to your 8-year-old daughter?" However, the

---

Sibling relationship aggression behavior of one sibling toward another sibling that is intended to induce social harm or psychic pain in the sibling.

Megan's Law law requiring that communities be notified of a neighbor's previous sex convictions.

In addition to physical and sexual sibling abuse, there is **sibling relationship aggression**. This is behavior of one sibling toward another sibling that is intended to induce social harm or psychic pain in the sibling. Examples include social alienation or exclusion (for example, not asking the sibling to go to a movie when a group is going), telling secrets or spreading rumors (for example, revealing a sibling's sexual or drug past), and withholding support or acceptance (for example, not acknowledging a sibling's achievements in school and sports). A team of researchers studied 185 sibling pairs—both younger siblings (mean age, 13.47 years) and older siblings (mean age, 15.95 years)—and their respective parents (married an average of 19.5 years). They found that sibling relationship aggression was more likely to occur when intimate feelings between siblings were low and their negativity toward each other was high. Family contexts in which parents are warm and involved with their children are conducive to lower levels of relationship aggression (Updegraff et al. 2005).

Elder Abuse

As increasing numbers of the elderly end up in the care of their children, abuse, though infrequent, is likely to increase.

National **Data**

About 6% of elderly Americans are victims of abuse each month (Cooper et al. 2008). Annually, 2.1 million elderly are abused, but only one in fourteen incidents come to the attention of the authorities (Plitnick 2008). When abuse is suspected, a nurse can ask simple questions of the patient to elicit information. This should be done in private.

Various examples of abuse include:

1. *Neglect.* Failing to buy or give the elderly needed medicine, failing to take them to receive necessary medical care, or failing to provide adequate food, clean clothes, and a clean bed are examples of neglect. Neglect is the most frequent type of domestic elder abuse.

reality is that little notification is afforded parents in most states. Rather, the issue is tied up in court and will likely remain so until the Supreme Court decides it. A group of concerned parents (Parents for Megan's Law) are trying to implement Megan's Law nationwide. See the Web Links section at the end of this chapter for the group's website address.

Parents for Megan's Law have also sought to enact legislation for the "civil commitment for a specific group of the highest-risk sexual predators who freely roam our streets, unwilling or unable to obtain proper treatment. This kind of predator commitment follows a criminal sentence and generally targets repeat sex offenders who then remain in a sexual predator treatment facility until it is safe to release them to a less restrictive environment or into the commu-

nity." Indeed, this group not only wants parents to be notified of criminal sex offenders but also wants the offenders to be housed in treatment centers on release from prison.

Your Opinion?

1. To what degree do you believe the Supreme Court should uphold Megan's Law?

2. To what degree do you believe that convicted child molesters who have served their prison sentence should be free to live wherever they like without neighbors being aware of their past?

3. Independent of Megan's Law, how can parents protect their children from sex abuse?

2. *Physical abuse.* This includes inflicting injury or physical pain or sexual assault.

3. *Psychological abuse.* Examples of psychological abuse include verbal abuse, deprivation of mental health services, harassment, and deception.

4. *Social abuse.* Unreasonable confinement and isolation, lack of supervision, and abandonment are examples of social abuse.

5. *Legal abuse.* Improper or illegal use of an elder's resources is considered legal abuse.

Lee (2008) studied elder abuse among older adults with disabilities being cared for by their family caregivers. The sample of 1,000 primary family caregivers revealed that elder abuse was more common in those cases where the elderly was cognitively impaired, had functional disabilities, and where there was limited money. The findings suggested that psychosocial support services and programs for family caregivers are needed to prevent and reduce the prevalence of elder abuse. Brozowski and Hall (2010) studied elder abuse among Canadians and found that women, the divorced, and those living in urban areas or with low income were more vulnerable to abuse.

Another type of elder abuse that has received recent media attention is **granny dumping**. Adult children or grandchildren who feel burdened with the care of their elderly parent or grandparent leave an elder at the entrance of a hospital with no identification. If the hospital cannot identify responsible relatives, it is required by state law to take care of the abandoned elder or transfer the person to a nursing home facility, which is paid for by state funds. Relatives of the dumped elder, hiding from financial responsibility, never visit or see the elder again.

Adult children who are most likely to dump or abuse their parents tend to be under a great deal of stress and to use alcohol or other drugs. In some cases, parent abusers are getting back at their parents for mistreating them as children. In other cases, the children are frustrated with the burden of having to care for their elderly parents. Such frustration is likely to increase. As baby boomers age, they will drain already limited resources for the elderly, and their children will be forced to care for them with little governmental support.

Granny dumping refers to adult children or grandchildren, burdened with the care of their elderly parent or grandparent, leaving the elder at the entrance of a hospital with no identification.

The ultimate form of elder abuse is murder. Karch and Nunn (2011) provided data on 68 homicides of dependent elderly by a caregiver and found that they were mostly (97%) non-Hispanic or white (88%) women (63%) killed in their own homes (92.6%) with a firearm (35.3%) or by intentional neglect (25.0%) by a husband (30.9%) or a son (22.1%). Nearly half of the victims were aged 80 years and older (48.5%). "Many homicide by caregiver incidents are precipitated by physical illness of the victim or caregiver, opportunity for perpetrator financial gain, mental illness of the caregiver, substance use by the caregiver, or an impending crisis in the life of the caregiver not related to illness."

The Future of Violence and Abuse in Relationships

Violence and abuse will continue to occur behind closed doors, in private contexts where social constraint is minimal. Reducing such violence and abuse will depend on prevention strategies focused at three levels: the general population, specific groups thought to be at high risk for abuse, and families who have already experienced abuse. Public education and media campaigns aimed at the general population will continue to convey the criminal nature of domestic assault, suggest ways the abuser might learn to prevent abuse (seek therapy for anger, jealousy, or dependency), and identify where abuse victims and perpetrators can get help.

Preventing or reducing family violence through education necessarily involves altering aspects of American culture that contribute to such violence. For example, violence in the media must be curbed (not easy, with nightly news clips of bombing assaults in other countries, violent films, etc.). In addition, traditional gender roles and views of women and children as property must be replaced with egalitarian gender roles and respect for women and children.

Another important cultural change is to reduce violence-provoking stress by reducing poverty and unemployment and by providing adequate housing, nutrition, medical care, and educational opportunities for everyone. Integrating families into networks of community and kin would also enhance family well-being and provide support for families under stress.

Summary

What is the nature of relationship abuse?

Violence or physical abuse may be defined as the intentional infliction of physical harm by either partner on the other. Emotional abuse (also known as psychological abuse, verbal abuse, or symbolic aggression) is designed to denigrate the partner, reduce the partner's status, and make the partner vulnerable, thereby giving the abuser more control. The most frequent emotionally abusive behavior is not talking to the partner. Spiritual abuse exists where the partner denigrates one's practice of religion or prevents or interferes with its expression.

What are explanations for violence in relationships?

Cultural explanations for violence include violence in the media, corporal punishment in childhood, gender inequality, and stress. Community explanations involve social isolation of individuals and spouses from extended family, poverty, inaccessible community services, and lack of violence prevention programs. Individual factors include psychopathology of the person (antisocial), personality (dependency or jealousy), and alcohol abuse. Family factors include child abuse by one's parents, observing parents who abuse each other, and not having a father in the home.

How does sexual abuse in undergraduate relationships manifest itself?

Violence in dating relationships begins as early as grade school, is mutual, and escalates with emotional involvement. Violence occurs more often among couples in which the partners disagree about each other's level of emotional commitment and when the perpetrator has been drinking alcohol or using drugs. Acquaintance rape is defined as nonconsensual sex between adults who know each other.

What are the effects of abuse?

The effects of IPV include physical harm, mental harm (depression, anxiety, low self-esteem, lost of trust in others, sexual dysfunctions), unintended pregnancy, and multiple abortions. High levels of anxiety and depression often lead to alcohol and drug abuse. Violence on pregnant women significantly increased the risk for infants of low birth weight, preterm delivery, and neonatal death.

What is the cycle of abuse and why do people stay in an abusive relationship?

The cycle of abuse begins when a person is abused and the perpetrator feels regret, asks for forgiveness, and starts acting nice (for example, gives flowers). The victim, who perceives few options and feels guilty terminating the relationship with the partner who asks for forgiveness, feels hope for the relationship at the contriteness of the abuser and does not call the police or file charges. The couple usually experiences a period of making up or honeymooning, during which the victim feels good again about the abusing partner. However, tensions mount again and are released in the form of violence. Such violence is followed by the familiar sense of regret and pleadings for forgiveness, accompanied by being nice (a new bouquet of flowers, and so on).

The reasons people stay in abusive relationships include love, emotional dependency, commitment to the relationship, hope, view of violence as legitimate, guilt, fear, economic dependency, and isolation. The catalyst for breaking free combines the sustained aversiveness of staying, the perception that they and their children will be harmed by doing so, and the awareness of an alternative path or of help in seeking one.

What is child abuse and what factors contribute to it?

Child abuse includes physical abuse, such as beating and burning; verbal abuse, such as insulting or demeaning the children; and neglect, such as failing to provide adequate food, hygiene, medical care, or adult supervision for children. Children can also experience emotional neglect by their parents.

Some of the factors that contribute to child abuse include parental psychopathology, a history of abuse, displacement of aggression, and social isolation.

What are the effects of general child abuse?

The negative effects of child abuse include impaired social relationships, difficulty in trusting others, aggression, low self-esteem, depression, low academic achievement, and posttraumatic stress disorder (PTSD). Physical injuries may result in disfigurement, physical disability, and even death.

One of the most devastating types of child abuse is child sexual abuse, which has serious negative long-term consequences. The effects include being withdrawn, anxious, and depressed; delinquency; suicide attempts; and substance abuse. PTSD, which means recurrent experiencing of the event, is common, as is heavy alcohol or substance abuse.

What is the nature of parent, sibling, and elder abuse?

Parent abuse is the deliberate harm (physical or verbal) of parents by their children. Sibling abuse is the most severe prevalent form of abuse. What passes for "normal," "acceptable," or "typical" behavior between siblings would often be regarded as violent and abusive behavior outside the family context.

Elder abuse is another form of abuse in relationships. Granny dumping is a new form of abuse in which children or grandchildren who feel burdened with the care of their elderly parents or grandparents leave them at the emergency entrance of a hospital. If the relatives of the elderly patient cannot be identified, the hospital will put the patient in a nursing home at state expense.

Key Terms

Acquaintance rape	Granny dumping	Red zone
Battered-woman syndrome	Honor crime/Honor killing	Rophypnol
Child abuse	Intimate partner homicide	Siblicide
Corporal punishment	Intimate partner violence (IPV)	Sibling relationship aggression
Cybercontrol	Intimate terrorism (IT)	Situational couple violence (SCV)
Cybervictimization	Intrafamilial	Spiritual abuse
Date rape	Marital rape	Stalking
Emotional abuse	Megan's Law	Uxoricide
Entrapped	Obsessive relational intrusion (ORI)	Violence
Extrafamilial	Parricide	
Filicide	Rape myths	

Web Links

Childabuse.com
http://www.childabuse.com/

Male Survivor
http://www.malesurvivor.org/

Minnesota Center against Violence and Abuse
http://www.mincava.umn.edu

National Sex Offender Data Base
http://www.nationalalertregistry.com

Parents for Megan's Law and the Crime Victims Center
http://www.parentsformeganslaw.com

Rape, Abuse & Incest National Network (RAINN)
http://www.rainn.org/

Stop and Recover from Abuse
www.chooserespect.org
www.safeplace.org

Stop It Now! The Campaign to Prevent Child Sexual Abuse
http://www.stopitnow.com/

V-Day (movement to stop violence against women and girls)
http://www.vday.org/

Stress and Crisis in Relationships

Life is not about waiting for the storm to pass... it is about learning to dance in the rain.

Unknown

Chelsea E. Curry

Learning Objectives

Review the definitions of stress, resilience, and the family stress model.

Identify positive stress management strategies (and what doesn't work).

Summarize five major crisis events.

Review the training of marriage/family therapists.

Know the caveats about becoming involved in marriage and family therapy.

Predict the future of stress and crisis in the family.

1. When a husband gets depressed, his wife is more likely to do so, too.

2. Unemployment is associated with both spouse and child abuse.

3. Volunteering while one is unemployed is associated with improved self-esteem, contributes to the community, improves one's skills, and increases one's network for finding a job.

4. Undergraduate males are more likely than undergraduate females to believe that if you love someone enough you will be able to solve problems with that person.

5. Twenty intervention studies on marriage and family therapy concluded that a "close friend" was just as effective as a "therapist" in terms of perceived improvement.

Answers: 1. T   2. T   3. T   4. T   5. F

A married couple enjoyed their respective careers and looked forward to retirement and travel. Since they both had lucrative careers, they put off retirement. Suddenly when the husband reached 55 he developed MS and was in a wheelchair by age 60. Their traveling days came to an abrupt end; he died at age 63. The wife noted that a personal, marital, or family crisis can come at any time and that one should not put off living since there is no guarantee of tomorrow.

Ann Landers was once asked what she would consider the single most useful bit of advice all people could profit from. She replied, "Expect trouble as an inevitable part of life, and when it comes, hold your head high, look it squarely in the eye and say, 'I will be bigger than you.'" Life indeed brings both triumphs and tragedies. No one is immune. This chapter is about experiencing and coping with crisis events and the day-to-day stress we feel during the in-between time.

Stress is an ignorant state. It believes that everything is an emergency.

Natalie Goldberg, *Wild Mind*

Personal Stress and Crisis Events

In this section, we review the definitions of crisis and stressful events, the characteristics of resilient families, and a framework for viewing a family's reaction to a crisis event.

Definitions of Stress and Crisis Events

Stress reaction of the body to substantial or unusual demands (physical, environmental, or interpersonal).

Stress is a reaction of the body to substantial or unusual demands (physical, environmental, or interpersonal). Stress is often accompanied by irritability, high blood pressure, and depression. Stress also has an effect on a person's relationships and sex life. Bodenmann et al. (2010) found that stress in daily life was associated with a decrease in relationship satisfaction and lower levels of sexual activity.

Stress is a process rather than a state. For example, a person will experience different levels of stress throughout a divorce—the stress involved in acknowledging that one's marriage is over, telling the children, leaving the family residence, getting the final decree, and seeing one's ex at the mall will all result in varying levels of stress.

Crisis a crucial situation that requires change in one's normal pattern of behavior.

A **crisis** is a crucial situation that requires changes in normal patterns of behavior. A family crisis is a situation that upsets the normal functioning of the family and requires a new set of responses to the stressor. Sources of stress and crises can be external, such as hurricanes which annually hit our coasts or the tsunami which devastated Japan in 2011. Other examples of an external crisis are economic recession, tornados, downsizing, or military deployment to Afghanistan. Stress and crisis events may also be induced internally (for example, alcoholism, an extramarital affair, or Alzheimer's disease of spouse or parents). Peilian et. al. (2011) studied stressful life events of 1,749 participants in seven Chinese cities and found that crisis events reduced marital satisfaction.

Stressors or crises may also be categorized as expected or unexpected. Examples of expected family stressors include the need to care for aging parents and the death of one's parents. Unexpected stressors include contracting human immunodeficiency virus (HIV), a miscarriage, or the suicide of one's teenager.

Both stress and crises are normal parts of family life and sometimes reflect a developmental sequence. Pregnancy, childbirth, job changes or loss, children leaving home, retirement, and widowhood are all stressful and predictable for most couples and families. Crisis events may have a cumulative effect: the greater the number in rapid succession, the greater the stress.

Resilient Families

Just as the types of stress and crisis events vary, individuals and families vary in their abilities to respond successfully to crisis events. **Resiliency** refers to a family's strengths and ability to respond to a crisis in a positive way. Black and Lobo (2008) defined family resilience as the successful coping of family members under adversity that enables them to flourish with warmth, support, and cohesion. The key factors that promote **family resiliency** include positive outlook, spirituality, flexibility, communication, financial management, shared family recreation, routines or rituals, and support networks. A family's ability to bounce back from a crisis (from loss of one's job to the death of a family member) reflects its level of resiliency. Resiliency may also be related to individuals' perceptions of the degree to which they are in control of their destiny. The Self-Assessment section measures this perception.

A Family Stress Model

Various theorists have explained how individuals and families experience and respond to stressors. Kahl et al. (2007) reviewed the ABCX model of family stress, developed by Reuben Hill in the 1950s. The model can be explained as follows:

A = stressor event

B = family's management strategies, coping skills

C = family's perception, definition of the situation

X = family's adaptation to the event

A is the stressor event, which interacts with B, the family's coping ability or crisis-meeting resources. Both A and B interact with C, the family's appraisal or perception of the stressor event. X is the family's adaptation to the crisis. Thus, a family that experiences a major stressor (for example, a spouse with a spinal cord injury) but has great coping skills (for example, religion or spirituality, love, communication, and commitment) and perceives the event to be manageable will experience a moderate crisis. However, a family that experiences a less critical stressor event (for example, their child makes Cs and Ds in school) but has minimal coping skills (for example, everyone blames everyone else) and perceives the event to be catastrophic will experience an extreme crisis. Johnson and Johnson (2010) also emphasized the importance of one's conceptualization of an event in regard to whether stress results.

A problem is a chance for you to do your best.

Duke Ellington,
American composer,
pianist, and big band leader

I have learned to seek my happiness by limiting my desires, rather than in attempting to satisfy them.

John Stuart Mill, British
philosopher

Resiliency a family's strength and ability to respond to a crisis in a positive way.

Family resiliency the successful coping of family members under adversity that enables them to flourish with warmth, support, and cohesion.

Finish each day and be done with it. You have done what you could. Some blunders and absurdities no doubt crept in, forget them as soon as you can. Tomorrow is a new day, you shall begin it well and serenely . . .

Ralph Waldo Emerson, poet

People have different feelings about their vulnerability to crisis events. The following scale addresses the degree to which you feel you have control, you feel others have control, or you feel that chance has control of what happens to you.

Directions

To assess the degree to which you believe that you have control over your own life (I = Internality), the degree to which you believe that other people control events in your life (P = Powerful Others), and the degree to which you believe that chance affects your experiences or outcomes (C = Chance), read each of the following statements and write a number from −3 to +3.

−3	−2	1	+2	+3
Strongly Disagree	Slightly Disagree	Neither Disagree/ Agree	Slightly Agree	Strongly Agree

Subscale

I _____ 1. Whether or not I get to be a leader depends mostly on my ability.

C _____ 2. To a great extent, my life is controlled by accidental happenings.

P _____ 3. I feel like what happens in my life is mostly determined by powerful people.

I _____ 4. Whether or not I get into a car accident depends mostly on how good a driver I am.

I _____ 5. When I make plans, I am almost certain to make them work.

C _____ 6. Often there is no chance of protecting my personal interests from bad luck happenings.

C _____ 7. When I get what I want, it's usually because I'm lucky.

P _____ 8. Although I might have good ability, I will not be given leadership responsibility without appealing to those in positions of power.

I _____ 9. How many friends I have depends on how nice a person I am.

C _____ 10. I have often found that what is going to happen will happen.

P _____ 11. My life is chiefly controlled by powerful others.

C _____ 12. Whether or not I get into a car accident is mostly a matter of luck.

P _____ 13. People like myself have very little chance of protecting our personal interests when they conflict with those of strong pressure groups.

C _____ 14. It's not always wise for me to plan too far ahead because many things turn out to be a matter of good or bad fortune.

P _____ 15. Getting what I want requires pleasing those people above me.

C _____ 16. Whether or not I get to be a leader depends on whether I'm lucky enough to be in the right place at the right time.

P _____ 17. If important people were to decide they didn't like me, I probably wouldn't make many friends.

I _____ 18. I can pretty much determine what will happen in my life.

I _____ 19. I am usually able to protect my personal interests.

P _____ 20. Whether or not I get into a car accident depends mostly on the other driver.

I _____ 21. When I get what I want, it's usually because I worked hard for it.

P _____ 22. To have my plans work, I make sure that they fit in with the desires of other people who have power over me.

I _____ 23. My life is determined by my own actions.

C _____ 24. Whether or not I have a few friends or many friends is chiefly a matter of fate.

Scoring

Each of the subscales of Internality, Powerful Others, and Chance is scored on a six-point Likert format from −3 to +3. For example, the eight Internality items are 1, 4, 5, 9, 18, 19, 21, 23. A person who has strong agreement with all eight items would score +24; strong disagreement, −24. After adding and subtracting the item scores, add 24 to the total score to eliminate negative scores. Scores for Powerful Others and Chance are similarly derived.

Norms

For the Internality subscale, means range from the low 30s to the low 40s, with 35 being the modal mean (SD values approximating 7). The Powerful Others subscale has produced means ranging from 18 through 26, with 20 being characteristic of normal college student subjects (SD = 8.5). The Chance subscale produces means between 17 and 25, with 18 being a common mean among undergraduates (SD = 8).

Source

This was published in *Research with the Locus of Control Construct* by Hervert M. Lefcourt, pp. 57–59. Copyright 1981 by Elsevier. Reprinted with permission from Elsevier.

The importance of how an event is perceived is crucial. In response to the recession of 2008–2009, Adolf Merckle threw himself in front of a train. He was a billionaire who had lost hundreds of millions of Euros on Volkswagen shares. How many people do you know who would commit suicide if they had only a few million left?

Positive Stress-Management Strategies

Researchers Burr and Klein (1994) administered an eighty-item questionnaire to seventy-eight adults to assess how families experiencing various stressors such as bankruptcy, infertility, a disabled child, and a troubled teen used various coping strategies and how useful they evaluated these strategies. In the following, we detail some helpful stress-management strategies.

The way we see the problem is the problem.

Stephen R. Covey,
best selling-author

Scaling Back and Restructuring Family Roles

Higgins et al. (2010b) studied 1,404 men and 1,623 women in dual-earner families. Spouses evidenced two stress reducing strategies: scaling back and restructuring family roles. Men were more likely to scale back than women.

Changing Basic Values and Perspective

The strategy that most respondents reported as being helpful in coping with a crisis was changing basic values as a result of the crisis situation. Survivors of hurricanes, tornados, and earthquakes often focus on the sparing of their lives and those close to them rather than the loss of their home or material possessions. A positive view of crisis conceptualizes it as positive change in the making. Buddhists have the saying, "Pain is inevitable; suffering is not." This is another way of emphasizing that how one views a situation, not the situation itself, determines its impact on you. Regardless of the crisis event, one can view the crisis as a positive. A betrayal can be seen as an opportunity for forgiveness, unemployment as a time to spend time with one's family, and ill health to appreciate one's inner life. Waller (2008) studied new parents at two time frames (when the child was 1 and 4) and noted that whether they survived the crisis of having a child was a function of their perception. She found that "parents in stable unions framed tensions as manageable within the context of a relationship they perceived to be moving forward, whereas those in unstable unions viewed tensions as intolerable in relationships they considered volatile."

The only way you can hurt your body is not use it. Inactivity is the killer—it is never too late.

Jack LaLanne, fitness
guru, died at 96

Exercise

The Centers for Disease Control and Prevention (CDC) and the American College of Sports Medicine (ACSM) recommend that people aged 6 years and older engage regularly, preferably daily, in light to moderate physical activity for at least thirty minutes at a time. These recommendations are based on research that has shown the physical, emotional, and cognitive benefits of exercise. Tetlie et al. (2008) confirmed that a structured exercise program lasting eight to twelve weeks was associated with individuals reporting improved feelings of well-being. In addition, Taliaferro et al. (2008) noted that vigorous exercise and involvement in sports are associated with lower rates of suicide among adolescents.

Friends and Relatives

A network of relationships is associated with successful coping with various life transitions. Women are more likely to feel connected to others and to feel that they can count on others in times of need (Weckwerth and Flynn 2006). News media covering hurricanes, earthquakes, and tsunamis emphasized that, as long as one's family is still together, individuals are less affected by the loss of material possessions.

Love

A love relationship also helps individuals cope with stress. Being emotionally involved with another and sharing the experience with that person helps to insulate individuals from being devastated by a crisis event. Love is also viewed as helping resolve relationship problems. Over 85% (85.9%) of undergraduate males and 72.5% of undergraduate females agreed that, "If you love someone enough, you will be able to resolve your problems with that person" (Dotson-Blake et al. 2010).

Religion and Spirituality

Religion may be helpful in adjusting to a crisis. Not only does religion provide a rationale for one's plight ("It is God's will"), but it also offers a mechanism to ask for help. Also, religion is a social institution which connects one to others who may offer both empathy and help. Green and Elliott (2010) analyzed national data and found that controlling for job satisfaction, relationship satisfaction, and financial status, people who identify as religious tend to report better health and happiness, regardless of religious affiliation or religious activities.

Laughter

A sense of humor is related to lower anxiety and a happier mood. Indeed, a team of researchers compared the effects of humor, aerobic exercise, and listening to music on the reduction of anxiety and found humor to be the most effective. Just sitting quietly seemed to have no effect on reducing one's anxiety (Szabo et al. 2005). Smedema et al. (2010) confirmed that humor was a positive coping mechanism.

Sleep

Getting an adequate amount of sleep is also associated with lower stress levels. Such sleep, including midday naps, is associated with positive functioning, particularly memory performance. Pietrzak et al. (2010) emphasized that sleep is important for adaptive cognitive functioning.

Pets

Hughes (2011) emphasized that animals are associated with reducing blood pressure and stress, preventing heart disease, and fighting depression. Veterinary practices are encouraged to increase the visibility of this connection so that more individuals might benefit.

Deep Muscle Relaxation

Tensing and relaxing one's muscles have been associated with an improved state of relaxation. Calling it abbreviated progressive muscle relaxation (APMR), Termini (2006) found that the cognitive benefits were particularly evident; people who tensed and relaxed various muscle groups noticed a mental relaxation more than a physical relaxation.

Education

Sometimes becoming informed about a family problem helps to cope with the problem. Friedrich et al. (2008) studied how siblings cope with the fact that a brother or sister is schizophrenic. Education and family support were the primary mechanisms. Just knowing that the "parents were not to blame" resulted from becoming informed about schizophrenia.

Counseling for Children

Baggerly and Exum (2008) reviewed the value of counseling for children experiencing natural disasters such as hurricanes. Such counseling involves providing a safe context for children and having them remind themselves that they are safe (cognitive behavior therapy).

Some Harmful Strategies

Some coping strategies not only are ineffective for resolving family problems but also add to the family's stress by making the problems worse. Respondents in

The hatred we bear our enemies injures their happiness less than our own.

J. Petit-Senn, French poet

But do not distress yourself with dark imaginings.

Many fears are born of fatigue and loneliness

Max Ehrmann, from *Desiderata*

How few there are who have courage enough to own their faults, or resolution enough to mend them.

Benjamin Franklin, Founding Father of the United States

the Burr and Klein (1994) research identified several strategies they regarded as harmful to overall family functioning. These included keeping feelings inside, taking out frustrations on or blaming others, and denying or avoiding the problem.

Burr and Klein's research also suggests that women and men differ in their perceptions of the usefulness of various coping strategies. Women were more likely than men to view as helpful such strategies as sharing concerns with relatives and friends, becoming more involved in religion, and expressing emotions. Men were more likely than women to use potentially harmful strategies such as using alcohol, keeping feelings inside, or keeping others from knowing how bad the situation was.

Family Crisis Examples

Some of the more common crisis events that spouses and families face include physical illness, mental challenges, an extramarital affair, unemployment, substance abuse, and death.

Physical Illness and Disability

While most spouses are healthy most of the time, when one partner has a debilitating illness there are profound changes in the roles of the respective partners and their relationship. Mutch (2010) interviewed eight partners who took care of a spouse with multiple sclerosis after 20 years of marriage. The partners reported experiencing a range of feelings, including a sense of duty and a sense of loss, as they prioritized the health and needs of their spouse above their own. In addition, partners reported losing their sense of identity as a spouse as they became "the caretaker." Some partners also felt out of control due to the unpredictable and progressive nature of MS and because it consumed their life 24 hours every day. In addition, some felt guilt at not being satisfied with their life and wanting some independence.

While 24 million adults in the United States report that they have a disability, only 6.8 million of these use a visible assistive device. Hence, individuals may have a "hidden disability" such a chronic back pain, multiple sclerosis, rheumatoid arthritis, and chronic fatigue syndrome (Lipscomb 2009). These illnesses are particularly invasive in that conventional medicine has little to offer besides pain medication. For example, spouses with chronic fatigue syndrome may experience financial consequences ("I could no longer meet the demands of my job so I quit"), gender role loss ("I couldn't cook for my family" or "I was no longer a provider"), and changed perceptions by their children ("They have seen me sick for so long they no longer ask me to do anything").

Some couples cope with cancer. For men, prostate cancer is a medical issue that can rock the foundation of their personal and marital well-being. The follow section details the experience of a husband coping with prostate cancer.

Since my father had prostate cancer, I was warned that it is genetic and to be alert as I reached age 50. At the age of 56, I noticed that I was getting up more frequently at night to urinate. My doc said it was probably just one of my usual prostate infections, and he prescribed the usual antibiotic. When the infection did not subside, a urologist did a transrectal ultrasound (TRUS) needle biopsy. The TRUS gives the urologist an image of the prostate while he takes about 10 tissue samples from the prostate with a thin, hollow needle.

Treatment Options   *Two weeks later I learned the bad news (I had prostate cancer) and the good news (it had not spread). Since the prostate is very close to the spinal column, failure to act quickly can allow time for cancer cells to spread from the prostate to the bones. My urologist outlined a number of treatment options, including traditional surgery, laparoscopic surgery, radiation treatment, implantation of radio-active "seeds," cryotherapy (in which liquid nitrogen is used to freeze and kill prostate cancer cells), and hormone therapy, which blocks production of the male sex hormones that stimulate growth of prostate cancer cells. He recommended traditional surgery ("radical retro pubic prostatectomy"), in which an incision is made between the belly button and the pubic bone to remove the prostate gland and nearby lymph nodes in the pelvis. This surgery is generally considered the "gold standard" when the disease is detected early. Within three weeks I had the surgery.*

Physical Effects Following Surgery *Every patient awakes from radical prostate surgery with urinary incontinence and impotence—which can continue for a year, two years, or forever. Such patients also awake from surgery hoping that they are cancer-free. This is determined by laboratory analysis of tissue samples taken during surgery. The patient waits for a period of about two weeks, hoping to hear the medical term "negative margins" from his doctor. That finding means that the cancer cells were confined to the prostate and did not spread past the margins of the prostate. A finding of negative margins should be accompanied by a PSA [prostate specific antigen] score of zero, confirming that the body no longer detects the presence of cancer cells. I cannot describe the feeling of relief that accompanies such a report, and I am very fortunate to have heard those words used in my case.*

Psychological Effects Following Surgery *The psychological effects have been devastating—more for me than for my partner. I have only been intimate with one woman—my wife—and having intercourse with her was one of the greatest pleasures in my life. For a year, I was left with no erection and an inability to have an orgasm. Afterwards, I was able to have an erection (via a self-injection of Alprostadil) and an attenuated (weakened) orgasm. Dealing with urinary incontinence (I refer to myself as Mr. Drippy) is a "wish it were otherwise" on my psyche. [Author note— Bronner et al. (2010) studied husbands who had had prostate surgery and found that all who met a sex therapist with their partner reported an improved sex life, even if erectile dysfunction wasn't resolved.]*

Evaluation *When faced with the decision to live or die, the choice for most of us is clear. In my case, I am alive, cancer free, and enjoying the love of my life (now in our 44th year together).*

In those cases in which the illness is fatal, **palliative care** is helpful. This term describes the health care for the individual who has a life-threatening illness (focusing on relief of pain and suffering) and support for them and their loved ones. Such care may involve the person's physician or a palliative care specialist who works with the physician, nurse, social worker, and chaplain. Pharmacists or rehabilitation specialists may also be involved. The effects of such care are to approach the end of life with planning (how long should life be sustained on machines?) and forethought to relieve pain and provide closure.

Mental Illness

Mental illness is defined as "alternations in thinking, mood, or behavior that are associated with distress and impaired function" (Marshall et al. 2010). It is estimated that one in four adults and one in five families experience mental illness (Marshall et. al. 2010). Pratt and Brody (2010) found a link between depression and smoking—43% of all adults 20 and over who report being depressed also report smoking. Depression is not only associated with smoking but also with poor diet, lack of exercise, diabetes, and cardiovascular disease. The latter is a leading cause of death for men and women. Kouros and Cummings (2010) studied 296 couples and found that higher levels of husbands' depressive symptoms predicted subsequent elevations in wives' depressive symptoms over time. Hence when one spouse gets depressed, the other is likely to mirror this.

You have to make taking care of yourself a priority.

Christina Maslach, psychologist

Chapter 14 Stress and Crisis in Relationships

Insel et al. (2008) noted the enormous economic costs of serious mental illness that involve a high rate of emergency room care (for example, suicide attempts), high prevalence of pulmonary disease (people with serious mental illness smoke 44% of all cigarettes in the United States), and early mortality (a loss of thirteen to thirty-two years). The toll of mental illness on a relationship can be immense. A major initial attraction of partners to each other includes intellectual and emotional qualities. Butterworth and Rodgers (2008) surveyed 3,230 couples to assess the degree to which mental illness of a spouse or spouses affects divorce and found that couples in which either men or women reported mental health problems had higher rates of marital disruption than couples in which neither spouse experienced mental health problems. For couples in which both spouses reported mental health problems, rates of marital disruption reflected the additive combination of each spouse's separate risk.

Children may also be mentally ill. Examples include autism (three to four times more common in boys), attention-deficit/hyperactivity disorder (11% of children, ages 12 to 17), and antisocial behavior (*Statistical Abstract of the United States 2011*, Table 184). These difficulties can stress spouses to the limit of their coping capacity, which may put an enormous strain on their marriage.

The reverse is also true; children must learn how to cope with the mental illness of their parents. Mordoch and Hall (2008) studied twenty-two children between 6 and 16 years of age, who were living part- or full-time with a parent with depression, schizophrenia, or bipolar illness. They found that the children learned to maintain connections with their parents by creating and keeping a safe distance between themselves and their parents so as not to be engulfed by their parents' mental illnesses.

Midlife Crisis (Middle Age Crazy)

The stereotypical explanation for 45-year-old people who buy convertible sports cars, have affairs, marry 20-year-olds, or adopt a baby is that they are "having a midlife crisis." The label conveys that they feel old, think that life is passing them by, and seize one last great chance to do what they have always wanted. Indeed, one father (William Feather) noted, "Setting a good example for your children takes all the fun out of middle age."

However, a ten-year study of close to 8,000 U.S. adults aged 25 to 74 by the MacArthur Foundation Research Network on Successful Midlife Development revealed that, for most respondents, the middle years brought no crisis at all but a time of good health, productive activity, and community involvement. Less than a quarter (23%) reported a "crisis" in their lives. Those who did experience a crisis were going through a divorce. Two-thirds were accepting of getting older; one-third did feel some personal turmoil related to the fact that they were aging (Goode 1999).

Of those who initiated a divorce in midlife, 70% had no regrets and were confident that they did the right thing. This is the result of a study of 1,147 respondents aged 40 to 79 who experienced a divorce in their forties, fifties, or sixties. Indeed, midlife divorcers' levels of happiness or contentment were similar to those of single individuals their own age and those who remarried (Enright 2004).

Some people embrace middle age. The Red Hat Society (http://www.redhatsociety.com/) is a group of women who have decided to "greet middle age with verve, humor, and elan. We believe silliness is the comedy relief of life [and] share a bond of affection, forged by common life experiences and a genuine enthusiasm for wherever life takes us next." The society traces its beginning to when Sue Ellen Cooper bought a bright red hat because of a poem Jenny Joseph wrote in 1961, titled the "Warning Poem." The poem says the following:

When I am an old woman I shall wear purple
With a red hat which doesn't go and doesn't suit me.

Seldom, or perhaps never, does a marriage develop into an individual relationship smoothly and without crises; there is no coming to consciousness without pain.

Carl Jung, Swiss psychiatrist

Perhaps we would bear our sadness with greater trust than we have in our joys. For they are the moments when something new has entered us, something unknown; our feelings grow mute in shy embarrassment, everything in us withdraws, a silence arises, and the new experience, which no one knows, stands in the midst of it all and says nothing.

Ranier Maria Rike,
Letters to a Young Poet

Cooper then gave red hats to friends as they turned 50. The group then wore their red hats and purple dresses out to tea, and that's how it got started. Now there are over 1 million members worldwide.

In the rest of this chapter, we examine how spouses cope with the crisis events of an extramarital affair, unemployment, drug abuse, and death. Each of these events can be viewed either as devastating and the end of meaning in one's life or as an opportunity and challenge to rise above.

Extramarital Affair (and Successful Recovery)

The media pounces on celebrities who have sex with someone outside their marriage. No less a reality are noncelebrity spouses who cope with this personal, marriage, and family crisis.

> *Now there is somebody new, These dreams I've been dreaming have all fallen through.*
>
> Bonnie Raitt, *Too Soon to Tell*

Extramarital affair refers to a spouse's sexual involvement with someone outside the marriage.

National **Data**

Of spouses in the United States, 23.2% of husbands and 13.18% of wives reported ever having had intercourse with someone to whom they were not married (Djamba et al. 2005).

Extramarital affair refers to a spouse's sexual involvement with someone outside the marriage. Arnold Schwarzenegger, Tiger Woods and John Edwards gave cultural visibility to this very private marital phenomenon. Affairs are of different types, which may include the following:

1. *Brief encounter.* A spouse meets and hooks up with a stranger. In this case, the spouse is usually out of town, and alcohol is often involved. Tiger Woods reported that he was in Las Vegas, alone, when he met several of his "13 mistresses." John Edwards was also out of town, away from his wife, when he met the woman who became the mother of his love child.

2. *Periodic sexual encounters.* A spouse is sexually unsatisfied in the marriage and seeks external sex, often with a hooker (for example, former New York governor Eliot Spitzer). A married person of bisexual orientation may also seek a periodic encounter outside the marriage.

3. *Instrumental or utilitarian affair.* This is sex in exchange for a job or promotion, to get back at a spouse, to evoke jealousy, or to transition out of a marriage.

4. *Coping mechanism.* Sex can be used to enhance one's self-concept or feeling of sexual inadequacy, compensate for failure in business, cope with the death of a family member, test out one's sexual orientation, and so on.

5. *Paraphiliac affairs.* In these, spouses act out sexual fantasies that most people would consider to be bizarre or abnormal sexual practices, such as sexual masochism, sexual sadism, or transvestite fetishism.

6. *New love.* A spouse may be in love with the new partner and may plan marriage after divorce.

International **Data**

Fifteen percent of the husbands and 5% of the wives in China have engaged in extramarital sex (Zhang 2010).

In addition, the computer or Internet affair is another type of affair. Although legally an extramarital affair does not exist unless two people (one being married) have intercourse, an online computer affair can be just as disruptive to a marriage or a couple's relationship. Computer friendships may move to feelings of intimacy, involve secrecy (one's partner does not know the level of involvement), include sexual tension (even though there is no overt sex), and take time, attention, energy, and affection away from one's partner. Cavaglion and

Infidelity can be defined as unfaithfulness in a committed monogamous relationship. Infidelity can affect anyone, regardless of race, color, or creed; it does not matter whether you are rich or attractive, where you live, or how old you are. The purpose of this survey is to gain a better understanding of what people think and feel about issues associated with infidelity. There are no right or wrong answers to any of these statements; we are interested in your honest reactions and opinions. Please read each statement carefully, and respond by using the following scale:

1	2	3	4	5	6	7
Strongly Disagree						Strongly Agree

_____ 1. Being unfaithful never hurt anyone.

_____ 2. Infidelity in a marital relationship is grounds for divorce.

_____ 3. Infidelity is acceptable for retaliation of infidelity.

_____ 4. It is natural for people to be unfaithful.

_____ 5. Online/Internet behavior (for example, visiting sex chat rooms, porn sites) is an act of infidelity.

_____ 6. Infidelity is morally wrong in all circumstances, regardless of the situation.

_____ 7. Being unfaithful in a relationship is one of the most dishonorable things a person can do.

_____ 8. Infidelity is unacceptable under any circumstances if the couple is married.

_____ 9. I would not mind if my significant other had an affair as long as I did not know about it.

_____ 10. It would be acceptable for me to have an affair, but not my significant other.

_____ 11. I would have an affair if I knew my significant other would never find out.

_____ 12. If I knew my significant other was guilty of infidelity, I would confront him/her.

Scoring

Selecting a 1 reflects the least acceptance of infidelity; selecting a 7 reflects the greatest acceptance of infidelity. Before adding the numbers you selected, reverse the scores for item numbers 2, 5, 6, 7, 8, and 12. For example, if you responded to item 2 with a "6," change this number to a "2"; if you responded with a "3," change this number to "5," and so on. After making these changes, add the numbers. The lower your total score (12 is the lowest possible), the less accepting you are of infidelity; the higher your total score (84 is the highest possible), the greater your acceptance of infidelity. A score of 48 places you at the midpoint between being very disapproving and very accepting of infidelity.

Scores of Other Students Who Completed the Scale

The scale was completed by 150 male and 136 female student volunteers at Valdosta State University. The average score on the scale was 27.85 (SD = 12.02). Their ages ranged from 18 to 49, with a mean age of 23.36 (SD = 5.13). The ethnic backgrounds of the sample included 60.8% white, 28.3% African American, 2.4% Hispanic, 3.8% Asian, 0.3% American Indian, and 4.2% other. The college classification level of the sample included 11.5% freshmen, 18.2% sophomores, 20.6% juniors, 37.8% seniors, 7.7% graduate students, and 4.2% post baccalaureate. Male participants reported more positive attitudes toward infidelity (mean = 31.53; SD = 11.86) than did female participants (mean = 23.78; SD = 10.86; p < .05). White participants had more negative attitudes toward infidelity (mean = 25.36; SD = 11.17) than did nonwhite participants (mean = 31.71; SD = 12.32; p <.05). There were no significant differences in regard to college classification.

Source

"Attitudes toward Infidelity Scale" 2006 by Mark Whatley, Ph.D., Department of Psychology, Valdosta State University, Valdosta, Georgia 31698-0100. Used by permission. Other uses of this scale by written permission of Dr. Whatley only (mwhatley@valdosta.edu). Information on the reliability and validity of this scale is available from Dr. Whatley.

Rashty (2010) noted the anguish embedded in 1,130 messages on self-help chat boards from female partners of males involved in cybersex relationships and pornographic websites. The females reported distress and feelings of ambivalent loss that had an individual, couple, and sexual relationship impact. Cramer et al. (2008) also noted that women become more upset when their man is emotionally unfaithful with another woman (although men become more upset when their partner is sexually unfaithful with another man). The Self-Assessment section allows you to measure your attitude toward infidelity.

Extradyadic involvement or extrarelational involvement refers to sexual involvement of a pair-bonded individual with someone other than the partner. Extradyadic involvements are not uncommon. Of 2,922 undergraduates, 24% agreed, "I have cheated on a partner I was involved with." Forty-one percent agreed, "A partner I was involved with cheated on me" (Knox and Hall 2010). What constitutes cheating is ambiguous. Wilson et al. (2011) reported a Perceptions of Dating Infidelity Scale (PDIS) which assesses attitudes toward specific behaviors that constitute various types of infidelity (ambiguous, deceptive, and

Extradyadic involvement
refers to sexual involvement of a pair-bonded individual with someone other than the partner; also called extrarelational involvement.

explicit behaviors) in romantic relationships. Those scoring high on deception and explicit behaviors also scored high on guilt. Marriage and family therapists rank an extramarital affair as the second most stressful crisis event for a couple (physical abuse is number one) (Olson et al. 2002). Men are more upset if their wife has a heterosexual than a homosexual affair while women are equally upset if their spouse has a heterosexual or homosexual affair (Confer and Cloud 2011). Characteristics associated with spouses who are more likely to have extramarital sex include male gender, a strong interest in sex, permissive sexual values, low subjective satisfaction in the existing relationship, employment outside the home, low church attendance, greater sexual opportunities, higher social status (power and money), and alcohol abuse (Hall et al. 2008; Olson et al. 2002). Elmslie and Tebaldi (2008) noted that women's infidelity behavior is influenced by religiosity (less religious = more likely), city size (urban = more likely), and happiness (less happy = more likely), whereas men's infidelity behavior is affected by race (white = less likely), happiness status (happy = less likely), city size (same for women), and employment status (employed = less likely).

Reasons for Extramarital Affair   Spouses report a number of reasons why they become involved in a sexual encounter outside their marriage:

1. *Variety, novelty, and excitement.* Extradyadic sexual involvement may be motivated by the desire for variety, novelty, and excitement. One of the characteristics of sex in long-term committed relationships is the tendency for it to become routine. Early in a relationship, the partners cannot seem to have sex often enough. However, with constant availability, partners may achieve a level of satiation, and the attractiveness and excitement of sex with the primary partner seem to wane. A high-end call girl said the following of the Eliot Spitzer affair:

> Almost all of my clients are married. I would say easily over 90 percent. I'm not trying to justify this business, but these are men looking for companionship. They are generally not men that couldn't have an affair [if they wanted to], but men who want this tryst with no stings attached. They're men who want to keep their lives at home intact. (Kottke 2008)

Coolidge effect term used to describe waning of sexual excitement and the effect of novelty and variety on increasing sexual arousal.

The **Coolidge effect** is a term used to describe this waning of sexual excitement and the effect of novelty and variety on sexual arousal:

> One day President and Mrs. Coolidge were visiting a government farm. Soon after their arrival, they were taken off on separate tours. When Mrs. Coolidge passed the chicken pens, she paused to ask the man in charge if the rooster copulated more than once each day. "Dozens of times," was the reply. "Please tell that to the President," Mrs. Coolidge requested. When the President passed the pens and was told about the rooster, he asked, "Same hen every time?" "Oh no, Mr. President, a different one each time." The President nodded slowly and then said, "Tell that to Mrs. Coolidge." (Bermant 1976, 76–77)

A man snatches the first kiss, pleads for the second, demands the third, takes the fourth, accepts the fifth—and endures all the rest.

Helen Rowland, journalist and humorist

Whether or not individuals are biologically wired for monogamy continues to be debated. Monogamy among mammals is rare (from 3% to 10%), and monogamy tends to be the exception more often than the rule (Morell 1998). Even if such biological wiring for plurality of partners does exist, it is equally debated whether such wiring justifies nonmonogamous behavior—that individuals are responsible for their decisions.

2. *Workplace friendships.* A common place for extramarital involvements to develop is the workplace (Merrill and Knox 2010). Neuman (2008) noted that

Penguins are serially monogamous. They have only one mate per year during the mating season and are faithful to that mate. After the season is over they seek a new mate.

four in ten of the affairs men reported began with a woman they met at work. Coworkers share the same world eight to ten hours a day and, over a period of time, may develop good feelings for each other that eventually lead to a sexual relationship. Tabloid reports regularly reflect that romances develop between married actors making a movie together (for example, Brad Pitt and Angelina Jolie, who cohabit/have children together and met at work).

3. Relationship dissatisfaction. It is commonly believed that people who have affairs are not happy in their marriage. Spouses who feel misunderstood, unloved, and ignored sometimes turn to another who offers understanding, love, and attention. Neuman (2008) confirmed that being emotionally dissatisfied in one's relationship is the primary culprit leading to an affair.

One source of relationship dissatisfaction is an unfulfilling sexual relationship. Some spouses engage in extramarital sex because their partner is not interested in sex. Others may go outside the relationship because their partners will not engage in the sexual behaviors they want and enjoy. The unwillingness of the spouse to engage in oral sex, anal intercourse, or a variety of sexual positions sometimes results in the other spouse's looking elsewhere for a more cooperative and willing sexual partner.

4. Revenge. Some extramarital sexual involvements are acts of revenge against one's spouse for having an affair. When partners find out that their mate has had or is having an affair, they are often hurt and angry. One response to this hurt and anger is to have an affair to get even with the unfaithful partner.

5. Homosexual relationship. Some individuals marry as a front for their homosexuality. Cole Porter, known for "I've Got You under My Skin," "Night and Day," and "Easy to Love," was a homosexual who feared no one would buy or publish his music if his sexual orientation were known. He married Linda Lee Porter (alleged to be a lesbian), and they had an enduring marriage for thirty years.

Other gay individuals marry as a way of denying their homosexuality. These individuals are likely to feel unfulfilled in their marriage and may seek involvement in an extramarital homosexual relationship. Other individuals may marry and then discover later in life that they desire a homosexual relationship. Such individuals may feel that (1) they have been homosexual or bisexual all along,

(2) their sexual orientation has changed from heterosexual to homosexual or bisexual, (3) they are unsure of their sexual orientation and want to explore a homosexual relationship, or (4) they are predominately heterosexual but wish to experience a homosexual relationship for variety.

6. *Aging.* A frequent motive for intercourse outside marriage is the desire to return to the feeling of youth. Ageism, which is discrimination against the elderly, promotes the idea that being young is good and being old is bad. Sexual attractiveness is equated with youth, and having an affair may confirm to older partners that they are still sexually desirable. Also, people may try to recapture the love, excitement, adventure, and romance associated with youth by having an affair.

7. *Absence from partner.* One factor that may predispose a spouse to an affair is prolonged separation from the partner. Some wives whose husbands are away for military service report that the loneliness can become unbearable. Some husbands who are away say that remaining faithful is difficult. Partners in commuter relationships may also be vulnerable to extradyadic sexual relationships.

8. *Delayed negative consequences.* Individuals are primarily influenced by short-term consequences. The allure of a new sexual encounter provides an immediate consequence to saying "yes" and moving forward. Months go by before the indiscretion is discovered and the negative consequences tumble out of life's barrel. Tiger Woods was confronted with numerous attractive females who offered companionship and sex. Engaging in the new sexual encounters was immediately reinforcing. Months went by before his divorce from Elin Woods (who reported that she was blindsided), disapproval from fans/colleagues, and being dropped by sponsors. The worst thing that can happen to a person who has an affair is that he or she will get away with it, since doing so ensures that they will repeat the behavior and eventually get caught.

Effects of an Affair Reactions to the knowledge that one's spouse has been unfaithful vary. For most, the revelation is difficult. The following is an example of a wife's reaction to her husband's affairs:

> *My husband began to have affairs within six months of our being married. Some of the feelings I experienced were disbelief, doubt, humiliation and outright heart-wrenching pain! When I confronted my husband he denied any such affair and said that I was suspicious, jealous and had no faith in him. In effect, I had the problem. He said that I should not listen to what others said because they did not want to see us happy but only wanted to cause trouble in our marriage. I was deeply in love with my husband and knew in my heart that he was guilty as sin; I lived in denial so I could continue our marriage.*

WHAT IF?

What if You Are Tempted to Cheat on Your Partner?

The costs of cheating on your partner will rarely be worth the anticipated gains in adventure or sex. Ask Democratic presidential candidate John Edwards if his indiscretion was worth it (his marriage ended and his political career was devastated). To control one's temptation, it is important to control words, alcohol, and context. Flirting, alcohol, and being alone with the other person increase the chance of cheating. Being polite but scripted, avoiding alcohol or drugs, and always being in a public place with the other person will ensure that nothing sexual will happen (and negative consequences will be avoided). If you are already drifting toward an affair, it is never too late to stop and reverse your behavior.

Of course, my husband continued to have affairs. Some of the effects on me included:

1. *I lost the ability to trust my husband and, after my divorce, other men.*
2. *I developed a negative self-concept—the reason he was having affairs is that something was wrong with me.*
3. *He robbed me of the innocence and my "VIRGINITY"—clearly he did not value the opportunity to be the only man to have experienced intimacy with me.*
4. *I developed an intense hatred for my husband.*

It took years for me to recover from this crisis. I feel that through faith and religion I have emerged "whole" again. Years after the divorce my husband made a point of apologizing and letting me know that there was nothing wrong with me, that he was just young and stupid and not ready to be serious and committed to the marriage.

Theiss and Nagy (2010) studied 220 married couples and found that relationships with those outside the marital dyad tend to introduce relationship uncertainty inside the dyad and that this had negative consequences for the sexual satisfaction reported by the spouses. Indeed, interference from others was associated with negative thoughts and emotional reactions to sex.

Seven states recognize **alienation of affection** lawsuits which give a spouse the right to sue a third party for taking the affections of a spouse away. Alienation of affection claims evolved from common law, which considered women property of their husbands. The reasoning was if another man was accused of stealing his "property," a husband could sue him for damages. The law applies to both women and men so a woman who steals another woman's man can be sued for taking her "property" away. Such was the case of Cynthia Shackelford who sued Anne Lundquist in 2010 for "alienating" her husband from her and breaking up her 33 year marriage. A jury awarded Cynthia Shackelford $4 million in punitive damages and $5 million in compensatory damages. The decision has been appealed.

As noted above, loss of trust is a major consequence of an extramarital affair. Such loss of trust is fed by the perception that the partner is withholding information, not disclosing. The perception that one's partner is hiding something may lead to snooping by the partner.

Snooping, also known as covert intrusive behavior, is defined as investigating (without the partner's knowledge or permission) a romantic partner's private communication (such as text messages, e-mail, and cell phone use) motivated by concern that the partner may be "hiding" something. Vinkers et al. (2011) noted that snooping is functional for the snooper since it provides information that is supposed to be a secret and gives power over the partner who cannot deny certain facts (e.g., "I know you have been texting your old girlfriend because I read the messages"). Snooping typically has negative consequences in that it is associated with increased conflict, decreased trust, and stained/conflictual interaction.

High levels of personal disclosure are expected in close relationships. When partners feel that their partner is not open/disclosing, they feel hurt and devalued. Individuals who feel anxious and uncertain about the partner's lack of disclosure are motivated to gain increased information which results in snooping. Hence low disclosure creates the context in which persons snoop to gain increased information about the partner's behavior to better predict the future of the relationship.

Vinkers et al. (2011) studied 188 couples married less than three years and found that lower levels of perceived partner disclosure were associated with higher levels of intrusive behavior (snooping). In addition, perceived disclosure was negatively associated with intrusive behavior at lower levels of trust in one's partner, but not at higher levels of trust. Hence, if one partner did not trust another, even though there was high disclosure, intrusive behavior still occurred. And, if the trust level was high, even though disclosure was low, snooping was less likely. Also, the couples in the study believed that wives engaged in more intrusive behavior than husbands (women more verbally demanding, have

Alienation of affection law which gives a spouse the right to sue a third party for taking the affections of a spouse away.

I don't know if you can ever mend something like this, in the sense of repair to the canvas so that you never see the tear in the fabric.

Eliot Spitzer, former governor of New York, of his affair

Snooping investigating (without the partner's knowledge or permission) a romantic partner's private communication (such as text messages, e-mail, and cell phone use) motivated by concern that the partner may be "hiding" something.

higher need for emotional involvement, and men have a greater need to control their privacy than women). The researchers did not find an actual gender difference in snooping behavior.

Successful Recovery from Infidelity   Olson et al. (2002) identified three phases of successful recovery from the discovery of a partner's affair. The "roller-coaster" phase involves agony at the initial discovery, which elicits an array of feelings including rage or anger, self-blame, the desire to give up, and the desire to work on the marital relationship. The second phase, "moratorium," involves less emotionality and a decision to work it out. The partners settle into a focused though tenuous commitment to get beyond the current crisis. The third phase, "trust building," involves taking responsibility for the infidelity, reassurance of commitment, increased communication, and forgiveness. The latter means letting go of one's resentment, anger, and hurt; accepting that we all need forgiveness; and moving forward (Hill 2010). Couples in this phase "reengage," "open up," and focus on problems leading up to the infidelity. Working through a discovered affair takes time, commitment to new behavior in the primary relationship, and forgiveness. Dean (2011) noted that part of the recovery a spouse experiences when a partner has an affair is the grieving over what the relationship was and what it represented and still moving forward in recovering from the event. Bagarozzi (2008) defined forgiveness as a conscious decision on the part of the offended spouse to grant a pardon to the offending spouse, to give up feeling angry, and to relinquish the right to retaliate against the offending spouse. In exchange, offending spouses take responsibility for the affair, agree not to repeat the behavior, and grant their partner the right to check up on them to regain trust. The Personal Choices section focuses on the decision to end a relationship rather than work through it when a partner has an affair.

> *My friend Linda is leaving her husband just because he is unfaithful to her. That is no reason to leave the person. I feel like, after that, you should stay with the person and make sure that the rest of their life is sheer hell.*
>
> Roseanne Barr, comedian

PERSONAL CHOICES

Should You Seek a Divorce If Your Partner Has an Affair?

About 20% of spouses face the decision of whether to stay with a mate who has had an extramarital affair. Such an affair may be physical or online. In a study of 123 committed couples, both men and women reported more distress when there was hypothetical emotional as compared to sexual online infidelity (Henline et al. 2007). Indeed, Amy Taylor, an Englishwoman, filed for divorce in 2008, after discovering that her husband had been having an affair in the online role-playing game, *Second Life.*

Regardless of whether the indiscretion is physical or emotional, one alternative for the partner is to end the relationship immediately on the premise that trust has been broken and can never be mended. People who take this position regard fidelity as a core element of the marriage that, if violated, necessitates a divorce. College students disapprove of an extramarital affair. In a sample of 2,922 undergraduates, 54% said that they would "divorce a spouse who had an affair" (Knox and Hall 2010). Americans in general tend to be unforgiving about an affair. In a national survey, 64% would not forgive their spouse for having an extramarital affair and almost as high a percentage (62%) say they would leave their spouse and get a divorce if they found out their spouse was having an affair; 31% would not (Jones 2008). For some, emotional betrayal is equal to sexual betrayal. Other couples build into their relationship the fact that each will have external relationships. In Chapter 2, we discussed polyamory, in which partners are open and encouraging of multiple relationships at the same time. The term *infidelity* does not exist for polyamorous couples.

Even for traditional couples, infidelity need not be the end of a couple's marriage but the beginning of a new, enhanced, and more understanding relationship. Healing takes time; the relationship may require a commitment on the part of the straying partner not to repeat the behavior, forgiveness by the partner (and not bringing it up again), and a new focus on improving the relationship. In spite of the difficulty of adjusting to an affair, most spouses are reluctant to end a marriage. Not one of fifty "successful couples" said that they would automatically end their marriage over adultery (Wallerstein and Blakeslee 1995).

The spouse who chooses to have an affair is often judged as being unfaithful to the vows of the marriage, as being deceitful to the partner, and as inflicting enormous pain on the partner (and children). When an affair is defined in terms of giving emotional energy, time, and economic resources to something or someone outside the primary relationship, other types of "affairs" may be equally as devastating to a relationship. Spouses who choose to devote their lives to their careers, parents, friends, or recreational interests may deprive the partner of significant amounts of emotional energy, time, and money and create a context in which the partner may choose to become involved with a person who provides more attention and interest.

Another issue in deciding whether to take the spouse back following extramarital sex is the concern over HIV. One spouse noted that, though he was willing to forgive and try to forget his partner's indiscretion, he required that she be tested for HIV and that they use a condom for six months. She tested negative, but their use of a condom was a reminder, he said, that sex outside one's bonded relationship in today's world has a life-or-death meaning. Related to this issue is that wives are more likely to develop cervical cancer if their husbands have other sexual partners.

There is no single way to respond to a partner who has an extramarital relationship. Most partners are hurt and think of ways to work through the crisis. Some succeed.

Sources

Henline, B. H., L. K. Lamke, and M. D. Howard. 2007. Exploring perceptions of online infidelity. *Personal Relationships* 14: 113–128.

Jones, J. M. 2008. Most Americans not willing to forgive unfaithful spouse. *The Gallup Poll Briefing,* March, 83. Washington.

Knox, D., and S. Hall. 2010 Relationship and sexual behaviors of a sample of 2,922 university students. Unpublished data collected for this text. Department of Sociology, East Carolina University, and Department of Family and Consumer Sciences, Ball State University.

Wallerstein, J. S., and S. Blakeslee. 1995. *The good marriage.* Boston: Houghton Mifflin.

Positive outcomes of having experienced and worked through infidelity include a closer marital relationship, increased assertiveness, placing higher value on each other, and realizing the importance of good marital communication (Olson et al. 2002).

Spouses who remain faithful to their partners have decided to do so. They avoid intimate conversations with members of the other sex and a context (for example, where alcohol drinking and/or being alone are involved) that are conducive to physical involvement. The best antidote is to be direct, simply telling the person you are not interested.

Prevention of Infidelity   Allen et al. (2008b) identified the premarital factors predictive of future infidelity. The primary factor for both partners was a negative pattern of interaction. Partners who end up being unfaithful were in relationships where they did not connect, argued, and criticized each other. Hence, spouses least vulnerable were in loving, nurturing, communicative relationships where each affirms the other. Neuman (2008) also noted that avoiding friends who have affairs and establishing close relationships with married couples who value fidelity further insulates one from having an affair.

Some individuals who have cheated and feel "out of control" seek help for their "addiction" by entering "rehab" (e.g., Tiger Woods). Whether their attending a "sex addiction clinic" is a marital ploy to be forgiven, do public penance, and resume the marriage or a call for help is unknown. Levine (2010) studied a sample of 30 married men who had been discovered to have violated monogamy rules with their wife via pornography, cybersex, commercial sex involvement, paraphilic pursuits, or affairs and who went to a "sex addiction clinic." According to Levine (2010), only 25% of the sample had issues (spiraling psychological deterioration) which could reasonably be described as having a sexual addiction, which is a primary criterion for being labeled as a "sex addict."

Unemployment

National **Data**

Of the 25.8 million married couples with children under 18, about 6% of husbands and 4% of wives were unemployed in 2009 (U.S. Census Bureau 2010).

America continues to downsize and outsource jobs to India and Mexico. The result is massive layoffs and insecurity in the lives of American workers. In recent years, unemployment in America has hovered around 10% in the general adult population and even higher among minorities. Forced unemployment or the threatened loss of one's job is a major stressor for individuals, couples, and families, sometimes leading to homelessness (Gould and Williams 2010). Also, when spouses or parents lose their jobs as a result of physical illness or disability, a family experiences a double blow—loss of income combined with higher medical bills. Unless an unemployed spouse is covered by the partner's medical insurance, unemployment can also result in loss of health insurance for the family. Insurance for both health care and disability is very important to help protect a family from an economic disaster. One reaction of unemployment on the part of the husband is for the wife to become employed or to seek work (Mattingly et al. 2010).

The effects of unemployment may be more severe for men than for women. Our society expects men to be the primary breadwinners in their families and equates masculine self-worth and identity with job and income. Stress, depression, alcohol abuse, and lowered self-esteem are all associated with unemployment. Unemployment is also associated with child maltreatment (Euser et al. 2010).

Women are not burdened with the cultural expectation of the provider role, and their identity is less tied to their work role. Hence, women may view unemployment as an opportunity to spend more time with their families; many enjoy doing so. For both women and men, Formichelli (2010) recommended that the unemployed consider volunteering. Doing so increases their self-esteem, contributes to the community, and provides additional skills/contacts for a subject. Some volunteer jobs do not require one to leave home (see volunteermatch.org).

Substance Abuse

Jeff Bridges won best actor (2010) for his role as Bad Blake in *Crazy Heart*, the story of an alcoholic country western singer. The film revealed the insidious consequences of alcohol on a relationship. Spouses, parents, and children who abuse alcohol/drugs contribute to the stress and conflict experienced in their respective marriages and families. Although some individuals abuse drugs to escape from unhappy relationships or the stress of family problems, substance abuse inevitably adds to the individual's marital and family problems when it results in health and medical problems, legal problems, loss of employment, financial ruin, school failure, divorce, and even death (due to accidents or poor health) (Dethier et al. 2011). The following reflects the experience of the wife of a man addicted to crack:

> Marriage needs a foundation of trust and open communication, but when you are married to an addict, you won't find either. When I first met "Nate," I thought he was everything I wanted in a partner—good looks, a great sense of humor, intelligence, and ambition. It didn't take me long to realize that I was extremely attracted to this man. Since we knew some of the same people, we ended up at several parties to-

Alcohol abuse sometimes goes undetected. This woman is home alone.

gether. We both drank and occasionally used cocaine but it wasn't a problem. I loved being with him and whenever I wasn't with him, I was thinking about him.

After only four months of dating, we realized we were falling madly in love with each other and decided to get married. Neither of us had ever been married before but I had a five-year-old son from a previous relationship. Nate really took to my son and it seemed like we had the perfect marriage—that changed drastically and quickly. I didn't know it but my husband had been using crack cocaine the entire time we dated. I wasn't extremely familiar with this new drug and I knew even less about its addictive power. Of course, I soon found out more about it than I cared to know.

Less than three months after we married, Nate went on his first binge. After leaving work one Friday night, he never came home. I was really worried and was calling all of his friends trying to find out where he was and what was wrong. I talked to the girlfriend of one of his best friends and found out everything. She told me that Nate had been smoking crack off and on for a long time and that he was probably out using again. I didn't know what to do or where to turn. I just stayed home all weekend waiting to hear from my husband.

Finally, on Sunday afternoon, Nate came home. He looked like hell and I was mad as hell. I sent my son to play with his friends next door and as soon as he was out of hearing range, I lost it. I began screaming and crying, asking why he never came home over the weekend. He just hung his head in absolute shame. I found out later that he had spent his entire paycheck, pawned his wedding ring, and had written checks off of a closed checking account. He went through over $1,000 of "our" money in less than three days.

I was devastated and in shock. After I calmed down and my anger subsided, we discussed his addiction. He told me he loved me with all his heart and that he was so sorry for what he had done. He also swore he would never do it again. Well, this is when I became an enabler and I continued to enable my husband for three years. He would stay clean for a while and things would be great between us, then he would go on another binge. It was the roller-coaster ride from hell. Nate was on a downward spiral and he was dragging my son and me down along with him.

By the end of our third year together, we were more like roommates. Our once wonderful sex life was virtually nonexistent and what love I still had for him was quickly fading away. I knew I had to leave before my love turned to hate. It was obvious that

Alcohol is the drug that college students most frequently use. Even in a private conservative religious college, 53.6% of a sample of undergraduates reported drinking an average of five or more beers at a time (Coll et al. 2008), and 43.1% reported drinking beer at least once a week. Seven percent of 2,922 undergraduates reported that "I have a problem with alcohol" (Knox and Hall 2010). Bulmer et al. (2010) reported that, in general, alcohol consumption patterns have remained stable. However, patterns vary by campus. At a northeastern university, increases in both frequency and volume were noted over a six year period—particularly by females, those over 21, those living off campus, and those performing well academically.

College students typically do not learn from negative consequences to their drinking. Mallett et al. (2006) found that students who threw up, made unwise sexual decisions, or experienced a hangover or blackout underestimated the amount of alcohol they could drink before they experienced negative consequences. Continuing to drink was a given.

Campus policies throughout the United States include alcohol-free dorms; alcohol bans, enforcement, and sanctions; peer support; and education. Most colleges and universities do not ban alcohol or its possession on campus. Administrators fear that students will attend other colleges where they are allowed to drink. Alumni may want to drink at football games and view such university banning as intrusive. Some attorneys think colleges and universities can be held liable for not stopping dangerous drinking patterns, but others argue that college is a place for students to learn how to behave responsibly. Should sanctions be used (for example, expelling a student or closing down a fraternity)? If police are too restrictive, drinking may go underground, where detecting may be more difficult.

For the purposes of peer education, students at the University of Buffalo in New York created a video illustrating responsible drinking and discussions by students of their own alcohol poisoning. Some universities have hired full-time alcohol education coordinators. Providing nonalcoholic ways to meet others and to socialize is now offered on

my son had been pulling away from Nathaniel emotionally so it was a good time to end the nightmare. I packed our belongings and moved in with my sister.

Less than three months after I left, Nate went into a rehabilitation program. I was happy for him but I knew I could never go back; it was too little, too late. That was nine years ago and the last I heard, he was still struggling with his addiction, living a life of misery. I have no regrets about leaving but I am saddened by what crack cocaine had done to my once wonderful husband—it turned him into a thief and a liar and ended our marriage.

Although getting married is associated with significant reductions in cigarette smoking, heavy drinking, and marijuana use for both men and women (Merline et al. 2008), family crises involving alcohol and/or drugs are not unusual. An additional concern is when it is the children, rather than the parents, who are involved in the drug abuse that can cause stress in the family. In 2010, Cameron Douglas (son of actor Michael Douglas) was sentenced to five years in prison for heroin possession and dealing cocaine and meth. The social policy section deals with alcohol abuse on campus.

As indicated in Table 14.1, drug use is most prevalent among 18- to 25-year-olds. [Note that this table does not reflect the abuse of prescription drugs such as Adderall.] Drug use among teenagers under age 18 is also high. Because teenage drug use is common, it may compound the challenge parents may have with their teenagers.

The Self-Assessment feature allows you to identify your motives for drinking alcohol.

Drug Abuse Support Groups Although treatments for alcohol abuse are varied, a combination of medications and behavioral interventions (for example, control for social context) may be used. Ruff et al. (2010) outlined a behavioral couples therapy program which involves the partners contracting to maintain

many campuses. But Wei et al. (2010) found attendance at alcohol-free parties and alcohol use indicates both heavy and light drinkers attend these parties.

Policies will continue to shift in reference to parental pressure to address the issue. Hustad et al. (2010) found that web-based alcohol prevention for freshmen students was effective in reducing the amount of alcohol freshmen drink. Some universities require that freshmen take the course in order to register for classes. Other new approaches include banning "beer pong" games (Keegan 2008) and informing parents of their sons' or daughters' alcohol or drug abuse on campus. An alcohol or drug infraction is reported to the parents, who may intervene early in curbing abuse.

Your Opinion?

1. To what degree do you believe drinking on your campus is an issue and policies should be developed to control it?

2. What policies might universities develop to reduce binge drinking?

3. Under what conditions should a student who abuses alcohol be expelled?

Sources

Bulmer, S. M., S. Irfan, R. Mugno, B. Barton, and L. Ackerman. 2010. Trends in alcohol consumption among undergraduate students at a northeastern public university, 2002–2008. *Journal of American College Health* 58: 383–390.

Coll, J. E., P. R. Draves, and M. E. Major. 2008. An examination of underage drinking in a sample of private university students. *College Student Journal* 42: 982–985.

Hustad, J. T. P., N.P. Barnett, B. Borsari, and K. M. Jackson. 2010. Web-based alcohol prevention for incoming college students: A randomized controlled trial *Addictive Behaviors* 35: 183–189.

Keegan, R. W. 2008. Beer pong's big splash. *Time*, August 18, 46–47.

Mallett, K. A., C. M. Lee, C. Neighbors, M. E. Larimer, and R. Turrisi. 2006. Do we learn from our mistakes? An examination of the impact of negative alcohol-related consequences on college students' drinking patterns and perceptions. *Journal of Studies on Alcohol* 67: 269–276.

Wei, J., N. P. Barnett, and M. Clark 2010 Attendance at alcohol-free and alcohol-service parties and alcohol consumption among college students. *Addictive Behaviors* 35: 572–579.

Alcohol poisoning can be lethal and happens on college campuses. This undergraduate has passed out and his friends wrote on him.

Chelsea E. Curry

TABLE 14.1	Current Drug Use by Type of Drug and Age Group		
Type of Drug Used	**Age 12 to 17**	**Age 18 to 25**	**Age 26 to 34**
Marijuana and hashish	7.0%	17.0%	9.0%
Cocaine	.4%	1.5%	1.5%
Alcohol	14.6%	61.2%	no data
Cigarettes	9.1%	35.7%	no data

Source: Adapted from *Statistical Abstract of the United States 2011*, 130th ed. Washington, DC: U.S. Bureau of the Census, Table 203.

Read the list of reasons people sometimes give for drinking alcohol. Thinking of all the times you drink, how often would you say you drink for each of the following reasons? (Write the appropriate number after reading each item.)

1 = almost never/never
2 = some of the time
3 = half of the time
4 = most of the time
5 = almost always

_____ 1. Because it helps me to enjoy a party
_____ 2. To forget my worries
_____ 3. Because the effects of alcohol feel good
_____ 4. Because of pressure from friends
_____ 5. To be sociable
_____ 6. To help my depression or nervousness
_____ 7. Because it is exciting
_____ 8. To avoid disapproval for not drinking
_____ 9. To make social gatherings more fun
_____ 10. To cheer me up
_____ 11. To get high
_____ 12. To fit in with the group
_____ 13. To improve parties and celebrations
_____ 14. To make me feel more self-confident
_____ 15. Because it gives me a pleasant feeling
_____ 16. To be liked
_____ 17. To celebrate a special occasion with friends
_____ 18. To help me forget about problems
_____ 19. Because it is fun
_____ 20. To avoid feeling left out

Scoring

The four basic drinking motives are social, coping, enhancement, and conformity. The items for these and the average scores from 1,243 respondents follow. The lowest score reflecting each motive would be 1 = never; the highest score reflecting each motive would be 5 = always. The most frequent motive for women and men is to be sociable. The least frequent motive for women and men is to conform. To compare your score with other respondents, add the numbers you wrote for each of the social, coping, enhancement, and conformity reasons identified and divide by five. For example, to ascertain the degree to which your motivation for drinking alcohol is to be sociable, add the numbers you wrote for items 1, 5, 9, 13, and 17 and divide that number by five.

Basic Reasons for drinking are social, coping, enhancement, and conformity. Items 1, 5, 9, 13, 17 pertain to social drinking; items 2, 6, 10, 14, 18 pertain to coping drinking; items 3, 7, 11, 15, 19 pertain to enhancement drinking; and items 4, 8, 12, 16, 20 pertain to conformity drinking. In this study, female respondents scored 2.29 for social drinking; 1.61 for coping drinking; 1.99 for enhancement drinking; and 1.34 for conformity drinking. Male respondents scored 2.63 for social drinking; 1.59 for coping drinking; 2.33 for enhancement drinking; and 1.43 for conformity drinking.

Source

© 1994 by the American Psychological Association. Adapted with permission. Cooper, M. Lynne. 1994 Motivations for Alcohol Use Among Adolescents: Development and Validation of a four-factor model. *Psychological Assessment* 6:117–128.

sobriety, involvement in a 12 step program, and regular monitoring of their behavior. This therapy does not focus solely on the alcohol abuser, but includes the spouse as an important context of support. Positive outcomes have been observed.

Abraham and Roman (2010) noted that in addition to Disulfiram, the U.S. Food and Drug Administration has approved three AUD (alcohol use disorder) medications—tablet naltrexone (Revia), acamprosate (Campral), and an extended-release injectable suspension of naltrexone (Vivitrol). The latter is being used more frequently with some success.

The support group Alcoholics Anonymous (AA; www.alcoholics-anonymous.org) has also been helpful. There are over 15,000 AA chapters nationwide; one in your community can be found through the Yellow Pages. The only requirement for membership is the desire to stop drinking.

Former abusers of drugs (other than alcohol) also meet regularly in local chapters of Narcotics Anonymous (NA), to help each other continue to be drug-free. Patterned after Alcoholics Anonymous, the premise of NA is that the best person to help someone stop abusing drugs is someone who once abused drugs. NA members of all ages, social classes, and educational levels provide a sense of support for each other to remain drug-free.

WHAT IF?

What if Your Partner is an Alcoholic?

If your partner is an alcoholic, it is important to recognize that you can do nothing to stop your partner from drinking. Your partner must "hit bottom" or have an epiphany and make a personal decision to stop. In the meantime, you can make choices to join Al-Anon to help you cope with your partner's alcoholism, regard the alcoholism as a "sickness" and stay with your partner, or end the relationship. Most individuals stay with their partner because there is an enormous emotional cost of leaving. However, some pay the price. Only you can decide.

Al-Anon is an organization that provides support for family members and friends of alcohol abusers. Spouses and parents of substance abusers learn how to live with and react to living with a substance abuser.

Parents who abuse drugs may also benefit from the Strengthening Families Program, which provides specific social skills training for both parents and children. After families attend a five-hour retreat, parents and children are involved in face-to-face skills training over a four-month period. A twelve-month follow-up has revealed that parenting skills remained improved and that reported heroin and cocaine use had declined.

Al-Anon organization that provides support for family members and friends of alcohol abusers.

Death of Family Member

Even more devastating than drug abuse are family crises involving death—of one's child, parent, or loved one (we discuss the death of one's spouse in Chapter 16 on Relationships in the Later Years). The crisis is particularly acute when the death is a suicide.

Death of One's Child   A parent's worst fear is the death of a child. Most people expect the death of their parents but not the death of their children. Sue Molhan (2010) experienced the death of her 20-year-old son, Stephen, the victim of a random shooting. The perpetrator was "high on drugs & alcohol" and began shooting people at random, one of whom was Stephen, who was in his car stopped at a stop sign. The shooter served 20 years in prison during which time Molhan met with him for "answers only he could provide and the possibility of closure." She found him contrite and could not forgive him, yet tutored him in prison to complete his GED to ensure that his lack of education would no longer keep him on the streets, into drugs, and repeating his former pattern.

But her anguish never ends. While Molhan has moved on with her life, she thinks of her son daily and laments the life he was not allowed to have. In a spiritual way, she "keeps in touch."

Mothers and fathers sometimes respond to the death of their child in different ways. When they do, the respective partners may interpret these differences in negative ways, leading to relationship conflict and

Chelsea E. Curry

This mother's son was shot at age 21 through a random act of violence. She reports that she thinks about her son every day.

Family Crisis Examples   427

unhappiness. To deal with these differences, spouses need to be patient and practice tolerance in allowing both to grieve in their own way.

Death of One's Parent　Terminally ill parents may be taken care of by their children. Such care over a period of years can be emotionally stressful, financially draining, and exhausting. Hence, by the time the parent dies, a crisis has already occurred.

Reactions to the death of a loved one is not something one "gets over." Burke et al. (1999) noted that grief is not a one-time experience that people adjust to and move on. Rather, for some, there is "**chronic sorrow**," where grief-related feelings occur periodically throughout the lives of those left behind. The late Paul Newman was asked how he got over the death of his son who overdosed. He replied, "You never get over it." Grief feelings may be particularly acute on the anniversary of the death or when the bereaved individual thinks of what might have been had the person lived. Burke et al. (1999) noted that 97% of the individuals in one study who had experienced the death of a loved one two to twenty years earlier met the criteria for chronic sorrow. Field et al. (2003) also observed bereavement-related distress five years after the death of a spouse.

Suicide of Family Member　In 2010, Michael, the 18-year-old adopted son of Marie Osmond, leaped to his death from an eight floor apartment. Suicide is a devastating crisis event for families, and is not that unusual. Annually there are 31,000 suicides (750,000 attempts), and each suicide immediately affects at least six other people in that person's life. These effects include depression (for example, grief), physical disorders (for example, shingles due to stress), and social stigma (for example, the person is viewed as weak and the family as a failure in not being able to help with the precipitating emotional problems) (De Castro and Guterman 2008). Schum (2007) identified family members who experience the suicide of a family member as having "the worst day of their lives."

Joan Rivers talked of the suicide of her husband in 2000 and the anger she felt in the aftermath. "He took the easy way out and left me and Melissa to pick up the pieces." She said she coped with his suicide by using it in her comedic routine. "He left a note telling me to visit him every day. So I had him cremated and his ashes sprinkled in Neiman Marcus" (NPR 2010).

People between 15 and 19 years old; homosexuals; males; those with a family history of suicide, mood disorders, or substance abuse; and those with past history of child abuse and parental sex abuse are more vulnerable to suicide than others (Melhem et al. 2007). As noted earlier, Taliaferro et al. (2008) found that vigorous exercise and involvement in sports are associated with lower rates of suicide among adolescents.

Suicide is viewed as a "rational act" in that the person feels that suicide is the best option available at the time. Therapists view suicide as a "permanent solution to a temporary problem" and routinely call 911 to have people hospitalized or restrained who threaten suicide or who have been involved in an attempt.

Adjustment to the suicide of a family member takes time. The son of physician T. Schum committed suicide, which set in motion a painful adjustment for Dr. Schum. "You will never get over this but you can get through it," was a phrase Dr. Schum found helpful (Schum 2007). Survivors of Suicide is also a helpful support group. Part of the recovery process is accepting that one cannot stop the suicide of those who are adamant about taking their own life and that one is not responsible for the suicide of another. Indeed, family members often harbor the belief that they could have done something to prevent the suicide. Singer Judy Collins lost her son to suicide and began to attend a support group for people who had lost a loved one to suicide. At a group she attended, one of the members in the group answered the question of whether there was something she could have done with a resounding *no*:

Chronic sorrow grief-related feelings that occur periodically throughout the lives of those left behind.

I was sitting on his bed saying, "I love you, Jim. Don't do this. How can you do this?" I had my hand on his hand, my cheek on his cheek. He said excuse me, reached his other hand around, took the gun from under the pillow, and blew his head off. My face was inches from his. If somebody wants to kill himself or herself, there is nothing you can do to stop them. (Collins 1998, 210)

Marriage and Family Therapy

University students have a positive view of marriage therapy. In a study of 288 undergraduate and graduate students, 93% of the females and 82% of the males agreed that, "I would be willing to see a marriage counselor before I got a divorce" (Dotson-Blake et al. 2010).

Couples might consider marriage and family therapy rather than continuing to drag through dissatisfaction related to a crisis event. Signs to look for in your own relationship that suggest you might consider seeing a therapist include feeling distant and not wanting or being unable to communicate with your partner, avoiding each other, feeling depressed, drifting into a relationship with someone else or having an affair, increased drinking, and privately contemplating separation or divorce.

If you are experiencing one or more of these symptoms, it may be wise to consider seeing a counselor with your partner to identify what behaviors each of you need to engage in to improve your relationship. Tilden et al. (2010) emphasized the relationship between a spouse being depressed and the couple being unhappy. As a result of therapeutic intervention, the depression decreased from the initial evaluation to the three year follow up. And, as the depression abated, the relationship improved.

A relationship is like a boat. A small leak unattended can become a major problem and sink the boat. Marriage therapy sometimes serves to reverse relationship issues early by helping the partners to sort out values, make decisions, and begin new behaviors so that spouses can start feeling better about each other.

Availability of Marriage and Family Therapists

There are around 50,000 marriage and family therapists in the United States. Their professional roles include medical doctors (MDs), nurses, psychologists, social workers, professional counselors, and those who focus exclusively on marriage and family therapy. Moore et al. (2011) compared the outcome of couples and clients who are treated by the various professionals and found the lowest dropout rates and recidivism among marriage/family therapists (who were viewed as more cost effective than the other professionals).

About 40% of "marriage counselors" are clinical members of the American Association for Marriage and Family Therapy (AAMFT). Currently there are fifty-seven masters, nineteen doctoral, and sixteen postgraduate programs accredited by the AAMFT. Forty-eight states require a license to practice marriage and family therapy. Del Rio (2010) recommended renaming the organization American Association for Couples and Family Therapy to be more inclusive.

Therapists holding membership in AAMFT have had graduate training in marriage and family therapy and a thousand hours of direct client contact (with 200 hours of supervision). Lee and Nichols (2010) emphasized that one only needs a master's degree to practice marriage and family therapy. Clients are customers and should feel comfortable with their therapists and the progress they are making. If they don't, they should switch therapists.

Marriage therapy is expensive—$100 to $125 an hour is not uncommon. Managed care has resulted in some private therapists lowering their fees so as to compete with what insurance companies will pay. Marriage therapists generally see both the husband and the wife together (called conjoint therapy).

Effectiveness of Behavioral Couple Therapy versus IBCT

Members of AAMFT use more than twenty different treatment approaches (Northey 2002). Most therapists (31%) report that they use either a behavioral or "cognitive-behavioral" approach. A behavioral approach (also referred to as **behavioral couple therapy** or BCT) means that the therapist focuses on behaviors the respective spouses want increased or decreased, initiated or terminated, and negotiate behavioral exchanges between the partners. Birditt et al. (2010) examined self-conflict behaviors in 373 couples over 16 years of marriage and found that destructive behavior the first year was associated with subsequent divorce. Destructive behaviors included four items: "I yelled and shouted at my spouse," "I insulted my spouse or called him or her names," "I brought up things that happened long ago," and "I had to have the last word." Behavioral therapeutic intervention was recommended.

The following is an example of a behavioral contract as a "homework assignment for the following week" that behavior therapists give to distressed couples (Crisp and Knox, 2009). The contract assumes that the partners argue frequently, never compliment each other, no longer touch each other, and do not spend time together. The contract calls for each partner to make no negative statements to the other, give two compliments per day to the other, hug or hold each other at least once a day, and allocate Saturday night to go out to dinner alone with each other. On the contract, under each day of the week, the partners would check that they did what they agreed and return to the therapist. Partners who change their behavior toward each other often discover that the partner changes also and there is a new basis for each to feel better about each other and their relationship.

Behavior Contract for Partners*

Name of Partners _____ Date:_Week of *June 8–14*
Behaviors each partner agrees to engage in and Days of Week

	Mon.	Tues.	Wed.	Thurs.	Fri.	Sat.	Sun.
1. No negative statements to partner	☐	☐	☐	☐	☐	☐	☐
2. Compliment partner twice each day	☐	☐	☐	☐	☐	☐	☐
3. Hug or hold partner once a day	☐	☐	☐	☐	☐	☐	☐
4. Out to dinner Saturday night	☐	☐	☐	☐	☐	☐	☐

*From B. Crisp and D. Knox. 2009. *Behavioral family therapy*. Durham, NC: Carolina Academic Press.

Sometimes clients do not like behavior contracts and say to the behavior therapist, "I want my partner to compliment me and hug me because my partner wants to, not because you wrote it down on one of these silly contracts." The behavior therapist acknowledges the desire for the behavior to come from the heart of the partner. Indeed it is, in that the partner has a choice about whether to please the partner. The partner could certainly say, "I don't care what my partner thinks or feels and I'm not doing anything to make things better. If you have to work at a relationship, it's not worth having." Of course, this position means the relationship is over (but not cooperating is unlikely because couples who come for marriage counseling are usually motivated to do things to improve the relationship).

Behavioral couple therapy (BCT) therapeutic focus on behaviors the respective spouses want increased or decreased, initiated or terminated.

Chapter 14 Stress and Crisis in Relationships

Cognitive-behavioral therapy, also referred to as **integrative behavioral couple therapy (IBCT)**, emphasizes a focus on the cognitions. South et al. (2010) found that spouses were most happy when they perceived the positive and negative behaviors of their spouse occurred at the desirable frequency—high frequency of positives and low frequency of negatives.

Christensen et al. (2006) provided follow-up data after two years on 130 couples who were part of a study that compared traditional couple behavior therapy (TCBT) with integrative behavioral couple therapy (IBCT). The researchers found that 60% of the couples who received TCBT reported significant improvement two years later compared to 69% of couples who received IBCT. They determined behavioral therapy involving cognitions is more effective.

Whether a couple in therapy remain together will depend on their motivation to do so, how long they have been in conflict, the severity of the problem, and whether one or both partners are involved in an extramarital affair. Two moderately motivated partners with numerous conflicts over several years are less likely to work out their problems than a highly motivated couple with minor conflicts of short duration. Severe depression or alcoholism on the part of either spouse is a factor that will limit positive marital and family gains. In general, these issues must be resolved individually before the spouses can profit from marital therapy.

Some couples come to therapy with the goal of separating amicably. The therapist then discusses the couple's feelings about the impending separation, the definition of the separation (temporary or permanent), the "rules" for their interaction during the period of separation (for example, see each other, date others), and whether to begin discussions with a divorce mediator or attorneys.

One alternative (to therapy) for enhancing one's relationship is the Association for Couples in Marriage Enrichment (ACME). This organization provides conferences for couples to enrich their marriage.

Computerized Internet Therapy

Stallard et al. (2010) reviewed a number of computerized therapeutic interventions available on the Internet, particularly those based on cognitive behavior therapy (CBT). These include "Beating the Blues," "COPE," "Overcoming Depression," and "MoodGym." While the results of these computer interventions have been generally positive, they were designed to treat depression and anxiety disorders rather than interpersonal conflict. A gap in intervention offerings is that of offering computerized help to couples online. Potential advantages are accessibility, convenience, privacy, and reduced embarrassment from meeting a marriage and family therapist.

Telerelationship Therapy

An alternative to mechanized therapeutic programs is **telerelationship therapy**, a variation of face to face therapy which uses Skype. Both therapist and couple log on to Skype, where each can see and hear the other so that the session is conducted online. Terms related to telerelationship therapy are telepsychology, telepsychiatry, and virtual therapy (Magaziner 2010). Telerelationship therapy allows couples to connect to a marriage/family therapist independent of where they live (e.g., isolated rural areas), the availability of transportation, and time (i.e., sessions can be scheduled outside the 9 to 5 block). While the efficacy of telerelationship therapy compared with face to face therapy continues to be researched, evidence regarding its value in individual therapy has been documented (Nelson and Bui 2010; Rees and Hawthornthwaite 2004).

Some Caveats about Marriage and Family Therapy

In spite of the potential benefits of marriage therapy, some valid reasons exist for not becoming involved in such therapy. Not all spouses who become involved in marriage therapy regard the experience positively. Some feel that their marriage

Integrative behavioral couple therapy (IBCT) therapy which focuses on the cognitions or assumptions of the spouses, which impact the way spouses feel and interpret each other's behavior.

Telerelationship therapy therapy sessions conducted online, often through Skype, where both therapist and couple can see and hear each other.

This therapist is conducting a session with a couple over the Internet (referred to as telerelationship therapy or Skype therapy).

David Knox

is worse as a result. Reasons some spouses cite for negative outcomes include saying things a spouse can't forget, feeling hopeless at not being able to resolve a problem "even with a counselor," and feeling resentment over new demands a spouse makes in therapy.

Therapists may also give clients an unrealistic picture of loving, cooperative, and growing relationships in which partners always treat each other with respect and understanding, share intimacy, and help each other become whomever each wants to be. In creating this idealistic image of the perfect relationship, therapists may inadvertently encourage clients to focus on the shortcomings in their relationship and to expect more of their marriage than is realistic. Dr. Robert Sammons (2010) calls this his first law of therapy: "That spouses always focus on what is missing rather than what they have . . . indeed the only thing that is important to couples in therapy is that which is missing. A new focus of what the person does that pleases them rather than what they are missing is needed."

Couples and families in therapy must also guard against assuming that therapy will be a quick and easy fix. Changing one's way of viewing a situation (cognitions) and behavior requires a deliberate, consistent, relentless commitment to make things better. Without it, couples are wasting their time and money.

As well, couples who become involved in marriage counseling may also miss work, have to pay for child care, and be "exposed" at work if they use their employer's insurance policy to cover the cost of therapy. Though these are not reasons to decide against seeing a counselor, they are issues that concern some couples.

The Future of Stress and Crisis in Relationships

Stress and crisis will continue to be a part of relationships. No spouse, partner, marriage, or family is immune. A major source of stress will be economic—the difficulty in securing and maintaining employment and sufficient income to take care of the needs of the family.

432   **Chapter 14** Stress and Crisis in Relationships

Relationship partners will also show resilience to rise above whatever crises happens. The motivation to do so is strong, and having a partner to share one's difficulties reduces the sting. As noted, it is always one's perception of an event, not the event itself, which will determine the severity of a crisis and the capacity to cope with and overcome it.

Summary

What is stress and what is a crisis event?

Stress is a reaction of the body to substantial or unusual demands (physical, environmental, or interpersonal). Stress is a process rather than a state. A crisis is a situation that requires changes in normal patterns of behavior. A family crisis is a situation that upsets the normal functioning of the family and requires a new set of responses to the stressor. Sources of stress and crises can be external (for example, hurricane, tornado, downsizing, military separation) or internal (for example, alcoholism, extramarital affair, Alzheimer's disease).

Family resilience is when family members successfully cope under adversity, which enables them to flourish with warmth, support, and cohesion. Key factors include positive outlook, spirituality, flexibility, communication, financial management, family shared recreation, routines or rituals, and support networks.

The ABCX model is a theoretical model of looking at family stress. *A* is the stressor event (for example, a hurricane), which interacts with *B*, the family's coping ability, or crisis-meeting resources (such as money, connections, spirituality). Both *A* and *B* interact with *C*, the family's appraisal or perception of the stressor event ("we can get through this and be better for it"). *X* is the family's adaptation to the crisis (for example, the family survives the hurricane).

What are positive stress management strategies?

Changing one's basic values and perspective is the most helpful strategy in reacting to a crisis. Viewing ill health as a challenge, bankruptcy as an opportunity to spend time with one's family, and infidelity as an opportunity to improve communication are examples. Other positive coping strategies are exercise, adequate sleep, love, religion, friends or relatives, humor, education, and counseling. Still other strategies include intervening early in a crisis, not blaming each other, keeping destructive impulses in check, and seeking opportunities for fun. Some harmful strategies include keeping feelings inside, taking out frustrations on others, and denying or avoiding the problem.

What are five of the major family crisis events?

Some of the more common crisis events that spouses and families face include physical illness, an extramarital affair, unemployment, alcohol or drug abuse, and the death of one's spouse or children. An extramarital affair is the second most stressful crisis event for a family (abuse is number one). Surviving an affair involves forgiveness on the part of the offended spouse to grant a pardon to the offending spouse, to give up feeling angry, and to relinquish the right to retaliate against the offending spouse. In exchange, an offending spouse must take responsibility for the affair, agree not to repeat the behavior, and grant the partner the right to check up on the offending partner to regain trust. The best affair prevention is a happy and fulfilling marriage as well as avoiding intimate conversations with members of the other sex and a context (for example, where alcohol is consumed and being alone), which are conducive to physical involvement. The occurrence of a "midlife crisis" is reported by less than a quarter of adults in the middle years. Those who did experience a crisis were going through a divorce.

What help is available from marriage and family therapists?

There are over 50,000 marriage and family therapists in the United States. About 40% are clinical members of the American Association for Marriage and Family Therapy (AAMFT). Whether a couple in therapy remain together will depend on their motivation to do so, how long they have been in conflict, and the severity of the problem. Two moderately motivated partners with numerous conflicts over several years are less likely to work out their problems than a highly motivated couple with minor conflicts of short duration. Couples tend to benefit from early intervention. About 70% of couples involved in integrative couple behavior therapy (ICBT) (which involves behavior change and cognitions) report continued significant gains two years after the end of treatment. Telerelationship therapy is an alternative to face to face therapy.

What caveats should be kept in mind for becoming involved in marriage/family therapy?

Not all spouses who become involved in marriage therapy regard the experience positively. Some feel that their marriage is worse as a result. Reasons some spouses cite for negative outcomes include saying things a spouse can't forget, feeling hopeless at not being able to resolve a problem "even with a counselor," and feeling resentment over new demands a spouse makes in therapy. Therapists may also give clients an unrealistic picture of loving, cooperative, and growing relationships in which partners always treat each other with respect and understanding, share intimacy, and help each other become whoever each wants to be. In creating this idealistic image of the perfect relationship, therapists may inadvertently encourage clients to focus on the shortcomings in their relationship and to expect more of their marriage than is realistic.

Key Terms

Al-Anon

Alienation of affection

Behavioral couple therapy (BCT)

Chronic sorrow

Coolidge effect

Crisis

Extradyadic involvement

Extramarital affair

Family resiliency

Integrative behavioral couple therapy (IBCT)

Palliative care

Resiliency

Snooping

Stress

Telerelationship therapy

Web Links

American Association for Marriage and Family Therapy
http://www.aamft.org

Association for Applied and Therapeutic Humor
http://www.aath.org

Association for Applied Psychophysiology and Biofeedback
http://www.aapb.org

Better Marriages
http://www.bettermarriages.org/

Dear Peggy.com Extramarital Affairs Resource Center
http://www.vaughan-vaughan.com/

Infidelity
http://www.infidelity.com/

Mental Health
http://www.bringchange2mind.org/

Red Hat Society
http://www.redhatsociety.com/

Volunteering
http://www.volunteermatch.org/

CHAPTER 15

Divorce and Remarriage

Families break up when people take hints you don't intend and miss hints you do intend.

Robert Frost, American poet

Mike Stoneman

Learning Objectives

Specify the definition of divorce, its prevalence, and low-risk divorce occupations.

Discuss the macro and micro factors of divorce.

Review the emotional and economic consequences of divorce for spouses.

Identify the criteria for deciding joint custody.

Summarize the factors associated with minimizing negative divorce consequences.

Review how one can have a "successful" divorce.

Know the issues involved in a remarriage.

Be aware of the unique aspects of stepfamilies.

List the developmental tasks of stepfamilies.

Predict the future of divorce and remarriage.

1.   One of the highest risk occupations for divorce is that of an attorney.

2.   By keeping income stable, children suffer less when moving from a home in which they live with two parents to a single parent home.

3.   Over 80% of college graduates who wait until they are age 26 or older to marry are still together 20 years after the wedding.

4.   Stepfamily relationship education workshops do NOT work because spouses are too hostile to the idea.

5.   Typical self-help stepfamily books are rich with research and very helpful for new stepfamily members.

Answers: 1. F   2. T   3. T   4. F   5. F

Being divorced is like being hit by a Mack truck. If you live through it, you start looking very carefully to the right and to the left.

Jean Kerr, playwright

Divorce the legal ending of a valid marriage contract.

The photo which begins this chapter is that of a couple who were married for 13 years, had two children, divorced, were apart for 10 years (during which time they had other lovers), and remarried each other. Even during the ten years they were divorced, they remained friends and involved parents for their two children. Their experience is a reminder that not all divorces result in sustained conflict.

Many college students have parents who are divorced. In a sample of 2,922 undergraduates, 26% reported that their parents were divorced (Knox and Hall 2010). In this chapter, we look at the social and individual causes of divorce, the consequences for the spouses and children, and ways to make the ending of one's marriage as painless and life-enhancing as possible. We begin by defining divorce and looking at its prevalence.

Divorce

Divorce is the legal ending of a valid marriage contract. This definition does not include the "informal divorces" which occur when long-term cohabitation relationships end (Amato 2010).

Prevalence

A common way of estimating the prevalence of divorce is to identify the percentage of married people who eventually get divorced. The problem with this statistic is the period of time considered in identifying those who divorced. The percentage of those who divorced within five years would be lower than the percentage of those who divorced within a ten-year period. Current estimates should include the demographics of a couple, since they are related to divorce probability (Parker-Pope 2010). For college graduates who wed in the 1980s at age 26 or older, 82% were still married 20 years later. For college grads who married when the partners were less than age 26, 65% were still married 20 years later. If the couple were high school graduates and married when the partners were below age 26, 49% were still married. Hence, education and age at marriage influence one's chance of divorce. In addition, race (whites less likely to divorce), religion (devout less likely to divorce), and previous marriage (if married before more likely to divorce) also influence one's chance of divorce.

National **Data**

The lifetime probability of divorce for couples getting married today is between 40 and 50% (Cherlin 2010).

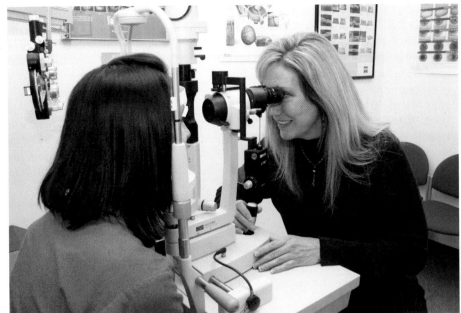
Chelsea E. Curry

Of all occupations, optometrists have the lowest risk of divorce.

Divorce rates have been relatively stable in recent years. The principal factor for the lack of increase is that people are delaying marriage so that they are older at the time of marriage. Indeed, the older a person at the time of marriage, the less likely the person is to divorce. In addition, divorce rates have dropped as cohabitation rates have increased because people who live together and break up do not add to the divorce rate (Clarke-Stewart and Brentano 2006).

National **Data**

About 1 million divorces occur each year (*Statistical Abstract of the United States 2011*, Table 78).

High- and Low-Risk Occupations for Divorce

Shawn and Aamodt (2009) identified high- and low-risk occupations associated with divorce. The highest risk occupations are those of being a dancer/choreographer (43%) or bartender (38%); among the lowest are clergy and optometrists at 5% and 4%, respectively. Other examples include nurses (29%), sociologists (23%), psychologists (19%), authors/teachers (15%), and lawyers (12%). Regardless of one's occupation, there are both macro and micro factors which help to explain divorce in U.S. society.

End an Unsatisfactory Relationship?

Breaking up is never easy. Almost a third of 279 undergraduates reported that they "sometimes" remained in relationships they thought should end (31%) or that became "unhappy" (32.3%). This finding reflects the ambivalence students sometimes feel in ending an unsatisfactory relationship and their reluctance to do so (Knox et al. 2002). All relationships have difficulties, and all necessitate careful consideration of various issues before they are ended. Light and Ahn (2010) noted that getting divorced is a risky behavior and that those who do so have a considerable tolerance for taking a risk (also that women have more to lose).

Before you pull the plug on your relationship, consider the following:

1. *Consider reviving and improving the relationship.* In some cases, people end relationships and later regret having done so. Setting unrealistically high standards may eliminate an array of individuals who might be superb partners, companions,

Ah, yes, divorce . . . from the Latin word, meaning to rip out a man's genitals through his wallet.

Robin Williams, comedian

Get your tongue outta my mouth 'cause I'm kissing you goodbye.

Waylon Jennings, country western singer

Divorce   437

and mates. If the reason for ending a relationship is conflict over an issue or set of issues, an alternative to ending the relationship is to attempt to resolve the issues through negotiating differences, compromising, and giving the relationship more time. (We do not recommend giving an abusive relationship more time, as abuse, once started, tends to increase in frequency and intensity. Nor do we recommend staying in a relationship in which there are great differences in levels of attachment.)

2. *Acknowledge and accept that terminating a relationship will be difficult and painful.* Yazedjian and Toews (2010) studied college breakups and found that they were associated with depression and a drop in grades. Both partners are usually hurt, although the person with the least interest in maintaining the relationship will suffer less.

3. *In talking with your partner, blame yourself for the end.* One way to end a relationship is to blame yourself by giving a reason that is specific to you ("I need more freedom," "I want to go to graduate school in another state," "I'm not ready to settle down," and so on). If you blame your partner or give your partner a way to make things better, the relationship may continue because your partner may promise to make the changes and *you* may feel obligated to give your partner a second chance.

4. *Cut off the relationship completely.* If you are the person ending the relationship, it will probably be easier for you to continue to see the other person without feeling too hurt. However, the other person will probably have a more difficult time and will heal faster if you stay away completely. Alternatively, some people are skilled at ending love relationships and turning them into friendships. Though this is difficult and infrequent, it can be rewarding across time for the respective partners.

5. *Learn from the terminated relationship.* Included in the reasons a relationship may end are behaviors of being too controlling; being oversensitive, jealous, or too picky; cheating; fearing commitment; and being unable to compromise and negotiate conflict. Some of the benefits of terminating a relationship are recognizing one's own contribution to the breakup and working on any characteristics that might be a source of problems. Otherwise, one might repeat the process.

6. *Allow time to grieve over the end of the relationship.* Ending a love relationship is painful. It is okay to feel this pain, to hurt, to cry. Allowing yourself time to experience such grief will help you heal for the next relationship. Recovering from a serious relationship can take twelve to eighteen months.

7. *Clean your Facebook page.* Angry spouses sometimes post nasty notes about their ex on their Facebook page which can be viewed by the ex's lawyer. Also, if there are any incriminating photos of indiscreet encounters, drug use, wild parties or the like, these can be used in court and should be purged. Twitter and MySpace postings should also be scrutinized.

Remain Unhappily Married or Divorce?

Suppose you are unhappily married. Is your best choice to stay in the relationship or take your chances and divorce? Hawkins and Booth (2005) analyzed longitudinal data of spouses in unhappy marriages over a twelve-year period and found that those who stayed unhappily married had lower life satisfaction, self-esteem, and overall health compared to those who divorced, whether or not they remarried. Similarly, Gardner and Oswald (2006) found that the psychological functioning and happiness of spouses going through divorce improved after the divorce—hence divorce was good for them. However, Waite et al. (2009) compared those who were unhappily married (over a five-year period) but who stayed married against those who divorced and remarried. The researchers did not find more positive outcomes for the latter. Hence, the data are unclear if remaining unhappily married or divorcing and remarrying has a more positive outcome for the spouses.

The Al and Tipper Gore Divorce—a Surprise?

When Al and Tipper Gore announced in 2010 that their marriage of 40 years was ending, the reaction of the American public ranged from sadness to shock. What is of note is not that this couple divorced, but that more couples of 40 years or more do not generally divorce. Reasons why we should not be surprised include:

1. *Marriage is a socially constructed relationship.* Societies create marriage to bond men and women together legally for the care and protection of children. Among lower animals, most (over 90%) go their merry way after sex. But because human infants have a sustained period of dependence, societies attempt to lock in the parents to take care of their offspring. There is no biological reason for two people to live in the same nest for 40 years or even 4 years. Social programming is much weaker than biological programming.

2. *Marital satisfaction declines over time.* Some studies on marital satisfaction reflect a diminution of reported marital happiness over time. We should not be surprised that 40 years of Gore togetherness had an eroding effect on some of their enjoyment of being together.

3. *Satiation is like the law of gravity.* Referred to in detail later in this chapter, satiation means that individuals tire of everything with repeated exposure. Whether food, a new house, or each other, nothing can remain "new," and it is the "new" which revives a person's wake from boredom. We are not suggesting that the Gores did not love each other or enjoy their companionship; we are suggesting that after 40 years together, the principle of satiation will be operative and take its toll for all couples.

4. *Social disapproval was minimal.* With both sets of parents dead, children out of the house and into their own lives, there was no longer the fear of abject disapproval (from people who mattered) for the Gores getting a divorce. Primary group relationships (those with one's own parents and children) exercise the most social control over one's behavior, not secondary group relationships (such as those with friends and the American public). The Gores knew their marriage would be on the cover of *USA Today* and *People Magazine* once, then, within a month, nothing. Hence, no sustained disapproval from primary group members and the public losing interest—fast—created an environment for divorce.

Mark Wilson/Getty Images News/Getty Images

The divorce of Al and Tipper Gore should not be a surprise.

5. *Each spouse had alternatives.* Both Al and Tipper are articulate attractive people that others would be interested in having a personal/romantic relationship with. The chance for a new love before old age and death could have its appeal.

6. *Money: The Gores could afford a divorce.* While some stay together out of economic survival, the $8.8 million house the Gores lived in reflects that they are well enough off to withstand splitting everything with neither suffering economically. Al will pay Tipper spousal support since he has the higher income… but he can afford it.

Divorce is a sad time for the couple ending their relationship. But the Gores made a private decision to make public what had already occurred behind closed doors—to go their separate ways. We should not be surprised.

Macro Factors Contributing to Divorce

Sociologists emphasize that social context creates outcomes. This is best illustrated in the statistic that the Puritans in Massachusetts, from 1639 to 1760, averaged only one divorce per year (Morgan 1944). The social context of that era involved strong pro-family values and strict divorce laws, with the result that divorce was almost nonexistent. In contrast, divorce occurs more frequently today as a result of various structural and cultural factors, also known as macro factors.

Increased Economic Independence of Women

In the past, an unemployed wife was dependent on her husband for food and shelter. No matter how unhappy her marriage was, she stayed married because she was economically dependent on her husband. Her husband literally represented her lifeline. Finding gainful employment outside the home made it possible for a wife to afford to leave her husband if she wanted to. Now that about three-fourths of wives are employed, fewer wives are economically trapped in unhappy marriage relationships. As we noted earlier, a wife's employment does not increase the risk of divorce in a happy marriage. However, it does provide an avenue of escape for women in unhappy or abusive marriages (Kesselring and Bremmer 2006).

Employed wives are also more likely to require an egalitarian relationship; although some husbands prefer this role relationship, others are unsettled by it. Another effect of a wife's employment is that she may meet someone new in the workplace so that she becomes aware of an alternative to her current partner. Finally, unhappy husbands may be more likely to divorce if their wives are employed and able to be financially independent (less alimony and child support).

Changing Family Functions and Structure

Many of the protective, religious, educational, and recreational functions of the family have been largely taken over by outside agencies. Family members may now look to the police for protection, the church or synagogue for meaning, the school for education, and commercial recreational facilities for fun rather than to each other within the family for fulfilling these needs. The result is that, although meeting emotional needs remains a primary function of the family, fewer reasons exist to keep a family together.

In addition to the changing functions of the family brought on by the Industrial Revolution, the family structure has changed from that of larger extended families in rural communities to smaller nuclear families in urban communities. In the former, individuals could turn to a lot of people in times of stress; in the latter, more stress necessarily falls on fewer shoulders.

Liberal Divorce Laws

All states recognize some form of **no-fault divorce**—where neither party is identified as the guilty party or the cause of the divorce (for example, due to adultery). In effect, divorce is granted after a period of separation ranging from six

No-fault divorce neither party is identified as the guilty party or the cause of the divorce.

weeks to twelve months. Nevada requires the short-est waiting period of six weeks. Most other states re-quire from six to twelve months. The goal of no-fault divorce has been to try to make divorce less acrimoni-ous. However, this has not always been successful as spouses who divorce may still fight over custody of the children, child support, spouse support, and division of property.

Fewer Moral and Religious Sanctions

Many priests and clergy recognize that divorce may be the best alternative in particular marital relationships and attempt to minimize the guilt that congrega-tional members may feel at the failure of their marriage. Churches increasingly embrace single and divorced or separated individuals, as evidenced by "divorce adjustment groups."

More Divorce Models

As the number of divorced individuals in our society increases, the probability increases that a person's friends, parents, siblings, or children will be divorced. The more divorced people a person knows, the more normal divorce will seem to that person. The less deviant the person perceives divorce to be, the greater the probability the person will divorce if that person's own marriage becomes strained. Divorce has become so common that numerous websites exclusively for divorced individuals are available (for example, www.heartchoice.com/divorce).

Mobility and Anonymity

When individuals are highly mobile, they have fewer roots in a community and greater anonymity. Spouses who move away from their respective family and friends often discover that they are surrounded by strangers who don't care if they stay married or not. Divorce thrives when pro-marriage social expecta-tions are not operative. In addition, the factors of mo-bility and anonymity also result in the removal of a consistent support system to help spouses deal with the difficulties they may encounter in marriage.

Ethnicity and Culture

Asian Americans and Mexican Americans have lower divorce rates than European Americans or African Americans because they consider the family unit to be of greater value (familism) than their individual interests (individualism). Unlike familistic values in Asian cultures, individualistic values in American culture em-phasize the goal of personal happiness in marriage. When spouses stop having fun (when individualistic goals are no longer met), they sometimes feel no reason to stay married. In effect, staying married in America is driven by individualistic val-ues focused on personal happiness. When such happiness turns to unhappiness, spouses bolt. In two national samples, 36% agreed that "The personal happiness of an individual is more important than putting up with a bad marriage" (Amato et al. 2007). Reflecting an individualistic philosophy, Geraldo Rivera asked of his divorces, "Who cares if I've been married five times?"

Larry King has been married seven times to six women. Journalist Belinda Luscombe (2010c) of *Time Magazine*, suggested that in light of King's relentless divorces that maybe our society should create a "Revoking the Marriage License" policy. She wrote:

In no other area of life can grown people flame out so often and so badly and still get official permission to go ahead and do the same thing again. If your driving is hazardous to those around you, your license is suspended. Fail too many courses at college and you'll get kicked out. You can lose your medical or law license for a single infraction. Stock analyst Henry Blodget was prohibited from trading securities forever for publicly saying things he knew weren't true. So why do people who are committed vows abusers keep getting handed marriage licenses at city hall? If batters and violent offenders get only three strikes, why should bad spouses get more? (p. 64)

Micro Factors Contributing to Divorce

Although macro factors may make divorce a viable cultural alternative to marital unhappiness, they are not sufficient to "cause" a divorce. One spouse must choose to divorce and initiate proceedings. Such a view is micro in that it focuses on the individual decisions and interactions within specific family units. The following subsections discuss some of the micro factors that may be operative in influencing a couple toward divorce.

Differences

As noted in the chapter on mate selection, the greater the differences between spouses, the more likely they are to divorce. Clarkwest (2007) noted that African Americans are more likely than non-Hispanic white individuals to marry someone different from themselves. Specifically, African American females are more likely to marry a man who is less supportive of gender equality, who is less religious, who wants fewer children, and who is less supportive of maternal employment than non-Hispanic white females. And the divorce rate is higher.

Falling Out of Love

Benjamin et al. (2010) noted that the absence of love in a relationship is associated with an increased chance of divorce. Indeed, almost half (48%) of 2,922 undergraduates reported that they would divorce a spouse they no longer loved (Knox and Hall 2010). No couple is immune to falling out of love and getting divorced. Lavner and Bradbury (2010) studied 464 newlyweds over a 4-year period and found that, even in those cases of reported satisfaction across the four years, some couples abruptly divorced. Whether the divorce was triggered by a personal indiscretion (e.g., infidelity) or crisis (e.g., death of child) is unknown, the point is that years of satisfaction do not make a couple immune to divorce.

Limited Time Together

Some spouses do not make time to be together. Their careers are so demanding that they rarely see each other. Barbara Walters reported that this happened with her husband Lee, who was busy producing Broadway plays. She was career-focused during the weekdays and he was career-focused at night and on weekends, so they drifted apart. She said in her autobiography: *I began to write good-bye letters to Lee, and then I would tear them up. How could I leave a man who had done me no harm? When our biggest mistake was that we had no life in common anymore? I didn't feel I wanted to reduce my workload. In fact, I had increased it by taking on Not for Women Only. Lee couldn't change his lifestyle or ambitions either* (Walters 2008, 213–214).

Decrease in Positive Behavior

People marry because they anticipate greater rewards from being married than from being single. During courtship, each partner engages in a high frequency of positive verbal (compliments) and nonverbal (eye contact, physical affection) behavior toward the other. The good feelings the partners experience as a result of these positive behaviors encourage them to marry to "lock in" these feelings across time. Mitchell (2010) interviewed 390 married couples and found that intimacy and doing things together were associated with marital happiness. Just as

My husband and I divorced over religious differences. He thought he was God and I didn't.

Unknown

Getting divorced just because you don't love a man is almost as silly as getting married just because you do.

Zsa Zsa Gabor, actress

Chapter 15 Divorce and Remarriage

love feelings are based on positive behavior from a partner, negative feelings are created when these behaviors stop and the partner engages in a high frequency of negative behavior. Thoughts of divorce then begin (to escape the negative behavior).

Affair

Some extramarital affairs result in a divorce. Involvement in an affair may result in both love and sex with the new partner and speed the spouse toward divorce. Alternatively, the spouse at home may become indignant and demand that the partner leave. In an Oprah.com survey of 6,069 adults, an affair is the reason most respondents said they would seek a divorce (Healy and Salazar 2010). Thirty-three percent said an affair would be the dealbreaker. Other top responses were chronic fighting (28%), no longer being in love (24%), boredom (8%), and sexual incompatibility (4%).

Lack of Conflict Resolution Skills

Managing differences and conflict in a relationship helps to reduce the negative feelings that develop in a relationship. Some partners respond to conflict by withdrawing emotionally from their relationship; others respond by attacking, blaming, and failing to listen to their partner's point of view. Ways to negotiate differences and reduce conflict were discussed in Chapter 4 on communication in relationships.

Value Changes

Both spouses change throughout the marriage. "He's not the person I married" is a frequent observation of people contemplating divorce. People may undergo radical value changes after marriage. Jane Fonda noted that she experienced a change in her assertiveness, that she was no longer willing to just go along with her mate but to specify what her needs were and to negotiate their fulfillment:

> *The very thing that I feared the most—that I would gain my voice and lose my man—was actually happening. . . . The problem comes when what you need and what you see isn't seen or needed by your partner. It doesn't mean your partner is bad; it just means that he or she wants something else in life. . . . I could see Ted [Turner] withering before my eyes. Clearly he wasn't going to be able (or willing) to make the journey with me. We agreed to separate. (Fonda 2005, 545)*

Because people change throughout their lives, the person selected at one point in life may not be the same partner one would select at another point. Margaret Mead, the famous anthropologist, noted that her first marriage was a student marriage; her second, a professional partnership; and her third, an intellectual marriage to her soul mate, with whom she had her only child. At each of several stages in her life, she experienced a different set of needs and selected a mate who fulfilled those needs.

Satiation

Satiation, also referred to as habituation, refers to the state in which a stimulus loses its value with repeated exposure. Spouses may tire of each other. Their stories are no longer new, their sex is repetitive, and their presence no longer stimulates excitement as it did at the beginning of the relationship. Some people who feel trapped by the boredom of constancy decide to divorce and seek what they believe to be more excitement by returning to singlehood and, potentially, new partners. One man said, "I traded something good for something new." A developmental task of marriage is for couples to enjoy being together and not demand a constant state of excitement (which is difficult over a fifty-year period). The late comedian George Carlin said, "If all of your needs are not being met, drop some of your needs." If spouses did not expect so much of marriage, maybe they would not be disappointed.

Satiation a stimulus loses its value with repeated exposure—also called habituation.

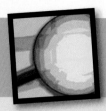
What advice do spouses who are in the process of divorce have for spouses unhappy in their marriage and who are contemplating divorce?

Sample and Methods

To find out, 59 respondents (47 females and 12 males) (average age, 35) completed a 29-item Internet survey designed to reveal what they (as people going through divorce) would suggest to those still married and contemplating divorce.

Selected Findings and Conclusions

Over two-thirds of the females (68%) and males (67%) recommended that others who are considering divorce not give up on their marriage but "work it out" or "see a counselor." One interpretation of this advice is regret on the part

Perception that One Would Be Happier if Divorced

Brinig and Allen (2000) noted that women file two-thirds of divorce applications. Their doing so may be encouraged by their view that they will achieve greater power over their own life. They feel that by getting a divorce they will have their own money (in the form of child support and/or alimony) without having a man in the house they don't want. In addition, they will have greater control over their children, since they are awarded custody in 80% of cases. The researchers argue that "who gets the children" is by far the greatest predictor of who files for divorce, and they contend that *if* the law were to presume that spouses would have joint custody, there would be fewer women filing for divorce because they would have less to gain.

Top Thirty Factors Associated with Divorce

Researchers have identified the characteristics of those most likely to divorce (Amato 2010; Nunley and Alan Seals 2010; Chiu and Busby 2010). Some of the more significant associations include the following:

1. Courting less than two years (partners know less about each other)
2. Having little in common (similar interests serve as a bond between people)
3. Marrying at age 17 and younger (associated with low education and income and lack of maturity)
4. Being different in race, education, religion, social class, age, values, and libido (widens the gap between spouses)
5. Not being religiously devout (less bound by traditional values)
6. Having a cohabitation history with different partners (pattern of establishing and breaking relationships)
7. Having been previously married (less fearful of divorce)
8. Having no children (less reason to stay married)
9. Having limited education (associated with lower income, more stress, less happiness)
10. Living in an urban residence (more anonymity, less social control in urban environment)
11. Being unfaithful (broken trust, emotional reason to leave relationship)
12. Growing up with divorced parents (models for ending rather than repairing relationship; may have inherited traits such as alcoholism that are detrimental to staying married)
13. Having poor communication skills (issues go unresolved and accumulated)

of these spouses now on the divorce path. Females were more likely to recommend seeing a counselor than males, and males were more likely than females to recommend a "cooling-off" period before beginning divorce proceedings. Clearly, one effect of involvement in the process of separation for both genders was a reevaluation of the desirability of seeking a divorce, to the degree that they would alert others contemplating separation to rethink their situation and to attempt reconciliation.

Source

Adapted from "Work it out/See a counselor, Advice from spouses in the separation process" by D. Knox and U. Corte. *Journal of Divorce & Remarriage* 48:1, 2007, 79–90, Taylor & Francis. Reprinted by permission of Taylor & Francis Group, http://www.informaworld.com).

14. Husband experiencing unemployment (his self-esteem is lost, loss of respect by wife who expects breadwinner)

15. Wife being employed (creates strife over division of labor at home)

16. Having mental (depression, anxiety, alcoholism) or physical disability (chronic fatigue syndrome)

17. Having seriously ill child (impacts stress, finances, couple time)

18. Having low self-esteem (associated with higher jealousy, less ability to love and accept love)

19. Being African American (live under more oppressive conditions, which increases stress in the marital relationship)

20. Lack of commitment (for nontraditional spouses, divorce is seen as an option if the marriage does not work out)

21. Experiencing rape (a spouse who has been raped before marriage is more likely to have sexual dysfunction, low self-esteem)

22. Having premarital pregnancy or unwanted child (spouses may feel pressure to get married; stress of parenting unwanted child)

23. Having stepchildren in the household (greater conflict in household)

24. Having high debt or sudden loss of income (stress of limited income)

25. Experiencing violence or abuse (reasons to leave marriage)

26. Having parents who never married (being unmarried is normative)

27. Marrying someone who has been divorced (escape from relationship problems rather than resolve them)

28. Wife earns higher income than husband (wife can afford to leave an unhappy marriage; husband feels threatened)

29. Falling out of love (spouses have less reason to stay married)

30. Bankruptcy (spouses conflict over limited economic resources)

The more of these factors that exist in a marriage, the more vulnerable a couple is to divorce. Regardless of the various factors associated with divorce, there is debate about the character of people who divorce. Are they selfish, amoral people who are incapable of making good on a commitment to each other and who wreck the lives of their children? Or are they individuals who care a great deal about relationships and won't settle for a bad marriage? Indeed, they may divorce precisely because they value marriage and want to rescue their children from being reared in an unhappy home.

Caryl Rusbult's investment model of commitment may also be used to identify why a relationship ends. When the emotional satisfaction for being in the relationship decreases, an attractive alternative presents itself, and a willingness to give up one's investments (e.g., material goods, friendships, etc.) converge, the person is ready to leave (Finkenauer 2010, 162).

Consequences of Divorce for Spouses/Parents

For both women and men, divorce is often an emotional and financial disaster. In addition to the death of a spouse, separation and divorce are among the most difficult of life's crisis events.

Recovering from a Broken Heart

A sample of 410 freshmen and sophomores at a large southeastern university completed a confidential survey revealing their recovery from a previous love relationship (Knox et al. 2000). Some of the findings were as follows:

1. *Sex differences in relationship termination.* Women were significantly more likely than men to report that they initiated the breakup (50% versus 40%). Sociologists suggest that women terminating relationships more often is related to their desire to select a better father for their offspring. One female student recalled, "I got tired of his lack of ambition—I just thought I could do better. He's a nice guy but his playing video games and our living in a trailer won't work."

2. *Sex differences in relationship recovery.* Though recovery was not traumatic for either men or women, men reported more difficulty than women did in adjusting to a breakup. When respondents were asked to rate their level of difficulty from "no problem" (0) to "complete devastation" (10), women scored 4.35 and men scored 4.96. In explaining why men might have more difficulty adjusting to terminated relationships, some of the female students said, "Men have such inflated egos, they can't believe that a woman would actually dump them." Others said, "Men are oblivious to what is happening in a relationship and may not have a clue that it is heading toward an abrupt end. When it does end, they are in shock."

Siegler and Costa (2000) also noted that women fare better emotionally after separation or divorce than do men. They note that women are more likely than men not only to have a stronger network of supportive relationships but also to profit from divorce by developing a new sense of self-esteem and confidence, because they are thrust into more independent roles. Women may also view divorce as a learning opportunity. Ellen Burstyn noted of her three divorces—"I've learned from all my failures, not all, but most of my learning has come from failures. I don't know if you can learn much from success or happiness" (Rountree 1993, 15).

3. *Time and new partner as factors in recovery.* The passage of time and involvement with a new partner were identified as the most helpful factors in getting over a love relationship that ended. Though the difference was not statistically significant, men more than women reported "a new partner" was more helpful in relationship recovery (34% versus 29%). Similarly, women more than men reported that "time" was more helpful in relationship recovery (34% versus 29%). However, a study of Jewish women who were adjusting to divorce emphasized that "a new romantic partner" was helpful in their adjustment (Kulik and Heine-Cohen 2011).

4. *Other findings.* Other factors associated with recovery for women and men were "moving to a new location" (13% versus 10%) and recalling that "the previous partner lied to me" (7% versus 5%). Men were much more likely than women to use alcohol to help them get over a previous partner (9% versus 2%). Neither men nor women reported using therapy to help them get over a partner (1% versus 2%). These data suggest that breaking up was not terribly difficult for these undergraduates (but more difficult for men than women) and that both time and a new partner enabled their recovery. These undergraduates are also young and may not view relationships at this age as permanent, making moving on easier.

Chapter 15 Divorce and Remarriage

Sometimes divorcing individuals who are church members seek solace through the church. Jenkins (2010) analyzed the experiences of 41 individuals who ended life partnerships while they were active in their congregation. The experience was mixed. While they felt tacit support as they were going through divorce, they also reported a "marked sense of aloneness, resulting from individual shame and congregational silence" (p. 278). The divorced who are able to forgive not only report being less angry but less depressed (Rohde-Brown and Rudestam 2011).

Financial Consequences

Getting divorced affects one's finances. Both women and men experience a drop in income following divorce, but women may suffer more. Because men usually have greater financial resources, they may take all they can with them when they leave. The only money they may continue to give to an ex-wife is court-ordered child support or spousal support (alimony).

Although 56% of custodial mothers are awarded child support, the amount is usually inadequate, infrequent, and not dependable, and women are forced to work (sometimes at more than one job) to take financial care of their children.

Remarriage generally restores a woman's economic stability. When remarried mothers and fathers are compared, there is a "matrilineal tilt" in terms of money transfers to children (Clark and Kenny 2010). In effect, while single divorced fathers give more money to their children than single divorced mothers, remarried mothers give more money to their children than remarried men. According to their data, 21% of divorced mothers gave financial transfers to their biological children over the past two years as compared to 16% of divorced fathers doing so.

Explanations include that remarried mothers will prefer to give their resources to their biological children with whom they are more likely to have maintained a relationship since the divorce. In contrast, remarried fathers are less likely to make money transfers since they have a less close relationship to their children since the divorce and they have a new wife monitoring their spending behavior. The new wife of a remarried father said to him when his biological son asked him for money, "We don't want to encourage his economic dependence...do we?" The son did not get the money.

How money is divided depends on whether the couple had a prenuptial agreement or a **postnuptial agreement**. We discussed prenuptial agreements in detail in Chapter 6. Appendix C at the end of this text provides an example of a prenuptial agreement for a couple getting remarried. Such agreements are most likely to be upheld if an attorney insists on four conditions—full disclosure by both parties, independent representation by separate counsel, absence of coercion or duress, and terms that are fair and equitable.

Fathers' Separation from Children

According to Finley (2004, F9), "... divorce transforms family power from intact patriarchy to post-divorce matriarchy," where women are typically given custody of the children and child support. Trinder (2008) emphasized that these women serve as gatekeepers for the relationship their husbands have with their children. Their patterns range from being proactive whereby they encourage such relationships and involvement to closing the gate and attempting to destroy the relationship. Since mothers end up with custody more often than fathers, fathers and paternal grandparents more often end up being shut off from their children and grandchildren (Doyle et al. 2010).

As a result, about 5 million divorced dads wake up every morning in an apartment or home while their children wake up with their separated or divorced mother. These are noncustodial fathers who may find the gate to their children

Loneliness and the feeling of being unwanted is the most terrible.

Mother Teresa, Nobel Peace Prize winner

Divorce is the one human tragedy that reduces everything to cash.

Rita Mae Brown, writer

A divorcée is a woman who got married so she didn't have to work, but now works so she doesn't have to get married.

Anna Magnani, Italian stage and film actress

Postnuptial agreement an agreement about how money is to be divided should a couple later divorce, which is made after the couple marry.

You're best off if you can put the emotions off to the side, and really realize that the end of a marriage is a business transaction.

Financial advisor Jean Chatzy

One father is worth more than a hundred schoolmasters.

George Herbert, Welsh poet, orator, and Anglican priest

Divorced fathers who stay involved in the lives of their children provide enormous emotional well being and economic support for them. This father and daughter are reunited after a six year blackout.

Chelsea E. Curry

shut so that they are allowed to see their children only at specified times (for example, two weekends a month).

Ahrons and Tanner (2003) found that low father involvement, not divorce, has a negative impact on the father-child relationship. Fathers who stay involved in the lives of their children emotionally, physically, and economically (in spite of being a noncustodial parent, having an adversarial former spouse, and a remarriage) mitigate any negative effects on the relationship with their children. Indeed, some relationships with the father may improve because a father may spend more one-on-one time with his children. Leidy et al. (2011) investigated how fathering behaviors (acceptance, rejection, monitoring, consistent discipline, and involvement) are related to preadolescent adjustment in Mexican American and European American stepfamilies and intact families. They concluded that in all contexts, fathering had significant positive effects on adolescent adjustment. Nielsen (2011) emphasized that, after divorce, a father's relationship with his daughter is often far more damaged than his relationship with his son. She noted the ways in which the daughter pays an ongoing price for this damaged—or destroyed—relationship with her father and suggests ways to reduce the damage.

Shared Parenting Dysfunction

Shared parenting dysfunction refers to the set of behaviors on the part of each parent (embroiled in a divorce) that are focused on hurting the other parent and are counterproductive for the well-being of the children. Turkat (2002) identified the following examples:

Shared parenting dysfunction behaviors on the part of both partners focused on hurting the other parent that are counterproductive for the well-being of the children.

- *A parent who forced the children to sleep in a car to prove the other parent had bankrupted them*
- *After losing a court battle over custody of the children, a noncustodial parent burned down the house of the primary residential parent*
- *One divorcing parent bought a cat for the children because the other divorcing parent was highly allergic to cats*

Finally, some of the most destructive displays of shared parenting dysfunction may include kidnapping, physical abuse, and murder (p. 390).

Chapter 15 Divorce and Remarriage

What if Your Former Spouse Tries to Turn Your Children against You?

The best antidote to parental alienation syndrome is to spend time with your children so that they can discover for themselves who you are as a parent, how you feel about them, and how much you love and value them. Regardless what your former spouse says to the children about you, the reality of how you treat them is what will determine how your children feel about you. However, should your spouse not allow you to see your children, it is imperative to hire a lawyer (and go to court if necessary) to ensure that you are awarded time with your children. You cannot have a relationship with your children if you don't spend time with them.

Parental Alienation

Shared parenting dysfunction may lead to parental alienation syndrome. **Parental alienation syndrome** is an alleged disturbance in which children are obsessively preoccupied with deprecation and/or criticism of a parent, denigration that is unjustified and/or exaggerated (Gardner 1998). Meier (2009) reviewed the history of parental alienation syndrome (PAS) and **parental alienation (PA)** and found the former to be questioned by most researchers (e.g., PAS is not a medical psychosis with specific criteria). However, parental alienation, a separate but related concept, denoting estrangement of a child from a parent, has experienced more credibility and acceptance (Baker and Chambers 2011). Examples of behavior that either parent may engage in to alienate a child from the other parent include the following (Schacht 2000; Teich 2007; and Baker and Darnall 2007):

1. Minimizing the importance of contact and the relationship with the other parent, including moving far away with the child to make regular contact difficult

2. Exhibiting excessively rigid boundaries; rudeness or refusal to speak to or inability to tolerate the presence of the other parent, even at events important to the child; refusal to allow the other parent near the home for drop-off or pick-up visitations

3. Having no concern about missed visits with the other parent

4. Showing no positive interest in the child's activities or experiences during visits with the other parent and withholding affection if the child expresses positive feelings about the absent parent

5. Granting autonomy to the point of apparent indifference ("It's up to you if you want to see your dad. I don't care.")

The most telling sign of children who have been alienated from a parent is the irrational behavior of the children, who for no properly explained reason say that they want nothing further to do with one of the parents. Indeed such children have a lack of ambivalence toward the alienation, lack of guilt or remorse about the alienation, and always take the alienating parent's side in the conflict (Baker and Darnall 2007).

Parental alienation syndrome an alleged disturbance in which children are obsessively preoccupied with deprecation and/or criticism of a parent; denigration that is unjustified and/or exaggerated.

Parental alienation estrangement of a child from a parent due to one parent turning the child against the other.

Consequences of Divorce for Children

Over a million children annually experience the divorce of their parents.

The well-being of children who grow up in homes where their parents remain in low-conflict marriages is clearly higher than children who have divorced parents. Children who grow up in homes where their parents are married but

Divorce is like two lions in a den attacking each other. You know somebody is going to get hurt real bad. All kids can do is sit behind a window and watch it happen.

Anonymous nine-year-old boy

who constantly fight also clearly suffer in terms of their subjective well-being (Sobolewski and Amato 2007). Gordon (2005) noted that divorce may actually benefit children. When parental conflict is very high prior to divorce, children benefit by no longer being subjected to the relentless anger and emotional abuse they observe between their parents. Indeed, when comparing children whose conflicted parents divorced with children whose parents were still together, the children were very similar. In self-reports of 158 Israeli young adults whose parents divorced when they were adolescents, Sever et al. (2007) determined that, although the children had painful feelings, almost half the participants reported more positive than negative outcomes. These included maturity and growth, empowerment, empathy, and relationship savvy. Lambert (2007) identified other advantages for children whose parents divorced as learning to be resilient, developing closer relationships with siblings, having happier parents, learning lessons about what not to do in a relationship, and receiving more attention. Knox et al. (2004) found that although 26% of the undergraduates in their study reported that the divorce of their parents had a negative effect, 32.9% reported a *positive* effect. Finally, when children of divorced parents become involved in their own marriage to a supportive, well-adjusted partner, the negative effects of parental divorce are mitigated (Hetherington 2003). In addition, when income is held constant, the overall effects of divorce are negligible (Biblarz and Stacey 2010).

Kelly and Emery (2003) reviewed the literature on the effect of divorce on children and concluded the following:

> . . . [I]t is important to emphasize that approximately 75–80% of children and young adults do not suffer from major psychological problems, including depression; have achieved their education and career goals; and retain close ties to their families. They enjoy intimate relationships, have not divorced, and do not appear to be scarred with immutable negative effects from divorce. (p. 357–58)

Indeed, Coltrane and Adams (2003) emphasized that claiming that divorce seriously damages children is a "symbolic tool used to defend a specific moral vision for families and gender roles within them" (p. 369). They go on to state that "Understanding this allows us to see divorce not as the universal moral evil depicted by divorce reformers, but as a highly individualized process that engenders different experiences and reactions among various family members . . ." (p. 370).

Nevertheless, divorce is often a difficult experience for children. D. Potter (2010) studied a sample of 20,000 kindergarteners through fifth graders and found that those who had parents who divorced experienced a decrease in psychological well-being which had a concomitant negative effect on grades. Indeed, children seem to benefit when the family structure is two biological married parents or adoptive parents in terms of positive mental health, lower drug use, later sexual behavior, and better physical health (Carr and Springer 2010).

Although the primary factor that determines the effect of divorce on children is the degree to which the divorcing parents are civil, legal and physical custody are important issues. The following section details how judges go about making this decision.

Who Gets the Children?

National **Data**

Twenty-three percent of children live with single mothers, 3% with single fathers (Carr and Springer 2010).

Legal custody judicial decision regarding whether one or both parents will have decisional authority on major issues affecting the children.

Judges who are assigned to hear initial child custody cases must make a judicial determination regarding whether one or both parents will have decisional authority on major issues affecting the children (called "**legal custody**"), and the distribution of parenting time (called "visitation" or "**physical custody**"). Toward this end, judges in all states are guided by the statutory dictum called "best interests of the child."

In a highly contested custody case, a judge will often appoint a mental health professional to conduct a custody evaluation to assist the judge in determining what will be the best future arrangement for the child. Of course, each custody case is different because the circumstances of the children are different, but some of the frequently employed custody factors include the following:

1. The child's age, maturity, sex, and activities, including culture and religion—all relevant information about the child's life—keeping the focus of custody on the child's best interests

2. The wishes expressed by the child, particularly the older child (judges will often interview children 6 years old or older in chambers)

3. Each parent's capacity to care for and provide for the emotional, intellectual, financial, religious, and other needs of the child (including the work schedules of the parents)

4. The parents' ability to agree, communicate, and cooperate in matters relating to the child

5. The nature of the child's relationship with the parents, which considers the child's relationship with other significant people, such as members of the child's extended family

6. The need to protect the child from physical and psychological harm caused by abuse or ill treatment (focuses on the issue of domestic violence)

7. The past and present parental attitudes and behavior (dealing with issues of parenting skills and personalities)

8. The proposed plans for caring for the child (Custody is not ownership of the child, so the judge will want to know how each parent proposes to raise the child, including proposed parenting times for the other parent.) (Lewis 2008)

These and other custody factors, whether presented by the custody evaluator or by testimony, will become the basis for the judge's custody determination. In some cases joint custody may be awarded.

Joint Custody?

Traditionally, sole custody to the mother was the only option courts considered for divorcing parents. The presumption was made that the "best interests of the child" were served if they were with their mother (the parent presumed to be more involved and more caring). As men have become more involved in the nurturing of children, the courts no longer assume that "parent" means "mother." Indeed, a new **family relations doctrine** has emerged which suggests that even nonbiological parents (such as stepparents) may be awarded custody or visitation rights if they have been economically and emotionally involved in the life of the child (Holtzman 2002).

Given that fathers are no longer routinely excluded from custody considerations, over half of the states in the United States have enacted legislation authorizing joint custody. About 16% of separated and divorced couples actually have a joint custody arrangement. In a typical joint physical custody arrangement, the parents continue to live in close proximity to each other. The children may spend part of each week with each parent or may spend alternating weeks with each parent.

New terminology is being introduced in the lives of divorcing spouses and in the courts. The term *joint custody*, which implies ownership, is being replaced with *shared parenting*, which implies cooperation in taking care of children.

There are several advantages of joint custody or shared parenting. Ex-spouses may fight less if they have joint custody because there is no inequity in their involvement in their children's lives. Children will benefit from the resultant decrease in hostility between parents who have both "won" them. Unlike sole-parent custody, in which one parent wins (usually the mother) and the other parent loses, joint custody allows children to continue to benefit from the love and attention of both parents. Children in homes where joint custody

Physical custody the distribution of parenting time between divorced spouses.

Family relations doctrine belief that even nonbiological parents may be awarded custody or visitation rights if they have been economically and emotionally involved in the life of the child.

has been awarded might also have greater financial resources available to them than children in sole-custody homes. Spruijt and Duindam (2010) studied divorced spouses who had joint custody and found positive outcomes for the children, mother, and father.

Joint physical custody may also be advantageous in that the stress of parenting does not fall on one parent but, rather, is shared. One mother who has a joint custody arrangement with her ex-husband said, "When my kids are with their dad, I get a break from the parenting role, and I have a chance to do things for myself. I love my kids, but I also love having time away from them." Another joint-parenting father said, "When you live with your kids every day, you can get very frustrated and are not always happy to be with them. But after you haven't seen them for three days, it feels good to see them again."

A disadvantage of joint custody is that it tends to put hostile ex-spouses in more frequent contact with each other, and the marital war continues. In a study (Markham and Coleman 2010) of 20 divorced women with shared physical custody, almost half (45%) were categorized as "continuously contentious", reflecting continuous anger. Factors fueling such anger/contentiousness were irresponsible fathers (e.g., they did not show up, drank irresponsibly, and/or did not provide structure or physical care like bathing the children), lack of money such as child support, and trying to control the ex-spouse.

Depending on the level of hostility between the ex-partners, their motivations for seeking sole or joint custody, and their relationship with their children, any arrangement can have positive or negative consequences for the ex-spouses as well as for the children. In those cases in which the spouses exhibit minimal hostility toward each other, have strong emotional attachments to their children, and want to remain an active influence in their children's lives, joint custody may be the best of all possible choices. In the Markham and Coleman (2010) study noted previously, the women who had positive outcomes ("always amicable" – 20%) with shared custody reported responsible fathers, money not being a source of conflict, and wanting shared custody. And, if things are bad, they can improve, as they did for 35% of the Markham and Coleman sample (referred to as "bad to better").

Minimizing Negative Effects of Divorce on Children

Researchers have identified the following conditions under which a divorce has the fewest negative consequences for children:

1. *Healthy parental psychological functioning.* Children of divorced parents benefit to the degree that the parents remain psychologically fit and positive, and socialize their children to view the divorce as a "challenge to learn from." Parents who nurture self-pity, abuse alcohol or drugs, and socialize their children to view the divorce as a tragedy from which they will never recover create negative outcomes for their children. Some divorcing parents can benefit from therapy as a method for coping with their anger or depression and for making choices in the best interests of their children.

Other parents become involved in divorce education programs. Forty-two percent of states mandate such programs for divorcing spouses (Riffe et al. 2010). However, these are meager programs, 70% of which typically require only one to two and a half hours. About half the courts provide such programs. Arizona mandates that all divorcing parents with young children attend a divorce education program (Amato 2010).

2. *A cooperative relationship between the parents.* The most important variable in a child's positive adjustment to divorce is when the child's parents continue to maintain a cooperative relationship throughout the separation, divorce, and post-divorce period. In contrast, bitter parental conflict places the children in the middle. One daughter of divorced parents said, "My father told me, 'If you love me, you would come visit me,' but my mom told me, 'If you love me, you won't visit him'" (see Personal Choices feature).

Choosing to Maintain a Civil Relationship with Your Former Spouse

Spouses who divorce must choose the kind of new relationship they will have with each other. This choice is crucial in that it affects not only their own but also their children's lives. What kind of relationship do former spouses have? Drapeau et al. (2009) interviewed children (ages 8–11) and their parents at 1 and 2.5 years after the separation and identified various patterns.

The typical pattern was for a high degree of conflict between the parents that dropped progressively over time. About a quarter displayed a low level of conflict that remained stable over time. Sometimes referred to as "sustained adjustment," these couples were committed to get along as well as they could, given that they were divorced. Another 25% of the former spouses were successful in achieving a level of cooperation after separation. This pattern was the most beneficial to children. A fourth pattern (a third of the divorces) reflected a sustained conflict that lasted for years after the divorce. This pattern was the most devastating for children since it continually exposed them to conflict between the parents despite their physical separation. We might call these four patterns "typical," "sustainers," "cooperatives," and "conflictors."

A final pattern that characterized 10% of the sample was that the early low level of conflict that set the tone for the relationship after divorce tended to rise over time. We might call these the "slow ragers." A factor associated with low-conflict was high income—if there was plenty of money after the divorce, the ex-spouses tended to get along. In addition, couples who had a high degree of initial agreement tended to fare better. Clearly, spouses make a choice in regard to how they relate to each other following their divorce.

---

3. *Parental attention to the children and allowing them to grieve.* Children benefit when both the custodial and the noncustodial parent continue to spend time with them and to communicate to them that they love them and are interested in them. Parents also need to be aware that their children do not want the divorce and to allow them to grieve over the loss of their family as they knew it. Indeed, children do not want their parents to separate.

4. *Encouragement to see noncustodial parent.* Children benefit when custodial parents (usually mothers) encourage them to maintain regular and stable visitation schedules with the noncustodial parent following divorce.

5. *Attention from the noncustodial parent.* Children benefit when they receive frequent and consistent attention from noncustodial parents, usually the fathers. Noncustodial parents who do not show up at regular intervals exacerbate their children's emotional insecurity by teaching them, once again, that parents cannot be depended on. Parents who show up often consistently teach their children to feel loved and secure.

6. *Assertion of parental authority.* Children benefit when both parents continue to assert their parental authority and continue to support the discipline practices of each other.

7. *Regular and consistent child support payments.* Support payments (usually from the father to the mother) are associated with economic stability for the child.

8. *Stability.* Moving to a new location causes children to be cut off from their friends, neighbors, and teachers. It is important to keep their life as stable as possible during a divorce.

9. *Children in a new marriage.* Manning and Smock (2000) found that divorced noncustodial fathers who remarried and who had children in the new marriages were more likely to shift their emotional and economic resources to the new family unit than were fathers who did not have new biological children. Fathers might be alert to this potential and consider each child, regardless of

ivorce mediation is a process in which spouses who have decided to separate or divorce meet with a neutral third party (mediator) to negotiate the following issues: (1) how they will parent their children, which is referred to as child custody and visitation; (2) how they are going to financially support their children, referred to as child support; (3) how they are going to divide their property, known as property settlement; and (4) how each one is going to meet their financial obligations, referred to as spousal support.

Benefits of Mediation

There are enormous benefits from avoiding litigation and mediating one's divorce:

1. **Better relationship.** Spouses who choose to mediate their divorce have a better chance for a more civil relationship because they cooperate in specifying the conditions of their separation or divorce. Mediation emphasizes negotiation and cooperation between the divorcing partners. Such cooperation is particularly important if the couple has children, in that it provides a positive basis for discussing issues in reference to the children and how they will be parented across time.

2. **Economic benefits.** Mediation is less expensive than litigation. The combined cost of hiring attorneys and going to court over issues of child custody and division of property is around $30,000. A mediated divorce typically costs less than $5,000. A couple cannot keep as assets to later divide what they spend in legal fees.

3. **Less time-consuming process.** Whereas a litigated divorce can take two to three years, a mediated divorce takes two to three months; a "mediated settlement conference" "... can take place in one session from 8:00 A.M. until both parties are satisfied with the terms" (ibid.).

4. **Avoidance of public exposure.** Some spouses do not want to discuss their private lives and finances in open court. Mediation occurs in a private and confidential setting.

5. **Greater overall satisfaction.** Mediation results in an agreement developed by the spouses, not one imposed by a judge or the court system. A comparison of couples who chose mediation with couples who chose litigation found that those who mediated their own settlement were much more satisfied with the conditions of their agreement. In addition, children of mediated divorces are exposed to less marital conflict, which may facilitate their long-term adjustment to divorce.

Basic Mediation Guidelines

Divorce mediators conduct mediation sessions with certain principles in mind:

1. **Children.** What is best for a couple's children should be the major concern of the parents because they know their children far better than a judge or the mediator. Children of divorced parents adjust best under three conditions: (1) that both parents have regular and frequent access to the children; (2) that the children see the parents relating in a polite and positive way; and (3) that each parent talks positively about the other parent and neither parent talks negatively about the other to the children. Sometimes children are included in the mediation. They may be interviewed without the parents present to provide information to the mediator about their perceptions and preferences. Such involvement of the children has superior outcomes for both the parents and the children (McIntosh et al. 2008).

2. **Fairness.** It is important that the agreement be fair, with neither party being exploited or punished. It is fair for both parents to contribute financially to the children and to have regular access to their children.

3. **Open disclosure.** The spouses will be asked to disclose all facts, records, and documents to ensure an informed and fair agreement regarding property, assets, and debts.

4. **Other professionals.** During mediation, spouses may be asked to consult an accountant regarding tax laws. In

---

> *For some reason, we see divorce as a signal of failure, despite the fact that each of us has a right, and an obligation, to rectify any other mistake we make in life.*
>
> Joyce Brothers, psychologist

when or with whom the child was born, as worthy of a father's continued love, time, and support.

   10. Age and reflection on the part of children of divorce. Sometimes children whose parents are divorced benefit from growing older and reflecting on their parents' divorce as an adult rather than a child. Nielsen (2004) emphasized that daughters who feel distant from their fathers can benefit from examining the divorce from the viewpoint of the father (Was he alienated by the mother?), the cultural bias against fathers (they are maligned as "deadbeat dads" who "abandon their families for a younger woman"), and the facts about divorced dads (they are more likely to be depressed and suicidal following divorce than mothers).

addition, spouses are encouraged to consult an attorney throughout the mediation and to have the attorney review the written agreements that result from the mediation. However, during the mediation sessions, all forms of legal action by the spouses against each other should be stopped.

Another term for involving a range of professionals in a divorce is **collaborative practice,** a process that brings a team of professionals (lawyer, psychologist, mediator, social worker, financial counselor) together to help a couple separate and divorce in a humane and cost-effective way.

5. *Confidentiality.* The mediator will not divulge anything spouses say during the mediation sessions without their permission. The spouses are asked to sign a document stating that, should they not complete mediation, they agree not to empower any attorney to subpoena the mediator or any records resulting from the mediation for use in any legal action.

Such an agreement is necessary for spouses to feel free to talk about all aspects of their relationship without fear of legal action against them for such disclosures.

Divorce mediation is not for every couple. It does not work where there is a history of spouse abuse, where the parties do not disclose their financial information, where one party is controlled by someone else (for example, a parent), where there is the desire for revenge, or where the mediator is biased. Mediation should be differentiated from **negotiation** (where spouses discuss and resolve the issues themselves), **arbitration** (where a third party listens to both spouses and makes a decision about custody, division of property, and so on), and **litigation** (where a judge hears arguments from lawyers representing the respective spouses and decides issues of custody, child support, division of property, and so on).

Haswell (2006) noted a continuum of consequences from negotiation to litigation.

Negotiation	Mediation	Arbitration	Litigation
Cooperative			Competitive
Low Cost			High Cost
Private			Public
Protects Relationships			Damages Relationships
Focus on the Future			Focus on the Past
Parties in Control			Parties Lose Control

Your Opinion?

1. To what degree do you believe the government should be involved in mandating divorce mediation?

2. How can divorce mediation go wrong? Why should a couple not want to mediate their divorce?

3. What are the advantages for children when parents mediate their divorce?

Source

*Appreciation is expressed to Mike Haswell for contributing to this section. See http://www.haswellmediation.com/ Costs of litigation versus mediation have been updated for 2011.

http://www.divorcewizards.com/Information-on-Divorce-Mediation-and-Divorce-Litigation.html

Conditions of a "Successful" Divorce

Although acknowledging that divorce is usually an emotional and economic disaster, it is possible to have a "successful" divorce. Indeed, most people are resilient and "are able to adapt constructively to their new life situation within two to three years following divorce, a minority being defeated by the marital breakup, and a substantial group of women being enhanced" (Hetherington 2003, 318). The following are some of the behaviors spouses can engage in to achieve this:

A lawyer is never entirely comfortable with a friendly divorce, anymore than a good mortician wants to finish his job and then have the patient sit up on the table.

Jean Kerr, playwright

Collaborative practice process involving a team of professionals (lawyer, psychologist, mediator, social worker, financial counselor) helping a couple separate and divorce in a humane and cost-effective way.

Negotiation spouses discuss and resolve the issues of custody, child support, and division of property themselves.

Arbitration third party listens to both spouses and makes a decision about custody, division of property, child support, and alimony.

Litigation a judge hears arguments from lawyers representing the respective spouses and decides issues of custody, child support, division of property, etc.

Divorce mediation meeting with a neutral professional who negotiates child custody, division of property, child support, and alimony directly with the divorcing spouses.

> *When one door of happiness closes, another opens; but often we look so long at the closed door that we do not see the one which has been opened for us.*
>
> Helen Keller, blind and deaf American author, political activist, and lecturer

1. *Mediate rather than litigate the divorce.* Divorce mediators encourage a civil, cooperative, compromising relationship while moving the couple toward an agreement on the division of property, custody, and child support. By contrast, attorneys make their money by encouraging hostility so that spouses will prolong the conflict, thus running up higher legal bills. In addition, the couple cannot divide money spent on divorce attorneys (average is $15,000 for *each* side so a litigated divorce cost will start at $30,000). Benton (2008) noted that the worse thing divorcing spouses can do is to respectively hire the "meanest, nastiest, most expensive yard dog lawyer in town" because doing so will only result in a protracted expensive divorce where neither spouse will "win." Because the greatest damage to children from a divorce is a continuing hostile and bitter relationship between their parents, some states require **divorce mediation** as a mechanism to encourage civility in working out differences and to clear the court calendar from protracted court battles. Research confirms there are enormous benefits of mediation versus litigation (Amato 2010). The preceding Social Policy section on pages 454–455 focuses on divorce mediation.

2. *Co-parent with your ex-spouse.* Setting aside negative feelings about your ex-spouse so as to cooperatively co-parent not only facilitates parental adjustment but also takes children out of the line of fire. Such co-parenting translates into being cooperative when one parent needs to change a child care schedule, sitting together during a performance by the children, and showing appreciation for the other parent's skill in responding to a crisis with the children. Brotherson et al. (2010) evaluated a divorce education program (Parents Forever) which revealed that participants learned valuable skills on how to co-parent effectively.

3. *Take some responsibility for the divorce.* Because marriage is an interaction between spouses, one person is seldom totally to blame for a divorce. Rather, both spouses share reasons for the demise of the relationship. Take some responsibility for what went wrong.

4. *Create positive thoughts.* Divorced people are susceptible to feeling as though they are failures. They see themselves as Divorced people with a capital D, a situation sometimes referred to as "hardening of the categories" disease. Improving self-esteem is important for divorced people. They can do this by systematically thinking positive thoughts about themselves.

One technique (called the stop-think technique) is to write down twenty-one positive statements about yourself ("I am honest," "I have strong family values," "I am a good parent," and so on) and transfer them to three-by-five cards, each containing three statements. Take one of the cards with you each day and read the thoughts at three regularly spaced intervals (for example, 7:00 A.M., 1:00 P.M., and 7:00 P.M.). This ensures that you are thinking positive thoughts about yourself and are not allowing yourself to drift into a negative set of thoughts (for example, "I am a failure" or "no one wants to be with me."). Webb et al. (2010) also noted the positive effect of conservative religion (Seventh Day Adventist) on reducing symptoms of depression following divorce. "Seeking support from God" and "working with God as partners to deal with problems" were most effective.

5. *Avoid alcohol and other drugs.* The stress and despair that some people feel following a divorce make them particularly vulnerable to the use of alcohol or other drugs. These should be avoided because they produce an endless negative cycle. For example, stress is relieved by alcohol; alcohol produces a hangover and negative feelings; the negative feelings are relieved by more alcohol, producing more negative feelings, and so on.

6. *Engage in aerobic exercise.* Exercise helps one to not only counteract stress but also to avoid it. Jogging, swimming, riding an exercise bike, or other similar exercise for thirty minutes every day increases the oxygen to the brain and helps facilitate clear thinking. In addition, aerobic exercise produces endorphins in the brain, which create a sense of euphoria ("runner's high").

7. Continue interpersonal connections. Adjustment to divorce is facilitated when continuing relationships with friends and family. These individuals provide emotional support and help buffer the feeling of isolation and aloneness. First Wives World (www.firstwivesworld.com) is a new interactive site to provide an Internet social network for women transitioning through divorce.

8. Let go of the anger for your ex-partner. Former spouses who stay negatively attached to an ex by harboring resentment and trying to get back at the ex prolong their adjustment to divorce. The old adage that you can't get ahead by getting even is relevant to divorce adjustment.

9. Allow time to heal. Because self-esteem usually drops after divorce, a person is often vulnerable to making commitments before working through feelings about the divorce. The time period most people need to adjust to divorce is between twelve and eighteen months. Although being available to others may help to repair one's self-esteem, getting remarried during this time should be considered cautiously. Two years between marriages is recommended.

Remarriage

Divorced spouses usually waste little time getting involved in a new relationship. Indeed, one-fourth of divorcées date someone new before the divorce is final. Those who do not have children have a higher percentage of remarrying. When comparing divorced individuals who have remarried against divorced individuals who have not remarried, the remarried individuals report greater personal and relationship happiness.

The majority of divorced people remarry for many of the same reasons as for a first marriage—love, companionship, emotional security, and a regular sex partner. Other reasons are unique to remarriage and include financial security (particularly for a wife with children), help in rearing one's children, the desire to provide a "social" father or mother for one's children, escape from the stigma associated with the label "divorced person," and legal threats regarding the custody of one's children. With regard to the latter, the courts view a parent seeking custody of a child more favorably if the parent is married.

National **Data**

Two thirds of divorced females and three fourths of divorced males remarry. Of the widowed, 5% of widowed females and 12% of widowed males remarry (Sweeney 2010).

Preparation for Remarriage

Higginbotham and Skogrand (2010) studied 356 adults who attended a 12-hour stepfamily relationship education course. These participants were either remarried, cohabitating, or seriously dating someone who had children from a previous relationship. Results revealed that regardless of their race or marital status, the individuals benefitted from the stepfamily relationship education—commitment to the relationship, agreement on finances, relationships with ex-partners, and parenting all improved over time.

Children also benefit when parents participate in stepfamily education programs. Higginbotham et al. (2010) interviewed 40 parents and 20 facilitators who were part of a stepfamily program. They found that children were perceived to benefit by their parents' increased empathy, engagement in family time, and enhanced relationship skills.

Persons getting remarried and becoming involved in a stepfamily might be aware that self-help books on the subject are lacking. One researcher (Shafer 2010) reviewed the book *Yours, Mine, and Hours: Relationship Skills for Blended*

Love is a feeling, marriage is a contract, and relationships are work.

Lori Gordon, artist

I don't care if I'm your first love, I just want to be your last.

Gretchen Wilson, country western singer

Men who have a pierced ear are better prepared for marriage. They've experienced pain & bought jewelry.

Rita Rudner, comedian

Negative commitment spouses who continue to be emotionally attached to and have difficulty breaking away from ex-spouses.

Families by John Penton and Shona Welsh and noted, "Typically, these books are poorly researched and unable to distill complex theories and empirical evidence into practical solutions for blended families. *Yours, Mine, and Hours* falls into the same trap."

Trust is a major issue for people getting remarried. Brimhall et al. (2008) interviewed sixteen remarried individuals and found that most reported that their first marriage ended over trust issues—the partner betrayed them by having an affair or by hiding or spending money without the partner's knowledge. Each was intent on ensuring a foundation of trust in the new marriage.

Issues of Remarriage for the Divorced

Several issues challenge people who remarry (Kim 2010; Ganong and Coleman 1999; Goetting 1982):

Boundary Maintenance   Movement from divorce to remarriage is not a static event that is over after a brief ceremony. Rather, ghosts of the first marriage—in terms of the ex-spouse and, possibly, the children—must be dealt with. A parent must decide how to relate to an ex-spouse to maintain a good parenting relationship for the biological children while keeping an emotional distance to prevent problems from developing with a new partner. Some spouses continue to be emotionally attached to and have difficulty breaking away from an ex-spouse. These former spouses have what Masheter (1999) terms a **"negative commitment."** Masheter says such individuals "have decided to remain [emotionally] in this relationship and to invest considerable amounts of time, money, and effort in it . . . [T]hese individuals do not take responsibility for their own feelings and actions, and often remain 'stuck,' unable to move forward in their lives" (p. 297).

Emotional Remarriage   Remarriage involves beginning to trust and love another person in a new relationship. Such feelings may come slowly as a result of negative experiences in a previous marriage.

Psychic Remarriage   Divorced individuals considering remarriage may find it difficult to give up the freedom and autonomy of being single and to develop a mental set conducive to pairing. This transition may be particularly difficult for people who sought a divorce as a means to personal growth and autonomy. These individuals may fear that getting remarried will put unwanted constraints on them.

WHAT IF?

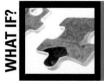

What if Your Ex-Spouse or Ex-Partner Wants to Get Back Together?

The dilemma of an ex-spouse or ex-partner wanting to get back together is not unusual and has been the subject of classic novels. Rhett Butler said of getting back with Scarlett O'Hara in *Gone with the Wind*, "I'd rather remember it as best it was than try and remend it and look at the broken pieces as long as I live." His decision is worthy of duplicating. Once an intense love relationship has been seriously broken (divorce), mending it and returning it to a durable, happy relationship is not likely (but it is possible as evidenced by the couple whose photo begins this chapter). The reason is that the factors that ended the relationship earlier may resurface.

Community Remarriage This stage involves a change in focus from single friends to a new mate and other couples with whom the new pair will interact. The bonds of friendship established during the divorce period may be particularly valuable because they have given support at a time of personal crisis. Care should be taken not to drop these friendships.

Parental Remarriage Because most remarriages involve children, people must work out the nuances of living with someone else's children. Mothers are usually awarded primary physical custody, and this translates into a new stepfather adjusting to the mother's children and vice versa. For individuals who have children from a previous marriage who do not live primarily with them, a new spouse must adjust to these children on weekends, holidays, and vacations or at other visitation times. Kim (2010) studied remarried families with adolescents and emphasized that having clear roles and understandings between the new stepparents/stepchildren was essential to smooth family functioning.

Economic and Legal Remarriage A second marriage may begin with economic responsibilities to a first marriage. Alimony and child support often threaten the harmony and sometimes even the economic survival of second marriages. Although the income of a new wife is not used legally to decide the amount her new husband is required to pay in child support for his children of a former marriage, his ex-wife may petition the court for more child support. The ex-wife may do so, however, on the premise that his living expenses are reduced with a new wife and that, therefore, he should be able to afford to pay more child support. Although an ex-wife is not likely to win, she can force the new wife to court and a disclosure of her income (all with considerable investment of time and legal fees for a newly remarried couple).

Economic issues in a remarriage may become evident in another way. A remarried woman who receives inadequate child support from an ex-spouse and needs money for her child's braces, for instance, might wrestle with how much money to ask her new husband for.

There may also be a need for a marriage contract to be drawn up before the wedding. Suppose a wife moves into the home of her new husband. If he has a will stating that his house goes to his children from a former marriage at his death and no marriage contract that either gives his wife the house or allows her to stay in the house rent free until her death, his children can legally throw her out of the house. The same is true for their beach house which he brought into the marriage. If his will gives the beach house to his children, his wife may have no place to live.

I'd marry again if I found a man who had fifteen million dollars, would sign over half to me, and guarantee that he'd be dead within a year.

Bette Davis, actress

Remarriage for Widowed Individuals

Only 10% of remarriages involve widows or widowers. Nevertheless, remarriage for widowed individuals is usually very different from remarriage for divorced people. Unlike divorced individuals, widowed individuals are usually much older and their children are grown.

Brimhall and Engblom-Deglmann (2011) interviewd 24 remarried individuals about the death of a previous spouse, either theirs or their partner's, and how this was affecting their marriage. Participants were interviewed individually and as a couple. Several themes emerged including past spouse on pedestal, current/past comparison, insecurity of the current spouse, curiosity about past spouse/relationship, partner's response to curiosity, and impact on the current relationship. Best new relationship outcomes seemed to come when the spouse of a deceased partner talked openly about the past relationship and reassured the current partner of his or her love for the partner and the current relationship.

A widow or widower may marry someone of similar age or someone who is considerably older or younger. Marriages in which one spouse is considerably older than the other are referred to as May-December marriages (discussed in

Chapter 7, Marriage Relationships). Here we will discuss only **December marriages**, in which both spouses are elderly.

A study of twenty-four elderly couples found that the primary motivation for remarriage was the need to escape loneliness or the need for companionship (Vinick 1978). Men reported a greater need to remarry than did the women.

Most of the spouses (75%) met through a mutual friend or relative and married less than a year after their partner's death (63%). Increasingly, elderly individuals are meeting online. Some sites cater to older individuals seeking partners, including seniorfriendfinder.com and thirdage.com.

The children of the couples in Vinick's study had mixed reactions to their parent's remarriage. Most of the children were happy that their parent was happy and felt relieved that someone would now meet the companionship needs of their elderly parent on a more regular basis. However, some children disapproved of the marriage out of concern for their inheritance rights. "If that woman marries Dad," said a woman with two children, "she'll get everything when he dies. I love him and hope he lives forever, but when he's gone, I want the house I grew up in." Though children may be less than approving of the remarriage of their widowed parent, adult friends of the couple, including the kin of the deceased spouses, are usually very approving (Ganong and Coleman 1999).

Stages of Involvement with a New Partner

After a legal separation or divorce (or being widowed), a parent who becomes involved in a new relationship passes through various transitions (see Table 15.1). These not only affect the individuals and their relationship but any children and/or extended family.

Stability of Remarriages

National data reflect that remarriages are more likely than first marriages to end in divorce in the early years of remarriage (Sweeney 2010). Remarriages most vulnerable to divorce are those that involve a woman bringing a child into the new marriage. Teachman (2008) analyzed data on women (N = 655) from National Survey of Family Growth to examine the correlates of second marital dissolution. He found that women who brought stepchildren into their second

TABLE 15.1 Stages of Parental Repartnering

Relationship Transition	Definition
Dating initiation	The parent begins to date.
Child introduction	The children and new dating partner meet.
Serious involvement	The parent begins to present the relationship as "serious" to the children.
Sleepover	The parent and the partner begin to spend nights together when the children are in the home.
Cohabitation	The parent and the partner combine households.
Breakup	The relationship experiences a temporary or permanent disruption.
Pregnancy in the new relationship	A planned or unexpected pregnancy occurs.
Engagement	The parent announces plans to remarry.
Remarriage	The parent and partner create a legal or civil union.

Source: Stages of Parental Repartnering by E. R. Anderson and S. M. Greene. Transitions in parental repartnering after divorce. *Journal of Divorce & Remarriage.* January 8, 2007, 43:49. Reprinted by permission of the publisher (Taylor & Francis Group, http://www.informaworld.com).

Newly remarried couples often have a baby of their own that results in giving greater stability to the new marriage.

marriage experienced an elevated risk of marital disruption. Premarital cohabitation or having a birth while cohabiting with a second husband did not raise the risk of marital dissolution, however. In addition, marrying a man who brought a child to the marriage did not increase the risk of marital disruption. One possible explanation for why a woman bringing a child into a second marriage is related to greater instability is that she may be less attentive to the new husband and more of a mother than a wife.

That second marriages, in general, are more susceptible to divorce than first marriages is because divorced individuals are less fearful of divorce than individuals who have never divorced. So, rather than stay in an unhappy second marriage, the spouses know they can survive a divorce and leave.

Though remarried people are more vulnerable to divorce in the early years of their subsequent marriage, they are less likely to divorce after fifteen years of staying in the second marriage than those in first marriages (Clarke and Wilson 1994). Hence, these spouses are likely to remain married because they want to, not because they fear divorce.

National **Data**

Of 2,691 adults, 42% say they have at least one step relative. Three-in-ten have a step- or half-sibling, 18% have a living stepparent, and 13% have at least one stepchild (Parker 2011).

Love and magic have a great deal in common. They enrich the soul, delight the heart. And they both take practice.

Nora Roberts, American author

Divorce is a journey that the children involved do not ask to take. They are forced along for a ride where the results are dictated by the road their parents decide to travel.

Diane Greene

Stepfamilies

Stepfamilies, also known as blended, binuclear, remarried, or reconstituted families, represent the fastest-growing type of family in the United States. A **blended family** is one in which spouses in a new marriage relationship blend their children from at least one other spouse from a previous marriage. The term

Blended family family wherein spouses in a remarriage bring their children to live with the new partner.

Binuclear family family that lives in two households as when parents live in separate households following a divorce.

Stepfamily family in which spouses in a new marriage bring children from previous relationships into the new relationship.

binuclear refers to a family that spans two households; when a married couple with children divorce, their family unit typically spreads into two households. There is a movement away from the use of the term *blended* because stepfamilies really do not blend. The term **stepfamily** (sometimes referred to as step relationships) is the term currently in vogue. This section examines how stepfamilies differ from nuclear families; how they are experienced from the viewpoints of women, men, and children; and the developmental tasks that must be accomplished to make a successful stepfamily.

Types of Stepfamilies

Although there are various types of stepfamilies (Sweeney 2010), the most common is a family in which the partners bring children from previous relationships into the new relationship (Sweeney, 2010). The couple may be married or living together, heterosexual or homosexual, and of all races. Although a stepfamily can be created when an individual who has never married or a widowed parent with children marries a person with or without children, most stepfamilies today are composed of spouses who are divorced and who bring children into a new marriage. This is different from stepfamilies characteristic of the early twentieth century, which more often were composed of spouses who had been widowed.

Stepfamilies may be both heterosexual and homosexual. Lesbian stepfamilies model gender flexibility in that a lesbian biological mother and a lesbian stepmother tend to share parenting (in contrast to a traditional family, in which the mother may take primary responsibility for parenting and the father is less involved). This allows a biological mother some freedom from motherhood as well as support in it. In gay male stepfamilies, the gay men may also share equally in the work of parenting.

Myths of Stepfamilies

Various myths abound regarding stepfamilies, including that new family members will instantly bond emotionally, that children in stepfamilies are damaged and do not recover, that stepmothers are "wicked" and "home-wreckers," that stepfathers are uninvolved with their stepchildren, and that stepfamilies are not "real" families. Regarding the latter, Sweeney (2010) reported that almost half of adult children chose a middle-range category—"quite a bit" or "a little" rather than "fully" or "not at all"—when asked to identify the extent to which they perceived a stepparent as being a family member.

Unique Aspects of Stepfamilies

Stepfamilies differ from nuclear families in a number of ways. These are identified in Table 15.2. These changes impact the parents, their children, and their stepchildren and require adjustment on the part of each member. In addition, stepfamilies are stigmatized. **Stepism** is the assumption that stepfamilies are inferior to biological families. Stepism, like racism, heterosexism, sexism, and ageism, involves prejudice and discrimination.

Adler-Baeder et al. (2010) focused on the economic problems of stepfamilies. Not only may divorce have reduced the disposable income available to the adults, one may be paying child support or alimony to a previous spouse, further reducing income to the unit. Low income groups are hit even harder by divorce.

Other matters with regard to nuclear families versus stepfamilies involve inheritance rights and child custody. Stepchildren do not automatically inherit from their stepparents, and courts have been reluctant to give stepparents legal access to stepchildren in the event of a divorce. In general, U.S. law does not consistently recognize stepparents' roles, rights, and obligations regarding their stepchildren. Without legal support to ensure such access, these relationships

Stepism the assumption that stepfamilies are inferior to biological families.

TABLE 15.2 Differences between Nuclear Families and Stepfamilies

Nuclear Families	Stepfamilies
1. Children are (usually) biologically related to both parents.	1. Children are biologically related to only one parent.
2. Both biological parents live together with children.	2. As a result of divorce or death, one biological parent does not live with the children. In the case of joint physical custody, children may live with both parents, alternating between them.
3. Beliefs and values of members tend to be similar.	3. Beliefs and values of members are more likely to be different because of different backgrounds.
4. The relationship between adults has existed longer than relationship between children and parents.	4. The relationship between children and parents has existed longer than the relationship between adults.
5. Children have one home they regard as theirs.	5. Children may have two homes they regard as theirs.
6. The family's economic resources come from within the family unit.	6. Some economic resources may come from an ex-spouse.
7. All money generated stays in the family.	7. Some money generated may leave the family in the form of alimony or child support.
8. Relationships are relatively stable.	8. Relationships are in flux: new adults adjusting to each other; children adjusting to a stepparent; a stepparent adjusting to stepchildren; stepchildren adjusting to each other.
9. No stigma is attached to nuclear family.	9. Stepfamilies are stigmatized.
10. Spouses had a childfree period.	10. Spouses had no childfree period.
11. Inheritance rights are automatic.	11. Stepchildren do not automatically inherit from stepparents.
12. Rights to custody of children are assumed if divorce occurs.	12. Rights to custody of stepchildren are usually not considered.
13. Extended family networks are smooth and comfortable.	13. Extended family networks become complex and strained.
14. Nuclear family may not have experienced loss.	14. Stepfamily has experienced loss.
15. Families experience a range of problems.	15. Stepchildren tend to be a major problem.

tend to become more distant and nonfunctional. In case of divorce, stepparents have little or no rights to the biological child (Sweeney 2010).

Stepfamilies in Theoretical Perspective

Structural functionalists, conflict theorists, and symbolic interactionists view stepfamilies from the following different points of view:

1. *Structural-functional perspective.* To the structural functionalist, integration or stability of the system is highly valued. The very structure of the stepfamily system can be a threat to the integration and stability of a family system. The social structure of stepfamilies consists of a stepparent, a biological parent, biological children, and stepchildren. Functionalists view the stepfamily system as vulnerable to an alliance between the biological parent and the biological children who have a history together.

In 75% of the cases, the mother and children create an alliance. The stepfather, as an outsider, may view this alliance between the mother and her children as the mother giving the children too much status or power in the family. Whereas a mother may relate to her children as equals, a stepfather may relate to the children as unequals whom he attempts to discipline. The result is a fragmented parental subsystem whereby the stepfather accuses the mother of being too soft and she accuses him of being too harsh.

Structural family therapists suggest that parents should have more power than children and that they should align themselves with each other. Not to do so is to give children family power, which they may use to splinter the parents off from each other and create another divorce.

2. *Conflict perspective.* Conflict theorists view conflict as normal, natural, and inevitable as well as functional in that it leads to change. Conflict in a stepfamily system is seen as desirable in that it leads to equality and individual autonomy.

Conflict is a normal part of stepfamily living. The spouses, parents, children, and stepchildren are constantly in conflict for the limited resources of space, time, and money. Space refers to territory (rooms) or property (television, CD player, or electronic games) in the house that the stepchildren may fight over. Time refers to the amount of time that the parents will spend with each other, with their biological children, and with their stepchildren. Money must be allocated in a reasonably equitable way so that each member of the family has a sense of being treated fairly.

Problems arise when space, time, and money are limited. Two new spouses who each bring a child from a former marriage into the house have a situation fraught with potential conflict. Who sleeps in which room? Who gets to watch which channel on television?

To further complicate the situation, suppose the couple have a baby. Where does the baby sleep? Because both parents may have full-time jobs, the time they have for the three children is scarce, not to speak of the fact that a baby will require a major portion of their available time. As for money, the cost of the baby's needs, such as formula and disposable diapers, will compete with the economic needs of the older children. Meanwhile, the spouses may need to spend time alone and may want to spend money as they wish. All these conflicts are functional because they increase the chance that a greater range of needs will be met within the stepfamily.

3. *Interactionist perspective.* Symbolic interactionists emphasize the meanings and interpretations that members of a stepfamily develop for events and interactions in the family. Children may blame themselves for their parents' divorce and feel that they and their stepfamily are stigmatized; parents may view stepchildren as spoiled.

Stepfamily members also nurture certain myths. Stepchildren sometimes hope that their parents will reconcile and that their nightmare of divorce and stepfamily living will end. This is the myth of reconciliation. Another is the myth of instant love, usually held by stepparents, who hope that the new partner's children will instantly love them. Although this does happen, particularly if the child is young and has no negative influences from the other parent, it is unlikely.

Stages in Becoming a Stepfamily

Just as a person must pass through various developmental stages in becoming an adult, a stepfamily goes through a number of stages as it overcomes various obstacles. Researchers such as Bray and Kelly (1998) and Papernow (1988) have identified various stages of development in stepfamilies. These stages include the following:

Stage 1: *Fantasy.* Both spouses and children bring rich fantasies into a new marriage. Spouses fantasize that their new marriage will be better than the previous one. If the new spouse has adult children, they assume that these children will be open to a rewarding relationship with them. Young children have their own fantasy—they hope that their biological parents will somehow get back together and that the stepfamily will be temporary.

Stage 2: *Reality.* Instead of realizing their fantasies, new spouses may find that stepchildren ignore or are rude to them. Indeed, stepparents may feel that they are outsiders in an already-functioning unit (the biological parent and child).

Stage 3: *Being Assertive.* Initially a stepparent assumes a passive role and accepts the frustrations and tensions of stepfamily life. Eventually, however, resentment can reach a level where the stepparent is driven to make changes. The stepparent may make the partner aware of the frustrations and suggest that the marital relationship should have priority some of the time. The stepparent may also make specific requests, such as reducing the number of conversations the partner has with the ex-spouse, not allowing the dog on the furniture, or requiring the stepchildren to use better table manners. This stage is successful to the degree that the partner supports the recommendations for change. A crisis may ensue.

Stage 4: *Strengthening Pair Ties.* During this stage, the remarried couple solidifies their relationship by making it a priority. At the same time, the biological parent must back away somewhat from the parent-child relationship so that the new partner can have the opportunity to establish a relationship with the stepchildren.

This relationship is the product of small units of interaction and develops slowly across time. Many day-to-day activities, such as watching television, eating meals, and riding in the car together, provide opportunities for the stepparent-stepchild relationship to develop. It is important that the stepparent not attempt to replace the relationship that the stepchildren have with their biological parents. The Self-Assessment feature provides a way to assess the degree to which a stepchild accepts and bonds with a stepfather.

Ganong et al. (2011) noted the influence of the relationship with one's biological parent as impacting the stepparent relationship. One respondent noted, "If my dad had a better relationship with my mom . . . that would have changed my relationships with my dad and stepmom. That was the one thing that kept me from getting close to my dad and stepmom."

Stage 5: *Recurring Change* A hallmark of all families is change, but this is even more true of stepfamilies. Bray and Kelly (1998) note that, even though a stepfamily may function well when the children are preadolescent, a new era can begin when the children become teenagers and begin to question how the family is organized and run. Such questioning by adolescents is not unique to stepchildren.

> *Holding everything inside will only result in your stomach keeping score.*
>
> Christy Borgeld, founder of National Stepfamily Day

David Knox

For some, it is never too late to start over. At fifty this man divorced, remarried, and had a daughter with his new wife (shown in photo). The man has three other daughters from his previous marriage and is now in his nineties.

The Parental Status Inventory (PSI) is a fourteen-item inventory that measures the degree to which respondents consider their stepfather to be a parent on an 11-point scale from 0% to 100%. Read each of the following statements and circle the percentage indicating the degree to which you regard the statement as true.

1. I think of my stepfather as my father. (0%, 10, 20, 30, 40, 50%, 60, 70, 80, 90, 100%)

2. I am comfortable when someone else refers to my stepfather as my father or dad. (0%, 10, 20, 30, 40, 50%, 60, 70, 80, 90, 100%)

3. I think of myself as his daughter/son. (0%, 10, 20, 30, 40, 50%, 60, 70, 80, 90, 100%)

4. I refer to him as my father or dad. (0%, 10, 20, 30, 40, 50%, 60, 70, 80, 90, 100%)

5. He introduces me as his son/daughter. (0%, 10, 20, 30, 40, 50%, 60, 70, 80, 90, 100%)

6. I introduce my mother and him as my parents. (0%, 10, 20, 30, 40, 50%, 60, 70, 80, 90, 100%)

7. He and I are just like father and son/daughter. (0%, 10, 20, 30, 40, 50%, 60, 70, 80, 90, 100%)

8. I introduce him as "my father" or "my dad." (0%, 10, 20, 30, 40, 50%, 60, 70, 80, 90, 100%)

9. I would feel comfortable if he and I were to attend a father-daughter/father-son function, such as a banquet, baseball game, or cookout, alone together. (0%, 10, 20, 30, 40, 50%, 60, 70, 80, 90, 100%)

10. I introduce him as "my mother's husband" or "my mother's partner." (0%, 10, 20, 30, 40, 50%, 60, 70, 80, 90, 100%)

11. When I think of my mother's house, I consider him and my mother to be parents to the same degree. (0%, 10, 20, 30, 40, 50%, 60, 70, 80, 90, 100%)

12. I consider him to be a father to me. (0%, 10, 20, 30, 40, 50%, 60, 70, 80, 90, 100%)

13. I address him by his first name. (0%, 10, 20, 30, 40, 50%, 60, 70, 80, 90, 100%)

14. If I were choosing a greeting card for him, the inclusion of the words *father* or *dad* in the inscription would prevent me from choosing the card. (0%, 10, 20, 30, 40, 50%, 60, 70, 80, 90, 100%)

Scoring

First, reverse the scores for items 10, 13, and 14. For example, if you circled a 90, change the number to 10; if you circled a 60, change the number to 40; and so on. Add the percentages and divide by 14. A 0% reflects that you do not regard your stepdad as your parent at all. A 100% reflects that you totally regard your stepdad as your parent. The percentages between 0% and 100% show the gradations from no regard to total regard of your stepdad as parent.

Norms

Respondents in two studies (one in Canada and one in America) completed the scale. The numbers of respondents in the studies were 159 and 156, respectively, and the average score in the respective studies was 45.66%. Between 40% and 50% of both Canadians and Americans viewed their stepfather as their parent.

*Developed by Dr. Susan Gamache. 2000. Hycroft Medical Centre, #217, 3195 Granville Street, Vancouver, B.C., Canada, V6H 3K2 gamache@interchange.ubc.ca. Details on construction of the scale including validity and reliability are available from Dr. Gamache. The PSI Scale is used in this text by permission of Dr. Gamache and may not be used otherwise (except as in class student exercises) without written permission.

Michaels (2000) noted that spouses who become aware of the stages through which stepfamilies pass report that they feel less isolated and unique. Involvement in stepfamily discussion groups such as the Stepfamily Enrichment Program provides enormous benefits.

Children in Stepfamilies

National **Data**

More than half of all children will spend some time in another family arrangement than the traditional family (Crosnoe and Cavanagh 2010). At any given time, 2% of children age 18 and below (5.3 million) are living with a biological parent and a married or cohabiting stepparent (Sweeney 2010).

Russell and Weaver (2010) noted that building step relationships is fluid and that patience and flexibility are needed. While step relationships are most easily developed when the children are in infancy and early development, older children can benefit if they are able to see their stepparents not as intrusive into their lives but as bringing positives (e.g., their parent is happier).

Stepfamily living is a stressful time for children and that stress may continue into adulthood. While the occurrence was low, Lizardi et al. (2010) found a relationship for women between having lived in a stepfamily and attempting suicide. The researchers analyzed data on 4,895 adults whose parents had divorced before age 13 and who had remarried. Seven percent of females who had lived with a stepparent had attempted suicide compared with 4.1% who had not lived with a stepparent. The respective percents for males were 2.5 and 2%.

Brown and Rinelli (2010) found that a cohabiting stepfamily was the family structure where adolescents were most like to be found smoking and drinking. Such a structure is likely to exercise the least strict social control on the children. However, Mackay (2005) compared children from intact and nonintact homes and found that, while the latter were "worse off," the differences between the groups were not large and were moderated by loss of income following the divorce, conflict between the parents, declines in mental health of custodial mothers, and compromised parenting (reduced attentiveness of parents due to their coping with their own adult issues). Similarly, Ruschena et al. (2005) compared longitudinal data on adolescents whose parents divorced or remarried versus those whose parents stayed married and found no significant group differences with regard to behavioral and emotional adjustment concurrently or across time, nor on academic outcomes and social competence. The researchers commented on the amazing resiliency of children as they transition through different family contexts.

A primary source of problems in stepfamilies is the continued conflict between the parents and their former spouses. Children caught in the conflict between their parents find being loyal to both parents impossible. Children temporarily resolve conflicts by siding with one parent at the expense of the relationship with the other parent. No one wins—children feel bad for abandoning a loving parent; the parent who has been tossed aside feels deprived of the opportunity for a close parent-child relationship; and the custodial parent runs the risk that the children, as adults, may resent being prevented from developing or continuing a relationship with the other parent.

Feelings of Abandonment, Divided Loyalties, New Discipline, and Stepsiblings

Children in stepfamilies often experience problems revolving around feeling abandoned, having divided loyalties, discipline, and stepsiblings. Some stepchildren feel that they have been abandoned twice—once when their parents got divorced and again when the parents turned their attention to new marital partners. One adolescent explained:

> It hurt me when my parents got divorced and my dad moved out. I really missed him and felt he really didn't care about me. But my sister and me adjusted with just my mom, and when everything was going right again, she got involved with this new guy and we were left with baby-sitters all the time. My dad also got involved with a new woman. I feel my sister and I have lost both parents in two years.

Coping with feelings of abandonment is not easy. It is best if the parents assure the children that the divorce was not the children's fault and that both parents love them a great deal. In addition, parents should be careful to find a balance between spending time with their new partner and spending time with their children. This translates into spending some alone time with their children.

Some children experience abandonment yet again if their parents' second marriage ends in divorce. They may have established a close relationship with the new stepparent only to find the relationship disrupted. Relationships with the stepgrandparents may also become strained. Stepgrandparents may be enormous sources of emotional support for children, but a divorce can interrupt this support.

Divided loyalties represent another issue children must deal with in step-families. Sometimes children develop an attachment to a stepparent that is more positive than the relationship with the natural parent of the same sex. When these feelings develop, children may feel they are in a bind. One adolescent boy explained:

> *My real dad left my mother when I was 6, and my mom remarried. My stepdad has always been good to me, and I really prefer to be with him. When my dad comes to pick me up on weekends, I have to avoid talking about my stepdad because my dad doesn't like him. I guess I love my dad, but I have a better relationship with my stepdad.*

For some adolescents, the more they care for the stepparent, the more guilty they feel, so they may try to hide their attachment. The stepparent may be aware of both positive and negative feelings coming from the child. Ideally, both the biological parent and the stepparent should encourage the child to have a close relationship with the other parent.

Discipline is another issue for stepchildren. Not only must they cope with new rules the stepparent may establish, they may feel no inclination to follow the rules of a virtual stranger. Therapists who work with stepfamilies encourage stepparents to discipline *their own* children only.

Siblings can also be a problem for stepchildren as they compete for parental approval, space, and shared materials (for example, the TV). Children who are already in a house may feel imposed upon and threatened. Children who are entering may feel out of place and that they do not belong. According to Pew Research, most adults who have step relatives feel a stronger sense of obligation to their biological family members than they do to their step kin. Adults are more inclined to come to the aid of their biological siblings than they are to assist their step- or half-siblings. Among those who have both biological siblings

This stepgrandparent is fishing with her stepgrandson.

David Knox

and step- or half-siblings, 64% say they would feel very obligated to a sibling who was in serious trouble. Only 42% say they would feel very obligated to provide assistance to a step- or half-sibling (Parker 2011).

Ambiguity of the Extended Family

A final issue for children in stepfamilies is their ambiguous place in the extended family system of the new stepparent. Although some parents and siblings of the new stepparent welcome a new stepchild into the extended family system, others may ignore the child because they are busy enough with their own grandchildren and children. In other cases, a biological parent may socialize the children to not accept the new stepgrandparents and stepparents or uncles: "My mom told us not to be fooled by dad's new wife's parents and siblings—that they really didn't care about us. I now know that she was just trying to get back at my dad. She knew that our not being open to a new family would make him sad. In effect my mom deprived us of getting to know some really great people."

Stepfamily living is often difficult for everyone involved: remarried spouses, children, even ex-spouses, grandparents, and in-laws. Though some of the problems begin to level out after a few years, others may take longer. Many couples become impatient with the unanticipated problems that are slow to abate, and they divorce.

Developmental Tasks for Stepfamilies

A **developmental task** is a skill that, if mastered, allows a family to grow as a cohesive unit. Developmental tasks that are not mastered will edge the family closer to the point of disintegration. Some of the more important developmental tasks for stepfamilies are discussed in this section.

Developmental task a skill that, if mastered, allows a family to grow as a cohesive unit.

Nurture the New Marriage Relationship

It is critical to the healthy functioning of a new stepfamily that the new spouses nurture each other and form a strong unit (Kim 2010). Indeed, the adult dyad is vulnerable because the couple had no childfree time upon which to build a common base. Once a couple develops a core relationship, they can communicate, cooperate, and compromise with regard to the various issues in their new blended family. Too often spouses become child-focused and neglect the relationship on which the rest of the family depends. Such nurturing translates into spending time alone with each other, sharing each other's lives, and having fun with each other. One remarried couple goes out to dinner at least once a week without the children. "If you don't spend time alone with your spouse, you won't have one," says one stepparent.

Allow Time for the Relationship between Partner and Children to Develop

In an effort to escape single parenthood and to live with one's beloved, some individuals rush into remarriage without getting to know each other. Not only do they have limited information about each other, but their respective children may also have spent little or no time with their future stepparent. One stepdaughter remarked, "I came home one afternoon to find a bunch of plastic bags in the living room with my soon-to-be stepdad's clothes in them. I had no idea he was moving in. It hasn't been easy." Both adults and children should have had meals together and spent some time in the same house before becoming bonded by marriage as a family.

The only things worth learning are the things you learn after you know it all.

Harry Truman, 33rd president of the U.S.

What if Your Children Will Not Accept Your New Spouse or Partner?

It never occurs to new lovers that their children will not embrace their new spouse or partner. After all, the new spouse or partner represents only joy and happiness for the recently divorced person. The children often see it otherwise. They see the new spouse or partner as the final ending of their safe and secure traditional family (mom and dad and the children in one house). Now they must be disrupted every other weekend and, when they are with their nonresident parent (usually the father), the "new woman is there."

Although most children mature and grow to accept their new stepparent, this may take years. In the meantime, the biological parent may need to lower expectations and settle for the children just being polite. The following is a script for a divorced father to give children who are barely civil: "I know the divorce is not what you wanted and you would prefer that your mom and I get back together and live in the same house again. That won't happen and we are both moving on. This means a new spouse for me and stepmother for you. Again, I know you do not want this and did not choose it. I don't require you to love and accept your new stepmother. I do require you to be polite. She will be kind to you and I expect you to be kind to her. That means "thank you" and "please." If you can't be at least polite, it is the end of your cell phone, computer/TV in your room, no friends over and no visiting friends. I will always be polite to your friends and I expect the same from you."

Have Realistic Expectations

Because of the complexity of meshing the numerous relationships involved in a stepfamily, it is important to be realistic. Dreams of one big happy family often set up stepparents for disappointment, bitterness, jealousy, and guilt. Some stepfamily members never bond. And others do so with a great deal of effort on the part of the divorcing spouses.

Accept Your Stepchildren

Rather than wishing your stepchildren were different, accepting them is more productive. All children have positive qualities; find them and make them the focus of your thinking. Stepparents may communicate acceptance of their stepchildren through verbal praise and positive or affectionate statements and gestures. In addition, stepparents may communicate acceptance by engaging in activities with their stepchildren and participating in daily activities such as homework, bedtime preparation, and transportation to after-school activities.

Establish Your Own Family Rituals

Don't marry someone who you can't see yourself being divorced from.

Kela Price, stepfamily counselor

Rituals are one of the bonding elements of nuclear families. Stepfamilies may integrate the various family members by establishing common rituals, such as summer vacations, visits to and from extended kin, and religious celebrations. These rituals are most effective if they are new and unique, not mirrors of rituals in the previous marriages and families.

Support the Children's Relationship with Their Absent Parent

A continued relationship with both biological parents is critical to the emotional well-being of children. Ex-spouses and stepparents should encourage children to have a positive relationship with both biological parents. In addition, if there is a stepparent, such access should also be provided.

Cooperate with the Children's Biological Parent and Co-Parent

A cooperative, supportive, and amicable co-parenting relationship between the biological parents and stepparents is a win-win situation for the children and parents. Otherwise, children are continually caught between the cross fire of the conflicted parental sets.

Structural Solution to Problems of Stepfamily Living

Living apart together (LAT) is a structural solution to many of the problems of stepfamily living (see Chapter 5 for a more thorough discussion). By getting two condos (side by side or one on top of the other), a duplex, or two small houses and having the respective biological parents live in each of the respective units with their respective children, everyone wins. The children and biological parent will experience minimal disruption as they transition to the new marriage. This arrangement is particularly useful where both spouses have children ranging in age from 10 to 18. Situations where only one spouse has children from a former relationship or those in which the children are very young will have limited benefit. The new spouses can still spend plenty of time together to nurture their relationship without spending all of their time trying to manage the various issues that come up with the stepchildren.

Stepfamily Education Web-based Program

Gelatt et al. (2010) studied the effect of a self-administered, interactive, and web-based stepfamily parent education program (*Parenting Toolkit: Skills for Stepfamilies* http://stepfamily.orcasinc.com) on a sample of 300 parents/stepparents of children, age 11–15, who were randomized into either treatment or delayed-access control groups. Results revealed that the stepfamily education program positively influenced several key areas of parenting and family functioning at post program and follow-up. Examples of positive change included reports of greater adjustment, harmony, life satisfaction, and a reduction of parent-child conflict.

The Future of Divorce and Remarriage

Divorce remains stigmatized in our society, as evidenced by the term **divorcism**— the belief that divorce is a disaster. In view of this cultural attitude, a number of attempts will continue to be made to reduce divorce rates. Marriage education workshops provide an opportunity for couples to meet with other couples and a leader who provides instruction in communication, conflict resolution, and parenting skills. The goal is to reduce divorce.

Another attempt at divorce prevention is **covenant marriage** (now available in Louisiana, Arizona, and Arkansas), which emphasizes the importance of staying married. Covenant marriage is discussed in more detail in Chapter 7. Although most of a sample of 1,324 adults in a telephone survey in Louisiana, Arizona, and Minnesota were positive about covenant marriage (Hawkins et al. 2002), fewer than 3% of marrying couples elected covenant marriages when given the opportunity to do so (Licata 2002). Although already married couples can convert their standard marriages to covenant marriages, there are no data on how many have done so (Hawkins et al. 2002).

Although most marriages begin with love and hope for a bright future, the reality is that over 40% will end in divorce. S. L. Brown (2010) noted the need for stable family structures for children and how researchers might become involved in informing federal policy debates to such stable structures. At the micro level, to protect themselves against a financial disaster should a marriage end in

Divorcism the belief that divorce is a disaster.

Covenant marriage attempt at divorce prevention which emphasizes the importance of staying married.

divorce, some individuals purchase "divorce insurance" (*www.divorceinsurance.com*). The cost is $16 a month for every $1,250 of coverage. To discourage couples from getting divorced just to collect the insurance money, policy holders must pay premiums for four years before they can collect on a divorce. The idea is the brainchild of John Logan, who watched his "wealth follow his divorce down the drain" (Luscombe 2010b). Persons who are getting remarried have the same hopes and dreams as those getting married for the first time. Since most first marrieds and remarrieds think their love is unique and immune to divorce, they will not be interested in divorce insurance.

Summary

What is divorce, how prevalent is it, and is it better to stay in an unhappy marriage?

Divorce is the legal ending of a valid marriage contract. About 43% of married couples eventually divorce after 15 years (this percentage lowers if the couple is college educated and wait until age 26 to marry). Divorce frequency has stabilized—the primary reason being that individuals are waiting until their mid-to-late twenties to marry. In general, the older a couple is at marriage, the less likely they are to divorce.

What are macro factors contributing to divorce?

Macro factors contributing to divorce include increased economic independence of women (women can afford to leave), changing family functions (companionship is the only remaining function), liberal divorce laws (it's easier to leave), fewer religious sanctions (churches embrace single individuals), more divorce models (Hollywood models abound), and individualism (rather than familism as a cultural goal of happiness).

What are micro factors contributing to divorce?

Micro factors include having numerous differences, falling out of love, negative behavior, lack of conflict resolution skills or satiation, value changes, and extramarital relationships. Those who are separated and going through a divorce say that they wished they had worked harder on their marriage.

What are the consequences of divorce for spouses/parents?

Women tend to fare better emotionally after separation and divorce than do men. Women are more likely than men not only to have a stronger network of supportive relationships but also to profit from divorce by developing a new sense of self-esteem and confidence, because they are thrust into a more independent role.

Both women and men experience a drop in income following divorce, but women may suffer more. Although 56% of custodial mothers are awarded child support, the amount is usually inadequate, infrequent, and not dependable, and women are forced to work (sometimes at more than one job) to take financial care of their children.

What are the effects of divorce on children?

Although researchers agree that a civil, cooperative, co-parenting relationship between ex-spouses is the greatest predictor of a positive outcome for children, researchers disagree on the long-term negative effects of divorce on children. However, there is no disagreement that most children do not experience long-term negative effects. Factors associated with minimizing the negative effects of divorce for children include healthy parental psychological functioning, a cooperative relationship between the parents, parental attention to the children, allowing them to grieve, encouragement to see noncustodial parent, attention from the noncustodial parent, assertion of parental authority, regular/consistent child support payments, stability (e.g., housing/neighborhood), and parents not having new baby, which drains off time from existing children.

What is the nature of remarriage in the United States?

Two thirds of divorced women and three fourths of divorced men remarry. When comparing divorced individuals who have remarried with divorced individuals who have not remarried, the remarried sometimes report greater personal and relationship happiness. Trust is a major issue for people getting remarried because they have already had at least one marriage fall apart. This is especially true if the partner had an affair or hid or spent money without the partner's knowledge.

Issues to confront/negotiate in a remarriage include boundary maintenance (not getting entangled with ex), emotional remarriage (loving/trusting again), psychic remarriage (give up freedom of singlehood), community remarriage (shift from single friends to new mate), parental remarriage (adjusting to stepfamily), and economic/legal remarriage (responsibilities to previous and current family).

What is the nature of stepfamilies in the United States?

Stepfamilies represent the fastest-growing type of family in the United States. Stepfamilies differ from nuclear families: the children in nuclear families are biologically related to both parents, whereas the children in stepfamilies are biologically related to only one parent. Also, in nuclear families, both biological parents live with their children, whereas only one biological parent in stepfamilies lives with the children. In some cases, the children alternate living with each parent. Stepism is the assumption that stepfamilies are inferior to biological families. Stepism, like racism, heterosexism, sexism, and ageism, involves prejudice and discrimination.

Stepfamilies go through a set of stages. New remarried couples often expect instant bonding between the new members of the stepfamily, but it does not often happen. The stages are: fantasy (everyone will love everyone), reality (possible bitter conflict), assertiveness (parents speak their mind), strengthening pair ties (spouses nurture their relationship), and recurring change (stepfamily members know there will continue to be change).

What are the developmental tasks of stepfamilies?

Developmental tasks for stepfamilies include nurturing the new marriage relationship, allowing time for partners and children to get to know each other, deciding whose money will be spent on whose children, deciding who will discipline the children and how, and supporting the children's relationship with both parents and natural grandparents. Both sets of parents and stepparents should form a parenting coalition in which they cooperate and actively participate in child rearing.

Key Terms

Arbitration	Divorce mediation	Parental alienation
Binuclear family	Divorcism	Parental alienation syndrome
Blended family	Family relations doctrine	Physical custody
Collaborative practice	Legal custody	Postnuptial agreement
Covenant marriage	Litigation	Satiation
December marriages	Negative commitment	Shared parenting dysfunction
Developmental task	Negotiation	Stepfamily
Divorce	No-fault divorce	Stepism

Web Links

Association for Conflict Resolution
http://www.acrnet.org

Center for Divorce Education
http://www.divorce-education.com/

Collaborative Divorce
http://www.collaborativepractice.com/default.asp

Dating after Divorce
http://www.heartchoice.com/hc/divorce/index.php

Divorce Laws by State
http://www.totaldivorce.com/state-laws/default.aspx

Divorce Source (a legal resource for divorce, custody, alimony, and support)
http://www.divorcesource.com/

Divorce Support Page
http://www.divorcesupport.com/index.html

Divorce 360 (divorce advice, news, blogs, and community)
http://www.divorce360.com/

Surviving Divorce
http://www.heartchoice.com/hc/divorce/index.php
http://www.divorceinfo.com/

DivorceBusting (solve marriage problems)
http://divorcebusting.com/

North Carolina Association of Professional Family Mediators
http://familymediators.org

The New Face of Divorce: First Wives World
http://firstwivesworld.com/

A Guide to the Parental Alienation Syndrome
http://www.coeffic.demon.co.uk/pas.htm

Positive Parenting Through Divorce Online Course
http://www.positiveparentingthroughdivorce.com/

National Center for Health Statistics (marriage and divorce data)
http://www.cdc.gov/nchs/

16 CHAPTER

Relationships in the Later Years

*Our wills and fates do so contrary run
That our devices still are overthrown,
Our thoughts are ours, their ends none of our own.*

Shakespeare, The Player King, in *Hamlet*

Sara DeGaetano

Learning Objectives

Specify the meanings of age and ageism.

Be aware of the challenges faced by the "sandwich" generation.

Discuss the primary issues (income, health, retirement) confronting the elderly.

Review relationships of the elderly.

Summarize grandparenting for the elderly.

Review end of life issues.

Predict the future of the elderly in the U.S.

A truism of relationships is that they are not the same in all of life's innings. The romance of youth, the exhilaration of early marriage, the discipline of parenting, the relief/sadness when the children leave home, the transition to retirement, and the uncertainty of the sunset years are typically experienced in a relationship context.

In 2015, about 14.4% of the 325 million individuals in the United States will be age 65 and older (*Statistical Abstract of the United States, 2011*, Table 8). This represents about 46 million "elderly" individuals. By 2030, this percentage will grow to 20% of the population in the United States. By 2050, there will be almost 90 million people over 65 (J. F. Potter 2010).

In this chapter, we focus on the factors that confront individuals and couples as they age and the dilemma of how to care for aging parents. We begin by looking at the concept of age.

Age and Ageism

All societies have a way to categorize their members by age. And all societies provide social definitions for particular ages.

The Concept of Age

A person's **age** may be defined chronologically, physiologically, psychologically, sociologically, and culturally. Chronologically, an "old" person is defined as one who has lived a certain number of years. How many years it takes to be regarded as old varies with one's own age. Children of 12 may regard siblings of 18 as old—and their parents as "ancient." Teenagers and parents may regard themselves as "young" and reserve the label "old" for their grandparents' generation.

Chronological age has obvious practical significance in everyday life. Bureaucratic organizations and social programs identify chronological age as a criterion of certain social rights and responsibilities. One's age determines the right to drive, vote, buy alcohol or cigarettes, and receive Social Security and Medicare benefits.

Age has meaning in reference to the society and culture of the individual. In ancient Greece and Rome, where the average life expectancy was 20 years, one was old at 18; similarly, one was old at 30 in medieval Europe and at age 40 in the United States in 1850. In the United States today, however, people are usually not considered old until they reach age 65. However, our society is moving toward new chronological definitions of "old." Three groups of the elderly are the "young-old," the "middle-old," and the "old-old." The young-old are typically

The years teach much which the days never knew.

Ralph Waldo Emerson

Age is an issue of mind over matter. If you don't mind, it doesn't matter.

Mark Twain, humorist

Age term which may be defined chronologically (number of years), physiologically (physical decline), psychologically (self-concept), sociologically (roles for the elderly/retired), and culturally (meaning of age in one's society).

TABLE 16.1	Life Expectancy			
Year	White Males	Black Males	White Females	Black Females
2010	76.5	70.2	81.3	77.2
2015	77.1	71.4	81.8	78.2
2020	77.7	72.6	82.4	79.2

Source: *Statistical Abstract of the United States, 2011,* 130th ed. Washington, DC: U.S. Bureau of the Census, Table 102.

between the ages of 65 and 74; the middle-old, 75 to 84, and the old-old, 85 and beyond. Current life expectancy is shown in Table 16.1.

The age of a person influences that person's view of when a person becomes "old." Individuals aged 18 to 35 identify 50 as the age when the average man or woman becomes "old." However, those between the ages of 65 and 74 define "old" as 80 (Cutler 2002). Hostetler (2011) noted that some senior citizens may avoid senior centers due to an "image" problem—they don't want to be associated with being labeled as one of the old folks.

Views of the elderly are changing from people who are needy and dependent to people who are active and resourceful. Research is underway to extend life. Dr. Aubrey de Grey of the Department of Genetics, University of Cambridge, predicted that continued research will make adding *hundreds* of years to one's life possible. Maher and Mercer (2009) noted that the most optimistic predictions are that we are two or three decades away from significant breakthroughs (Ray Kurzweil says we are forty-nine years away). Maher and Mercer (2009) also note the paths are known via genetic engineering (we must learn how to switch off the genes responsible for aging), tissue or organ replacement (knees and kidneys are already being replaced), and the merging of computer technology with human biology (hearing and seeing are now being improved). Should the human lifespan be extended hundreds of years, imagine the impact on marriage—"till death do us part"?

Some individuals have been successful in delaying the aging process. Clint Eastwood, now in his eighties, continues to make Academy Award winning movies. Actress Betty White, at age 90, is still hosting Saturday Night Live, making movies, and appearing in commercials. Fitness guru Jack LaLanne, was still active until his death at age 96.

Physiologically, people are old when their auditory, visual, respiratory, and cognitive capabilities decline significantly. At age 80 and over, 45% are hearing impaired and 25% have visual impairment (Dillion et al. 2010). Indeed, Siedlecki (2007) confirmed that increased age is associated with increased difficulty in retrieving information. Becoming "disabled" is associated with being "old." Individuals tend to see themselves as disabled when their driver's licenses are taken away and when home health care workers come to their home to care for them (Kelley-Moore et al. 2006). Sleep changes occur for the elderly including going to bed earlier, waking up during the night, and waking up earlier in the morning, as well as issues like restless legs syndrome, snoring, and obstructive sleep apnea (Wolkove et al. 2007).

People who need full-time nursing care for eating, bathing, and taking medication properly and who are placed in nursing homes are thought of as being old. Failing health is the criterion the elderly use to define themselves as old (O'Reilly 1997), and successful aging is culturally defined as maintaining one's health, independence, and cognitive ability. Jorm et al. (1998) observed that the prevalence of successful aging declines steeply from age 70 to age 80. Garrett

In youth the days are short and the years are long; in old age the years are short and the days long.

Nikita Ivanovich Panin

To be young is all there is in the world... [Adults] talk so beautifully about work and having a family and a home (and I do, too, sometimes)—but it's all worry and headaches and respectable poverty and forced gushing.... Telling people how nice it is, when, in reality, you would rather give all of your last thirty years for one hour of your first thirty. Old people are tremendous frauds.

Walter Stevens

and Martini (2007) noted the impact on the U.S. health care system as increasing numbers of baby boomers age.

People who have certain diseases are also regarded as old. Although younger individuals may suffer from Alzheimer's, arthritis, and heart problems, these ailments are more often associated with aging. As medical science conquers more diseases, the physiological definition of aging changes so that it takes longer for people to be defined as "old."

Psychologically, a person's self-concept is important in defining how old that person is. As individuals begin to fulfill the roles associated with the elderly—retiree, grandparent, nursing home resident—they begin to see themselves as aging. Sociologically, once they occupy these roles, others begin to see them as "old." Culturally, the society in which an individual lives defines when and if a person becomes old and what being old means. In U.S. society, the period from age 18 through 64 is generally subdivided into young adulthood, adulthood, and middle age. Cultures also differ in terms of how they view and take care of their elderly. Spain is particularly noteworthy in terms of care for the elderly, with eight of ten elderly people receiving care from family members and other relatives. The elderly in Spain report very high levels of satisfaction in the relationships with their children, grandchildren, and friends (Fernandez-Ballesteros 2003).

Ageism

Every society has some form of **ageism**—the systematic persecution and degradation of people because they are old. Ageism is similar to sexism, racism, and heterosexism. The elderly are shunned, discriminated against in employment, and sometimes victims of abuse. Media portrayals contribute to the negative image of the elderly. They are portrayed as difficult, complaining, and burdensome and are often underrepresented in commercials and comic strips.

Negative stereotypes and media images of the elderly engender **gerontophobia**—a shared fear or dread of the elderly, which may create a self-fulfilling prophecy. For example, an elderly person forgets something and attributes the behavior to age. A younger person, however, engaging in the same behavior, is unlikely to attribute forgetfulness to age, given cultural definitions surrounding the age of the onset of senility. Individuals are also thought to become more inflexible and more conservative as they age. Analysis of national data suggests this is not true. In fact, the elderly become more tolerant (Danigelis et al. 2007).

The negative meanings associated with aging underlie the obsession of many Americans to conceal their age by altering their appearance. With the hope of holding on to youth a little bit longer, aging Americans spend billions of dollars each year on plastic surgery, exercise equipment, hair products, facial creams, and Botox injections.

Theories of Aging

Gerontology is the study of aging. Table 16.2 identifies several theories, the level (macro or micro) of the theory, the theorists typically associated with the theory, assumptions, and criticisms. As noted, there are diverse ways of conceptualizing the elderly. Currently

Ageism the systematic persecution and degradation of people because they are old.

Gerontophobia fear or dread of the elderly, which may create a self-fulfilling prophecy.

Gerontology the study of aging.

Filial piety love and respect toward parents including bringing no dishonor to parents and taking care of elderly parents.

Filial responsibility emphasizes duty, protection, care, and financial support for one's parents.

Everyone is the age of their heart.

Guatemalan Proverb

DIVERSITY IN OTHER COUNTRIES

Among Asians, the high status of the elderly in the extended family derives from religion. Confucian philosophy, for example, prescribes that all relationships are of the subordinate-superordinate type—husband-wife, parent-child, and teacher-pupil. For traditional Asians to abandon their elderly rather than include them in larger family units would be unthinkable. However, commitment to the elderly may be changing as a result of the Westernization of Asian countries such as China, Japan, and Korea.

DIVERSITY IN OTHER COUNTRIES

Whereas female children in the United States have more frequent contact with and are more involved in the caregiving of their elderly parents than male children, the daughters-in-law in Japan offer the most help to elderly individuals. The female child who is married gives her attention to the parents of her husband (Ikegami 1998).

Eastern cultures emphasize filial piety, which is love and respect toward their parents. **Filial piety** involves respecting parents, bringing no dishonor to parents, and taking good care of parents. Western cultures are characterized by **filial responsibility** emphasizing duty, protection, care, and financial support to one's parents.

TABLE 16.2   Theories of Aging

Name of Theory	Level of Theory	Theorists	Basic Assumptions	Criticisms
Disengagement	Macro	Elaine Cumming William Henry	The gradual and mutual withdrawal of the elderly and society from each other is a natural process. It is also necessary and functional for society that the elderly disengage so that new people can be phased in to replace them in an orderly transition.	Not all people want to disengage; some want to stay active and involved. Disengagement does not specify what happens when the elderly stay involved.
Activity	Macro	Robert Havighurst	People continue the level of activity they had in middle age into their later years. Though high levels of activity are unrelated to living longer, they are related to reporting high levels of life satisfaction.	Ill health may force people to curtail their level of activity. The older a person, the more likely the person is to curtail activity.
Conflict	Macro	Karl Marx Max Weber	The elderly compete with youth for jobs and social resources such as government programs (Medicare).	The elderly are presented as disadvantaged. Their power to organize and mobilize political resources such as the American Association of Retired Persons is underestimated.
Age stratification	Macro	M. W. Riley	The elderly represent a powerful cohort of individuals passing through the social system that both affect and are affected by social change.	Too much emphasis is put on age, and little recognition is given to other variables within a cohort such as gender, race, and socioeconomic differences.
Modernization	Macro	Donald Cowgill	The status of the elderly is in reference to the evolution of the society toward modernization. The elderly in premodern societies have more status because what they have to offer in the form of cultural wisdom is more valued. The elderly in modern technologically advanced societies have low status because they have little to offer.	Cultural values for the elderly, not level of modernization, dictate the status of the elderly. Japan has high respect for the elderly and yet is highly technological and modernized.
Symbolic	Micro	Arlie Hochschild	The elderly socially construct meaning in their interactions with others and society. Developing social bonds with other elderly can ward off being isolated and abandoned. Meaning is in the interpretation, not in the event.	The power of the larger social system and larger social structures to affect the lives of the elderly is minimized.
Continuity	Micro	Bernice Neugarten	The earlier habit patterns, values, and attitudes of the individual are carried forward as a person ages. The only personality change that occurs with aging is the tendency to turn one's attention and interest on the self.	Other factors than one's personality affect aging outcomes. The social structure influences the life of the elderly rather than vice versa.

popular in sociology is the life-course perspective (Willson 2007). This approach examines differences in aging across cohorts by emphasizing that "individual biography is situated within the context of social structure and historical circumstance" (p. 150).

Caregiving for the Frail Elderly—the "Sandwich Generation"

Elderly people are defined as **frail** if they have difficulty with at least one personal care activity or other activity related to independent living; the severely disabled are unable to complete three or more personal care activities. These personal care activities include bathing, dressing, getting in and out of bed, shopping for groceries, and taking medications.

Only 6.8% of the frail elderly have long-term health care insurance (Johnson and Wiener 2006). While they may have other resources to pay for elder care, their children often end up taking care of their parents. The term *children* typically means female adult children. Indeed, women account for about two-thirds of all unpaid caregivers (ibid.). The adults who provide **family caregiving** to these elderly parents (and their own children simultaneously) are known as the "**sandwich generation**" because they are in the middle of taking care of the needs of both their parents and children.

Caregiving for an elderly parent has two meanings. One, caregiving refers to providing personal help with the basics of daily living such as helping the parent get in and out of bed, bathing, toileting, and eating. A second form of caregiving refers to instrumental activities such as grocery shopping, money management (including paying bills), and driving the parent to the doctor.

The typical caregiver is a middle-aged married woman who works outside the home. High levels of stress and fatigue may accompany caring for one's elders. Lee et al. (2010a) studied family members (average age = 46) providing care for elderly parents as well as children and found that both responsibilities created time and stress problems. When the two roles were compared, taking care of the parents often translated into missing more work days. Kingsberry et al. (2010) noted additional stress of African-American caregivers due to the effect of limited economic resources, multiple caregiving roles, and dwindling social support.

The number of individuals in the sandwich generation will increase for the following reasons:

1. *Longevity.* The over-85 age group, the segment of the population most in need of care, is the fastest-growing segment of our population.

2. *Chronic disease.* In the past, diseases took the elderly quickly. Today, diseases such as arthritis and Alzheimer's are associated not with an immediate death sentence but a lifetime of managing the illness and being cared for by others. Family caregivers of parents with Alzheimer's note the difficulty of the role: "He's not the man I married," lamented one wife. Hilgeman et al. (2007) noted that having a positive view of taking care of an Alzheimer's patient was associated with experiencing less stress in doing so.

3. *Fewer siblings to help.* The current generation of elderly had fewer children than the elderly in previous generations. Hence, the number of adult siblings to help look after parents is more limited. Only children are more likely to feel the weight of caring for elderly parents alone.

4. *Commitment to parental care.* Contrary to the myth that adult children in the United States abrogate responsibility for taking care of their elderly parents, most children institutionalize their parents only as a last resort. Asian children, specifically Chinese children, are socialized to expect to take care of their elderly in the home.

In a world there are no people so piteous and forlorn as those who are forced to eat the bitter bread of dependency in their old age, and find how steep are the stairs of another man's house.

Dorothy Dix, mental health advocate

Frail term used to define elderly people if they have difficulty with at least one personal care activity (feeding, bathing, toileting).

Family caregiving adult children providing care for their elderly parents.

Sandwich generation generation of adults who are "sandwiched" between caring for their elderly parents and their own children.

5. *Lack of support for the caregiver.* Caring for a dependent, aging parent requires a great deal of effort, sacrifice, and decision making on the part of more than 14 million adults in the United States who are challenged with this situation. The emotional toll on the caregiver may be heavy. Guilt (over not doing enough), resentment (over feeling burdened), and exhaustion (over the relentless care demands) are common feelings that are sometimes mixed. One caregiver adult child said, "I must be an awful person to begrudge taking my mother supper, but I feel that my life is consumed by the demands she makes on me, and I have no time for myself, my children, or my husband." Sheehy (2010) cared for her husband who had terminal cancer. She noted the importance of NOT trying to be the sole caretaker but getting help. Marks et al. (2002) noted an increase in symptoms of depression among a national sample of caregivers (of a child, parent, or spouse). Caregiving can also be expensive and can devastate a family budget.

Some reduce the strain of caring for an elderly parent by arranging for home health care. This involves having a nurse go to the home of a parent and provide such services as bathing the parent and giving medication. Other services may include taking meals to the elderly (for example, through Meals on Wheels). The National Family Caregiver Support Program (see web link at end of chapter) provides support services for individuals (including grandparents) who provide family caregiving services. Such services might include eldercare resource and referral services, caregiver support groups, and classes on how to care for an aging parent. In addition, states are increasingly providing family caregivers a tax credit or deduction.

Offspring who have no help may become overwhelmed and frustrated. Elder abuse, an expression of such frustration, is not unheard of (we discussed elder abuse in Chapter 13). Many wrestle with the decision to put their parents in a nursing home or other long-term care facility. We discuss this issue in the following Personal Choices section.

> *If you are yearning for the good old days, just turn off the air conditioning.*
>
> Griff Niblack, author

Fred Johnson

This daughter made the painful decision to put her mother in a nursing home when the mother developed Alzheimer's and was no longer safe to be by herself.

Should I Put My Parents in a Long-Term Care Facility?

While 80% of long-term care for the elderly is met by families, increasingly the elderly are being cared for by nursing homes (Potter 2010). Over 1.8 million individuals are in a nursing home. Twenty-eight percent of these are 75–84 years old (*Statistical Abstract of the United States, 2011,* Table 73). Factors relevant in deciding whether to care for an elderly parent at home, arrange for nursing home care, or provide another form of long-term care include the following.

1. *Level of care needed.* As parents age, the level of care that they need increases.

An elderly parent who cannot bathe, dress, prepare meals, or be depended on to take medication responsibly needs either full-time in-home care or a skilled nursing facility that provides 24-hour nursing supervision by registered or licensed vocational nurses. Commonly referred to as "nursing homes" or "convalescent hospitals," these facilities provide medical, nursing, dietary, pharmacy, and activity services.

An intermediate-care facility provides eight hours of nursing supervision per day. Intermediate care is less extensive and expensive and generally serves patients who are ambulatory and who do not need care throughout the night.

A skilled nursing facility for special disabilities provides a "protective" or "security" environment to people with mental disabilities. Many of these facilities have locked areas where patients reside for their own protection.

An assisted living facility is for individuals who are no longer able to live independently but who do not need the level of care that a nursing home provides. Although nurses and other health care providers are available, assistance is more typically in the form of meals and housekeeping.

Retirement communities involve a range of options, from apartments where residents live independently to skilled nursing care. These communities allow older adults to remain in one place and still receive the care they need as they age.

2. *Temperament of parent.* Some elderly parents have become paranoid, accusatory, and angry with their caregivers. Family members no longer capable of coping with the abuse may arrange for their parents to be taken care of in a nursing home or other facility.

3. *Philosophy of adult child.* Most children feel a sense of filial responsibility—a sense of personal obligation for the well-being of aging parents. Theoretical explanations for such responsibility include the norm of reciprocity (adult children reciprocate the care they received from their parents), attachment theory (caring results from positive emotions for one's parents), and a moral imperative (caring for one's elderly parents is the right thing to do).

One only child promised his dying father that he would take care of the father's spouse (the child's mother) and not put her in a nursing home. When the mother became 82 and unable to care for herself, the son bought a bed and made the living room of his home the place for his mother to spend the last days of her life. "No nursing home for my mother," he said. This man also had the same philosophy for his mother-in-law and moved her into the home with his wife when she was 94.

4. *Siblings.* Most adult children have siblings with whom they can share the work and expense of caring for their elderly parents. Shared elder parental care may be an enormous source of pride or conflict for siblings. While some cooperate, others disagree over who is to do what, often resulting in only one sibling doing the work/spending the money, which can destroy the relationship between the siblings. One sister who took care of her aging mother with zero help from her sister said in retrospect, "I am shocked at the absence of my sister during the declining days of our mother. Our relationship is done and I'll never see her again." Russo (2010) identified the issues siblings confront with the care of their aging parents.

5. *Length of time for providing care.* Offspring must also consider how long they will be in the role of caring for an aging parent. The duration of caregiving can last from

less than a year to more than forty years. About 40% of caregivers provide assistance for five or more years; nearly a fifth provides assistance for ten or more years (Family Caregiver Alliance 2006). A team of researchers (Walz and Mitchell 2007) noted that both adult offspring and their parents underestimated the length of time both they and their parents would need full-time care as an older person.

6. *Privacy needs of caregivers.* Some spouses take care of their elderly at home but note the effect on their own marital privacy. A wife who took care of her husband's mother for twelve years in her home said that it was a relief for his mother to die and for them to get their privacy back.

7. *Cost.* For private full-time nursing home care, including room, board, medical care, and so on, count on spending $1,000 to $1,500 a week. Because women live longer than men and represent 70% of the older population living in poverty, they are more likely to have the need for eldercare and will not have the resources for such care unless provided by their children (Willson 2007). **Medicare**, a federal health insurance program for people 65 and older, was developed for short-term acute hospital care. Medicare generally does not pay for long-term nursing care. In practice, adult children who arrange for their aging parent to be cared for in a nursing home end up paying for it out of the elder's own funds. After all of these economic resources are depleted, **Medicaid**, a state welfare program for low-income individuals, will pay for the cost of care. A federal law prohibits offspring from shifting the assets of an elderly parent so as to become eligible for Medicaid.

A crisis in care for the elderly is looming. In the past, women have taken care of their elderly parents. However, these were women who lived in traditional families where one paycheck took care of a family's economic needs. Women today work out of economic necessity, and quitting work to take care of an elderly parent is becoming less of an option. As more women enter the labor force, less free labor is available to take care of the elderly. Government programs are not in place to take care of the legions of elderly Americans. Who will care for them when both spouses are working full-time? China is facing a similar crisis. Zhan et al. (2008) noted that, due to the unavailability of adult children (for example, China has had its one-child policy), more and more elderly parents are ending up in nursing homes.

8. *Chain nursing home.* Lucas et al. (2007) found that residents of nursing homes reported higher satisfaction if they were not in a "chain nursing home." Offspring might keep this in mind when selecting a long-term care facility for their parents.

9. *Sexual orientation.* Homosexual elders may be resistant to go to a nursing home because they fear prejudice and discrimination from workers and patients at the facility.

10. *Wishes or readiness of the elderly.* The elderly should be included in the decision to be cared for in a long-term care facility. Wielink et al. (1997) found that the more frail the elderly person was, the more willing he or she was to go to a nursing home. Cohen-Mansfield and Wirtz (2007) also noted that characteristics of those "ready" to go into a nursing home included advanced age, depression, and a higher number of psychiatric diagnoses. Once a decision is made for nursing home care, it is important to assess several facilities. Taking a tour of the facility, eating a meal at the facility, and meeting staff are also helpful in making a decision and a smooth transition. Because there is an acute nursing shortage, it is important to find out the ratio of registered nurses per patient. Some elderly adapt well to living in a residential facility, such as a nursing home. They enjoy the community of others of similar age, enjoy visiting others in the nursing home, and reach out to others to make new friends. In essence, they find positive meaning in the nursing home experience.

11. *Other issues.* Whether or not deciding to put one's parent (or spouse) in a nursing home, the elderly person or those with power of attorney should complete a document called an **advance directive** (also known as a **living will**), detailing the conditions under which life support measures should be used (do they want to be sustained on a respirator?). These decisions, made ahead of time, spare the adult children the responsibility of making them in crisis contexts and give clear directives to the medical staff in charge of the elderly person. For example, elderly people can direct that a feeding tube

Medicare a federal health insurance program for people 65 and older.

Medicaid a state welfare program for low-income individuals.

Advance directive (living will) details for medical care personnel the conditions under which life support measures should be used for one's partner.

should not be used if they become unable to feed themselves. Hence, by making this decision, the children are spared the decision regarding a feeding tube. A **durable power of attorney**, which gives adult children complete authority to act on behalf of the elderly, is also advised. These documents also help to save countless legal hours, time, and money for those responsible for the elderly. Appendixes D and E present examples of the living will and durable power of attorney.

Finally, adult children may consider buying long-term care insurance (LTCI) to cover what Medicare and many private health care plans do not cover "nonmedical" day-to-day care such as bathing or eating for an Alzheimer's parent, as well as nursing home costs. Costs of LTCI begin at about $1,000 a year but vary a great deal (including over a thousand a month) depending on the age and health of the insured individual.

As an aside, the first author's mother was 90 when she died. With a husband who had died forty-five years earlier, two adult sons who lived in other states, and no grandchildren or great-grandchildren near, she lived out the last twelve years of her life in a nursing home. There is considerable stigma about children "putting their parents in a nursing home." However, this woman's last years and days were happy ones. Both her mother and younger brother had lived out their last days in a nursing home. Life in a nursing home had become normative. She never complained, always smiled, and was grateful for each day. She loved the food, enjoyed the visits and phone calls, and commented on the helpful care she received from nurses and doctors. Her oldest son was present when she died.

Sources

Cohen-Mansfield, J., and P. W. Wirtz. 2007. Characteristics of adult day care participants who enter a nursing home. *Psychology and Aging* 22:354–360.

Lucas, J. A., C. A. Levin, T. J. Lowe, and B. Robertson. 2007. The relationship between organizational factors and resident satisfaction with nursing home care and life. *Journal of Aging & Social Policy* 19:125–135.

Potter, J. F. 2010. Aging in America: essential considerations in shaping senior care policy. *Aging Health* 6.3: 289–300.

Statistical Abstract of the United States, 2011. 130th ed. Washington, DC: U.S. Bureau of the Census.

Walz, H. S., and T. E. Mitchell. 2007. Adult children and their parents' expectations of future elder care needs. *Journal of Aging and Health* 19:482–491.

Wielink, G., R. Huijsman, and J. McDonnell. 1997. A study of the elders living independently in the Netherlands. *Research on Aging* 19:174–198.

I think I would never worry about age if I knew I could go on being loved or having the possibility of love.

Audrey Hepburn

Issues Confronting the Elderly

Numerous issues become concerns as people age. In middle age, the issues are early retirement (sometimes forced), job layoffs (recession-related cutbacks), **age discrimination** (older people are often not hired and younger workers are hired to take their place), separation or divorce from a spouse, and adjustment to children leaving home. For some in middle age, grandparenting is an issue if they become the primary caregiver for their grandchildren. As couples move from the middle to the later years, the issues become more focused on income, health, retirement, and sexuality.

Middle age—when you are sitting at home on Saturday night and the phone rings and you hope it is not for you.

Ring Lardner, American humorist

Income

For most individuals, the end of life is characterized by reduced income. Social Security and pension benefits, when they exist, are rarely equal to the income a retired person formerly earned.

National **Data**

The median annual income of men aged 65 and older is $25,503; women, $14,559 (*Statistical Abstract of the United States, 2011*, Table 701).

The HBO movie *You Don't Know Jack* (2010) reviewed the life of Jack Kevorkian and revived the debate on physician assisted suicide. **Euthanasia** is from the Greek words meaning "good death," or dying without suffering. Euthanasia may be passive, where medical treatment is withdrawn and nothing is done to prolong the life of the patient, or active, which involves deliberate actions to end a person's life.

Adult children or spouses are often asked their recommendations about withdrawing life support (food, water, or mechanical ventilation), starting medications to end life (intravenous vasopressors), or withholding certain procedures that would prolong life (cardiopulmonary resuscitation). In the United States, 60% of adults in a Gallup poll reported that they approved of physician-assisted suicide (Carroll 2006). The debate includes not only the terminally ill but those who find that their life is intolerable (Sullivan 2010).

One's aging parents may experience a significant drop in **quality of life**—defined in terms of physical functioning, independence, economic resources, social relationships, and spirituality (Willson 2007). These parents may also not want to be a burden to their children (Schaffer 2007), and may ask for death. The top reasons patients cited for wanting to end their lives are losing autonomy (84%), decreasing ability to participate in activities they enjoyed (84%), and losing control of bodily functions (47%) (Chan 2003).

Physician-assisted suicide (PAS) is legal in the Netherlands. Georges (2008) found that nearly half of a group of general practitioners wanted to avoid physician-assisted suicide because it was against their own personal values or because it was an emotional burden for them to confront the issue. Douglas (2008) observed a "double effect" of sedatives and analgesics administered at the end of life—not only do these relieve pain but can also hasten death. Some physicians find that "slow euthanasia" is more psychologically acceptable to doctors than active voluntary euthanasia by injection.

All fifty states now have laws for living wills, permitting individuals to decide (or family members to decide on their behalf) to withhold artificial nutrition (food) and hydration (water) from a patient who is wasting away. In practice, this means not putting in a feeding tube. For elderly, frail patients who may have a stroke or heart attack, do-not-resuscitate (DNR) orders may also be put in place (Cardozo 2006). The Supreme Court has ruled that state law will apply in regard to physician-assisted suicide. In January 2006, the Supreme Court ruled that Oregon has a right to physician-assisted suicide. Its Death with Dignity Act requires that two physicians must agree that the patient is terminally ill and is expected to die within six months, the patient must ask three times for death both orally and in writing, and the patient must swallow the barbiturates themselves rather than be injected with a drug by the physician. The number of physician-assisted suicides increased from twenty-one in 2001, to thirty-eight in 2002 (an 81% increase). Most patients had cancer and were more likely to be white, male, and well-educated (Chan 2003).

Euthanasia from the Greek meaning "good death" or dying without suffering; either passively or actively ending the life of a patient.

Quality of life one's physical functioning, independence, economic resources, social relationships, and spirituality.

Financial planning to provide end-of-life income is important. Sometimes only major changes—health issues, death of a spouse, divorce, or remarriage—jolt a person into end of life planning. Some adults buy long-term health care insurance. Such insurance can be costly and does not always cover needed expenses.

To be caught at the end of life without adequate resources is not unusual. And surviving a major health crisis if there is no insurance can be a catastrophe (Cook et al. 2010). Women are particularly disadvantaged because their work history has often been discontinuous, part-time, and low-paying. Social Security and private pension plans favor those with continuous full-time work histories. Even so, Social Security benefits amount to only 42% of the worker's preretirement wage (J. F. Potter 2010).

Physical Health

Good physical health is the single most important determinant of an elderly person's reported happiness (Smith et al. 2002). Franks et al. (2010) noted that one elderly spouse with diabetes affects both spouses in terms of depressive symptoms. Weight also has an effect on one's health as one ages. Gadalla (2010) studied the characteristics of those over 65 who had limitations in IADL

The official position of the American Medical Association (AMA) is that physicians must respect the patient's decision to forgo life-sustaining treatment but that they should not participate in patient-assisted suicide: "PAS is fundamentally incompatible with the physician's role as healer." Rather, the AMA affirms physicians who support life. Arguments against PAS emphasize that people who want to end the life of those they feel burdened by (or worse, for money) can abuse the practice and that, because physicians make mistakes, what is diagnosed as "terminal" may not in fact be terminal.

Physician-assisted suicide has been legal in Holland for almost 20 years. A concern has been the potential to misuse the law. However, a Dutch study of 5,000 requests per year of euthanasia and physician-assisted suicide in general practice over twenty-five years concluded, "Some people feared that the lives of increasing numbers of patients would end through medical intervention, without their consent and before all palliative options were exhausted. Our results, albeit based on requests only, suggest that this fear is not justified" (Marquet et al. 2003, 202).

Your Opinion?

1. Suppose your father has Alzheimer's disease and is in a nursing home. He is 88 and no longer recognizes you. He has stopped eating. Would you have a feeding tube inserted to keep him alive?

2. To what degree do you agree with the Death with Dignity policy operative in Oregon?

3. What do you think the position of the government should be in regard to physician-assisted suicide?

Sources

Carroll, J. 2006. Public continues to support right-to-die for terminally ill patients. Gallop Poll, June 19. www.galluppoll.com/content/CI=23356 (retrieved June 21).

Chan, S. 2003. Rates of assisted suicides rise sharply in Oregon. *Student BMJ* 11:137–138.

Douglas, C., I. Kerridge, and R. Ankeny. 2008. Managing intentions: End of life administration of analgesics and sedatives, and the possibility of slow euthanasia. *Bioethics* 22:388–402.

Georges, J. J. 2008. Dealing with requests for euthanasia: a qualitative study investigating the experience of general practitioners. *Journal of Medical Ethics* 34:150–163.

Marquet, R., A. Bartelds, G. J. Visser, P. Spreeuwenberg, and I. Peters. 2003. Twenty-five years of requests for euthanasia and physician assisted suicide in Dutch practice: Trend analysis. *British Medical Journal* 327: 201–202.

Schaffer, M. 2007. Ethical problems in end of life decisions for elderly Norwegians. *Nursing Ethids* 14:242–257.

Sullivan, S. 2011. The right to die: a discussion of 'rational suicide'. *Mental Health Practice.* 14:32–34.

Willson, A. E. 2007. The sociology of aging. In *21st century sociology: A reference handbook*, ed. Clifton D. Bryant and Dennis L. Peck, 148–155. Thousand Oaks, California: Sage.

(instrumental activities of daily living) due to body weight and found that older women and those not in a relationship were more vulnerable. Weight may also affect one's mental health. Carroll et al. (2010) found that, for women, becoming obese was associated with depressive symptoms.

Perceived health status varies by race/ethnicity. Liang et al. (2010) compared the self-rated health of 18,486 Americans 50 or above in age and found that White Americans rated their health most positively,

DIVERSITY IN OTHER COUNTRIES

In a cross-cultural study of 21,000 adults, ages 40 to 80, in 21 countries, key ingredients associated with reported happiness were good health, decent standard of living, genes that predispose individuals to being optimistic, having happy parents, and continued involvement in a meaningful activity (for example, taking care of a grandchild, volunteering, and employment) (Coombes 2007).

followed by Black Americans, with Hispanics rating their health least positively. Regardless of racial/ethnic background, most elderly individuals, even those of advanced years, continue to define themselves as being in good health. Ostbye et al. (2006) studied an elderly population in Cache County, Utah, and found that 80% to 90% of those aged 65 to 75 were healthy on ten dimensions of health (independent living, vision, hearing, activities of daily living, instrumental activities of daily living, absence of physical illness, cognition, healthy mood, social support and participation, and religious participation and spirituality). Prevalence of excellent and good self-reported health decreased with age, to approximately 60%

What if Your Spouse Says "No" to Your Mother Living with You?

As parents age and one spouse dies (usually the father), an older adult married child sometimes needs to take the remaining parent into his or her home. Seventeen percent of elderly women (and 7% of elderly men) live with their adult children (J. F. Potter 2010). When a spouse is adamantly against such a change in living space, some alternatives should be explored: get siblings involved so that the widowed parent spends some time with each adult child, share expenses with siblings for cost of nursing home facility, or share the cost with siblings of hiring a full-time person to live with the widowed parent. If none of these alternatives are acceptable, a marriage therapist may be helpful in examining the various options and moving beyond this impasse.

The key to successful aging is to pay as little attention to it as possible.

Judith Regan

Two crucial factors in how happy an older person is—having a purpose in life and maintaining quality relationships with others.

Jimmy Carter, former President

The young have aspirations that never come to pass, the old have reminiscences of what never happened.

H.H. Munro (Saki), British writer

among those aged 85 and older. Although most (over 90%) of the elderly do not exercise, Morey et al. (2008) studied the elderly ages 65 to 94 and found that the greater their physical activity, the greater their ability to function physically.

Some elderly become so physically debilitated that questions about the quality of life sometimes lead to consideration of physician-assisted suicide. Useda et al. (2007) compared a group of adults over 50 who attempted suicide versus those who had been successful in killing themselves and found that those in the latter group were more focused, planned, and organized in their movement toward ending their own life. Some debilitated elderly ask their physicians to end their lives. The debilitated elderly may also ask spouses or adult children for help in ending their life. Physician-assisted suicide is addressed in the Social Policy section on pages 484–485.

Mental Health

Aging also affects mental processes. Elderly people (particularly those 85 and older) more often have a reduced capacity for processing information quickly, for cognitive attention to a specific task, for retention, and for motivation to focus on a task. However, judgment may not be affected, and experience and perspective are benefits to decision making. Mental health may worsen for some elderly. Mood disorders, with depression being the most frequent, are more common among the elderly. Ryan et al. (2008) analyzed detailed reproductive histories of 1,013 women aged 65 years and over and found that the prevalence of depressive symptoms was 17%. Women who reported menopause at an earlier age had an increased risk. Women who had taken the oral contraceptive pill for at least ten years were less likely to report depression. The elderly who abuse alcohol and who do not exercise are also more likely to report being depressed (Van Gool et al. 2007).

National **Data**

Depression among the elderly may be linked to suicide. Suicide rates are among the highest for white males aged 85 or older. In 2004, 16% of all suicides were among individuals aged 65 and over. Non-Hispanic whites have the highest suicide rates—15.8 per 100,000; Non-Hispanic blacks have the lowest suicide rate—5.0 per 100,000 (National Institute of Mental Health 2010).

	True	False	Don't Know
1. Alzheimer's disease can be contagious.	___	___	___
2. People will almost certainly get Alzheimer's disease if they live long enough.	___	___	___
3. Alzheimer's disease is a form of insanity.	___	___	___
4. Alzheimer's disease is a normal part of getting older, like gray hair or wrinkles.	___	___	___
5. There is no cure for Alzheimer's disease at present.	___	___	___
6. A person who has Alzheimer's disease will experience both mental and physical decline.	___	___	___
7. The primary symptom of Alzheimer's disease is memory loss.	___	___	___
8. Among people older than age 75, forgetfulness most likely indicates the beginning of Alzheimer's disease.	___	___	___
9. When the husband or wife of an older person dies, the surviving spouse may suffer from a kind of depression that looks like Alzheimer's disease.	___	___	___
10. Stuttering is an inevitable part of Alzheimer's disease.	___	___	___

	True	False	Don't Know
11. An older man is more likely to develop Alzheimer's disease than an older woman.	___	___	___
12. Alzheimer's disease is usually fatal.	___	___	___
13. The vast majority of people suffering from Alzheimer's disease live in nursing homes.	___	___	___
14. Aluminum has been identified as a significant cause of Alzheimer's disease.	___	___	___
15. Alzheimer's disease can be diagnosed by a blood test.	___	___	___
16. Nursing-home expenses for Alzheimer's disease patients are covered by Medicare.	___	___	___
17. Medicine taken for high blood pressure can cause symptoms that look like Alzheimer's disease.	___	___	___

Answers: 1– 4, 8, 10, 11, 13–16 are False; remaining items are True.

Source

Neal E. Cutler, Boettner/Gregg Professor of Financial Gerontology, Widener University. Originally published in 1987, in *Psychology Today*, 20th Anniversary Issue, "Life Flow: A Special Report—The Alzheimer's Quiz," 21(5):89, 93. Reprinted with permission from *Psychology Today* magazine, © 1987 Sussex Publishers, LLC. The scale was completed by sixty-nine undergraduates at East Carolina University in 1998. Of the respondents, 40% identified less than 50% of the items correctly.

Dementia, which includes Alzheimer's disease, is the mental disorder most associated with aging. In spite of the association, only 3% of the aged population experience severe cognitive impairment—the most common symptom is loss of memory. It can be devastating to an individual and the partner. An 87-year-old woman, who was caring for her 97-year-old demented husband, said, "After 56 years of marriage, I am waiting for him to die, so I can follow him. At this point, I feel like he'd be better off dead. I can't go before him and abandon him" (Johnson and Barer 1997, 47). The Self-Assessment feature reflects some of the misconceptions about Alzheimer's disease.

Dementia a disorder of the mental processes marked by memory loss, personality changes, and impaired reasoning.

Retirement

Retirement represents a rite of passage through which most elderly pass. Feldman and Beehr (2011) identified three stages of thinking about retirement—imagining the possibility of retirement, assessing when it is time to let go of long-held jobs, and putting concrete plans for retirement into action at present. Pond et al. (2010) interviewed 60 individuals from 55 to 70 who revealed their reasons for retirement. These included poor health and the "maximization of life."

The latter referred to retiring while they were healthy and could enjoy/fulfill other life goals. Another reason was "health protection"—decisions motivated by health protection and promotion. Being able to retire in terms of financial security was also in the mix.

Individuals in the United States can take early retirement at age 62, with reduced benefits. Retirement affects an individual's status, income, privileges, power, and prestige. Fabian (2007) noted that, for most of our history, the concept of retirement did not exist—older individuals were viewed as a source of

WHAT IF?

What if Your Spouse with Alzheimer's Disease Says "No" to a Nursing Home?

McLennon et al. (2010) interviewed caregivers of elderly with Alzheimer's disease and identified two themes in their decision to institutionalize their loved one—anticipating the inevitable and reaching the limit. The caregivers recognized 3 to 4 months before institutionalization that they would not be able to continue caring for their relative. Advanced-stage Alzheimer's disease often results in affected people being a danger to themselves and others. Some of the symptoms of advanced Alzheimer's disease include leaving something on the stove, getting completely lost while driving or in a department store, and becoming paranoid and aggressive. When a spouse refuses to consider an alternative living arrangement, the healthy spouse may need to resort to deception. One spouse took her husband with Alzheimer's disease out to eat at a nursing home facility and then had security restrain her husband while she left. The experience was traumatic for both the husband and wife, but both adjusted within a month. Indeed, the husband fell in love with another patient and the wife often had dinner with him and his new love.

wisdom and they continued to work. In the twentieth century, retirement was developed in reference to the economy, which was faced with an aging population and surplus labor (Willson 2007). Indeed, retirement is a socially programmed stage of life that is being reevaluated. The retirement age for those born after 1960 is age 67.

People least likely to retire are unmarried, widowed, single-parent women who need to continue working because they have no pension or even Social Security benefits—if they don't work or continue to work, they will have no income, so retirement is not an option. Some workers experience what is called **blurred retirement** rather than a clear-cut one. A blurred retirement means the individual works part-time before completely retiring or takes a "bridge job" that provides a transition between a lifelong career and full retirement. Others may plan a **phased retirement** whereby an employee agrees to a reduced work load in exchange for reduced income. This arrangement is beneficial to both the employee and employer (Rappaport 2009).

Wang et al. (2011) identified five variables associated with enjoying retirement—individual attributes (e.g., physical and mental health/financial stability), preretirement job-related variables (e.g., escaped from work stress/job demands), family-related variables (e.g., being happily married), retirement-transition-related issues (e.g., planned voluntary retirement), and postretirement activities (e.g., bridge employment, volunteer work). Indeed, individuals who have a positive attitude toward retirement are those who have a pension waiting for them, are married (and thus have social support for the

Blurred retirement an individual working part-time before completely retiring or taking a "bridge job" that provides a transition between a lifelong career and full retirement.

Phased retirement an employee agreeing to a reduced work load in exchange for reduced income.

David Knox

Fred Johnson

During the sunset years, fishing is one of the hobbies of the retired.

transition), have planned for retirement, are in good health, and have high self-esteem. Those who regard retirement negatively have no pension waiting for them, have no spouse, gave no thought to retirement, have bad health, and have negative self-esteem (Mutran et al. 1997). Dew and Yorgason (2010) found that retirement seemed to insulate couples from marital distress during difficult economic times.

Largier (2010) identified several factors related to a happy retirement. These include good health (the single most important factor), involvement with a significant other (and even better if that person is also retired), friends, lots of interests/activities (other than television watching), reading (or other brain activity), enough money not to worry, and a willingness to drop out of the achievement ("I must be the best") value promoted by society. Cozijnsen et al. (2010) noted that once individuals retire, they are likely to maintain at least one relationship with someone they knew at work. Hence, some social ties are maintained. Paul Yelsma is a retired university professor who noted several recommendations for a successful retirement:

- Just like going to high school, the military, college, or your first job, know that the learning curve is very sharp. Don't think retirement is a continuation of what you did a few years earlier. Learn fast or you will be bored and burnt out.
- For every 10 phone calls you make expect about 1 back, your status has dwindled and you have less to offer those who want something from you.
- Have at least three to five hobbies that you can do almost any time and any place. Feed your hobbies or they will die.
- Having several things to look forward to is so much more exciting than looking backward. If you looked backward to high school while in college, your life would be boring.
- Develop a new physical exercise program that you want to do. Don't expect others to support your activities. The "pay off" in retirement is vastly different than what colleagues expected of you.
- Find new friends who have time to share (colleagues are often too busy).
- Retirement is a NEW time to live and NEW things have to happen.
- Travel—we have traveled to 7 new countries in the past few years.

Some retired individuals volunteer—giving back their time and money to attack poverty, illiteracy, oppression, crime, and so on. Examples of volunteer organizations are the Service Corps of Retired Executives, known as SCORE (www.score.org), Experience Works (www.experienceworks.org), and Generations United (*www.gu.org*).

Social Relationships

Luong et al. (2011) observed that older adults typically report higher levels of satisfaction with their social relationships than younger adults. Specifically, older adults engage in strategies that optimize positive social experiences and minimize negative ones by avoiding conflicts. In addition social partners of the elderly often treat older adults more positively and with greater forgiveness than they do younger adults.

Sexuality

There are numerous changes in the sexuality of women and men as they age (Vickers 2010). Frequency of intercourse drops from about once a week for those 40 to 59 to once every six weeks for those 70 and older. Changes in men include a decrease in the size of the penis from an average of 5 to about 4 ½ inches. Elderly men also become more easily aroused by touch rather than visual stimulation, which was arousing when they were younger. Erections take longer to achieve, are less rigid, and it takes longer for the man to recover so that he can have another erection. "It now takes me all night to do what I use to do all night" is the motto of the aging male.

Levitra, Cialis, and Viagra (prescription drugs that help a man obtain and maintain an erection) are helpful for about 50% of men. Others with erectile dysfunction may benefit from a pump that inflates two small banana-shaped tubes that have been surgically implanted into the penis. Still others benefit from devices placed over the penis to trap blood so as to create an erection.

Women also experience changes including menopause, which is associated with a surge of sexual libido, an interest in initiating sex with her partner, and greater orgasmic capacity. Not only are they free from worry about getting pregnant, estrogen levels drop and testosterone levels increase. Her vaginal walls become thinner and less lubricating; the latter issue can be resolved by lubricants like KY Jelly.

Table 16.3 describes these and other physiological sexual changes that occur as individuals age.

I regret to say that when you retire, and you can "do what you want to every day!"—what you actually do IS PASS INTO OBLIVION.

Fred Johnson

For me as an individual, aging has also brought freedom from romance, freedom from the ways in which your hormones distort your judgment and make you do things that aren't right for you.

Gloria Steinem

TABLE 16.3   Sexuality and Aging

Physical Changes in Sexuality as Men Age

1. Takes longer to get erection; erection less firm.
2. More direct stimulation needed for erection.
3. Extended refractory period (12 to 24 hours before second erection can occur)
4. Fewer expulsive contractions during orgasm.
5. Less forceful expulsion of seminal fluid and a reduced volume of ejaculate.
6. Rapid loss of erection after ejaculation.
7. Loss of ability to maintain an erection for a long period.
8. Decrease in size and firmness of the testes.

Physical Changes in Sexuality as Women Age

1. Reduced or increased sexual interest.
2. Possible painful intercourse due to menopausal changes.
3. Decreased volume of vaginal lubrication.
4. Decreased expansive ability of the vagina.
5. Possible pain during orgasm due to less flexibility.
6. Thinning of the vaginal walls.
7. Shortening of vaginal width and length.
8. Decreased sex flush, reduced increase in breast volume, and longer post orgasmic nipple erection.

Source: Adapted from Janell L. Carroll. *Sexuality Now: Embracing Diversity*, 3e, p. 273. © 2007. Wadsworth, a part of Cengage Learning, Inc. Reproduced by permission. www.cengage.com/permissions.

The National Institute on Aging surveyed 3,005 men and women ages 57 to 85 and found that sexual activity decreases with age (Lindau et al. 2007). Almost three-fourths (73%) of those 57 to 64 reported being sexually active in the last twelve months. This percentage declined to about half (53%) for those 65 to 74 and to about a fourth (26%) for those ages 75 to 80. An easy way to remember these percentages is three-fourths of those around 60, a half of those about 70, and a fourth of those around 80 report being sexually active.

Those most sexually active were also in good health. Diabetes and hypertension were major causes of sexual dysfunction. Incontinence (leaking of urine) was particularly an issue for older women and can be a source of embarrassment. The most frequent sexual problem for men was erectile dysfunction; for women, the most frequent sexual problem was the lack of a partner. Women also reported low sexual desire (43%), less vaginal lubrication (39%), and inability to climax (34%). Beckman et al. (2006) also reported that sexual interest continued for 95% of their sample of over 500 people in their seventies. Alford-Cooper (2006) studied the sexuality of couples married for more than fifty years and found that sexual interest, activity, and capacity declined with age but that marital satisfaction did not decrease with these changes.

Although sex may occur less often, it remains satisfying for most elderly. Winterich (2003) found that, in spite of vaginal, libido, and orgasm changes past menopause, women of both sexual orientations reported that they continued to enjoy active, enjoyable sex lives due to open communication with their partners.

The debate continues about whether menopausal women should become involved in estrogen replacement therapy and estrogen-progestin replacement therapy (collectively referred to as HRT—hormone replacement therapy). Schairer et al. (2000) studied 2,082 cases of breast cancer and concluded that the estrogen-progestin regimen increased the risk of breast cancer beyond that associated with estrogen alone. Beginning in 2003, women were no longer routinely encouraged to take HRT, and the effects of their not doing so on their physical and emotional health were minimal. A major study (Hays et al. 2003) of more than 16,608 postmenopausal women aged 50 to 79 found no significant benefits from HRT in terms of quality of life. Those with severe symptoms (for example, hot flashes, sleep disturbances, irritability) do seem to benefit without negative outcomes. Indeed, one woman reported, "I'd rather be on estrogen so my husband can stand me (and I can stand myself)." New data suggest that women who have had a hysterectomy can benefit from estrogen-alone therapy without raising their breast cancer risk.

Successful Aging

There is considerable debate about the meaning of successful aging. Torres and Hammarström (2009) noted that at least three definitions have been used: absence of physical problems/disability (no diseases or cognitive impairment), presence of strategies to cope with aging that results in positive emotions, sense of well-being, and not feeling lonely, and the definition of successful aging used by the elderly such as pride in viewing one's aging positively, coping with changing, having friends, and maintaining financial health.

Torres and Hammarström (2009) interviewed 16 elderly, ages 77 to 86, to identify their definitions of successful aging. They identified three factors:

1. Resources—physical: being physically healthy, mental: being alert/remembering things, social: having friends/relationships, and financial: not worrying about money

2. Attitude—positive approach to/view of life, learning new things, being curious about life. An example of attitude was expressed by one of the interviewees:

> *To be able to remember, to be able to do things in a calmer way, whether it is reading, or writing or taking a walk . . . there is a sort of calmness, a sort of relaxation that I don't think one can reach as a younger person and which I don't think a younger person would appreciate . . . young people prefer that which is stressful and exciting and not that which is calm and relaxed. And I have to say that it is fun to be able to*

One aspect of successful aging is staying mentally alert. This woman, in her mid eighties, surfs the Internet and plays games on the computer daily.

David Knox

appreciate this calmness. And that is one of the things that one could count amongst the pros of growing old . . . that one can become more relaxed. (Hans, 79 years old, without home-help care) (p. 43)

3. Continuity—being able to be healthy and continuing to be involved in and to enjoy life. One of the respondents noted that successful aging means not aging.

Schieman et al. (2010) studied individuals over the age of 65 and found that a "sense of mattering" (which was defined as belief in items such as "you are important to the people you know," "your well being is important to the people you know," and "people count on you when they are down and blue") was associated with a belief in a "divine control" (e.g., "When good or bad things happen, God has a plan for you"). This study went beyond church attendance and membership and focused on beliefs as the important variable in "mattering."

Researchers who worked on the Landmark Harvard Study of Adult Development (Valliant 2002) followed 824 men and women from their teens into their eighties and identified those factors associated with successful aging. These include not smoking (or quitting early), developing a positive view of life and life's crises, avoiding alcohol and substance abuse, maintaining healthy weight, exercising daily, continuing to educate oneself, and having a happy marriage. Indeed, those who were identified as "happy and well" were six times more likely to be in a good marriage than those who were identified as "sad and sick."

Not smoking is "probably the single most significant factor in terms of health" according to Valliant (2002), of the Landmark Harvard Study of Adult Development. Smokers who quit before age 50 were as healthy at 70 as those who had never smoked.

Exercise is one of the most beneficial activities the elderly can engage in to help them maintain good health. Takata et al. (2010) studied the association between physical fitness and quality of life in a sample of 207 men and women 85 years of age and found that increases in their physical fitness level can contribute to improvements in quality of life. Hence, exercise, which enhances physical fitness, can have payoffs to those even in their mid 80s and beyond. John Keller, for example, ran his 60th Boston Marathon at age 83.

In spite of the benefits of exercise for the elderly, the occurrence of exercise activities significantly decreases as a person ages. Wrosch et al. (2007) studied the exercise behavior of 172 elderly adults and found that the greatest predictor of whether elderly people exercised was their having done so previously. Hence, people who exercised at year 2 of the study were likely to still be exercising at year 5 of the study. Stephenson et al. (2007) noted that community-sponsored walking programs at malls were beneficial in getting the elderly to exercise. Jack LaLanne, the health fitness guru, remarked at age 90, "I hate to exercise; I love the results." At his 90th birthday party, he challenged his well-wishers to be at his 115th birthday party.

Agahi (2008) also noted a strong positive effect of involvement in social leisure activities and successful aging among a representative sample of 1,246 men and women ages 65 to 96. Participating in only a few activities doubled mortality risk compared to those with the highest participation levels, even after controlling for age, education, walking ability, and other health indicators. Strongest benefits were found for engagement in organizational activities and study circles among women and hobby activities and gardening among men.

Use of the Internet is also helpful in keeping the elderly socially connected. Hogeboom et al. (2010) examined associations between Internet use and the social networks of adults over 50 years of age (sample of 2284) and found that frequency of contact with friends, the frequency of contact with family, and the attendance at organizational meetings (not including religious services) were positively associated with Internet use for adults over 50.

One key to successful aging is to stop paying attention to how old you are.

Ray Vickers, retired geriatric physician

Relationships and the Elderly

Relationships continue into old age.

Use of Technology to Maintain Relationships

Youth regularly text friends and romantic partners throughout the day. Madden (2010) noted that older adults and senior citizens are increasing their use of technology to stay connected as well. Over 40% of adults over the age of 50 use e-mail. And almost half (47%) of Internet users 50–64 and 25% of users 65 and older use social networking sites such as Facebook. Indeed these older adults and seniors view themselves among the Facebooking and LinkedIn masses. These figures have doubled since 2009.

Use of Twitter has not caught on among older adults and senior citizens, with only 11% of online adults ages 50–64 reporting such usage. E-mail and on-line news are the most frequent Internet behaviors of this age group.

Relationships among Elderly Spouses

Marriages that survive into late life are characterized by little conflict, considerable companionship, and mutual supportiveness. All but one of the thirty-one spouses over age 85 in the Johnson and Barer (1997) study reported "high expressive rewards" from their mate. Walker and Luszcz (2009) reviewed the literature on elderly couples and found marital satisfaction related to equality of roles and marital communication. Health may be both improved by positive relationships and decreased by negative relationships.

Field and Weishaus (1992) reported interview data on seventeen couples who had been married an average of fifty-nine years and found that the husbands and wives viewed their marriages very differently. Men tended to report more marital satisfaction, more pleasure in the way their relationships had been across time, more pleasure in shared activities, and closer affectional ties. Sex was also more important to the husbands. Every man in the study reported that sex was always an important part of the relationship with his wife, but only four of the seventeen wives reported the same.

Only a small percentage (8%) of individuals older than 100 are married. Most married centenarians are men in their second or third marriage. Many have outlived some of their children. Marital satisfaction in these elderly marriages is related to a high frequency of expressing love feelings to one's partner. Though it is assumed that spouses who have been married for a long time should know how their partners feel, this is often not the case. Telling each other "I love you" is very important to these elderly spouses.

Relationship with One's Own Children at Age 85 and Beyond

In regard to relationships of the elderly with their children, emotional and expressive rewards are high. Actual caregiving is rare. Only 12% of the Johnson and Barer (1997) sample of adults older than 85 lived with their children. Most preferred to be independent and to live in their own residence. "This independent stance is carried over to social supports; many prefer to hire help rather than bother their children. When hired help is used, children function more as mediators than regular helpers, but most are very attentive in filling the gaps in the service network" (Johnson and Barer 1997, 86).

Relationships among multiple generations will increase. Whereas three-generation families have been the norm, four- and five-generation families will increasingly become the norm. These changes have already become evident.

This 50-year-old son and his 82-year-old mother share a wonderful relationship.

Grandparenthood

Another significant role for the elderly is grandparenting. Among adults aged 40 and older who had children, close to 95% are grandparents and most have, on average, five or six grandchildren. Grandparents may actively take care of their grandchildren full-time, provide supplemental help in a multigenerational family, help on an occasional part-time basis, or occasionally visit their grandchildren. Grandparents see themselves as caretakers, emotional and/or economic resources, teachers, and historical connections on the family tree.

Households headed by grandparents are increasing. Six million grandparents are living in homes with grandchildren; 2.5 million are rearing grandchildren alone because the parents are unavailable (for example, due to illness, drug addiction, AIDS, incarceration, wars in the Middle East, divorce, or poverty) (Ackerman and Banks 2007).

Motherhood gets in the way of being able to enjoy your children.

Carol Barnes, therapist

Grandparents in a Full-Time Role of Raising Grandchildren

National **Data**

About 4% of children live with their grandparents or other relatives (Carr and Springer 2010).

While grandmothers provide the most care for grandchildren, grandfathers may also be involved. Bates and Taylor (2010) studied 351 grandfathers who reported the details of their relationship with one of their grandchildren under the age of 26. Findings revealed that grandchildren were a significant source of social interaction and grandfathers who connected to their grandchildren experienced fewer depressive symptoms.

Styles of Grandparenting

Grandparents also have different styles of relating to grandchildren. Whereas some grandparents are formal and rigid, others are informal and playful, and authority lines are irrelevant. Still others are surrogate parents providing considerable care for working mothers and/or single parents. Baker et al. (2010) surveyed grandmothers who were caring for their grandchildren while the mother of the children was in prison (such care benefited both the grandmother and the children).

Some grandparents have regular contact with their grandchildren. Amicable relationships between children and their grandparents (particularly paternal grandparents) is associated with parents staying together when the child is young (Hogns and Carlson 2010). In addition, Barnett et al. (2010a) emphasized the benefit to young children of time with the grandmother in terms of social development. Grandmother involvement also served as a buffer for harsh parenting of the mother on subsequent externalizing behavior of the child. While some grandparents are close with their grandchildren, others are distant and show up only for special events like birthdays and holidays. E-mail and Skype are helping grandparents to stay connected to their grandchildren.

Some grandparents, grandmothers particularly, take their grandchildren to the college they attended and stay with them on campus during a week in the summer. Grandparent Universities began over ten years ago at the University of Wisconsin-Madison and has become more common on college campuses during the summer. The program allows for grandmothers and teens to bond while providing exposure to a real college campus (Hunt 2010).

Age seems to be a factor in determining how grandparents relate to their grandchildren. Grandparents over the age of 65 are less likely to be playful and fun-seeking than those under 65. This may be because the older grandparents are less physically able to engage in playful activities with their grandchildren. Indeed, according to Johnson and Barer (1997):

> . . . *members of the oldest generation in our study place more emphasis on their relationship with their own children over their grandchildren. This situation could stem from the fact that as grandchildren reach adulthood, they become more independent from their own parent. That parent then is freed up to strengthen the relationship with their oldest parent, at the same time they, as the middle generation, maintain a lineage bridge linking their parent to their child. (p. 89)*

Finally, the quality of the grandparent-grandchild relationship can be affected by the parents' relationship to their own parents. If a child's parents are estranged from their parents, it is unlikely that the child will have an opportunity to develop a relationship with the grandparents. Divorce often shatters the relationship grandparents have with their grandchildren (Doyle et al. 2010).

The "Myth" of the Happy Grandmother

The stereotype of "grandmother" is that of a white-haired lady with glasses who wears an apron and bakes an apple pie for her grandchildren. The reality can be a woman who wears a stylish windbreaker as she jogs around the block or a

The role of most grandfathers is to play with their grandchildren.

Lisa Knox

woman who jets to New York for a corporate meeting. Although some women love their role of grandmother, the reality is sometimes different. Kulik (2007) noted the following negatives that grandmothers identify:

1. *Conflict.* Taking care of grandchildren may interfere with what grandmothers want to do. "I have a life and am involved in a new relationship," reported one grandmother.

2. *Demanding children.* Full-time care of grandchildren can be exhausting. "They wear me out after three days," one woman said of her twin grandkids.

3. *Boredom.* Not all grandmothers enjoy the activities of young children. Coloring, drawing, and playing with toys are OK for a couple of hours for some grandmothers; others get bored silly.

4. *Exploitation.* Some grandmothers feel exploited by their children. They feel that their children "dump" the kids on them and "expect" them to be a full-time mother. "I don't appreciate it," said one grandmother, but she felt guilty at not wanting to be more helpful.

5. *End of childbearing capacity.* Some women view grandparenting as symbolic that they are getting old and are no longer capable of having more children of their own.

Effect of Divorce on the Grandparent-Child Relationship

Divorce often has a negative effect on grandparents seeing their grandchildren. The situation most likely to produce this outcome is when the children are young and the wife is granted primary custody. She may withhold the children from seeing the parents of her ex-husband so the grandparents may only see them when their son has "his weekend with them." On the other hand, the maternal grandmother may see her grandchildren more since her daughter (who has custody) may need her to take care of the children as she reenters the work force. Sometimes a custodial parent will purposefully try to sever the relationship in an attempt to seek revenge or vent hostility against a former spouse.

Indeed, when their children divorce, some grandparents are not allowed to see their grandchildren. In all fifty states, the role of the grandparent has limited legal and political support. By a vote of six to three in 2000, the Supreme Court *(Troxel v. Granville)* sided with the parents and virtually denied 60 million grandparents the right to see their grandchildren. The court viewed parents as having a fundamental right to make decisions about with whom their children could spend time—including keeping them from grandparents. Grandparents are often reluctant to seek legal access to their grandchildren for fear that doing so will anger the custodial parent who will further retreat with the children (LaPierre 2010). However, some courts have ruled in favor of grandparents. Stepgrandparents have no legal rights to their stepgrandchildren.

If you were going to die soon and had only one phone call you could make, who would you call and what would you say? And why are you waiting?

Stephen Levine, American poet

Benefits to Grandchildren

Of a sample of 2,922 university students, 72% agreed with the statement, "I have a loving relationship with my grandmother" (63% reported having had a loving relationship with their grandfather) (Knox and Hall 2010). Grandchildren report enormous benefits from having a close relationship with grandparents, including development of a sense of family ideals, moral beliefs, and a work ethic. Kennedy (1997) focused on the memories grandchildren had of their grandparents and found that "love and companionship" were identified most frequently by the grandchildren. In addition, a theme running through a fourth of the memories was that the grandchildren felt that they were regarded as "special" by a grandparent, either because of being the first or last grandchild, having personality characteristics similar to the grandparent's, or being the child of a favorite son or daughter of the grandparent.

The End of One's Life

Thanatology the examination of the social dimensions of death, dying, and bereavement.

Thanatology is the examination of the social dimensions of death, dying, and bereavement (Bryant 2007). The end of one's life sometimes involves the death of one's spouse.

Death of One's Spouse

The death of one's spouse is one of the most stressful life events a person ever experiences. One year after the death of Patrick Swayze, Niemi, his wife of 34 years, noted, "I get through the days putting one foot in front of the other…I know he's not here…but I like having him in my dreams" (Jordan 2010). Because women tend to live longer than men, and because women are often younger than their husbands, women are more likely than men to experience the death of their marital partner.

Although individual reactions and coping mechanisms for dealing with the death of a loved one vary, several reactions to death are common. These include shock, disbelief and denial, confusion and disorientation, grief and sadness, anger, numbness, physiological symptoms such as insomnia or lack of appetite, withdrawal from activities, immersion in activities, depression, and guilt. Eventually, surviving the death of a loved one involves the recognition that life must go on, the need to make sense out of the loss, and the establishment of a new identity.

Women's response to the death of their husbands may involve practical considerations. Johnson and Barer (1997) identified two major problems of widows—the economic effects of losing a spouse and the practical problems of maintaining a home alone. The latter involves such practical issues as cleaning the gutters, painting the house, and changing the air filters.

Women and men tend to have different ways of reacting to and coping with the death of a loved one. Women are more likely than men to express and share feelings with family and friends and are also more likely to seek and accept help, such as attending support groups of other grievers. Initial responses of men are often cognitive rather than emotional. From early childhood, males are taught to be in control, to be strong and courageous under adversity, and to be able to take charge and fix things. Showing emotions is labeled weak.

Men sometimes respond to the death of their spouse in behavioral rather than emotional ways. Sometimes they immerse themselves in work or become involved in physical action in response to the loss. For example, a widower immersed himself in repairing a beach cottage he and his wife had recently bought. Later, he described this activity as crucial to getting him through those first two months. Another coping mechanism for men is the increased use of alcohol and other drugs. They may also seek new sexual partners. Ball (2010) found that being a widow is associated with an increased risk of contracting a sexually transmitted infection. From six months to one year after the death of one's wife, widowed men had an increased use of Viagra and accompanying STIs.

Bennett (2010) studied 60 males whose wives had died and found that almost 40% (38%) demonstrated resilience in terms of participating in relationships and activities and returning to a life that had satisfaction and meaning. Having a positive outlook and moving beyond adversity were personal characteristics associated with resiliency. In addition, having social support was also beneficial.

Whether a spouse dies suddenly or after a prolonged illness has an impact on the reaction of the remaining spouse. The sudden rather than prolonged death of one's spouse is associated with being less at peace with death and being more angry. The suddenness of the death provides no cognitive preparation time.

The age at which one experiences the death of a spouse is also a factor in one's adjustment. People in their eighties may be so consumed with their own health and disability concerns that they have little emotional energy left to grieve. On the other hand, a team of researchers (Hansson et al. 1999) noted that death does not end the relationship with the deceased. Some widows and widowers report a feeling that their spouses are with them at times and are watching out for them a year after the death of their beloved. They may also dream of them, talk to their photographs, and remain interested in carrying out their wishes. Such continuation of the relationship may be adaptive by providing meaning and purpose for the living or maladaptive in that it may prevent one from establishing new relationships. Research is not consistent on the degree to which individuals are best served by continuing or relinquishing the emotional bonds with the deceased (Stroebe and Schut 2005).

Preparing for One's Own Death

What is it like for those near the end of life to think about death? To what degree do they go about actually "preparing" for death? Johnson and Barer (1997) interviewed forty-eight individuals with an average age of 93 to find out their perspective on death. Most interviewees were women (77%); of them 56% lived alone, but 73% had some sort of support in terms of children or one or more social support services. The following findings are specific to those who died within a year after the interview.

Thoughts in the Last Year of Life   Most had thought about death and saw their life as one that would soon end. Most did so without remorse or anxiety. With their spouses and friends dead and their health failing, they accepted death as the next stage in life. They felt like the last leaf on the tree. Some comments follow:

> If I die tomorrow, it would be all right. I've had a beautiful life, but I'm ready to go. My husband is gone, my children are gone, and my friends are gone.
> That's what is so wonderful about living to be so old. You know death is near and you don't even care.
> I've just been diagnosed with cancer, but it's no big deal. At my age, I have to die of something. (Johnson and Barer 1997, 205)

The major fear these respondents expressed was not the fear of death but of the dying process. Dying in a nursing home after a long illness is a dreaded fear. Sadly, almost 60% of the respondents died after a long, progressive illness. They had become frail, fatigued, and burdened by living. They identified dying in their sleep as the ideal way to die. Some hasten their death by no longer taking their medications or wish they could terminate their own life. "I'm feeling kind of useless. I don't enjoy anything anymore. . . . What the heck am I living for? (Johnson and Barer, 1997, 204). Some move to Oregon where "death with dignity" is legal- a person can swallow drugs prescribed by a physician and be dead in 90 seconds.

Behaviors in the Last Year of Life   Aware that they are going to die, most simplify their life, disengage from social relationships, and leave final instructions. In simplifying their life, they sell their home and belongings and move to smaller quarters. One 81-year-old woman sold her home, gave her car away to a friend, and moved into a nursing home. The extent of her belongings became a chair, lamp, and TV.

Some divorce just before they die. Dennis Hopper, upon learning he had terminal cancer, filed for divorce from his 5th wife. One explanation for this behavior is that divorce removes a spouse from automatically getting part of a deceased spouse's estate and allows control of dispensing one's assets (often to one's own children) while one is alive. "A spouse is pretty much the only person you can't cut out of a will" says attorney Wynne Whitman, an estate planning lawyer in New Jersey (Luscombe 2010a).

A long marriage is two people trying to dance a duet and two solos at the same time.

Anne Taylor Fleming

I feel like the last lobster in the tank and the waiter is rolling up his sleeve.

Johnny Carson

*We hope to grow old; we fear old age;
That is to say we love life and flee death.*

Jean de la Bruyere

We are all terminal.

Jack Kevorkian

Disengaging from social relationships is selective. Some maintain close relationships with children and friends but others "let go." They may no longer send out Christmas cards, stop sending letters, and phone calls become the source of social connections. Some leave final instructions in the form of a will or handwritten note expressing wishes of where to be buried, handling costs associated with disposal of the body, and what to do about pets. One of Johnson and Barer's (1997) respondents left $30,000 to specific caregivers to take care of each of several pets (p. 204).

Elderly who may have counted on children to take care of them may discover that they have too few children who may have scattered because of job changes, divorce, or both. Some children simply walk away from their parents or leave their care to their siblings. The result is that the elderly may have to fend for themselves.

One of the last legal acts of the elderly is to have a will drawn up. Stone (2008) emphasized how wills may stir up sibling rivalry (for example, one sibling may be left more than another), be used as a weapon against a second spouse (for example, by leaving all of one's possessions to one's children), or reveal a toxic secret (for example, reveal a mistress and several children who are left money).

Some live full lives up until the very end. Ted Kennedy died of brain cancer at age 77. Up until the last week of his life, he was active in politics (he had sponsored over 300 bills in his Senate career). His words resonate…"For all those whose cares have been our concern, the work goes on, the cause endures, the hope still lives, and the dream shall never die."

End of Life Preferences

Researchers (Carr and Moorman 2009) identified the end of life preferences of 5,106 adults for one of two scenarios—severe physical pain with no cognitive impairment or severe cognitive impairment with no physical pain. Those who were nonreligious, mainline Protestants, and those who had made end of life preparations were more likely to reject prolonging life in either scenario. Women who had witnessed the painful death of a loved one were more likely to reject prolonging life, particularly if there was cognitive impairment. The theory of reasoned action was used to explain the findings—persons act consistently with perceived positive outcomes.

The Final Days

Sometimes family members provide care the final days. Carlander et al. (2011) confirmed that caregiving for an elderly dying family member affects the health and daily lives of family caregivers. Ten family caregivers who cared for a dying family member at home with support from an advanced home care team were interviewed 6-12 months after the death of the family member. The interviews revealed experiences such as 'forbidden thoughts', intimacy, and decreasing personal space. This study emphasizes the need for creating a context in which caregivers can talk about caregiving experiences.

Alternatively, Hirano et al. (2011) reported on a unique end of life context. They studied 31 home care nurses in Japan of an elderly individual who helped family members by (a) discussing the possibility of having the individual die at home, (b) explaining what was about to happen as the individual neared death and what could be done, (c) proposing where and how the family could say goodbye, (d) building family consensus, (e) coordinating resources, and (f) offering psychological support for end-of-life care. Rather than let these various issues unfold without warning, the nurses were trained to take the family members through the process.

The Future of the Elderly in the U.S.

The elderly will increase in number and political clout. By 2030, 30% of the U.S. population will be over the age of 55 (now 21%). The challenges of old age will be the same—coping with dwindling income, declining health, and the death of loved ones. On the positive side, greater attention will be paid to the health needs of the elderly. Medicare will help pay for some of the medical needs of the elderly. However, it alone will be inadequate and other private sources will be needed.

Summary

What is meant by the terms age and ageism?

Age is defined chronologically (by time), physiologically (by capacity to see, hear, and so on), psychologically (by self-concept), sociologically (by social roles), and culturally (by the value placed on elderly). Ageism is the denigration of the elderly, and gerontophobia is the dreaded fear of being elderly. Theories of aging range from disengagement (individuals and societies mutually disengage from each other) to continuity (the habit patterns of youth are continued in old age). Life course is currently the popular aging theory.

What is the "sandwich generation"?

Eldercare combined with child care is becoming common among the sandwich generation—adult children responsible for the needs of both their parents and their children. Two levels of eldercare include help with personal needs such as feeding, bathing, and toileting as well as instrumental care such as going to the grocery store, managing bank records, and so on. Members of the sandwich generation report feelings of exhaustion over the relentless demands, guilt over not doing enough, and resentment over feeling burdened.

Deciding whether to arrange for an elderly parent's care in a nursing home requires attention to a number of factors, including the level of care the parent needs, the philosophy and time availability of the adult child, and the resources of the adult children and other siblings. Full-time nursing care (not including medication) can be between $1,000–$1,500 a week.

Elderly parents who are dying from terminal illnesses incur enormous medical bills. Some want to die and ask for help. Our society continues to wrestle with physician-assisted suicide and euthanasia. Only Oregon currently has a death with dignity policy.

What issues confront the elderly?

Issues of concern to the elderly include income, health, retirement, and sexuality. One's health becomes a primary focus for the elderly. Good health is the single most important factor associated with an elderly person's perceived life satisfaction. Mental problems may also occur with mood disorders; depression is the most common.

Though the elderly are thought to be wealthy and living in luxury, most are not. The median household income of people over the age of 65 is less than half of what they earned in the prime of their lives. Social Security provides only 42% of preretirement wages. The most impoverished elderly are those who have lived the longest, who are widowed, and who live alone. Women are particularly disadvantaged because their work history may have been discontinuous, part-time, and low-paying. Social Security and private pension plans favor those with continuous, full-time work histories.

For most elderly women and men, sexuality involves lower reported interest, activity, and capacity. Inability to have an erection and the absence of a sexual partner are the primary sexual problems of elderly men and women.

What factors are associated with successful aging?

There is debate about the definition of successful aging. Three factors include: resources, attitude, and continuity. Not smoking is "probably the single most significant factor in terms of health." Exercise is one of the most beneficial activities the elderly can engage in to help them maintain good health. Other factors associated with successful aging include a positive view of life and life's crises, avoiding alcohol and substance abuse, continuing to educate oneself, and having a happy marriage.

What are relationships like for the elderly?

Marriages that survive into old age (beyond age 85) tend to have limited conflict, considerable companionship, and mutual supportiveness. Relationships with siblings are primarily emotional rather than functional. In regard to relationships of the elderly with their children, emotional and expressive rewards are high. Caregiving help is available but rare. Only 12% of one sample of adults older than 85 lived with their children.

What is grandparenthood like?

Among adults aged 40 and older who had children, close to 95% are grandparents. There is considerable variation in role definition and involvement. Whereas some delight in seeing their lineage carried forward in their grandchildren and provide emotional and economic support, others are focused on their own lives or on their own children and relate formally and at a distance to their grandchildren. When grandparents are involved in their lives, grandchildren benefit in terms of positive psychological and economic benefits.

What are some end of life issues?

Coping with the death of one's spouse and preparing for one's own death are end of life issues for the elderly. While the death of one's mate results in emotional free-fall, one's own death, if not protracted with ill health, is smoother than might be anticipated.

Key Terms

Advance directive (living will)	Family caregiving	Medicare
Age	Filial piety	Phased retirement
Age discrimination	Filial responsibility	Quality of life
Ageism	Frail	Sandwich generation
Blurred retirement	Gerontology	Thanatology
Durable power of attorney	Gerontophobia	
Euthanasia	Medicaid	

Web Links

AARP (American Association of Retired Persons)
   http://www.aarp.org

Calculate Your Life Expectancy
   www.livingto100.com

ElderSpirit Community
   www.elderspirit.net

ElderWeb
   http://www.elderweb.com

ExperienceCorps
   http://www.experiencecorps.org/index.cfm

Family Caregiver Alliance
   http://caregiver.org/caregiver/jsp/home.jsp

Foundation for Grandparenting
   http://www.gu.org

GROWW (Grief Recovery Online)
   http://www.groww.com

Nolo: Law for All (wills and legal issues)
   http://www.nolo.com

Senior Corps
   http://www.seniorcorps.gov/

Senior Sex
   http://www.holisticwisdom.com/senior-sex.htm

Silver Sage Village
   http://www.silversagevillage.com

SPECIAL TOPIC 1

Careers in Marriage and the Family

Let the beauty we love, be what we do.

Rumi, philosopher

David Knox

Students who take courses in marriage and the family sometimes express an interest in working with people and ask what careers are available if they major in marriage and family studies. In this Special Topics section, we review some of these career alternatives, including family life education, marriage and family therapy, child and family services, and family mediation. These careers often overlap, so you might engage in more than one of these at the same time. For example, you may work in family services but participate in family life education as part of your job responsibilities.

For all the careers discussed in this section, having a bachelor's degree in a family-related field such as family science, sociology, or social work is helpful. Family science programs are the only academic programs that focus specifically on families and approach working with people from a family systems perspective. These programs have many different names, including child and family studies, human development and family studies, child development and family relations, and family and consumer sciences. Marriage and family programs are offered through sociology departments; family service programs are typically offered through departments of social work as well as through family science departments. Whereas some jobs are available at the bachelor's level, others require a master's or PhD degree. More details on the various careers available to you in marriage and the family follow.

Family Life Education

Family life education (FLE) is an educational process that focuses on prevention and on strengthening and enriching individuals and families. Family life educators empower family members by providing them with information that will help prevent problems and enrich their family well-being. This education may be offered to families in different ways: a newsletter, one-on-one, online, or through a class or workshop. Examples of family life education programs include parent education for parents of toddlers through a child-care center, a brown-bag lunch series on balancing work and family in a local business, a premarital or marriage enrichment program at the local church, a class on sexuality education in a high school classroom, or a workshop on family finance and budgeting at a local community center. The role of family life educator involves making presentations in a variety of settings, including schools, churches, and even prisons. Family life educators may also work with military families on military bases, within the business world with human resources or employee assistance programs, and within social service agencies or cooperative extension programs. Some family life educators develop their own business providing family life education workshops and presentations.

To become a family life educator, you need a minimum of a bachelor's degree in a family-related field such as family science, sociology, or social work. You can become a certified family life educator (CFLE) through the National Council on Family Relations (NCFR). The CFLE credential offers you credibility in the field and shows that you have competence in conducting programs in all areas of family life education. These areas are individuals and families in societal contexts, internal dynamics of the family, human growth and development, interpersonal relationships, human sexuality, parent education and guidance, family resource management, family law and public policy, and ethics. In addition, you must show competence in planning, developing, and implementing family life education programs.

Your academic program at your college or university may be approved for provisional certification. In other words, if you follow a specified program of study at your school, you may be eligible for a provisional CFLE certification. Once you gain work experience, you can then apply for full certification.

Marriage and Family Therapy

Whereas family life educators help prevent the development of problems, marriage and family therapists help spouses, parents, and family members resolve existing interpersonal conflicts and problems. They treat a range of problems including communication, emotional and physical abuse, substance abuse, and sexual dysfunctions. Marriage and family therapists work in a variety of contexts, including mental health clinics, social service agencies, schools, and private practice. There are about 50,000 marriage and family therapists in the United States and Canada (about 40 percent are members of the American Association for Marriage and Family Therapy).

Currently, all fifty states and the District of Columbia license or certify marriage and family therapists. Although an undergraduate degree in sociology, family studies, or social work is a good basis for becoming a marriage and family therapist, a master's degree and two years of post-graduate supervised clinical work is required to become a licensed marriage and family therapist. Some universities offer accredited master's degree programs specific to marriage and family therapy; these involve courses in marriage and family relationships, family systems, and human sexuality, as well as numerous hours of clinical contact with couples and families under supervision. A list of graduate programs in marriage and family therapy is available at http://www.aamft.org/cgi-shl/twserver. exe?run:COALIST.

Full certification involves clinical experience, with 1,000 hours of direct client, couple, and family contact; 200 of these hours must be under the direction of a supervisor approved by the American Association of Marriage and Family Therapists (AAMFT). In addition, most states require a licensure examination. The AAMFT is the organization that certifies marriage and family therapists. A marriage and family therapist can be found at http://family-marriage-counseling. com/therapists-counselors.htm.

Child and Family Services

In addition to family life educators and marriage and family therapists, careers are available in agencies and organizations that work with families, often referred to as social service agencies. The job titles within these agencies include family interventionist, family specialist, and family services coordinator. Your job responsibilities in these roles might involve helping your clients over the telephone, coordinating services for families, conducting intake evaluations, performing home visits, facilitating a support group, or participating in grant writing activities. In addition, family life education is often a large component of child and family services. You may develop a monthly newsletter, conduct workshops or seminars on particular topics, or facilitate regular educational groups.

Some agencies or organizations focus on helping a particular group of people. If you are interested in working with children, youth, or adolescents, you might find a position with Head Start, youth development programs such as the Boys and Girls Club, after-school programs (for example, pregnant or parenting teens), child-care resource or referral agencies, or early intervention services. Child-care resource and referral agencies assist parents in finding child care, provide training for child-care workers, and serve as a general resource for parents and for child-care providers. Early intervention services focus on children with special needs. If you work in this area, you might work directly with children or you might work with the families and help to coordinate services for them.

Other agencies focus more on specific issues that confront adults or families as a whole. Domestic violence shelters, family crisis centers, and employee-assistance programs are examples of employment opportunities. In many of these positions, you will function in multiple roles. For example, at a family crisis center, you might take calls on a crisis hotline, work one-on-one with clients to help them find resources and services, and offer classes on sexual assault or dating violence to high school students.

Another focus area in which jobs are available is gerontology. Opportunities include those within residential facilities such as assisted-living facilities or nursing homes, senior centers, organizations such as the Alzheimer's Association, or agencies such as National Association of Area Agencies on Aging. There is also a need for eldercare resource and referral, as more and more families find that they have caregiving responsibilities for an aging family member. These families have a need for resources, support, and assistance in finding residential facilities or other services for their aging family member. Many of the available positions with these types of agencies are open to individuals with bachelor's degrees. However, if you get your master's degree in a program emphasizing the elderly, you might have increased opportunity and will be in a position to compete for various administrative positions.

Family Mediation

In Chapter 15 on Divorce and Remarriage, we emphasized the value of divorce mediation. This is also known as family mediation and involves a neutral third party who negotiates with divorcing spouses on the issues of child custody, child support, spousal support, and division of property. The purpose of mediation is not to reconcile the partners but to help the couple make decisions about children, money, and property as amicably as possible. A mediator does not make decisions for the couple but supervises communication between the partners, offering possible solutions.

Although some family and divorce mediators are attorneys, family life professionals are becoming more common. Specific training is required that may include numerous workshops or a master's degree, offered at some universities (for example, University of Maryland). Most practitioners conduct mediation in conjunction with their role as a family life educator, marriage and family therapist, or other professional. In effect, you would be in business for yourself as a family or divorce mediator.

Students interested in any of these career paths can profit from obtaining initial experience in working with people through volunteer or internship agencies. Most communities have crisis centers, mediation centers, and domestic abuse centers that permit students to work for them and gain experience. Not only can you provide a service, but you can also assess your suitability for the "helping professions," as well as discover new interests. Talking with people already in the profession is also a good way to gain new insights. Your instructor may already be in the marriage and family profession you would like to pursue or be able to refer you to someone who is.

Note: Appreciation is expressed to Sharon Ballard, PhD, CFLE, for the development of this Special Topic section. Dr. Ballard is an associate professor of Child Development and Family Relations at East Carolina University. She is also a certified family life educator through the National Council on Family Relations.

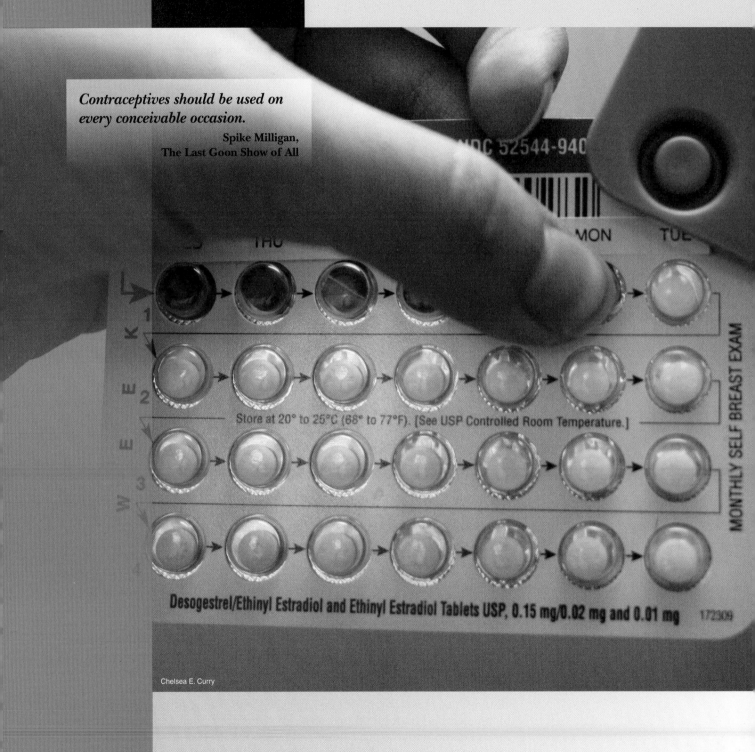

Contraceptives should be used on every conceivable occasion.

**Spike Milligan,
The Last Goon Show of All**

Store at 20° to 25°C (68° to 77°F). [See USP Controlled Room Temperature.]

MONTHLY SELF BREAST EXAM

Desogestrel/Ethinyl Estradiol and Ethinyl Estradiol Tablets USP, 0.15 mg/0.02 mg and 0.01 mg

172309

Chelsea E. Curry

507

O nce individuals have decided on whether and when they want children, they need to make a choice about contraception. All contraceptive practices have one of two common purposes: to prevent the male sperm from fertilizing the female egg or to keep the fertilized egg from implanting itself in the uterus. About five to seven days after fertilization, pregnancy begins. Although the fertilized egg will not develop into a human unless it implants on the uterine wall, pro-life supporters believe that conception has already occurred.

In selecting a method of contraception, the important issues to consider are pregnancy prevention, STI prevention, opinion of the partner, ease of use, and cost. Often sexual partners use no contraception and live in denial that "this one time won't end in a pregnancy." In a study of 2,922 undergraduates, only half reported that the last time they had sexual intercourse, they used a form of birth control (other than withdrawal). Thirty percent used a condom to prevent contracting an STI (Knox and Hall 2010). Even when contraception is used, Speidel et al. (2008) noted that 9% of pill users and 17% of condom users become pregnant during the first year of typical use.

Hormonal Contraceptives

Hormonal methods of contraception currently available to women include "the pill," Norplant, Depo-Provera, NuvaRing, and Ortho Evra.

Oral Contraceptive Agents (Birth Control Pills)

Birth control pills are the most commonly used method of all the nonsurgical forms of contraception. Although almost 10% of women who take the pill still become pregnant in the first year of use, it remains a desirable birth control option.

Oral contraceptives are available in basically two types: the combination pill, which contains varying levels of estrogen and progestin, and the minipill, which is progestin only. Combination pills work by raising the natural level of hormones in a woman's body, inhibiting ovulation, creating an environment where sperm cannot easily reach the egg, and hampering implantation of a fertilized egg.

The second type of birth control pill, the minipill, contains the same progesterone-like hormone found in the combination pill but does not contain estrogen. Progestin-only pills are taken every day, with no hormone-free interval. As with the combination pill, the progestin in the minipill provides a hostile environment for sperm and does not allow implantation of a fertilized egg in the uterus, but unlike the combination pill, the minipill does not always inhibit ovulation. For this reason, the minipill is somewhat less effective than other types of birth control pills. The minipill has also been associated with a higher incidence of irregular bleeding. Neither the combination pill nor the minipill should be taken unless prescribed by a health care provider who has detailed information about the woman's medical history.

Contraindications—reasons for not prescribing birth control pills—include hypertension, impaired liver function, known or suspected tumors that are estrogen-dependent, undiagnosed abnormal genital bleeding, pregnancy at the time of the examination, and history of poor blood circulation or blood clotting. The major complications associated with taking oral contraceptives are blood clots and high blood pressure. Also, the risk of heart attack is increased for those who smoke or have other risk factors for heart disease. The risk of cancer from using hormonal contraception is actually lower (Walling 2008).

If they smoke, women older than 35 should generally use other forms of contraception. Although the long-term negative consequences of taking birth control pills are still the subject of research, short-term negative effects are experienced by 25% of all women who use them. These side effects include increased susceptibility to vaginal infections, nausea, slight weight gain, vaginal bleeding between periods, breast tenderness, headaches, and mood changes (some women become depressed and experience a loss of sexual desire). Women should also be aware of situations in which the pill is not effective, such as the first month of use, with certain prescription medications (antibiotics), and when pills are missed. On the positive side, pill use reduces the incidence of ectopic pregnancy and offers noncontraceptive benefits, such as reduced incidence of ovarian and endometrial cancers, pelvic inflammatory disease, anemia, and benign breast disease.

Finally, women should be aware that pill use is associated with an increased incidence of chlamydia and gonorrhea. One reason for the association of pill use and a higher incidence of STIs is that sexually active women who use the pill sometimes erroneously feel that they are also protected from contracting STIs because they are protected from becoming pregnant. The pill provides no protection against STIs; the only methods that provide some protection against STIs are the male and female condoms.

Despite the widespread use of birth control pills, many women prefer a method that is longer acting and does not require daily action. Research continues toward identifying safe, effective hormonal contraceptive delivery methods that are more convenient. One recent hormonal contraceptive is Seasonale®, which reduces the number of periods a woman experiences from thirteen to four per year. This hormonal contraceptive manages the menstrual cycles by skipping the hormone-free week and limiting women's menstrual periods to once every three months. However, unexpected menstrual bleeding occurs four times as often with this extended-cycle regimen, and women who use Seasonale are four times as likely to discontinue use in the first twenty-six weeks as a result (Wilson and Kudis 2005). Two other similar hormonal contraceptives are Seasonique® (also an extended-cycle regime) and Lybrel® (a continuous-cycle regime) that is taken orally every day with no hormone-free week.

Contraceptive Implant—Implanon

Implanon® is a newer type of implant that was approved by the Food and Drug Administration in 2006. It is a progestin-only method that provides three years of pregnancy protection. A flexible plastic rod about the size of a match is inserted under the skin of the upper inner arm. The most common side effect associated with Implanon® is irregular menstrual bleeding.

Depo-Provera®

Also known as "Depo" and "the shot," Depo-Provera is a synthetic compound similar to progesterone that is injected into the woman's arm or buttock and protects against pregnancy for three months, either by preventing ovulation, thickening cervical mucus, and/or changing the lining of the uterus which prevents implantation. Side effects of Depo-Provera include menstrual spotting, irregular bleeding, and some heavy bleeding the first few months of use, although eight of ten women using Depo-Provera will eventually experience amenorrhea, or the absence of a menstrual period. Mood changes, headaches, dizziness, and fatigue have also been observed. Some women report a weight gain of five to ten pounds.

Also, after the injections are stopped, it takes an average of six to eighteen months before the woman will become pregnant at the same rate as women who have not used Depo-Provera. Slightly fewer than 3% of American women report using the injectable contraceptive. Most fear the side effects or are just satisfied with their current contraceptive method.

Recent research has shown that prolonged use of Depo-Provera is associated with significant loss of bone density, which may not be completely reversible after discontinuing use. A "black box" warning highlights this information. Depo-Provera should be used only as a long-term contraceptive method (longer than two years) if other methods are inadequate.

Vaginal Rings

NuvaRing, which is a soft, flexible, transparent ring approximately two inches in diameter that is worn inside the vagina, provides month-long pregnancy protection. NuvaRing has two major advantages. First, because the hormones are delivered locally rather than systemically, very low levels are administered (the lowest of any of the hormonal contraceptives). Second, unlike oral contraceptives, with which the hormone levels rise and fall depending on when the pill is taken, the hormone level from the NuvaRing remains constant. The NuvaRing is a highly effective contraceptive when used according to the labeling. Out of a hundred women using NuvaRing for a year, one or two will become pregnant. This method is self-administered.

NuvaRing is inserted into the vagina and is designed to release hormones that are absorbed by the woman's body for three weeks. The ring is then removed for a week, at which time the menstrual cycle will occur; afterward, the ring is replaced with a new ring. Side effects of NuvaRing are similar to those of the birth control pill. Barreios et al. (2010) assessed use of the vaginal ring in 75 women (18–37 years of age). Compared to pre-enrollment status, at the end of the study period, the patients reported significantly less dysmenorrhea and irritability. In addition, although weight and body mass index increased significantly among users, they remained within the expected limits.

Transdermal Applications

Ortho Evra is a contraceptive transdermal patch that delivers hormones to a woman's body through skin absorption. The contraceptive patch is worn on the buttocks, abdomen, upper torso (excluding the breasts), or on the outside of the upper arm for three weeks and is changed on a weekly basis. The fourth week is patch-free and the time when the menstrual cycle will occur. Under no circumstances should a woman allow more than seven days to lapse without wearing a patch.

Ortho Evra provides pregnancy protection and has side effects similar to those of the pill. Ortho Evra simply offers a different delivery method of the hormones needed to prevent pregnancy from occurring. In clinical trials, the contraceptive patch was found to keep its adhesiveness even through showers, workouts, and water activities, such as swimming. However, the FDA approved updating the labeling of Ortho Evra to warn users and health care providers that this product exposes women to about 60% more estrogen than most birth control pills and it may be less effective for women who weigh more than 198 pounds. In general, increased estrogen exposure may increase the risk of blood clots.

Male Hormonal Methods

Because of dissatisfaction with the few contraceptive options available to men, research and development are occurring in this area. Numerous studies have found that administration of testosterone to men markedly reduces sperm count and is a very efficient and well-tolerated method of contraception. Combinations with progestogens or with gonadotropin-releasing hormone antagonists are even more effective and suggest that hormonal contraception in men is feasible and may be as effective as the currently used methods (Amory et al. 2006). Several promising male products under development rely on MENT, a synthetic steroid that resembles testosterone. In contrast to testosterone, however, MENT does not have the effect of enlarging the prostate. A MENT implant and MENT transdermal gel and patch formulation are being developed for contraception.

Male Condom

Batar and Sivin (2010) noted that worldwide use of male condoms is about 6%. They remain the only proven preventive tool against STIs. Condoms are thin sheaths made of latex, polyurethane, or natural membranes. Latex condoms, which can be used only with water-based lubricants, historically have been more popular. They are also less likely to slip off and have a much lower chance of breakage. However, the polyurethane condom, which is thinner but just as durable as latex, is growing in popularity.

Polyurethane condoms can be used with oil-based lubricants, are an option for some people who have latex-sensitive allergies, provide some protection against HIV and other STIs, and allow for greater sensitivity during intercourse. Condoms made of natural membranes (sheep intestinal lining) are not recommended because they are not effective in preventing transmission of HIV or other STIs.

The condom works by being rolled over and down the shaft of the erect penis before intercourse. When the man ejaculates, sperm are caught inside the condom. When used in combination with a spermicidal lubricant that is placed on the inside of the reservoir tip of the condom as well as a spermicidal (sperm-killing) agent that the woman inserts into her vagina, the condom is a highly effective contraceptive. Care should be taken to avoid using nonoxynol-9 (N-9) as a contraceptive lubricant because it has been shown to provide no protection against STIs or HIV. N-9 products, such as condoms that have N-9 as a lubricant, should not be used rectally because doing so could increase one's risk of getting HIV or other STIs.

Like any contraceptive, the condom is effective only when used properly. Practice putting the male condom on reduces the chance of the condom breaking or coming off during intercourse.

The condom should be placed on the penis early enough to avoid any seminal leakage into the vagina. In addition, polyurethane or latex condoms with a reservoir tip are preferable, as they are less likely to break. Even when a condom has a reservoir tip, air should be squeezed out of the tip as it is being placed on the penis to reduce the chance of breaking during ejaculation. However, such breakage does occur. Finally, the penis should be withdrawn from the vagina immediately after ejaculation, before it returns to its flaccid state. If the penis is not withdrawn and the erection subsides, semen may leak from the base of the condom into the vaginal lips. Alternatively, when the erection subsides, the condom will come off when the man withdraws his penis if he does not hold onto the condom. Either way, the sperm will begin to travel up the vagina to the uterus and fertilization may occur.

In addition to furnishing extra protection, spermicides also provide lubrication, which permits easy entrance of the condom-covered penis into the vagina. If no spermicide is used and the condom is not of the prelubricated variety, a sterile lubricant (such as K-Y Jelly) may be needed. Vaseline or other kinds of oil-based products should not be used with condoms because vaginal infections and/or condom breakage may result. Though condoms should also be checked for visible damage and for the date of expiration, this is rarely done. Three-fourths of the respondents in Lane's (2003) study did not check for damage, and 61% did not check the date of expiration.

Female Condom

The female condom resembles the male condom except that it fits in the woman's vagina to protect her from pregnancy, HIV infection, and other STIs. The female condom is a lubricated, polyurethane adaptation of the male version. It is about six inches long and has flexible rings at both ends. It is inserted like a diaphragm, with the inner ring fitting behind the pubic bone against the cervix; the outer ring remains outside the body and encircles the labial area. Like the male version, the female condom is not reusable. Female condoms have been approved by the FDA and are being marketed under the brand names Femidom and Reality. The one-size-fits-all device is available without a prescription. The female condom has not "caught on" and is used by less than 1% of respondents (Batar and Sivin 2010)

The female condom is durable and may not tear as easily as latex male condoms. Some women may encounter some difficulty with first attempts to insert the female condom. A major advantage of the female condom is that, like the male counterpart, it helps protect against transmission of HIV and other STIs, giving women an option for protection if their partner refuses to wear a condom.

Placement may occur up to eight hours before use, allowing greater spontaneity. Women who have had an STI are more likely to use the female condom. If women have instruction and training in the use of the female condom, including a chance to practice the method, this increases its use. Women who had a chance to practice skills on a pelvic model were more likely to rate the method favorably, use it, and use it correctly.

Problems with the female condom occur when there is a large disparity between the size of the woman's vagina and the size of her partner's penis and when intercourse is very active. The female condom is also very expensive in comparison with other nonreusable barrier methods.

Vaginal Spermicides

A spermicide is a chemical that kills sperm. Vaginal spermicides come in several forms, including foam, cream, jelly, film, and suppository. Only spermicides without N-9 should be used (Battar 2010). Spermicidal creams or gels should be used with a diaphragm. Spermicidal foams, creams, gels, suppositories, and films may be used alone or with a condom. Spermicides must be applied before the penis enters the vagina no more than twenty minutes before intercourse (appropriate applicators are included when the product is purchased). Foam is effective immediately, but suppositories, creams, and jellies require a few minutes to allow the product to melt and spread inside the vagina (package instructions describe the exact time required). Each time intercourse is repeated, more spermicide must be applied. Spermicide must be left in place for at least six to eight hours after intercourse; douching or rinsing the vagina should not be done during this period.

Spermicides are advantageous in that they are available without a prescription or medical examination. They also do not manipulate the woman's hormonal system and have few side effects such as urinary tract infections that young women may develop. It was believed that a major noncontraceptive benefit of some spermicides is that they offer some protection against STIs. However, spermicides are not recommended for STI or HIV protection.

Contraceptive Sponge

The Today contraceptive sponge is a disk-shaped polyurethane device containing the spermicide N-9. This small device, which has a 72% to 86% effectiveness rate, is dampened with water to activate the spermicide and then inserted into the vagina before intercourse.

The sponge protects for repeated acts of intercourse for twenty-four hours without the need for supplying additional spermicide. It cannot be removed for at least six hours after intercourse, but it should not be left in place for more than thirty hours. Possible side effects that may occur with use include irritation, allergic reactions, or difficulty with removal; the risk of toxic shock syndrome, a rare but serious infection, is greater when the device is kept in place longer than recommended. The sponge provides no protection from STIs. The Today sponge was taken off the market for eleven years due to manufacturing concerns. National distribution has resumed (http://www.todaysponge.com/).

Intrauterine Device (IUD)

Although not technically a barrier method, the intrauterine device (IUD) is a structural device that prevents implantation. A physician inserts it into the uterus to prevent the fertilized egg from implanting on the uterine wall or to dislodge the fertilized egg if it has already implanted. Two common IUDs sold in the United States are ParaGard Copper T 380A and the Mirena® Intrauterine System (IUS). The Copper T is partly wrapped in copper and can remain in the uterus for up to ten years. The Mirena® IUS contains a supply of progestin, which it continuously releases into the uterus in small amounts for up to five years.

Although there is risk of perforation of the uterus, Hollander (2008) noted that the American College of Obstetricians and Gynecologists' Committee on Adolescent Health Care has come out strongly in favor of providing IUDs to adolescents. The organization states, "health care providers should strongly encourage young women who are appropriate candidates to use this method." Speidel et al. (2008) also noted that long-acting reversible contraception (LARC) methods, including intrauterine contraceptives and implants, have a proven record of very high effectiveness, many years of effectiveness, convenience, cost effectiveness, suitability for a wide variety of women and, in general, high user satisfaction.

Diaphragm

The diaphragm is a shallow rubber dome attached to a flexible, circular steel spring. Varying in diameter from two to four inches, the diaphragm covers the cervix and prevents sperm from moving beyond the vagina into the uterus. This device should always be used with a spermicidal jelly or cream.

To obtain a diaphragm, a woman must have an internal pelvic examination by a physician or nurse practitioner, who will select the appropriate size and instruct the woman on how to insert the diaphragm. The woman will be told to apply spermicidal cream or jelly on the inside of the diaphragm and around the rim before inserting it into the vagina (no more than two hours before intercourse).

The diaphragm must also be left in place for six to eight hours after intercourse to permit any lingering sperm to be killed by the spermicidal agent. After the birth of a child, a miscarriage, abdominal surgery, or the gain or loss of ten pounds, a woman who uses a diaphragm should consult her physician or health practitioner to ensure a continued good fit. In any case, the diaphragm should be checked every two years for fit.

A major advantage of the diaphragm is that it does not interfere with the woman's hormonal system and has few, if any, side effects. Also, for those couples who feel that menstruation diminishes their capacity to enjoy intercourse, the diaphragm may be used to catch the menstrual flow for a brief time. On the negative side, some women feel that using the diaphragm with spermicidal gel is messy and a nuisance, and using a spermicide may possibly produce an allergic reaction. Furthermore, some partners feel that spermicides make oral genital contact less enjoyable. Finally, if the diaphragm does not fit properly or is left in place too long (more than twenty-four hours), pregnancy or toxic shock syndrome can result.

Cervical Cap

The cervical cap is a thimble-shaped contraceptive device made of rubber or polyethylene that fits tightly over the cervix and is held in place by suction. Like the diaphragm, the cervical cap, which is used in conjunction with spermicidal cream or jelly, prevents sperm from entering the uterus. The cervical cap cannot be used during menstruation, because the suction cannot be maintained. The effectiveness, problems, risks, and advantages are similar to those of the diaphragm. Use of the cervical cap is rare—less than 1% of respondents (Batar and Sivin 2010).

Natural Family Planning

Also referred to as periodic abstinence, rhythm method, and fertility awareness, natural family planning involves refraining from sexual intercourse during the seven to ten days each month when the woman is thought to be fertile. Women who use natural family planning must know their time of ovulation and avoid intercourse just before, during, and immediately after that time. Calculating the fertile period involves knowing when ovulation has occurred. This is usually the fourteenth day (plus or minus two days) before the onset of the next menstrual period. Numerous "home ovulation kits" are available without a prescription in a pharmacy; these allow women to identify twelve to twenty-six hours in advance when they will ovulate (by testing their urine). Freund et al. (2010) noted that about 4% of all couples of reproductive age use this method.

Another method of identifying when ovulation has occurred is by observing an increase in the basal body temperature and a sticky cervical mucus. Calculating the time of ovulation may also be used as a method of becoming pregnant because it helps the couple to know when the woman is fertile.

Nonmethods: Withdrawal and Douching

Because withdrawal and douching are not effective in preventing pregnancy, we call them "nonmethods" of birth control. Also known as coitus interruptus, withdrawal is the practice whereby the man withdraws his penis from the vagina before he ejaculates. Freund et al. (2010) noted that about 3% of all couples of reproductive age use this method.

The advantages of coitus interruptus are that it requires no devices or chemicals, and it is always available. The disadvantages of withdrawal are that it does not provide protection from STIs, it may interrupt the sexual response cycle and diminish the pleasure for the couple, and it is very ineffective in preventing pregnancy.

Withdrawal is not a reliable form of contraception for two reasons. First, a man can unknowingly emit a small amount of pre-ejaculatory fluid, which may contain sperm. One drop can contain millions of sperm. In addition, the man may lack the self-control to withdraw his penis before ejaculation, or he may delay his withdrawal too long and inadvertently ejaculate some semen near the vaginal opening of his partner. Sperm deposited there can live in the moist vaginal lips and make their way up the vagina.

Though some women believe that douching is an effective form of contraception, it is not. Douching refers to rinsing or cleansing the vaginal canal. After intercourse, the woman fills a syringe with water, any of a variety of solutions that can be purchased over the counter, or a spermicidal agent and flushes (so she assumes) the sperm from her vagina. In some cases, however, the fluid will actually force sperm up through the cervix. In other cases, a large number of sperm may already have passed through the cervix to the uterus, so the douche may do little good.

Sperm may be found in the cervical mucus within ninety seconds after ejaculation. In effect, douching does little to deter conception and may even encourage it. In addition, douching is associated with an increased risk for pelvic inflammatory disease and ectopic pregnancy.

Emergency Contraception

National **Data**

Based on a national sample of 7,643 women aged 15 to 44, 4% of those who had ever had sex with a man reported having used emergency contraception (Kavanaugh and Schwarz 2008).

Also called postcoital contraception, **emergency contraception** (EC) refers to various types of morning-after pills that are used primarily in three circumstances: when a woman has unprotected intercourse, when a contraceptive method fails (such as condom breakage or slippage), and when a woman is raped. EC methods should be used in emergencies for those times when unprotected intercourse has occurred, and medication can be taken within seventy-two hours of exposure. Plan B, an emergency contraceptive, is now available over the counter for someone who is age 18 or above. Trussell et al. (2010) noted that in spite of the availability of emergency contraception pills, unintended pregnancy rates have not decreased. The authors suggest a need for highly effective reversible (HER) contraceptives (e.g., IUDs containing copper) to be constantly in place since the act of having sexual intercourse without immediate planning is likely to recur.

We consider emergency contraception something of a staple that ought to be in every woman's medicine cabinet just in case...

Kate Looby, South Dakota State Director for Planned Parenthood Minnesota, North Dakota, South Dakota

Combined Estrogen-Progesterone

The most common morning-after pills are the combined estrogen-progesterone oral contraceptives routinely taken to prevent pregnancy. In higher doses, they serve to prevent ovulation, fertilization of the egg, or transportation of the egg to the uterus. They may also make the uterine lining inhospitable to implantation. Known as the "Yuzpe method" after the physician who proposed it, this method involves ingesting a certain number of tablets of combined estrogen-progesterone. These pills must be taken within seventy-two hours of unprotected intercourse to be effective. A new pill, ulipristal acetate, is effective up to five days (120 hours) after unprotected sex. It received FDA approval in 2010 and became available in the United States in 2011. A prescription is required.

Common side effects of combined estrogen-progesterone EC pills (sold under the trade names Plan B and One-Step) include nausea, vomiting, headaches, and breast tenderness, although some women also experience abdominal pain, headache, and dizziness. Side effects subside within a day or two after treatment is completed.

Postcoital IUD

Insertion of a copper IUD within five to seven days after ovulation in a cycle when unprotected intercourse has occurred is very effective for preventing pregnancy. This option, however, is used much less frequently than hormonal treatment because women who need EC often are not appropriate IUD candidates. Women who have chlamydia or gonorrhea when an IUD is inserted are at higher risk of developing pelvic inflammatory disease; therefore, testing is recommended prior to insertion.

Mifepristone (RU-486)

Mifepristone, also known as RU-486, is a synthetic steroid that effectively inhibits implantation of a fertilized egg by making the endometrium unsuitable for implantation. The so-called abortion pill, approved by the FDA in the United States in 2000, is marketed under the name Mifeprex and can be given to induce abortion within seven weeks of pregnancy. Side effects of RU-486 are usually very severe and may include cramping, nausea, vomiting, and breast tenderness. Potential serious adverse effects include the possibility of hemorrhage and infection. The pregnancy rate associated with RU-486 is low, which suggests that RU-486 is an effective means of EC. More than 90% of U.S. women who tried RU-486 would recommend it to others and choose it over surgery again. Its use remains controversial. To be clear, an abortion is not a method of contraception because conception has already occurred. An abortion simply prevents conception, fertilization, and development to progressing toward a normal pregnancy.

Effectiveness of Various Contraceptives

Over half (54%) of women who have an abortion were using a contraceptive method during the month they became pregnant (Frantz 2010). In Table ST2.1, we present data on the effectiveness of various contraceptive methods in preventing pregnancy and protecting against STIs. Table ST2.1 also describes the benefits, disadvantages, and costs of various methods of contraception. Also included in the chart is the obvious and most effective form of birth control: abstinence. Its cost is the lowest, it is 100% effective for pregnancy prevention, and it also eliminates the risk of HIV and other STIs from intercourse. Abstinence can be practiced for a week, a month, several years, until marriage, or until someone finds the "right" sexual partner.

Note: Appreciation is expressed to Tywanna Purkett, MA, Health Education Specialist, Campus Recreation and Wellness, East Carolina University for her assistance in the development of this section.

TABLE ST2.1 Methods of Contraception and Sexually Transmitted Infection Protection

Method	Typical Use[1] Effectiveness Rates	STI Protection	Benefits	Disadvantages	Cost[2]
Oral contraceptive, combined or progestin-only pills ("the pill")	92%	No	High effectiveness rate; 24-hour protection; menstrual regulation	Daily administration; side effects possible; medication interactions	$10–60 per month
Implanon® (one rod, three-year implant)	99.95%	No	High effectiveness rate; long-term protection	Menstrual changes; side effects possible	$150–200, not including implantation fees
Depo-Provera® (three-month injection)	97%	No	High effectiveness rate; no estrogen; privacy of use	Can seriously impact bone density; short-term use recommended unless other methods are inadequate; side effects likely	$45–75 per injection
Ortho Evra® (transdermal patch)	92%	No	Same as oral contraceptives except maintenance is weekly, not daily	Visibility; side effects possible; 60% more hormone exposure than pills	$15–32 per month
NuvaRing® (vaginal ring)	92%	No	Lower dosage than other hormonal methods	Must be comfortable with body for insertion	$15–48 per month
Male condom	85%	Yes	Few or no side effects; easy to purchase and use	Can interrupt spontaneity	$2–10 a box
Female condom	79%	Yes	Few or no side effects; easy to purchase	Decreased sensation; insertion takes practice	$4–10 a box
Spermicide	71%	No	Many forms to choose; easy to purchase and use	Can cause irritation and be messy	$8–18 per box, tube, or can
Today® Sponge[3]	68–84%	No	Few side effects; effective for 24 hours after insertion	Spermicide irritation possible	$3–5 per sponge
Diaphragm, Lea's Shield, and cervical cap[3]	68–84%	No	Few side effects; can be inserted within two hours before intercourse	Can be messy; increased risk of vaginal or urinary tract infections	$50–200 plus spermicide
Intrauterine device (IUD): Mirena	98.2–99%	No	Little maintenance; longer-term protection	Risk of pelvic inflammatory disease increased; chance of expulsion	$200–400
Withdrawal	73%	No	Requires little planning; always available	Pre-ejaculatory fluid can contain sperm	$0
Periodic abstinence (also called natural family planning or fertility awareness)	75%	No	No side effects; accepted in all religions or cultures	Requires a lot of planning; need ability to interpret fertility signs	$0

(continued)

Method	Typical Use[1] Effectiveness Rates	STI Protection	Benefits	Disadvantages	Cost[2]
Emergency contraception	75%	No	Provides an option after intercourse has occurred	Must be taken within 72 hours; side effects likely	$10–32
Abstinence	100%	Yes	No risk of pregnancy or STDs	Partners both have to agree to abstain	$0
Sterilization	99.5% (Female) 99.85% (Male)	No	Long-term; complications are rare	May not be reversible	$250–5,000

[1]Effectiveness rates are listed as percentages of women not experiencing an unintended pregnancy during the first year of typical use. Typical use refers to use under real-life conditions. Perfect use effectiveness rates are higher.

[2]Costs may vary.

[3]Lower percentages apply to parous women (women who have given birth). Higher rates apply to nulliparous women (women who have never given birth).

Source: Tywanna Purkett, MA, a health education specialist, Campus Recreation and Wellness, East Carolina University.

References

Amory, J., S. Page, and W. Bremner. 2006. Drug insight: recent advances in male hormonal contraception. *Nature Clinical Practice Endocrinology & Metabolism* 2: 32–41.

Barreios, F. A., C. A. F. Guazzelli, R. Barbosa, F. de Assis, and F. F. de Araujo. 2010. Extended regimens of the contraceptive vaginal ring: evaluation of clinical aspects. *Contraception* 81: 223–225.

Batar, I., and I. Sivin. 2010. State-of-the-art of non-hormonal methods of contraception: I. Mechanical barrier contraception. *European Journal of Contraception & Reproductive Health Care* 15: 67–88.

Batar, I. 2010. State-of-the-art of non-hormonal methods of contraception: II. Chemical barrier contraceptives. *European Journal of Contraception & Reproductive Health Care* 15: 89–95.

Frantz, K. 2010. A questionable pro-choice strategy (in 140 characters or less). A woman who used the Twitter online service to broadcast her abortion. *The Humanist* 70: 7–10.

Freund, G., I. Sivin, and I. Batar. 2010. State-of-the-art of non-hormonal methods of contraception: IV. Natural family planning. *European Journal of Contraception & Reproductive Health Care* 15: 113–123.

Hollander, D. 2008. IUDs for teenagers? *Perspectives on Sexual and Reproductive Health* 40: 5–6.

Kavanaugh, M. L., and E. B. Schwarz. 2008. Counseling about and use of emergency contraception in the United States. *Perspectives on Sexual and Reproductive Health* 40: 81–87.

Knox, D., and S. Hall. 2010 Relationship and sexual behaviors of a sample of 2,922 university students. Unpublished data collected for this text. Department of Sociology, East Carolina University, and Department of Family and Consumer Sciences, Ball State University.

Lane, T. 2003. High proportion of college men using condoms report errors and problems. *Perspectives on Sexual and Reproductive Health* 35: 50–52.

Speidel, J. J., C. C. Harper, and W. C. Shields. 2008. The potential of long-acting reversible contraception to decrease unintended pregnancy. *Contraception* 78: 197–270.

Trussell, J., E. B. Schwarz, and K. Guthrie. 2010. Research priorities for preventing unintended pregnancy: Moving beyond emergency contraceptive pills. *Perspectives on Sexual & Reproductive Health* 42: 8–9.

Walling, A. D. 2008. Oral contraception is associated with reduced overall cancer risk. *American Family Physician* 77: 1597–1599.

See Also Social &
ARAWAN SANDY M
1035 A Directors Ct 27858
CARENET COUNSELING EA
3219 Landmark St 27834
CHRISTIAN COUNSELING MINISTR

Counselors - Marriage, Family & Individual

See Also Social & Human Svce Organizations
CLINICAL & SUBSTANCE
COUNSELING

Chelsea E. Curry

What's right about America is that although we have a mess of problems, we have great capacity-intellect and resources-to do some thing about them.

Henry Ford

Abortion—Pro-Choice

Religious Coalition for Reproductive Choice
1413 K Street NW
14th Floor
Washington, DC 20005
Phone: 202-628-7700
    www.rcrc.org

Abortion—Pro-Life

National Right to Life Committee
512 10th Street NW
Washington, DC 20004
Phone: 202-626-8800
    www.nrlc.org/Unborn_Victims/

Adoption

Dave Thomas Foundation for Adoption
525 Metro Place North
Suite 220
Dublin, Ohio 43017
Phone: 800-275-3832
    www.davethomasfoundationforadoption.org/

Evan B. Donaldson Adoption Institute
120 East 38th Street
New York, NY 10016
Phone: 212-925-4089
    www.adoptioninstitute.org/

Al-Anon Family Groups

1600 Corporate Landing Parkway
Virginia Beach, VA 23454
Phone: 757-563-1600
Fax: 757-563-1655
    www.al-anon.org

Child Abuse

Prevent Child Abuse America
228 South Wabash Avenue
10th Floor
Chicago, IL 60604
Phone: 312-939-8962
    www.preventchildabuse.org/index.shtml

Children

National Association for the Education of Young Children
1313 L St. N.W. Suite 500
Washington, DC 20005
Phone: 800-424-2460
    www.naeyc.org

Communes/Intentional Communities

Twin Oaks
138 Twin Oaks Road
Louisa, VA 23093
Phone: 540-894-5798
    www.twinoaks.org/

Divorce—Custody of Children

Child Custody Evaluation Services of Philadelphia, Inc.
Dr. Ken Lewis
P. O. Box 202
Glenside, PA 19038
Phone: 215-576-0177

Domestic Attorney
Leslie Fritscher
Wsrd and Smith, PA
120 West Fire Tower Road
Winterville, NC 28590
252 215 4002

Family Law Specialist/Divorce Mediator
Shelby Duffy Benton
130 S. John St.
P. O. Box 947
Goldsboro, NC 27533
Phone: 919-736-1830

Divorce Recovery

The Divorce Room
    http://www.heartchoice.com/hc/divorce/index.php

American Coalition for Fathers and Children
1718 M St. NW. #187
Washington, DC 20036
Phone: 800-978-3237
    www.acfc.org

Domestic Violence

National Coalition against Domestic Violence
1120 Lincoln Street
Suite 1603
Denver, CO 80203
Phone: 303-839-1852
    www.ncadv.org/

National Toll Free Number for Domestic Violence
Phone: 800-799-7233

Family Planning

Planned Parenthood Federation of America
434 West 33rd Street
New York, NY 10001
Phone: 212-541-7800
    www.plannedparenthood.org

Grandparenting

The Foundation for Grandparenting
108 Farnham Road
Ojai, CA 93023
    www.grandparenting.org

AARP (grandparent information center)
601 E Street NW
Washington, DC 20049
Phone: 888-687-2277
www.aarp.org/grandparents/

Healthy Baby
National Healthy Mothers, Healthy Babies Coalition
2000 N. Beauregard Street
6th Floor
Alexandria, VA 22311
Phone: 703-837-4792
www.hmhb.org

Homosexuality
National Gay and Lesbian Task Force
1325 Massachusetts Ave. NW
Washington, DC 20005
Phone: 202-393-5177
www.thetaskforce.org/

Parents, Families and Friends of Lesbians and Gays (PFLAG)
1828 L Street, NW
Suite 660
Washington, DC 20036
Phone: 202-467-8180
www.pflag.org

Infertility
American Fertility Association
315 Madison Avenue
Suite 901
New York, NY 10017
Phone: 888-917-3777
www.theafa.org/

Marriage
National Marriage Project, Rutgers, The State University of New Jersey
25 Bishop Place
New Brunswick, NJ 08901-1181
Phone: 732-932-2722
E-mail: marriage@rci.rutgers.edu

Marriage: Keeping It Healthy and Happy
http://www.heartchoice.com/hc/marriage/marriage.php

Coalition for Marriage, Family and Couples Education
5310 Belt Road NW
Washington, DC 20015-1961
Phone: 202-362-3332
www.smartmarriages.com/

National Healthy Marriage Resource Center
9300 Lee Highway
Fairfax, VA 22031-6050
Phone: 866-916-4672 or 866-91-NHMRC
http://www.healthymarriageinfo.org/

Marriage and Family Therapy
American Association for Marriage and Family Therapy
112 South Alfred Street
Alexanderia, VA 22314
Phone: 703-838-9805
www.aamft.org/

Marriage Enrichment
ACME (Association for Couples in Marriage Enrichment)
56 Windsor Court
New Brighton, MN 55112
Phone: 800-634-8325
E-mail: hamb1001@tc.umn.edu
www.bettermarriages.org

Mate Selection
Right Mate
http://www.heartchoice.com/hc/rightmate/index.php

Men's Awareness
American Men's Studies Association
1507 Pebble Drive
Greensboro, NC 27410
Phone: 336-323-2672
www.mensstudies.org/

Motherhood
MOTHERS
c/o National Association of Mothers' Centers
1740 Old Jericho Turnpike
Jericho, NY 11753
Phone: 877-939-6667
www.mothersoughttohaveequalrights.org/

Parental Alienation Syndrome
http://www.deltabravo.net/custody/rand01.php

Parenting Education
Family Education Network
501 Boylston St,
Suite 900
Bosta, MA 02116
Phone: 617-671-2000
www.familyeducation.com/home/

Relationships
http://www.heartchoice.com/hc/

Reproductive Health
Association of Reproductive Health Professionals
1901 L Street, NW
Suite 300
Washington, DC 20036-1718
Phone: 202-466 3825
www.arhp.org

Sex Abuse

VOICES in Action, Inc. (Victims of Incest Can Emerge Survivors)
P.O. Box 148309
Chicago, Illinois 60614
Phone: 800-7-VOICE-8

National Clearinghouse on Marital and Date Rape
2325 Oak Street
Berkeley, California 94708-1697
Phone: 510-524-1582
www.ncmdr.org

Sex Education

Sexuality Information and Education Council of the United States
90 John St.
Suite 402
New York, New York 10038
Phone: 212-819-9776
www.siecus.org

Sexual Intimacy
http://www.heartchoice.com/hc/intimacy/index.php

Sex Fantasy Questionnaire-Glenn Wilson
http://www.cymeon.com/sfq.asp

Sexually Transmissible Diseases

American Social Health Association (Herpes Resource Center and HPV Support Program)
P.O. Box 13827
Research Triangle Park, NC 27709
Phone: 919-361-8400
www.ashastd.org

National AIDS Hotline
Phone: 800-232-4636
www.thebody.com/hotlines/national.html

National Herpes Hotline
Phone: 919-361-8488 or 800-230-6039
www.ashastd.org/herpes/herpes_overview.cfm

National STD Hotline
Phone: 800-227-8922
www.ashastd.org/

STD/AIDS
Phone: 919-361-8400

Single Parenthood

Parents without Partners
1650 South Dixie Highway,
Suite 402
Boca Raton, Fl 33432
Phone: 800-637-7974
www.parentswithoutpartners.org

Single Mothers by Choice
P.O Box 1642
New York, NY 10028
Phone: 212-988-0993
   www.singlemothersbychoice.com

Singlehood

Alternatives to Marriage Project
PMB 131
358 7th Avenue
Brooklyn NY, 11215
Tel: 347-987-1068
   www.unmarried.org/

Stepfamilies

National Stepfamily Resource Center
   http://www.stepfamilies.info/

Stepfamily Associates
1368 Beacon Street
Suite 108
Brookline, MA 02446
Phone: 617-731-5767
   www.stepfamilyboston.com

Stepfamily Foundation
310 West 85th St. # 1B
New York, NY 10024
Phone: 212-877-3244
   www.stepfamily.org

Transgender

Tri-Ess: The Society for the Second Self, Inc.
P.O. Box 980638
Houston, TX 77098-0638
Phone: 713-349-9910
   www.tri-ess.org

Widowhood

Widowed Persons Service, American Association of Retired Persons
601 E St. NW
Washington, DC 20049
Phone: 202-434-2260
   http://seniors-site.com/widowm/wps.html

Women's Awareness

National Organization for Women
1100 H Street NW,
3rd Floor
Washington, DC 20005
Phone: 202-331-9002
   www.now.org

Individual Autobiography Outline

Your instructor may ask you to write a paper that reflects the individual choices you have made that have contributed to you becoming who you are. Check with your instructor to determine what credit (if any) is available for completing this autobiography. Use the following outline to develop your paper. Some topics may be too personal, and you may choose to avoid writing about them. Your emotional comfort is important, so skip any questions you want and answer only those questions you feel comfortable responding to.

I. Choices: Free Will Versus Determinism

Specify the degree to which you feel that you are free to make your own interpersonal choices, versus the degree to which social constraints influence and determine your choices. Give an example of social influences being primarily responsible for a relationship choice and an example of you making a relationship choice where you acted contrary to the pressure you were getting from parents, siblings, and peers.

II. Relationship Beginnings

A. *Interpersonal context into which you were born.* If your parents are married, how long had they been married before you were born? How many other children had been born into your family? How many followed your birth? Describe how these and other parental choices affected you before you were born and the choices you will make in regard to family planning that will affect the lives of your children.

B. *Early relationships.* What was your relationship with your mother, father, and siblings when you were growing up? What is your relationship with each of them today? Who took care of you as a baby? If this person was other than your parents or siblings (for example, a grandparent), what is your relationship with that person today? How often did your mother or father tell you that they loved you? How often did they embrace or hug you? How has this closeness or distance influenced your pattern with others today? Give other examples of how your experiences in the family in which you were reared have influenced who you are and how you behave today. How easy or difficult is it for you to make decisions and how have your parents influenced this capacity?

C. *Early self-concept.* How did you feel about yourself as a child, an adolescent, and a young adult? What significant experiences helped to shape your self-concept? How do you feel about yourself today? What choices have you made that have resulted in you feeling good about yourself? What choices have you made that have resulted in you feeling negatively about yourself?

III. Subsequent Relationships

A. *First love.* When was your first love relationship with someone outside your family? What kind of love was it? Who initiated the relationship? How long did it last? How did it end? How did it affect you and your subsequent relationships? What choices did you make in this first love relationship that you are glad you made? What choices did you make that you now feel were a mistake?

B. *Subsequent love relationships.* What other significant love relationships (if any) have you had? How long did they last and how did they end? What choices did you make in these relationships that you are glad you made? What choices did you make that you now feel were a mistake? On a ten-point scale (0 = very distant and 10 = very close), how emotionally close do you want to be to a romantic partner?

C. *Subsequent relationship choices.* What has been the best relationship choice you have made? What relationship choices have you regretted?

D. *Lifestyle preferences.* What are your preferences for remaining single, being married, or living with someone? How would you feel about living in a commune? What do you believe is the ideal lifestyle? Why?

IV. Communication Issues

A. *Parental models.* Describe your parents' relationship and their manner of communicating with each other. How are your interpersonal communication patterns similar to and different from theirs?

B. *Relationship communication.* How comfortable do you feel talking about relationship issues with your partner? How comfortable do you feel telling your partner what you like and don't like about his or her behavior? To what degree have you told your partner your feelings for him or her? To what degree have you told your partner your desires for the future?

C. *Sexual communication.* How comfortable do you feel giving your partner feedback about how to please you sexually? How comfortable are you discussing the need to use a condom with a potential sex partner? How would you approach this topic?

D. *Sexual past.* How much have you disclosed to a partner about your previous relationships? How honest were you? Do you think you made the right decision to disclose or withhold? Why?

V. Sexual Choices

A. *Sex education.* What did you learn about sex from your parents, peers, and teachers (both academic and religious)? What choices did your parents, peers, and teachers make about your sex education that had positive consequences? What decisions did they make that had negative consequences for you?

B. *Sexual experiences.* What choices have you made about your sexual experiences that had positive outcomes? What choices have you made that resulted in sexual regret?

C. *Sexual values.* To what degree are your sexual values absolutist, legalist, or relativistic? How have your sexual values changed since you were an adolescent?

D. *Safer sex.* What is the riskiest choice you have made with regard to your sexual behavior? What is the safest choice you have made with regard to your sexual behavior? What is your policy about asking your partner about previous sex history and requiring that both of you be tested for HIV and STIs before having sex? How comfortable are you buying and using condoms?

VI. Violence and Abuse Issues

A. *Violent or abusive relationship.* Have you been involved in a relationship in which your partner was violent or abusive toward you? Give examples of the violence or abuse (verbal and nonverbal) if you have been involved in such a relationship. How many times did you leave and return to the relationship before you left permanently?

What was the event that triggered your decision to leave the first and last time? Describe the context of your actually leaving (for example, did you leave when the partner was at work?). To what degree have you been violent or abusive toward a partner in a romantic relationship?

B. *Family or sibling abuse.* Have you been involved in a relationship where your parents or siblings were violent or abusive toward you? Give examples of any violence or abuse (verbal or nonverbal) if such experiences were part of your growing up. How have these experiences affected the relationship you have with the abuser today?

C. *Forced sex.* Have you been pressured or forced to participate in sexual activity against your will by a parent, sibling, partner, or stranger? How did you react at the time and how do you feel today? Have you pressured or forced others to participate in sexual experiences against their will?

VII. Reproductive Choices

A. *Contraception.* What is your choice for type of contraception? How comfortable do you feel discussing the need for contraception with a potential partner? In what percentage of your first-time intercourse experiences with a new partner did you use a condom? (If you have not had intercourse, this question will not apply.)

B. *Children.* How many children (if any) do you want and at what intervals? How important is it to you that your partner wants the same number of children as you do?

How do you feel about artificial insemination, sterilization, abortion, and adoption? How important is it to you that your partner feels the same way?

VIII. Childrearing Choices

A. *Discipline.* What are your preferences for your use of "time-out" or "spanking" as a way of disciplining your children? How important is it to you that your partner feel the same as you do on this issue?

B. *Day care.* What are your preferences for whether your children grow up in day care or whether one parent will stay home with and rear the children? How important is it to you that your partner feel the same as do you on this issue?

C. *Education.* What are your preferences for whether your children attend public or private school or are "homeschooled"? What are your preferences for whether your child attends a religious school? To what degree do you feel it is your responsibility as parents to pay for the college education of your children? How important is it to you that your partner feels the same as do you on these issues?

IX. Education or Career Choices

A. *Own educational or career choices.* What is your major in school? How important is it to you that you finish undergraduate school? How important is it to you that you earn a master's degree, PhD, MD, or law degree? In what? To what extent do you want to be a stay-at-home mom or stay-at-home dad? How important is it to you that your partner be completely supportive of your educational, career, and family aspirations?

B. *Expectations of partner.* How important is it to you that your partner has the same level of education that you have? To what degree are you willing to be supportive of your partner's educational and career aspirations? What are the career goals of you and your partner?

Family Autobiography Outline

Your instructor may want you to write a paper that reflects the influence of your family on your development. Check with your instructor to determine what credit (if any) is available for completing this family autobiography. Use the following outline to develop your paper. Some topics may be too personal, and you may choose to avoid writing about them. Your emotional comfort is important, so skip any questions you like and answer only those you feel comfortable responding to.

I. Family Background

A. Describe yourself, including age, gender, place of birth, and additional information that helps to identify you. On a scale of 0 to 10 (10 is highest), how happy are you? Explain this number in reference to the satisfaction you experience in the various roles you currently occupy (for example, offspring, sibling, partner in a relationship, employee, student, parent, roommate, friend).

B. Identify your birth position; give the names and ages of children younger and older than you. How did you feel about your "place" in the family? How do you feel now?

C. What was your relationship with and how did you feel about each parent and sibling when you were growing up?

D. What is your relationship with and how do you feel today about each of these family members?

E. Which parental figure or sibling are you most like? How? Why?

F. Who else lived in your family (for example, grandparent, spouse of sibling), and how did they impact family living?

G. Discuss the choice you made before attending college that you regard as the wisest choice you made during this time period. Discuss the choice you made prior to college that you regret.

H. Discuss the one choice you made since you began college that you regard as the wisest choice you made during this time period. Discuss the one choice you made since you began college that you regret.

II. Religion and Values

A. In what religion were you socialized as a child? Discuss the impact of religion on yourself as a child and as an adult. To what degree will you choose to teach your own children similar religious values? Why?

B. Explain what you were taught in your family in regard to each of the following values: intercourse outside of marriage, the need for economic independence, manners, honesty, importance of being married, qualities of a desirable spouse, importance of having children, alcohol and drugs, safety, elderly family members, people of other races and religions, people with disabilities, people of alternative sexual orientation, people with less or more education, people with and without "wealth," and occupational role (in regard to the latter, what occupational role were you encouraged to pursue?). To what degree will you choose to teach your own children similar values? Why?

C. What was the role relationship between your parents in terms of dominance, division of labor, communication, affection, and so on? How has your observation of the parent of the same sex and opposite sex influenced the role you display in your current relationships with intimate partners? To what degree will you choose to have a similar relationship with your own partner that your parents had with each other?

D. How close were your parents emotionally? How emotionally close were/are you with your parents? To what degree did you and your parents discuss feelings? On a scale of 0 to 10, how well do your parents know how you think and feel? To what degree will you choose to have a similar level of closeness or distance with your own partner and children?

E. Did your parents have a pet name for you? What was this name and how did you feel about it?

F. How did your parents resolve conflict between themselves? How do you resolve conflict with partners in your own relationships?

III. Economics and Social Class

A. Identify your social class (lower, middle, upper); the education, jobs, or careers of your respective parents; and the economic resources of your family. How did your social class and economic well-being affect you as a child? How has the economic situation in which you were reared influence your own choices of what you want for yourself? To what degree are you economically self-sufficient?

B. How have the career choices of your parents influenced your own?

IV. Parental Plusses and Minuses

A. What is the single most important thing your mother and father, respectively, said or did that has affected your life in a positive way?

B. What is the single biggest mistake your mother and father, respectively, made in rearing you? Discuss how this impacted you negatively. To what degree does this choice still affect you?

V. Personal Crisis Events

A. Everyone experiences one or more crisis events that have a dramatic impact on life. Identify and discuss each event you have experienced and your reaction and adjustment to them.

B. How did your parents react to this crisis event you were experiencing? To what degree did their reaction help or hinder your adjustment? What different choice or choices (if any) could they have made to assist you in ways you would have regarded as more beneficial?

VI. Family Crisis Events

Identify and discuss each crisis event your family has experienced. How did each member of your family react and adjust to each event? An example of a family crisis event would be unemployment of a primary breadwinner, prolonged illness of a family member, aging parent coming to live with the family, death of a sibling, or alcoholism.

VII. Family Secrets

Most families have secrets. What secrets are you aware of in your family and kinship system? To what degree has it been difficult for you to be aware of this family secret?

VIII. Future

Describe yourself two, five, and ten years from now. What are your educational, occupational, marital, and family goals? How has the family in which you were reared influenced each of these goals? What choices might your parents make to assist you in achieving your goals?

Prenuptial Agreement of a Remarried Couple

Pam and Mark are of sound mind and body and have a clear understanding of the terms of this contract and of the binding nature of the agreements contained herein; they freely and in good faith choose to enter into the PRENUPTIAL AGREEMENT and MARRIAGE CONTRACT and fully intend it to be binding upon themselves.

Now, therefore, in consideration of their love and esteem for each other and in consideration of the mutual promises herein expressed, the sufficiency of which is hereby acknowledged, Pam and Mark agree as follows:

Names

Pam and Mark affirm their individuality and equality in this relationship. The parties believe in and accept the convention of the wife's accepting the husband's name, while rejecting any implied ownership.

Therefore, the parties agree that they will be known as husband and wife and will henceforth employ the titles of address Mr. and Mrs. Mark Stafford and will use the full names of Pam Hayes Stafford and Mark Robert Stafford.

Relationships with Others

Pam and Mark believe that their commitment to each other is strong enough that no restrictions are necessary with regard to relationships with others.

Therefore, the parties agree to allow each other freedom to choose and define their relationships outside this contract, and the parties further agree to maintain sexual fidelity each to the other.

Religion

Pam and Mark reaffirm their belief in God and recognize He is the source of their love. Each of the parties has his/her own religious beliefs.

Therefore, the parties agree to respect their individual preferences with respect to religion and to make no demands on each other to change such preferences.

Children

Pam and Mark both have children. Although no minor children will be involved, there are two (2) children still at home and in school and in need of financial and emotional support.

Therefore, the parties agree that they will maintain a home for and support these children as long as is needed and reasonable. They further agree that all children of both parties will be treated as one family unit, and each will be given emotional and financial support to the extent feasible and necessary as determined mutually by both parties.

Careers and Domicile

Pam and Mark value the importance and integrity of their respective careers and acknowledge the demands that their jobs place on them as individuals and on their partnership.

Both parties are well established in their respective careers and do not foresee any change or move in the future.

The parties agree, however, that if the need or desire for a move should arise, the decision to move shall be mutual and based on the following factors:

1. The overall advantage gained by one of the parties in pursuing a new opportunity shall be weighed against the disadvantages, economic and otherwise, incurred by the other.

2. The amount of income or other incentive derived from the move shall not be controlling.

3. Short-term separations as a result of such moves may be necessary.

Mark hereby waives whatever right he might have to solely determine the legal domicile of the parties.

Care and Use of Living Spaces

Pam and Mark recognize the need for autonomy and equality within the home in terms of the use of available space and allocation of household tasks. The parties reject the concept that the responsibility for housework rests with the woman in a marriage relationship whereas the duties of home maintenance and repair rest with the man.

Therefore, the parties agree to share equally in the performance of all household tasks, taking into consideration individual schedules, preferences, and abilities.

The parties agree that decisions about the use of living space in the home shall be mutually made, regardless of the parties' relative financial interests in the ownership or rental of the home, and the parties further agree to honor all requests for privacy from the other party.

Property, Debts, Living Expenses

Pam and Mark intend that the individual autonomy sought in the partnership shall be reflected in the ownership of existing and future-acquired property, in the characterization and control of income, and in the responsibility for living expenses. Pam and Mark also recognize the right of patrimony of children of their previous marriages.

Therefore, the parties agree that all things of value now held singly and/or acquired singly in the future shall be the property of the party making such acquisition. In the event that one party to this agreement shall predecease the other, property and/or other valuables shall be disposed of in accordance with an existing will or other instrument of disposal that reflects the intent of the deceased party.

Property or valuables acquired jointly shall be the property of the partnership and shall be divided, if necessary, according to the contribution of each party. If one party shall predecease the other, jointly owned property or valuables shall become the property of the surviving spouse.

Pam and Mark feel that each of the parties to this agreement should have access to monies that are not accountable to the partnership.

Therefore, the parties agree that each shall retain a mutually agreeable portion of their total income and the remainder shall be deposited in a mutually agreeable banking institution and shall be used to satisfy all jointly acquired expenses and debts.

The parties agree that beneficiaries of life insurance policies they now own shall remain as named on each policy. Future changes in beneficiaries shall be mutually agreed on after the dependency of the children of each party has been terminated. Any other benefits of any retirement plan or insurance benefits that accrue to a spouse only shall not be affected by the foregoing.

The parties recognize that in the absence of income by one of the parties, resulting from any reason, living expenses may become the sole responsibility of the employed party, and in such a situation, the employed party shall assume responsibility for the personal expenses of the other.

Both Pam and Mark intend their marriage to last as long as both shall live.

Therefore, the parties agree that, should it become necessary, due to the death of either party, the surviving spouse shall assume any last expenses in the event that no insurance exists for that purpose.

Pam hereby waives whatever right she might have to rely on Mark to provide the sole economic support for the family unit.

Evaluation of the Partnership

Pam and Mark recognize the importance of change in their relationship and intend that this CONTRACT shall be a living document and a focus for periodic evaluations of the partnership.

The parties agree that either party can initiate a review of any article of the CONTRACT at any time for amendment to reflect changes in the relationship. The parties agree to honor such requests for review with negotiations and discussions at a mutually convenient time.

The parties agree that, in any event, there shall be an annual reaffirmation of the CONTRACT on or about the anniversary date of the CONTRACT.

The parties agree that, in the case of unresolved conflicts between them over any provisions of the CONTRACT, they will seek mediation, professional or otherwise, by a third party.

Termination of the Contract

Pam and Mark believe in the sanctity of marriage; however, in the unlikely event of a decision to terminate this CONTRACT, the parties agree that neither shall contest the application for a divorce decree or the entry of such decree in the county in which the parties are both residing at the time of such application.

In the event of termination of the CONTRACT and divorce of the parties, the provisions of this and the section on "Property, Debts, Living Expenses" of the CONTRACT as amended shall serve as the final property settlement agreement between the parties. In such event, this CONTRACT is intended to effect a complete settlement of any and all claims that either party may have against the other, and a complete settlement of their respective rights as to property rights, homestead rights, inheritance rights, and all other rights of property otherwise arising out of their partnership. The parties further agree that, in the event of termination of this CONTRACT and divorce of the parties, neither party shall require the other to pay maintenance costs or alimony.

Decision Making

Pam and Mark share a commitment to a process of negotiations and compromise that will strengthen their equality in the partnership. Decisions will be made with respect for individual needs. The parties hope to maintain such mutual decision making so that the daily decisions affecting their lives will not become a struggle between the parties for power, authority, and dominance. The parties agree that such a process, although sometimes time-consuming and fatiguing, is a good investment in the future of their relationship and their continued esteem for each other.

Now, therefore, Pam and Mark make the following declarations:

1. They are responsible adults.
2. They freely adopt the spirit and the material terms of this prenuptial and marriage contract.

3. The marriage contract, entered into in conjunction with a marriage license of the State of Illinois, County of Wayne, on this 12th day of June 2004, hereby manifests their intent to define the rights and obligations of their marriage relationship as distinct from those rights and obligations defined by the laws of the State of Illinois, and affirms their right to do so.

4. They intend to be bound by this prenuptial and marriage contract and to uphold its provisions before any Court of Law in the Land.

Therefore, comes now, Pam Hayes Stafford, who applauds her development that allows her to enter into this partnership of trust, and she agrees to go forward with this marriage in the spirit of the foregoing PRENUPTIAL and MARRIAGE CONTRACT.

Therefore, comes now, Mark Robert Stafford, who celebrates his growth and independence with the sign-

ing of this contract, and he agrees to accept the responsibilities of this marriage as set forth in the foregoing PRENUPTIAL and MARRIAGE CONTRACT.

This CONTRACT AND COVENANT has been received and reviewed by the Reverend Ray Brannon, officiating.

Finally, come Vicki Whitfield and Rodney Whitfield, who certify that Pam and Mark did freely read and sign this MARRIAGE CONTRACT in their presence, on the occasion of their entry into a marriage relationship by the signing of a marriage license in the State of Illinois, County of Wayne, at which they acted as official witnesses. Further, they declare that the marriage license of the parties bears the date of the signing of this PRENUPTIAL and MARRIAGE CONTRACT. (Although this document is real, the names are fictitious to protect the real parties.)

Living Will

I, [Declarant], ("Declarant" herein), being of sound mind, and after careful consideration and thought, freely and intentionally make this revocable declaration to state that, if I should become unable to make and communicate my own decisions on life-sustaining or life-support procedures, then my dying shall not be delayed, prolonged, or extended artificially by medical science or life-sustaining medical procedures, all according to the choices and decisions I have made and which are stated here in my Living Will.

It is my intent, hope, and request that my instructions be honored and carried out by my physicians, family, and friends, as my legal right.

If I am unable to make and communicate my own decisions regarding the use of medical life-sustaining or life-support systems and/or procedures, and if I have a sickness, illness, disease, injury, or condition that has been diagnosed by two (2) licensed medical doctors or physicians who have personally examined me (or more than two (2) if required by applicable law), one of whom shall be my attending physician, as being either (1) terminal or incurable certified to be terminal, or (2) a condition from which there is no reasonable hope of my recovery to a meaningful quality of life, which may reasonably be referred to as hopeless, although not necessarily "terminal" in the medical sense, or (3) has rendered me in a persistent vegetative state, or (4) a condition of extreme mental deterioration, or (5) permanently unconscious, then in the absence of my revoking this Living Will, all medical life-sustaining or life-support systems and procedures shall be withdrawn, unless I state otherwise in the following provisions.

Unless otherwise provided in this Living Will, nothing herein shall prohibit the administering of pain-relieving drugs to me, or any other types of care purely for my comfort, even though such drugs or treatment may shorten my life, or be habit forming, or have other adverse side effects.

I am also stating the following additional instructions so that my Living Will is as clear as possible: (1) resuscitation (CPR)—I do not want to be resuscitated; (2) intravenous and tube feeding—I do not want to be kept alive via intravenous means, and I do not want a feeding tube installed if I am unable to consume food/liquids naturally; (3) Life-Sustaining Surgery—I do not want to be kept alive by any new or old life-sustaining surgery; (4) new medical developments—I do not want to become a participant in any new medical developments that would prolong my life; (5) home or hospital—I want to die wherever my family chooses, including a hospital or nursing home.

In the event that any terms or provisions of my Living Will are not enforceable or are not valid under the laws of the state of my residence, or the laws of the state where I may be located at the time, then all other provisions which are enforceable or valid shall remain in full force and effect, and all terms and provisions herein are severable.

IN WITNESS WHEREOF, I have read and understand this Living Will, and I am freely and voluntarily signing it on this the day of (month), (year) in the presence of witnesses.

Signed: [Declarant]

Street Address:

County:

City and State:

Witness:

We, the undersigned witnesses, certify by our signatures below, that we are adult (at least 18 years old), mentally competent persons; that we are not related to the Declarant by blood, marriage, or adoption; that we do not stand to inherit anything from the Declarant by any means, including will, trust, operation of law or the laws of intestate succession, or by beneficiary designation, nor do we stand to benefit in any way from the death of the Declarant; that we are not directly responsible for the health or medical care, or general welfare of the Declarant; that neither of us signed the Declarant's signature on this document; and that the Declarant is known to us.

We hereby further certify that the Declarant is over the age of 18; that the Declarant signed this document freely and voluntarily, not under any duress or coercion; and that we were both present together, and in the presence of the Declarant to witness the signing of this Living Will on this the day of (month), (year).

Witness signature:
Residing at:
Witness signature:
Residing at:
Notary Acknowledgment:
This instrument was acknowledged before me on this the day of (month), (year) by [Declarant], the Declarant herein, on oath stating that the Declarant is over the age of 18, has fully read and understands the above and foregoing Living Will, and that the Declarant's signing and execution of same is voluntary, without coercion, and is intentional.
Notary Public
My commission or appointment expires:

Durable Power of Attorney

KNOW ALL MEN BY THESE PRESENTS That I, _____ , as principal ("Principal"), a resident of the State and County aforesaid, have made, constituted, appointed and by these present do make, constitute, and appoint and, either or both of them, as my true and lawful agent or attorney-in-fact ("Agent") to do and perform each and every act, deed, matter, and thing whatsoever in and about my estate, property, and affairs as fully and effectually to all intents and purposes as I might or could do in my own proper person, if personally present, including, without limiting the generality of the foregoing, the following specifically enumerated powers which are granted in aid and exemplification of the full, complete and general power herein granted and not in limitation or definition thereof:

1. To forgive, request, demand, sue for, recover, elect, receive, hold all sums of money, debts due, commercial paper, checks, drafts, accounts, deposits, legacies, bequests, devises, notes, interest, stock of deposit, annuities, pension, profit sharing, retirement, Social Security, insurance, and all other contractual benefits and proceeds, all documents of title, all property and all property rights, and demands whatsoever, liquidated or unliquidated, now or hereafter owned by me, or due, owing, payable, or belonging to me or in which I have or may hereafter acquire an interest to have, use and take all lawful means and equitable and legal remedies and proceedings in my name for collection and recovery thereof, and to adjust, sell, compromise, and agree for the same, and to execute and deliver for me, on my behalf, and in my name all endorsements, releases, receipts, or other sufficient discharges for the same.

2. To buy, receive, lease as lessor, accept, or otherwise acquire; to sell, convey, mortgage, grant options upon, hypothecate, pledge, transfer, exchange, quitclaim, or otherwise encumber or dispose of; or to contract or agree for the acquisition, disposal, or encumbrance of any property whatsoever or any custody, possession, interest, or right therein for cash or credit and upon such terms, considerations, and conditions as Agent shall think proper, and no person dealing with Agent shall be bound to see to the application of any monies paid.

3. To take, hold, possess, invest, or otherwise manage any or all of the property or any interest therein; to eject, remove, or relieve tenants or other persons from, and recover possession of, such property by all lawful means; and to maintain, protect, preserve, insure, remove, store, transport, repair, build on, raze, rebuild, alter, modify, or improve the same or any part thereof, and/or to lease any property for me or my benefit, as lessee without option to renew, to collect, and receive any receipt for rents, issues, and profits of my property.

4. To invest and reinvest all or any part of my property in any property and undivided interest in property, wherever located, including bonds, debentures, notes secured or unsecured, stock of corporations regardless of class, interests in limited partnerships, real estate or any interest in real estate whether or not productive at the time of the investment, interest in trusts, investment trusts, whether of the open and/or closed funds types, and participation in common, collective, or pooled trust funds or annuity contracts without being limited by any statute or rule of law concerning investment by fiduciaries.

5. To make, receive, and endorse checks and drafts; deposit and withdraw funds; acquire and redeem certificates of deposit in banks, savings and loan associations, or other institutions; and execute or release such deeds of trust or other security agreements as may be necessary or proper in the exercise of the rights and powers herein granted.

6. To pay any and all indebtedness of mine in such manner and at such times as Agent may deem appropriate.

7. To borrow money for any purpose, with or without security or on mortgage or pledge of any property.

8. To conduct or participate in any lawful business of whatsoever nature for me and in my name; execute partnership agreements and amendments thereto; incorporate, reorganize, merge, consolidate, recapitalize, sell, liquidate, or dissolve any business; elect or employ officers, directors, and agents; carry out the provisions of any agreement for the sale of any business interest or stock therein; and exercise voting rights with respect to stock either in person or by proxy, and to exercise stock options.

9. To prepare, sign, and file joint or separate income tax returns or declarations of estimated tax for any year or years; to prepare, sign, and file gift tax returns with respect to gifts made by me for any year or years; to consent to any gift and to utilize any gift-splitting provision or other tax election; and to prepare, sign, and file any claims for refund of tax.

10. To have access at any time or times to any safe-deposit box rented by me, wheresoever located, and to remove all or any part of the contents thereof, and to surrender or relinquish said safety deposit box in any institution in which such safe-deposit box may be located shall not incur any liability to me or my estate as a result of permitting Agent to exercise this power.

11. To execute any and all contracts of every kind or nature.

As used herein, the term "property" includes any property, real or personal, tangible or intangible, wheresoever situated.

The execution and delivery by Agent of any conveyance paper instrument or document in my name and behalf shall be conclusive evidence of Agent's approval of the consideration therefore, and of the form and contents thereof, and that Agent deems the execution thereof in my behalf necessary or desirable.

Any person, firm, or corporation dealing with Agent under the authority of this instrument is authorized to deliver to Agent all considerations of every kind or character with respect to any transactions so entered into by Agent and shall be under no duty or obligation to see to or examine into the disposition thereof.

Third parties may rely upon the representation of Agent as to all matters relating to any power granted to Agent, and no person who may act in reliance upon the representation of Agent or the authority granted to Agent shall incur liability to me or my estate as a result of permitting Agent to exercise any power. Agent shall be entitled to reimbursement for all reasonable costs and expenses incurred and paid by Agent on my behalf pursuant to any provisions of this durable power of attorney, but Agent shall not be entitled to compensation for services rendered hereunder.

Notwithstanding any provision herein to the contrary, Agent shall not satisfy any legal obligation of Agent out of any property subject to this power of attorney, nor may Agent exercise this power in favor of Agent, Agent's estate, Agent's creditors, or the creditors of Agent's estate.

Notwithstanding any provision hereto to the contrary, Agent shall have no power or authority whatever with respect to (a) any policy of insurance owned by me on the life of Agent, and (b) any trust created by Agent as to which I am Trustee.

When used herein, the singular shall include the plural and the masculine shall include the feminine.

This power of attorney shall become effective immediately upon the execution hereof.

This is a durable power of attorney made in accordance with and pursuant to Section (this section to be completed in reference to laws of the state in which the document is executed).

This power of attorney shall not be affected by disability, incompetency, or incapacity of the principal.

Principal may revoke this durable power of attorney at any time by written instrument delivered to Agent. The guardian of Principal may revoke this instrument by written instrument delivered to Agent.

IN WITNESS WHEREOF, I have executed this durable power of attorney in three (3) counterparts, and I have directed that photostatic copies of this power be made, which shall have the same force and effect as an original.

DATED THIS THE day of (month), (year).

WITNESS:

Name of Principal here

THE STATE OF and (County)

I, a Notary Public in and for said County, in said State, hereby certify that _____, whose name is signed to the foregoing durable power of attorney, and who is known to me, acknowledged before me on this day that being informed of the contents of the durable power, executed the same voluntarily on the day the same bears date.

GIVEN under my hand and seal this day of (month), (year).

NOTARY PUBLIC

My Commission Expires:

Glossary

A

Abortion rate—the number of abortions per thousand women aged 15 to 44

Abortion ratio—the number of abortions per thousand live births

Absolutism—a sexual value system that is based on unconditional allegiance to tradition or religion (e.g., waiting until marriage to have sexual intercourse)

Acquaintance rape—nonconsensual sex between adults (of same or other sex) who know each other

Advance directive (living will)—details for medical care personnel identifying the conditions under which life support measures should be used for one's partner

Agape love style— characterized by a focus on the well-being of the love object, with little regard for reciprocation (e.g., love of parent for a child)

Age—term which may be defined chronologically (number of years), physiologically (physical decline), psychologically (self-concept), sociologically (roles for the elderly—retired), and culturally (meaning of age in one's society)

Age discrimination—a situation where older people are often not hired and younger workers are hired to take their place

Ageism—the systematic persecution and degradation of people because they are old

Al-Anon—organization that provides support for family members and friends of alcohol abusers

Alcohol-exposed pregnancy—sexual intercourse that occurs in the context of alcohol, which involves little consideration for using contraception

Alienation of affection—law which gives a spouse the right to sue a third party for taking the affections of a spouse away

Androgyny—a blend of traits that are stereotypically associated with masculinity and femininity

Anodyspareunia—frequent, severe pain during receptive anal sex

Antinatalism—opposition to children

Anxious jealousy—obsessive ruminations about the partner's alleged infidelity that can make one's life a miserable emotional torment

Arbitration—third party listens to both spouses and makes a decision about custody, division of property, child support, and alimony

Arranged marriage—mate selection pattern whereby parents select the spouse of their offspring. A matchmaker may be used but the selection is someone of whom the parents approve.

Artifact—concrete symbol that reflects the existence of a cultural belief or activity (e.g., wedding ring)

Asceticism—sexual belief system, which emphasizes that giving in to carnal lust, is unnecessary and one should attempt to rise above the pursuit of sensual pleasure into a life of self-discipline and self-denial

Asexual—the absence of sexual behavior with a partner as well as oneself (masturbation)

Attachment theory of mate selection—developed early in reference to one's parents, the drive toward an intimate, social/emotional connection

Authentic—being who one is and saying what one feels

B

Baby blues—transitory symptoms of depression twenty-four to forty-eight hours after the baby is born

Battered-woman syndrome—the general pattern of battering that a woman is subjected to, defined in terms of the frequency, severity, and injury she experiences

Behavioral couple therapy (BCT)—therapeutic focus on behaviors the respective spouses want increased or decreased, initiated, or terminated

Benevolent sexism—the belief that women are innocent creatures who should be protected and supported

Binuclear family—a family in which the members live in two households, as when parents live in separate households following a divorce

Biosocial theory—also referred to as sociobiology, social behaviors (for example, gender roles) are biologically based and have an evolutionary survival function

Biphobia— negative attitudes toward bisexuality and those identified as bisexual

Birth control sabotage—partner interference with contraception

Bisexuality—cognitive, emotional, and sexual attraction to members of both sexes

Blended family—a family created when two individuals marry and at least one of them brings a child or children from a previous relationship or marriage into the unit. Also referred to as a stepfamily.

Blurred retirement—an individual working part-time before completely retiring or taking a "bridge job" that provides a transition between a lifelong career and full retirement

Brainstorming—suggesting as many alternatives as possible without evaluating them

Branching—in communication, going out on different limbs of an issue rather than staying focused on the issue

Bride wealth—also known as bride price or bride payment, the amount of money or goods paid by the groom or his family to the bride's family for giving her up

C

Child abuse—any interaction or lack of interaction between children and their parents or caregivers that results in nonaccidental harm to the children's physical or psychological well-being

Child marriage—marriages in which females as young as 8 to 12 are required by their parents to marry an older man

Chronic sorrow—grief-related feelings that occur periodically throughout the lives of those left behind

Civil union—a pair-bonded relationship given legal significance in terms of rights and privileges

Closed-ended question—question that allows for a one-word answer and does not elicit much information

Cohabitation—two adults, unrelated by blood or by law, involved in an emotional and sexual relationship who sleep in the same residence at least four nights a week for three months

Cohabitation effect—those who have multiple cohabitation experiences prior to marriage are more likely to end up in marriages characterized by violence, lower levels of happiness, lower levels of positive communication, and depression

Coitus—the sexual union of a man and woman by insertion of the penis into the vagina

Collaborative practice—process involving a team of professionals (lawyer, psychologist, mediator, social worker, financial counselor) helping a couple separate and divorce in a humane and cost-effective way

Collectivism—pattern in which one regards group values and goals as more important than one's own values and goals

Coming out—being open to others about one's sexual orientation and identity

Commitment—the intent to maintain a relationship

Common-law marriage—a heterosexual cohabiting couple presenting themselves as married

Communication—the process of exchanging information and feelings between two or more people

Compersion—sometimes thought of as the opposite of jealousy, the approval/support of a partner's emotional and sexual involvement with another person

Competitive birthing—having the same number (or more) of children in reference to one's peers

Complementary-needs theory—states that we tend to select mates whose needs are opposite and complementary to our own needs

Comprehensive sex education program—learning experience which recommends abstinence but also discusses contraception and other means of pregnancy protection

Conception—refers to the fusion of the egg and sperm; also known as fertilization

Concurrent sexual partnership—relationship in which the partners have sex with several individuals concurrently

Conflict—the context in which the perceptions or behavior of one person are in contrast to or interfere with the other

Conflict framework—the view that individuals in relationships compete for valuable resources

Congruent messages—message in which the verbal and nonverbal behaviors are the same

Conjugal love—the love between married people characterized by companionship, calmness, comfort, and security

Connection rituals—rituals which occur daily in which the couple share time and attention.

Consumerism—economic societal value that one must buy everything and have everything now

Conversion therapy—also called reparative therapy, focused on changing the sexual orientation of homosexuals

Coolidge effect—term used to describe waning of sexual excitement and the effect of novelty and variety on increasing sexual arousal

Corporal punishment—the use of physical force with the intention of causing a child to experience pain, but not injury, for the purpose of correction or control of the child's behavior

Correct rejections—both partners are aware that one of them chose not to engage in a negative comment

Cougar—a woman, usually in her thirties and forties, who is financially stable and mentally independent and looking for a younger man with whom to have fun

Covenant marriage—type of marriage whereby the spouses agree to have marriage counseling before getting married, to have a cooling off period of two years after children are born, and to divorce only for serious faults (such as abuse, adultery, or imprisonment for a felony)

Crisis—a crucial situation that requires change in one's normal pattern of behavior

Cross-dresser—individuals who dress or present themselves in the gender of the other sex

Cryopreservation—the freezing of fertilized eggs for implantation at a later stage

Cybercontrol—use of communication technology, such as cell phones, e-mail, and social networking sites, to monitor or control a partner with whom one has been in an intimate relationship

Cybersex—any consensual, computer mediated, participatory sexual experience involving two or more individuals

Cybervictimization—harassing behavior, which includes being sent threatening e-mail, unsolicited obscene e-mail, computer viruses, or junk mail (spamming); can also include flaming (online verbal abuse) and leaving improper messages on message boards

D

Date rape—nonconsensual sex between people who are dating or are on a date

December marriages—both spouses are elderly

Defense mechanism—individual unconscious cognitive techniques that function to protect a person from anxiety and to minimize emotional hurt (e.g., rationalization)

Defense of Marriage Act—legislation which says that marriage is a "legal union between one man and one woman" and denies federal recognition of same-sex marriage

Demandingness—the manner in which parents place demands on children in regard to expectations and discipline

Developmental task—a skill that, if mastered, allows a family to grow as a cohesive unit

Dementia—a disorder of the mental processes marked by memory loss, personality changes, and impaired reasoning

Disenchantment—the transition from a state of newness and high expectation to a state of mundaneness tempered by reality

Displacement—shifting one's feelings, thoughts, or behaviors from the person who evokes them onto someone else (e.g., anger at boss becomes directed to spouse)

Divorce—the legal ending of a valid marriage contract

Divorce mediation—meeting with a neutral professional who negotiates child custody, division of property, child support, and alimony directly with the divorcing spouses

Divorcism—the belief that divorce is a disaster

Domestic partnership—two adults who have chosen to share each other's lives in an intimate and committed relationship. These relationships are given official recognition by a city or corporation so as to receive partner benefits (for example, health insurance).

Double Dutch—strategy of using both the pill and condom by sexually active youth in the Netherlands

Dual-career marriage—a marriage in which both spouses pursue careers

Durable power of attorney—gives adult children complete authority to act on behalf of the elderly (e.g., parent)

E

Emergency contraception—refers to various types of morning-after pills

Emotional abuse—abuse designed to make the partner feel bad—to denigrate the partner, reduce the partner's status, and to make the partner vulnerable to being controlled by the abuser

Endogamy—cultural expectation to select a marriage partner within one's own social group

Engagement—time in which the romantic partners are sexually monogamous, committed to marry, and focused on wedding preparations

Entrapped—stuck in an abusive relationship and unable to extricate oneself from the abusive partner

Eros love style—also known as romantic love; the love of passion and sexual desire

Escapism—the simultaneous denial of and avoidance of dealing with a problem

Euthanasia—from the Greek meaning "good death" or dying without suffering; either passively or actively ending the life of a patient

Exchange theory—theory that emphasizes that relations are formed and maintained based on which partner offers the greatest rewards at the lowest costs

Exogamy—the cultural pressure to marry outside the family group

Extended family—the nuclear family or parts of it plus other relatives such as grandparents, aunts, uncles, and cousins

Extradyadic involvement—refers to sexual involvement of a pair-bonded individual with someone other than the partner; also called extrarelational involvement

Extrafamilial—child sexual abuse in which the perpetrator is someone outside the family who is not related to the child

Extramarital affair—refers to a spouse's sexual involvement with someone outside the marriage

F

Familism—value that decisions are made in reference to what is best for the family

Family—a group of two or more people related by blood, marriage, or adoption

Family caregiving—adult children providing care of their elderly parents

Family life course development—the stages and process of how families change over time

Family life cycle—stages that identify the various developmental tasks family members face across time

Family of orientation—also known as the family of origin; the family into which a person is born

Family of procreation—the family a person begins typically by getting married and having children

Family relations doctrine—belief that even nonbiological parents may be awarded custody or visitation rights if they have been economically and emotionally involved in the life of the child

Family resiliency—the successful coping with adversity by family members that enables them to flourish with warmth, support, and cohesion

Family systems framework—views each member of the family as part of a system and the family as a unit that develops norms of interaction

Female genital alteration—cutting off the clitoris or excising (partially or totally) the labia minora

Feminist framework—views marriage and family as contexts of inequality and oppression for women

Feminization of poverty—the idea that women (particularly those who live alone or with their children) disproportionately experience poverty

Feral children—"wild, not domesticated" children thought to have been reared by animals

Filial piety—love and respect toward parents including bringing no dishonor to parents and taking care of elderly parents

Filial responsibility—emphasizes duty, protection, care, and financial support to one's parents

Filicide—murder of one's child by a parent

Flirting—to show interest without serious intent

Foster parent—neither a biological nor an adoptive parent but a person who takes care of and fosters a child taken into custody by the courts

Frail—term used to define elderly people if they have difficulty with at least one personal care activity (feeding, bathing, toileting)

Friends with benefits—a relationship of nonromantic friends who also have a sexual relationship

G

Gatekeeper role—term used to refer to the influence/control of the mother on the father's involvement/relationship with his children

Gay—term which refers to women or men who prefer same-sex partners

Gender—social construct which refers to the social and psychological characteristics associated with being female or male

Gender identity—the psychological state of viewing oneself as a girl or a boy, and later as a woman or a man

Gender role ideology—the typical role relationships between women and men in a society

Gender role transcendence—abandoning gender frameworks and looking at phenomena independent of traditional gender categories

Gender roles—social norms which specify the socially appropriate behavior for females and males in a society

Generation Y—children of the baby boomers (typically born between 1979 and 1984)

Gerontology—the study of aging

Gerontophobia—fear or dread of the elderly, which may create a self-fulfilling prophecy

GLBT—general term which refers to gay, lesbian, bisexual, and transgender individuals

Granny dumping—refers to adult children or grandchildren burdened with the care of their elderly parent or grandparent leaving the elder at the entrance of a hospital with no identification

H

Hanging out—refers to going out in groups where the agenda is to meet others and have fun

Hedonism—the belief that the ultimate value and motivation for human actions lie in the pursuit of pleasure and the avoidance of pain

HER/his career—dual career marriage in which wife's career takes precedence

Heterosexism—the institutional and societal reinforcement of heterosexuality as the privileged and powerful norm; assumes that homosexuality is "bad"

Heterosexuality—emotional and sexual attraction to individuals of the other sex

HIS/her career—dual career marriage in which husband's career takes precedence

HIS/HER career marriage—dual career marriage in which both careers are treated as equal

Homogamy—the tendency for an individual to seek a mate who has similar characteristics

Homonegativity—attaching negative connotations to homosexuality

Homophobia—negative attitudes and emotions toward homosexuality and those who engage in homosexual behavior

Homosexuality—refers to the predominance of cognitive, emotional, and sexual attraction to individuals of the same sex

Honeymoon—the time following the wedding whereby the couple becomes isolated to recover from the wedding and to solidify their new status change from lovers to spouses

Honor crime/Honor killing—refers to unmarried women who are killed because they bring shame on their parents and siblings; occurs in Middle Eastern countries such as Jordan

Hooking up—a sexual encounter that occurs between individuals who have no relationship commitment

Hypothesis—a suggested explanation for a phenomenon

Hysterectomy—form of female sterilization whereby the woman's uterus is removed

I

"I" statements—statements which focus on the feelings and thoughts of the communicator without making a judgment on others

Individualism—making decisions that serve the individual's rather than the family's interests

Induced abortion—the deliberate termination of a pregnancy through chemical or surgical means

Infatuation—intense emotional feelings based on little actual exposure to the love object

Infertility—the inability to achieve a pregnancy after at least one year of regular sexual intercourse without birth control, or the inability to carry a pregnancy to a live birth

Institution—established and enduring patterns of social relationships (e.g., the family)

Integrative behavioral couple therapy (IBCT)—therapy which focuses on the cognitions or assumptions of the spouses, which impact the way spouses feel and interpret each other's behavior

Internalized homophobia—a sense of personal failure and self-hatred among lesbians and gay men resulting from social rejection and stigmatization of being gay

Intersexed individuals—those with mixed or ambiguous genitals

Intimate partner homicide—murder of a spouse

Intimate partner violence (IPV)—an all-inclusive term that refers to crimes committed against current or former spouses, boyfriends, or girlfriends

Intimate terrorism (IT)—behavior designed to control the partner

Intrafamilial—child sexual abuse referring to exploitive sexual contact or

attempted sexual contact between relatives before the victim is 18

J

Jealousy—an emotional response to a perceived or real threat to an important or valued relationship

L

Laparoscopy—a form of tubal ligation that involves a small incision through the woman's abdominal wall just below the navel

Legal custody—judicial decision regarding whether one or both parents will have decisional authority on major issues affecting the children

Leisure—the use of time to engage in freely chosen activities perceived as enjoyable and satisfying

Lesbian—a woman who prefers same-sex partners

Litigation—a judge hears arguments from lawyers representing the respective spouses and decides issues of custody, child support, division of property, etc.

Living apart together (LAT)—committed couple who does not live in the same home (and others such as children or elderly parents may live in the respective homes of the partners)

Lose-lose solution—a solution to a conflict in which neither partner benefits

Ludic love style—views love as a game and the player has no intention of getting involved

Lust—sexual desire

M

Mania love style—the out-of-control love whereby the person "must have" the love object. Obsessive jealousy and controlling behavior are symptoms of manic love.

Marital rape—forcible rape by one's spouse—a crime in all states

Marital success—refers to the quality of the marriage relationship measured in terms of marital stability and marital happiness

Marriage—traditionally, a legal relationship that binds a couple together for the reproduction, physical care, and socialization of children

Marriage benefit—when compared to being single, married persons are healthier, happier, live longer, less drug use, etc.

Marriage rituals—deliberate repeated social interactions that reflect emotional meaning to the couple

Marriage squeeze—the imbalance of the ratio of marriageable-aged men to marriageable-aged women

Masturbation—stimulating one's own body with the goal of experiencing pleasurable sexual sensations

Mating gradient—the tendency for husbands to be more advanced than their wives with regard to age, education, and occupational success

May-December marriage—age dissimilar marriage (ADM) in which the woman is typically in the spring of her life (May) and her husband is in the later years (December)

Medicaid—a state welfare program for low-income individuals

Medicare—a federal health insurance program for people 65 and older

Megan's Law—law requiring that communities be notified of a neighbor's previous sex convictions

Millenials—workers born between 1980 and 1995

Mommy track—mothers who stop paid employment to spend time with young children

Momprenuer—a woman who has a successful at-home business

N

Negative commitment—spouses who continue to be emotionally attached to and have difficulty breaking away from ex-spouses

Negotiation—spouses discuss and resolve the issues of custody, child support, division of property themselves

New relationship energy (NRE)—the euphoria of a new emotional/sexual relationship, which dissipates over time

No fault divorce—divorce whereby neither party is identified as the guilty party

Nonverbal communication—the "message about the message," using gestures, eye contact, body posture, tone, volume, and rapidity of speech

Nuclear family—consists of you, your parents, and your siblings or you, your spouse, and your children

O

Obsessive relational intrusion (ORI)—the relentless pursuit of intimacy with someone who does not want it

Occupational sex segregation—the concentration of women and men in different occupations

Oophorectomy—form of female sterilization whereby the woman's ovaries are removed

Open-ended question—question which elicits a lot of information

Open-minded—an openness to understanding alternative points of view, values, and behaviors

Open relationship—relationship in which the partners agree that each may have emotional and sexual relationships with those outside the dyad

Oppositional defiant disorder—disorder in which children fail to comply with requests of authority figures

Opting out—when women leave their career and return home to take care of their children

Overindulgence—defined as more than just giving children too much, includes overnurturing and providing too little structure

Ovum transfer—a form of in vitro fertilization wherein a fertilized egg is implanted in the uterine wall

Oxytocin—a hormone released from the pituitary gland during the expulsive stage of labor that has been associated with the onset of maternal behavior in lower animals

P

Palimony—refers to the amount of money one "pal" who lives with another "pal" may have to pay if the partners end their relationship

Palliative care—health care for the individual who has a life threatening illness which focuses on relief of pain/suffering and support for the individual

Pantagamy—a group marriage in which each member of the group is "married" to the others

Parental alienation—estrangement of a child from a parent due to one parent turning the child against the other

Parental alienation syndrome—an alleged disturbance in which children are obsessively preoccupied with deprecation and/or criticism of a parent, denigration that is unjustified and/or exaggerated

Parental consent—a woman needs permission from a parent to get an abortion if under a certain age, usually 18

Parental investment—any investment by a parent that increases the offspring's chance of surviving and thus increases reproductive success

Parental notification—a woman has to tell a parent she is getting an abortion if she is under a certain age, usually 18, but she doesn't need parental permission

Parenting—defined in terms of roles including caregiver, emotional

resource, teacher, and economic resource

Parenting self-efficiency—feeling competent as a parent

Parricide—murder of a parent by an offspring

Phased retirement—an employee agrees to a reduced workload in exchange for reduced income

Physical custody—the distribution of parenting time between divorced spouses

Polyamory—a lifestyle in which two lovers embrace the idea of having multiple lovers. By agreement, each partner may have numerous emotional and sexual relationships.

Polyandry—a type of marriage in which one wife has two or more husbands

Polygamy—generic term for a marriage involving more than two spouses

Polygyny—type of marriage involving one husband and two or more wives

Pool of eligibles—the population from which a person selects a mate

Positive androgyny—a view of androgyny that is devoid of the negative traits associated with masculinity (e.g., aggression) and femininity (e.g., being passive)

Positive assortative personality mating—individuals who sort each other out on the basis of similar personality characteristics

Possessive jealousy—involves attacking the partner or the alleged person to whom the partner is showing attention

Postnuptial agreement—an agreement about how money is to be divided should a couple later divorce that is made after the couple marry

Postpartum depression—a severe reaction following the birth of a baby which occurs in reference to a complicated delivery as well as numerous physiological and psychological changes; usually in the first month after birth but can be experienced after a couple of years have passed

Postpartum psychosis—a reaction in which a woman wants to harm her baby

Power—the ability to impose one's will on the partner and to avoid being influenced by the partner

Pragma love style—love style that is logical and rational. The love partner is evaluated in terms of pluses and minuses and is regarded as a good or bad "deal."

Pregnancy—when the fertilized egg is implanted (typically in the uterine wall)

Pregnancy coercion—coercion by a male partner for the woman to become pregnant

Prenuptial agreement—a contract between intended spouses specifying which assets will belong to whom and who will be responsible for paying what in the event of a divorce or when the marriage ends by the death of one spouse

Primary groups—small numbers of individuals whereby the interaction is intimate and informal

Principle of least interest—the person who has the least interest in a relationship controls the relationship

Procreative liberty—the freedom to decide to have children or not

Pronatalism—cultural attitude which encourages having children

Projection—attributing one's own thoughts, feelings, and desires to someone else while avoiding recognition that these are one's own thoughts, feelings, and desires

Q

Quality of life—refers to one's physical functioning, independence, economic resources, social relationships, and spirituality

R

Rape myths—beliefs that deny victim injury or cast blame on the woman for her own rape

Rationalization—the cognitive justification for one's own behavior that unconsciously conceals one's true motives

Reactive attachment disorder—common among children who were taught as infants that no one cared about them; these children have no capacity to bond emotionally with others since they have no learning history of the experience and do not trust adults, caretakers, or parents

Reactive jealousy—jealous feelings that are a reaction to something the partner is doing

Red zone—the first month of the first year of college when women are alleged to be most likely to be victims of sexual abuse

Reflective listening—paraphrasing or restating what the person has said to you while being sensitive to what the partner is feeling

Relativism—value system emphasizing that sexual decisions should be made in the context of a particular relationship

Resiliency—a family's strength and ability to respond to a crisis in a positive way

Responsiveness—refers to the extent to which parents respond to and meet the needs of their children

Rite of passage—an event that marks the transition from one social status to another (e.g., wedding ceremony)

Role compartmentalization—strategy used to separate the roles of work and home so that an individual does not think about or dwell on the problems of one when he or she is at the physical place of the other

Role overload—not having the time or energy to meet the demands or responsibilities in the roles of wife, parent, and worker

Role strain—the anxiety that results from being able to fulfill only a limited number of role obligations

Role theory of mate selection—theory which focuses on the social learning of roles. A son or daughter models after the parent of the same sex by selecting a partner similar to the one the parent selected.

Romantic love—passionate love whereby the lover believes in love at first sight, only one true love, and love conquers all

Rophypnol—causes profound, prolonged sedation and short-term memory loss; also known as the date rape drug, roofies, Mexican Valium, or the "forget (me) pill"

S

Salpingectomy—type of female sterilization whereby the fallopian tubes are cut and the ends are tied

Sandwich generation—generation of adults who are "sandwiched" between caring for their elderly parents and their own children

Satiation—a stimulus loses its value with repeated exposure; also called habituation

Second shift—the housework and child care that employed women engage when they return home in the evening-from their jobs

Second-parent adoption—legal procedure that allows individuals to adopt their partner's biological or adoptive child without terminating the first parent's legal status as parent

Secondary groups—groups in which the interaction is impersonal and formal

Secondary virginity—a sexually initiated person's deliberate decision to refrain from intimate encounters for a set period of time and to refer to that decision as a kind of virginity (rather than "mere" abstinence)

Sex—the biological distinction between females and males

Sex roles—roles defined by biological constraints and can be enacted by members of one biological sex only—for example, wet nurse, sperm donor, child-bearer

Sexism—an attitude, action, or institutional structure that subordinates or discriminates against individuals or groups because of their biological sex

Sexting—sending erotic text or photo images via a cell phone

Sextortion—online sexual extortion

Sexual compliance—an individual willingly agrees to participate in sexual behavior without having the desire to do so

Sexual double standard—the view that encourages and accepts sexual expression of men more than women

Sexual identity—term used synonymously with sexual orientation

Sexual orientation—classification of individuals as heterosexual, bisexual, or homosexual, based on their emotional, cognitive, sexual attractions, and self-identity

Sexual readiness—factors such as autonomy of decision (not influenced by alcohol or peers), consensuality (both partners equally willing), and absence of regret (the right time for me) are more important than age as meaningful criteria for determining when a person in ready for first intercourse

Sexual values—moral guidelines for making sexual choices in nonmarital, marital, heterosexual, and homosexual relationships

Shared parenting dysfunction—behaviors on the part of both partners focused on hurting the other parent and that are counterproductive for the well-being of the children

Siblicide—murder of a sibling

Sibling relationship aggression—behavior of one sibling toward another sibling that is intended to induce social harm or psychic pain in the sibling

Single-parent family—family in which there is only one parent and the other parent is completely out of the child's life through death, sperm donation, or abandonment and no contact is made with the other parent

Single parent household—one parent has primary custody of the child/children with the other parent living outside of the house but still being a part of the child's family; also called binuclear family

Situational couple violence (SCV)—conflict escalates over an issue and one or both partners lose control

Snooping—investigating (without the partner's knowledge or permission) a romantic partner's private communication (e.g., text messages, cell phone calls) motivated by concern that the partner may be "hiding" something

Social allergy—being annoyed and disgusted by a repeated behavior on the part of the partner

Social exchange framework—views interaction and choices in terms of cost and profit

Socialization—the process through which we learn attitudes, values, beliefs, and behaviors appropriate to the social positions we occupy

Sociobiology—theory which emphasizes the biological basis for all social behavior, including mate selection

Sociological imagination—the influence of social structure and culture on interpersonal decisions

Spectatoring—involves mentally observing your sexual performance and that of your partner

Spiritual abuse—defined as any attempt to impair the woman's spiritual life, spiritual self, or spiritual well-being

Spontaneous abortion (miscarriage)—the unintended termination of a pregnancy

Stalking—unwanted following or harassment that induces fear in a target person

Stepfamily—family in which spouses in a new marriage bring children from previous relationships into the new home

Stepism—the assumption that stepfamilies are inferior to biological families

Sterilization—a permanent surgical procedure that prevents reproduction

STI (sexually transmitted infection)—refers to the general category of sexually transmitted infections such as chlamydia, genital herpes, gonorrhea, and syphilis

Storge love style—also known as companionate love; a calm, soothing, non-sexual love devoid of intense passion

Stress—reaction of the body to substantial or unusual demands (physical, environmental, or interpersonal)

Structure-function framework—emphasizes how marriage and family contribute to society

Superwoman/supermom—a cultural label that allows a woman to regard herself as very efficient, bright and confident; usually a cultural cover-up for an overworked and frustrated woman

Symbolic interaction framework—views marriages and families as symbolic worlds in which the various members give meaning to each other's behavior

T

Telerelationship therapy—therapy sessions conducted online, often through Skype, where both therapist and couple can see and hear each other

Texting—short typewritten messages sent via a cell phone that are used to "commence, advance, and maintain" interpersonal relationships

Thanatology—the examination of the social dimensions of death, dying, and bereavement

THEIR career marriage—dual career marriage in which spouses share a career or work together

Theoretical frameworks—a set of interrelated principles designed to explain a particular phenomenon

Therapeutic abortion—abortions performed to protect the life or health of the woman

Third shift—the expenditure of emotional energy by a spouse or parent in dealing with various emotional issues in family living (e.g., parent helps an adolescent through a first love breakup)

Time-out—a noncorporal form of punishment that involves removing the child from a context of reinforcement to a place of isolation

Transgender—a generic term for a person of one biological sex who displays characteristics of the other sex

Transition to parenthood—period from the beginning of pregnancy through the first few months after the birth of a baby during which the mother and father undergo changes

Transracial adoption—adopting children of a race different from that of the parents

Transsexual—an individual with the biological and anatomical sex of one gender (for example, male) but the self-concept of the opposite sex (that is, female)

U

Unrequited love—love that is not returned

Utilitarianism—individuals rationally weigh the rewards and costs associated with behavioral choices

Uxoricide—the murder of a woman by a romantic partner

V

Vasectomy—form of male sterilization whereby the vas deferens is cut so that sperm cannot continue to travel outside the body via the penis

Violence—intentional infliction of physical harm by either partner on the other

W

Weed dating—another form of speed dating where people in a rural area meet on a farm and get to know each other while weeding between rows

Win-lose solution—outcome of a conflict in which one partner wins and the other loses

Win-win relationships—relationship in which conflict is resolved so that each partner derives benefits from the resolution

Y

"You" statement—statement that blames or criticizes the listener and often results in increasing negative feelings and behavior in the relationship

References

A

ABC Television. 20/20. 2010. "Ancient Ritual, Human Sacrifice, Stolen Children," February 28.

Abowitz, D., D. Knox, and K. Berner. 2011. Traditional and non-traditional husband preference among college women. Annual Meeting Eastern Sociological Society, March. Philadelphia, Pennsylvania.

Abowitz, D. A., D. Knox, and M. Zusman. 2010. Emotional abuse among undergraduates romantic relationships. *International Journal of Sociology of the Family* 36: 117–139.

Abowitz, D. A., D. Knox, M. Zusman, and A. McNeely. 2009. Beliefs about romantic relationships: Gender differences among undergraduates. *College Student Journal* 43: 276–284.

Abraham, A. J., and P. M. Roman. 2010. Early adoption of injectable naltrexone for alcohol-use disorders: findings in the private-treatment sector. *Journal of Studies on Alcohol and Drugs* 71: 460–467.

Ackbar, S., and C. Y. Senn. 2010. What's the confusion about fusion?—Differentiating positive and negative closeness in lesbian relationships. *Journal of Marital & Family Therapy* 36: 416–430.

Ackerman, D. 1994. *A natural history of love.* New York: Random House.

Ackerman, R. J., and M. E. Banks. 2007. Women over 50: Caregiving issues. In *Women over fifty: Psychological perspectives,* ed. Varda Muhlbauer and Joan C. Chrisler, 147–163. New York: Springer Science and Business Media.

Adams, M. et al. 2008. Rhetoric of alternative dating: Investment versus exploration. Southern Sociological Society, Richmond, VA, April.

Adams, M., L. A. and P. Griffin. 2007. *Teaching for diversity and social justice.* (2nd ed.). New York: Routledge.

Adler-Baeder, F., A. Robertson, and D. G. Schramm. 2010. Conceptual framework for marriage education programs for stepfamily couples with considerations for socioeconomic context. *Marriage & Family Review* 46: 300–322.

Adolfsen, A., J. Iedema, and S. Keuzenkamp. 2010. Multiple dimensions of attitudes about homosexuality: Development of a multifaceted scale measuring attitudes toward homosexuality. *Journal of Homosexuality* 57: 1237–1257.

Agahi, N. 2008. Leisure activities and mortality: Does gender matter? *Journal of Aging and Health* 20: 855–871.

Agate, J. R., R. Zabriskie, S. T. Agate, and R. Poff. 2009. Family leisure satisfaction and satisfaction with family life. *Journal of Leisure Research* 41: 205–223.

Ahnert, L., and M. E. Lamb. 2003. Shared care: Establishing a balance between home and child care settings. *Child Development* 74: 1044–1049.

Ahrons, C. R., and J. L. Tanner. 2003. Adult children and their fathers: Relationship changes 20 years after parental divorce. *Family Relations* 52: 340–351.

Akmatov, M. K. 2011. Child abuse in 28 developing and transitional countries—results from the Multiple Indicator Cluster Surveys. *International Journal of Epidemiology* 40: 219–227.

Albright, J. M. 2007. How do I love thee and thee and thee?: Self-presentation, deception, and multiple relationships online. In *Online Matchmaking,* ed. M. T. Whitty, A. J. Baker, and J. A. Inman, 81–93. New York: Palgrave Macmillan.

Alexander, P. C., A. Tracy, M. Radek, and C. Koverola 2009. Predicting stages of change in battered women. *Journal of Interpersonal Violence* 24: 1652–1672.

Alford-Cooper, F. 2006. Where has all the sex gone? Sexual activity in lifetime marriage. Paper presented at the Southern Sociological Society, New Orleans, March 23–26.

Algoe, S. B., S. L. Gable, and N. C. Maisel. 2010. It's the little things: Everyday gratitude as a booster shot for romantic relationships. *Personal Relationships* 17: 217–233.

Ali, M. M., and D. S. Dwyer. 2010. Social network effects in alcohol consumption among adolescents. *Addictive Behaviors* 35: 337–342.

Allen, D. W. 2010. A better method for assessing the value of "housewife" services. *American Journal of Family Law* 23: 219–223.

Allen, E. S., G. K. Rhoades, S. M. Stanley, H. J. Markman, et al. 2008b. Premarital precursors of marital infidelity. *Family Process* 47: 243–260.

Allen, K. N., and D. Wozniak. 2011. The language of healing: Women's voices in healing and recovering from domestic violence. *Social Work in Mental Health* 9: 37–55.

Allen, K. R., E. K. Husser, D. J. Stone, and C. E. Jordal. 2008a. Agency and error in young adults' stories of sexual decision making. *Family Relations* 57: 517–529.

Allgood, S., and A. Bakker. 2009. Connection rituals and marital satisfaction. Poster, National Council on Family Relations, San Francisco, November.

Allgood, S. M., and J. Gordon. 2010. Premarital advice" Do engaged couples listen? Poster, National Council on Family Relations annual meeting, November 3–5. Minneapolis, MN.

Allport, G. W. 1954. *The nature of prejudice.* Cambridge, MA: Addison-Wesley.

Amato, P. R. 2004. Tension between institutional and individual views of marriage. *Journal of Marriage and Family* 66: 959–965.

Amato, P. R. 2010. Research on divorce: Continuing trends and new developments. *Journal of Marriage and Family* 72: 650–666.

Amato, P. R., and J. B. Kane. 2011. Life-course pathways and the psychosocial adjustment of young adult women. *Journal of Marriage & Family* 73: 279–295.

Amato, P. R., A. Booth, D. R. Johnson, and S. F. Rogers. 2007. *Alone together: How marriage in America is changing.* Cambridge, Massachusetts: Harvard University Press.

Ambwani, S., and J. Strauss. 2007. Love thyself before loving others? A qualitative and quantitative analysis of gender differences in body image and romantic love. *Sex Roles* 56: 13–22.

American Psychological Association. 2004. *Sexual Orientation, Parents, & Children.* APA Policy Statement on Sexual Orientation, Parents, & Children. APA Online. http://www.apa.org.

Anderson, J. R., M. J. Van Ryzin, and W. J. Doherty. 2010. Developmental trajectories of marital happiness in continuously married individuals: A group-based modeling approach. *Journal of Family Psychology* 24: 587–596.

Anderson, K. L. 2010a. Conflict, power, and violence in families *Journal of Marriage and Family* 72: 726–742.

Anderson, S. 2010b. The polygamists: An exclusive look inside the FLDS. *National Geographic* (February): 38–51.

Anderson, P. L., J. A. Tiro, A. W. Price, M. A. Bender, and N. J. Kaslow. 2003. Additive impact of childhood emotional, physical, and sexual abuse on suicide attempts among low-income African American woman. *Suicide and Life-Threatening Behavior* 32: 131–138.

Aponte, R., and R. Pessagno. 2010. The communications revolution and its impact on the family: Significant, growing, but skewed and limited in scope. *Marriage & Family Review* 45: 576–586.

544

Asch-Goodkin, J. 2006. An unsuccessful attempt to adopt a constitutional amendment that bans gay marriage. *Contemporary Pediatrics* 23: 14–15.

Aulette, J. R. 2010. *Changing American families.* Boston: Allyn and Bacon.

Azziz-Baumgartner, E., L. McKeown, P. Melvin, D. Quynh, and J. Reed. 2011. Rates of femicide in women of different races, ethnicities, and places of birth: Massachusetts, 1993–2007. *Journal of Interpersonal Violence* 26: 1077–1090.

B

Baden, A. L., and M. O. Wiley. 2007. Counseling adopted persons in adulthood: Integrating practice and research. *Counseling Psychologist* 35: 868–879.

Baek, E. and S. A. DeVaney. 2010. How do families manage their economic hardship? *Family Relations* 59: 358–368.

Baggerly, J., and H. A Exum. 2008. Counseling children after natural disasters: Guidance for family therapists. *The American Journal of Family Therapy* 36: 79–93.

Bagarozzi, D. A. 2008. Understanding and treating marital infidelity: A multidimensional model. *The American Journal of Family Therapy* 36: 1–17.

Baker, A. 2010. Afghan women and the return of the Taliban. *Time Magazine,* July 9. 20–28.

Baker, A. J. 2007. Expressing emotion in text: Email communication of online couples. In *Online Matchmaking,* ed. M. T. Whitty, A. J. Baker, and J. A. Inman, 97–111. New York: Palgrave Macmillan.

Baker, A. J. L., and D. Darnall. 2007. A construct study of the eight symptoms of severe parental alienation syndrome: A survey of parental experiences. *Journal of Divorce & Remarriage* 47: 55–62.

Baker, A. J. L., and J. Chambers. 2011. Adult recall of childhood exposure to parental conflict: Unpacking the black box of parental alienation. *Journal of Divorce & Remarriage* 52: 55–76.

Baker, J., J. McHale, A. Strozier, and D. Cecil. 2010. Mother–grandmother coparenting relationships in families with incarcerated mothers: A pilot investigation. *Family Process* 49: 165–184.

Ball, H. 2010. Death of a spouse may be associated with increased STD diagnosis among older men. *Perspectives on Sexual & Reproductive Health* 42: 64–64.

Baly, A. R. 2010. Leaving abusive relationships: Constructions of self and situation by abused Women. *Journal of Interpersonal Violence* 25: 2297–2315.

Barelds, D. P., and P. Barelds-Dijkstra. 2007. Love at first sight or friends first? Ties among partner personality trait similarity, relationship onset, relationship quality, and love. *Journal of Social and Personal Relationships* 24: 479–496.

Barelds-Dijkstra, D. P. H., and P. Barelds. 2007. Relations between different types of jealousy and self and partner perceptions of relationship quality. *Clinical Psychology & Psychotherapy* 14: 176–188.

Barnes, K., and J. Patrick. 2004. Examining age-congruency and marital satisfaction. *The Gerontologist* 44: 185–187.

Barnett, M. A., L. V. Scaramella, T. K. Neppl, L. L. Ontai, and R. D. Conger. 2010a. Grandmother involvement as a protective factor for early childhood social adjustment. *Journal of Family Psychology* 24: 635–645.

Barnett, R. C., K. C. Gareis, L. Sabattini, and N. M. Carter. 2010b. Parental concerns about after-school time: Antecedents and correlates among dual-earner parents. *Journal of Family Issues* 31: 606–625.

Bartle-Haring, S. 2010. Using Bowen theory to examine progress in couple therapy. *The Family Journal* 18: 106–115.

Basow, S. A. 2010. Changes in psychology of women and psychology of gender textbooks (1975–2010) *Sex Roles* 62: 151–291.

Bates, J. S., and A. C. Taylor. 2010. Variations in grandfathering: Innovations in conceptualizing grandfather styles. Paper, Annual Meeting of the National Council on Family Relations, November 3–6.

Bauerlein, M. 2010. Literary learning in the hyperdigital age. *Futurist* 44: 24–25.

Baumrind, D. 1966. Effects of authoritative parental control on child behavior. *Child Development* 37: 887–907.

Becker, K. 2010. Risk behavior in adolescents: What we need to know. *European Psychiatry* 25: 85–85.

Becker, K. D., J. Stuewig, and L. A. McCloskey. 2010. Traumatic stress symptoms of women exposed to different forms of childhood victimization and intimate partner violence. *Journal of Interpersonal Violence* 25: 1699–1715.

Beekman, D. 1977. *The mechanical baby: A popular history of the theory and practice of child rearing.* Westport, Ct: Lawrence Hill & Company

Beckman, N. M., M. Waern, I. Skoog, and The Sahlgrenska Academy at Goteborg University, Sweden. 2006. Determinants of sexuality in 70 year olds. *The Journal of Sex Research* 43: 2–3.

Beerthuizen, R. 2010. State-of-the-art of non-hormonal methods of contraception: V. Female sterilisation. *European Journal of Contraception & Reproductive Health Care* 15: 124–135.

Begue, L. 2001. Social judgment of abortion: A black-sheep effect in a Catholic sheepfold. *Journal of Social Psychology* 141: 640–650.

Benjamin, Le., N. L. Dove, C. R. Agnew, M. S. Korn, and A.A. Musso. 2010. Predicting nonmarital romantic relationship dissolution: A meta-analytic synthesis. *Personal Relationships* 17: 377–390.

Bennett, K. M. 2010. How to achieve resilience as an older widower: turning points or gradual change? *Ageing & Society* 30: 369–382.

Bennetts, L. 2007. *The feminine mistake: Are we giving up too much?* New York: Voice/ Hyperion.

Benton, S. D. 2008. Divorce mediation. Lecture, East Carolina University, November 10.

Berg, N., and D. Lein. 2006. Same-sex behaviour: U.S. frequency estimates from survey data with simultaneous misreporting and non-response. *Applied Economics* 39: 757–770.

Berger, R., and M. Paul. 2008, Family secrets and family functioning: The case of donor assistance. *Family Process* 47: 553–566.

Bergman, K., R. J. Rubio, R. J. Green, and E. Padron. 2010. Gay men who become fathers via surrogacy: The transition to parenthood. *Journal of GLBT Family Studies* 6: 111–141.

Bermant, G. 1976. Sexual behavior: Hard times with the Coolidge Effect. In *Psychological research: The inside story,* ed. M. H. Siegel and H. P. Zeigler. New York: Harper and Row.

Berle, W., with B. Lewis. 1999. *My father uncle Miltie.* New York: Barricade Books, Inc.

Berscheid, E. 2010. Love in the fourth dimension. *Annual Review of Psychology* 61: 1–25.

Berzenski, S. R., and T. Yates. 2010. A developmental process analysis of the contribution of childhood emotional abuse to relationship violence. *Journal of Aggression, Maltreatment & Trauma* 19: 180–203.

Biblarz, T. J., and J. Stacey. 2010. How does the gender of parents matter? *Journal of Marriage and Family* 72: 3–22.

Biblarz, T. J., and E. Savci. Lesbian, gay, bisexual, and transgender families. 2010 *Journal of Marriage and Family* 72: 480–497.

Birditt, K. S., E. Brown, T. L. Orbuch and J. J. McIlvane. 2010. Marital conflict behaviors and implications for divorce over 16 years. *Journal of Marriage and Family* 72: 1188–1204.

Bisakha, S. 2010. The relationship between frequency of family dinner and adolescent problem behaviors after adjusting for other family characteristics. *Journal of Adolescence* 33: 187–196.

Black, K., and M. Lobo. 2008. A conceptual review of family resilience factors. *Journal of Family Nursing* 14: 1–33.

Blair, S. L. 2010. The influence of risk-taking behaviors on the transition into marriage: An examination of the long-term consequences of adolescent behavior. *Marriage & Family Review* 46: 126–146.

Blakely, K. 2008. Busy brides and the business of family life. *Journal of Family Issues* 29: 639–643.

Block, S. 2008. Adopting domestically can lower hurdles to claiming tax credit. *USA Today,* August 19, 3B.

Blumer, H. G. 1969. The methodological position of symbolic interaction. In *Symbolic interactionism: Perspective and method.* Englewood Cliffs, NJ: Prentice-Hall.

Bobbe, J. 2002. Treatment with lesbian alcoholics: Healing shame and internalized homophobia for ongoing sobriety. *Health and Social Work* 27: 218–223.

Bock, J. D. 2000. Doing the right thing? Single mothers by choice and the struggle for legitimacy. *Gender and Society* 14: 62–86.

Bodenmann, G., D. C. Atkins, M. Schär, and V. Poffet. 2010. The association between daily stress and sexual activity. *Journal of Family Psychology* 24: 271–279.

Boehnke, M. 2011. Gender role attitudes around the globe: Egalitarian vs. traditional views. *Asian Journal of Social Science* 39: 57–74.

Bogle, K. A. 2008. *Hooking up: Sex, dating, and relationships on campus.* New York: New York University Press.

Boislard P., and F. Poulin. 2011. Individual, familial, friends-related and contextual predictors of early sexual intercourse. *Journal of Adolescence* 34: 289–300.

Bonello, K., and M. C. Cross. 2010. Gay monogamy: I love you but I can't have sex with only you. *Journal of Homosexuality* 57: 117–139.

Bontempo, D. E., and A. R. D'Augelli. 2002. Effects of at-school victimization and sexual orientation on lesbian, gay, or bisexual youths' health risk behavior. *Journal of Adolescent Health* 30: 364–374.

Booth, C. L., K. A. Clarke-Stewart, D. L. Vandell, K. McCartney, and M. T. Owen. 2002. Child-care usage and mother-infant "quality time." *Journal of Marriage and the Family* 64: 16–26.

Bos, E. H., A. L. Bouhuys, E. Geerts, T.W. D. P. Van Os, and J. Ormel. 2007. Stressful life events as a link between problems in nonverbal communication and recurrence of depression *Journal of Affective Disorders* 97: 161–169.

Bos, H., and T. G. M. Sandfort. 2010. Children's gender identity in lesbian and heterosexual two-parent families. *Journal Sex Roles* 62: 114–126.

Bos, H., and N. Gartrell. 2010. Adolescents of the USA National Longitudinal Lesbian Family Study: Can family characteristics counteract the negative effects of stigmatization? *Family Process* 49: 559–572.

Boylan, J. 2004. *She's not there: A life in two genders.* New York: Broadway Books.

Brackett, A., D. Knox, and B. Easterling. 2010. Secrets in romantic relationships: Sexual orientation and other correlates. Poster, Southern Sociological Society, Atlanta, April.

Bradford, J., K. Barrett, and J. A. Honnold. 2002. *The 2000 census and same-sex households: A user's guide.* New York: National Gay and Lesbian Task Force Policy Institute, Survey and Evaluation Research Laboratory, and Fenway Institute. http://www.thetaskforce.org.

Bradshaw, C., A. S. Kahn, and B. K. Saville. 2010. To hook up or date: Which gender benefits? *Journal Sex Roles* 62: 661–669.

Brandes, M., C. Hamilton, J. van der Steen, J. de Bruin, R. Bots, W. Nelen, and J. Kremer. 2011. Unexplained infertility: overall ongoing pregnancy rate and mode of conception. *Human Reproduction* 26: 360–368.

Brantley, A., D. Knox, and M. E. Zusman. 2002. When and why gender differences in saying "I love you" among college students. *College Student Journal* 36: 614–615.

Bratter, J. L., and R. B. King. 2008. "But Will It Last?": Marital instability among interracial and same-race couples. *Family Relations* 57: 160–171.

Braun, M., K. Mura, M. Peter-Wright, R. Hornhung, and U. Scholz. 2010. Toward a better understanding of psychological well-being in dementia caregivers: The link between marital communication and depression. *Family Process* 49: 185–203.

Bray, J. H. and J. Kelly. 1998. *Stepfamilies: Love, marriage and parenting in the first decade.* New York: Broadway Books.

Brecklin, L. R., and S. E. Ullman. 2005. Self-defense or assertiveness training and women's responses to sexual attacks. *Journal of Interpersonal Violence* 20: 738–762.

Bredow, C. A., T. L. Huston, and G. Noval. 2011. Market value, quality of the pool of potential mates, and singles' confidence about marrying. *Personal Relationships* 18: 39–57.

Breheny, M. and C. Stephens, 2010. Youth or disadvantage? The construction of teenage mothers in medical journals. *Health & Sexuality* 12: 307–322.

Brimhall, A., K. Wampler, and T. Kimball. 2008. A warning from the past, altering the future: A tentative theory of the effect of past relationships on couples who remarry. *Family Process* 47: 373–387.

Brimhall, A. S., and M. L. Engblom-Deglmann. 2011. Starting over: A tentative theory exploring the effects of past relationships on postbereavement remarried couples. *Family Process* 50: 47–62.

Brinig, M. F., and D. W. Allen. 2000. "These boots are made for walking": Why most divorce filers are women. *American Law and Economic Association* 2: 126–169.

Bristol, K., and B. Farmer. 2005. *Sexuality among Southeastern university students: A survey.* Unpublished data. Greenville, NC: East Carolina University.

British Columbia Reproductive Mental Health Program. 2005. Reproductive mental health: Psychosis. http://www.bcrmh.com/disorders/psychosis.htm (retrieved June 15, 2005).

Brizendine, L. 2010. *The male brain.* New York: Broadway Books.

Bronner, G., S. Shai, and G. Raviv. 2010. Sexual dysfunction after radical prostatectomy: Treatment failure or treatment delay? *Journal of Sex & Marital Therapy* 36: 421–429.

Bronte-Tinkew, J., J. Carrano, A. Horowitz, and A. Kinukawa. 2008. Involvement among resident fathers and links to infant cognitive outcomes. *Journal of Family Issues* 29: 1211–1231.

Brotherson, S. E., J. White, and C. Masich. 2010. Parents forever: An assessment of the perceived value of a brief divorce education program *Journal of Divorce & Remarriage* 51: 465–490.

Brown, A. 2010. How to accurately interpret a peer's social class: Symbols of class status and presentation of self in college students. Paper, Southern Sociological Society, Atlanta, April.

Brown, J. D. 2008. Foster parents' perceptions of factors needed for successful foster placements. *Journal of Child and Family Studies* 17: 538–555.

Brown, G. L., S. C. Mangelsdorf, C. Neff, S. J. Schoppe-Sullivan, and C. A. Frosch. 2009. Young children's self-concepts: Associations with child temperament, mothers' and fathers' parenting, and triadic family interaction. *Merrill-Palmer Quarterly* 55: 207.

Brown, S. and K. Guthrie. 2010. Why don't teenagers use contraception? A qualitative interview study. *European Journal of Contraception & Reproductive Health Care* 15: 197–204.

Brown, S. L. 2010. Marriage and child well-being: Research and policy perspectives. *Journal of Marriage and Family* 72: 1059–1077.

Brown, S. L., and L. N. Rinelli. 2010. Family structure, family processes, and adolescent smoking and drinking. *Journal of Research on Adolescence* 20: 259–273.

Brownridge, D. A. 2010. Does the situational couple violence- Intimate terrorism typology explain cohabitors' high risk of intimate partner violence? *Journal of Interpersonal Violence* 25: 1264–1283.

Brozowski, K. and D. R. Hall. 2010. Aging and risk: Physical and sexual abuse of elders in Canada. *Journal of Interpersonal Violence* 25: 1183–1199.

Brucker, H., and P. Bearman. 2005. After the promise: The STD consequences of adolescent virginity pledges. *Journal of Adolescent Health* 36: 271–278.

Brumbaugh, C. C., and R. C. Fraley. 2010. Adult attachment and dating strategies: How do insecure people attract mates?. *Personal Relationships* 17: 599–614.

Bryant, C. D. 2007. The sociology of death and dying. In *21st century sociology: A reference handbook*, ed. Clifton D. Bryant and Dennis L. Peck, 156–66. Thousand Oaks, California: Sage.

Buchler, S., J. Baxter, M. Haynes, and M. Western. 2009. The social and demographic characteristics of cohabiters in Australia: Towards a typology of cohabiting couples. *Family Matters* 2009: 22–29.

Bulanda, J. R. 2011. Gender, marital power, and marital quality in later life. *Journal of Women & Aging* 23: 3–22.

Bulanda, J. R., and S. L. Brown. 2007. Race-ethnic differences in marital quality and divorce *Social Science Research* 36: 945–959.

Burke, S., M. Wallen, K. Vail-Smith and D. Knox. 2011. Using technology to control intimate partners: An exploratory study of college undergraduates. *Computers in Human Behavior* 27: 1162–1167.

Burke, M. L., G. G. Eakes, and M. A. Hainsworth. 1999. Milestones of chronic sorrow: Perspectives of chronically ill and bereaved persons and family caregivers. *Journal of Family Nursing* 5: 384–387.

Burr, W. R., and S. R. Klein. 1994. *Reexamining family stress: New theory and research.* Thousand Oaks, CA: Sage.

Burton, L. M., E. Bonilla-Silva, V. Ray, R. Buckelew, and E. H. Freeman. 2010. Critical race theories, colorism, and the decade's research on families of color. *Journal of Marriage and Family* 72: 440–459.

Busby, D. M., T. B. Holman, and E. Walker. 2008. Pathways to relationship aggression between adult partners. *Family Relations* 57: 72–83.

Busse, P., M. Fishbein, A. Bleakley, and M. Hennessy. 2010. The role of communication with friends in sexual initiation. *Communication Research* 37: 239–255.

Butterworth, P., and B. Rodgers. 2008. Mental health problems and marital disruption: Is it the combination of husbands and wives' mental health problems that predicts later divorce? *Social Psychiatry and Psychiatric Epidemiology* 43: 758–764.

Buunk, A. P., J. Goor, and A. C. Solano. 2010. Intrasexual competition at work: Sex differences in the jealousy-evoking effect of rival characteristics in work settings. *Journal of Social and Personal Relationships* 27: 671–684.

Buxton, A. P. 2004. Paths and pitfalls: How heterosexual spouses cope when their husbands or wives come out. *Journal of Couple and Relationship Therapy* 3: 95–105.

Buxton, A. 2005. A family matter: When a spouse comes out as gay, lesbian, or bisexual. *Journal of GLBT Family Studies* 1: 49–70.

C

Cade, R. 2010. Covenant marriage. *Family Journal* 18: 230–233.

Cadden, M., and D Merrill 2007. What married people miss most (based on *Reader's Digest* study of 1,001 married adults). *USA Today*, D1.

Cahill, S., and S. Slater. 2004. *Marriage: Legal protections for families and children. Policy brief.* Washington, DC: National Gay and Lesbian Task Force Policy Institute.

Calligas, A., F. Adler-Baeder, M. Keiley, T. Smith, and S. Ketring. 2010. Examining change in parenting dimensions in relation to change in couple dimensions. Poster, National Council on Family Relations annual meeting, November 3–5. Minneapolis, MN.

Cantón-Cortés, D., and J. Cantóna. 2010. Coping with child sexual abuse among college students and post-traumatic stress disorder: The role of continuity of abuse and relationship with the perpetrator. *Child Abuse & Neglect.* 34: 496–506.

Capizzano, J., G. Adams, and J. Ost. 2006. The child care patterns of white, black and Hispanic children. Urban Institute. http://www.urban.org/url.cfm?ID=311285 (retrieved April 2006).

Cardozo, M. 2006. What is a good death? Issues to examine in critical care. *British Journal of Nursing* 14: 1056–1060.

CareerBuilder.com. 2010. "Forty Percent of Workers Have Dated a Co-Worker, Finds Annual CareerBuilder.com Valentine's Day Survey" Posted on their website, Feb 15, 2010. Retrieved February 15, 2010 at http://www.careerbuilder.com/share/aboutus/pressreleasesdetail.aspx?id=pr553&sd=2%2f9%2f2010&ed=12%2f31 %2f 2010&siteid=cbpr&sc_cmp1=cb_pr553_&cbRecursionCnt=1&cbsid=9354b8fc6d164b7cad8a6ed186882567-319524618-VM-4

Carey, A. R., and K. Gellers. 2010. Do you believe in the idea of soulmates? *USA Today*, August 31. 1A.

Carey, A. R., and P. Trap. 2010a. How honest are you on your social networking sites? *USA Today* Jan 3, A1.

Carey, A, and P. Trap. 2010b. Wired vacations. *USA Today* June 3, A1.

Carey, A. , and V. Salazar. 2011. Women talk and text more. *USA Today* Feb 1, 2011. P. 1.

Carlander, I., E. Sahlberg-Blom, I. Hellström, and B. Ternestedt. 2011. The modified self: Family caregivers' experiences of caring for a dying family member at home. *Journal of Clinical Nursing* 20: 1097–1105.

Carpenter, C., and G. J. Gates. 2008. Gay and lesbian partnership: Evidence from California. *Demography* 45: 573–691.

Carpenter, L. 2009. Virginity loss in reel/real life: Using popular movies to navigate sexual initiation. *Sociological Forum* 24: 804–827.

Carpenter, L. M. 2010. Like a Virgin… again?: Secondary virginity as an ongoing gendered social Construction. *Sexuality and Culture* 14: 253–270.

Carr, D., and K. W. Springer. 2010. Advances in families and health research in the 21st Century. *Journal of Marriage and Family* 72: 743–761.

Carr, D., and S. M. Moorman. 2009. End-of-life treatment preferences among older adults: An assessment of psychosocial influences. *Sociological Forum* 24: 754–778.

Carroll, D. D., H. M. Blanck, M. K. Serdula, and D. R. Brown. 2010. Obesity, physical activity, and depressive symptoms in a cohort of adults aged 51 to 61. *Journal of Aging and Health* 22: 384–398.

Cassidy, M. L., and G. Lee. 1989. The study of polyandry: A critique and synthesis. *Journal of Comparative Family Studies* 20: 1–11.

Castrucci, B. C., J. F. Culhane, E. K. Chung, I. Bennett, and K. F. McCollum. 2006. Smoking in pregnancy: Patient and provider risk reduction behavior. *Journal of Public Health Management & Practice* 12: 68–76.

Caughlin, J. P., and M. B. Ramey. 2005. The demand/withdraw pattern of communication in parent-adolescent dyads. *Personal Relationships* 12: 337–355.

Cavaglion, G., and E. Rashty. 2010. Narratives of suffering among Italian female partners of cybersex and cyber-porn. *Sexual Addiction & Compulsivity* 17: 270–287.

Cavazos-Rehg, P. A. E. L. Spitznagel, K. K. Bucholz, J. Nurnberger Jr., H. J. Edenberg, J. R. Kramer, S. Kuperman, V. Hesselbrock, and L. J. Bierut. 2010. Predictors of sexual debut at age 16 or younger. *Journal Archives of Sexual Behavior* 39: 664–673.

Chaney, C., and K. Marsh. 2009. Factors that facilitate relationship entry among married and cohabiting African Americans. *Marriage & Family Review* 45: 26–51.

Chapleau, K. M., D. L. Oswald, and B. L. Russell 2008. Male rape myths: The role of gender, violence, and sexism. *Journal of Interpersonal Violence* 23: 600–615.

Chartiera, M. J., J. R. Walkerb and B. Naimarkc. 2010. Separate and cumulative effects of adverse childhood experiences in predicting adult health and health care utilization *Child Abuse & Neglect* 34: 454–464.

Chase, B. 2000. NEA president Bob Chase's historic speech from 2000 GLSEN Conference. http://www.glsen.org.

Cherlin, A. J. 2009. *The Marriage-go-round: The State of marriage and the family in America today.* New York: Alfred a Knopf Inc.

Cherlin, A. J. 2010 Demographic trends in the United States: A review of research in the 2000s. *Journal of Marriage and Family* 72: 403–419.

Chich-Hsiu, H., L. Chia-Ju, J. Stocker, and Y. Ching-Yun. 2011. Predictors of postpartum stress. *Journal of Clinical Nursing* 20: 666–674.

Chih,-Chien, W., and C. Ya-Ting. 2010. Cyber relationship motives: Scale development and motivation. *Social Behavior & Personality: An International Journal* 38: 289–300.

Chiu, H., and D. Busby. 2010. Parental influence in adult children's marital relationship. Poster, National Council on Family Relations annual meeting, November 3–5. Minneapolis, MN.

Christensen, A. J. Yi, Atkins, D. C., D. H. Baucom, and W. I. George. 2006 Couple and individual adjustment for 2 years following a randomized clinical trial comparing traditional versus integrative behavioral couple therapy. *Journal of Consulting & Clinical Psychology* 74: 1180–1191.

Chu, K. 2007. As higher education costs rise, so do debt loads. *USA Today* May 25, 3B.

Cianciotto, J. 2005. Hispanic and Latino same-sex couple households in the U.S.: A report from the 2000 Census. National Gay and Lesbian Task Force Policy Institute. http:www.ngltf.org.

Cianciotto, J., and S. Cahill. 2006. Youth in the crosshairs: The third wave of ex-gay activism. National Gay and Lesbian Task Force Policy Institute. http:www.the-taskforce.org.

Cinamon, R. G. 2006. Anticipated work-family conflict: effects of gender, self-efficacy, and family background. *Career Development Quarterly* 54: 202–216.

Claffey, S. T., and K. D. Mickelson. 2010. Division of household labor and distress: The role of perceived fairness for employed mothers. *Sex Roles* 60: 819–831.

Clark, S. and C. Kenney. 2010. Is the United States experiencing a "matrilineal tilt": Gender, family structures and financial transfers to adult children. *Social Forces* 88: 1753–1776.

Clarke, S. C., and B. F. Wilson. 1994. The relative stability of remarriages: A cohort approach using vital statistics. *Family Relations* 43: 305–310.

Clarke, J. I. 2004. The overindulgence research literature: Implications for family life educators. Poster at the National Council on Family Relations, Annual Meeting, November. Orlando, Florida.

Clarke-Stewart, A., and C. Brentano. 2006. *Divorce: Causes and Consequences.* New Haven: Yale University Press.

Clarkwest, A. 2007. Spousal dissimilarity, race, and marital dissolution. *Journal of Marriage and the Family* 69: 639–653.

Claxton, A., and M. Perry-Jenkins. 2008. No fun anymore: Leisure and marital quality across the transition to parenthood. *Journal of Marriage and the Family* 70: 28–43.

Clements, C. M., and R. L. Ogle. 2009 Does acknowledgment as an assault victim impact post assault psychological symptoms and coping? *Journal of Interpersonal Violence* 24: 1595–1614.

Cleverley, K., and M. H. Boyle. 2010. The individual as a moderating agent of the long-term impact of sexual abuse. *Journal of Interpersonal Violence* 25: 274–290.

Cline, C., and K. Marie. 2010. Psychological effects of dog ownership: Role strain, role enhancement, and depression. *Journal of Social Psychology* 150: 117–131.

Clunis, D. M., and G. Dorsey Green. 2003. *The lesbian parenting book*, 2nd ed. Emeryville, CA: Seal Press.

Cobey, K. D., T. V. Pollet, S. C. Roberts, and A. P. Buunk. 2011. Hormonal birth control use and relationship jealousy: Evidence for estrogen dosage effects. *Personality & Individual Differences* 50: 315–317.

Cohen-Kettenis, P. T. 2005. Gender change in 46, XY persons with 5[alpha]-reductase-2 deficiency and 17[beta]-hydroxysteroid dehydrogenase-3 deficiency. *Archives of Sexual Behavior* 34: 399–411.

Cohen Westbrook, R. A. 2011. Masturbation and sex positive/shameful sexual attitudes. Poster, Easter Sociological Society, Philadelphia, Feb 25.

Cokes, C., and W. Kornblum. 2010. Experiences of mental distress by individuals during an economic downturn: the story of an urban city. *The Western Journal of Black Studies* 34: 24–36.

Colapinto, J. 2000. *As nature made him: The boy who was raised as a girl.* New York: Harper Collins.

Collins, J. 1998. *Singing lessons: A memoir of love, loss, hope, and healing.* New York: Pocket Books.

Colson, M., A. Lemaire, P. Pinton, K. Hamidi, and P. Klein. 2006. Sexual behaviors and mental perception, satisfaction, and expectations of sex life in men and women in France. *The Journal of Sexual Medicine* 3: 121–131.

Coltrane, S., and M. Adams. 2003. The social construction of the divorce "problem": Morality, child victims, and the politics of gender. *Family Relations* 52: 363–372.

Confer, J. C., and M. D. Cloud. 2011. Sex differences in response to imagining a partner's heterosexual or homosexual affair. *Personality & Individual Difference* 50: 129–134.

Congressional Digest. 2010a. Legislative background on don't ask, don't tell recent action in Congress. *Congressional Digest* April: 115.

Congressional Digest. 2010b. Homosexuals in the military evolution of the don't ask, don't tell." *Congressional Digest.* April: 103–107.

Consolatore, D. 2002. What next for the women of Afghanistan? *The Humanist* 62: 10–15.

Cook, K., D. Dranove, and A. Sfekas. 2010. Does major illness cause financial catastrophe? *Health Services Research* 45: 418–436.

Cooley, C. H. 1964. *Human nature and the social order.* New York: Schocken.

Coombes, A. 2007. What are the key ingredients to a happier old age? It's easier than you think. http://www.marketwatch.com/news/story/here-key-ingredients-staying-happy/story.aspx?guid=%7BFD3

F3A18-7DF3-4BF5-A3A3-B6438317E 693%7D (retrieved August 24).

Coontz, S. 2000. Marriage: Then and now. *Phi Kappa Phi Journal* 80:16–20.

Cooper, C., A. Selwood, and G. Livingston. 2008. The prevalence of elder abuse and neglect: a systematic review. *Age and Ageing* 37: 151–161.

Copeland, L. 2010. Parent-teen driving contracts. *USA Today.* Oct 19, A1.

Cora, M., S. Carter, J. Scott Carter, and D. Knox. 2009 Trends in marital happiness by sex and race, 1973–2006. *Journal of Family Issues* 30: 1379–1404.

Cornelius-Cozzi, T. 2002. Effects of parenthood on the relationships of lesbian couples. *PROGRESS: Family Systems Research and Therapy* 11: 85–94.

Cousins, G., H. McGee, and R. Layte. 2010. Suppression effects of partner type on the alcohol-risky sex relationship in young Irish adults. *Journal of Studies on Alcohol and Drugs* 7: 357–366.

Cowburn, M. 2010. Invisible men: Social reactions to male sexual coercion—bringing men and masculinities into community safety and public policy. *Critical Social Policy* 30: 225–244.

Coyne, S. M., L. Stockdale, D. Busby, B. Iverson, and D. M. Grant. 2011. "I luv u :)!": A descriptive study of the media use of individuals in romantic relationships. *Family Relations* 60: 150–162.

Cozijnsen, R., N. L. Stevens, and T. G. Van Tilburg. 2010. Maintaining work-related personal ties following retirement. *Personal Relationships,* 17: 345–356.

Craig, L., and K. Mullan 2010. Parenthood, gender and work-family time in the United States, Australia, Italy, France, and Denmark. *Journal of Marriage and Family* 72: 1344–1361.

Cramer, R. E., R. E. Lipinski, J. D. Meteer, and J. A. Houska. 2008. Sex differences in subjective distress to unfaithfulness: Testing competing evolutionary and violation of infidelity expectations hypotheses. *The Journal of Social Psychology* 148: 389–406.

Crawley, S. L., L. J. Foley, and C. L. Shehan. 2008. *Gendering bodies.* Boston: Rowman and Littlefield.

Crisp, B., and D. Knox. 2009. *Behavioral family therapy: An evidence based approach.* Chapel Hill, NC: Carolina Academic Press. http://www.cap-press.com/books/1870.

Crosnoe, R., and S. E. Cavanagh. 2010. Families With Children and Adolescents: A Review, Critique, and Future Agenda. *Journal of Marriage and Family* 72: 594–611.

Crowell, J. A., D. Treboux, and E. Waters. 2002. Stability of attachment representations: The transition to marriage. *Developmental Psychology* 38: 467–79.

Crowl, A., S. Ahn, and J. Baker. 2008. A meta-analysis of developmental outcomes for children of same-sex and heterosexual parents. *Journal of GLBT Family Studies* 4: 385–407.

Cui, M., and F. D. Fincham. 2010. The differential effects of parental divorce and marital conflict on young adult romantic relationships. *Personal Relationships* 17: 331–343.

Cui, M., F. D. Fincham, and B. Kay Pasley. 2008. Young adult romantic relationships: The role of parents' marital problems and relationship efficacy. *Personality and Social Psychology Bulletin* 34: 1226–1235.

Cunningham, M. R., S. R. Shamblen, A. P. Barbee, and L. K. Ault. 2005. Social allergies in romantic relationships: Behavioral repetition, emotional sensitization, and dissatisfaction in dating couples. *Personal Relationships* 12: 273–295.

Curran, M. A., E. A. Utley, and J. A. Muraco. 2010. An exploratory study of the meaning of marriage for African Americans. *Marriage & Family Review* 46: 346–365.

Curtis, C. 2003. Poll: U.S. public is 50–50 on gay marriage. *PlanetOut* (October 7). http:// www.planetout.com.

Curtis, C. 2004. Poll: 1 in 20 High school students is gay. *PlanetOut.* http://www.planetout.com.

Cutler, N. E. 2002. *Advising mature clients.* New York: Wiley.

D

Dabbous, Y., and A. Ladley. 2010. A spine of steel and a heart of gold: newspaper coverage of the first female Speaker of the House. *Journal of Gender Studies* 19: 181–194.

Daigle, L. E., B. S. Fisher, and F. T. Cullen. 2008. The violent and sexual victimization of college women: Is repeat victimization a problem? *Journal of Interpersonal Violence* 23: 1296–1313.

Danigelis, N. L., M. Hardy, and S. J. Cutler. 2007. Population aging, intracohort aging, and sociopolitical attitudes. *American Sociological Review* 72: 812–830.

Darlow, S., and M. Lobel. 2010. Who is beholding my beauty? Thinness ideals, weight, and women's responses to appearance evaluation. *Sex Roles* 63: 833–843.

Davidson, J. K., Sr., N. B. Moore, J. R. Earle, and R. Davis. 2008. Sexual attitudes and behavior at four universities: Do region, race, and/or religion matter? *Adolescence* 43: 189–223.

Dawson, C., D. Bredehoft, and J. I. Clarke. 2003. How much is enough? Boston: De Capo Press.

de Anda, D. 2006. Baby think it over: Simulation intervention for adolescent pregnancy prevention *Health and Social Work.* 31: 26–35.

De Castro, S., and J. T. Guterman. 2008. Solution-focused therapy for families with suicide. *Journal of Marital and Family Therapy* 34: 93–107.

De Schipper, J. C., L. W. C. Tavecchio, and M. H Van IJzendoorn. 2008. Children's attachment relationships with day care caregivers: Associations with positive caregiving and the child's temperament. *Social Development* 17: 454–465.

de Vries, B. 2007. LGBT couples in later life: a study in diversity. *Couples in Later Life* 31(3): 18–23.

de Vries, J. M. A., L. Swenson, and R. P. Walsh. 2007. Hot picture or great self-description: Predicting mediated dating success with parental investment theory. *Marriage & Family Review* 42: 7–23.

Deacon, S. A., L. Peinke, and D. Viers. 1996. Cognitive-behavioral therapy for bisexual couples: expanding the realms of therapy.

The American Journal of Family Therapy 24(3): 242–258.

Dean, C. J. 2011. Psychoeducation: A first step to understanding infidelity-related systemic trauma and grieving. *Family Journal* 19: 15–21.

DeCuzzi, A., D. Knox, and M. Zusman. 2006. Racial differences in perceptions of women and men. *College Student Journal* 40: 343–349.

Dehan, N., and Z. Levi. 2009. Spiritual abuse: An additional dimension of abuse experienced by abused Haredi (Ultraorthodox) Jewish wives. *Violence Against Women* 15: 1294–1310.

Del Rio, C. M. 2010. "Marriage" misnames "couples" and familial therapies. *The Family Journal* 18: 169–177.

DeLamater, J., and M. Hasday. 2007. The sociology of sexuality. In *21st century sociology: A reference handbook*, ed. Clifton D. Bryant and Dennis L. Peck, 254–264. Thousand Oaks, California: Sage.

Dema-Moreno, S. and C. Díaz-Martínez. 2010. Gender inequalities and the role of money in Spanish dual-income couples. *European Societies* 12: 65–84.

DeMaris, D. 2010. The 20-year trajectory of marital quality in enduring marriages: Does equity matter? *Journal of Social and Personal Relationships* 27: 449–471.

Demir, M. 2008. Sweetheart, you really make me happy: Romantic relationship quality and personality as predictors of happiness among emerging adults. *Journal of Happiness Studies* 9: 257–277.

Denney, J. T. 2010. Family and household formations and suicide in the United States. *Journal of Marriage and the Family* 72: 202–213.

DeOllos, I. Y. 2005. Predicting marital success or failure: Burgess and beyond. In *Sourcebook of family theory and research,* ed. Vern L. Bengtson, Alan C. Acock, Katherine R. Allen, Peggye Dilworth-Anderson, and David M. Klein, 134–136. Thousand Oaks, CA: Sage Publications.

Department of Commerce etc. 2011. *Women in America: Indicators of social and economic well-being*. Retrieved March 15. http://www.whitehouse.gov/sites/default/files/rss_viewer/Women_in_America.pdf

DePaulo, B. 2006. *Singled out: How singles are stereotyped, stigmatized, and ignored, and still live happily ever after*. New York: St. Martin's Press.

Dethier, M, C. Counerotte, and S. Blairy. 2011. Marital satisfaction in couples with an alcoholic husband. *Journal of Family Violence* 26: 151–162.

Deutsch, F. M., A. P. Kokot, and K. S. Binder. 2007. College women's plans for different types of egalitarian marriages. *Journal of Marriage and Family* 69: 916–629.

Devall, E., M. Montanez, and D. Vanleeuwen. 2009. Effectiveness of relationship education with single, cohabiting, and married parents. Poster, National Council on Family Relations, November.

Dew, J. 2008. Debt change and marital satisfaction change in recently married couples. *Family Relations* 57: 60–71.

Dew, J. 2011. Financial issues and relationship outcomes among cohabiting individuals. *Family Relations* 60: 178–190.

Dew, J., and J. Yorgason. 2010. Economic pressure and marital conflict in retirement-aged couples. *Journal of Family Issues* 31: 164–188.

Dewaele, A., N. Cox, W. Van den Berghe, and J. Vincke. 2011. Families of choice? Exploring the supportive networks of lesbians, gay men, and bisexuals. *Journal of Applied Social Psychology* 41: 312–331.

Diamond, L. M. 2003. What does sexual orientation orient? A biobehavioral model distinguishing romantic love and sexual desire. *Psychological Review* 110: 173–192.

Diamond, A., J. Bowes, and G. Robertson. 2006. Mothers' safety intervention strategies with toddlers and their relationship to child characteristics. *Early Child Development and Care* 176: 271–284.

DiDonato, M. D., S. A. Berenbaum. 2011. The benefits and drawbacks of gender typing: How different dimensions are related to psychological adjustment. *Archives of Sexual Behavior* 40: 457–463.

Diem, C., and J. M. Pizarro. 2010. Social structure and family homicides. *Journal of Family Violence*. 25: 521–532.

DiLillo, D. S. A. Hayes-Skeltona, M. A. Fortier, A. R. Perrya, S. E. Evans, T. L. Messman Moore, K. Walsha, C. Nasha and A. Fauchierc. 2010. Development and initial Psychometric properties of the Computer Assisted Maltreatment Inventory (CAMI): A comprehensive self-report measure of child maltreatment history *Child Abuse & Neglect* 34: 305–317.

Dillion C. F., Q. Gu, H. Hoffman, and CW Ko. 2010. Vision, hearing, balance, and sensory impairment in Americans aged 70 years and over: United States, 1999–2006. NCHS data brief, no 31. Hyattsville, MD: National Center for Health Statistics.

Dixon, L., C. Hamilton-Giachritsis, and K. Browne. 2005. Attributions and behaviors of parents abused as children: a mediational analysis of the intergenerational continuity of child maltreatment (Part II). *Journal of Child Psychology and Psychiatry and Allied Disciplines* 46: 58–73.

Djamba, Y. K., M. J. Crump, and A. G. Jackson. 2005. Levels and determinants of extramarital sex. Paper presented at the Southern Sociological Society, March. Charlotte, NC.

Dotson-Blake, K., D. Knox, and A. Holman 2008. College student attitudes toward marriage, family, and sex therapy. Unpublished data from 288 undergraduate/graduate students. East Carolina University, Greenville, NC.

Dotson-Blake, K., D. Knox, and A. R. Holman. 2010. Reaching out: College student perceptions of counseling. *Professional Issues in Counseling* (Fall). Retrieved from http://www.shsu.edu/~piic/CollegeStudentPerceptions.htm.

Dotson-Blake, K., D. Knox, and M. Zusman. 2009. Oral sex and still a virgin?: A profile of undergraduates who agree. Data collected for this text.

Dowd, D. A., M. J. Means, J. F. Pope, and J. H. Humphries. 2005. Attributions and marital satisfaction: The mediated effects of self-disclosure. *Journal of Family and Consumer Sciences* 97: 22–27.

Dowd, D. A. 2009. The how and why of religion and marital commitment: Multiple dimensions of religiosity and their relationship to components of marital commitment. Poster, National Council on Family Relations, San Francisco, November.

Doyle, M., C. O'Dywer, and V. Timonen. 2010. "How can you just cut off a whole side of the family and say move on?" The reshaping of paternal grandparent-grandchild relationships following divorce or separation in the middle generation. *Family Relations* 59: 587

Dozetos, B. 2001. School shooter taunted as 'gay.' *PlanetOut* (March 7). http:www.planetout.com.

Drapeau, A. M. Gagn, M. Saint-Jacques, R. Lpine, and H. Ivers. 2009. Post-separation conflict trajectories: A longitudinal study. *Marriage & Family Review* 45: 353–373.

Drefahl, S. 2010. How does the age gap between partners affect their survival? *Demography* 47: 313–326.

Drucker, D. J. 2010. Male sexuality and Alfred Kinsey's 0–6 Scale: Toward "A sound understanding of the realities of sex". *Journal of Homosexuality* 57: 1105–1123.

Druckerman, P. 2007. *Lust in translation*. New York: Penguin Group.

Dubbs, S. L., and A. P. Buunk. 2010. Sex differences in parental preferences over a child's mate choice: A daughter's perspective. *Journal of Social and Personal Relationships* 27: 1051–1059.

Dugan, L., D. S. Nagin, and R. Rosenfeld. 2003. Exposure reduction or retaliation? The effects of domestic violence resources on intimate-partner homicide. *Law & Society Review* 37: 169–198.

Dumka, L. E., N. A. Gonzales, L.A. Wheeler, and R. E. Millsap. 2010. Parenting self-efficacy and parenting practices over time in Mexican American families. *Journal of Family Psychology* 24: 522–531.

Duncan, S. F., T. B. Holman, and C. Yang. 2007. Factors associated with involvement in marriage preparation programs. *Family Relations* 56: 270–278.

Dunn, K. M., P. R. Croft, and G. I. Hackett. 2000. Satisfaction in the sex life of a general population sample. *Journal of Sex and Marital Therapy* 26: 141–151.

Dysart-Gale, D. 2010. Social justice and social determinants of health: Lesbian, gay, bisexual, transgendered, intersexed, and queer youth in Canada. *Journal of Child & Adolescent Psychiatric Nursing* 23: 23–28.

E

East, L., D. Jackson, L. O'Brien, and K. Peters. 2007. Use of the male condom by heterosexual adolescents and young people: Literature review. *Journal of Advanced Nursing* 59: 103–110.

East, L., D. Jackson, L. O'Brien and K. Peters. 2011. Condom negotiation: Experiences of sexually active young women. *Journal of Advanced Nursing* 67: 77–85.

Easterling, B. A. 2005. *The Invisible Side of Military Careers: An Examination of Employment and Well-Being Among Military Spouses.* Master's Thesis, University of North Florida.

Easterling, B. A., and D. Knox. 2010. Left behind: How military wives experience the deployment of their husbands. *Journal of Family Life.* Published July 20, 2010 online http://www.journaloffamilylife.org/militarywives

Eckstein, D., M. Sperber, and S. McRae. 2009. Forgiveness: Another relationship ``F Word''—A couple's dialogue. *The Family Journal* 17: 256–262.

Edin, K., and R. J. Kissane 2010 Poverty and the American family: A decade in review *Journal of Marriage and Family* 72: 460–479.

Edser, S.J., and J.D. Shea. 2002. An exploratory investigation of bisexual men in monogamous, heterosexual marriages. *Journal of Bisexuality* 2(4): 5–43.

Edwards, T. M. 2000. Flying solo. *Time* August 28, 47–53.

Eggebeen, D. J., J. Dew, and L. E. Dumka, N. A. Gonzales, L. A. Wheeler, R. E. Millsap. 2010. Parenting self-efficacy and parenting practices over time in Mexican American families. *Journal of Family Psychology.* 24: 522–531.

Eke, A., N. Hilton, G. Harris, M. Rice, and R. Houghton. 2011. Intimate partner homicide: Risk assessment and prospects for prediction. *Journal of Family Violence* 26: 211–216.

Elgar, F. J., J. Knight, G. J. Worrall, and G. Sherman. 2003. Attachment characteristics and behavioral problems in rural and urban juvenile delinquents. *Child Psychiatry and Human Development* 34: 35–48.

Ellison, C. G., A. M. Burdette, and W. B. Wilcox. 2010. The couple that prays together: Race and ethnicity, religion, and relationship quality among working-age adults. *Journal of Marriage and Family* 72: 963–975.

Elmslie, B., and E. Tebaldi. 2008. So, What did you do last night? The economics of infidelity *Kyklos* 61: 391–406.

Else-Quest, N. M., J. S. Hyde, and J. D. DeLamater. 2005. Context counts: Long-term sequelae of premarital intercourse of abstinence. *Journal of Sex Research* 42: 102–112.

Emons, P., F. Wester, and P. Scheepers. 2010. "He Works Outside the Home; She Drinks Coffee and Does the Dishes": Gender roles in fiction programs on Dutch television. *Journal of Broadcasting & Electronic Media* 54: 40–53.

England, P. 2010. The Gender Revolution: Uneven and Stalled. *Gender & Society* 24: 149–166.

England, P. and R. J. Thomas. 2006. The decline of the date and the rise of the college hook up. In *Family in transition*, 14th ed., ed. A. S. Skolnick and J. H. Skolnick, 151–62. Boston: Pearson Allyn & Bacon.

Ennis, E., A. Vrij, and C. Chance. 2008. Individual differences and lying in everyday life. *Journal of Social and Personal Relationships* 25: 105–118.

Enright, E. 2004. A house divided. *AARP The Magazine,* July/August, 60.

Epstein, M., J. P. Calzo, A. P. Smiler, and L M. Ward. 2009. "Anything from making out to having sex": Men's negotiations of hooking up and friends with benefits scripts. *Journal of Sex Research* 46: 414–424.

Erich, S., P. Leung, and P. Kindle. 2005. A Comparative analysis of adoptive family functioning with gay, lesbian, and heterosexual parents and their children. *Journal of GLBT Family Studies* 1: 43–60.

Eshbaugh, E. M., and G. Gute. 2008. Hookups and sexual regret among college women. *The Journal of Social Psychology* 148: 77–87.

Esmaila, A. 2010. Negotiating fairness": A study on how lesbian family members evaluate, construct, and maintain "fairness" with the division of household labor. *Journal of Homosexuality* 57: 591–609.

Euser, E. M., M. H. van Ijzendoorn, P. Prinzie, M. J. Bakermans-Kranenburg. 2010. Prevalence of child maltreatment in the Netherlands. *Child Maltreatment* 15: 5–17.

F

Fabian, N. 2007. Rethinking retirement—And a footnote on diversity. *Journal of Environmental Health* 69: 85–86.

Facer, J., and R. Day. 2004. Explaining diminished marital satisfaction when parenting adolescents. Poster at an Annual Meeting National Council on Family Relations, Orlando, Florida.

Fairlie, R. W., D. O. Beltran, and K. K. K. Das. 2010. Home computers and educational outcomes: Evidence from the NLSY97 and CPS. *Economic Inquiry* 48: 771–793.

Falcon, M., F. Valero, M. Pellegrini, M. Rotolo, G. Scaravelli, J. Joya, and O. Vall. et al. 2010. Exposure to psychoactive substances in women who request voluntary termination of pregnancy assessed by serum and hair testing. *Forensic Science International* 196: 22–26.

Fallona, B., N. Trocméb, J. Flukec, B. MacLaurind, L. Tonmyre, and Ying-Ying Yuanf. 2010. Methodological challenges in measuring child maltreatment *Child Abuse & Neglect* 34: 70–79.

Family Caregiver Alliance. http://caregiver.org/caregiver/jsp/content_node.jsp?nodeid=439 (retrieved April 10, 2006).

Farris, C., R. J. Viken, and T. Treat. 2010. Alcohol alters men's perceptual and decisional processing of women' s sexual interest. *Journal of Abnormal Psychology* 119: 427–432.

Feldman, D. C., and T. A. Beehr. 2011. A three-phase model of retirement decision making. *American Psychologist* 66: 193–203.

Fernandez-Ballesteros, R. 2003. Social support and quality of life among older people in Spain. *Journal of Social Issues* 58: 645–660.

Few, A. L., and K. H. Rosen 2005. Victims of chronic dating violence: How women's vulnerabilities link to their decisions to stay *Family Relations* 54: 265–279.

Field, D., and S. Weishaus. 1992. Marriage over half a century: A longitudinal study. In *Changing lives*, ed. M. Bloom, 269–73. Columbia, SC: University of South Carolina Press.

Field, N. P., E. Gal-Oz, and G. A. Bananno. 2003. Continuing bonds and adjustment at 5 years after the death of a spouse. *Journal of Consulting and Clinical Psychology* 71: 110–117.

Fieldera, R. L., and M. P. Careya 2010. Prevalence and characteristics of sexual hookups among first-semester female college students. *Journal of Sex & Marital Therapy* 36: 346–359.

Filippi, V., S. Goufodji, S. C. Sismanidis, L. Kanhonou, E. Fottrell, C. Ronsmans, E. Alihonou, and V. Patel. 2010. Effects of severe obstetric complications on women's health and infant mortality in Benin. *Tropical Medicine & International Health* 15: 733–742.

Fincham, F. D., and S. R. H. Beach. 2010. Marriage in the new millennium: A decade in review. *Journal of Marriage and Family* 72: 630–649.

Finer, L. B., L. F. Frohwirth, L. A. Dauphinne, S. Singh, and A. M. Moore. 2005. Reasons U.S. women have abortions: quantitative and qualitative reasons. *Perspectives on Sexual and Reproductive Health* 37: 110–118.

Finkenauer, C. 2010. Although it helps, love is not all you need: How Caryl Rusbult made me discover what relationships are all about. *Personal Relationships* 17: 161–163.

Finkenauer, C., and H. Hazam. 2000. Disclosure and secrecy in marriage: Do both contribute to marital satisfaction? *Journal of Social and Personal Relationships* 17: 245–263.

Finkenauer, C., L. Wijngaards-De Meij, H. T. Reis, and C. E. Rusbult. 2010. The importance of seeing what is not there: A quasi-signal detection analysis of positive and negative behavior in newlywed couples. *Personal Relationships* 17: 615–633.

Finley, G. E. 2004. Divorce inequities. *NCFR Family Focus Report* 49(3):F7.

Fisher, C. M. 2009b. Queer youth experiences with abstinence-only-until marriage sexuality education: "I can't get married so where does that leave me?" *Journal of LBGT Youth* 6: 61–79.

Fisher, H. 2009a. *Why him? Why her?* New York: Henry Holt and Company.

Fisher, H. 2010. The new monogamy: forward to the past: an author and anthropologist looks at the future of love. *The Futurist* 44: 26–29.

Fisher, H. et al. 2006 Romantic love: A mammalian brain system for mate choice, *Philosophical Transactions of the Royal Society B* 361: 2173–2186.

Fisher, T. D., and J. K. McNulty. 2008. Neuroticism and marital satisfaction: The mediating role played by the sexual relationship. *Journal of Family Psychology* 22: 112–123. Fortune 500 in 2008.

Fisher, M., and A. Cox. 2011. Four strategies used during intrasexual competition for mates. *Personal Relationships* 18: 20–38.

Flack, Jr., W. F., M. L. Caron, S. J. Leinen, K. G. Breitenbach, A. M. Barber, E. N. Brown, C. T. Gilbert, T. F. Harchak, M. M. Hendricks, C. E. Rector, H. T. Schatten, and H. C. Stein. 2008. "The Red Zone": Temporal risk for unwanted sex among college students. *Journal of Interpersonal Violence* 23: 1177–1196.

Flood, M. 2009. The harms of pornography exposure among children and young people. *Child Abuse Review* 18: 384–400.

Flouri, E., and A. Buchanan. 2003. The role of father involvement and mother involvement in adolescents' psychological well-being. *British Journal of Social Work* 33: 399–406.

Follingstad, D. R., and M. Edmundson. 2010. Is psychological abuse reciprocal in intimate relationships? Data from a national sample of American adults. *Journal of Family Violence* 25: 495–508.

Fonda, J. 2005. *Jane Fonda: My life, so far.* New York: Random House.

Fone, B. 2000. *Homophobia: A history.* New York: Henry Holt.

Formichelli, L. 2010. Surviving unemployment. *Momentum* 3: 27–28.

Foster, J. D. 2008. Incorporating personality into the investment model: Probing commitment processes across individual differences in narcissism. *Journal of Social and Personal Relationships* 25: 211–223.

Foster, J. 2010. How love and sex can influence recognition of faces and words: A processing model account. *European Journal of Social Psychology* 40: 524–535.

Foubert, J. D., E. E. Godin, and J. L. Tatum. 2010. In their own words: Sophomore college men describe attitude and behavior changes resulting from a rape-prevention program 2 years after their participation. *Journal of Interpersonal Violence* 25: 2237–2257. http://money.cnn.com/galleries/2008/fortune/0804/gallery.500_women_ceos.fortune/index.html.

Fram, A. 2007. A fifth vacation with laptops. *Associated Press,* June 1.

Franks, M. M., T. Lucas, M. A. Stephens, K.S. Rook, and Gonzalez, R. 2010. Diabetes distress and depressive symptoms: A dyadic investigation of older patients and their spouses. *Family Relations* 59: 599–610.

Freeman, D., and J. Temple. 2010. Social factors associated with history of sexual assault among ethnically diverse adolescents. *Journal of Family Violence* 25: 349–356.

Freud, S. 1905/1938. Three contributions to the theory of sex. In *The basic writings of Sigmund Freud,* ed. A. A. Brill. New York: Random House.

Friedrich, R. M., S. Lively, and L. M Rubenstein. 2008. Siblings' coping strategies and mental health services: A national study of siblings of persons with schizophrenia *Psychiatric Services* 59: 261–273.

Fu, X. 2006. Impact of socioeconomic status on inter-racial mate selection and divorce. *Social Science Journal* 43: 239–258.

Fulkerson, J. A., K E. Pasch, M. H. Stigler, K. Farbakhsh, C. L. Perry, and K. A. Komro 2010. Longitudinal associations between family dinner and adolescent perceptions of parent–child communication among racially diverse urban youth. *Journal of Family Psychology* 24: 261–270.

Fuller, J. A., and R. M. Warner. 2000. Family stressors as predictors of codependency. *Genetic, Social, and General Psychology Monographs* 126: 5–22.

Fusco, R. A., M. E. Rauktis, J. S. McCrae, M. A. Cunningham, and C. K. Bradley-King 2010. Aren't they just black kids? Biracial children in the child welfare system. *Child & Family Social Work* 15: 441–451.

Futris, T. G., A. W. Barton, T. M. Aholou, and D. M. Seponski. 2011. The impact of PREPARE on engaged couples: Variations by delivery format. *Journal of Couple & Relationship Therapy* 10: 69–86.

G

Gable, S. L., H. T. Reis, and G. Downey. 2003. He said, she said: A Quasi-Signal detection analysis of daily interactions between close relationship partners. *Psychological Science* 14: 100–105.

Gadalla, T. M. 2010. Relative body weight and disability in older adults: Results from a national survey. *Journal of Aging and Health* 22: 403–418.

Gager, C. T. and S. T. Yabiku. 2010. Who has the time? The relationship between household labor time and sexual frequency. *Journal of Family Issues* 31: 135–163.

Gaines, S. O., Jr., and J. Leaver. 2002. Interracial relationships. In *Inappropriate relationships: The unconventional, the disapproved, and the forbidden,* ed. R. Goodwin and D. Cramer, 65–78. Mahwah, NJ: Lawrence Erlbaum.

Galambos, N. L., E. T. Barker, and D. M. Almeida. 2003. Parents do matter: Trajectories of change in externalizing and internalizing problems in early adolescent. *Child Development* 74: 578–595.

Galinsky, E. 2010. *Mind in the making.* New York: Harper Collins.

Gallmeier, C. P., M. E. Zusman, D. Knox, and L. Gibson. 1997. Can we talk? Gender differences in disclosure patterns and expectations. *Free Inquiry in Creative Sociology* 25: 219–225.

Gameiro, S., M. C. Canavarro, J. Boivin, and I. Soares. 2010. Social nesting: Changes in social network and support across the transition to parenthood in couples that conceived spontaneously or through assisted reproductive technologies. *Journal of Family Psychology* 24: 175–187.

Ganong, L. H., and M. Coleman. 1999. *Changing families, changing responsibilities: Family obligations following divorce and remarriage.* New York: Lawrence Erlbaum Assoc. Inc.

Ganong, L. H., M. Coleman, and T. Jamison. 2011. Patterns of stepchild–stepparent relationship development. *Journal of Marriage and Family* 73: 396–413.

Gardner, J., and A. J. Oswald. 2006. Do divorcing couples become happier by breaking up? *Journal of the Royal Statistical Society: Series A (Statistics and Society)* 169: 319–336.

Gardner, R. A. 1998. *The parental alienation syndrome.* 2d ed. Cresskill, N.J.: Creative Therapeutics.

Garfield, R. 2010. Male emotional intimacy: How therapeutic men's groups can enhance couples therapy. *Family Process* 49: 109–122.

Garnets, L., G. M. Herek, and B. Levy. 1990. Violence and victimization of lesbians and gay men: Mental health consequences. *Journal of Interpersonal Violence* 5: 366–383.

Garrett, N., and E. M. Martini. 2007. The boomers are coming: A total cost of care model of the impact of population aging on the cost of chronic conditions in the United States. *Disease Management* 10: 51–60.

Garrett, T. M., H. W. Baillie, and R. M. Garrett. 2001. *Health care ethics,* 4th ed. Upper Saddle River, NJ: Prentice Hall.

Gates, Gary. 2009 (September). "Same-Sex Couples in the 2008 American Community Survey." Los Angeles, CA: The Williams Institute on Sexual Orientation Law and Public Policy.

Gatzeva, M., and A. Paik. 2011. Emotional and physical satisfaction in noncohabiting, cohabiting, and marital relationships: The importance of jealous conflict. *Journal of Sex Research* 48: 29–42.

Gavin, J. 2003. *Deep in a dream: The long night of Chet Baker.* New York: Welcome Rain.

Gavin, L. E., M. M. Black, S. Minor, Y. Abel, and M. E. Bentley. 2002. Young, disadvantaged fathers' involvement with their infants: An ecological perspective. *Journal of Adolescent Health* 31: 266–276.

Gay, Lesbian, and Straight Education Network. 2000. Homophobia 101: Teaching respect for all. Gay, Lesbian, and Straight Education Network. http://www.glsen.org.

Ge, X., M. N. Natsuaki, D. Martin, J. M. Neiderhiser, D. S. Shaw, L. Scaramella, J. B. Reid, and D. Reiss. 2008. Bridging the divide: Openness in adoption and post adoption psychosocial adjustment among birth and adoptive parents issue: Public health perspectives on family interventions. American Psychological Association. Sage Periodicals Press.

Gelatt, V. A., F. Adler-Baeder, and J. R. Seeley. 2010. An interactive web-based program for stepfamilies: Development and evaluation of efficacy. *Family Relations* 59: 572–586.

Geller, P., C. Psaros, and S. L. Kornfield. 2010. Satisfaction with pregnancy loss aftercare: are women getting what they want? *Archives of Women's Mental Health* 13: 111–124.

Gesell, A., F. L. Ilg, and L. B. Ames. 1995. *Infant and child in the culture of today.* Northvale, NJ: Jason Aronson.

Gibbs, N. 2008. The pursuit of purity. *Time,* July 28, 46–49.

Gibson, V. 2002. *Cougar: A guide for older women dating younger men.* Boston, MA: Firefly Books.

Gidycz, C. A., A. V. Wynsberghe, and K. M. Edwards. 2008. Prediction of women's utilization of resistance strategies in a sexual assault situation: A prospective study. *Journal of Interpersonal Violence* 23: 571–588.

Gilla, D. L., R. G. Morrow, K. E. Collinsc, A. B. Lucey, and A. M. Schultze. 2010. Perceived climate in physical activity settings. *Journal of Homosexuality* 57: 895–913.

Gillath, O., M. Mikulincer, G. E. Birnbaum, and P. R. Shaver. 2008. When sex primes love: Subliminal sexual priming motivates relationship goal pursuit. *Personality and Social Psychology Bulletin* 34: 1057–1073.

Gilman, S. E., S. D. Cochran, V. M. Mays, M. Hughes, D. Ostrow, and R. C. Kessler. 2001. Risk of psychiatric disorders among individuals reporting same-sex sexual partners in the National Comorbidity Survey. *American Journal of Public Health* 91: 933–939.

Giordano, P. C., W. D. Manning, and M. A. Longmore. 2005. The romantic relationships of African-American and white adolescents. *The Sociological Quarterly* 46: 545–568.

Girgis, S., R. P. George, and R. Anderson. 2011. What is marriage? *Harvard Journal of Law & Public Policy* 34: 245–287.

Goetting, A. 1982. The six stations of remarriage: The developmental tasks of remarriage after divorce. *The Family Coordinator* 31: 213–222.

Goldberg, A. E., J. Z. Smith, and D. A. Kashy. 2010. Preadoptive factors predicting lesbian, gay, and heterosexual couples' relationship quality across the transition to adoptive parenthood. *Journal of Family Psychology* 24: 221–232.

Goldberg, A. E., L. A. Kinkler, and D. A. Hines. 2011a. Perception and internalization of adoption stigma among gay, lesbian, and heterosexual adoptive parents. *Journal of GLBT Family Studies* 7: 132–154.

Goldberg, A. E., L. A. Kinkler, H. B. Richardson, and J. B. Downing. 2011b. Lesbian, gay, and heterosexual couples in open adoption arrangements: A Qualitative Study. *Journal of Marriage and Family* 73: 502–518.

Golombok, S., B. Perry, A. Burston, C. Murray, J. Money-Somers, M. Stevens, and J. Golding. 2003. Children with lesbian parents: A community study. *Developmental Psychology* 39: 20–33.

Gonzaga, G. C., S. Carter, and J. Galen Buckwalter. 2010. Assortative mating, convergence, and satisfaction in married couples. *Personal Relationships* 17: 634–644.

Goodall, N. 2010. *The secret world of Johnny Depp*. London: John Blake Publishing.

Goode, E. 1999. New study finds middle age is prime of life. *New York Times*, July 17, D6.

Gordon, R. M. 2005. The doom and gloom of divorce research- Comment on Wallerstein and Lewis (2004) *Psychoanalytic Psychology* 22: 450–451.

Gordon, T. 2000. *Parent effectiveness training: The parents' program for raising responsible children*. New York: Random House.

Gordon, J. R., and K. S. Whelan-Berry. 2005. Contributions to family and household activities by the husbands of midlife professional women. *Journal of Family Studies* 26: 899–923.

Gordon, R. A., and R. S. Högnäs. 2006. The best laid plans: Expectations, prefer-

ences, and stability of child-care arrangements. *Journal of Marriage and Family* 68: 373–393.

Gormley, B. and F. G. Lopez. 2010. Psychological abuse perpetration in college dating Relationships: Contributions of gender, stress, and adult attachment orientations. *Journal of Interpersonal Violence* 25: 204–218. http://www.edge.org/3rd_culture/gottman05/gottman05_index.html (accessed August 23).

Gottman, J. 1994. *Why marriages succeed or fail*. New York: Simon & Schuster.

Gottman, J., and S. Carrere. 2000. Welcome to the love lab. *Psychology Today*, September/October, 42.

Gould, T. E., and A. Williams. 2010. Family homelessness: An investigation of structural effects. *Journal of Human Behavior in the Social Environment* 20: 170–192.

Graham, C. A., S. A. Sanders, R. R. Milhausen, and K. R. McBride. 2004. Turning on and turning off: A focus group study of the factors that affect women's sexual arousal. *Archives of Sexual Behavior* 33: 527–538.

Grandey, A. A., B. L. Cordeiro, and A. C. Crouter. 2005. A longitudinal and multisource test of the work-family conflict and job satisfaction. *Journal of Occupational and Organizational Psychology* 78: 305–323.

Granie, M. 2010. Gender stereotype conformity and age as determinants of preschoolers' injury-risk behaviors. *Accident Analysis & Prevention* 42: 726–733.

Green, J. C. 2004. The American religious landscape and political attitudes: a baseline for 2004. Pew Forum on Religion and Public Life. http://pewforum.org.

Green, M., and M. Elliott. 2010. Religion, health, and psychological well-being. *Journal of Religion & Health* 49: 149–163.

Green, R. J., J. Bettinger, and E. Sacks. 1996. Are lesbian couples fused and gay male couples disengaged? In *Lesbians and gays in couples and families*, ed. J. Laird and R. J. Green, 185–230. San Francisco: Jossey-Bass.

Greene, K. and S. Faulkner. 2005. Gender, belief in the sexual double standard, and sexual talk in heterosexual dating relationships. *Sex Roles* 53: 239–251.

Greenfield, E. A. and N. F. Marks. 2010. Identifying experiences of physical and psychological violence in childhood that jeopardize mental health in adulthood. *Child Abuse & Neglect* 34: 161–171.

Gregory, J. D. 2010. Pet custody: Distorting language and the law. *Family Law Quarterly* 44: 35–64.

Greitemeyer, T. 2010. Effects of reciprocity on attraction: The role of a partner's physical attractiveness. *Personal Relationships* 17: 317–330.

Grekin, E. R., K. J. Sher, and J. L Krull. 2007. College spring break and alcohol use: Effects of spring break activity. *Journal of Studies on Alcohol and Drugs* 68: 681–693.

Grief, G. L. 2006. Male friendships: Implications from research for family therapy. *Family Therapy* 33:1–15.

Grogan, S. 2010. Promoting positive body image in males and females:

Contemporary issues and future directions. *Sex Roles* 63: 757–765.

Gross, J., and W. Connors. 2007. Ethopia, Open doors for foreign adoptions. *The New York Times*, June 4, A1.

Gross, K. 2006. Teenage kissing may increase meningococcal risk. *Youth Studies Australia* 25: 5–6.

Grov, C., D. S. Bimbi, J. E. Nanin, and J. T. Parsons. 2006. Race, ethnicity, gender and generational factors associated with the coming-out process among gay, lesbian, and bisexual individuals. *The Journal of Sex Research* 43: 115–22.

Grusec, J. E., and M. Davidov 2010. Integrating different perspectives on socialization theory and research: A domain-specific perspective. *Child Development* 81: 687–709.

Guldner, G. T. 2003. *Long distance relationships: The complete guide*. Corona, CA: JFMilne Publications.

Gunnar, M. R., E. Kryzer, M. J. Van Ryzin, and D. A. Phillips. 2010. The rise in cortisol in family day care: Associations with aspects of care quality, child behavior, and child sex *Child Development* 81: 851–869.

Gutter, M. S., S. Garrison, and Z. Copur. 2010. Social learning opportunities and the financial behaviors of college students. *Family and Consumer Sciences Research Journal* 38: 387–404.

H

Ha, J. H. 2008. Changes in support from confidants, children, and friends following widowhood. *Journal of Marriage and Family* 70: 306–329.

Haandrikman, K. 2011. Spatial homogamy: The geographical dimensions of partner choice. *Journal of Economic and Social Geography* 102: 100–110

Haas, A. P., M. Eliason, V. M. Mays, R. M. Mathy, S. D. Cochran, A. R. D'Augelli, M. M. Silverman, P. W. Fisher, T. Hughes, M. Rosario, S. T. Russell, E. Malley, R. Reed, D. A Litts, E. Haller, R. L. et al. 2011. Suicide and suicide risk in lesbian, gay, bisexual, and transgender populations: Review and recommendations. *Journal of Homosexuality* 58: 10–51.

Hall, J. A., N. Park, M. J. Cody, and H. Song. 2010. Strategic misrepresentation in online dating: The effects of gender, self-monitoring, and personality traits. *Journal of Social and Personal Relationships* 27: 117–135.

Hall, J. H., W. Fals-Stewart, and F. D. Fincham. 2008. Risky sexual behavior among married alcoholic men. *Journal of Family Psychology* 22: 287–299.

Hall, S. J. 2010b. Gauging the gatekeepers: How do adoption workers assess the suitability of gay, lesbian, or bisexual prospective parents? *GLBT Family Studies* 6: 265–293.

Hall, S. 2010a. Implicit theories of the marital institution: Origins and implications. Poster, National Council on Family Relations annual meeting, November 3–5. Minneapolis, MN

Halpern, C. T., R. B. King, S. G. Oslak, and J. R. Udry. 2005. Body mass index, dieting, romance, and sexual activity in adolescent girls: Relationships over time. *Journal of Research on Adolescence* 15: 535–559.

Halpern-Meekin. 2011. High school relationship and marriage education: A comparison of mandated and self-selected treatment. Detail *Journal of Family Issues*. 32: 394–419.

Hamman, R. 2007. Cyberorgasms: Ten years on and not enough learned. In *Online Matchmaking* ed. M. T. Whitty, A. J. Baker, and J. A. Inman, 31–39. New Work: Paulgrave Macmillan.

Hammarberg, K., J. R. Fisher, and K. H. Wynter. 2008a. Psychological and social aspects of pregnancy, childbirth and early parenting after assisted conception: A systematic review *Human Reproduction* 14: 395–415.

Hammarberg, K., J. R. W. Fisher, and H. J. Rowe. 2008b. Women's experiences of childbirth and post-natal healthcare after assisted conception. *Human Reproduction* 23: 1567–1574.

Hannon, P. A., C. Rusbult, E. Finkel, and M. Kamashiro. 2010. In the wake of betrayal: Amends, forgiveness, and the resolution of betrayal. *Personal Relationships* 17: 253–278.

Hansen, T., T. Moum, and A. Shapiro. 2007. Relational and individual well-being among cohabitors and married individuals in midlife. *Journal of Family Issues* 28: 910–933.

Hansson, R. O., J. O. Berry, and M. E. Berry. 1999. The bereavement experience: Continuing commitment after the loss of a loved one. In *Handbook of interpersonal commitment and relationship stability*, ed. J. M. Adams and W. H. Jones, 281–91. New York: Academic/Plenum Publishers.

Hardie, J. H. and A. Lucas. 2010. Economic factors and relationship quality among young couples: Comparing cohabitation and marriage. *Journal of Marriage and the Family* 72: 1141–1154.

Haring, M., P. L. Hewitt, and G. L. Flett. 2003. Perfectionism, coping, and quality of relationships. *Journal of Marriage and the Family* 65: 143–159.

Harris Interactive and GLSEN. 2005. *From teasing to torment: School climate in America.* New York: GLSEN (Gay, Lesbian, and Straight Education Network).

Harris Poll. 2007. Pets as family members. www.harrispollonline.com (retrieved December 11, 2007).

Hart, K. 2007. Love by arrangement: The ambiguity of "spousal choice" in a Turkish village. *Journal of the Royal Anthropological Institute* 13: 345–363.

Harvey, C. D. H. 2005. Families in Canada. In *Handbook of world families*, ed Bert N. Adams and Jan Trost, 539–559. Thousand Oaks, CA: Sage Publications.

Hawkins, A. J. and T. A. Fackrell. 2009. Review of *Covenant Marriage. Journal of Marriage and the Family* 71: 804–806.

Haswell, W. M. 2006. Mediation: An alternative to litigation. Presentation for the Department of Sociology, East Carolina University, April 5. www.HaswellMediation.com.

Hawes, Z. C., K. Wellings, and J. Stephenson. 2010. First heterosexual intercourse in the United Kingdom: A review of the literature. *Journal of Sex Research* 47: 137–152.

Hawkins, A. J., S. L. Nock, J. C. Wilson, L. Sanchez, and J. D. Wright. 2002. Attitudes about covenant marriage and divorce: Policy implications from a three state comparison. *Family Relations* 51: 166–175.

Hawkins, D. N., and A. Booth. 2005. Unhappily ever after: Effects of long-term, low-quality marriages on well-being. *Social Forces* 84: 445–465.

Haynie, D. L., and D. W. Osgood. 2005. Reconsidering peers and delinquency: How do peers matters? *Social Forces* 84: 1109–1130.

Haynie, D. L., P. C. Giordano, W. D. Manning, and M. A. Longmore. 2005. Adolescent romantic relationships and delinquency involvement. *Criminology* 43: 177–210.

Hays, J., J. K. Ockene, R. L. Brunner, J. M. Kotchen, J. E. Manson, R. E. Patterson, A. K. Aragki, M. S., S. A. Shumaker, R. G. Bryzyski, et al. 2003. Effects of estrogen plus progestin on health-related quality of life. *The New England Journal of Medicine* 348: 1839–1854.

Healy, M., and V. Salazar. 2010. Dealbreakers. *USA Today*. April 27. D 1.

Heatherington, L., and J. A. Lavner. 2008. Coming to terms with coming out: Review and recommendations for family systems-focused research. *Journal of Family Psychology* 22: 329–343.

Hegi, K. E., and R. M. Bergner 2010. What is love? An empirically-based essentialist account. *Journal of Social and Personal Relationships* 27: 620–636.

Heino, R. D., N. B. Ellison, and J. L. Gibbs. 2010. Relationshopping: Investigating the market metaphor in online dating. *Journal of Social and Personal Relationships* 27: 427–447.

Held, M. 2005. Mix it up: T-shirts and activism. (March 16). http://www.tolerance.org/teens.

Heller, N. 2008. Will the transgender dad be a father? What goes on the birth certificate? http://www.slate.com/id/2193475/ (accessed June 13, 2008).

Helms, H. M., J. K. Walls, A. C. Crouter, and S. M. McHale. 2010. Provider role attitudes, marital satisfaction, role overload, and housework: A dyadic approach. *Journal of Family Psychology* 24: 568–577.

Henning, K. and J. Connor-Smith. 2010. Why doesn't he leave? Relationship continuity and satisfaction among male domestic violence offenders. *Journal of Interpersonal Violence*. Published online June 28, 2010 (address will not be valid when textbook is printed so not provided here. The article will be available in the Journal).

Henry, R. G., R. B. Miller, and R. Giarrusso. 2005. Difficulties, disagreements, and disappointments in late-life marriages. *International Journal of Aging & Human Development* 61: 243–265.

Herbenick, D., M. Reece, S. A. Sanders, B. Dodge, A. Ghassemi, and D. Fortenberry. 2010. Women's vibrator use in sexual partnerships: Results from a nationally representative survey in the United States. *Journal of Sex & Marital Therapy* 36: 49–65.

Hertenstein, M. J., M. J. Hertenstein, J. M. Verkamp, A. M. Kerestes, and R. M. Holmes. 2007. The communicative functions of touch in humans, nonhuman primates, and rats: A review and synthesis of the empirical research. *Genetic Social and General Psychology Monographs* 132: 5–94.

Hess, J. 2009. Personal communication. Appreciation is expressed to Judye Hess for the development of this section. For more information about Judye Hess, see http://www.psychotherapist.com/judyehess/.

Hetherington, E. M. 2003. Intimate pathways: Changing patterns in close personal relationships across time. *Family Relations* 52: 318–331.

Higginbotham, B. H., L. Skogrand, and E. Torres. 2010. Stepfamily education: Perceived benefits for children. *Journal of Divorce & Remarriage* 51: 36–49.

Higginbotham, B. J. and L. Skogrand, L. 2010. Relationship education with both married and unmarried stepcouples: An exploratory study. *Journal of Couple & Relationship Therapy* 9: 133–148.

Higgins, C. A., L. E. Duxbury, and S. T. Lyons. 2010a. Coping with overload and stress: Men and women in dual-earner families. *Journal of Marriage and Family* 72: 847–859.

Higgins, J. A., J. Trussell, N. B. Moore, and J. K. Davison. 2010b. Virginity lost, satisfaction gained? Physiological and psychological sexual satisfaction at heterosexual debut. *Journal of Sex Research* 47: 384–394.

Hilgeman, M. M., R. S. Allen, J. DeCoster, and L. D. Burgio. 2007. Positive aspects of caregiving as a moderator of treatment outcome over 12 months. *Psychology and Aging* 22: 361–371.

Hill, D. B. 2007. Differences and similarities in men's and women's sexual self-schemas. *Journal of Sex Research* 44: 135–144.

Hill, E. W. 2010. Discovering forgiveness through empathy: implications for couple and family Therapy. *Journal of Family Therapy* 32: 169–185.

Hill, M. R., and V. Thomas. 2000. Strategies for racial identity development: Narratives of black and white women in interracial partner relationships. *Family Relations* 49: 193–200.

Hira, N. A. 2007. The baby boomers' kids are marching into the workplace and look out. This crop of twentysomethings really is different. *Fortune Magazine*, May.

Hirano, Y., N. Yamamoto-Mitani, M. Ueno, S. Takemori, M. Kashiwagi, I. Sato, N. Miyata, M. Kimata, H. Fukahori, and M. Yamada. 2011. Home care nurses' provision of support to families of the elderly at the end of life. *Qualitative Health Research* 21: 199–213.

Hochschild, A. R. 1989. *The second shift*. New York: Viking.

———. 1997. *The time bind*. New York: Metropolitan Books.

Hoff, C. C., and S. C. Beougher. 2010. Sexual agreements among gay male couples. *Journal Archives of Sexual Behavior* 39: 774–787.

Hofferth, S., and F. Goldscheider. 2010. Family structure and the transition to early parenthood. *Demography* 47: 415–437.

Hogeboom, D. L., R. J. McDermott, K. Perrin, H. Osman, and B. Bell-Ellison. 2010. Internet use and social networking among middle aged and older adults. *Educational Gerontology* 36: 93–111.

Högns, R. S., and Carlson, M. J. 2010, Intergenerational relationships and union stability in fragile families. *Journal of Marriage and Family* 72: 1220–1233.

Hogue, M., C. L. Z. Dubois, and L. Fox-Cardamone. 2010. Gender differences in pay expectations: The roles of job intention and self-view. *Psychology of Women Quarterly* 34: 215–227.

Hohmann-Marriott, B. E., and P. Amato. 2008. Relationship quality in interethnic marriages and cohabitation. *Social Forces* 87: 825–855.

Hollander, D. 2006. Many teenagers who say they have taken a Virginity Pledge retract that statement after having intercourse. *Perspectives on Sexual and Reproductive Health* 38: 168–173.

Hollander, D. 2010. Body image predicts some risky sexual behaviors among teenage women. *Perspectives on Sexual & Reproductive Health* 42: 67–67.

Hollingsworth, L. D. 1997. Same race adoption among African Americans: A ten-year empirical review. *African American Research Perspectives* 13: 44–49.

Holt-Lunstad J., T. B. Smith, and J. B. Layton. 2010. Social relationships and mortality risk: A meta-analytic review. *PLoS Medicine* 7: 316–333.

Holthouse, D. 2005. Curious cures. *Intelligence Report* 117 (Spring):14.

Holtzman, M. 2002. The "family relations" doctrine: Extending Supreme Court precedent to custody disputes between biological and nonbiological parents. *Family Relations* 51: 335–343.

Hostetler, A. J. 2009. Single by choice? Assessing and understanding voluntary singlehood among mature gay men. *Journal of Homosexuality* 56: 499–531.

Hostetler, A. J. 2011. Senior centers in the era of the "Third Age:" Country clubs, community centers, or something else? *Journal of Aging Studies* 25: 166–176.

Howard, J. 2006. *Expanding resources for children: Is adoption by gays and lesbians part of the answer for boys and girls who need homes?* New York: Evan B. Donaldson Adoption Institute.

Howard, M. 2003. *A life in letters: Ann Landers's letters to her only child*. New York: Warner.

Hsueh, A. C., K. R. Morrison, and B. D. Doss. 2009. Qualitative reports of problems in cohabiting relationships: Comparisons to married and dating relationships. *Journal of Family Psychology* 23: 236–246.

Huang, H., and L. Leung. 2010. Instant messaging addiction among teenagers in China: Shyness,alienation and academic performance decrement. *CyberPsychology & Behavior* 12: 675–679.

Hughes, M. 2011. Hey, we love animals! *Industrial Engineer* 43: 6–6.

Hughes, M., K. Morrison, and K. J. Asada. 2005. What's love got to do with it? Exploring the impact of maintenance rules, love attitudes, and network support on friends with benefits relationships. *Western Journal of Communication* 69: 49–66.

Huh, N. S., and W. J. Reid. 2000. Intercountry, transracial adoption and ethnic identity: a Korean example. *International Social Work* 43: 75–87.

Human Rights Campaign. 2000. *Feeling free: Personal stories How love and self-acceptance saved us from "ex-gay" ministries*. Washington, DC: Human Rights Campaign Foundation.

Human Rights Campaign. 2004. *Resource guide to coming out*. Washington, DC: Human Rights Campaign Foundation.

Human Rights Campaign. 2005. *The state of the workplace for lesbian, gay, bisexual, and transgendered Americans, 2004*. Washington, DC: Human Rights Campaign. http://www.hrc.org.

Hunt, A. 2010. Enrollment is up at Grandparent's University. *USA Today*, June 30. 2D.

Hunter, D. J., M. J. Khoury, and J. M. Drazen. 2008. Letting the genome out of the bottle—Will we get our wish? *The New England Journal of Medicine* 358: 105–108.

Huston, T. L., J. P. Caughlin, R. M. Houts, S. E. Smith, and L. J. George. 2001. The connubial crucible: Newlywed years as predictors of marital delight, distress, and divorce. *Journal of Personality and Social Psychology* 80: 237–252.

Huyck, M. H., and D. L. Gutmann. 1992. Thirty something years of marriage: Understanding experiences of women and men in enduring family relationships. *Family Perspective* 26: 249–265.

Hymowitz, K. S. 2011. *Manning up: How the rise of women has turned men into boys*. New York: Basic Books.

I

Ikegami, N. 1998. Growing old in Japan. *Age and Ageing* 27: 277–278.

Impett, E. A., J. B. Breines, and A. Strachman. 2010. Keeping it real: Young adult women's authenticity in relationships and daily condom use. *Personal Relationships* 17: 573–584.

Impett, E. A., L. A. Peplau, and S. L. Gable. 2005. Approach and avoidance sexual motives: Implications for personal and interpersonal well-being. *Personal Relationships* 12: 465–482.

Ingoldsby, B., P. Schvaneveldt, and C. Uribe. 2003. Perceptions of acceptable mate attributes in Ecuador. *Journal of Comparative Family Studies* 34: 171–186.

Insel, T. R. 2008. Assessing the economic costs of serious mental illness. *The American Journal of Psychiatry* 165: 663–666.

Israel, T., and J. J. Mohr. 2004. Attitudes toward bisexual women and men: Current research, future directions. In *Current research on bisexuality*, ed. R. C. Fox, 117–134. New York: Harrington Park Press.

J

Jaegar, M. M. 2011. "A thing of beauty is a joy forever"? Returns to physical attractiveness over the life course. *Social Forces* 89: 983–1004.

Jalovaara, M. 2003. The joint effects of marriage partners' socioeconomic positions on the risk of divorce. *Demography* 40: 67–81.

James, S. D. 2008. Wild child speechless after tortured life. *ABC News,* May 7.

Jaudesa, P. K., and L. Mackey-Bilaverb. 2008. Do chronic conditions increase young children's risk of being maltreated? *Child Abuse and Neglect* 32: 671–681.

Jayson, S. 2011. Is dating dead? *USA Today* March 30, A1.

Jefson, C. 2006. Candy hearts: Messages about love, lust, and infatuation. *Journal of School Health* 76: 117–122.

Jenkins, K. E. 2010. In concert and alone: Divorce and congregational experience. *Journal for the Scientific Study of Religion* 49: 278–292.

Jenkins, M., E. G. Lambert, and D. N. Baker. 2009. The attitudes of Black and White college students toward gays and lesbians. *Journal of Black Studies* 39: 589–601.

Jerin, R. A., and B. Dolinsky. 2007. Cyber victimization and online dating. In *Online Matchmaking*, ed. M. T. Whitty, A. J. Baker, and J. A. Inman, 147–156. New Work: Palgrave Macmillan.

Johnson, C. L., and B. M. Barer. 1997. *Life beyond 85 years: The aura of survivorship*. New York: Springer Publishing.

Johnson, D. W., C. Zlotnick, and S. Perez. 2008. The relative contribution of abuse severity and PTSD severity on the psychiatric and social morbidity of battered women in shelters. *Behavior Therapy* 39: 232–247.

Johnson, D. C. and L. B. Johnson. 2010. Reinventing the stress concept. *Ethical Human Psychology & Psychiatry* 12: 218–231.

Johnson, M. D., J. R. Anderson, and C. J. Aducci. 2010. Reasons for marriage and the association with life satisfaction. Poster, National Council on Family Relations annual meeting, November 3–5. Minneapolis, MN.

Johnson, M. D., J. R. Anderson, and C. Aducci. 2011. Understanding the decision to marry versus cohabit: The role of interpersonal dedication and constraints and the impact on life satisfaction. *Marriage & Family Review* 47: 73–89.

Jonason, P. K., and P. Kavanagh. 2010. The dark side of love: Love styles and the Dark Triad. *Personality & Individual Differences* 49: 606–610.

Johnson, R. W., and J. M. Wiener. 2006. A profile of frail older Americans and their caregivers. Urban Institute Report.

http://www.urban.org/url.
cfm?ID=311284 (posted March 1).

Jones, D. 2006a. One of USA's exports: Love, American style. *USA Today*, February 14, 1B.

Jones, J. 2006b. Marriage is for white people. *The Washington Post*, March 26, B3–B4.

Jones, K. 2010. Personal communication.

Jones, R. K., and K. Kooistra. 2011. Abortion incidence and access to services in the United States, 2008. *Perspectives on Sexual & Reproductive Health* 43: 41–50.

Jones, R. K., A. M. Moore, and L. F. Frohwirth. 2011. Perceptions of male knowledge and support among U.S. women obtaining abortions. *Women's Health Issues* 21: 117–123.

Jordan, J. 2010. "I miss Patrick so much" *People*, September 20, 78–85.

Jorm, A. F., H. Christensen, A. S. Henderson, P. A. Jacomb, A. E. Korten, and A. Mackinnon. 1998. Factors associated with successful ageing. *Australian Journal of Ageing* 17:33–37.

Juffer, F., M. van Ijzendoorn, and J. Palacios. 2011. Children's recovery after adoption. *Infancia y Aprendizaje* 34: 3–18.

K

Kahl, S. F., L. C. Steelman, L. M. Mulkey and P. R. Koch, et al. 2007. Revisiting Reuben Hill's theory of familial response to stressors: The mediating role of mental outlook for offspring of divorce. *Family and Consumer Sciences Research Journal* 36: 5–25.

Kahneman, D. and A. Deaton. 2010. High income improves evaluation of life but not emotional well-being. Proceedings of the National Academy of Sciences, Early Edition, September 6, 2010.

Kaiser Family Foundation. 2010. Media use among teens. Retrieved Feb 7, 2010. http://www.kff.org/entmedia/entmedia012010nr.cfm

Kalish, R., and M. Kimmel. 2011. Hooking up. *Australian Feminist Studies* 26: 137–151.

Kalmijn, M. 2007. Gender differences in the effects of divorce, widowhood and remarriage on intergenerational support: Does marriage protect fathers? *Social Forces* 85: 1079–1085.

Kalmijn, M., and H. Flap. 2001. Assortative meeting and mating: Unintended consequences of organized settings for partner choices. *Social Forces* 79: 289–312.

Kalush, W., and L. Sloman. 2006. *The secret life of Houdini*. New York: Altria Books.

Kaplan, M. S., and R. B. Krueger. 2010. Diagnosis, assessment, and treatment of hypersexuality. *Journal of Sex Research* 47: 181–198.

Karch, D., and K. C. Nunn. 2011. Characteristics of elderly and other vulnerable adult victims of homicide by a caregiver: National violent death reporting system—17 U.S. States, 2003–2007. *Journal of Interpersonal Violence* 26: 137–157.

Karten, E. Y., and J. C. Wade. 2010. Sexual orientation change efforts in men: A client perspective. *Journal of Men's Studies* 18: 84–102.

Kase_ru, K. 2010. Intending to marry.... students' behavioral intention towards family forming. *TRAMES: A Journal of the Humanities & Social Sciences* 14: 3–20.

Katz, J., and L. Myhr. 2008. Perceived conflict patterns and relationship quality associated with verbal sexual coercion by male dating partners. *Journal of Interpersonal Violence* 23: 798–804.

Katz, J., J. Moore, and P. May. 2008. Physical and sexual co victimization from dating partners: A distinct type of intimate abuse? *Violence Against Women* 14: 961–973.

Katz, P., J. Showstack, J. F. Smith, R. D. Nachtigall, S. G. Millstein, H. Wing, M. L. Eisenberg, L. Pasch, M. S. Croughan, and N Adler. 2011. Costs of infertility treatment: results from an 18-month prospective cohort study. *Fertility & Sterility* 95: 915–921.

Kaye, K., K. A. Moore, E. C. Hair, A. M. Hadley, R. D. Day, and D. K. Orthner. 2009. Parent marital quality and the parent-adolescent relationship: Effects on sexual activity among adolescents and youth. *Marriage & Family Review* 45: 270–288.

Keller, E. G. 2008b. *The comeback: Seven stories of women who went from career to family and back again.* New York: Bloomsbury.

Keller, G. 2008a. French businesses loath to end 35-hour work week. *Associated Press.* http://www.wtop.com/?nid=105&sid=1471646.

Kelley-Moore, J. A., J. G. Schumacher, E. Kahana, and B. Kahana. 2006. When do older adults become 'disabled'? Social and health antecedents of perceived disability in a panel study of the oldest old. *Journal of Health and Social Behavior* 47: 126–142.

Kelly, J. B., and R. E. Emery. 2003. Children's adjustment following divorce: Risk and resilience perspectives. *Family Relations* 52: 352–362.

Kem, J. 2010. Fatal lovesickness in Marguerite de Navarre's Heptaméron. *Sixteenth Century Journal* 41: 355–370.

Kennedy, D. P., J. S. Tucker, M. S. Pollard, M. Go, and H. D. Green. 2011. Adolescent romantic relationships and change in smoking status. *Addictive Behaviors* 36: 320–326.

Kennedy, G. E. 1997. Grandchildren's memories: A window into relationship meaning. Paper presented at the Annual Conference of the National Council on Family Relations, Crystal City, Virginia.

Kennedy, M. 2007. Gender role observations of East Africa. Written exclusively for this text.

Kerley, K. R., X Xu, B. Sirisunyaluck, and J. M. Alley. 2010. Exposure to family violence in childhood and intimate partner perpetration or victimization in adulthood: Exploring intergenerational transmission in urban Thailand. *Journal of Family Violence* 25: 337–347.

Kennedy, R. 2003. *Interracial intimacies.* New York: Pantheon.

Kero, A., and A. Lalos. 2004. Reactions and reflections in men, 4 and 12 months post-abortion. *Journal of Psychosomatic Obstetrics and Gynecology* 25: 135–143.

Kesselring, R. G., and D. Bremmer. 2006. Female income and the divorce decision: Evidence from micro data. *Applied Economics* 38: 1605–1617.

Khanchandani, L. 2005. Jealousy during dating among college women. Paper presented at Third Annual East Carolina University Undergraduate Research and Creative Activities Symposium, Greenville, NC, April 8.

Kiecolt, K. J. 2003. Satisfaction with work and family life: No evidence of a cultural reversal. *Journal of Marriage and the Family* 65: 23–35.

Kiernan, K. 2000. European perspectives on union formation. In *The ties that bind*, ed. L. J. Waite, 40–58. New York: Aldine de Gruyter.

Kim, H. 2011. Exploratory study on the factors affecting marital satisfaction among remarried Korean couples. *Families in Society* 91: 193–200.

Kim, J. L., C. L. Sorsoli, K. Collins, B. A. Zylbergold, D. Schooler, and D. L. Tolman. 2007. From sex to sexuality: Exposing the heterosexual script on primetime network television. *Journal of Sex Research* 44: 145–157.

Kim-Cohen, J., T. E. Moffitt, A. Taylor, S. J. Pawlby, and A. Caspi. 2005. Maternal depression and children's antisocial behavior: Nature and nurture effects. *Archives of General Psychiatry* 62: 173–182.

Kimmel, M. 2008. *Guyland: The perilous world where boys become men.* New York: Harper Collins.

King, A. 2010. Clarifying the term "queer". e-mail June 1.

King, P. A. 2003. Stalking: A control factor. Paper presented at 73rd Annual Meeting of the Eastern Sociological Society, Philadelphia, February 28.

Kingsberry, S. Q., M. A. Saunders, and A. Richardson. 2010. The effect of psychosocial stressors on the mental health status of African American caregivers of the elderly. *Families in Society* 91: 408–414.

Kinsey, A. C., W. B. Pomeroy, and C. E. Martin. 1948. *Sexual behavior in the human male.* Philadelphia: Saunders.

Kinsey, A. C., W. B. Pomeroy, C. E. Martin, and P. H. Gebhard. 1953. *Sexual behavior in the human female.* Philadelphia: Saunders.

Kirkpatrick, R. C. 2000. The evolution of human sexual behavior. *Current Anthropology* 41: 385–414.

Kirn, W. 2007. Vacation, all I never wanted. *The New York Times Magazine*, August 5, 11–12.

Kline, S. L., and S. Zhang. 2009. The role of relational communication characteristics and filial piety in mate preferences: Cross-cultural comparisons of Chinese and US college students. *Journal of Comparative Family Studies* 40: 325–353.

Knox, D., C. Schacht, J. Turner, and P. Norris. 1995. College students' preference for win-win relationships. *College Student Journal* 29: 44–46.

Knox, D., and K. Leggett. 2000. *The divorced dad's survival book: How to stay connected with your kids.* Reading, MA: Perseus Books.

Knox, D., M. E. Zusman, and A. Decuzzi. 2004. Effects of divorce on romantic relationships of college students. *College Student Journal* 38: 597–601.

Knox, D., M. E. Zusman, L. Mabon, and L. Shivar. 1999. Jealousy in college student relationships. *College Student Journal* 33: 328–329.

Knox, D., and M. Zusman. 2007. Traditional wife? Characteristics of college men who want one. *Journal of Indiana Academy of Social Sciences* 11: 27–32.

Knox, D., and M. Zusman. 2009. Become involved with someone on the rebound? How fast should you run? *College Student Journal* 43: 99–104.

Knox, D., M. Zusman, and W. Nieves. 1998. What I did for love: Risky behavior of college students in love. *College Student Journal* 32: 203–205.

Knox, D., M. E. Zusman, and H. R. Thompson. 2004. Emotional perceptions of self and others: Stereotypes and data. *College Student Journal* 38: 130–142.

Knox, D., M. E. Zusman, and W. Nieves. 1997. College students' homogamous preferences for a date and mate. *College Student Journal* 31: 445–448.

Knox, D., M. E. Zusman, K. McGinty, and B. Davis. 2002. College student attitudes and behaviors toward ending an unsatisfactory relationship. *College Student Journal* 36: 630–634.

Knox, D., M. E. Zusman, M. Kaluzny, and C. Cooper. 2000 College student recovery from a broken heart. *College Student Journal* 34: 322–324.

Knox, D., and S. Hall. 2010. Relationship and sexual behaviors of a sample of 2,922 university students. Unpublished data collected for this text. Department of Sociology, East Carolina University, and Department of Family and Consumer Sciences, Ball State University.

Knox, D., S. Hatfield, and M. E. Zusman. 1998. College student discussion of relationship problems. *College Student Journal* 32: 19–21.

Knox, D., and U. Corte. 2007. "Work it out/ See a counselor": Advice from spouses in the separation process. *Journal of Divorce and Remarriage* 48: 79–90.

Kohlberg, L. 1966. A cognitive-developmental analysis of children's sex-role concepts and attitudes. In *The development of sex differences*, ed. E. E. Macoby. Stanford, CA: Stanford University Press.

———. 1969. State and sequence: The cognitive developmental approach to socialization. In *Handbook of socialization theory and research*, ed. D. A. Goslin, 347–480. Chicago: Rand McNally.

Kolko, D. J., L. D. Dorn, O. Bukstein, and J. D. Burke. 2008. Clinically referred ODD children with or without CD and healthy controls: Comparisons across contextual domains. *Journal of Child and Family Studies* 17: 714–734.

Kornreich, J. L., K. D. Hern, G. Rodriguez, and L. F. O'Sullivan. 2003. Sibling influence, gender roles, and the sexual socialization of urban early adolescent girls. *Journal of Sex Research* 40: 101–110.

Koropeckyj-Cox, T., and G. Pendell. 2007a. Attitudes about childlessness in the United States: Correlates of positive, neutral, and negative responses. *Journal of Family Issues* 28: 1054–1082.

Koropeckyi-Cox, T., and G. Pendell. 2007b. The gender gap in attitudes about childlessness in the United States. *Journal of Marriage and Family* 69: 899–915.

Kottke, J. 2008. The Eliot Spitzer affair and the business of sex. http://kottke.org/08/03/the-eliot-spitzer-affair-and-the-business-of-sex (retrieved December 6, 2008).

Kouros, C. D. and E. M. Cummings. 2010. Longitudinal associations between husbands' and wives' depressive symptoms. *Journal of Marriage and the Family* 72: 135–147.

Kozloskia, M. J. 2010. Homosexual moral acceptance and social tolerance: Are the effects of education changing? *Journal of Homosexuality* 57: 1370–1383.

Krebs, C. P., C. H. Lindquist, T. Warner, B. Fisher, and S. Martin. 2009. The differential risk factors of physically forced and alcohol- or other drug-enabled sexual assault among university women. *Violence & Victims* 24: 302–321.

Kress, V. E., J. J. Protivnak, and L. Sadlak. 2008. Counseling clients involved with violent intimate partners: The mental health counselor's role in promoting client safety. *Journal of Mental Health Counseling* 30: 200–211.

Krumrei, E., C. Coit, S. Martin, W. Fogo, and A. Mahoney. 2007. Post-divorce adjustment and social relationships: A meta-analytic review. *Journal of Divorce & Remarriage* 46: 145–156.

Kulik, L. 2007. Contemporary midlife grandparenthood. In *Women over fifty: Psychological perspectives*, ed. Varda Muhlbauer and Joan C. Chrisler, 131–146. New York: Springer Science and Business Media.

Kulik, L., and E. Heine-Cohen. 2011. Coping resources, perceived stress and adjustment to divorce among Israeli women: Assessing effects. *Journal of Social Psychology* 151: 5–30.

Kulkarni, M., S. Graham-Bermann, S. Rauch, and J. Seng. 2011. Witnessing versus experiencing direct violence in childhood as correlates of adulthood PTSD. *Journal of Interpersonal Violence* 26: 1264–1281.

Kurdek, L. A. 1994a. Areas of conflict for gay, lesbian, and heterosexual couples: What couples argue about influences relationship satisfaction. *Journal of Marriage and the Family* 56: 923–934.

———. 1994b. Conflict resolution styles in gay, lesbian, heterosexual nonparent, and heterosexual parent couples. *Journal of Marriage and the Family* 56: 705–722.

———. 1995. Predicting change in marital satisfaction from husbands' and wives' conflict resolution styles. *Journal of Marriage and the Family* 57: 153–164.

———. 2004. Gay men and lesbians: The family context. In *Handbook of contemporary families: Considering the past, contemplating the future*, ed. M. Coleman and L. H. Ganong, 96–115. Thousand Oaks, CA: Sage Publications.

———. 2005. What do we know about gay and lesbian couples? *Current Directions in Psychological Science* 14: 251–254.

———. 2008. Change in relationship quality for partners from lesbian, gay male, and heterosexual couples. *Journal of Family Psychology* 22: 701–711.

L

Lacayo, R. 2010. Appreciation: J. D. Salinger. *Time* (February 15): 66.

Lacey, K. K., D. G. Saunders, and Z. Lingling. 2011. A comparison of women of color and non-Hispanic White women on factors related to leaving a violent relationship. *Journal of Interpersonal Violence* 26: 1036–1055.

Lachance-Grzela, L., and G. Bouchard 2010. Why do women do the Lion's Share of housework? A decade of research. *Sex Roles* 63: 767–780.

Lai, Y., and M. Hynie. 2011. A tale of two standards: An examination of young adults' endorsement of gendered and ageist sexual double standards. *Sex Roles* 64: 360–371.

Lalicha, J., and K. McLarena. 2010. Inside and outcast: Multifaceted stigma and redemption in the lives of gay and lesbian Jehovah's Witnesses. *Journal of Homosexuality* 57: 1303–1333.

Lambda Legal and Deloitte Financial Advisory Services LLP. 2006. 2005 Workplace Fairness Survey (April). http:www.lambdalegal.org.

Lambert, A. N. 2007. Perceptions of divorce-Advantages and disadvantages: A comparison of adult children experiencing one parental divorce versus multiple parental divorces. *Journal of Divorce & Remarriage* 48: 55–77.

Landale, N. S., R. Schoen, and K. Daniels. 2010. Early family formation among White, Black, and Mexican American women. *Journal of Family Issues* 31: 445–474.

Landale, N. S., and R. S. Oropesa. 2007. Hispanic families: Stability and change. *Annual Review of Sociology* 33: 381–405.

Linville, D., K. Chronister, T. Dishion, J. Todah, J. Miller, D. Shaw, F. Gardner and M. Wilson. 2010. A longitudinal analysis of parenting practices, couple satisfaction, and child behavior problems. *Journal of Marital & Family Therapy* 36: 244–255.

Landis, D. 1999. Mississippi Supreme Court made a tragic mistake in denying custody to gay father, experts say. American Civil Liberties Union News (February 17). http://www.aclu.org.

Landor, A., and L. G. Simons. 2010. The impact of virginity pledges on sexual attitudes and behaviors among college students. Paper, Annual Meeting of the National Council on Family Relations, November 3–6. Minneapolis, MN

LaPierre, T. A. 2009. Marital status and depressive symptoms over time: Age and gender variations. *Family Relations* 58: 404–416.

LaPierre, T. A. 2010. The legal context of grandfamilies: Meaning and role conflict. Poster, National Council on Family Relations annual meeting, November 3–5. Minneapolis, MN.

Lareau, A., and E. B. Weininger. 2008. Time, work, and family life: Reconceptualizing gendered time patterns through the case of children's organized activities. *Sociological Forum* 23: 419–454.

Largier, S. 2010. Seven secrets to a happy retirement. *U. S. News and World Report,* July 22.

Larson, J. H., R. W. Blick, J. B. Jackson, and T. B. Holman. 2010. Partner traits that predict relationship satisfaction for neurotic individuals in premarital relationships. *Journal of Sex & Marital Therapy* 36: 430–444.

Laumann, E. O., A. Paik, D. B. Glasser, J.-H. Kang, T. Wang, B. Levinson, E. D. Moreira, Jr., A. Nicolosi, and C. Gingell. 2006. A cross-national study of subjective sexual well-being among older women and men: Findings from the global study of sexual attitudes and behaviors. *Archives of Sexual Behavior* (April).

Lavee, Y., and A. Ben-Ari 2007. Relationship of dyadic closeness with work-related stress: A daily diary study. *Journal of Marriage and Family* 69: 1021–1035.

Lavner, J. A., and T. N. Bradbury. 2010. Patterns of change in marital satisfaction over the newlywed years. *Journal of Marriage and Family* 72: 1171–1187.

Lawrence, A. 2010. Societal individualism predicts prevalence of nonhomosexual orientation in male-to-female transsexualism. *Archives of Sexual Behavior* 39: 573–583.

Lawyer, S., H. Resnick, V. Bakanic, T. Burkett and D. Kilpatrick. 2010. Forcible, drug-facilitated, and incapacitated rape and sexual assault among undergraduate women. *Journal of American College Health* 58: 453–460.

Lease, S. H., A. B. Hampton, K. M. Fleming, L. R. Baggett, S. H. Montes, R. J. Sawyer. 2010. Masculinity and interpersonal competencies: Contrasting White and African American men *Psychology of Men & Masculinity* 11: 195–207.

LeCouteur, A., and M. Oxlad. 2011. Managing accountability for domestic violence: Identities, membership categories and morality in perpetrators' talk. *Feminism & Psychology.* 21: 5–28.

Lee, D. 2006 Device brings hope for fertility clinics. http://www.indystar.com/apps/pbcs.dll/article?AID=/20060221/BUSINESS/602210365/1003 (retrieved February 22, 2006).

Lee, J. A. 1973. *The colors of love: An exploration of the ways of loving.* Don Mills, Ontario: New Press.

———. 1988. Love-styles. In *The psychology of love,* ed. R. Sternberg and M. Barnes, 38–67. New Haven, CN: Yale University Press.

Lee, M. 2008. Caregiver stress and elder abuse among Korean family caregivers of older adults with disabilities. *Journal of Family Violence* 23: 707–713.

Lee, J. A., P. Foos, and C. Clow. 2010a. Caring for one's elders and family-to-work conflict. *Psychologist-Manager Journal* 13: 15–39.

Lee, J. T., C. L. Lin, G. H. Wan, and C. C. Liang. 2010b. Sexual positions and sexual satisfaction of pregnant women. *Journal of Sex & Marital Therapy* 36: 408–420.

Lee, R. E., and W. C. Nichols. 2010. The doctoral education of professional marriage and family therapists. *Journal of Marital and Family Therapy* 36: 259–269.

Lehman, A. D. 2010. Inappropriate injury: The case for barring consideration of a parent's homosexuality in custody actions. *Family Law Quarterly* 44: 115–133.

Lehmiller, J. J., and C. R. Agnew. 2007. Perceived marginalization and the prediction of romantic relationship stability. *Journal of Marriage and Family* 69: 1036–1049.

Leidy, M. S., T. J. Schofield, M. A. Miller, R. D. Parke, S. Coltrane, S. Braver, J. Cookston, W. Fabricius, D. Saenz, and M. Adams. 2011. Fathering and adolescent adjustment: Variations by family structure and ethnic background. *Fathering: A Journal of Theory, Research, & Practice about Men as Fathers* 9: 44–68.

Lemer, J. L., E. H. B. Salafia, and K. E. Benson. 2010. Women's body image, attitudes, and sexual frequency: A mediation analysis. Paper, Annual Meeting of the National Council on Family Relations, November 3–6. Minneapolis, MN.

Lengua, L. J., S. A. Wolchik, I. N. Sandler, and S. G. West. 2000. The additive and interactive effects of parenting and temperament in predicting problems of children of divorce. *Journal of Clinical Child Psychology* 29: 232–244.

Leno, J. 1996. *Leading with my chin.* New York: Harper Collins.

Lenton, A. P., and A. Bryan. 2005. An affair to remember: The role of sexual scripts in perceptions of sexual intent. *Personal Relationships* 12: 483–498.

Leopold, T., and M. Raab. 2011. Short-term reciprocity in late parent-child relationships. *Journal of Marriage & Family* 73: 105–119.

Letartea, M. J., S. Normandeaub and J. Allardb. 2010. Effectiveness of a parent training program "Incredible Years" in a child protection service. *Child Abuse & Neglect* 34: 253–261.

Leung, P., S. Erich, and H. Kanenberg. 2005. A comparison of family functioning in gay/lesbian, heterosexual and special needs adoptions. *Children and Youth Services Review* 27: 1031–1044.

Lever, J. 1994. The 1994 Advocate survey of sexuality and relationships: The men. *The Advocate* August 23, 16–24.

Levi, B. H., and S. G. Portwood. 2011. Reasonable suspicion of child abuse: Finding a common language. *Journal of Law, Medicine & Ethics* 39: 62–69.

Levine, S. B. 2010. What is sexual addiction? *Journal of Sex & Marital Therapy* 36: 261–275.

Levy, S. 2009. *Paul Newman: A life.* New York: Harmony Books.

Lewis, G. B. 2003. Black-white differences in attitudes toward homosexuality and gay rights. *Public Opinion Quarterly* 67: 59–78.

Lewis, K. 2008. Personal communication. Dr. Lewis is also the author of *Five Stages of Child Custody.* Glenside, PA: CCES Press.

Lewis, S. K., and V. K. Oppenheimer. 2000. Educational assortative mating across marriage markets: Non-Hispanic whites in the United States. *Demography* 37: 29–40.

Li, N. P., K. A. Valentine, and L. Patel. 2011. Mate preferences in the US and Singapore: A cross-cultural test of the mate preference priority model. *Personality & Individual Differences* 50: 291–294.

Liang, J., A. R. Quinones, J. M. Bennett, Y. Wen, X. Xiao, B. Shaw, and M. B. Ofstedal. 2010. Evolving self-rated health in middle and old age: How does it differ across Black, Hispanic, and White Americans? *Journal of Aging & Health* 22: 3–26.

Licata, N. 2002. Should premarital counseling be mandatory as a requisite to obtaining a marriage license? *Family Court Review* 40: 518–532.

Lichter, D. T., R. N. Turner, and S. Sassler. 2010. National estimates of the rise in serial cohabitation. *Social Science Research* 39: 754–765.

Light, A., and T. Ahn. 2010. Divorce as risky behavior. *Demography* 47: 895–921.

Lincoln, A. E. 2010. The shifting supply of men and women to occupations: Feminization in veterinary education. *Social Forces* 88: 1969–1998.

Lindau, S. T., L. P. Schumm, E. O. Laumann, W. Levinson, C. A. O'Muircheartaigh, and L. J. Waite. 2007. A study of sexuality and health among older adults in the United States. *The New England Journal of Medicine* 357: 762–774.

Lipman, E. L., K. Georgiades, and M. Boyle. 2011. Young adult outcomes of children born to teen mothers: Effects of being born during their teen or later years. *Journal of the American Academy of Child & Adolescent Psychiatry* 50: 232–241.

Lipscomb, R. 2009. Person-first practice: Treating patients with disabilities. *Journal of the American Dietetic Association* 109: 21–25.

Littleton, H., A. Grills-Taquechel, and D. Axsom. 2009. Impaired and incapacitated rape victims: Assault characteristics and post-assault experiences. *Violence & Victims* 24: 439–457

Liu, C. 2003. Does quality of marital sex decline with duration? *Archives of Sexual Behavior* 32: 55–60.

Livingston, G. and D'Vera Cohn. 2010. More women without children. Pew Research Center June 25 http://pewresearch.org/pubs/1642/more-women-without-children.

Livermore, M. M., and R. S. Powers. 2006. Unfulfilled plans and financial stress: unwed mothers and unemployment. *Journal of Human Behavior in the Social Environment* 13: 1–17.

Lizardi, D. M., R. G. Thompson, K. M. Keyes, and D. S. Hasin. 2010. The effect of parental remarriage following parental divorce on offspring suicide attempt. *Families in Society* 91: 186–192.

Looi, C., P. Seow, B. Zhang, H. So, W. Chen, and L. Wong. 2010 Leveraging mobile technology for sustainable seamless learning: a research agenda. *British Journal of Educational Technology* 41: 154–169.

López Turley, R. N., M. Desmond and S. K. Bruch 2010. Unanticipated educational consequences of a positive parent-child relationship. *Journal of Marriage and Family* 72: 1377–1390.

Lorber, J. 1998. *Gender inequality: Feminist theories and politics*. Los Angeles, CA: Roxbury.

Losada, A., A. Perez-Penaranda, E. Rodriguiz-Sanchez, M. Gomez-Marcos, C. Ballesteros-Rios, et al. 2010. Leisure and distress in caregivers for elderly patients. *Archives of Gerontology & Geriatrics* 50: 347–350.

Lubin, J. 2010. Attitudes towards interethnic relationships and interethnic relationships behavior. University of Surrey. Dissertation, Department of Psychology.

Lucier-Greer, M., and F. Adler-Baeder. 2010. Gender role attitudes during divorce & remarriage: Plastic or plaster? Poster, National Council on Family Relations annual meeting, November 3–5. Minneapolis, MN.

Lundquist, J. H. 2007. A comparison of civilian and enlisted divorce rates during the early all-volunteer force era. *Journal of Political and Military Sociology* 35(2), 199–217.

Lundstrom, J. N., and M. Jones-Gotman. 2009. Romantic love modulates women's identification of men's body odors. *Hormones and Behavior* 55: 280–95.

Luo, S. H., and E. C. Klohnen. 2005. Assortative mating and marital quality in newlyweds: A couple-centered approach. *Journal of Personality and Social Psychology* 88: 304–326.

Luong, G., S.T. Charles, and K. L Fingerman. 2011. Better with age: Social relationships across adulthood. *Journal of Social & Personal Relationships* 28: 9–23.

Luscombe, B. 2010a. Divorcing while dying. *Time* February 15, p. 49.

Luscombe, B. 2010b. Making divorce pay. *Time Magazine* Sept. 13.

Luscombe, B. 2010c. Revoking the marriage license. *Time Magazine* May 5: 64.

Lussier, P., E. Beauregard, J. Proulx, and N. Alexandre. 2005. Developmental factors related to deviant sexual preferences in child molesters. *Journal of Interpersonal Violence* 20: 999–1017.

Lydon, J., T. Pierce, and S. O'Regan. 1997. Coping with moral commitment to long-distance dating relationships. *Journal of Personality and Social Psychology* 73: 104–113.

M

MacArthur, S. 2010. Adolescent religiosity, religious affiliation, and premarital predictors of marital quality and stability. Poster, National Council on Family Relations annual meeting, November 3–5. Minneapolis, MN.

Mackay, R. 2005. The impact of family structure and family change on child outcomes: a personal reading of the research literature. *Social Policy Journal of New Zealand*, March 111–134.

MacLeod, C. 2010b. China may relax one child rule. *USA Today*, Sept 9. 1A.

Macleod, C. 2010a. China smitten with TV dating. *USA Today*, May 18, p. 8A.

Madathil, J., and J. M. Benshoff. 2008. Importance of marital characteristics and marital satisfaction: A comparison of Asian Indians in arranged marriages and Americans in marriages of choice. *Family Journal* 16: 222–232.

Madden, M. 2010. Older adults and social media. Pew Internet & American Life ProjectPosted and retrieved August 27 http://pewresearch.org/pubs/1711/older-adults-social-networking-facebook-twitter.

Madden, M., and A. Lenhart. 2006. *Online dating*. Washington, DC: Pew Internet & American Life Project.

Maddox, A. M., G. K. Rhoades, and H. J. Markman. 2011. Viewing sexually-explicit materials alone or together: Associations with relationship quality. *Archives of Sexual Behavior* 40: 441–448.

Magaziner, J. 2010 The new technologies of change. *Psychology Networker*, September/October: 42–47.

Maher, D. and C. Mercer (editors). 2009. Introduction. *Introduction to religion and the implications of radical life extension*. New York: Palgrave Macmillan.

Mahoney, S. 2006. The secret lives of single women—lifestyles, dating and romance: A study of midlife singles. *AARP: The Magazine* May/June, 62–69.

Maikovich-Fonga, A. K., and S. R. Jaffee. 2010. Sex differences in childhood sexual abuse characteristics and victims' emotional and behavioral problems: Findings from a national sample. *Child Abuse & Neglect* 34: 429–437.

Major, B., M. Appelbaum, and C. West. 2008. Report of the APA task force on mental health and abortion. August 13.

Malacad, B., and G. Hess. 2010. Oral sex: Behaviours and feelings of Canadian young women and implications for sex education. *European Journal of Contraception & Reproductive Health Care* 15: 177–185.

Malinen, K., U. Kinnunen, A.Tolvanen, A. Rönka, H. Wierda-Boer and J. Gerris. 2010. Happy spouses, happy parents? Family relationships among Finnish and Dutch dual earners. *Journal of Marriage and Family* 72: 293–306.

Maltby, L E., M. E. L. Hall, T. L. Anderson, and K. Edwards. 2010. Religion and sexism: The moderating role of participant gender. *Sex Roles* 62: 615–622.

Mandara, J., F. Varner, and S. Richman. 2010. Do African American mothers really "love" their sons and "raise" their daughters? *Journal of Family Psychology* 24: 41–50.

Manning, W. D., D. Trella, H. Lyons, and N. C. DuToit. 2010. Marriageable women: A focus on participants in a community Healthy Marriage Program. *Family Relations* 59: 87–102.

Manning, W. D., and P. J. Smock. 2000. "Swapping" families: Serial parenting and economic support for children. *Journal of Marriage and the Family* 62: 111–122.

Manning, W. D., J. A. Cohen, and P. J. Smock. 2011. The role of romantic partners, family, and peer networks in dating couples' views about cohabitation. *Journal of Adolescent Research* 26: 115–149.

Marano, H. E. 1992. The reinvention of marriage. *Psychology Today* January/February, 49.

Marano, H. E. 2010. The expectations trap. *Psychology Today* 43: 62–71.

Marklein, M. B. 2008. High mark for foreign students here. *USA Today* November 17, 4D.

Markham, M. S., and M. Coleman. 2010. The good, the bad, and the ugly: Divorced mothers' experiences with coparenting. Poster, National Council on Family Relations annual meeting, November 3–5. Minneapolis, MN.

Markman, H. J., G. K. Rhoades, S. M. Stanley, E. P. Ragan, and S. W. Whitton. 2010. The premarital communication roots of marital distress and divorce: The first five years of marriage. *Journal of Family Psychology* 24: 289–298.

Markman, H. J., S. M. Stanley, and S. L. Blumberg. 2010. *Fighting for your marriage* (3rd ed.). San Francisco, CA: Jossey-Bass.

Marks, L. D., D. C. Dollahite, and J. Baumgartner. 2010. In God we trust: Qualitative findings on finances, family, and faith from a diverse sample of U.S. families. *Family Relations* 59: 439–452.

Marks, N. F., J. D. Lambert, and H. Choi. 2002. Transitions to caregiving, gender, and psychological well-being: A prospective U.S. national study. *Journal of Marriage and Family* 64: 657–667.

Marquet, R., A. Bartelds, G. J. Visser, P. Spreeuwenberg, and I. Peters. 2003. Twenty-five years of requests for euthanasia and physician assisted suicide in Dutch practice: Trend analysis. *British Medical Journal* 327: 201–202.

Marshall, A., J. M. Bell, and N. J. Moules. 2010. Beliefs, suffering, and healing: A clinical practice model for families experiencing mental illness. *Perspectives in Psychiatric Care* 46: 197–208.

Marshall, N. L., and A. J. Tracy. 2009. After the baby: Work-family conflict and working mothers' psychological health. *Family Relations* 58: 380–391.

Marshall, T. C., K. Chuong, and A. Aikawa. 2011. Day-to-day experiences of amae in Japanese romantic relationships. *Asian Journal of Social Psychology* 14: 26–35.

Martin, B. A., and C. Dula. 2010. More than skin deep: Perceptions of, and stigma against, tattoos. *College Student Journal* 44: 200–206.

Masheter, C. 1999. Examples of commitment in postdivorce relationships between spouses. In *Handbook of interpersonal commitment and relationship stability*, edited by J. M. Adams and W. H. Jones. New York: Academic/Plenum Publishers, 293–306.

Mashoa, S. W., D. Chapmana, and M. Ashbya. 2010. The impact of paternity and marital status on low birth weight and preterm births. *Marriage & Family Review* 46: 243–256.

Mason, M. A., and M. Goulden. 2004. Do babies matter? The effect of family formation on the lifelong careers of academic men and women. Annual Conference of the National Council on Family Relations, November. Orlando, Florida.

Masters, W. H., and V. E. Johnson. 1970. *Human sexual inadequacy*. Boston: Little, Brown.

Mathy, R. M. 2007. Sexual orientation moderates online sexual activity. In *Online Matchmaking*, ed. M. T. Whitty, A. J. Baker, and J. A. Inman, 159–77. New Work: Paulgrave Macmillan.

Mattingly, M. J., and K. E. Smith. 2010. Changes in wives' employment when husbands stop working: A recession-prosperity comparison. *Family Relations* 59: 343–357.

Maume, D. J. 2006. Gender differences in taking vacation time. *Work and Occupations* 33: 161–190.

Maynard, E., A. Carballo-Diéguez, A. Ventuneac, T. Exner, and K. Mayer. 2009. Women's experiences with anal sex: Motivations and implications for STD prevention. *Perspectives on Sexual & Reproductive Health* 41: 142–149.

Mbopi-Keou, F. X, R. E. Mbu, H. Gonsu Kamga, G. C. M. Kalla, M. Monny Lobe, C. G. Teo, R. J. Leke, P. M. Ndumbe, and L. Belec. 2005. Interactions between human immunodeficiency virus and herpes viruses within the oral mucosa. *Clinical Microbiology and Infection* 11: 83–85.

McAlister, A. R., N. Pachana, and C. J. Jackson. 2005. Predictors of young dating adults' inclination to engage in extra dyadic sexual activities: A multi-perspective study. *British Journal of Psychology* 96: 331–350.

McBride, B. A., S. J. Schoppe, and T. R. Rane. 2002. Child characteristics, parenting stress, and parental involvement: Fathers versus mothers. *Journal of Marriage and Family* 64: 998–1011.

McBride, K., and J. D. Fortenberry. 2010. Heterosexual anal sexuality and anal sex behaviors: A review. *Journal of Sex Research* 47: 123–136.

McClintock, E. A. 2010. When does race matter? Race, sex, and dating at an elite university 2010. *Journal of Marriage and Family* 72: 45–72.

McGinty, C. 2010. Internet dating. Presentation to courtship and marriage class, East Carolina University, Greenville, NC.

McGinty, K., D. Knox, and M. Zusman. 2007. Friends with benefits: Women want "friends," men want "benefits." *College Student Journal* 41: 1128–1131.

McIntosh, J. E., Y. D. Wells, B. M. Smyth, and C. M. Long. 2008. Child-focused and child-inclusive divorce mediation: Comparative outcomes from a prospective study of post separation adjust. *Family Court Review* 46: 105–115.

Mcintosh,W.D., L. Locker, K. Briley, R. Ryan, and A. Scott. 2011. What do older adults seek in their potential romantic partners? Evidence from online personal ads. *International Journal of Aging & Human Development* 72: 67–82.

McKee, K. S., and A. J. Blow. 2010. Proposed model of decision making for genetic testing in families. Poster, National Council on Family Relations annual meeting, November 3–5. Minneapolis, MN.

McKinney, J. 2004. *The Christian case for gay marriage.* Pullen Memorial Baptist Church (February 8). http://www.pullen.org.

McKinney, C., and K. Renk 2008. Differential parenting between mothers and fathers: Implications for late adolescents. *Journal of Family Issues* 29: 806–827.

McLanahan, S. S. 1991. The long term effects of family dissolution. In *When families fail: The social costs,* ed. Brice J. Christensen, 5 26. New York: University Press of America for the Rockford Institute.

McLanahan, S. S., and K. Booth. 1989. Mother-only families: Problems, prospects, and politics. *Journal of Marriage and the Family* 51: 557–580.

McLean, K. 2004. Negotiating (non)monogamy: Bisexuality and intimate relationships. In *Current research on bisexuality,* ed. R. C. Fox, 82–97. New York: Harrington Park Press.

McLennon, S. M., B. Habermann, and L L. Davis. 2010. Deciding to institutionalize: why do family members cease caregiving at home? *Journal of Neuroscience Nursing* 42: 95–104.

McMahon, S. 2010. Rape myth beliefs and bystander attitudes among incoming college students. *Journal of American College Health* 59: 3–11.

McNulty, J. K. 2008. Forgiveness in marriage: Putting the benefits into context. *Journal of Family Psychology* 22: 171–183.

McPherson, M., L. Smith-Lovin, and M. E. Brashears. 2006. Social isolation in America, 1985–2004. *American Sociological Review* 71: 353–375.

McWey, L. M., M. Cui, and A. L. Pazdera. 2010. Changes in externalizing and internalizing problems of adolescents in foster care. *Journal of Marriage and the Family* 72: 1128–1140.

Mead, G. H. 1934. *Mind, self, and society.* Chicago: University of Chicago Press.

Mead, M. 1935. *Sex and temperament in three primitive societies.* New York: William Morrow.

Medora, N. P. 2003. Mate selection in contemporary India: Love marriages versus arranged marriages. In *Mate Selection across Cultures,* ed. R. R. Hamon and B. B. Ingoldsby, 209, 30. Thousand Oaks, California: Sage Publications.

Medora, N. P., J. H. Larson, N. Hortacsu, and P. Dave. 2002. Perceived attitudes towards romanticism: A cross-cultural study of American, Asian-Indian, and Turkish young adults. *Journal of Comparative Family Studies* 33: 155–178.

Meehan, D., and C. Negy. 2003. Undergraduate students' adaptation to college: Does being married make a difference? *Journal of College Student Development* 44: 670–690.

Meier, A., K. E. Hull, and T. A. Orty. 2009. Young adult relationship values at the intersection of gender and sexuality. *Journal of Marriage and the Family* 71: 510–525.

Meier, J. S. 2009. A historical perspective on parental alienation syndrome and parental alienation. *Journal of Child Custody* 6: 232–257.

Meinhold, J. L., A. Acock, and A. Walker. 2006. The influence of life transition statuses on sibling intimacy and contact in early adulthood. Paper presented at the Annual Meeting of the National Council on Family Relations in Orlando in November 2005.

Meisenbach, M. J. 2010. The female breadwinner: Phenomenological experience and gendered identity in work/family spaces. *Journal of Sex Roles* 62: 2–19.

Melhem, N. M., D. A. Brent, M. Ziegler, S. Iyengar, et al. 2007. Familial pathways to early-onset suicidal behavior: Familial and individual antecedents of suicidal behavior. *American Journal of Psychiatry* 164: 1364–1371.

Mellor, D., L. A. Ricciardelli, M. P. McCabe, J. Yeow, N. Hidayah bt Mamat, and N. Fizlee bt Mohd Hapidzal. 2010. Psychosocial correlates of body image and body change behaviors among malaysian adolescent boys and girls. *Sex Roles* 63: 386–398.

Meloy, J. R., and H. Fisher. 2005. Some thoughts on the neurobiology of stalking. *Journal of Forensic Science* 50: 1472–1480.

Meltzer, A. L., and J. K. McNulty. 2010. Body image and marital satisfaction: Evidence for the mediating role of sexual frequency and sexual satisfaction. *Journal of Family Psychology* 24: 156–164.

Meltzer, H., P. Bebbington, T. Brugha, R. Jenkins, S. McManus, and M. S. Dennis. 2011. Personal debt and suicidal ideation. *Psychological Medicine* 41: 771–778.

Merline, A. C., J. E. Schulenberg, P. M. O'Malley, J. G. Bachman, and L. D. Johnston. 2008. Substance use in marital dyads: Premarital assortment and change over time. *Journal of Studies on Alcohol and Drugs* 69: 352–365.

Merolla, A. J., and S. Zhang. 2011. In the wake of transgressions: Examining forgiveness communication in personal relationships. *Personal Relationships* 18: 79–95.

Merrill, J., and D. Knox. 2010. *Finding love from 9 to 5: Secrets of office romance.* Santa Barbra, California: Praeger.

Meston, C. M., and D. M.Buss. 2007. Why humans have sex. *Archives of Sexual Behavior* 36: 477–507.

Meyer, A. S., L. M. McWey, W.McKendrick, and T. L. Henderson. 2010. Substance using parents, foster care, and termination of parental rights: The importance of risk factors for legal outcomes. *Children & Youth Services Review* 32: 639–649.

Meyer, D. 2007. Selective serotonin reuptake inhibitors and their effects on relationship satisfaction. *The Family Journal* 15: 392–397.

Meyer, I. H., J. Dietrich, and S. Schwartz. 2008. Lifetime prevalence of mental disorders and suicide attempts in diverse lesbian, gay, and bisexual populations. *American Journal of Public Health* 98: 1003–1004.

Meyer, J. P., and S. Pepper. 1977. Need compatibility and marital adjustment in young married couples. *Journal of Personality and Social Psychology* 35: 331–342.

Meyer-Bahlburg, H. F. L. 2005. Introduction: Gender dysphoria and gender change in persons with intersexuality. *Archives of Sexual Behavior* 34: 371–374.

Michael, R. T., J. H. Gagnon, E. O. Laumann, and G. Kolata. 1994. *Sex in America: A definitive survey.* Boston: Little, Brown.

Michaels, M. L. 2000. The stepfamily enrichment program: A preliminary evaluation using focus groups. *American Journal of Family Therapy* 28: 61–73.

Michielsen, D., and R. Beerthuizen. 2010. State-of-the art of non-hormonal methods of contraception: VI. Male sterilisation. *European Journal of Contraception & Reproductive Health Care* 15: 136–149.

Miller, A. S., and R. Stark. 2002. Gender and religiousness: Can socialization explanations be saved? *American Journal of Sociology* 107: 1399–1423.

Miller, E., M. Decker, H. McCauley, D. J. Tancredi, R. Levenson, J. Waldman, P. Schoewald, and J. Silverman. 2010. Pregnancy coercion, intimate partner violence and unintended pregnancy. *Contraception* 81: 316–322.

Miller, S., A. Taylor, and D. Rappleyea. 2011. The influence of religion on young adult's attitudes of dating events. Poster, Fifth Annual Research & Creative Achievement Week, East Carolina University, April 4–8.

Miller, et al. 2002. Bisexual health: an introduction and model practices for HIV/ STD prevention programming. *Final Report*. National Gay and Lesbian Task force Policy Institute.

Minnottea, K. L., D. E. Pedersena, S. E. Mannonb, and G. Kiger 2010. Tending to the emotions of children: Predicting parental performance of emotion work with children. *Marriage & Family Review* 46: 224–241.

Mitchell, B. A. 2010. Midlife marital happiness and ethnic culture: A life course perspective. *Journal of Comparative Family Studies* 41: 167–183.

Moen, P. 2011. From 'work-family' to the 'gendered life course' and 'fit': five challenges to the field. *Community, Work & Family* 14: 81–96.

Moghadam, V. M. 2002. Patriarchy, the Taliban, and the politics of public space in Afghanistan. *Women's Studies International Forum* 25: 19–31.

Mohipp, C., and M. M. Morry. 2004. Relationship of symbolic beliefs and prior contact to heterosexuals' attitudes toward gay men and lesbian women. *Canadian Journal of Behavioral Science* 36: 36–44.

Mohr, J., R. Cook-Lyon, and M. R. Kolchakian. 2010 Love imagined: Working models of future romantic attachment in emerging adults. *Personal Relationships* 17: 457–473.

Monin, J. K., M. S. Clark, and E. P. Lemay. 2008. Communal responsiveness in relationships with female versus male family members. *Sex Roles* 59: 176–188.

Monro, S. 2000. Theorizing transgender diversity: Towards a social model of health. *Sexual and Relationship Therapy* 15: 33–42.

Monteoliva, A., J. Garcia-Martinez, and A. Miguel. 2005. Adult attachment style and its effect on the quality of romantic relationships in Spanish students. *Journal of Social Psychology* 145: 745–747.

Montemurro, B. 2006. *Something old, something bold*. New Brunswick, NJ: Rutgers University Press.

Mooney, L. A. 2009. ECU: GLBT campus climate assessment. Department of Sociology, East Carolina Universitiy. Unpublished manuscript.

Mooney, L., C. Reiser, and K. Wilson. 2010. Pet nation: Demographic correlates of the human-companion animal bond. Poster, Southern Sociological Society, Atlanta, April 23.

Moore, A., S. Hamilton, D. R. Crane, and D. Fawcett. 2011. The influence of professional license type on the outcome of family therapy. *American Journal of Family Therapy* 39: 149–161.

Moore, M. M. 2010. Human nonverbal courtship behavior—A Brief Historical Review. *Journal of Sex Research* 47: 171–180.

Morawska, A., and M. Sanders. 2011. Parental use of time out revisited: A useful or harmful parenting strategy? *Journal of Child & Family Studies* 20: 1–8.

Mordoch, E., and W. A. Hall. 2008. Children's perceptions of living with a parent with a mental illness: Finding the rhythm and maintaining the frame. *Qualitative Health Research* 18: 1127–1135.

Morell, V. 1998. A new look at monogamy. *Science* 281: 1982.

Morey, M. C., R. Sloane, C. F Pieper, and M. J. Peterson. 2008. Effect of physical activity guidelines on physical function in older adults. *Journal of the American Geriatrics Society* 56: 1873–1885.

Morgan, E. M., M. G. Steiner, and E. M. Thompson 2010. Processes of sexual orientation questioning among heterosexual men. *Men and Masculinities* 12: 425–443.

Morgan, E. S. 1944. *The Puritan family*. Boston: Public Library.

Morin, R., and D. Cohn. 2008. Women call the shots at home; Public mixed on gender roles in jobs, gender and power. Pew Research Center, September 25.

Morr Serewicz, M. C., and D. J. Canary. 2008. Assessments of disclosure from the in-laws: Links among disclosure topics, family privacy orientations, and relational quality. *Journal of Social and Personal Relationships* 25: 333–357.

Morrill, I., and M. Morrill 2010. Pathways between marriage and parenting for wives and husbands: The role of coparenting. *Family Process* 49: 59–73.

Morris, M. L., H. McMillan, S. D. Duncan and J. Larson. 2011. Who will attend? Characteristics of couples and individuals in marriage education. *Marriage & Family* 47: 1–22.

Moskowitz, D. A., G. Rieger, and M. E. Roloff. 2010. Heterosexual attitudes toward same-sex marriage. *Journal of Homosexuality* 57: 325–336.

Mosuo, 2010. http://en.wikipedia.org/wiki/Mosuo#General_Practice.

Murdock, G. P. 1949. *Social structure*. New York: Free Press.

Murphy, M. J., L. Deets, and M. Peterson. 2010. The effect of emotional labor and emotional abuse on relationship satisfaction. Poster, National Council on Family Relations annual meeting, November 3–5. Minneapolis, MN.

Murphy-Graham, E. 2010 And when she comes home? Education and women's empowerment in intimate relationships. *International Journal of Educational Development* 30: 320–331.

Murray, C. I., and N. Kimura. 2003. Multiplicity of paths to couple formation in Japan. In *Mate selection across cultures* ed. R. R. Hamon and B. B. Ingoldsby, 247–268. Thousand Oaks, CA: Sage Publications.

Mutch, K. 2010. In sickness and in health: experience of caring for a spouse with MS. *British Journal of Nursing* 19: 214–219.

Mutran, E. J., D. Reitzes, and M. E. Fernandez. 1997. Factors that influence attitudes toward retirement. *Research on Aging* 19: 251–273.

Myers, S. A. 2011. I have to love her, even if sometimes I may not like her: The reasons why adults maintain their sibling relationships. *North American Journal of Psychology* 13: 51–62.

N

National Center for Lesbian Rights. 2003. Second-parent adoptions: A snapshot of current law. http://www.nclrights.org.

National Coalition of Anti-Violence Programs. 2005. *2004 National hate crimes report: Anti-lesbian, gay, bisexual and transgender violence in 2004*. New York: National Coalition of Anti-Violence Programs.

National Gay and Lesbian Task Force. 2004. Anti-Gay parenting laws in the U.S. National Gay and Lesbian Task Force (June). http://www.thetaskforce.org.

National Gay and Lesbian Task Force. 2005. Second-parent adoption in the U.S. National Gay and Lesbian Task Force. http://www.thetaskforce.org.

National Gay and Lesbian Task Force. 2006. State nondiscrimination laws in the U.S. (as of March 2006). http:www.thetaskforce.com.

National Institute of Mental Health. 2010. Older adults: Depression and suicide facts. http://www.nimh.nih.gov/health/publications/older-adults-depression-and-suicide-facts-fact-sheet/index.shtml (retrieved on June 12, 2010).

National Runaway Switchboard. 2010. Annual survey. http://www.nrscrisisline.org/news_events/call_stats.html.

NCFR Policy Brief. 2004. *Building strong communities for military families*. Minneapolis, MN: National Council on Family Relations.

Nelson, M. C. 2010. *Parents out of control*. New York: New York Press.

Nelson, B. S., and K. S. Wampler. 2000. Systemic effects of trauma in clinic couples: An exploratory study of secondary trauma resulting from childhood abuse. *Journal of Marriage and Family Counseling* 26: 171–184.

Nelson, Eve-Lynn, and Thao Bui. 2010. Rural telepsychology services for children and adolescents. *Journal of Clinical Psychology: In Session* 66: 490–501.

Neuman, M. G. 2008. *The truth about cheating: Why men stray and what you can do to prevent it*. New York: John Wiley & Sons.

Newman, R. Six strains on your financial future. 2010. *US News and World Report*. May 26.

Newton, N. 2002. *Savage girls and wild boys: A history of feral children*. New York: Thomas Dunne Books/St. Martin's Press.

Nielsen, L. 2004. *Embracing your father: How to build the relationship you always wanted with your dad*. New York: McGraw-Hill.

Nielsen, L. 2011. Divorced fathers and their daughters: A review of recent research. *Journal of Divorce & Remarriage* 52: 77–93.

Niven, K. J. 2010. *The circle of life: The process of sexual recovery workbook* (available at Amazon.com).

Nock, S. L., L. A. Sanchez, and J. D. Wright. 2008. *Covenant Marriage: The movement to reclaim tradition in America*. New Brunswick, New Jersey: Rutgers University Press.

Nolla, J. G. 2008. Sexual abuse of children—Unique in its effects on development? *Child Abuse and Neglect* 32: 603–605.

Nomaaguchi, K. M., and M. A. Milkie. 2006. Maternal employment in childhood and adult's retrospective reports of parenting practices. *Journal of Marriage and the Family* 68: 573–591.

Noonan, M. C., and M. E. Corcoran. 2004. The mommy track and partnership: Temporary delay or dead end? *The Annals of the American Academy of Political and Social Science* 596: 130–150.

North, R. J., C. J. Holahan, R. H. Moos, and R. C. Cronkite. 2008. Family support, family income, and happiness: A 10-year perspective. *Journal of Family Psychology* 22: 475–483.

Northey, W. F. J. 2002. Characteristics and clinical practices of marriage and family therapists: A national survey. *Journal of Marriage and Family Therapy* 28: 487–494.

NPR -National Public Radio. 2010. Joan Rivers on her husband's suicide, June 12.

Nuner, J. E. 2004. A qualitative study of mother-in-law/daughter-in-law relationships. *Dissertation Abstracts International, A: The Humanities and Social Sciences*, 65(August): 712A–13A.

Nunley, J. M., and A. Seals. 2010. The effects of household income volatility on divorce. The *American Journal of Economics and Sociology* 69: 983–1011.

O

Ochs, R. 1996. Biphobia: It goes more than two ways. In *Bisexuality: The psychology and politics of an invisible minority*, ed. B. A. Firestein, 217–239. Thousand Oaks, CA: Sage.

O'Flaherty, K. M., and L. W. Eells. 1988. Courtship behavior of the remarried. *Journal of Marriage and the Family* 50: 499–506.

Ogilvie, G. S., D. L. Taylor, T. Trussler, R. Marchand, M. Gilbert, A. Moniruzzaman, and M. L. Rekart. 2008. Seeking sexual partners on the Internet: A marker for risky sexual behaviour in men who have sex with men. *Canadian Journal of Public Health* 99: 185–189.

Olayemi, O., D. Strobino, C. Aimakhy, K. Adedapo, A. Kehinde, A. Odukogbe, and B. Salako. 2010. Influence of duration of sexual cohabitation on the risk of hypertension in nulliparous parturients in Ibadan: A cohort study. *Australian & New Zealand Journal of Obstetrics & Gynaecology* 50: 40–44.

Olsen, K. M., and S. Dahl. 2010. Working time: implications for sickness absence and the work–family balance. *International Journal of Social Welfare* 19: 45–53.

Olson, M. M., C. S. Russell, M. Higgins-Kessler, and R. B. Miller. 2002. Emotional processes following disclosure of an extramarital affair. *Journal of Marital and Family Therapy* 28: 423–434.

O'Reilly, E. M. 1997. *Decoding the cultural stereotypes about aging: New perspectives on aging talk and aging issues*. New York: Garland.

O'Reilly, S., D. Knox, and M. Zusman. 2006. "I have never masturbated": 973 college students who said "yes" or "no." Paper presented at the annual Meeting Southern Sociological Society, March. New Orleans.

O'Reilly, S., D. Knox, and M. Zusman. 2007. College student attitudes toward pornography use. *College Student Journal* 41: 402–406.

Orthner, D. K., and R. Roderick. 2009. Work separation demands and spouse psychological well-being. *Family Relations* 58: 392–403.

Oscarsson, M. 2009. Mandatory sex education classes for 14-year-olds anger Muslim immigrants in Sweden. *Global Post*, June 29.

Ostbye, T., K. M. Krause, M. C. Norton, J. Tschanz, L. Sanders, K. Hayden, C. Pieper, and K. A. Welsh-Bohmer. 2006. Ten dimensions of health and their relationships with overall self-reported health and survival in a predominately religiously active elderly population: *The Cache County Memory Study. Journal of the American Geriatrics Society* 54: 199–209.

Ostenson, J., R. Funk, K. Hollist, K. Kearl, and L. Bythway. 2010. Defending marriage against individualism: An ethnographic perspective. Poster, National Council on Family Relations annual meeting, November 3–5. Minneapolis, MN.

Oswald, D. L., and B. L. Russell. 2006. Perceptions of sexual coercion in heterosexual dating relationships: the role of aggressor gender and tactics. *The Journal of Sex Research* 43: 87–98.

Otis, M. D., S. S. Rostosky, E. D. B. Riggle, and R. Hamrin. 2006. Stress and relationship quality in same-sex couples. *Journal of Social and Personal Relationships* 23: 81–99.

Owen, J., and F. D. Fincham. 2011. Effects of gender and psychosocial factors on 'Friends with Benefits' relationships among young adults. *Archives of Sexual Behavior* 40: 311–320.

Owen, J. J., G. K. Rhoades, S. M. Stanley, and F. D. Fincham. 2010. "Hooking Up" among college students: Demographic and psychosocial correlates. *Journal Archives of Sexual Behavior* 39: 653–663.

Oyamot Jr., C. M., P. T. Fuglestad, and M. Snyder. 2010. Balance of power and influence in relationships: The role of self-monitoring. *Journal of Social and Personal Relationships* 27: 23–46.

P

Padilla, Y. C., C. Crisp, and D. L.Rew. 2010. Parental acceptance and illegal drug use among gay, lesbian, and bisexual adoles-

cents: results from a national survey. *Social Work* 55: 265–276.

Paik, A. 2010. The contexts of sexual involvement and concurrent sexual partnerships *Perspectives on Sexual and Reproductive Health* 42: 33–43.

Paintal, S. 2007. Banning corporal punishment of children. *Childhood Education* 83: 410–421.

Palmer, L., D. L. Bliss, J. W. Goietz, and D. Moorman. 2010a. Improving financial awareness among college students: Assessment of a financial management project. *College Student Journal* 44: 659–676.

Palmer, R., and R. Bor. 2001. The challenges to intimacy and sexual relationships for gay men in HIV serodiscordant relationships: A pilot study. *Journal of Marital and Family Therapy* 27: 419–431.

Palmer, R. S., T. J. McMahon, B. J. Rounsaville, and S. A. Ball. 2010b. Coercive sexual experiences, protective behavioral strategies, alcohol expectancies and consumption among male and female college students. *Journal of Interpersonal Violence* 25: 1563–1578.

Palomares, N. A. 2009. Women are sort of more tentative than men, aren't they? How men and women use tentative language differently, similarly, and counter stereotypically as a function of gender salience. *Communication Research* 36: 538–560.

Paluscia, V. J., S. J. Wirtzb and T. M. Covington. 2010. Using capture-recapture methods to better ascertain the incidence of fatal child maltreatment. *Child Abuse & Neglect* 34: 396–402.

Papernow, P. L. 1988. Stepparent role development: From outsider to intimate. In *Relative strangers*, edited by William R. Beer. Lanham, MD: Rowman and Littlefield, 54–82.

Paradis, A., and S. Boucher. 2010. Child maltreatment history and interpersonal problems in adult couple. *Journal of Aggression, Maltreatment & Trauma* 19: 138–158.

Parelli, S. 2007. Why ex-gay therapy doesn't work. *The Gay & Lesbian Review Worldwide* 14: 29–32.

Parish, W. L., Y. Luo, E. O. Laumann, M. Ken, and Z. Yu. 2007. Unwanted sexual activity among married women in urban China. *Journal of Sex Research* 44: 158–171.

Parker, K. 2011. A portrait of stepfamilies. Pew Research Center. Putlished and retrieved January 13, 2011. http://pewsocialtrends.org/2011/01/13/a-portrait-of-stepfamilies/

Parker-Pope, T. 2010. *For better: The science of a good marriage*. New York: E. P. Dutton.

Parra-Cardona, J. R. , D. Córdova, Jr., K. Holtrop, F. A. Villarruel, and E. Wieling. 2008. Shared ancestry, evolving stories: Similar and contrasting life experiences described by foreign born and U.S. born Latino parents. *Family Process* 47: 157–173.

Pasanen, M., and A. Silverman. 2010. Weed dating. *USA Today* Oct 4. P. 5D.

Passel, J. S., W. Wang, and P. Taylor. 2010. Marrying out: One-in-seven new U.S. marriages is interracial or interethnic. Pew Research Center, June 4. http://pewresearch.org/pubs/1616/american-marriage-interracial-interethnic.

Patrick, M. E., and C. M. Lee. 2010. Sexual motivations and engagement in sexual behavior during the transition to college. *Journal Archives of Sexual Behavior* 39: 674–681.

Patrick, S., J. N. Sells, F. G. Giordano, and T. Tollerud. 2010. Intimacy, differentiation, and personality variables as predictors of marital satisfaction. *The Family Journal: Counseling and Therapy for Couples and Families* 20 (2): in press.

Patrick, T. 2005. Stay-at-home dads. *Futurist* 39: 12–13.

Paul, J. P. 1996. Bisexuality: Exploring/exploding the boundaries. In *The lives of lesbians, gays, and bisexuals: Children to adults*, ed. R. Savin-Williams and K. M. Cohen, 436–61. Fort Worth, TX: Harcourt Brace.

Pawelski, J. G., E. C. Perrin, J. M. Foy, C. E. Allen, J. E. Crawford, M. Del Monte, M. Kaufman, J. D. Klein, K. Smith, S. Springer, J. L. Tanner, and D. L. Vickers. 2006. The effects of marriage, civil union, and domestic partnership laws on the health and well-being of children *Pediatrics* 118: 349–355.

Payne, J. L., E. S. Fields, J. M. Meuchel, C. J. Jaffe, and M. Jha. 2010. Post adoption depression. *Archives of Women's Mental Health* 13: 147–151.

Pearcey, M. 2004. Gay and bisexual married men's attitudes and experiences: Homophobia, reasons for marriage, and self-identity. *Journal of GLBT Family Studies* 1: 21–42.

Peilian , C., S. K. M. Tsang, S. C. Kin, X Xiaoping, P. S. F. Yip, T. C. Yee, and Z. Xiulan. 2011. Marital satisfaction of Chinese under stress: Moderating effects of personal control and social support. *Asian Journal of Social Psychology* 14: 15–25.

Pelton, S. L., and K. M. Hertlein. 2011. A proposed life cycle for voluntary childfree couples. *Journal of Feminist Family Therapy* 23: 39–53.

Penhollow, T. M., A. Marx, and M. Young. 2010. Impact of recreational sex on sexual satisfaction and leisure satisfaction. *Electronic Journal of Human Sexuality* 13: March 31.

Peplau, L. A., R. C. Veniegas, and S. N. Campbell. 1996. Gay and lesbian relationships. In *The lives of lesbians, gays, and bisexuals: Children to adults*, ed. R. C. Savin-Williams and K. M. Cohen, 250–273. Fort Worth, TX: Harcourt Brace.

Pepper, T. 2006. Fatherhood: Trying to do it all. *Newsweek,* International ed., February 27.

Perilloux, C., D. S. Fleischman, and D. M. Buss. 2011. Meet the parents: Parent-offspring convergence and divergence in mate preferences. *Personality & Individual Differences* 50: 253–258.

Perry, S. L. 2010. The effects of race, religion, and religiosity on attitudes towards transracial adoption. *Journal of Comparative Family Studies* 41: 837–854.

Perry-Jenkins, M., R. L. Repetti, and A. C. Crouter. 2001. Work and family in the 1990s. In *Understanding families into the new millennium: A decade in review,* ed. R. M. Milardo, 200–217. Minneapolis: National Council on Family Relations.

Pescosolido, B. A., J. K. Martin, A. Lang, and S. Olafsdottir. 2008. Rethinking theoretical approaches to stigma: A Framework Integrating Normative Influences on Stigma (FINIS). *Social Science & Medicine* 67: 431–451.

Peterson, Z. D., E. K. Voller, M. Polusny, and M. Murdoch. 2011. Prevalence and consequences of adult sexual assault of men: Review of empirical findings and state of the literature. *Clinical Psychology Review* 31: 1–24.

Petrie, J., J. A. Giordano, and C. S. Roberts. 1992. Characteristics of women who love too much. *Affilia: Journal of Women and Social Work* 7: 7–20.

Petroni, S. 2011. Historical and current influences on United States international family planning policy. *Journal of Women, Politics & Policy* 32: 28–51.

Pettigrew, J. 2009. Text messaging and connectedness within close interpersonal relationships. *Marriage & Family Review* 45: 697–716.

Pew Research Center. 2006. Less opposition to gay marriage, adoption and military service (March 22). http://people-press.org.

Pew Research Center. 2008a. Gay Marriage Opposed. http://pewforum.org/docs/index.php?DocID=39 (accessed November 16).

Pew Research Center. 2008b. The U.S. religious landscape survey. Pew Forum on Religion & Public Life. http://pewresearch.org/pubs/743/united-states-religion.

Pew Research Center. 2010b. Religion among the Millennials. Pew Forum on Religion & Public Life. http://pewforum.org/Age/Religion-Among-the-Millennials.aspx

Pew Research Center. 2010a. Social & Demographic Trends: The decline of marriage and rise of new families. Published and retrieved on November 18 http://pewresearch.org/pubs/1802/decline-marriage-rise-new-families.

Pfeffer, C. A. 2010 "Women's work"? Women partners of transgender men doing housework and emotion work. *Journal of Marriage & Family* 72: 165–183.

Phillips, T. M., and J. D. Wilmoth. 2010. Keys to longevity: A study of enduring African American marriages. Poster, National Council on Family Relations annual meeting, November 3–5. Minneapolis, MN.

Pietrzak, R. H., C. A. Morgan, and S. M. Southwick. 2010. Sleep quality in treatment-seeking veterans of Operations Enduring Freedom and Iraqi Freedom: The role of cognitive coping strategies and unit cohesion. *Journal of Psychosomatic Research* 69: 441–448.

Pillemer, K. J. Suitor, S. Pardo, and C. Henderson, Jr. 2010. Mothers' differentiation and depressive symptoms among adult children. *Journal of Marriage and Family* 72: 333–345.

Pimentel, E. E. 2000. Just how do I love thee? Marital relations in urban China. *Journal of Marriage and the Family* 62: 32–47.

Pinello, D. R. 2008. Gay marriage: For better or for worse? What we've learned from the evidence. *Law & Society Review* 42: 227–230.

Pines, A. M. 1992. *Romantic jealousy: Understanding and conquering the shadow of love.* New York: St. Martin's Press.

Pinheiro, R. T., R. A. da Silva, P. V. S. Magalhaes, B. L. Hortam, and K. A. T. Pinheiro. 2008. Two studies on suicidality in the postpartum. *Acta Psychiatrica Scandinavica* 118: 160–162.

Plagnol, A. C., and R. A. Easterlin. 2008. Aspirations, attainments, and satisfaction: Life cycle differences between American women and men. *Journal of Happiness Studies.* Published online July 2008.

Plant, E. A., J. Kunstman, and J. K. Maner. 2010. You do not only hurt the one you love: Self-protective responses to attractive relationship alternatives. *Journal of Experimental Social Psychology,* 46: 474–477.

Platt, L. 2001. Not your father's high school club. *American Prospect* 12: A37–39.

Plitnick, K. R. 2008. Elder abuse. *Association of Operating Room Nurses Journal* 87: 422–432.

Plouffe, D. 2009. *The audacity to win.* New York: Viking.

Pompper, D. 2010. Masculinities, the metrosexual, and media images: Across dimensions of age and ethnicity. *Sex Roles* 63: 682–696. *Cache County Memory Study. Journal of the American Geriatrics Society* 54:199–209.

Pond, R. L., C. Stephens, and F. Alpass. 2010. How health affects retirement decisions: three pathways taken by middle-older aged New Zealanders. *Ageing & Society.* 30: 527–545.

Poortman, A. and T. Van der Lippe. 2009. Attitudes toward housework and child care and the gendered division of labor. *Journal of Marriage and the Family* 71: 526–541.

Potok, M. 2010. Anti-gay hate crimes: Doing the math. *Intelligence Report* 140 (Winter): 29.

Potter, J. F. 2010. Aging in America: essential considerations in shaping senior care policy. *Aging Health* 6.3: 289–300.

Potter, D. 2010. Psychosocial well-being and the relationship between divorce and children's academic achievement. *Journal of Marriage and Family* 72: 933–946.

Pratt, L. A., and D. J. Brody. 2010. Depression and smoking in the U.S. household population aged 20 and over. 2005–2008. NCHS National Center for Health Statistics Data Brief. No. 34. April.

Presser, H. B. 2000. Nonstandard work schedules and marital instability. *Journal of Marriage and the Family* 62: 93–110.

Proulx, C. M. and L. A. Snyder. 2009. Families and health: An empirical resource guide for researchers and practitioners. *Family Relations* 58: 489–504.

Pryor, J. H., S. Hurtado, L. DeAngelo, J. Sharkness, L. C. Romero, W. K. Korn, and S. Trans. 2008. *The American freshmen: National Norms for fall 2008.* Los Angeles: Higher Education Research Institute, UCLA.

Pryor, J. H., S. Hurtado, L. DeAngelo, L. P. Blake, and S. Tran. 2011. The American Freshman: National Norms Fall 2010. Los Angeles: Higher Education Research Institute. UCLA Graduate School of Education and Information Studies.

Puccinelli, M. 2005. Students support, decry gays with t-shirts. CBS 2 Chicago (April 19). http://cbs2chicago.com.

Puentes, J., D. Knox, and M. Zusman. 2008. Participants in "Friends with Benefits" relationships. *College Student Journal* 42: 176–180.

Puri, S. and R. D. Nachtigall. 2010. The ethics of sex selection: a comparison of the attitudes and experiences of primary care physicians and physician providers of clinical sex selection services. *Fertility & Sterility* 93: 2107–2114.

Purkett, T. 2010. Sexually transmitted infections. Presented to Courtship and Marriage class, Department of Sociology, Fall.

Q

Qing, M., Z. Li-xia, and S. Xiao-yin. 2011. A comparison of postnatal depression and related factors between Chinese new mothers and fathers. *Journal of Clinical Nursing* 20: 645–652.

R

Rafferty, A., and J. Wiggan. 2011. Choice and welfare reform: Lone parents' decision making around paid work and family life. *Journal of Social Policy* 40: 275–293.

Raj, A., C. Gomez, and J. G. Silverman. 2008. Driven to a fiery death: The tragedy of self-immolation in Afghanistan. *The New England Journal of Medicine* 358: 2201–2217.

Raley, R. K. 2000. Recent trends and differentials in marriage and cohabitation: The United States. In *The ties that bind,* ed. L. J. Waite, 19–39. New York: Aldine de Gruyter.

Raley, R. K., and M. K. Sullivan. 2010. Social-contextual influences on adolescent romantic involvement: Constraints of being a numerical minority. *Sociological Spectrum* 30: 65–89.

Randall, B. 2008. *Songman: The story of an Aboriginal elder of Uluru.* Sydney, Australia: ABC Books.

Randler, C., and S. Kretz 2011. Assortative mating in morningness-eveningness. *International Journal of Psychology* 46: 91–96.

Rankin, S. R. 2003. Campus climate for gay, lesbian, bisexual, and transgendered people: A national perspective. New York: National Gay and Lesbian Task Force Policy Institute. http://www.thetaskforce.org.

Rapoport, B., and C. Le Bourdais. 2008. Parental time and working schedules. *Journal of Population Economics* 21: 903–933.

Rappaport, A. 2009. Phased retirement- An important part of the evolving retirement scene. *Benefits Quarterly* 25: 38–50.

Ray, R., and J. A. Rosow. 2010. Getting off and getting intimate: How normative institutional arrangements structure Black and White fraternity men's approaches toward women. *Men and Masculinities* 12: 523–546.

Read, S., and E. Grundy. 2011. Mental health among older married couples: the role of gender and family life. *Social Psychiatry & Psychiatric Epidemiology* 46: 331–341.

Realinia, J. P., R. S. Buzib, P. B. Smithc, and M. Martinezd. 2010. Evaluation of "Big Decisions": An abstinence-plus sexuality curriculum. *Journal of Sex & Marital Therapy* 36: 313–326.

Reczek, C., H. Liu, and D. Umberson. 2010. Just the two of us? How parents influence adult children's marital quality. *Journal of Marriage and Family* 72: 1205–1219.

Reece, M., D. Herbenick, B. Dodge, S. A. Sanders, A. Ghassemi, and D. Fortenberry. 2010. Vibrator use among heterosexual men varies by partnership status: Results from a nationally representative study in the United States. *Journal of Sex & Marital Therapy* 36: 389–407.

Rees, C. S., and S. Hawthornthwaite. 2004. Telepsychology and videoconferencing: Issues, opportunities and guidelines for psychologists. *Australian Psychologist* 39: 212–219.

Regan, P. 2000. Love relationships. In *Psychological perspectives on human sexuality,* ed. L. T. Szuchman and F. Muscarella, 232–282. New York: Wiley.

Regan, P. C., and A. Joshi. 2003. Ideal partner preferences among adolescents. *Social Behavior and Personality* 31: 13–20.

Reiss, I. L. 1960. Toward a sociology of the heterosexual love relationship. *Journal of Marriage and Family Living* 22: 139–145.

Reynaud, M., L. Blecha, and A. Benyamina. 2011. Is love passion an addictive disorder? *American Journal of Drug & Alcohol Abuse* 36: 261–267.

Reyome, N. D. 2010. Childhood emotional maltreatment and later intimate relationships: Themes from the empirical literature. *Journal of Aggression, Maltreatment & Trauma* 19: 224–242.

Rhatigan, D. L., and A. M. Nathanson. 2010. The role of female behavior and attributions inpredicting behavioral responses to hypothetical male aggression. *Violence Against Women* 16: 621–637.

Rhoades, G. K., S. M. Stanley, and H. J. Markman. 2009. Couples' reasons for cohabitation: Associations with individual well-being and relationship quality. *Journal of Family Issues* 30: 233–246.

Rhoades, G. K., S. M. Stanley, and H. J. Markman. 2010. Should I stay or should I go? Predicting dating relationship stability from four aspects of commitment. *Journal of Family Psychology* 24: 543–550.

Rhodes, K V., C. Cerulli, M. E. Dichter, C. L. Kothari, and F. K. Barg. 2010. "I Didn't Want To Put Them Through That": The influence of children on victim decision-making in intimate partner violence cases. *Journal of Family Violence* 25: 485–493.

Richeimer, S. 2011. Love hurts. KABC, Los Angeles http://abclocal.go.com/kabc/story?section=news/health/your_health&id=8039618

Ridley, C. 2009. Personal communication with retired professor from University of Arizona.

Riela, S., G. Rodriguez, A. Aron, X. Xu, and B. P. Acevedo. 2010. Experiences of falling in love: Investigating culture, ethnicity, gender, and speed. *Journal of Social and Personal Relationships* 27: 473–493.

Riffe, J., D. Brandon, M. Mulroy, and A. Faulkner. 2010. Parent education for separating or divorcing families: Results of a national extension survey. Poster, National Council on Family Relations annual meeting, November 3–5. Minneapolis, MN.

Rill, L., E. Baiocchi, M. Hopper, K. Denker, and L. Olson. 2009. Exploration of the relationship between self-esteem, commitment, and verbal aggressiveness in romantic dating relationships. *Communication Reports* 22: 102–113.

Rinelli, L., and S. L. Brown. 2010. Race differences in union transitions among cohabitors: The role of relationship features. *Marriage & Family Review* 46: 22–40.

Rivadeneyraa, R., and M. J. Lebob. 2008. The association between television-viewing behaviors and adolescent dating role attitudes and behaviors. *Journal of Adolescence* 31: 291–305.

Rivers, D. 2010. "In the best interests of the child": lesbian and gay parenting custody cases, 1967–1985. *Journal of Social History* 43: 917–954.

Robbins, C. A. 2005. ADHD couple and family relationships: Enhancing communication and understanding through Imago Relationship Therapy. *Journal of Clinical Psychology* 61: 565–578.

Robnett, B. and C. Feliciano. 2011. Patterns of racial-ethnic exclusion by Internet daters. *Social Forces* 89: 807–828.

Robnett, R. D., and J. E. Susskind 2010. Who cares about being gentle? The impact of social identity and the gender of one's friends on children's display of same-gender favoritism. *Sex Roles* 63: 820–832.

Roby, J., and H. White. 2010. Adoption activities on the Internet: A call for regulation. *Social Work* 55: 203–212.

Rochman, B. 2009 The ethics of octuplets. *Time,* February 16, 43–44.

Rock, M., T. S. Carlson and C. R. McGeorge. 2010. Does affirmative training matter? Assessing CFT students' beliefs about sexual orientation and their level of affirmative training. *Journal of Marital & Family Therapy* 36: 171–184.

Rodriguez, Y., J. Su, and H. Helms. 2010. Sociocultural context, relationship quality and gender: "His" and "her" experiences in married and cohabiting couples of Mexican origin in the childbearing years. Paper, Annual Meeting of the National Council on Family Relations,November 3–5. Minneapolis, MN.

Roeters, A., and J. K. Treas. 2011. Parental work demands and parent-child, family, and couple leisure in Dutch families: What gives? *Journal of Family Issues* 32: 269–291.

Rohde-Brown, J., and K. E. Rudestam. 2011. The role of forgiveness in divorce adjustment and the impact of affect. *Journal of Divorce & Remarriage* 52: 109–124.

Rohrbaugh, J. B. 2006. Domestic violence in same-gender relationships. *Family Court Review* 44: 287–299.

Rolfe, A. 2008. "You've got to grow up when you've got a kid": Marginalized young women's accounts of motherhood. *Journal of Community & Applied Social Psychology* 18: 299–312.

Roopnarine, J. L., P. Bynoe, R. Singh, and R. Simon. 2005. Caribbean families in English-speaking countries. In *Families in global perspective*, ed. J. L. Roopnarine and U. P. Gielen, 311–329. Boston: Pearson Allyn & Bacon.

Rosario, M., E. W. Schrimshaw, J. Hunter, and L. Braun. 2006. Sexual identity development among lesbian, gay, and bisexual youth: Consistency and change over time. *Journal of Sex Research* 43: 46–58.

Rose, K., and K. J. Elicker. 2008 Parental Decision Making about Child Care. *Journal of Family Issues* 29: 1161–1179.

Rose, S. 2005. Going too far? Sex, sin and social policy. *Social Forces* 84: 1207–1232.

Rosen, E. 2006. Derailed on the mommy track? There's help to get going again. *New York Times* 155(10): 1–3.

Rosenberg, M. B. 2010. Who's your mommy, Who's your daddy? Legal complexities of ART and third-party reproduction. *American Journal of Family Law* 24: 95–98.

Rosin, H. 2010. The end of men. *The Atlantic*, August.

Rosin, H., and R. Morin. 1999. In one area, Americans still draw a line on acceptability. *Washington Post National Weekly Edition* 16(January 11): 8.

Ross, C. B. 2006. An exploration of eight dimensions of self-disclosure on relationship. Paper for Southern Sociological Society, New Orleans, LA. March 24.

Ross, C., D. Knox, and M. Zusman. 2008. "Hey Big Boy": Characteristics of university women who initiate relationships with men. Poster, Southern Sociology Society, Richmond.

Rossi, N. E. 2010 "'Coming out" Stories of gay and lesbian young adults *Journal of Homosexuality* 57: 1174–1191.

Rothman, E. F., D. Exner, and A. Baughman. 2011. The prevalence of sexual assault against people who identify as gay, lesbian, or bisexual in the United States: A systematic review. *Trauma, Violence & Abuse* 12: 55–66.

Rountree, C. 1993. *On women turning 50: Celebrating mid-life discoveries.* New York: Harper Collins.

Routt, G. G., and L. Anderson. 2010. Adolescent violence towards parents. *Journal of Aggression,Maltreatment & Trauma* 20: 1–18.

Royo-Vela, M., J. Aldas-Manzano, I. Kuster, and N. Vila. 2008. Adaptation of marketing activities to cultural and social context: Gender role portrayals and sexism in Spanish commercials. *Sex Roles* 58: 379–391.

Rubin, D. M., C. W. Christian, L. T. Bilaniuk, K. A. Zaxyczny, and D. R. Durbin. 2003. Occult head injury in high-risk abused children. *Pediatrics* 111: 1382–1386.

Rubinstein, G. 2010. Narcissism and self-esteem among homosexual and heterosexual male students. *Journal of Sex & Marital Therapy* 36: 24–34.

Ruff, S., J. L. McComb, C. J. Coker, and D.H. Sprenkle. 2010. Behavioral couples therapy for the treatment of substance abuse: A substantive and methodological review of O'Farrell, Fals-Stewart, and colleagues' program of research. *Family Process* 49: 439–456.

Ruschena, E., M. Prior, A. Sanson, and D. Smart. 2005. A longitudinal study of adolescent adjustment following family transitions. *Journal of Child Psychology & Psychiatry & Allied Disciplines* 46: 353–363.

Russell, L., and S. E. Weaver. 2010. Building steprelationships: Emerging adult stepchildrens' perceptions and experiences of their stepparents. Poster, National Council on Family Relations annual meeting, November 3–5. Minneapolis, MN.

Russella, D., K. W. Springerb, and E. A. Greenfieldd. 2010. Witnessing domestic abuse in childhood as an independent risk factor for depressive symptoms in young adulthood. *Child Abuse & Neglect* 34: 448–453.

Russo, F. 2010. *They're your parents too! How siblings can survive their parents' aging without driving each other crazy.* New York: Bantam Dell Publishing Group.

Rust, Paula. 1993. Neutralizing the political threat of the marginal woman: lesbians' beliefs about bisexual women. *Journal of Sex Research* 30(3): 214–218.

Rutherford, L. G., and W. S. Fox. 2010. Financial wellness of young adults age 18–30. *Family and Consumer Sciences Research Journal* 38: 468–484.

Ryan, J., I. Carrière, J. Scali, K. Ritchie, and M. Ancelin. 2008. Lifetime hormonal factors may predict late-life depression in women. *International Psychogeriatrics* 20: 1203–1229.

S

Saad, L. 2005. Gay rights attitudes a mixed bag. Gallup Organization (May 20). http://www.gallup.com.

Sack, K. 2008. Health benefits inspire rush to marry, or divorce. *The New York Times* August 12.

Saint, D. J. 1994. Complementarily in marital relationships. *Journal of Social Psychology* 134: 701–704.

Sammons, L. 2008. Personal communication, Grand Junction, Colorado.

Sammons, R. A., Jr. 2010. First law of therapy. Personal communication. Grand Junction, Colorado.

Samuels, G. M. 2010. Building kinship and community: Relational processes of bicultural identity among adult multiracial adoptees. *Family Process* 49: 26–42.

Samuels-Dennis, J., M. Ford-Gilboe and S. Ray. 2011. Single mother's adverse and traumatic experiences and post-traumatic stress symptoms. *Journal of Family Violence* 26: 9–20.

Sanday, P. R. 1995. Pulling train. In *Race, class, and gender in the United States*, 3rd ed., ed. P. S. Rothenberg, 396–402. New York: St. Martin's Press.

Sandler, L. 2010. One and done. *Time*, July19: 34–41.

Sanford, K., and A. J. Grace. 2011. Emotion and underlying concerns during couples' conflict: An investigation of within-person change. *Personal Relationships* 18: 96–109.

Sarkar, N. N. 2008. The impact of intimate partner violence on women's reproductive health and pregnancy outcome. *Journal of Obstetrics and Gynecology* 28: 266–278.

Sasaki, T., N. L. Hazen, and W. B. Swann Jr. 2010. The supermom trap: Do involved dads erode moms' self-competence? *Personal Relationships* 17: 71–79.

Sassler, S. 2010. Partnering across the life course: Sex, relationships, and mate selection. *Journal of Marriage and Family* 72: 557–575.

Sassler, S., and A. Cunningham. 2008. How cohabitors view childrearing. *Sociological Perspectives* 51: 3–29.

Sassler, S. and A. J. Miller. 2011b. Class differences in cohabitation processes. *Family Relations* 60: 163–177.

Sassler, S., and A. Miller. 2011a. Waiting to be asked: Gender, power, and relationship progression among cohabiting couples. *Journal of Family Issues* 32: 482–506.

Savin-Williams, R. C. 2006. Who's gay? Does it matter? Current Directions in *Psychological Science* 15: 40–44.

Sayare, S., and M. Ia De La Baume. 2010. In France, Civil Unions gain favor over marriage. *The New York Times* Dec 15.

Sayer, L. C., S. M. Bianchi, and J. P. Robinson. 2004. Are parents investing less in children? Trends in mothers' and father's time with children. *American Journal of Sociology* 110: 143.

Sayer, L. S., and L. Fine. 2011. Racial-ethnic differences in U.S. married women's and men's housework. *Social Indicators Research* 101: 259–265.

Schacht, T. E. 2000. Protection strategies to protect professionals and families involved in high-conflict divorce. *UALR Law Review* 22(3): 565–592.

Schairer, C., J. Lubin, R. Troisi, S. Sturgeon, L. Brinton, and R. Hoover. 2000. Menopausal estrogen and estrogen-progestin replacement therapy and breast cancer risk. *Journal of the American Medical Association* 283: 485–491.

Scheib, J. E., M. Riordan, and S. Rubin 2005. Adolescents with open-identity sperm donors: reports from 12–17 year olds. *Human Reproduction* 20: 239–252.

Scherrer, K. S. 2010. The intergenerational family relationships of grandparents and GLBQ Grandchildren. *Journal of GLBT Family Studies* 6: 229–264.

Schiappa, E., P. B. Gregg, and D. E. Hewes. 2005. The parasocial contact hypothesis. *Communication Monographs* 72: 92–115.

Schieman, S., C. G. Ellison, and A. Bierman. 2010. Religious involvement, beliefs about God, and the sense of mattering among older adults. *Journal for the Scientific Study of Religion* 49: 517–535.

Schindler, H. S. 2010. The importance of parenting and financial contributions in promoting fathers' psychological health. *Journal of Marriage and Family* 72: 318–332.

Schmeer, K. K. 2011. The child health disadvantage of parental cohabitation. *Journal of Marriage & Family* 73: 181–193.

Schoebi, D. 2008. The coregulation of daily affect in marital relationships. *Journal of Family Psychology* 22: 595–604.

Schoen, R., N. M. Astone, K. Rothert, N. J. Standish, and Y. J. Kim. 2002. Women's employment, marital happiness, and divorce. *Social Forces* 81: 643–662.

Schoen, R., N. S. Landale, and K. Daniels. 2007. Family transitions in young adulthood. *Demography* 44: 807–830.

Schoen, R., S. J. Rogers, and P. R. Amato. 2006. Wives' employment and spouses' marital happiness: Assessing the direction of influence using longitudinal couple data. *Journal of Family Issues* 27: 506–528.

Schoppe-Sullivan, S. J., G. L. Brown, E. A. Cannon, S. C. Mangelsdorf, and M. S. Sokolowski. 2008. Maternal gatekeeping, coparenting quality, and fathering behavior in families with infants. *Journal of Family Psychology* 22: 389–397.

Schum, T. R. 2007. Dave's dead! Personal tragedy leading a call to action in preventing suicide. *Ambulatory Pediatrics* 7: 410–412.

Schwartz, S J., B. L. Zamboanga, R. D. Ravert, S. Y. Kim, R. S. Weisskirch, M. K. Williams, M. Bersamin, and G. E. Finley. 2009. Perceived parental relationships and health-risk behaviors in college-attending emerging adults. *Journal of Marriage and the Family* 71: 727–740.

Schwarz, S., M. Hassebrauck, and R. Dorfler. 2010. Let us talk about sex: Prototype and personal templates. *Personal Relationship* 17: 533–555.

Scott, L. S. 2009. *Two is enough*. Berkeley, California: Seal Press.

Sears, W., and M. Sears. 1993. *The baby book*. Boston: Little, Brown.

Sefton, B. W. 1998. The market value of the stay-at-home mother. *Mothering* 86: 26–29.

Seguin-Levesque, C., M. L. N. Laliberte, L. G. Pelletier, C. Blanchard, and R. J. Vallerand. 2003. Harmonious and obsessive passion for the Internet: Their associations with the couple's relationship. *Journal of Applied Social Psychology* 33: 197–221.

Seifert, A. E, M. Polusny, and M. Murdoch. 2011. The association between childhood physical and sexual abuse and functioning and psychiatric symptoms in a sample of U.S. Army soldier. *Military Medicine* 176: 176–181.

Sener, I. P., R. B. Copperman, R. M. Pendyala, and C. R. Bhat. 2008. An analysis of children's leisure activity engagement: examining the day of week, location, physical activity level, and fixity dimensions. *Transportation* 35: 673–697.

Serido, J., S. Shim, A. Mishra, and C. Tang. 2010. Financial parenting, financial coping behaviors, and well-being of emerging adults. *Family Relations* 59: 453.

Serovich, J. M., S. M. Craft, P. Toviessi, R. Gangamma, et al. 2008. A systematic review of the research base on sexual reorientation therapies. *Journal of Marital and Family Therapy* 34: 227–239.

Sever, I., J. Guttmann, and A. Lazar. 2007. Positive consequences of parental divorce among Israeli young adults: A long-term effect model. *Marriage and Family Review* 42: 7–21.

Shackelford, T. K., A. T. Goetz, D. M. Buss, H. A. Euler, and S. Hoier. 2005. When we hurt the ones we love: Predicting violence against women from men's mate retention. *Personal Relationships* 12: 447–463.

Shafer, K. 2010. Yours, mine, and hours: Relationship skills for blended familiers. *Journal of Comparative Family Studies* 41: 185–186.

Shamai, M., and E. Buchbinder. 2010. Control of the self: Partner-violent men's experience of therapy. *Journal of Interpersonal Violence* 25: 1338–1362.

Sharp, E. A., and L. Ganong. 2007. Living in the gray: Women's experiences of missing the marital transition *Journal of Marriage and Family* 69: 831–844.

Shawn P. McCoy and M. G. Aamodt. 2009. A comparison of law enforcement divorce rates with those of other occupations. *Journal of Police and Criminal Psychology* Online Journal published Oct 20, 2009.

Sheehy, G. 2010. *Passages in caregiving*. New York: William Morrow.

Shelon, J. N., T. E. Trail, T. V. West, and H. B. Bergsieker. 2010. From strangers to friends: The interpersonal process model of intimacy in developing interracial friendships. *Journal of Social and Personal Relationships* 27: 71–90.

Sherif-Trask, B. 2003. Love, courtship, and marriage from a cross-cultural perspective: The upper middle class Egyptian example. In *Mate Selection Across Cultures* ed. R. R. Hamon and B. B. Ingoldsby, 121–136. Thousand Oaks, CA: Sage Publications.

Shin, K. H. Shin, J. A. Yang, and C. Edwards. 2010. Gender role identity among Korean and American college students: Links to gender and academic achievement. *Social Behavior & Personality: An International Journal* 38: 267–272.

Shinn, L. K., and M. O'Brien. 2008. Parent-child conversational styles in middle childhood: Gender and social class differences. *Sex Roles* 59: 61–69.

Shook, S. E., D. J. Jones, R. Forehand, S. Dorsey, and G. Brody. 2010. The mother–coparent relationship and youth adjustment: A study of African American single-mother families. *Journal of Family Psychology* 24: 243–251.

SIECUS (Sexuality Information and Education Council of the United States). 2009. Facts. Retrieved March 23, http://www.dianedew.com/siecus.htm.

Siedlecki, K. L. 2007. Investigating the structure and age invariance of episodic memory across the adult lifespan. *Psychology & Aging* 22: 251–268.

Siegler, I., and P. Costa. 2000. Divorce in midlife. Paper presented at the Annual Meeting of the American Psychological Association, Boston.

Silvergleid, C., and E. S. Mankowski. 2006. How batterer intervention programs work: Participant and facilitator accounts of processes of change. *Journal of Interpersonal Violence* 21: 139–159.

Silverstein, L. B., and C. F. Auerbach. 2005. (Post) modern families. In *Families in global perspective*, ed. Jaipaul L. Roopnarine and U. P. Gielen, 33–48. Boston, MA: Pearson Education.

Simon, R. J., and R. M. Roorda. 2000. *In their own voices: Transracial adoptees tell their stories*. New York: Columbia University Press.

Simon, R. W., and K. Lively. 2010. Sex, anger, and depression. *Social Forces* 88: 1543–1568.

Simonelli, C. J., T. Mullis, A. N. Elliott, and T. W. Pierce. 2002. Abuse by siblings and subsequent experiences of violence within the dating relationship. *Journal of Interpersonal Violence* 17: 103–121.

Simrns, D. C., and E. S. Byers. 2009. Interpersonal perceptions of desired frequency of sexual behaviours. *Canadian Journal of Human Sexuality* 18: 15–25.

Singer, E. 2010. The 'W.I.S.E. Up!' tool: Empowering adopted children to cope with questions and comments about adoption. *Pediatric Nursing* 36: 209–212.

Skaine, R. 2002. *The women of Afghanistan under the Taliban*. Jefferson, NC: McFarland.

Slinger, M. R., and D. J. Bredehoft. 2010. Relationships between childhood overindulgence and adult attitudes and behavior. Poster, National Council on Family Relations annual meeting, November 3–5. Minneapolis, MN

Smedema, S. M., D. Catalano, and D. J. Ebener. 2010. The relationship of coping, self-worth, and subjective well-being: A structural equation model. *Rehabilitation Counseling Bulletin* 53: 131–142.

Smith, T. 2007. Four kids is the new standard. National Public Radio, August 5.

Smith, L. H., and J. Ford. 2010. History of forced sex and recent sexual risk indicators among young adult males. *Perspectives on Sexual & Reproductive Health* 42: 87–92.

Smith, A., A. Lyons, J. Ferris, J. Richter, M. Pitts, and J. Shelley. 2010a. Are sexual problems more common in women who have had a tubal ligation? A population-based study of Australian women. *An International Journal of Obstetrics & Gynecology* 117: 463–468.

Smith, P. H., C. E. Murray, and A. L. Coker. 2010b. The Coping Window: A contextual understanding of the methods women use to cope with battering. *Violence and Victims* 25: 18–28.

Smith, J., M. Borchelt, H. Maier, and D. Jopp. 2002. Health and well-being in the young old and oldest old. *Journal of Social Issues* 58: 715–733.

Smock, P. J. 2000. Cohabitation in the United States: An appraisal of research themes, findings, and implications. *Annual Review of Sociology* 26: 1–20.

Smock, P. J., and F. R. Greenland. 2010. Diversity in pathways to parenthood: Patterns, implications, and emerging research directions. *Journal of Marriage and Family* 72: 576–593.

Snorton, R. 2005. GLSEN's 2004 State of the States Report is the first objective analysis of statewide safe schools policies (April 1). http://www.glsen.org.

Snyder, K. A. 2007. A vocabulary of motives: Understanding how parents define quality time *Journal of Marriage and Family* 69: 320–340.

Sobolewski, J. M., and P. R. Amato. 2007. Parents' discord and divorce, parent-child relationships and subjective well-being in early adulthood: Is feeling close to two parents always better than feeling close to one? *Social Forces* 1105–1125.

Society for the Advancement of Education. 2005. Mr. Mom nation reaches new peak. *USA Today* August 13.

Solem, M., K. Christophersen, and M. Martinussen. 2011. Predicting parenting stress: children's behavioral problems and parents' coping. *Infant & Child Development* 20: 162–180.

Solomon, Z., S. Debby-Aharon, G. Zerach, and D. Horesh. 2011. Marital adjustment, parental functioning, and emotional sharing in war veterans. *Journal of Family Issues* 32: 127–147.

Soltz, V., and R. Dreikurs. 1991. *Children: The challenge.* New York: Penguin.

Soo Jung, J., and A. Zippay. 2011. Juggling act: Managing work-life conflict and work-life balance. *Families in Society* 92: 84–90.

South, S. C., B. D. Doss, and A. Christensen, A. 2010. Through the eyes of the beholder: The mediating role of relationship acceptance in the impact of partner behavior. *Family Relations* 59: 611–622.

Spitzberg, B. H., and W. R. Cupach, eds. 1998. *The dark side of close relationships.* Mahway, N.J.: Erlbaum.

Spitzberg, B. H., and W. R. Cupach. 2007. Cyberstalking as (mis)matchmaking. In *Online matchmaking*, ed. M. T. Whitty, A. J. Baker, and J. A. Inman, 127–46. New York: Palgrave Macmillan.

Spoto, D. 2006. *Enchantment: The life of Audrey Hepburn.* New York: Harmony Books.

Sprecher, S. 2009. Relationship initiation and formation on the Internet. *Marriage & Family Review* 45: 761–782.

Sprecher, S. 2002. Sexual satisfaction in premarital relationships: Associations with satisfaction, love, commitment, and stability. *Journal of Sex Research* 39: 190–196.

Sprecher, S., and P. C. Regan. 2002. Liking some things (in some people) more than others: Partner preferences in romantic relationships and friendships. *Journal of Social and Personal Relationships* 19: 463–481.

Spruijt, E., and V. Duindam. 2010. Joint physical custody in The Netherlands and the well-being of children. *Journal of Divorce & Remarriage* 51: 65–82.

Stack, S., and J. R. Eshleman. 1998. Marital happiness: A 17-nation study. *Journal of Marriage and the Family* 60: 527–536.

Stallard, P., S. Velleman and T. Richardson 2010. Computer use and attitudes towards computerised therapy amongst young people and parents attending child and adolescent mental health services. *Child and Adolescent Mental Health* 15: 80–84.

Stanfield, J. B. 1998. Couples coping with dual careers: A description of flexible and rigid coping styles. *Social Science Journal* 35: 53–62.

Stanley, S. M., and H. J. Markman. 1992. Assessing commitment in personal relationships. *Journal of Marriage and the Family* 54: 595–608.

Stanley, S. M., and L. A. Einhorn. 2007. Hitting pay dirt: Comment on "Money: A therapeutic tool for couples therapy." *Family Process* 46: 293–299.

Statistical Abstract of the United States, 2011, 130th ed. Washington, DC: U.S. Census Bureau.

Stein, J. 2011. Can buy me love. *Time Magazine* (March 28), 82.

Steinberg, J. R., and N. F. Russo. 2008. Abortion and anxiety: What's the relationship? *Social Science & Medicine* 67: 238–42.

Stephenson, L. E., S. N. Culos-Reed, P. K. Doyle-Baker, J. A. Devonish, and J. A. Dickinson. 2007. Walking for wellness: Results from a mall walking program for the elderly. *Journal of Sport & Exercise Psychology* 29: 204–214.

Sternberg, R. J. 1986. A triangular theory of love. *Psychological Review* 93: 119–135.

Stone, A. 2006. Drives to ban gay adoption heat up. *USA Today*, February 21, A1.

Stone, E. 2008. The last will and testament in literature: Rupture, rivalry, and some-times Rapprochement from Middlemarch to Lemony Snikert. *Family Process* 47: 425–439.

Stone, N., B. Hatherall, R. Ingham, and J. McEachron. 2006. Oral sex and condom use among young people in the United Kingdom. *Perspectives on Sexual and Reproductive Health* 38: 6–13.

Stone, P. 2007. *Opting out?* Berkley: University of California Press.

Stott, P. 2010. No recession for workplace romance. Vault.Com Office Romance Survey 2010. Posted on Vault.com's website February 14 and retrieved February 18 at http://vaultcareers.wordpress.com/2010/02/17/no-recession-for-workplace-romance/

Straus, M. A. 2000. Corporal punishment and primary prevention of physical abuse. *Child Abuse and Neglect* 24: 1109–1114.

Strassberg, D. S., and S. Holty. 2003. An experimental study of women's Internet personal ads. *Archives of Sexual Behavior* 32: 253–261.

Strazdins, L., M. S. Clements, R. J. Korda, D. H. Broom, and M. Rennie. 2006. Unsociable work? Nonstandard work schedules, family relationships, and children's well-being. *Journal of Marriage and Family* 68: 394–410.

Strickland, S. M. 2008. Female sex offenders: Exploring issues of personality, trauma, and cognitive distortions. *Journal of Interpersonal Violence* 23: 474–489.

Strickler, B. L., and J. D. Hans (2010). Defining infidelity and identifying cheaters: An inductive approach with a factoral design. Poster, National Council on Family Relations annual meeting, November 3–5. Minneapolis, MN.

Stroebe, M., and H. Schut. 2005. To continue or relinquish bonds: A review of consequences for the bereaved. *Death Studies* 29: 477–495.

Sugarman, S. D. 2003. Single-parent families. In *All our families: New policies for a new century*, 2nd ed., ed. M. A. Mason, A. Skolnick, and S. D. Sugarman, 14–39. New York: Oxford University Press.

Suitor, J. J., and K. Pillemer. 2007. Mothers' favoritism in later life: The role of children's birth order. *Research on Aging* 29: 32–42.

Sullivan, A. 1997. The conservative case. In *Same sex marriage: Pro and con*, ed. A. Sullivan, 146–54. New York: Vintage Books.

Sulloway, F. J. 1996. *Born to rebel: Birth order, family dynamics, and creative lives.* New York: Vintage Books.

Sulloway, F. J. 2007. Birth order and intelligence. *Age and Intelligence* 316: 1711–1721.

Sutphina, S. T. 2010. Social exchange theory and the division of household labor in same-sex couples. *Marriage and Family Review* 46: 191–206.

Swan, S. C., L. J. Gambone, J. E. Caldwell, T. P. Sullivan, and D. L Snow. 2008. A review of research on women's use of violence with male intimate partners. *Violence and Victims* 23: 301–315.

Swartout, K. M. and J. W. White. 2010. The relationship between drug use and sexual aggression in men across time. *Journal of Interpersonal Violence* 25: 1716–1735.

Sweeney, M. M. 2010. Remarriage and stepfamilies: Strategic sites for family scholarship in the 21st century. *Journal of Marriage and the Family* 72: 667–684.

Swenson, D., J. G. Pankhurst, and S. K. Houseknecht. 2005. Links between families and religion. In *Sourcebook of family theory & research*, ed. V. L. Bengtson, A. C. Acock, K. R. Allen, P. Dilworth-Anderson, and D. M. Klein, 530 33. Thousand Oaks, California: Sage Publications.

Sylvestrea, A. and C. Mérettec. 2010. Language delay in severely neglected children: A cumulative or specific effect of risk factors? *Child Abuse & Neglect* 34: 414–428.

Szabo, A., S. E. Ainsworth, and P. K. Danks. 2005. Experimental comparison of the psychological benefits of aerobic exercise, humor, and music. *International Journal of Humor Research* 18: 235–246.

T

Taliaferro, L. A., B. A. Rienzo, M. D. Miller, R. M. Pigg, and V. J. Dodd. 2008. High school youth and suicide risk: Exploring protection afforded through physical activity and sport participation. *The Journal of School Health* 78:545–556.

Tabi, M. M., C. Doster, and T. Cheney. 2010 A qualitative study of women in polygamous marriages. *International Nursing Review* 57: 121–127.

Takata, Y., T. Ansai, I. Soh, S. Awano, Y. Yoshitake, Y. Kimura, et al. 2010. Quality of life and physical fitness in an 85-year-old population. *Archives of Gerontology & Geriatrics* 50: 272–276.

Tannen, D. 1990. *You just don't understand: Women and men in conversation.* London: Virago.

Tannen, D. 1998. *The argument culture* New York: Random House.

Tannen, D. 2006. *You're wearing that? Understanding mothers and daughters in conversation.* New York: Random House.

Tao, G. 2008. Sexual orientation and related viral sexually transmitted disease rates among U.S. women aged 15 to 44 years. *American Journal of Public Health* 98: 1007–1010.

Taubman-Ben-Ari, O., and A. Noy. 2011. Does the transition to parenthood influence driving?. *Accident Analysis & Prevention* 43: 1022–1035.

Taylor, R. L. 2002. Black American families. In *Minority families in the United States: A multicultural perspective*, ed. Ronald L. Taylor, 19–47. Upper Saddle River, NJ: Prentice Hall.

Taylor, A. C., and A. Bagd. 2005. The lack of explicit theory in family research: The case analysis of the *Journal of Marriage and the Family 1990–1999*. In *Sourcebook of family theory & research*, ed. Vern L. Bengtson, Alan C. Acock, Katherine R. Allen, Peggye Dilworth-Anderson, and David M. Klein, 22–25. Thousand Oaks, CA: Sage Publications.

Teachman, J. 2008. Complex life course patterns and the risk of divorce in second marriages *Journal of Marriage & Family* 70: 294–305.

Teich, M. 2007. A divided house. *Psychology Today* 40: 96–102.

Ter Gogt, T. F. M., R. C. M. E. Engles, S. Bogers, and M. Kloosterman. 2010. "Shake It Baby, Shake It": Media preferences, sexual attitudes and gender stereotypes among adolescents. *Sex Roles* 63: 844–859.

Termini, K. A. 2006. Reducing the negative psychological and physiological effects of chronic stress. 4th Annual ECU Research and Creative. Activities Symposium, April 21, East Carolina University, Greenville, NC.

Teten, A. L., B. Ball, L. A. Valle, R. Noonan, and B. Rosenbluth. 2009. Considerations for the definition, measurement, consequences, and prevention of dating violence victimization among adolescent girls. *Journal of Women's Health* 18: 923–927.

Teten, A. L., J. A. Schumacher, C. T. Taft, M. A. Stanley, T. A. Kent, S. D. Bailey, N. Jo Dunn, and D. L. White. 2010. Intimate partner aggression perpetrated and sustained by male Afghanistan, Iraq, and Vietnam veterans with and without post-traumatic stress disorder. *Journal of Interpersonal Violence* 25: 1612–1630.

Tetlie, T., N. Eik-Nes, T. Palmstierna, P. Callaghan, and J. A Nøttestad. 2008. The effect of exercise on psychological & physical health outcomes: Preliminary results from a Norwegian forensic hospital. *Journal of Psychosocial Nursing & Mental Health Services* 46: 38–44.

Textor, K. 2007. "The Millennials are Coming" *Sixty Minutes*, November 11. CBS Television.

The, N. S., and P. Gordon-Larsen 2009. Entry into romantic partnership is associated with obesity. *Obesity* 10: 97–112.

Theiss, J. A. and M. E. Nagy. 2010. Actor-partner effects in the associations between relationship characteristics and reactions to marital sexual intimacy. *Journal of Social and Personal Relationships* 27: 1089–1109.

Thomas, P. A., E. M. Krampe, and R. R. Newton. 2008. Father presence, family structure, and feelings of closeness to the father among adult African American children. *Journal of Black Studies* 38: 529–541.

Thomas, S. G. 2007. *Buy, buy baby: How consumer culture manipulates parents and harms young minds*. Houghton Mifflin Publisher.

Thomsen, D., and I. J. Chang. 2000. Predictors of satisfaction with first intercourse: A new perspective for sexuality education. Poster at the 62nd Annual Conference of the National Council on Family Relations, Minneapolis, November.

Thomson, E., and E. Bernhardt. 2010. Education, values, and cohabitation in Sweden. *Marriage & Family Review* 46: 1–21.

Thompson, M. 2008. America's medicated army. *Time* July 16, 38–42.

Thompson, E. H., and J. Hampton. 2011. The effect of relationship status on communicating emotions through touch. *Cognition & Emotion* 25: 295–306.

Tilden, T., T. Gude, A. Hoffart, and H. Sexton. 2010. Individual distress and dyadic adjustment over a three-year follow-up period in couple therapy: a bi-directional relationship? *Journal of Family Therapy* 32: 119–141.

Time Magazine. 2010 Geneva: Gender equality around the world. October 25, p. 19.

Tobias, S., and S. Cahill. 2003. School lunches, the Wright brothers, and gay families. National Gay and Lesbian Task Force. http://www.thetaskforce.org.

Toma, C. L., and J. T. Hancock. 2010. Looks and lies: The role of physical attractiveness in online dating self-presentation and deception. *Communication Research* 37: 335–351.

Tomkins, S. A., J. Rosa, and J. Benavente. 2010. A holistic approach to relationship enrichment. Poster, National Council on Family Relations annual meeting, November 3–5. Minneapolis, MN.

Torpy, J. M., C. Lynm, and R. M Glass. 2008. Intimate partner violence. *Journal of the American Medical Association* 300: 754–766.

Torres, S. and G. Hammarström. 2009. Successful aging as an oxymoron: older: people with and without home-help care: talk about what aging well means to them *International Journal of Ageing & Later Life* 4: 23–54.

Toufexis, A. 1993. The right chemistry. *Time* February 15, 49–51.

Trask, B. S. 2010. *Globalization and families: Accelerated systemic social change*. New York: Springer.

Treas, J., and D. Giesen. 2000. Sexual infidelity among married and cohabiting Americans. *Journal of Marriage and the Family* 62: 48–60.

Trinder, L. 2008. Maternal gate closing and gate opening in postdivorce families. *Journal of Family Issues* 29: 1298–2011.

Trotter, P. B. 2010. The influence of parental romantic relationships on college students' attitudes about romantic relationships. *College Student Journal* 44: 71–84.

True Love Waits. 2010. http://www.lifeway.com/tlw/students/join.asp (accessed May 21, 2010).

Tsai, M., and C. Tsai. 2010 Junior high school students' Internet usage and self-efficacy: A re-examination of the gender gap. *Computers & Education* 54: 1182–1192.

Tucker, P. 2005. Stay-at-home dads. *The Futurist* 39: 12–15.

Tucker, C. J., S. M. McHale, and A. C. Crouter. 2003. Dimensions of mothers' and fathers' differential treatment of siblings: Links with adolescents' sex-typed personal qualities. *Family Relations* 52: 82–89.

Tucker, P., A. Dahlgren, T. Akerstedt, and J. Waterhouse. 2008. The impact of free-time activities on sleep, recovery and well-being. *Applied Ergonomics* 39: 653–661.

Tumulty, K. 2010. America, the doctor will see you now. *Time*, April 8. 24–32.

Turkat, I. D. 2002. Shared parenting dysfunction. *The American Journal of Family Therapy* 30: 385–393.

Turla, A., C. Dundar, C. Ozkanli and C. Ozkanli. 2010. Prevalence of childhood physical abuse in a representative sample of college students in Samsun, Turkey. *Journal of Interpersonal Violence* 25: 1298–1308.

Turner, A. J. 2005. Communication basics. Personal communication. Huntsville, Alabama Turner, J. H. 2007. Self, emotions, and extreme violence: Extending symbolic interactionist theorizing. *Symbolic Interaction* 30: 501–531.

Twenge, J. M., W. K. Campbell, and C. A. Foster. 2003. Parenthood and marital satisfaction: A meta-analytic review. *Journal of Marriage and Family* 65: 574–583.

Tyagart, C. E. 2002. Legal rights to homosexuals in areas of domestic partnerships and marriages: Public support and genetic causation attribution. *Educational Research Quarterly* 25: 20–29.

Tyre, P., and D. McGinn. 2003. She works, he doesn't. *Newsweek*, May 12: 45–52.

Tzeng, O. C. S., K. Wooldridge, and K. Campbell. 2003. Faith love: A psychological construct in intimate relations. *Journal of the Indiana Academy of the Social Sciences* 7: 11–20.

U

Uecker, J., and M. Regnerus. 2011. *Premarital sex in America: How young Americans meet, mate, and think about marrying*. United Kingdom: Oxford University Press.

Ulman, A., and M. A. Straus. 2003. Violence by children against mothers in relation to violence between parents and corporal punishment by parents. *Journal of Comparative Family Studies* 34: 41–56.

Umminger, A., and S. Parker. 2005. What hinders vacation? *USA Today*, February 18, A1.

Updegraff, K. A., S. M. Thayer, S. D. Whiteman, D. J. Denning, and S. M. McHale. 2005. Relational aggression in adolescents' sibling relationships: Links to sibling and parent-adolescent relationship quality. *Family Relations* 54: 373–386.

U.S. Census Bureau. Report issued Jan 15, 2010 http://www.census.gov/newsroom/releases/archives/families_households/cb10-08.html.

Useda, J. D., K. R. Conner, A. Beckman, N. Franus, Z. Tu, and Y. Conwell. 2007. Personality differences in attempted suicide versus suicide in adults 50 years of age or older. *Journal of Consulting and Clinical Psychology* 75: 126–133.

Usher-Seriki, K. K., M. S. Bynum, and T. A. Callands. 2008. Mother daughter communication about sex and sexual intercourse among middle- to upper-class African American girls. *Journal of Family Issues* 29: 901–917.

V

Vail-Smith, K., D. Knox, and M. Zusman. 2007. The lonely college male. *International Journal of Men's Health* 6: 273–279.

Vail-Smith, K., L. MacKenzie, and D. Knox. 2010. The illusion of safety in "monogamous" undergraduates. *American Journal of Health Behavior* 34: 15–20.

Vaillant, C. O., and G. E. Vaillant. 1993. Is the U-curve of marital satisfaction an illusion? A 40 year study of marriage. *Journal of Marriage and the Family* 55: 230–239.

Valliant, G. E. 2002. *Aging well: Surprising guideposts to a happier life from the Landmark Harvard study on adult development.* New York: Little, Brown.

Van Gool, C. H., G. Kempen, H. Bosma, J. Van Eijk, M. P. J. Van Boxtel, and J. Jolles. 2007. Associations between lifestyle and depressed mood: longitudinal results from the Maastricht Aging Study. *American Journal of Public Health* 97: 887.

Van Laningham, J., and D. R. Johnson. 2009. Measuring marital quality: Comparing single and multiple item measures. Poster, National Council on Family Relations, Nov. San Francisco.

Vandell, D. L., J. Belsky, M. Burchinal, L. Steinberg, N. Vandergrift. and NICHD Early Child Care Research Network. 2010. Do effects of early child care extend to age 15 years? Results from the NICHD study of early child care and youth development, *Child Development* 81: 737–756.

Vandello, J. A., and D. Cohen. 2003. Male honor and female fidelity: Implicit cultural scripts that perpetuate domestic violence. *Journal of Personality and Social Psychology* 84: 997–1010.

Vanderkam, L. 2010. *168 hours: You have more time than you think to achieve your dreams.* Portfolio Press.

Vannier, S. A. and L. F. O'Sullivan. 2010. Sex without desire: Characteristics of occasions of sexual compliance in young adults' committed relationships. *Journal of Sex Research* 47: 429–439.

Vazonyi, A. I., and D. D. Jenkins. 2010. Religiosity, self control, and virginity status in college students from the "Bible belt": A research note. *Journal for the Scientific Study of Sex* 49: 561–568.

Veenhoven, R. 2007. Quality-of-life-research. In *21st century sociology: A reference handbook,* ed. Clifton D. Bryant and Dennis L. Peck, 54–62. Thousand Oaks, California: Sage Publications.

Verbakel, E., and T. A. Diprete. 2008. The value of non-work time in cross-national quality of life comparisons: The case of the United States versus the Netherlands. *Social Forces* 87: 679–712.

Vickers, R. 2010. Sexuality and the Elderly. Presentation, Sociology of Human Sexuality class, Department of Sociology, East Carolina University, March 22.

Vinick, B. 1978. Remarriage in old age. *The Family Coordinator* 27: 359–363.

Vinkers, C. D. W., C. Finkenauer, and S. T. Hawk. 2011. Why do close partners snoop? Predictors of intrusive behavior in newlywed couples. *Personal Relationships* 18: 110–124.

Vrangalova, Z., and R. C. Savin-Williams. 2010. Correlates of same-sex sexuality in heterosexually identified young adults. *Journal of Sex Research* 47: 92–102.

W

Wagner, C. G. 2006. Homosexual relationships. *Futurist* 40: 6.

Waite, L. J., Y. Luo, and A. C. Lewin. 2009. Marital happiness and marital stability: Consequences for psychological well-being. *Social Science Research* 28: 201–217.

Walch, S. E., P. M. Orlosky, K. A. Sinkanen, and H. R. Stevens. 2010. Demographic and social factors associated with homophobia and fear of AIDS in a community sample. *Journal of Homosexuality* 57: 310–324.

Walcheski, M. J. and D. J. Bredehoft. 2010. Exploring the relationship between overindulgence and parenting styles. Poster, National Council on Family Relations annual meeting, November 3–5. Minneapolis, MN.

Walker, R. B., and M. A. Luszcz. 2009. The health and relationship dynamics of late-life couples: a systematic review of the literature. *Ageing and Society* 29: 455–481.

Walker, K., M. Krehbiel, and L. Knoyer. 2009. "Hey you! Just stopping by to say hi!": Communicating with friends and family on MySpace. *Marriage & Family Review* 45: 677–696.

Walker, S. K. 2000. Making home work: Family factors related to stress in family child care providers. Poster at the Annual Conference of the National Council on Family Relations, November 3–5. Minneapolis.

Waller, M. R. 2008. How do disadvantaged parents view tensions in their relationships? Insights for relationship longevity among at-risk couples. *Family Relations* 57: 128–143.

Waller, W., and R. Hill. 1951. *The family: A dynamic interpretation.* New York: Holt, Rinehart and Winston.

Wallerstein, J., and S. Blakeslee. 1995. *The good marriage.* Boston: Houghton-Mifflin.

Wallis, C. 2011. Performing gender: A content analysis of gender display in music videos. *Sex Roles* 64: 160–172.

Walsh, J. L., and L. M. Ward. 2010. Magazine reading and involvement and young adults' sexual health knowledge, efficacy, and behaviors. *Journal of Sex Research* 47: 285–300.

Walsh, K. T. 2006. And now it's her turn. *U.S. News & World Report* (May 15): 27.

Walsh, T., M. Schembri, P. Turek, J. Chan, P. Carroll, J. F. Smith, M. L. Eisenberg, S. Van Den Eeden, K. Stephen, and M. S. Croughan. 2010. Increased risk of high-grade prostate cancer among infertile men *Cancer* 9: 2140–2147.

Walster, E., and G. W. Walster. 1978. *A new look at love.* Reading, MA: Addison-Wesley.

Walters, B. 2008. *Barbara Walters audition: A memoir.* New York: Alfred A. Knopf.

Walzer, S. 2008. Redoing gender through divorce. *Journal of Social and Personal Relationships* 25: 5–21.

Wang, M., K. Henkens, and H. Solinge. 2011. Retirement djustment: A review of theoretical and empirical advancements. *American Psychologist* 66: 204–213.

Wang, Y. A., D. Healy, D. Black, and E. A. Sullivan. 2008 Age-specific success rate for women undertaking their first assisted reproduction technology treatment using their own oocytes in Australia, 2002–2005. *Human Reproduction* 23: 1533–1639.

Wang, W., and P. Taylor. 2011. For millennials, parenthood trumps marriage. Pew Reseach Center Publications. March 9. (See PewSocialTrends.org for full report)

Warash, B. G., C. A. Markstrom, and B. Lucci. 2005. The early childhood environment rating scale-revised as a tool to improve child care centers. *Education* 126: 240–250.

Ward, R. A., and G. D. Spitze. 2007. Nestleaving and coresidence by young adult children: The role of family relations. *Research on Aging* 29: 257–271.

Webb, A. P., C. G. Ellison, M. J. McFarland, J. W. Lee, K. Morton, and J. Walters. 2010. Divorce, religious coping, and depressive symptoms in a conservative protestant religious group. *Family Relations* 59: 544–557.

Webb, F. J. 2005. The new demographics of families. In *Sourcebook of family theory & research,* ed. Vern L. Bengtson, Alan C. Acock, Katherine R. Allen, Peggye Dilworth-Anderson, and David M. Klein, Thousand Oaks, California: Sage Publications.

Weckwerth, A. C., and D. M. Flynn. 2006. Effect of sex on perceived support and burnout in university students. *College Student Journal* 40: 237–249.

Weden, M., and R. T. Kimbro. 2007. Racial and ethnic differences in the timing of first marriage and smoking cessation. *Journal of Marriage and Family* 69: 878–887.

Weigel, D. J. 2010. Mutuality of commitment in romantic relationships: Exploring a dyadic model. *Personal Relationships* 17: 495–513.

Weigel, D. J., and D. S. Ballard-Reisch. 2002. Investigating the behavioral indicators of relational commitment. *Journal of Social and Personal Relationships* 19: 403–423.

Weinstein, L., and R. Alexander. 2010. College students and their cats. *College Student Journal* 44: 626–628.

Weisgram, E. S., R. S. Bigler, and L. S. Liben 2010. Gender, values, and occupational interests among children, adolescents, and adults. *Child Development* 81: 778–796.

Weiss, K. G. 2010. Male sexual victimization: Examining men's experiences of rape and sexual assault *Men and Masculinities* 12: 275–298.

Welch, V. 2009. *Doctorate recipients from United States universities.* Chicago: National Opinion Research Center.

Wells, G. 2011. Making room for daddies: Male couples creating families through adoption. *Journal of GLBT Family Studies* 7: 155–181.

Wells, R. S., T. A. Seifert, R. D. Padgett, S. Park, and P. Umbach. 2011 Why do more women than men want to earn a four-year degree? Exploring the effects of gender, social origin, and social capital on educational expectations. *Journal of Higher Education* 82: 1–32.

West III, J. L. W. 2005. *The perfect hour.* New York: Random House.

West, P. C., and L. C. Merriam, Jr. 2009. Outdoor recreation and family cohesiveness: a research approach. *Journal of Leisure Research* 41: 351–360.

Wetherill, R. R., D. J. Neal, and K. Fromme. 2010. Parents, peers, and sexual values influence sexual behavior during the transition to college. *Journal Archives of Sexual Behavior* 39: 682–694.

Whatley, M. 2005. The effect of participant sex, victim dress, and traditional attitudes on casual judgments for marital rape victims. *Journal of Family Violence* 20: 191–201.

Wheeler, L. A., K. A. Updegraff, and S. M. Thayer. 2010. Conflict resolution in Mexican-origin couples: Culture, gender, and marital quality. *Journal of Marriage and the Family* 72: 991–1005.

Whisman, M. A., L. A. Uebelacker, and T. D. Settles. 2010. Marital distress and the metabolic syndrome: Linking social functioning with physical health. *Journal of Family Psychology* 24: 367–370.

White, J. M., and D. M. Klein. 2002. *Family theories,* 2d ed. Thousand Oaks, CA: Sage Publications.

White, M. 2005. A thorn in their side. *Intelligence Report* 117(Spring): 27–30.

White, S. S., N. El-Bassel, L. Gilbert, E. Wu, and M. Chang. 2010. Lack of awareness of partner STD risk among heterosexual couples. *Perspectives on Sexual & Reproductive Health* 42: 49–55.

Whitehead, B. D., and D. Popenoe. 2004. The state of our union: The social health of marriage in America. The National Marriage Project. Rutgers University. http://www.marriage.rutgers.edu/.

Whitley, B., C. Childs, and J. Collins. 2011. Differences in black and white American college students' attitudes toward lesbians and gay men. *Sex Roles* 64: 299–310.

Whitty, M. T. 2007. The art of selling one's 'self' on an online dating site: The BAR approach. In *Online matchmaking,* ed. M. T. Whitty, A. J. Baker, and J. A. Inman, 37–69. New York: Palgrave Macmillan.

Wickrama, K. A. S., C. M. Bryant, and T. K. Wickrama. 2010 Perceived community disorder, hostile marital interactions, and self-reported health of African American couples: An interdyadic process. *Personal Relationships* 17: 515–531.

Wielink, G., R. Huijsman, and J. McDonnell. 1997. A study of the elders living independently in the Netherlands. *Research on Aging* 19: 174–198.

Wienke, C., and G. J. Hill. 2009. Does the "Marriage Benefit" extend to partners in gay and lesbian relationships?: Evidence from a random sample of sexually active adults. *Journal of Family Issues* 30: 259–273.

Wilcox, C., and R. Wolpert. 2000. Gay rights in the public sphere: Public opinion on gay and lesbian equality. In *The Politics of Gay Rights,* ed. C. A. Rimmerman, K. D. Wald, and C. Wilcox, 409–32. Chicago: University of Chicago Press.

Wilcox, W. B. (editor) 2010. *The state of our unions: When marriage disappears.* Charlottesville, Virginia: National Marriage Project and The Center for Marriage and Families at the Institute for American Values.

Wilcox, W. B., and S. L. Nock. 2006. What's love got to do with it? Equality, equity, commitment and marital quality. *Social Forces* 84:1 321–345.

Wildsmith, E., K. B. Guzzo, and S. R. Hayford. 2010. Repeat unintended, unwanted and seriously mistimed childbearing in the United States. *Perspectives on Sexual and Reproductive Health* 42: 14–23.

Willer, E. K. and J. Soliz. 2010. Face needs, intragroup status, and women's reactions to socially aggressive face threats. *Personal Relationships* 17: 557–571.

Willson, A. E. 2007. The sociology of aging. In *21st century sociology: A reference handbook,* ed. Clifton D. Bryant and Dennis L. Peck, 148–55. Thousand Oaks, California: Sage Publications.

Wilmoth, J. D., S. L. Smyser, T. Staier, and T. M. Phillips. 2010. Influence of a statewide marriage initiative on clergy involvement in marriage preparation. *Marriage and Family Review* 46: 278–299.

Wilmoth, J., and G. Koso. 2002. Does marital history matter? Marital status and wealth outcomes among preretirement adults. *Journal of Marriage and the Family* 64: 254–268.

Wilson, C. and R. Rodrigous. 2010. Polyamory. Presentation, Sociology of Human Sexuality, Department of Sociology, East Carolina University, Spring.

Wilson, E. K., B. T. Dalberth, H. P. Koo, and J. C. Gard. 2010. Parents' perspectives on talking to preteenage children about sex. *Perspectives on Sexual and Reproductive Health* 42: 56–64.

Wilson, G. D., and J. M. Cousins. 2005. Measurement of partner compatibility: Further validation and refinement of the CQ test. *Sexual and Relationship Therapy* 20: 421–429.

Wilson, G. D., J. M. Cousins, and B. Fink. 2006. The CQ as a predictor of speed-date outcomes. *Sexual & Relationship Therapy* 21: 163–169.

Wilson, H., and A. Huntington. 2006. Deviant mothers: the construction of teenage motherhood in contemporary discourse. *Journal of Social Policy* 35: 59–76.

Wilson, K., B. A. Mattingly, E. M. Clark, D. J. Weidler, and A. W. Bequette. 2011. The gray area: Exploring attitudes toward infidelity and the development of the perceptions of dating infidelity scale. *Journal of Social Psychology* 151: 63–86.

Wilson, S. A. and H. A. Kudis. 2005. Ethinyl estradiol/levinorgestrol (Seasonale) for oral contraception. *American Family Physician* 71: 1581–1586.

Wilson, S. M., L. W. Ngige, and L. J. Trollinger. 2003. Kamba and Maasai paths to marriage in Kenya. In *Mate selection across cultures,* ed. R. R. Hamon and B. B. Ingoldsby, 95–117. Thousand Oaks, CA: Sage Publications.

Winch, R. F. 1955. The theory of complementary needs in mate selection: Final results on the test of the general hypothesis. *American Sociological Review* 20: 552–555.

Winterich, J. A. 2003. Sex, menopause, and culture: Sexual orientation and the meaning of menopause for women's sex lives. *Gender and Society* 17: 627–642.

Wirtberg I., A. Möller, L. Hogström, S. E. Tronstad, and A. Lalos. 2007. Life 20 years after unsuccessful infertility treatment. *Human Reproduction* 22: 598–604.

Witt, M. G., and W. Wood. 2010. Self-regulation of gendered behavior in everyday life. *Journal Sex Roles* 62: 9–10.

Witte, T. H. and R. Kendra. 2010. Risk recognition and intimate partner violence. *Journal of Interpersonal Violence* 25: 2199–2216.

Wolkove, N., O. Elkholy, M. Baltzan, and M. Palayew. 2007. Sleep and aging. *Canadian Medical Association Journal* 176: 1299–1304.

Woodhill, B. M., and C. A. Samuels. 2003. Positive and negative androgyny and their relationship with psychological health and well-being. *Sex Roles* 48: 555–565.

Wrosch, C., R. Schulz, G. E. Miller, S. Lupien, and E. Dunne. 2007. Physical health problems, depressive mood, and cortisol secretion in old age: Buffer effects of health engagement control strategies *Health Psychology* 26: 341–349.

Wu, P., and W. Chiou. 2009. More options lead to more searching and worse choices in finding partners for romantic relationships online: An experimental study. *CyberPsychology & Behavior* 12: 315–318.

X

Xia, Y. R., and Z. G. Zhou 2003. The transition of courtship, mate selection, and marriage in China. In *Mate Selection across Cultures,* ed. R. R. Hamon and B. B. Ingoldsby, 231–246. Thousand Oaks, CA: Sage Publications.

Xie, Y., J. M. Raymo, K. Govette, and A. Thornton. 2003. Economic potential and entry into marriage and cohabitation. *Demography* 40: 351–364.

Y

Yang, J. and S. Ward. 2010. Working wives/ Ernst & Young Survey. *USA Today.* December 7. B1.

Yarhouse, M. A., C. H. Gow, and E. B. Davis. 2009. Intact marriages in which one partner experiences same-sex attraction: A 5-year follow-up study. *The Family Journal* 17: 329–334.

Yazedjian, A., and M. L. Toews. 2010. Breakups, depression, and self-esteem as predictors of college adjustment. Poster, National Council on Family Relations annual meeting, November 3–5. Minneapolis, MN.

Yoo, H. C., M. F. Steger, and R. M. Lee. 2010. Validation of the subtle and blatant racism scale for Asian American college students (SABR-A²). *Cultural Diversity and Ethnic Minority Psychology* 16: 323–334.

Yoshimura, C. G. 2010. The experience and communication of envy among siblings, siblings-in-law, and Spouses. *Journal of Social and Personal Relationships* 27: 1075–1088.

Yu, L., and D. Zie. 2010. Multidimensional gender identity and psychological adjustment in middle childhood: A study in China. *Journal of Sex Roles* 62: 100–113.

Yvonne. 2004. Devout Christian finds a reason to stand up for equality. Human Rights Campaign (December 1). http://www.hrc.org.

Z

Zeitzen, M. K. 2008. *Polygamy: A cross-cultural analysis.* Oxford: Berg.

Zeki, S. 2007. The neurobiology of love. *Febs Letters* 581: 2575–2579.

Zhan, H. J., X. Feng, and B. Luo. 2008. Placing elderly parents in institutions in urban China: A reinterpretation of filial piety. *Research on Aging* 30: 543–558.

Zhang, Y. 2010. A mixed-methods analysis of extramarital sex in contemporary China. *Marriage & Family Review* 46: 170–190.

Zhang, Y. and G. Lee. 2011. Intercountry versus transracial adoption: Analysis of adoptive parents' motivations and preferences in adoption. *Journal of Family Issues* 32: 75–98.

Zinzow, H. M., H. S. Resnick, A. B. Amstadter, J. L. McCauley, K.J. Ruggiero, and D. G. Kilpatrick. 2010. Drug- or alcohol-facilitated, incapacitated, and forcible rape in relationship to mental health among a national sample of women. *Journal of Interpersonal Violence* 25: 2217–2236.

Zolotor, A., A. Theodore, D. Runyan, K. Desmond, J. J. Chang, A. Laskey, L. Antoinette, and L. Izvor. 2011. Corporal punishment and physical abuse: population-based trends for three-to-11-year-old children in the United States. *Child Abuse Review* 20: 57–66.

Zusman, M. 2011. Interview, January 28. on social control of choices. Dr. Zusman is retired from Indiana University Northwest. See also Zusman, M. E., D. Knox, and T. Gardner. 2009. *The social context view of sociology.* Durham, North Carolina: Carolina Academic Press.

Name Index

A

Aamodt, M.G., 437
Abdelhady, R., 56
Abowitz, D., 43, 161, 376
Abraham, A.J., 426
Ackbar, S., 231
Ackerman, D., 81
Ackerman, R.J., 495
Adams, M., 136, 218, 450
Adler, A., 337–338, 339
Adler-Baeder, F.A., 8, 61, 462
Adolfsen, A.J., 225
Afifi, T.D., 107
Agahi, N., 493
Agate, J.R., 367
Ahn, T., 437
Ahnert, L., 361
Ahrons, C.R., 448
Akmatov, M.K., 393
Albright, J.M., 135, 136
Alexander, P.C., 390
Alexander, R., 9
Alford-Cooper, F., 491
Algoe, S.B., 81
Ali, M.M., 322
Allen, D.W., 352, 444
Allen, E.S., 421
Allen, K.N., 392
Allen, K.R., 17
Allgood, S.M., 170, 211
Allport, G.W., 229
Amato, P.R., 14, 156, 158, 160,
    163, 179, 199, 210, 211, 212,
    239, 290, 351, 436, 441, 444,
    450, 452, 456
Ambwani, S., 78
Amory, J., 510
Anderson, E.R., 460
Anderson, J.R., 213
Anderson, K.L., 6, 381
Anderson, L., 399
Anderson, P.L., 398
Andrews, M., 324
Aponte, R., 115
Asch-Goodkin, J., 243
Auerbach, C.F., 12
Axelson, Lee, 14
Ayres, I., 108
Azziz-Baumgartner, E., 375

B

Baden, A.L., 296
Baek, E., 348
Bagarozzi, D., 420
Bagd, A., 27, 369
Baggerly, J., 410
Baker, A., 51, 52

Baker, A.J., 136
Baker, A.J.L., 449
Baker, J., 496
Baker, K., 108
Bakker, A., 211
Ballard-Reisch, S., 189
Balsam, K.F., 143
Baly, A.R., 389
Banks, M.E., 495
Barelds, D.P., 70, 87
Barelds-Dijkstra, D.P.H., 70, 87
Barer, B.M., 487, 494
Barnes, K., 208
Barnett, M.A., 496
Barnett, R.C., 359
Barreios, F.A., 510
Bartle-Haring, S., 189
Basow, S.A., 49
Batar, I., 511, 512, 514
Bates, J.S., 495
Bauerlein, M., 95
Baumgartner, S.E., 140
Baumrind, D., 324
Beach, S.R.H., 34
Bearman, P., 251
Becker, K., 332
Becker, K.D., 388
Beckman, N.M., 491
Beehr, T.A., 487
Beekman, D., 322
Beerthuizen, R., 299, 301
Begue, L., 304
Ben-Ari, A., 362
Benjamin, L., 442
Bennett, K.M., 498
Bennetts, L., 352
Benton, S.D., 456
Beougher, S.C., 233
Berenbaum, S.A., 60
Bergman, K., 242
Bergner, R.M., 66
Berle, W., 327
Bermant, G., 416
Berscheid, E., 66, 68, 83
Berzenski, S.R., 395
Biblarz, T.J., 242, 243, 450
Birditt, K.S., 429
Birkimer, J.C., 270
Bisakha, S., 331
Blair, S.L., 167
Blakely, K., 191
Blakeslee, S., 213, 421
Block, S., 299
Blumer, H., 30
Bobbe, J., 227
Bock, J.D., 341
Bodenmann, G., 406
Bogle, K.A., 132

Boislard, P., 251
Bontempo, D.E., 229
Booth, A., 145, 438
Booth, C.L., 311
Booth, K., 342
Bor, R., 233
Borgerhoff Mulder, M., 283
Bos, H., 93, 243
Bouchard, G., 42
Boucher, S., 395
Boyle, M.H., 398
Brackett, A., 227
Bradbury, T.N., 111, 210, 442
Bradford, J., 222
Bradshaw, C., 132
Brandes, M., 290
Brantley, A., 72
Bratter, J.L., 158
Braun, M., 92
Bray, J.H., 464, 465
Brecklin, L.R., 386
Bredehoft, D.J., 325, 327
Bredow, C., 165
Breheny, M., 289
Bremmer, D., 440
Brentano, C., 437
Bretherick, K., 280
Brimhall, A., 458, 459
Brinig, M.F., 444
Bristol, K., 304
Brizendine, L., 319
Brody, D.J., 412
Bronner, G., 412
Bronte-Tinkew, J., 319
Brotherson, S.E., 456
Brown, A., 159
Brown, G.L., 30
Brown, J.D., 299
Brown, M., 193
Brown, S., 301
Brown, S.L., 141, 467, 471
Brownridge, D.A., 374, 376, 377
Brozowski, K., 401
Brucker, H., 251
Brumbaugh, C.C., 167
Bryant, C.D., 498
Buchanan, A., 319
Buchbinder, E., 392
Buchler, S., 139
Bui, T., 431
Bulanda, J.R., 55
Bulmer, S.M., 424
Burke, M.L., 428
Burke, S., 381
Burr, W.R., 409, 411
Burton, L.M., 156, 204
Busby, D.M., 384, 444
Buss, D.M., 264

571

Busse, P., 25
Butterworth, P., 413
Buunk, A.P., 179
Buxton, A., 235, 242
Byers, E.S., 257, 264

C

Cadden, M., 197
Cade, R., 190
Cahill, S., 224, 239
Calligas, A.F., 214
Campbell, J.A., 375
Canary, D.J., 199
Cantón-Cortés, D., 397
Cantóna, J., 397
Capizzano, J., 358
Carey, A., 95, 105, 366
Careya, M.P., 131
Carlander, I., 500
Carlson, M.J., 496
Carpenter, C., 222, 252
Carpenter, L.M., 251
Carr, D., 5, 369, 450, 500
Carrere, S., 210
Carroll, D.D., 485
Carroll, J., 484
Carroll, J.L., 491
Carter, D., 140
Carter, P., 140
Cary, A.R., 68
Castrucci, B.C., 311
Caughlin, J.P., 113
Cavaglion, G., 414
Cavanagh, S.E., 327, 466
Cavazos-Rehg, P.A., 264
Chambers, J., 449
Chan, S., 484
Chaney, C., 16, 159
Chang, I.J., 263
Chapleau, K.M., 384
Chartiera, M.J., 397
Chase, B., 229
Chen, C., 252
Cherlin, A.J., 24, 27, 140, 190
Chich-Hsiu, H., 316
Chih-Chien, W., 134
Chiou, W., 135, 369
Chiu, H., 444
Christensen, A., 431
Chu, K., 162
Cianciotto, J., 224
Cinamon, R.G., 363
Claffey, S.T., 97
Clark, S., 447
Clarke, S.C., 461
Clarke-Stewart, A., 437
Clarkwest, A., 156, 163, 442
Clements, C.M., 386
Cleverley, K., 398
Cloud, M.D., 416
Clunis, D.M., 245
Cobey, K.D., 87
Cohen, D., 382
Cohen-Kettenis, P.T., 41
Cohen-Mansfield, J., 482
Cohn, D., 284, 350
Cokes, C., 348
Colapinto, J., 41
Coleman, M., 452, 458, 460
Coll, J.E., 424
Colson, M., 271
Coltrane, S., 450

Confer, J.C., 416
Connor-Smith, J., 387
Connors, W., 298
Consolatore, D., 51
Cook, K., 484
Cooley, C.H., 30
Coontz, S., 187, 188
Cooper, M.L., 140, 426
Copeland, L., 312
Corcoran, M.E., 352
Cornelius-Cozzi, T., 318
Corra, M., 55, 212
Corte, U., 119, 445
Costa, P., 446
Cousins, G., 84, 179
Cousins, J.M., 166
Cowburn, M., 386
Cox, A., 86
Coyne, S.M., 95
Cozijnsen, R., 489
Craig, L., 55
Crawley, S.L., 39
Crisp, B., 329, 334, 336, 339, 429
Crosnoe, R., 327, 466
Crowl, A.S., 242, 244
Cui, M., 311
Cummings, E.M., 412
Cunningham, M.R., 76, 144
Cupach, W.R., 134, 379, 380
Curran, M.A., 30
Curtis, C., 222
Cutler, N.E., 486

D

D'Augelli, A.R., 229
Dabbous, Y., 60
Dahl, S., 358
Daigle, L.E., 385
Danigelis, N.L., 477
Dantzler, T., 256
Darlow, S., 54
Darnall, D., 449
Davidov, M., 333
Dawson, C., 327
Day, R., 321
de Anda, D., 281
De Castro, S., 428
de Grey, Aubrey, 476
de Marneffe, D., 283
De Schipper, J.C., 361
de Vries, B., 234
Deacon, S.A., 234
Dean, C.J., 420
Deaton, A., 348
DeCuzzi, A.D., 24
Dehan, N., 378
Del Rio, C.M., 429
DeLamater, J., 261
Dema-Moreno, S., 350
DeMaris, D., 210, 211
Demir, M., 65
Denney, J.T., 121
DeOllos, I.Y., 211
DePaulo, B., 118
Dethier, M.C., 422
Deutsch, F.M., 354
Devall, E., 174, 176
DeVaney, S.A., 348
Dew, J., 139, 199, 489
Diamond, A., 73
Diamond, A.J., 312

Diamond, L.M., 221
Díaz-Martínez, C., 350
DiDonato, M.D., 60
Diem, C., 375
DiLillo, D.S., 388, 395
Dillion, C.F., 476
Diprete, T.A., 365
Dolinsky, B., 380
Dotson-Blake, K., 59, 69, 263, 410, 429
Douglas, C., 484
Dowd, D.A., 210
Doyle, M., 447, 496
Dozetos, B., 230–231
Drapeau, A., 453
Drefahl, S., 209
Dreikurs, R., 337–338
Drucker, D.J., 221
Druckerman, P., 416
Dubbs, S.L., 179
Dugan, L., 376
Duindam, V., 452
Dula, C., 33
Dumka, L.E., 326
Duncan, S.F., 174
Duncan, S.R., 8
Dunn, K.M., 268
Dush, C.K., 145
Dwyer, D.S., 322
Dysart-Gale, D., 39, 225

E

East, L.D., 84
Easterlin, R.A., 214
Easterling, B.A., 99, 203, 204
Edin, K., 347
Edmundson, M., 376–377
Edser, S.J., 234
Eells, L.W., 130
Eggebeen, D.J., 320
Einhorn, L.A., 348
Eke, A., 383
Elgar, F.J., 340
Elicker, K.J., 360
Elliott, M., 410
Ellison, C.G., 160
Elmslie, B., 416
Elnashar, A., 56
Else-Quest, N.M., 264
Emery, R.E., 450
Engblom-Deglmann, M.L., 459
England, P., 48, 131, 132, 257
Enright, E., 413
Epstein, M., 133
Erich, S., 244
Eshbaugh, E.M., 132
Eshleman, J.R., 213
Esmaila, A., 234
Euser, E.M., 422
Exum, H.A., 410

F

Fabian, N., 488
Facer, J., 321
Fairlie, R.W., 10
Fallona, B., 393, 395
Farmer, B., 304
Farris, C., 385
Faulkner, S., 258
Feldman, D.C., 487
Feliciano, C., 157
Few, A.L., 390

Subject Index

Note: Page number followed with t refers to tables.

Certified family life educators (CFLE),
  504
Cervical caps, 514
Chaplin, Charles, 209
Chaplin, Oona, 209
Cheating, 108–110
Cheney, Mary, 235
Child abuse, 383–384, 392–398
Childless by Choice Project, 285
Child marriage, 22
Children
  abuse of, 383–384, 392–398
  adult, 313–314, 479
  birth order effects, 323–324
  childfree marriages, 285–286
  cohabitation and, 144–145
  corporal punishment/spanking,
    328–330, 381
  death of, 427–428
  divorce and, 447–455
  of divorced parents, 130
  domestic abuse and, 388
  dual-income parents and, 358–362
  feral, 28
  historical views, 322
  influences on, 322–323
  marriage and, 4, 187
  planning for, 278, 305
  of same-sex relationships, 241–245
  sex selection, 288, 294
  sexual abuse of, 395–398
  small/large families, 286–288
  in stepfamilies, 466–469
  uniqueness of, 323
Child support, 147
Choices. *See* Relationship choices
Chronic sorrow, 428
Circadian preferences, 161
Civil unions, 7–8
Clemons, Roger, 106
Clinton, Bill, 249
Clinton, Chelsea, 139, 207
Clooney, George, 255
Closed-ended questions, 100
Codependency, 79
Cognitive behavior therapy (CBT), 431
Cognitive-developmental theory, 46–47
Cognitive restructuring, 363
Cohabitation effect, 145
Cohabitation relationships
  children and, 144–145
  consequences of, 144
  definition and overview, 138–139
  future of, 151
  legal aspects, 146–147
  marriage and, 145–146
  same-sex, 139
  types of, 139–141
Coitus, 264
Cold feet, 190
Collaborative practice, 456
Collectivism *vs.* individualism, 24
College Cost Calculator, 284
Collins, Judy, 428–429
Coming out, 235–238
Commitment, 18, 188–190
Common-law marriage, 3, 147
Communication
  conflict resolution, 97, 110–115
  conflicts, 96–97
  cultural differences, 103–104

future of, 115
gender differences, 103
honesty, 104–105
interpersonal, 92–96
lying and cheating, 106–109
negative, 178
secrets, 105–106
self-disclosure, 80–81, 104
social exchange theory and, 110
symbolic interactionism theory and, 110
techniques of, 97–102
Companionship, 187
Compersion, 88
Competitive birthing, 288
Complementary-needs theory, 163–164
Comprehensive sex education
  programs, 252–253
Conception, 290
Condoms, 511
Conflict framework
  age/aging and, 478t
  on marriage and families, 29–30
  stepfamilies, 463–464
Conflict resolution, 97, 110–115
Conflicts, 96–97
Congruent messages, 102
Conjugal love, 71
Connection rituals, 211
Consumerism, 346
Continuity theory, 478t
Contraception
  cervical caps, 514
  condoms, 511–512
  contraceptive sponges, 512–513
  diaphragms, 513–514
  effectiveness of, 516–518
  emergency, 301, 515–516
  hormonal, 508–510
  intrauterine devices (IUDs), 513
  natural family planning, 514
  sexual values and, 261
  talking with partner about, 301–302
  withdrawal/douching, 514–515
Conversion therapy, 223–224
Coolidge effect, 416
Cooper, Sue Ellen, 413–414
Corporal punishment, 328–329, 381
Correct rejections, 99
Cougar: A Guide for Older Women Dating
  Younger Men (Gibson), 209
Cougars, 209
Covenant marriages, 190, 471
Cowgill, Donald, 478t
Crazy Heart (film), 422
Crazy Love (film), 65
Crises. *See* Stressors/crises
Cross-dressers, 40
Cross-national marriages, 207–208
Cryopreservation, 294
Culture. *See also* African Americans;
    Asians; Hispanics
  Aboriginal, 13
  abortion and, 302
  abuse and, 381–382
  Afghans, 51–52
  African, 52
  age/aging and, 477
  beliefs and values, 23–24
  body image and, 54
  Caribbean, 52
  child abuse and, 393, 394

childbearing and, 281
child marriage, 22
cohabitation and, 142
collectivism *vs.* individualism, 24
communication and, 101, 103–104
divorce rates, 441
extended families, 13
extramarital affairs and, 416
family size and, 287
gender role development and, 50–52
homophobia and, 227
mate selection and, 154–155
nuclear families, 12
online relationships and, 134
romantic love and, 69–70
same-sex marriage and, 240
sexuality and, 256
sociological imagination and, 24
work weeks, 365
Cumming, Elaine, 478t
Cunnilingus, 263
Cupid Cowboy (film), 137
Cybercontrol, 380–381
Cybersex, 266, 414–415
Cybervictimization, 380

D

Dangerfield, Rodney, 342
Date rape, 385–386
Dating
  after divorce, 128–130
  changes in, 127–128
  functions of, 125–127
  hanging out, 131
  HIV and, 130–131
  hooking up, 131–133
  international, 136–137
  interracial, 206
  long distance relationships, 137–138
  online, 133–136
  Selective Search, 136
  sexual abuse and, 384–386
  speed-dating, 136
Day care, 360–362
Death and dying, 498–500
Death with Dignity Act, 484
Debt, 161–162, 199–200, 347
December marriages, 460
Deception, 104–108
Defense mechanisms, 114–115
Defense of Marriage Act (DOMA),
    238–239
Demandingness, 324
Dementia, 487
De Niro, Robert, 158
Depo-Provera, 509–510
Depp, Johnny, 27
Developmental-maturational approach,
    333–334
Developmental tasks, 469–471
Diaphragms, 513–514
Disenchantment, 197
Disengagement theory, 478t
Displacement, 114
Divorce. *See also* Stepfamilies
  in Canada, 201
  children and, 447–455
  contributing factors, 440–445
  dating after, 128–130
  extramarital affairs and, 420–421, 443
  financial consequences, 447

Hochschild, Arlie, 478t
Homicide, 374–376, 402
Homogamy, 155–163
Homonegativity, 225, 227. *See also* Sexual orientation diversity
Homophobia, 225, 227. *See also* Sexual orientation diversity
Homosexuality, 219, 221–222, 223–224. *See also* Sexual orientation diversity
Honesty, 104–105
Honeymoons, 192–193
Honor crimes/killings, 382
Hooking up, 131–133
Hopper, Dennis, 499
Hormonal contraceptives, 508–510
Hormone therapy, 291
Houdini, Harry, 92
Howard, Margo, 25
How Much is Enough? (Dawson), 327
Huguely, George, 374
Humor, sense of, 410
Hunter, Rielle, 106
Hypersexuality, 265
Hypotheses, 33
Hysterectomy, 300

I

Identification, 46
IM (instant messaging), 95–96
Implanon, 509
Incest, 396
Incontinence, 491
Individualism, 24
Individualism *vs.* collectivism, 24
Induced abortions, 302
Infatuation, 66
Infertility, 290–295
Infidelity, 414–422
Inheritance, 147
In-laws, 198–199
Instant messaging (IM), 95–96
Institutional Review Boards (IRBs), 34
Institutions, 21–22
Integrative behavioral couple therapy (IBCT), 431
Interactionist perspective, 464
Intercountry adoptions, 298
Intercourse, 264–265
Internalized homophobia, 227
International brides, 136–137
International marriages, 207–208
Internet content, 322–323, 324–325
Internet dating, 133–136
Internet Generation. *See* Generation Y
Interracial marriages, 204–206
Interreligious marriages, 206–207
Intersexed individuals, 39–40
Intimate-partner homocide, 375
Intimate-partner violence (IPV), 374
Intimate terrorism (IT), 374
Intrafamilial child abuse, 395–396
Intrauterine devices (IUDs), 513
In vitro fertilization (IVF), 293–294
Involved couple's inventory, 171–173
Islam. *See* Muslim-Americans
I statements, 100
It Gets Better Project, 230

J

Jacob Wetterling Act (1994), 400
James, Jesse, 163

Jealousy, 85–88, 383
Jessop, Joe, 6
Jewish families, 194, 378
Johnsen, Coco, 147
Joint custody, 451–452
Jolie, Angelina, 295, 298, 417
Jon and Kate Plus 8 (reality show), 340
Judgmental statements, 101t

K

Kanka, Megan, 400
Kennedy, Meredith, 52, 122, 125
Kennedy, Ted, 500
Kevorkian, Jack, 484
The Kids are All Right (film), 242
King, Alissa R., 219–220
King, Larry, 441
Kinsch, Amanda, 356
Kissing, 228, 261–262
Kurweil, Ray, 476
Kutcher, Ashton, 209

L

LaLanne, Jack, 476, 493
Lambert, Adam, 60
Landers, Ann, 25
Landmark Harvard Study of Adult Development, 493
Laparoscopy, 300
Latchkey children, 358–360
Latin Americans. *See* Hispanics
Latinos. *See* Hispanics
Learning theory, 75–76
Legal custody, 450–451
Leisure, 366–370
Leno, Jay, 164
Lesbians and gays. *See* Same-sex relationships; Sexual orientation diversity
Levine, Robert, 209
LGBITQ, 220
LGBT. *See* Sexual orientation diversity
Life experiences, 25
Listening skills, 100. *See also* Communication
Litigation, 456
Living apart together, 148–151, 471
Living wills, 482–483, 484
Logan, John, 472
Loneliness, 124–125
Long distance relationships, 137–138
Long-term care, 481–483
Looking-glass self, 30
Lose-lose situations, 113
Love
    development of, 78–81
    end of, 82–85
    future of, 89
    jealousy and, 85–88
    lasting, 72
    marriage and, 3, 186
    multiple partners and, 88–89
    origins of, 73–78
    romantic *vs.* realistic, 68–71
    social control of, 72–73
    stress management and, 409–410
    styles of, 66–68
    triangular view of, 71–72
    in the workplace, 74–75
Love, Yeardley, 374
Ludic love style, 66

Lundquist, Anne, 419
Luscombe, Belinda, 441
Lust, 66
Lust in Translation (Druckerman), 416
Lying, 104–108, 134–135

M

Machismo, 51
Madoff, Bernie, 106
Maher, Bill, 147
Male contraceptives, 510–511
Mania love style, 68
Marital rape, 388
Marital success, 210–214
Markman, Howard, 112
Marriage. *See also* Mate selection; Parenthood/parenting
    age differences and, 208–209
    arranged, 72–73
    benefits of, 4–5
    Canadian families, 201
    changes in, 14–15
    childfree, 285–286
    cohabitation and, 145–146
    cold feet, 190
    commitment and, 188–190
    common-law, 3, 147
    cross-national, 207–208
    defined, 2
    delay of, 121–123, 142, 176–182
    dual-career, 353–355, 358–362
    education programs, 8
    elements of, 3–4
    extramarital affairs and, 414–422
    functions of, 187–188
    future of, 34–35, 215
    Hispanic families and, 200–201
    honeymoons, 192–193
    interracial, 204–206
    interreligious, 206–207
    legal changes, 193–200
    licenses, 176
    military families, 202–204
    motivations for, 186–187
    Muslim-Americans, 201–202
    older adults and, 494
    remarriage, 457–461, 471–472
    same-sex, 238–241
    sex and, 266–267
    sexual abuse in, 386–388
    success/happiness in, 210–214
    therapy and, 429–432
    traditional *vs.* egalitarian, 188t
    types of, 5–7
    weddings, 191–192, 194–195
    widowhood, 121, 459–460, 498–499
    work and, 350–358
Marriage and family therapy, 429–432, 505
Marriage squeeze, 158–159
Marxism, 29
Marx, Karl, 478t
Mass media. *See* Media
Masturbation, 262–263
Matchmaking, 136
Mate selection
    attachment theory and, 165–166
    complementary-needs theory and, 163–164
    cultural aspects, 154–155
    engagement and, 169–176
    exchange theory and, 164–165
    future of, 182

Mate selection (*continued*)
  homogamy, 155–163
  marriage delay, 176–182
  personality characteristics and, 161, 166–168
  positive assortative personality mating, 166
  role theory and, 165
  sociobiological factors and, 168–169
Mating gradient, 22–23, 159
May-December marriages, 208–209
McCain, Cindy, 174
McCain, John, 174
McCartney, Paul, 174
McKinney, Jack, 240
Meals on Wheels, 480
Media
  gender role development and, 49
  relationship choices, 24–25
  sexual values and, 261
  violence in, 381
Mediation, 454–456
Medicaid, 482
Medicare, 482
Megan's Law, 398, 400–401
Men's traditional gender role socialization, 56–59, 161, 188t
Mental illness, 412–413, 486–487
Merckle, Adodlf, 408
Mezvinsky, Marc, 207
Middleton, Catherine, 28
Midlife crisis, 413–414
Mifepristone (RU-486), 516
Military families, 202–204, 225, 382
Milk, Harvey, 237
Millennial Generation. *See* Generation Y
Millennials, 367
Mills, Heather, 174
Mixed-orientation relationships, 234–235
Modernization theory, 478t
Molhan, Sue, 427
Mommy track, 352–353
Momprenuers, 351
Money. *See also* Work
  age/aging and, 483–484
  debt, 161–162, 199–200, 347
  divorce and, 447
  financial behaviors, 349–350
  management of, 161–162
  poverty, 53–54, 289, 347–349
  as power, 350–351
  recession and, 347
Money, John, 40
Monogamy, 4, 108–110, 232–233, 268
Moore, Demi, 209
Moore, Mary Tyler, 209
Mormons, 6, 22, 360
Motherhood, 289–290, 316–319.
    See also Parenthood/parenting
Mr. Mom, 354
Murder. *See* Homicide
Muslim-Americans, 201–202

N

Narcotics Anonymous, 426
National Council on Family Relations (NCFR), 504
National Domestic Violence Hotline, 391
National Family Caregiver Support Program, 480

National Health Marriage Resource Center, 8
Negative commitment, 458
Negotiation, 456
Neugarten, Bernice, 478t
Newman, J.H., 65
Newman, Paul, 309, 428
New relationship energy (NRE), 267
No-fault divorces, 440–441
Nonverbal communication, 92–93
Nordegren, Erin, 186
Nuclear families, 11–12

O

Obama, Barack, 163
Obama, Michelle, 163
Obesity, 84
Obsessive relational intrusion (ORI), 380
Occupational sex segregation, 48–49
Office romances, 74–75, 353, 354–355
Online relationships, 133–136
Oophorectomy, 300
Open-ended questions, 100
Open-mindedness, 159
Open relationships, 89, 234
Oppositional defiant disorder, 309
Opting out, 352–353
Opting Out (Stone), 352–353
Oral sex, 251–252, 263–264
Osmond, Marie, 428
Outlaw, Tracy E., 147
Overindulgence, 327
Ovum transfer, 294
Oxytocin, 76–77

P

Palimony, 147
Palliative care, 412
Pantagamy, 7
Parent abuse, 399
Parental alienation (PA), 449
Parental alienation syndrome, 449
Parental consent, 302
Parental investment, 45
Parental notification, 302
Parent effectiveness training (PET), 336–337
Parenthood/parenting. *See also* Children; Families
  abortion, 302–305
  adoption, 243–245, 295–299
  adult children, 313–314, 347, 479, 486, 494
  birth order, 323–324
  choices perspective, 309, 313–314
  control of Internet content and, 324–325
  corporal punishment/spanking, 328–330
  future of, 342
  genetic testing, 169, 288
  historical views of children, 322
  infertility, 290–295
  influences on children, 322–323
  lifestyle changes and economic costs, 281–284
  motivations for, 279–281
  planning for, 278, 305
  principles of, 326–332
  roles, 310–312
  shared, 451–452

shared parenting dysfunction, 448
  single parenting, 340–342
  small/large families, 286–288
  sterilization, 299–302
  styles of, 324–326
  teenaged, 289–290
  theories of, 333–340
  transition to, 315–321
  uniqueness of children, 323
Parenting self-efficiency, 326
Parents. *See also* Parenthood/parenting
  death of, 428
  disapproving, 84, 143, 179
  of partner, 173–174
  same-sex, 241–245
Parents, Families, and Friends of Lesbians and Gays (PFLAG), 236
Parricide, 375
Partner selection. *See* Mate selection
Partner's night out, 196
Patient Protection and Affordable Healthcare Act (2010), 349
Peers, 48
Pelosi, Nancy, 60
Personality, 25, 161, 166–168
Peter the Wild Boy, 28
Pets, 8–10
PFLAG. *See* Parents, Families, and Friends of Lesbians and Gays
Phased retirement, 488
Phenylethylamine (PEA), 77
Physical appearance, 159–160
Physical custody, 451
Physical health, 484–486
Physical illness/disability, 411–412
Physician-assisted suicide, 484–485
Pirates of the Caribbean (film), 27
Pitt, Brad, 295, 417
Plouffe, David, 163
Polyamory, 88, 255
Polyandry, 6
Polygamy, 5–6
Pool of eligibles, 155
Pornography, 258, 327
Porter, Cole, 417
Porter, Linda Lee, 417
Portman, Natalie, 25
Positive androgyny, 60
Positive assortative personality mating, 166
Postnuptial agreements, 447
Postpartum depression, 316–317
Postpartum psychosis, 317
Posttraumatic stress syndrome. *See* PTSD
Poverty, 53–54, 289, 347–349, 382
Power, 102, 350–351
Pragma love style, 66
Pregnancy, 180, 290
Pregnancy coercion, 278
Premarital counseling, 169–170
Premarital education, 8, 174
Premarital sex, 254
Prenuptial agreements, 174–176
Primary groups, 23
Prince William, 28
Principle of least interest, 165
Procreative liberty, 284
Projection, 114
Pro-life positions, 304
Pronatalism, 279
Psychological abuse, 376–377
Psychological blackmail, 180–181
Psychosexual theory, 76

Stepfamilies
  children in, 466–469
  definitions, 461–462
  developmental tasks, 469–471
  stages, 464–466
  theoretical perspectives, 463–464
  types of, 462
  unique aspects, 462–463
Stepism, 462
Sterilization, 299–302
STIs. *See* Sexually transmitted infections
Storge love style, 68
Straight Spouse Network, 235, 238
Strengthing Families Program, 427
Stress management, 409–411
Stressors/crises
  child abuse and, 382
  death of a family member, 427–429
  definitions, 406–407
  extramarital affairs, 414–422
  family resiliency and, 407
  family stress model, 407–408
  future of, 432–433
  management of, 409–411
  marriage and family therapy, 429–432
  mental illness, 412–413
  midlife crisis, 413–414
  physical illness/disability, 411–412
  substance abuse, 422–427
  unemployment, 422
Strong, Charlene, 143
Structure-function framework, 27–29, 463–464
Substance abuse, 422–427
Suicide, 229–230, 428, 484–485, 486
Suleman, Nadya, 293
Supermon/superperson, 363
Surrogate mothers, 292–293
Survivors of Suicide, 428
Symbolic aggression, 376–377
Symbolic interaction framework, 30–31, 110
Symbolic theory, 478t

T

Taliban, 51–52
Teenagers, 289–290, 332–333, 348–349
Telerelationship therapy, 431
Texting, 95–96, 494
Thanatology, 498
THEIR career marriage, 353
Theories and theoretical frameworks
  activity theory, 478t
  attachment parenting, 338–340
  attachment theory, 77–78, 165–166
  behavioral approach, 334–336
  biochemical theory, 76–77
  biosocial theory, 45–46

cognitive-developmental theory, 46–47
complementary-needs theory, 163–164
conflict framework, 29–30
continuity theory, 478t
developmental-maturational approach, 333–334
disengagement theory, 478t
evolutionary theory, 73–74
exchange theory, 164–165
family life course development, 27
family systems, 31
feminist, 31–32
identification, 46
learning theory, 75–76
modernization theory, 478t
parent effectiveness training (PET), 336–337
psychosexual theory, 76
queer theory, 219–220
role theory, 165
social exchange, 26–27, 110
social learning, 46
sociological theory, 76
socioteleological approach, 337–338
structure-function, 27–29
symbolic interaction, 30–31, 110
symbolic theory, 478t
Therapeutic abortions, 303
Third shift, 363
Thompson, Ivan, 136–137
Thoreau, Henry David, 124
Time management, 364–365
Time outs, 328, 330
Timmendeqas, Jesse, 400
Transdermal patches, 510
Transgender individuals, 40
Transition to parenthood, 315–321
Transracial adoption, 296–297
Transsexuals, 40–41
Troxel v. Granville (2000), 497
True Love Waits, 251
Trump, Donald, 175
Trust, 80
Turkish families, 195
Turner, Ted, 161
Twin Oaks Community, 31, 88

U

Unemployment, 422
United Church of Christ (UCC), 239–240
Unrequited love, 84
Up in the Air (film), 255
Utilitarianism, 26–27
Uxorcide, 375

V

Vacations, 366–370
Vaginal intercourse, 264
Vaginal rings, 510
Values, 24
Vasectomy, 301
Verbal abuse, 376–377
Vibrators, 262, 273
Video games, 368–369
Violence, 374–376. *See also* Abusive relationships
Virginity, 250–252

W

Walters, Barbara, 442
Weber, Max, 478t
Weddings, 191–192, 194–195
Weed dating, 136
Welles, Orson, 30
West, Mae, 128
White, Betty, 476
White, Mel, 235
Whitman, Wynne, 499
Widowhood, 121, 459–460, 498–499
Williams, Charles Andrew, 230–231
Win-lose situations, 113
Win-win situations, 112–113
Withdrawal/douching, 514–515
Wives, 43–45
Women's traditional gender role socialization, 53–56, 161, 188t
Woods, Tiger, 106, 157, 186, 414, 421
Woodward, Joanne, 309
Work
  divorce and, 437
  dual-career marriages, 353–355, 358–362
  family life and, 362–365
  government and corporate policies/ programs, 364–365
  leisure time and, 366–370
  mommy track, 352–353
  money as power, 350–351
  office romances, 74–75, 353, 354–355
  unemployment, 422
  working wives, 350–352, 355–358

Y

Yelsma, Paul, 490
You Don't Know Jack (film), 484
You statements, 100–101

Z

Zuma, Jacob, 6